Yeasts in the Production of Wine

Patrizia Romano • Maurizio Ciani
Graham H. Fleet
Editors

Yeasts in the Production of Wine

Springer

Editors
Patrizia Romano
University of Basilicata
School of Agriculture, Forest, Food
and Environmental Sciences
Potenza, Italy

Maurizio Ciani
Department of Life and Environmental
Sciences
Polytechnic University of Marche
Ancona, Italy

Graham H. Fleet
Food Science and Technology
University of New South Wales
School of Chemical Engineering
Sydney, NSW, Australia

ISBN 978-1-4939-9780-0 ISBN 978-1-4939-9782-4 (eBook)
https://doi.org/10.1007/978-1-4939-9782-4

This Springer imprint is published by the registered company Springer Science+Business Media, LLC, part of Springer Nature.
The registered company address is: 233 Spring Street, New York, NY 10013, U.S.A.

In Memory of Graham

The idea of this book originated several years ago from the desire of two friends (Graham Fleet and I) to share a project, focused on the role of yeasts in winemaking. The project was planned to involve colleagues from all over the world, experts in wine yeasts. Our aim was to publish an updated book covering the occurrence and activities of yeasts in wine production, with contributions from an impressive cohort of international authors. I was well aware of Graham's scientific production on wine yeasts. His work has accompanied many students and researchers over the years, always stimulating new and innovative topics. I personally met Graham in Perugia in 1988 at the General Congress of the International Yeast Commission. It was a great emotion for me to finally meet this great scientist known all over the world for his works, in particular concerning yeasts in the wine sector. Since then, a scientific bond based on the common interest on food yeasts has grown between us year after year discussing yeast topics while attending the symposia of the International Commission on Yeasts, of which we were both commissioners.

The project of the book *Yeasts in the Production of Wine* arose from this knowledge and friendship. We were bound by a great enthusiasm and the desire to organize an updated and useful book that would cover all the topics of interest. However, several drawbacks have delayed the work that Graham and I were doing together. The last time we saw each other was in Perugia for the ISSY32, where we discussed the book's progress. Unfortunately, a short time later, Graham suddenly passed, and his painful loss blocked the book. On the one hand, it was difficult and painful to continue this work without him; on the other hand, I felt strongly the feeling of having to complete this book to dedicate it to this great scientist of yeasts. Therefore, I shared this final stage of the book with our colleague and friend, Maurizio Ciani.

For me, he was not only a dear colleague but primarily a benchmark for research on food yeasts. Over the years, our friendship was firmly established, and despite living in such distant countries, we were in touch via email, talking not only about yeasts but also about everyday life. Beyond the great esteem I felt for him, "the scientist", I admired his depth as a generous person and very caring towards others. I have many fond moments and memories of Graham talking about yeasts, and I had the pleasure and honour on numerous occasions of sharing with him many wine tastings while exchanging opinions. He was an educator and a great scientist who shared with all of us his love and passion for yeast and the lifetime he spent in research with great commitment, generosity and joy, trying to understand yeast contributions in foods and beverages.

It was hard to complete the book without him, but thanks to the availability of Maurizio Ciani, who has shared this project, and of all the authors of the various chapters, we were able to complete the book. The greatest satisfaction is to dedicate it to this great researcher and friend who was Graham.

Patrizia Romano

Preface

The main role of yeasts in the bioconversion of grape juice into wine is well established even if other microorganisms may affect the composition of the final product. In the last few decades, research on winemaking process has made more clarity on the complex interactions of many microbial species and the complex ecological and biochemical processes, highlighting the fundamental impact of yeasts on wine production. Researchers across the world have demonstrated the great diversity of yeasts at the species and strain level, which is expressed through different biochemical, physiological and molecular mechanisms, which are the basis of the many roles of yeasts in wine production.

The aim of this book is to collect the new recent developments on the key role of yeast in wine production, evaluating the ecological, genetic and metabolic aspects. The book contains 16 chapters, written by international contributors who are recognized authorities in their field, which cover the most important topics concerning yeasts and their biotechnological applications in wine production. Chapter 1 describes the recent developments and new approaches in yeast ecological distribution from grapes to cellar during the fermentation process and its role in wine production. Chapter 2 gives an exhaustive description of the metabolic impact of yeasts on wine. Chapters 3, 4, 5 and 6 focus a molecular approach on the monitoring and quality assurance of yeasts (Chap. 3), on yeast diversities (Chap. 4), on gene expression during fermentation (Chap. 5) and on synthetic genome engineering in wine yeasts (Chap. 6). Chapter 7 examines the presence and the role of yeasts in a specific environment and fermentation process such as botrytized wines, while Chapter 8 describes the modalities of production of commercial starter cultures in large scale. Chapter 9 describes kinetics and control of alcoholic fermentation, while Chapter 10 is focused on the description of the recent development of strategies for the genetic improvement of wine yeasts. Chapter 11 provides an updated overview on the role of wine yeasts in determining the content of different compounds affecting wine consumer health. Chapter 12 reports the diversity and significance of yeast spoilage species in wine production, analysing the appropriate methodologies for their control. In the next chapters, the occurrence and role of yeasts in specific and peculiar wines are described. Chapter 13 describes the role and influence of yeasts in primary and secondary fermentation during sparkling wine production, while in Chapter 14, the peculiar characteristics and the fundamental role of yeasts associated with fortified wines are discussed. Chapter 15 reports the recent development on fruit wines which are an increasingly widespread alternative to winemaking. Chapter 16, the last chapter, provides an in-depth review on yeasts associated with the production of distilled alcoholic beverages.

In summary, this book provides a comprehensive account on the occurrence, role and biotechnological use of wine yeasts written by a group of expert scientists from key wine production countries and experts in their field. We believe that each chapter contains information which should be valuable to students of winemaking courses, PhD students and researchers who study or work with yeasts. The content of this book can also be useful for wineries and wine yeast companies.

Potenza, Italy Patrizia Romano
Ancona, Italy Maurizio Ciani

Contents

Contributors

Warren Albertin Institut des Sciences de la Vigne et du Vin, Unité de recherche Oenologie, Villenave d'Ornon Cedex, France

Hervé Alexandre Institut Universitaire de la Vigne et du Vin Jules Guyot, Université de Bourgogne, Dijon, France

Melchor Arellano-Plaza Unidad de Biotecnología Industrial, Centro de Investigación y Asistencia en Tecnología y Diseño del Estado de Jalisco A.C. (CIATEJ), Zapopan, Jalisco, México

Tahía Benítez Department of Genetics, Faculty of Biology, University of Seville, Sevilla, Spain

Linda F. Bisson Department of Viticulture and Enology, University of California, Davis, CA, USA

Marilena Budroni Department of Agriculture, University of Sassari, Sassari, Italy

Angela Capece University of Basilicata, School of Agriculture, Forest, Food and Environmental Sciences, Potenza, Italy

Charoen Charoenchai Department of Food Science and Technology, Faculty of Agricultural Technology, Rajamangala University of Technology Thanyaburi, Thanyaburi, Pathumthani, Thailand

Maurizio Ciani Department of Life and Environmental Sciences, Polytechnic University of Marche, Ancona, Italy

Antonio C. Codón Department of Genetics, Faculty of Biology, University of Seville, Sevilla, Spain

Francesca Comitini Department of Life and Environmental Sciences, Polytechnic University of Marche, Ancona, Italy

Richard Degré Lallemand Inc., Montréal, Québec, Canada

Virginie Galeote Sciences pour l'œnologie (SPO), University of Montpellier, INRA, Montpellier SupAgro, Montpellier Cedex 2, France

Ramon Gonzalez Instituto de Ciencias de la Vid y del Vino (ICVV), Logroño, Spain

Lisa Granchi Department of Agriculture, Food, Environment and Forestry, University of Florence, Florence, Italy

Anne Christine Gschaedler-Mathis Unidad de Biotecnología Industrial, Centro de Investigación y Asistencia en Tecnología y Diseño del Estado de Jalisco A.C. (CIATEJ), Zapopan, Jalisco, México

Manuel Kirchmayr Unidad de Biotecnología Industrial, Centro de Investigación y Asistencia en Tecnología y Diseño del Estado de Jalisco A.C. (CIATEJ), Zapopan, Jalisco, México

Patricia Lappe-Oliveras Departamento de Botánica, Instituto de Biología, Universidad Nacional Autónoma de México, Cd. de México, México

Jean-Luc Legras Sciences pour l'œnologie (SPO), University of Montpellier, INRA, Montpellier SupAgro, Montpellier Cedex 2, France

Manuel Malfeito-Ferreira Linking Landscape Environment Agriculture and Food Research Centre (LEAF), Instituto Superior de Agronomia, University of Lisbon, Lisbon, Portugal

Ilaria Mannazzu Department of Agriculture, University of Sassari, Sassari, Italy

Isabelle Masneuf-Pomarede Institut des Sciences de la Vigne et du Vin, Unité de recherche Oenologie, Villenave d'Ornon Cedex, France

Rubén Moreno-Terrazas C. Departamento de Ingenierías Química, Industrial y de Alimentos, Universidad Iberoamericana, Cd. de Mexico, México

Anne Ortiz-Julien Lallemand S.A.S., Blagnac Cedex, France

Isak S. Pretorius Chancellery, Macquarie University, Sydney, NSW, Australia

Doris Rauhut Department of Microbiology and Biochemistry, Hochschule Geisenheim University, Geisenheim, Germany

Ana M. Rincón Department of Genetics, Centro Andaluz de Biología Molecular y Medicina Regenerativa (CABIMER), University of Seville, Sevilla, Spain

Patrizia Romano University of Basilicata, School of Agriculture, Forest, Food and Environmental Sciences, Potenza, Italy

Jean-Marie Sablayrolles UMR Sciences for Enology, INRA, Montpellier Cedex 2, France

Ana Carla Silva Linking Landscape Environment Agriculture and Food Research Centre (LEAF), Instituto Superior de Agronomia, University of Lisbon, Lisbon, Portugal

Matthias Sipiczki Department of Genetics and Appled Micobiology, University of Debrecen, Debrecen, Hungary

Jordi Tronchoni Instituto de Ciencias de la Vid y del Vino (ICVV), Logroño, Spain

Ileana Vigentini Department of Food, Environmental and Nutritional Sciences, University of Milan, Milan, Italy

Graeme M. Walker School of Applied Sciences, Abertay University, Dundee, Scotland, UK

Forbes Wardrop Lallemand Inc., Montréal, Québec, Canada

Giacomo Zara Department of Agriculture, University of Sassari, Sassari, Italy

Severino Zara Department of Agriculture, University of Sassari, Sassari, Italy

Zhigen Zhang Lallemand Inc., Montréal, Québec, Canada

About the Editors

Patrizia Romano graduated in Biological Sciences at the University of Bologna (Italy), where she developed teaching and research activities in the area of agro-food microbiology. Since 1994, she started to work at the University of Basilicata (Potenza, Italy) as Full Professor in Agricultural Microbiology. During her career at the University of Basilicata, she has held various academic positions, as President of the degree course of Food Technology; Vice-Dean of the Agriculture Faculty and successively Vice-Director of the School of Agricultural, Forestry, Food, and Environmental Sciences; Member of the Academic Senate; and actually Vice-Rector for International Education. She is Member of the doctorate boarding in food technology sector, responsible of Erasmus mobility programs with different European countries, and Tutor of research activity of students and researchers from different countries. She is Member of the Italian Academy of Vine and Wine and of International Commission on Yeasts (ICY); Delegate and Expert of Italian Ministry of Agricultural, Food, and Forest Politics for the OIV (International Organisation of Vine and Wine) in the Microbiology Group (Oenology Commission); and Coordinator of the Italian Wine Microbiology Group (GMV). Her research activity is focused on yeast biotechnology, related to different fermentative processes. She is Author and Coauthor of about 230 publications, comprising chapters in books, publications on Italian and international journals, posters, and oral presentations in Italian and international conferences. The main research lines deal with ecology of natural yeasts and their characterization for technological parameters, yeast characterization for secondary compound production related to organoleptic and healthy quality of fermented beverages, and formulation of yeast starters for application at industrial level. She is Co-inventor of two patents on yeasts and Member of an academic spin-off.

Maurizio Ciani is Full Professor of Biotechnology of Microorganisms in the Department of Life and Environmental Sciences at the Polytechnic University of Marche. He graduated in Agricultural Sciences at the University of Perugia, where he was Researcher from 1990 to 2000. In 1993, with a Research Fellow of the

National Research Council (NRC), he spent a period at Viticulture and Enology Department of the University of California, Davis, USA. He is Associate Professor from 2000 at Polytechnic University of Marche and Full Professor from October 2006 in the same university. He is the Department Coordinator for International Mobility at the Faculty of Science and Delegate of the Rector for the orientation of students. He is Member of Italian Academy of Vine and Wine from 2006 and is Expert of Italian Ministry of Agricultural, Food and Forest Politics for the OIV (International Organisation of Vine and Wine) in the Microbiology Group (Oenology Commission). He has been Scientific Responsible of several research projects financially supported by public funds or private institutions and industries. The scientific activity is focused on yeasts: isolation, selection, and characterization of yeasts from various environments. The investigations concerned the new fermentation biotechnology in wine-making (immobilization, multistarter fermentations), the use of nonconventional yeasts in wine and beer production to improve complexity and aroma profile as well as increase/reduce specific compounds, the physiology of wine yeasts (metabolic characterization of yeasts for food and industrial application), and the molecular and biochemical characterization of antimicrobial compounds from yeasts for food and industrial application. He is Author or Coauthor of more than 180 publications including 18 book chapters and a monograph.

Graham H. Fleet had completed his MSc degrees (1969) in Microbiology and Biochemistry from the University of Queensland. He joined the research group of Prof. Herman Phaff at the University of California Davis, where he developed a pioneer project on the biochemistry of yeast cell walls. In 1973, he received his PhD in Microbiology, Food Science, and Technology from the University of California, Davis. After completing his studies in the USA, he went for postdoctoral studies at the Department of Brewery and Biological Sciences, Heriot-Watt University, Edinburgh, UK. In 1975, he returned to Sidney, where he developed a productive academic carrier at the University of New South Wales (UNSW) as Lecturer, Associate Professor (1982), and Professor (1996) of Food Science and Technology, School of Chemical Science and Engineering, until retirement in 2007, when he continued his academic activities at the UNSW as Emeritus Professor. He was a talented Mentor to students from several countries. He left a remarkable legacy in the literature on food science and technology, microbiology, food safety, and other areas, demonstrating his leading role. He was the Editor/Coeditor of 9 books and Author/Coauthor of over 30 book chapters and more than 100 papers in prestigious journals. He was fascinated by the microbial communities associated with natural fermentations of foods. His last book, *Cocoa and Coffee Fermentations* (2015), is focused on this topic, and his book *Wine Microbiology and Biotechnology* (1993) is a known reference. He was Editor of the *International Journal of Food Microbiology* and served at the editorial board of several scientific journals. Since 2010, he served as the Executive Board of the International Committee on Food Microbiology and Hygiene. He was the Chair of the International Commission on Yeasts (ICY, 1996–2000), then ICY Vice-Chair (2000–2004), and later ICY Honorary Member. He was a Member of the Executive Board of the International Union of Microbiological Societies (IUMS) and Chairperson of the IUMS Mycological Division (2002–2008), where he acted as a passionate Representative of yeast researchers. He participated on several ICY meetings, often presenting conferences on yeasts in food and beverages. His last lecture, presented in 2015 during ISSY 32 in Perugia (Italy), was a memorable conference on "yeasts and the fermented food renaissance."

Chapter 1
Yeast Ecology of Wine Production

Maurizio Ciani and Francesca Comitini

Contents

1 Introduction

Wine fermentation is a complex biotechnological process in which yeasts play an essential role. In this context, the ecological distribution of yeasts through the production chain of wine production is a crucial factor the quality of wine. Although

M. Ciani (✉) · F. Comitini
Department of Life and Environmental Sciences, Polytechnic University of Marche, Ancona, Italy
e-mail: m.ciani@univpm.it; f.comitini@univpm.it

P. Romano et al. (eds.), *Yeasts in the Production of Wine*,
https://doi.org/10.1007/978-1-4939-9782-4_1

Saccharomyces cerevisiae is the main microorganism involved during the transformation of grape juice in wine, many other yeasts species occur in grape juice fermentation and may actively take part in the process. Nowadays, selected starter cultures of *S. cerevisiae* are usually added by oenologists to control the fermentative process and to achieve specific desired enological characters (inoculated fermentations). The aim is that to dominate indigenous yeasts belonging to the vineyard environment, winery facilities and cellar equipment. Indeed, it has been clearly demonstrated that the microbial population is a multi-comprehensive consortium that includes filamentous fungi, yeasts and bacteria with different physiological characteristics and different impact on the grape metabolome and final wine quality (Pinto et al. 2015; Verginer et al. 2010). The composition of grape microbiota can be influenced, in complexity and frequency, by various abiotic or biotic factors, including climatic conditions, temperature, UV exposure, rainfall, sunlight and winds, ripeness or variety of grapes and interaction within strains that co-habitat. The study and the monitoring of microbiota of grape barriers is important to recognize the evolution of yeasts and the relationship between the microorganisms, fundamental to predict the progress of fermentative process. The use of conventional and innovative molecular methods allow to analyse the microbial members of consortium from grape berries to wine. Indeed, spontaneous wine fermentation is typically carried out by a complex evolution of microorganisms extensively examined during the years. Now, it is well established that together with *S. cerevisiae*, non-*Saccharomyces* species actively participate during the alcoholic fermentation and their contribution was recently positively revaluated. Non-*Saccharomyces* yeasts, coming from grape berry and winery environment, if well managed, can positively impact on the analytical and sensory characteristics of wines. In this regard, growing interest on the use of controlled mixed fermentation with selected non-*Saccharomyces* and *S. cerevisiae* wine yeasts draw the applied research in oenological field.

2 Yeasts on Grapes

Grapes represent a complex ecological niche where filamentous fungi, yeasts and bacteria cohabit. The microbial community colonizing this ecological niche includes microbial species whose concentration depending on multiple factors; the most important are related to grape ripening and nutrients availability. Actually, the microbial ecology of grape berry is a wide concept including closed relations between the ecosystems and their microbial interactions, microbial vectors and sources of microorganisms. Herman Phaff, the pioneer of yeast ecology, described the concept of ecology as "where microbes live and why they live in one habitat and how yeasts interact with other microorganisms" (Lachance 2003). This comprehensive approach implies that microbial communities may be affected by many other variables in grapes, such as viticultural practices, pedoclimatic factors, diseases and pests that could modify grape integrity. In general, the yeast populations of mature grapes are comprised of between 10^3 and 10^5 cells/g (Fleet et al. 2002), but approximately one log higher values have often been found on damaged berries in presence of higher availability of sugar and nutrients (Barata et al. 2008). Over the last century many researchers have described the occurrence and association of yeasts with grape surface and the results were reviewed by Amerine and Kunkee 1968; Kunkee and Goswell 1977; Kunkee and Bisson 1993. More recently, the yeast ecology of wine grapes was reviewed by Fleet et al. 2002, Barata et al. 2012 and Jolly et al. 2014 evaluating the factors that affect their occurrence and quantitative presence.

2.1 Occurrence and Diversity of Yeasts

The composition, in terms of occurrence and amount, of indigenous microbiota naturally present on grape berry surfaces is crucial during winemaking process, as it can positively or negatively affect the quality of final wine. The presence and fitness of yeasts are essential in alcoholic fermentation, as promoters of transformation of grape sugars into principal products of fermentations: ethanol, carbon dioxide and hundreds of other metabolites responsible for aroma and flavours (Romano et al. 2003; Fleet 2003).

Kurtzman et al. (2011) already several years ago, ascribed overall yeasts potentially associated with grape/wine ecosystem in 15 different yeast genera, such as *Dekkera/Brettanomyces*, *Candida*, *Cryptococcus*, *Debaryomyces*, *Hanseniaspora/Kloeckera*, *Kluyveromyces*, *Metschnikowia*, *Pichia*, *Rhodotorula*, *Saccharomyces*, *Saccharomycodes*, *Schizosaccharomyces* and *Zygosaccharomyces*. On the other hand, the dynamic yeast taxonomy poses challenge on the nomenclature of wine microbiology (Bisson et al. 2017). The yeast *Hanseniaspora* and its anamorph counterpart *Kloeckera* are the numerically predominant genera present on the surface of grape, with more than 50% of the total yeast population (Fleet and Heard 1993). To a lesser extent, species belonging to *Candida*, *Starmerella*, *Cryptococcus*, *Pichia*, *Metschnikowia* and *Kluyveromyces (Lachancea)* genera are detected (Heard and Fleet 1988; Mills et al. 2002; Rosini et al. 1982). However, the variability may be reduced to few groups of similar physiological characteristics. For instance, the ubiquitous *Candida* spp. and *Pichia* spp. are highly heterogeneous, and new species are likely to be found in each new survey because the accuracy of molecular identifications is constantly increasing. A division of yeast biota of grape berries into three main groups with similar characteristics are proposed: (i) oxidative yeasts as basidiomycetous *Rhodotorula* and *Cryptococcus* along with the yeast-like fungus *Aerobasidium pullulans and some Candida* species; (ii) oxidative-fermentative ascomycetes *Hanseniaspora* spp., *Pichia* spp., and *Metschnikowia* spp. together with some *Candida* species; (iii) strongly fermentative yeasts with higher alcohol producing *Saccharomyces* spp., *Starmerella* spp. *Torulaspora* spp., *Zygosaccharomyces* spp., and *Lachancea* spp. In Table 1.1 are summarized the main yeast species colonizing wine making environment.

Table 1.1 Main yeasts genera found on grape berry surfaces and in winery

Yeast Genera	species	Presence in grapes and characters
Aerobasidium	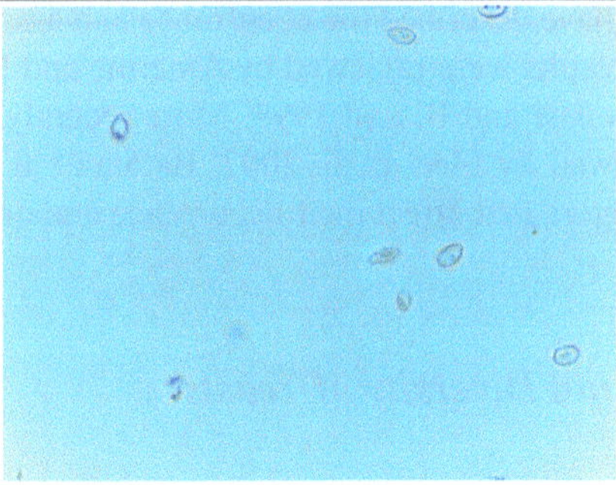 *A. pullulans*	Oxidative yeast-like fungus present on the berries surface still before this reach maturity and in the early stages of fermentation.
Dekkera/ Brettanomyces	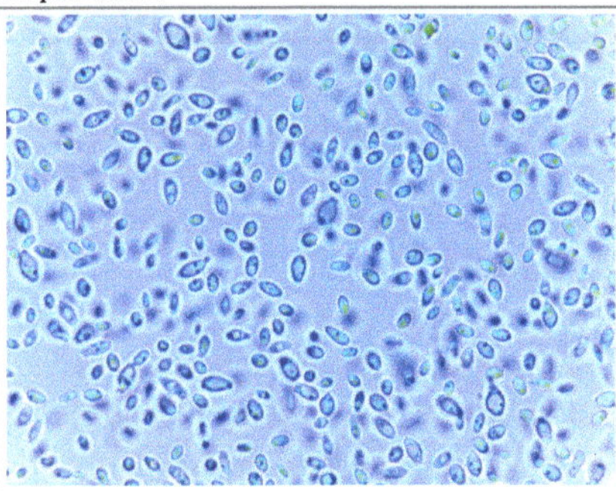 *D. bruxellenisis*	Vineries colonization, wine aging; rarely found on grapes; spoilage yeast.
Starmerella	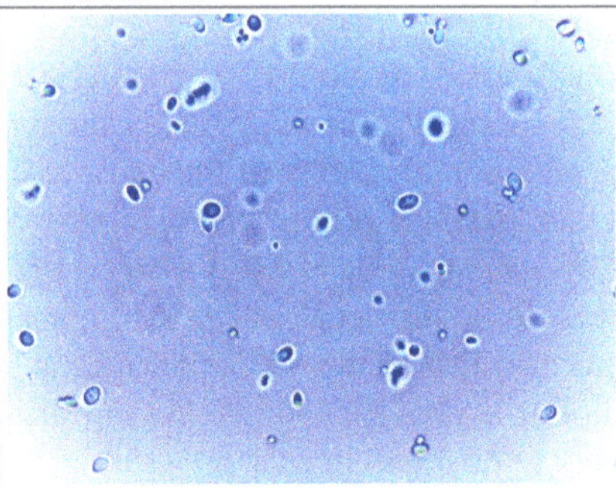 *S. bombicola (formerly Candida stellata)*	Occurrence on grapes surface at harvest time; low production of volatile acidity;strong fructofilic character;high amounts of glycerol production.

(continued)

Table 1.1 (continued)

Cryptococcus	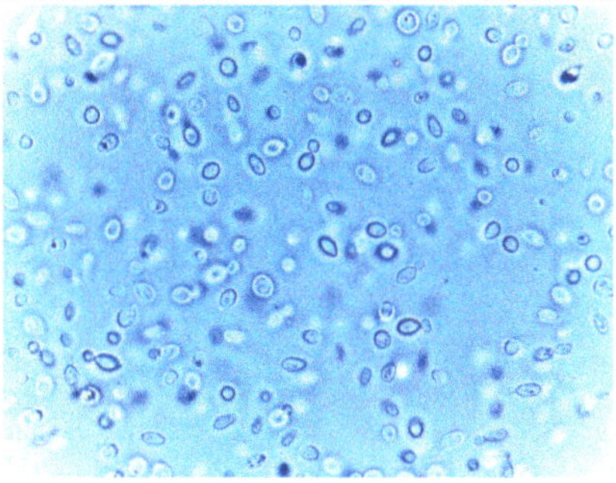 *C. flavescens*	Occurrence on grape berries before and during full ripeness.
Debaryomyces	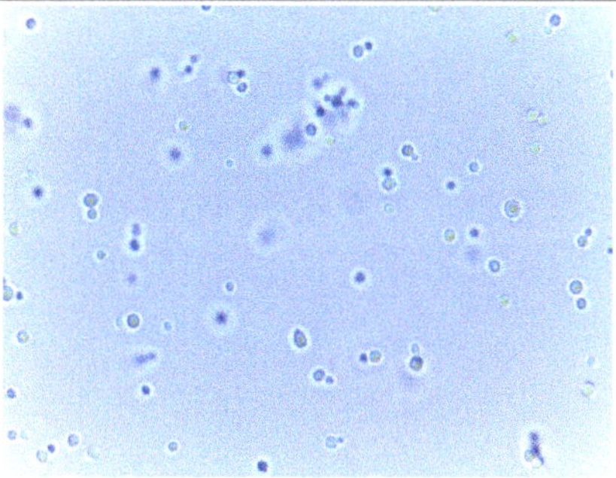 *D. hansenii*	Potential presence on grape surface.
Hanseniaspora/ Kloeckera	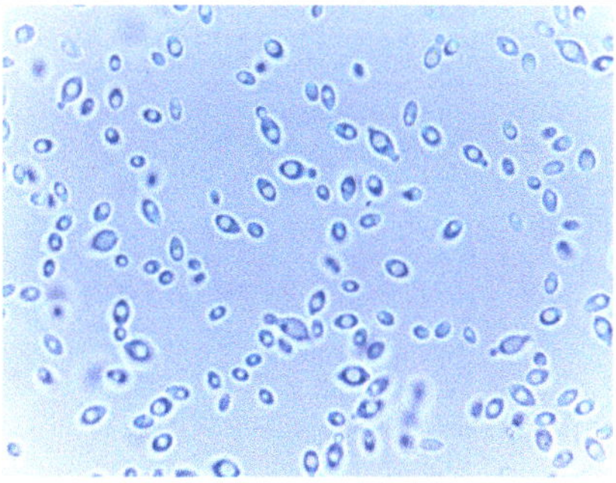 *H. uvarum*	Diffuse presence in grape berries surfaces;colonization of grape treated with organic and conventional treatment; occasionally present in cellar; spontaneous fermentation process; generally limited in the first few days of fermentation; weak ethanol tolerance; less efficient fermentation; highest ability for acetate formation.

(continued)

Table 1.1 (continued)

Kluyveromyces	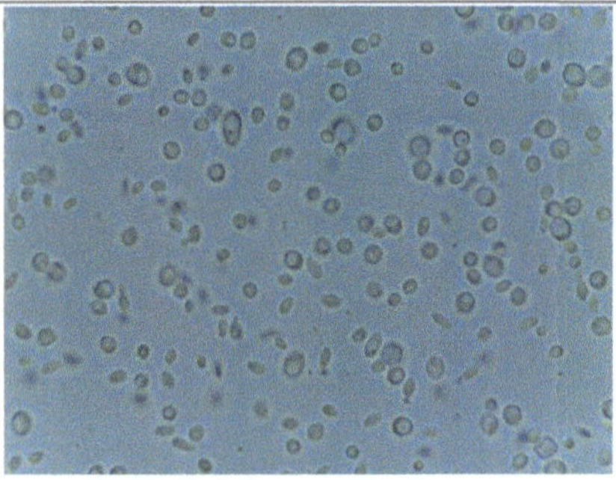 *K. marxianus*	Generally present in grape berries; found during various stage of fermentation.
Metschnikowia	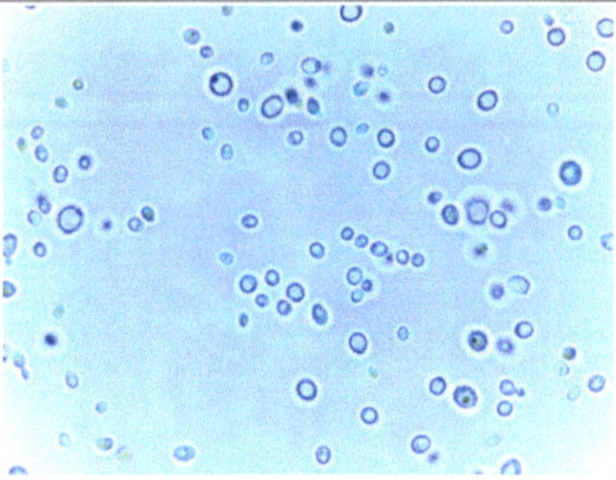 *M. pulcherrima*	Widely present in grapes and at harvest time; occurrence during various stage of fermentation (first few days); less efficient fermentation; High ability for acetate and acetate ester formation; antimicrobial activity; positive features to produce polysaccharides; glicosidase activity; ethanol reduction.
	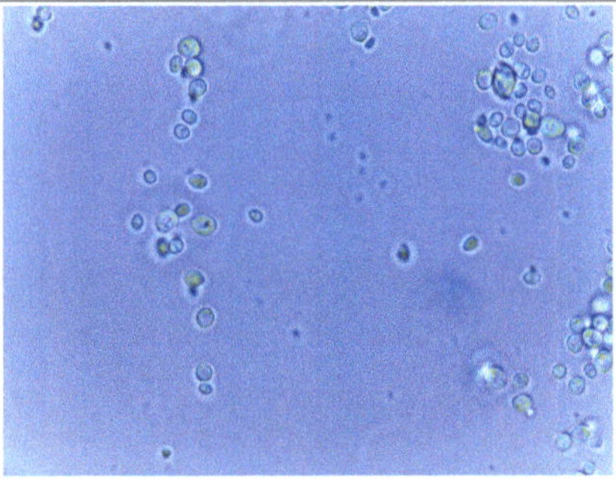 *M. chrysoperlae*	Present in grape berry; occasionally present in cellar.

(continued)

Table 1.1 (continued)

	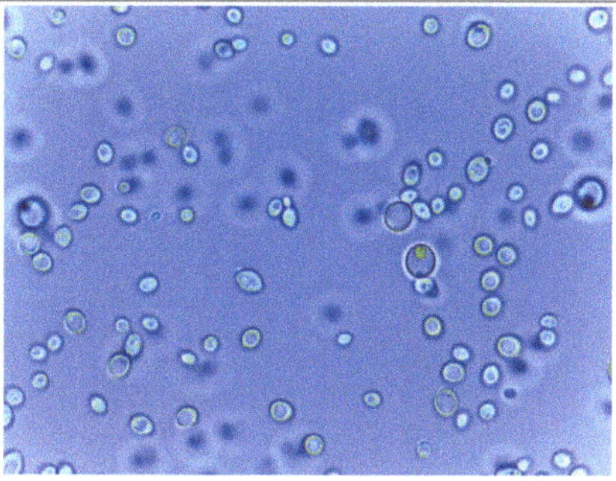 *M. fructicola*	It possesses divergent copy or rDNA GENE in comparison with other species;unambiguous identification.
Issakentia	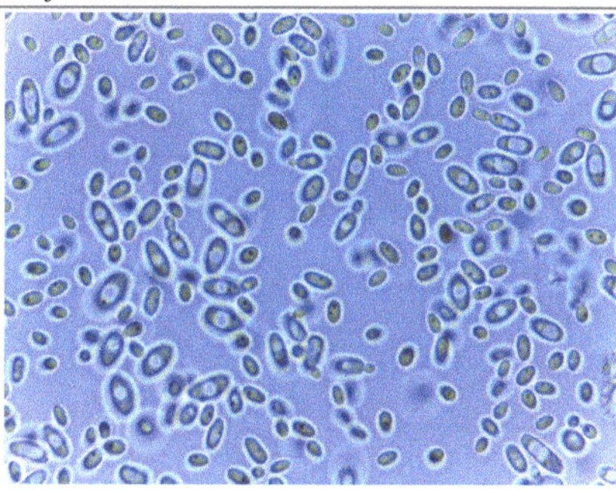 *I. terricola (Pichia terricola)*	Present in grape berry surface and in winery environment; present during various stage of fermentation; potential enzymatic activity.
Candida	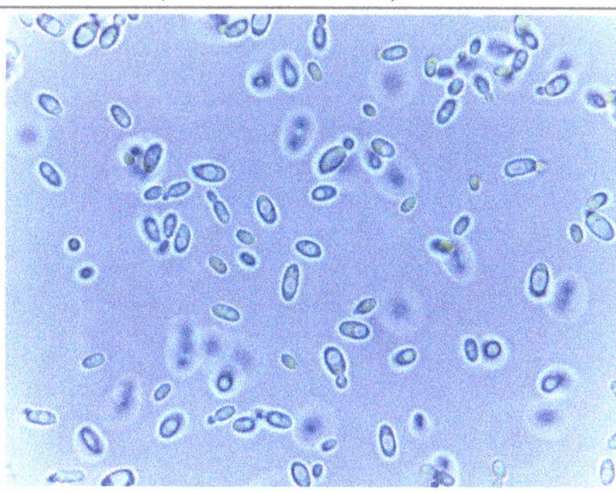 *C. californica*	Present in grape berry surface during various stage of ripening; fructophilic character.
Whickeramomyces	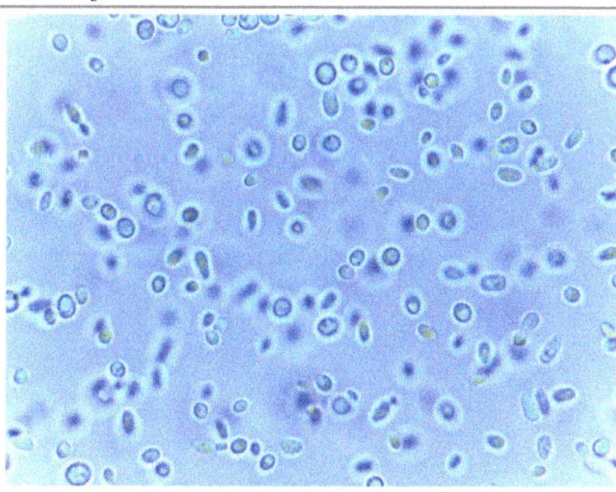 *W. anomalus (Pichia anomala)*	Less frequent in grape berry; Effect against *Dekkera/Brettanomyces* during fermentation or aging

(continued)

Table 1.1 (continued)

Rhodotorula	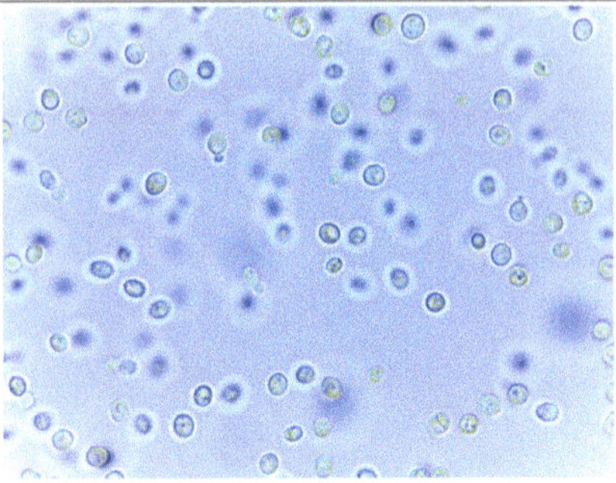 *R. mucillaginosa*	Oxidative yeast; colonize grape berries; ubiquitous yeast
Saccharomyces	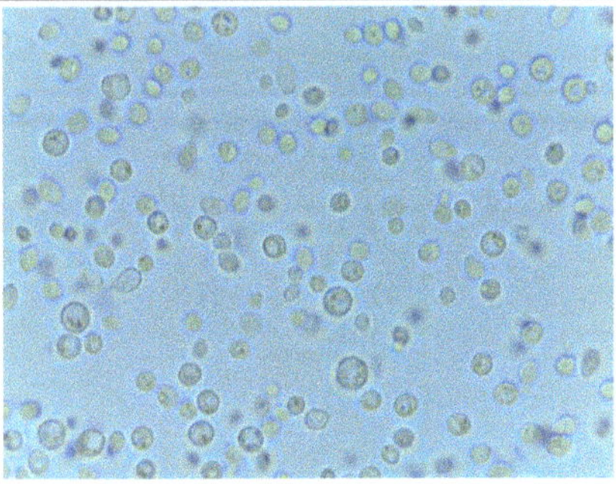 *S. cerevisiae*	Rarely present in grape berry; The main fermenting yeast involved in winemaking process; selected starter yeast; used to create new hybrid strains; strong cellar colonization; selective pressure by ethanol and SO_2 in winery is widely present (30-40% of total yeast population)
Schizosaccharomyces	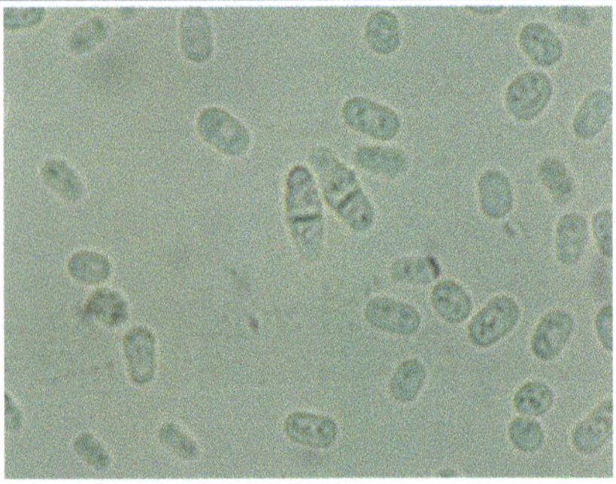 *S. pombe*	Present in grape/wine ecosystem; malo-alcoholic fermentation; increase in pigment production; High producer of polysaccharides
Zygosaccharomyces	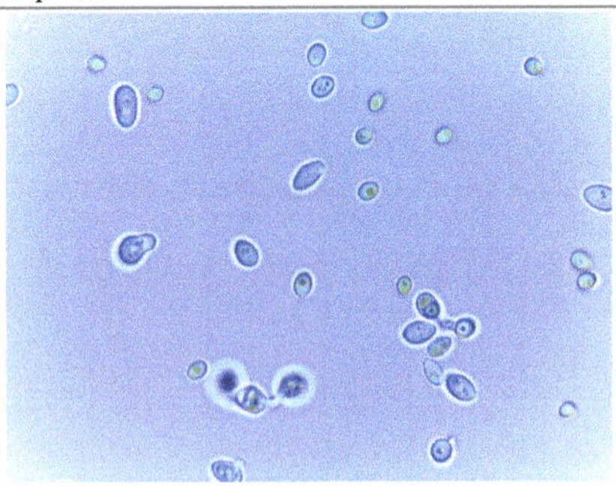 *Z. bailii*	Strongly fermentative yeast; occasionally present in unripe and overripe grapes; high ethanol tolerance

(continued)

Table 1.1 (continued)

Torulaspora	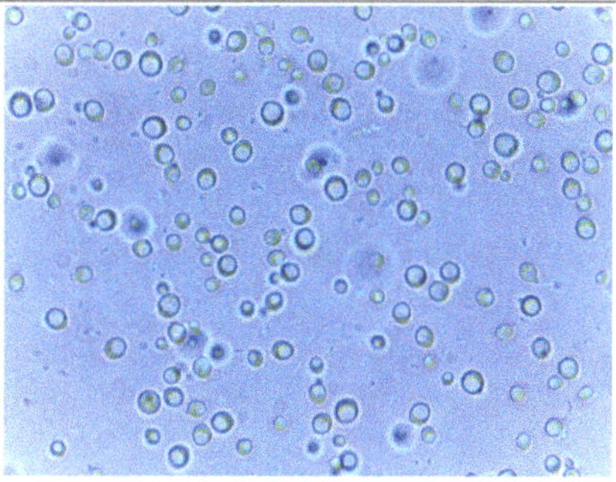 *T. delbrueckii*	Strongly fermentative yeast; increase of fruity aroma; high ethanol tolerant; low acetic acid production; high competitiveness with *S. cerevisiae*; frequently isolate in grape surface; impact of aromatic composition of wine; increase of acetate esters, thiols and terpenes and β-phenyl ethanol
Lachancea	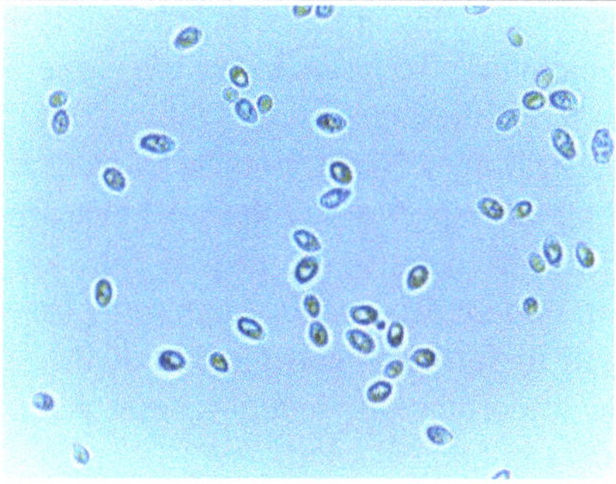 *L. thermotolerans*	Strongly fermentative yeast; most frequently isolated in grape; good competitiveness with *S. cerevisiae* with exception in limited oxygen concentration; large amount of acid lactic together with glycerol and β-phenyl ethanol

Oxidative Yeasts

Relatively to oxidative yeasts, these are present on the surface of the berries still before this reach maturity when there is a high sugar content and can be found in the early stages of fermentation. In the middle and last phase of grape ripeness, the oxidative yeasts decrease in concurrently to the detriment of nutrient availability due to the competition with other yeast species, but they are still widely present at harvest time depending on the agronomical practices (Fleet et al. 2002; Hernández et al. 2018).

Oxidative-Fermenting Yeasts

Hanseniaspora/Kloeckera species are the most abundant ascomycetes yeasts colonizing the grape surface of grape berries at harvest time. Regardless of the geographic distribution of winemaking areas, the presence and colonization of the yeasts *Hanseniaspora / Kloeckera* on grape surface is everywhere dominant over the other yeast species. Within the apiculate yeasts the species *Hanseniaspora uvarum (Kloeckera apiculata)* are the most frequent but other species such as *Hanseniaspora hosmophila* or *Hanseniaspora guilliermondii* can be found at lower concentration (Giorello et al. 2018). Other ascomycetes widely found at harvest time on grape surfaces are species belonging to *Pichia*, *Candida* and *Metschnikowia* genera. In this regard, several species have been described. Among the species described within *Pichia* genera, *Pichia membranifaciens*, *Pichia fermentans*, *Pichia kluyvery* and *Pichia kudriadzvii* (synonimum *Issatchenkia orientalis*) are the most widely isolated (del Monaco et al. 2014). In the *Candida* genus several fermenting and non-fermenting species were isolated from grapes. The most diffused fermenting species is *Candida stellata* that it was successively reclassified as *Candida zemplinina* and more recently enclosed in clade of *Starmerella* as *Starmerella bacillaris* (Duarte et al. 2012)

Within *Metschnikowia* genera, new species *Metschnikowia viticola* was recently isolated, studied and characterized from a Hungarian vineyard. From a genetic point of view *M. viticola* is well disconnected species within the genus *Metschnikowia*. However, very little is known about the ecological distribution of *M. viticola* and their frequency on grape berries (Peter et al. 2005; Brysch-Herzberg and Seidel 2015). Other many new species have recently been described in the *Metschnikowia* genus, including *Metschnikowia chrysoperlae* (Suh et al. 2004), *Metschnikowia fructicola* (Kurtzman and Droby 2001) and *Metschnikowia andauensis* (Molnar and Prillinger 2005). In these cases, there was a real difficult in the delimitation among new species and the well characterized *Metschnikowia pulcherrima* The experimental results obtained by Sipiczki et al. (2013) explain that the type strains of *M. andauensis* and *M. fructicola* possess divergent copies of the rDNA gene will lead to further investigations of the species concept in the clade. This support the importance of unambiguous yeast identification in any study of the yeast diversity in grape habitat.

Strongly Fermentative Yeasts
Regarding to the fermentative, higher alcohol tolerant yeasts, their colonization is related to the high nutrient availability resulting from grape damage that possess, besides much higher cell counts, wider species diversity than sound grapes (Barata et al. 2012). *S. bacillaris* may be present in higher numbers but its relative proportion also decreases in favour of higher fermentative yeasts such as *Zygosaccharomyces* spp., *Lachancea* spp. and *Torulaspora* spp., which, as mentioned above, may occasionally dominate the overall microbiota.

2.2 *Factors Affecting Yeast Community*

The composition and complexity of microbiota of grape berries depend on the interactions between individuals. The resulting consortium is generally stable over time and depending on several biotic and abiotic factors (Fig. 1.1). Relative to abiotic factors, the climatic and microclimatic conditions, including the effect of temperature, UV exposure, rainfall, sunlight and winds, can influence microbial populations.

Among biotic factors, microbial vectors, such as bees and wasps, can actively transfer yeasts on the grape surfaces (Francesca et al. 2012; Goddard et al. 2010; Stefanini et al. 2012). The microbial habitat associated with birds represents the

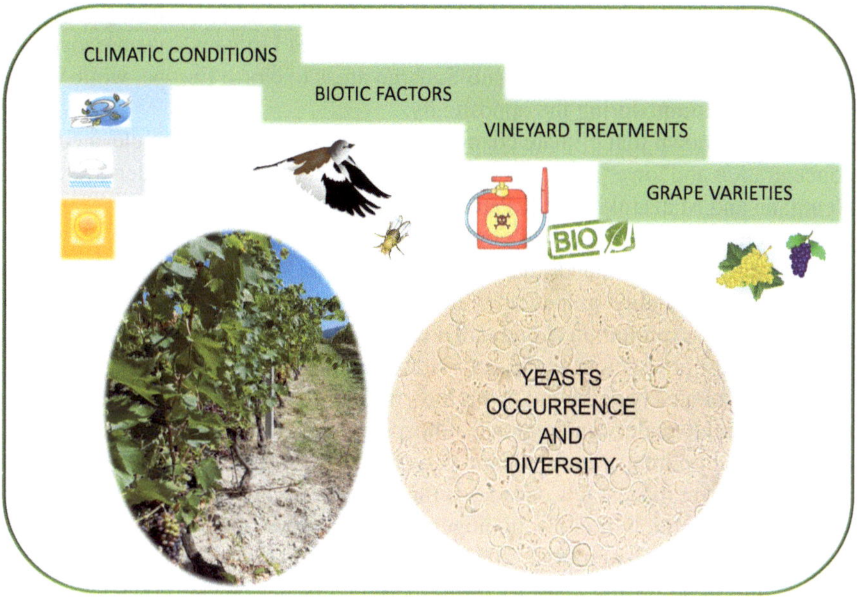

Fig. 1.1 Factors affecting yeast community in vineyard

object of several ecological surveys not only in applied food microbiology. Indeed, associations between wild birds and microorganisms have been studied mainly focusing on bacteria, whereas limited studies on yeasts are available (Cafarchia et al. 2006). The monitoring of bird movements allows the investigation about behavioural and demographical responses to a given environment (Riffell et al. 2006). The migration of birds includes a round trip to the resting areas and a return to the territories of nesting, which occur in autumn and spring respectively. These movements follow the seasonality of food resources. Since birds act as microorganism vectors, the analysis of the microflora they host may be important to evaluate the microbial diversity of the sites visited. From an applicative perspective, yeasts carried out by birds have not been deeply investigated. Nowadays, there is a growing interest of wine producers to perform winemaking employing 'autochthonous' strains which may ensure typical terroir characteristics (Capozzi et al., 2015). At this regard, it was recently reported the dissemination of oenological yeasts by vineyard inhabiting birds, mainly Black birds (*Turdus merula*), although no yeast with technological relevant traits was found in the few migratory birds analysed (Francesca et al. 2012). Those authors evidenced an issue related to the autochthonous status of yeasts, since they may not be indigenous in each environment. Yeasts may be moved at different distances depending on the vector type. Some studies provided evidences that insects such as honey bees disseminate *S. cerevisiae* strains until approximately 10 Km (Goddard et al. 2010), so that the investigation of migratory birds need clarifications for the associated movements of yeasts with the support of technological relevance. During migration, several sites are visited by birds because they represent important stop-over points. For example, during migration from Africa to Europe and vice versa, Lampedusa and Ustica islands are visited in spring when the direction is from sub-Saharan areas to North Europe, while Linosa is visited in autumn during the opposite fly.

It is well known that climate change is partly responsible for the elevated sugar concentration and lower acidity in grape berry and then in must (Godden et al. 2015; Neumann and Matzarakis 2014; Petrie and Sadras 2008; Teslić et al. 2018). This is strictly related to the microbial composition in grape berries. Vine sensitivity to weather properties (Holland and Smit 2014), narrow spatial surfaces suitable for producing high-quality grapes as wine industry raw material, and possibility of perennial plant exploitation (Lereboullet et al. 2014) are indicative of the need for a climate change assessment associated with winemaking. Despite the importance of the global climate change trend, from the vine grower/winemaker perspective, it is more essential to understand regional atmospheric conditions (Orlandini et al. 2009) and local microclimatic environment as well. Generally, increasing average global temperature over the last few decades is more than evident, as is the increasing temperature trend, although is not homogenous in every vine-growing region (Pielke et al. 2002; Van Leeuwen et al. 2013). For example, a significant growing season temperature trend for the majority of northern hemisphere wine-producing regions between 1950 and 1999, with an average increase of 1.26 °C, was demonstrated. However, there was also an insignificant trend in the majority of southern hemisphere wine regions, which emphasizes the necessity to focus study on smaller study areas.

Since climate modifications are vastly complex, examinations of simple temperature and precipitation values are insufficient to explain climate change trends. Therefore, several bioclimatic indices, such as Huglin index (Huglin 1978), Cool night index (Tonietto 1999), Winkler or growing degree day index (Winkler 1962), number of days with maximum temperatures higher than 30 °C (ND > 30 °C) (Ramos et al. 2008), number of days with precipitations <1 mm (Dry spell index, DSI) (Moisselin and Dubuisson 2006), etc. are commonly used in viticulture to provide an improved insight into climate change tendencies. However, the selected bioclimatic indices were mainly based only on-air temperature, as it has the strongest influence on overall growth, productivity, and berry ripening of the grapevine (Jones-Vaid et al. 2012).

Another important parameter influencing the grape microbiota is related to the water intended as rainfall. Indeed, moderate water stress may positively affect berry sugar accumulation during grape-growing season (Coombe et al. 1989), while increasing temperature advances phenological stages and speeds up sugar accumulation in grape berries (Jones-Vaid et al. 2012; Bonnefoy et al. 2013). The association of water stress together with increasing temperature later lead to the production of wines with higher alcohol content and other microbiological, technological, sensorial, and financial implications (Mira de Orduña et al. 2000). As direct consequence, increase of grape sugar content at harvest may cause slow or stuck alcoholic fermentations during hot years as well as alter sensory features due to the ethanol's tendency to increase bitterness perception (Sokolowsky and Fischer 2012), suppress the perception of sourness, and reduce astringency perception.

Concerning the total yeast counts, Combina et al. (2005) found that rainy years increased yeast presence. This climatic condition probably increases the berry volume and permits the release of juice in joint areas, such as the part between the pedicel and the berry, and higher exosmosis leads to nutrients on the grape surface. With careful and sound berry sampling, Čadež et al. (2010) found that colder harvests with higher rainfall lead to increased yeast counts. In contrast, Comitini and Ciani (2006) found ten-fold less total counts in years with high rainfall. In addition, the geographic location, grape variety and vineyard age and size can influence the composition and occurrence of microflora that are present on the surface of grape berries.

Another important aspect is related to vineyard chemical treatments. A lot of studies showed that agronomical practices, such as organic or biodynamic management can modify the microbiota of grape and must (Cordero-Bueso et al. 2011; Milanovic et al. 2013; Pretorius 2000; Mezzasalma et al. 2017). Some authors suggested that the occurrence of specific bacteria in must and wine influences wine characteristics and typicity (Belda et al. 2017a; Liu et al. 2017).

The main vineyard treatment studied is related with the use of pesticide treatments, mainly those against fungi (downy mildew, powdery mildew and grey rot). The studies are either based on analysing grapes after vineyard treatment, which do not exclude the influence of other factors, or from auto-enrichment fermentations which cannot be correctly extrapolated to evaluate the variations on berry microbiota. Conventional pesticides can produce a decrement in the yeast population and diversity in fermenting musts. Ganga and Martínez (2004) detected less diversity of non-*Saccharomyces* species, which was explained using fungicides against *Botritys cinerea*. Differently, there are discordant results on the effect of chemical treatments on *S. cerevisiae* presence on grapes. Ganga and Martínez (2004) did not find reduced *S. cerevisiae* occurrence after fungicide use while other investigations recovered lower numbers of this species (Regueiro et al. 1993; Van der Westhuizen et al. 2000). It is quite evident that the influence of chemical pesticides on microbiota of grape berry is related to other factors, such as climatic conditions or grape variety, which cannot be correctly extrapolated to evaluate the single effect on berry microbiota. About this concern, Ganga and Martínez (2004) detected less diversity of non-*Saccharomyces* species, which was explained using fungicides against *B. cinerea*, while Regueiro et al. (1993) and recovered lower numbers of these species. Milanovic et al. (2013) found that *Candida zemplinina (synonimus S. bacillaris)* and *Hanseniaspora* species colonised surface of grapes treated with both organic and conventional treatment, while *M. pulcherrima* was widely found in conventional samples and only occasionally in organic grapes.

A specific influence of grape varieties on indigenous yeast community of grape berries was found. Clavijo and collegues (2010) carried out an ecological survey of wine yeasts present on grapes growing in two vineyards located in the southern Spain (Serranía de Ronda region). They found that, although *Kluyveromyces (Lachancea) thermotolerans*, *H. guilliermondii*, *H. uvarum and Issatchenkia orientalis* (*Pichia kudriavzevii*) are the most frequent species, a specific distribution of strains was found in the three grape varieties studied. The influence of grape varieties on the indigenous yeast community of grape berries was also evaluated by Raspor et al. (2006) The frequency of occurrence of yeast species showed their preferences for certain grape varieties. The white grape variety mostly attracted pigmented *Basidiomycetous* yeasts belonging to the genera *Rhodotorula*, *Sporobolomyces* and *Cryptococcus* that dominated on all sampling locations. Differently, yeast populations isolated from the red grape surfaces belonged both to *Ascomycetous* and *Basidiomycetous* yeasts in the ratio of 1: 1.

In the last 10 years, due to the advances in metagenomics, it has become clearer and clearer that in general, plants host a wide array of bacteria and yeasts most of which are not cultivable and therefore are almost unknown at the taxonomic and metabolic levels. Such microorganisms interact with the plant organs and can influence plant nutrition, development, productivity, and stress responses (Bacon and White 2016).

Another important question regards the influence of grapevine cultivar on the grape microbiota. Recently, it was shown that some epiphytic bacteria were shared by aerial plant portions and the soil (Martins et al. 2013). This finding led them to propose that the physical proximity between soil and the plant might facilitate microbial migration through rain splash, winds, pollinators and other foragers, and parasites.

Moreover, any grapevine cultivar shows peculiar secondary metabolites, and most of these are concentrated in the fruit. Some of these metabolites have antimicrobial properties (Chong et al. 2009; Katalinić et al. 2010) and could influence the composition of grape microbiota both quantitatively and qualitatively. Based on these assumptions, it was hypothesized that each cultivar could have an active and specific role in the interaction with and selection of its microbial community (Mezzasalma et al. 2017).

2.3 Recent Methodologies for Detecting the Presence of Yeasts on the Grape Berry

To know the microbial composition in grape barriers and to further monitor their evolution during wine fermentation understanding the relationship between the microorganisms is of relevant importance in applied studies (Bokulich et al. 2014; Piao et al. 2015; Stefanini and Cavalieri 2018). The use of conventional methods including culture-dependent techniques, allow to analyze culturable fungi, yeasts, acetic acid- and lactic acid-bacteria associated with grape berries and wine. As well as in many other natural habitats, there are several viable but non-culturable wine microorganisms, that could not be studied under conventional laboratory microbial conditions, leaving an incomplete knowledge about the occurrence and dynamics of the microbial community involved in winemaking (Cocolin et al. 2013; Piao et al. 2015). Recent advances in sequencing technologies based on culture-independent techniques allow to capture a large proportion of cells (culturable, non-culturable and slight represented species) finding a complete microbial ecology picture (Bokulich et al. 2013; De Filippis et al. 2013; Abdelfattah et al. 2016). The beginning of massively parallel, high-throughput sequencing approach (sometimes referred to as next-generation sequencing) represent a revolution in applied microbial ecology research. Several platforms and chemistries exist, such as Illumina, 454/pyrosequencing, ion semiconductor, and nanopore sequencing, also if all employ nanotechnology to tether individual strands of DNA and detect the incorporation of individual nucleotides into each strand during polymerization events. Each system has its strengths and weaknesses, including different sequence read lengths, number of strands sequenced, and error rates – but each has been a stepping stone in advancing the ability to investigate the inner workings of the microbial community. This approach is appropriate for the food sciences, bringing manifold improvements over earlier mixed-microbial detection techniques (Bokulich and Mills 2012).

These new sequencing strategies rely on the analysis of a single core molecule DNA (and by transcription RNA) yet possess many applications for microbial ecology analysis. The first is amplicon sequencing whereby marker-genes are amplified from mixed genomic DNA by PCR, sequenced directly, and aligned against a reference dataset to identify the taxonomic composition of whole microbial communities. This same process can also be applied to RNA, by reverse-transcribed to cDNA, to profile the actively transcribing community within a sample. The taxonomic information provided by amplicon sequencing is frequently lower-resolution than that delivered by metagenome sequencing (which enables reconstruction of full-length marker genes) but is substantially higher throughput, facilitating exploration of massive numbers of unique microbial communities. With the availability of new metagenomic approaches the monitoring and composition of microbial populations can be better and faster described. In this regard, the relation between complexity of microbial community and geographical wine producing area represent a very interesting current topic. Using the metagenomic approaches, several studies showed the variation of the microbial community of grapes in function of regional distribution, (Gilbert et al. 2014b; Taylor et al. 2014; Morrison-Whittle and Goddard 2015; Pinto et al. 2015; Belda et al. 2017b). Moreover, the correlation among microbial complexity and organoleptic characteristics of wine was studied (Knight et al. 2015; Bokulich et al. 2016). An unambiguously explanation to the diversity of microbial communities among geographic locations is not currently known. In addition, recent studies showed that microbial populations found in musts may originate also from the environment surrounding the vineyard (Morrison-Whittle and Goddard 2018). Because of the observation of a putative microbial "terroir", the role and persistence of environmental microbial species in the wine fermentative process gained a renewed interest.

In this regard, from the application point of view, studies on indigenous yeasts strongly adapted to specific grape musts are growing, both to study the biodiversity associated to different geographic area and to select new indigenous strains associated with "terroir" (Capozzi et al. 2016; Zarraonaindia et al. 2015). These new concepts of microbial colonization and effectiveness showed that microbiomes associated with grapes and with the earlier stage of fermentation are biogeographically defined, illustrating that different regional wine profiles are related with specific microbial communities.

3 Yeast in Winery

In addition to natural habitats such as woods and agricultural areas near the vineyard, vineyard soil, vines and grapes, a relevant and consistent yeast community have found niches in man-made environments such as wine cellars. During the vinification process, grape juice and wine encounter a large area of equipment surfaces within wineries which may serve as important reservoirs of microorganisms that influence and contribute the final composition of the wines. For these reasons the surfaces of winery equipments become locations for the developments of so-called residential or winery microflora (Peynaud and Domercq 1959; Pretorius et al. 1999; Rosini 1984).

3.1 Diversity of Yeasts in Winery Environment

The role of winery environments in shaping the microbiota of wine fermentations and vectoring wine spoilage organisms is poorly understood at the systems level. Indeed, although the presence and importance of winery yeasts have been known or surmised for a long-time year (Peynaud and Domercq 1959) their actual contribution to must fermentations has been poor investigated and somewhat ignored (Pretorius et al. 1999). However there are several factors that potentially determine a stable colonization of yeasts in this anthropized environment (Fig. 1.2).

Winery equipment, including crush/press equipment, valves, collectors and barrels, for its difficult to clean, become useful for microbial adhesion and biofilm production and consequent potential sources of contamination. In this regard, one of the most important features that characterize the winery microbiota is the survival and the modality of colonization over the course of harvest campaign (before, during and after grape harvest). Therefore, to track the occurrence of equipment microbiota and evaluate the fluctuation of yeast population, samplings before, during, and after grape harvest are an important aspect to be investigated.

The pre-harvest yeast communities represent the resting state of the winery and play an important role since these is the first population encountered by fresh grape juice prior to fermentation. The composition of this microbial community may play an important role and can impact on wine fermentation qualities downstream.

Most of the studies on the occurrence and yeasts colonization of cellar were carried out on *S. cerevisiae* the main agent involved in alcoholic fermentation. Indeed, colonization of winery surfaces by *Saccharomyces* has been widely reported and it

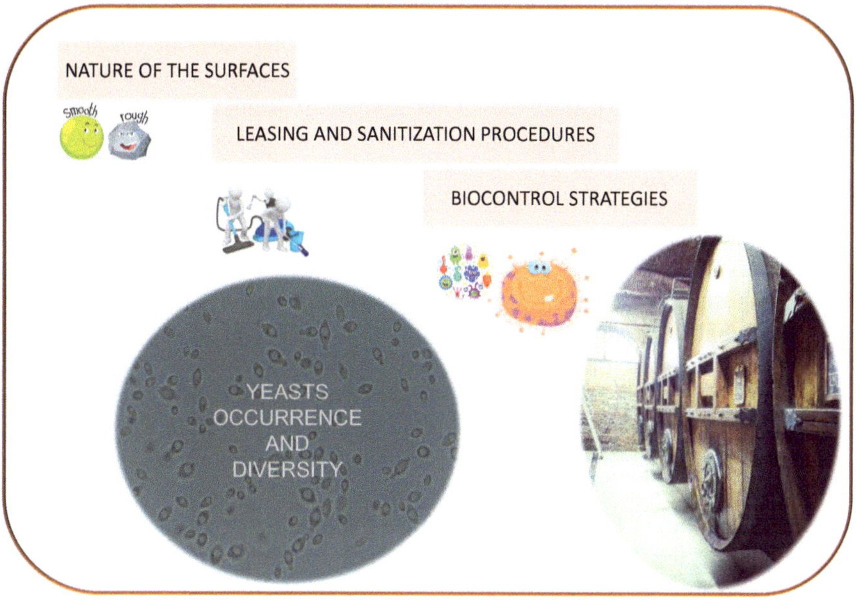

Fig. 1.2 Factors affecting yeast community in winery

is probably an important source of this yeast in wine fermentations, particularly in non-inoculated wines (Bokulich et al. 2013). Studies, investigating on the yeast biota of winery, showed the constant presence of *S. cerevisiae* and had identified this species as dominant on winery surfaces at pre-harvest time (Ciani and Rosini 1986; Ocón et al. 2010). Indeed, the winery environment is colonized by many cells of *S. cerevisiae*, which go through innumerable generations during fermentation for each vintage (Rosini 1984; Ciani and Rosini 1990; Ocón et al. 2010). Here may exist a significant selective pressure on the *S. cerevisiae* population of the winery by factors such as ethanol and SO_2 (Cocolin et al. 2004). However, even if *S. cerevisiae* is the most abundant species in winery environments may account only 30–40% of the total yeast population, other non-*Saccharomyces* yeast species may colonize the winery surfaces and equipments depending on the spatial variation in the winery surface. Indeed, some crush equipment (hopper, elevator, crusher, and press samples) that entering almost exclusively in contact with the grapes are colonized by yeast-like (e.g., *Aureobasidium pullulans*) and yeast genera that colonize the grape surfaces such as *Hanseniaspora*, *Candida* or *Metschnikowia* (Bokulich et al. 2013). The occurrence and persistence of non-*Saccharomyces* in the cellar environment was well documented (Ciani and Rosini 1986; Ocón et al. 2010). A more recently work, using identification methods at the strain level, found a large number of isolates belong to *S. bacillaris*, *H. guilliermondii* and *H. uvarum* demonstrating the persistence of non-*Saccharomyces* yeast strains from year to year in the cellar.

Indeed, some strains of these three non-*Saccharomyces* species were found in the must for two consecutive years and found also in cellar environment before the second harvest indicating the persistence of these yeasts in this environment (Grangeteau et al. 2016). In fermentation equipment samples (fermentors, hoses, filters, and pumps) that all deal strictly with fermenting and fermented wine, *S. cerevisiae* is largely present, together with other fermenting yeasts such as of *Pichia kudiavzevii* (synonimum *Issatchenkia orientalis*), *Torulopsis bacillaris* (synonymus of *Candida zemplinina*) and *Pichia* spp. These non-*Saccharomyces* yeasts can explain for 60–70% of total yeast biota which colonizes the winery surfaces and their role in this context has been little investigated (Ciani and Rosini 1986; Ocón et al. 2010; Sabate et al. 2002; Bokulich et al. 2013). Some oenological practices such as cold maceration prior to fermentation may affect the yeast ecology during wine fermentation in favour of non-*Saccharomyces* species. Hierro et al. (2005) found that cold maceration favoured the presence of *H. osmophila*, *Candida tropicalis* and *Zygosaccharomyces bisporus*, the only species isolated from the unripe and overripe fermentations after cold maceration.

Another important feature of winery biota that involves ecological and technological aspect is the flux of specific *S. cerevisiae* strain from winery surface to must fermentation and vice versa.

A series of studies have found an effective flow of strains of *S. cerevisiae* from the cellar equipments and surfaces to fermentation musts. Rosini in 1984 in a new pilot scale winery demonstrated the occurrence of a flow of *S. cerevisiae* cells from the winery surfaces to freshly pressed grape musts, and vice versa. These results were then confirmed by Costantì et al. (1997) that found a competition of resident winery yeasts and pure *S. cerevisiae* starter cultures. A 6 year follow-up study carried out in a new built winery showed that indigenous winery resident *S. cerevisiae* strains competed with commercial strains inoculated in other fermentation tanks of the cellar (Beltran et al. 2002). The contribution of winery-resident *S. cerevisiae* strains to spontaneous grape must fermentation was shown under real vinification conditions. The *S. cerevisiae* strains colonizing the winery surfaces were the ones that conducted the natural must fermentation (Ciani et al. 2004). Other investigations found that specific *Saccharomyces* strains become established on winery surfaces, resulted in repeatable detection over multiple years in uninoculated wines (Santamarıa et al. 2008; Blanco et al. 2011; Ciani et al. 2004).These results support the role of winery as a man-made niche of *S. cerevisiae* and a possible reproducibility, as well as regionality, of wine sensory characteristics produced at a given winery. In this regard the selectivity of the winery environment (winery effect) may have a selective pressure towards some enological characters as maximum ethanol production (ethanol resistance) fermentation rate and SO_2 resistance of *S. cerevisiae* population (Cocolin et al. 2004). On the other hand, the role of *S. cerevisiae* of winery environment on the specificity of wine sensory profile at regional level remains unclear.

3.2 *Factors Affecting Yeast Community in Winery*

The extended of the development of a residential microflora will depends by several factors such as nature of the surfaces (irregular, unpolished surfaces, cracks and welds) and cleaning and sanitization procedures and possible biocontrol procedures (Fig. 1.2). The nature of the surfaces may strongly influence the colonization of yeast species and determines their persistence from one harvest to another.

On the other hand, the modalities of cleaning and sanitization procedures also influence the quantitative presence and relative abundance of the different yeasts species. However, several works reported that also in well cleaned wineries, the widely presence of microorganisms and specifically of yeast biota, was found. However, under normal correct procedures of cleaning and sanitization the presence of spoilage-related microorganisms (e.g. *Brettanomyces* spp.) was undetected or detected at very low levels (Bokulich et al., 2013; Ocòn et al., 2013).

Classical studies on spoilage yeasts by Van der Walt and van der Kerken (1961) on *Brettanomyces* spp., Rankine and Pilone (1973) and on *Zygosaccharomyces bailii*, Peynaud and Domercq (1955) on *Saccharomycodes ludwigii* have demonstrated that they may be winery contaminants, even if most results from literature suggest that their prevalence is low. Chatonnet et al. (1992) were the first authors to identify oak barrels as an ecological niche for *Dekkera/Brettanomyces* spp., which become more dangerous with repeated use. This suggests that barrel sanitation and sulfite utilization (sulfur burning in empty barrels) is not enough to eliminate *Dekkera/ Brettanomyces* spp., which develop during the lifetime of the barrel. Connell et al. (2002) also recovered *Dekkera bruxellensis* from air samples of crush, tank, barrel, and bottling line areas using BSM medium (Millipore) followed by a filter-based chemiluminescent in situ hybridization technique. However, the primary source of these yeasts remains obscure. A recent study investigating on the occurrence of *Brettanomyces bruxellensis* found a flux of isolates form grapes to winery (Comitini et al. 2019). Currently, some of the procedures that applied to limit the risks of *Brettanomyces/Dekkera* colonization in wineries and wines are not particularly appropriate for use during wine ageing. This has led to increased interest in the exploration of yeasts that can counteract the activities of these undesired microorganisms in wine (Comitini et al. 2004). Investigations on biocontrol topic, relative to killer yeasts as producers of mycocins that can neutralize the activities of undesired microorganisms in wines represent a valid strategy for the control of these undesired yeasts (Druvefors et al. 2005).

4 Alcoholic Fermentation

Wine fermentation is typically carried out by a complex evolution of microorganisms involving both yeasts and bacteria. During the years, a lot of studies extensively examined the succession of yeasts that occurs during the spontaneous fermentation in must as non-sterile source. Now, it is well established that together with *S. cerevisiae*, non-*Saccharomyces* species actively participate during the alcoholic fermentation. In the past these non-*Saccharomyces* wine yeasts were negatively considered because of reduced fermentation power, high production of undesired products that affects the aromatic profile of wines. For these reasons, the use of selected *S. cerevisiae* as starter culture was a common and widely diffused winemaking practice to control the fermentation process and give the desired characteristics to the wines. More recently, several studies have been revaluated the role of non-*Saccharomyces* yeasts during alcoholic fermentation and their metabolic impact on the analytical sensory characteristics of white and red wines (Benito et al., 2014). In this regard, there are a growing interest on the use of controlled mixed fermentation with selected non-*Saccharomyces* wine yeasts tin co-culture or sequential inoculation.

4.1 Spontaneous Fermentation

Grape bunches, the primary substrate of winemaking process, are perhaps the most obvious potential source of microbial diversity of spontaneous grape juice fermentation. However, the winery, as previously indicated as a man-made ecological niche, may play also an important role in spontaneous grape juice fermentation particularly regarding to the fermenting yeasts. Indeed, a serial of ecological surveys of the yeast flora associated with spontaneous fermentation of grape juice in almost all the geographical winemaking areas revealed a sequential occupation of the substrate: initially apiculate yeasts (*Hanseniaspora*, *Kloeckera*) take over, after 3–4 days they are replaced by *S. cerevisiae* (Martini 1993; Pretorius 2000). While the first one yeasts are abundant on the grape surfaces at harvesting time, *S. cerevisiae* (*Saccharomyces uvarum*) species colonize the winery surface where resulted the most widely diffused species during the different stages of winemaking (before during and after the fermentation). With ethanol increasing *S. cerevisiae* the higher resistant to alcohol is the first explanation to this substitution but other contributing factors may be involved.

In this sequential occupation of the grape juice by apiculate-elliptical yeasts, during the various stages of fermentation it is possible to isolate other yeast genera, such as *Starmerella*, *Candida*, *Pichia*, *Zygosaccharomyces*, *Schizosaccharomyces*, *Torulaspora, Lachancea* (*Kluyveromyces*) and *Metschnikowia* (Fleet et al. 1984; Pardo and Serrano 1989; Belda et al. 2015, del Carmen Portillo and Mas 2016; Garofalo et al. 2016). The indigenous non-*Saccharomyces* yeasts, are present in the grape juice in high numbers in active growth state, which gives them a competitive edge (Cray et al. 2013).

The growth of non-*Saccharomyces* species belonging to the genera *Kloeckera/Hanseniaspora*, *Starmerella* and *Metschnikowia* is generally limited to the first few days of fermentation, because of their weak ethanol tolerance and less efficient fermentation. Other more ethanol tolerant fermenting yeasts such as

Torulaspora delbrueckii, *Lachancea theromotolerans* and *Zygosaccharomyces* spp. are generally present less frequently but occasionally they were found at higher concentration. (Clavijo et al. 2010; Zott et al. 2008; Jolly et al. 2003; Garofalo et al. 2016).

Therefore, spontaneous fermentation contains a mixture of yeast species but, as the fermentation progresses, the environment becomes more selective and the dominance of *S. cerevisiae* is expected and desired. However, *S. cerevisiae* populations showed a genetic variability with the presence of more than one strain. Indeed, several studies carried out in various winemaking areas showed that different strains of *S. cerevisiae* are involved during the fermentation process. In this regard, some *S. cerevisiae* strains are present at high relative amount and were able to dominate the alcoholic fermentation and are denominated "dominant" or "predominant" strains while other strains occur at lower relative amounts and are defined "secondary" strains. The association of the dominant *S. cerevisiae* strains in spontaneous fermentation with the environment is not well defined and it was linked both winery (Frezier and Dubourdieu 1992; Guillamón et al. 1996; Ganucci et al. 2018) or grape variety (Blanco et al. 2006; Schuller et al. 2012).

As reported above, in a spontaneous fermentation several yeast species and strains coexist interacting with each other and environment factors. The specific fitness of yeast species and the evolution occurring in grape must during alcoholic fermentation toward more selective conditions determined the sequential occupation of the substrate and the progressive dominant presence of *S. cerevisiae*. Similarly, within *S. cerevisiae* strains the dominant strains were selected on the bases of its fitness toward the specific environmental factors. Currently, there are two lines of research that investigate the origin and occurrence of *S. cerevisiae* dominating spontaneous fermentation and that characterize the analytical sensory characteristics of the wine of a given territory or winery.

As previously reported, several recent investigations (Bokulich et al. 2014; Gilbert et al. 2014a; Taylor et al. 2014; Morrison-Whittle and Goddard 2015; Pinto et al. 2015; Belda et al. 2017a, b), showed evidences for a relation between microbial community of grapes and geographic distribution and organoleptic characteristics of fermenting musts (Knight et al. 2015; Bokulich et al. 2016). This adaptation, however, could be due to the selective pressure performed during the winemaking process in the winery. Some factors such as ethanol, temperature, SO_2 and others can play a fundamental role in the yeast species and strains selection during spontaneous fermentation and determining the dominant *S. cerevisiae* strains that for these reasons, colonize the man-made winery environment.

4.2 Factors Affecting the Occurrence and Succession of Yeast During Spontaneous Fermentation

There are two principal factors determining the evolution of yeast community during the spontaneous fermentation: (i) the quantitative occupation of the substrate by yeast species; (ii) the progressive increase of ethanol concentration. These features play a key role on the yeast species succession in spontaneous grape juice fermentation. However, the dynamics among the yeast species present during fermentation are more complex and strongly influenced by their interactions and other environmental factors (Ciani and Comitini 2015).

Temperature of grape juice fermentation is one of the most influencing factors on yeast species dynamics. Indeed, the presence and permanence of non-*Saccharomyces* yeast species during fermentation is affected by temperature. Indeed low temperatures (10–15 °C). increased tolerance to ethanol of *K. apiculata* and *C. stellata* (formerly *S. bacillaris*) (Erten 2002; Gao and Fleet 1988: Ciani et al. 2006; Mendoza et al. 2007; Ciani et al. 2010). Such increase in ethanol tolerance of non-*Saccharomyces* yeasts at low temperatures appear to be the major factor that affects their stronger contribution in this condition.

Another factor that regulates the presence and occurrence of yeast species during spontaneous fermentation is the availability of oxygen. Reduced oxygen availability during fermentation may have a key role in yeast-yeast interactions. Indeed, the limited availability of oxygen could explain in part the reduced competitiveness showed by *K. thermotolerans* and *T. delbrueckii* toward *S. cerevisiae* (Hansen et al. 2001).

In the yeast species interactions during wine fermentation cell-to cell contact appears to be also involved. Indeed, in the presence of high concentrations of viable cells of *S. cerevisiae*, the growth of *T. delbrueckii* and *K. thermotolerans* is inhibited (Nissen and Arneborg 2003; Nissen et al. 2003; Arneborg et al. 2005).

The quorum-sensing-like phenomena could also be involved in some yeast interactions during spontaneous wine fermentation and the identification of active molecules and their influence on gene expression of yeast co-culture deserves to be investigated. In this regard, recent investigations on the putative quorum-sensing molecules as 2-phenylethanol, tryptophol, and tyrosol have begun to elucidate the mechanisms and role of quorum sensing in yeast under winemaking condition (high cell density or under low nutrient conditions) (Zupan et al. 2013; Williams et al. 2015). In addition to ethanol, acetic acid, medium chain fatty acids, acetaldehyde and the synergistic action of their combinations, could play an important role on the inhibitory mechanism that can occur in wine fermentation (Bisson 1999; Fleet 2003).

The production of toxic compounds from *S. cerevisiae* has also been hypothesized as a cause of the early death of *H. guilliermondi* in mixed fermentations (Pérez-Nevado et al. 2006). Indeed, several compounds produced by yeasts during must fermentation may become inhibitory to other yeast species or strains. Between them, secretion of killer toxins by specific yeasts represents an efficient tool to eliminate competitors without direct cell-to-cell contact. Yeasts with killer phenotype secrete protein or glycoprotein that generally kill sensitive cells in a two-step receptor-mediated manner: First, there is a specific bind to primary cell wall receptors; secondary the killer toxin translocate to the plasma membrane where they interact with secondary receptors or enter susceptible cells, thus exerting their cytocidal effect through different mode of actions. In addition to killer proteins other proteinacenous compounds have been found in the yeast–yeast and yeast–bacteria interactions in wine fermentations.

Indeed, it was found that certain *S. cerevisiae* strains produce proteinaceous compounds that are active against malolactic bacteria (Comitini et al. 2005; Osborne and Edwards 2007), peptides with molecular mass less than 10 kDa, that inhibit the growth of some non-*Saccharomyces* such as *H. guilliermondii*, *T. delbrueckii*, *K. marxianus* and *K. thermotolerans* (Albergaria et al. 2010; Peña and Ganga 2018). However, also if the identity of these antimicrobial peptides remained elusive and need some deepness, the possibility of using them as natural biopreservatives in alcoholic fermentations could be an interesting alternative for the microbial control of winemaking process.

4.3 **S. cerevisiae** *Inoculated Fermentation*

From the ecology surveys carried out in different winemaking environments *S. cerevisiae* is a minority species and it is difficult to isolate from vineyard soil or the surface of ripe grapes, while it is the dominant yeast species of winery and its equipment (Martini 1993). Indeed, *S. cerevisiae* is generally found in association with the production of alcoholic beverages and for this reason it is defined a "domesticated" species, strongly specialized for fermenting high sugars substrates. For their fermenting features and oenological aptitude, *S. cerevisiae* is the species that conduct and determine the rightness of the fermentation process characterizing the chemical and sensory profile of wine. However, for the long time the fermentation of grape juice was carried out without yeast starter strain inoculation and spontaneous must fermentation occurred. After 1960 scientific and technological improvements allowed the diffusion of active dry yeasts commercial preparation belonging to *S. cerevisiae*. The diffusion of commercial starters in active dry form was one of the most significant technological advances in winemaking. As direct result, the quality and quantity of wine production were highly improved, as the winemaking process was controlled and safe (Heard and Fleet 1985; Henick-Kling et al. 1998). The introduction in winemaking process of selected and efficient strains announced the concept of innovations revolutionizing the wine industry and market. Nowadays, the large-scale wine production, where rapid and reliable fermentations are essential for wine flavour and predictable quality, the practice of the inoculation of selected pure starter strains of known ability is a common practice. However, the current challenge of applied research in biotechnology is the producing new yeast strains with even more reliable performance, reducing processing inputs, and facilitating the production of peculiar and high-quality wines (Pretorius 2000). The forces of market and technology continue to challenge the tension between tradition and innovation. On the one hand, it is evident the tendency to use commercial strains that guarantee controlled processes, from another hand it is still recognized the potentially of native yeasts to obtain distinctive features. Indeed, despite the immense wealth of natural yeast diversity, the extremely selective and specific conditions of industrial fermentations sometimes require a combination of phenotypic traits that might not be commonly encountered in nature. In this picture, a lot studies focused on the isolation, manipulation and develop of novel *S. cerevisiae* strains tailored for a specific wine product that bring greater complexity to wine than strains currently available to the industry (Bellon et al. 2013). The most intuitive way to generate artificial diversity in yeasts is based on genetic manipulation, to artificially increase the already existing yeast diversity and generate variants that may perform better in industrial settings than the strains that are selected in natural environments. A specific approach is the genetic engineering reshuffle in selected strains, applied to modify single genes by introducing, disrupting or modulating enzymatic key-steps of metabolic networks (Santos and Stephanopoulos 2008). However, wine yeasts show complex and continuous variation for most industrially relevant traits, such as stress-related response, fermentative performance and profile of secondary metabolites and this approach is ineffective in modifying quantitative traits. Furthermore, genetically engineered strains are opposed by regulation No 1829/2003 of the European Parliament and of the Council, which prohibits GMOs in foodstuffs. In this picture, genome hybridization techniques, understood as natural and random rearrangement between strains exploiting the natural phenotypic variation within wild yeast populations, is a valid biotechnological tool to create genetically non-modified organisms (non-GMO) with improved phenotypes (Steensels et al. 2014). The hybridization (both sexual and asexual) produce random gene arrangements that are then tested and selected through screening procedures and technological simulations. *Saccharomyces* sensu stricto interspecific hybrids have been found in different fermentation processes: in addition to *Saccharomyces. pastorianus*, present in lager brewing, other hybrid strains have also been described from wine and cider (Masneuf et al. 1998; Groth et al. 1999; Naumova et al. 2005). For example, the type strain of *Saccharomyces bayanus*, originally isolated from beer, has recently been suggested to be a hybrid between *S. cerevisiae* and *S. bayanus* based on the presence of subtelomeric repeated sequences and genes (Nguyen et al. 2000; de Barros Lopes et al. 2002; Nguyen and Gaillardin 2005).

In general, the strength of this approach is the production of new fermenting strains that acquire physiological properties from both parents. The principal improved phenotype traits concern the ethanol and acetic acid tolerance, the copper resistance, the glycerol production, the high osmotic stress resistance, and the utilization of xylose (Brown and Oliver 1982; Aarnio et al. 1991; Adamo et al. 2012; Kutyna et al. 2012; Ekberg et al. 2013; Shen et al. 2012).

The prospects to obtain superior industrial wine yeasts are extremely bright. Yeasts offer unique advantages for strain improvement: they combine sexual and asexual life cycles, they can be easily cultivated in high numbers, and genetic transformation is often easy. Moreover, most strain improved by hybridization could be profitable involved in the rectification of fermentation disorders in spontaneous fermentations has been recently described in the literature (König and Claus 2018). Recent investigations have provided convincing evidence that fermentation problems can be overcome when must fermentations are successively performed with *S. bayanus* and the triple hybrid *S. cerevisiae* × *Saccharomyces kudriavzevii* × *S. bayanus*.

The triple hybrid uses amino acids as a nitrogen source in the absence of ammonium and it also exhibits a fructophilic character with an enhanced uptake of fructose in comparison to glucose. This applicative example revealed that hybrid strains could be a promising tool for winemakers not always for the creation of novel wine types with desired sensory characteristics under more challenging conditions, but also *ex post* to solve fermentation problems during spontaneous fermentation or especially when the composition of the must components is not optimal because of critical climatic or soil conditions.

Several yeast hybrid strains are already commercially available, such as the strain "Oenoferm® X-treme" that is a GMO-free hybrid yeast obtained from the protoplast fusion of two different *S. cerevisiae* strains; the strain "Cross Evolution" a natural cross hybrid between *S. cerevisiae* yeasts; strain NT 202 is a product of the yeast hybridization program; the strain S6U a hybrid of *S. cerevisiae* × *S. bayanus* and the strain VIN7 an allotriploid interspecific hybrid of a heterozygous diploid complement of *S. cerevisiae* chromosomes and a haploid *S. kudriavzevii* genomic contribution (Borneman et al. 2012, 2016; Hart et al. 2016; Pérez-Torrado et al. 2018).

4.4 Controlled Mixed Fermentation

Although most research on wine microbiology has focused on *Saccharomyces* yeasts, particularly *S. cerevisiae*, there is a growing interest in studying and characterising non-*Saccharomyces* yeasts for development of starter cultures.

As already reported above, in the past wine was produced through spontaneous non-controlled fermentation by microflora residing on grapes, vineyard and in winery. In this way, many yeast species, not always identified, contribute to wine fermentation to obtain not reproducible and determining sometime failed results. Afterwards, the use of pure *S. cerevisiae* starter cultures, which establishes a dominant yeast population from the beginning of fermentation, has enabled modern wineries to produce predictable and reliable wines with established quality standards.

The sensory profile of wines produced by monoculture-inoculated fermentations differ substantially from those that are spontaneously fermented, principally for the biochemical composition of un-inoculated wines, which are distinctly different from wine obtained by pure fermentations (Varela et al. 2009). Certainly, spontaneous fermentations imply a higher risk of sluggish and/or incomplete fermentation and spoilage trend if compared to pure processes characterized by many default desirable characteristics but less complex flavour profiles (Jolly et al. 2014; Ugliano and Henschke 2009).

On the basis of this view, in the last years wine researchers have explored the controlled use of non-*Saccharomyces* starter cultures in addition to commercial and conventional *S. cerevisiae* starters. It is certainly known that non-*Saccharo*myces yeasts are generally unable to complete alcoholic fermentation on their own, for this they are used in pairs with *S. cerevisiae* wine strain. This can be achieved by inoculating first with the non-*Saccharomyces* yeast followed by a wine strain of *S. cerevisiae* to finish the fermentation. This is known as sequential inoculation, as opposed to simultaneous inoculation, in which two or more yeasts are added at the same time (Ciani et al. 2010; Jolly et al. 2003). In this regard, The use of non-*Saccharomyces* yeast in winemaking has grown enormously in the last years and several investigations have been carried out to better understand the impact of non-*Saccharomyces* strains on the chemical and sensorial properties of wine (Ciani and Maccarelli 1998; Comitini et al. 2011; Renault et al. 2015; Swiegers et al. 2005). In this regard, it is well established a wide intraspecific variability of oenological characters, peculiar positive oenological traits and, above all, a different behaviour in co-culture due to interactions with *S. cerevisiae*. All these aspects have highlighted a significant role of these non-conventional yeasts in determining the analytical and sensory profile and the aromatic complexity of wine.

*4.5 Non-*Saccharomyces *Yeasts as Biotechnological Tool*

The controlled multistarter fermentation with *S. cerevisiae* is the most profitable modality to use of these selected non-conventional wine yeasts. Several objectives can be pursued with the use of controlled mixed cultures with non-conventional yeasts: (i) enhancement of flavour and aroma complexity; (ii) distinctive features; (iii) ethanol reduction; (iv) control of spoilage microflora.

4.5.1 Aroma Enhancement

The contribution of selected non-*Saccharomyces* yeasts during wine production will be provided focusing the attention on the principal features such as aromatic profile, the color stability and polysaccharides production, the modulation of acidity, the ethanol reduction and concerning about antimicrobial activity toward undesired strains.

Certainly, the aromatic profile is one of most important traits that contribute to the quality of wine. As in many foods, wine aroma is composed by 100 s of different compounds with concentrations that can vary between 10^{-1} and 10^{-10} mg/mL. The balance and interaction of all of them determine the wine aromatic quality (Padilla et al. 2016). In literature, several works investigated on the production of volatile aromas, such as esters by different non-*Saccharomyces* yeast species that positively contribute to enhance the aroma profile of wines. (Moreira et al. 2008; Rojas et al. 2003; Viana et al. 2008). Between them, ethyl acetate and isoamyl acetate is often produced by yeast strains in natural grape juice during fermentation. For example, *Kloeckera apiculata* exhibited the highest ability for acetate formation; *Hansenula subpelliculosa*, *Kluyveromyces marxianus*, *T. delbrueckii* and *S. cerevisiae* produced intermediate levels and *P. membranaefaciens* and *C. guilliermondii* very low levels of the two esters. In general, the high production of esters did not always negatively influence the aromatic profile of wines Moreira et al. (2008).

Several applied studies focused the attention on *T. delbrueckii*, a species low frequently isolated on grape surface but one of the most studied species to increase flavour and aroma complexity in alcoholic beverages. Indeed, *T. delbrueckii* possesses several positive features that could be profitable used. Several investigations agree that *T. delbrueckii* impact on aromatic composition and sensory attributes of wines in both simultaneous and sequential fermentation through an increase of acetate ester (Cordero-Bueso 2013), thiols (3-sulfanylhexan-1-ol and 3-sulfanylhexyl acetate (Renault et al. 2015; Zott et al. 2008), terpenes (α terpineol and linalool) (Čuš and Jenko 2013), 2 phenyl-ethanol (Comitini et al. 2011).

Another non-*Saccharomyces* yeast is *M. pulcherrima*, species, frequently present on the grape surface and often recovered during the initial stages of alcoholic fermentation. *M. pulcherrima* is a high producer of β-glucosidase (Rodriguez et al. 2010), and its presence in mixed cultures can provide significant enhancements in the wine of higher alcohols, esters and terpenoids. Its aromatic profile in mixed fermentation was characterized by "citrus/grape fruits" some smoky and flowery attributes in Risling and Macabeo grape varieties respectively González-Royo et al. 2015). Also *W. anomalus* (formerly *P. anomala*) resulted in positive contribution to aroma profile of wines in mixed fermentation determining an enhancement of isoamyl acetate and ethyl esters (Kurita 2008). Finally, an interesting non-*Saccharomyces* wine yeast to enhance complexity and overall aroma profile is *Zygotorulaspora florentina*, a yeast responsible of increased fruity and floral notes as well as lower perception of astringency (Lencioni et al. 2018). A wide and deepened information about the aroma enhancement of non-*Saccharomyces* yeasts in winemaking are dealt in the Chap. 2.

4.5.2 Distinctive Features

It has long been known the ability of some yeast species to metabolize malic acid. *Schizosaccharomyces* yeasts (*Schizosaccharomyces pombe, Schyzosaccharomyces japonicus*) are characterized by malo-alcoholic fermentation and they are capable to completely metabolize the malic acid present in grape must and wine (Magyar and Panyik 1989; Ciani 1995) and could be profitable used in winemaking. In addition, more recently works showed that these yeasts species in mixed fermentation determined and increase in the production of pigments and large amounts of polysaccharides (Domizio et al. 2017; Escott et al. 2018). On the other hand, biological acidification is a desired feature in grape juices deficient in acidity generally coming from wines of warm climates. In addition in the last years, there is an increasing interest due to a progressive reduction of the total acidity of wines caused by global climate change and variations in viticulture and oenology practices. In this context *Lachancea thermotolerans* showed a peculiar ability to produce large amounts of lactic acid, together with glycerol and 2-phenyl ethanol during fermentation of grape musts (Kapsopoulou et al. 2007; Comitini et al. 2011; Gobbi et al. 2013).

Glycerol production is another relevant feature of non - *Saccharomyces* yeasts. *S. bacillaris* (synonym *Candida zemplinina*) (Rantsiou et al. 2012; Duarte et al. 2012) and *Starmerella bombicola* (formerly *C. stellata*) exhibit strong fructophilic character and shows the ability to produce high amounts of glycerol (Ciani and Ferraro 1996). In mixed fermentation with *S. cerevisiae* these yeast species exhibited positive interactions in the production and degradation of metabolites (Ciani and Ferraro 1998). In addition to large glycerol production *M. pulcherrima* exhibited some positive features such as polysaccharides and glycosidase activity (Comitini et al. 2011). Various enzymatic activities important for enzymatic release of aromatic compounds in winemaking, were found in several other non *Saccharomyces* yeasts such as *Hanseniaspora, Pichia and Candida* genera (Rodríguez et al. 2007) and well as large production of polysaccharides (Domizio et al. 2011).

Another positive trait desired and pursued by non-*Saccharomyces* yeast is the low production of volatile acidity. This feature is one of the fundamental character to select strain for the oenological use. Some non-*Saccharomyces* species such as *T. delbrueckii* and *C. stellata* (now reclassified as *Starmerella bombicola*) exhibited a very low production of volatile acidity (Ciani and Maccarelli 1998). In mixed fermentation with *S. cerevisiae* both *T. delbrueckii* and *C. stellata* showed a consistent reduction of volatile acidity (Ciani and Ferraro 1998). Similarly a reduction of acetic acid production was obtained in sweet wine fermentations in mixed fermentations using *C. zemplinina* (now reclassified as *S. bacillaris*) (Rantsiou et al. 2012).

4.5.3 Ethanol Reduction in Wine

Nowadays, the progressive increase in alcohol levels in wine, is a growing problem affecting the winemaking industry. Indeed, over the last two decades, there has been a progressive increase in the ethanol content in wines of c.a. two degrees over the viticulture areas (Alston et al. 2011; Gonzalez et al. 2013). This increase is mainly due to two main concerns: global climate change and the new wine styles often associated with increased grape maturity. For example, in wine the harvest the grapes at complete phenolic maturation may determine a overripe grapes and consequently the production of wines with high ethanol content. On the other hand, global climate change has deeply influenced the vine phenology and the grape composition, resulting in grapes with lower acidity, phenolic maturation and tannin content modifying other wine sensory attributes.

In order to overcome these issues, the market focus is directed to wines with a moderate alcohol content. In addition, lowering ethanol content has an economic interest due to the high taxes imposed in some countries. In this context, there are a rising interest in ethanol reduction in wine. Microbiological approach for decreasing ethanol concentrations appears a promising way and there is a growing interest to evaluate the use of non-*Saccharomyces* wine yeasts. There are several features possessed by non-*Saccharomyces* wine yeast that are a potential tool for the reduction of alcohol content in wine: a wide variability in ethanol yield (Contreras et al.

2014; 2015; Gobbi et al. 2014; Magyar and Tóth 2011) and the differences in regulatory respiro-fermentative metabolism with *S. cerevisiae* (Gonzalez et al. 2013). Indeed, among non-*Saccharomyces* wine yeasts some strains/species showed and sugar consumption by respiration (Crabtree negative). Therefore, both of these features of non-*Saccharomyces* yeasts have indicated a promising way to limit ethanol production. The approach used to use non-*Saccharomyces* wine yeasts to limit the production of ethanol is the mixed culture (simultaneous or sequential) since the inability of these yeasts of completing alcoholic fermentation (Ciani et al. 2016).

4.5.4 Control of Spoilage Microflora

Another possible applicative use of non-*Saccharomyces* yeasts in winemaking regards the control of spoilage microorganisms. During different stages of fermentation, a punctual and timely control of potential spoilage microorganisms is needed. In particular, during fermentation and aging stages of wine, the most spoilage yeast is *B. bruxellensis* responsible of undesired odors and considered the current major concern for winemakers, since an effective method to control their growth has not yet been developed.

Dekkera/*Brettanomyces* are described in the literature as part of the microbiota of many fermented beverages including cider, some type of beer, kombucha and kefyr, etc. (Morrissey et al. 2004). *Dekkera*/*Brettanomyces* can grow during the wine aging and even after their bottling; on the contrary, these yeasts are rarely found during the alcoholic fermentation of grape must. A few studies have reported their presence on grapes due to the difficult cultivation while in winery, in particular in vats, pumps or equipments difficult to sanitizes, *Brettanomyces* yeasts are more easily found (Fugelsang and Zoecklein 2003; Pretorius 2000; Renouf and Lonvaud-Funel 2007).

Different strains of *Brettanomyces* can show great differences in their production of volatile phenols. The variety of grape used also affects the sensorial perception of ethylphenols. Phister and Mills (2004) indicate detection thresholds to be high in monovarietal Cabernet Sauvignon wines, and lower in Tempranillo wines. The treatments to reduce the negative effect caused by *Dekkera*/*Brettanomyces* are based on both preventive and curative actions. Certain additives can inhibit the growth of *Brettanomyces*, including sulphur dioxide (SO_2). The recommended molecular dose of SO_2 is highly variable, from 0.3 to 0.8 mg/L. But these doses do not consider differences of strain resistance to sulfites or yeast population levels. Moreover, SO_2 is known as a chemical stressor inducing a viable but nonculturable (VBNC) state of *B. bruxellensis* that are non-detectable by plate counting, can lead to new contamination when the amount of sulfite decreases over time (Capozzi et al. 2016). Moreover, the SO_2 preservative agent has been largely demonstrated to have negative effects in wine consumers, including allergic reactions, asthma and headaches. This led to the establishment of strict regulations governing its use in the wine industry (Guerrero and Cantos-Villar 2015) with a direct consequence that the industry is interested to new ways to reduce sulphur dioxide levels, without changing the sensory quality of the wine. On the basis of this, a valid and natural alternative could be represented by bioactive compound produced by yeasts (Muccilli and Restuccia 2015). Biopreservation or biocontrol refers to the use of natural or controlled microorganisms, or their antimicrobial products, to extend the shelf life and to enhance the safety of food and beverages. This can be achieved by the addition of antimicrobial metabolites, such as killer toxins, or the direct application of pro-technological killer strain. A number of microorganisms and other biological agents have been regarded to be crucial in the biopreservation of food and beverages. In this context, a large group of non-*Saccharomyces* killer yeasts, able to produce killer toxins, can counteract *Dekkera*/*Brettanomyces* spoilage yeasts in wine.

Yeast killer toxins, also named mycocins or zymocins were initially defined as extracellular proteins, glycoproteins or glycolipids that disrupt the cell membrane function in susceptible yeast bearing receptors for the compound, whose activity is directed primarily against yeast closely related to the producer strain, which has a protective factor. The first mycocins were identified in association with *S. cerevisiae* in the brewing industry, but several others have since been isolated, frequently where yeast populations occur in high density and in highly competitive conditions, as for example fermented olive brine and fermenting grape must. Biological control could have an important application during the maturation and wine ageing of wines. In this regard, killer toxins secreted by *W. anomalus* (Pikt) and *K. wickerhamii* (Kwkt) were tested to control *Dekkera/Brettanomyces* spoilage yeasts. The stability in wine and the fungicidal effect of these two zymocins were demonstrated (Comitini et al. 2004). Thus, a potential application for the two toxins as antimicrobial agents active on *Dekkera/Brettanomyces* during wine ageing and storage can be hypothesized. Also, another killer toxin produced by *Ustilago maydis* it was seen to have efficacy to control *B. bruxellensis*, in mixed cultures under winemaking conditions Santos et al. (2011).

Recently, two new killer toxins produced by *Candida pyralidae* with an antimicrobial effect against *B. bruxellensis*, was tested in wine (Mehlomakulu et al. 2014). The killer toxins were stable under winemaking conditions and the activity was not affected by the ethanol and sugar concentrations typically found in grape juice and wine. Another new killer toxin from *T. delbrueckii* was identified and partial characterized. This zymocin, showed also a potential biocontrol effect on *B. bruxellensis* and other spoilage non-*Saccharomyces* yeasts such as *Pichia* spp.

However, other biological methods besides killer yeasts, were evaluated to control *B. bruxellensis* using non-*Saccharomyces* specific strains. For example, *M. pulcherrima* secretes pulcherriminic acid that exhibits an effective inhibitory effect to the growth of *B. bruxellensis*. In this case, the antimicrobial activity of *M. pulcherrima* does not seem due to proteinaceous compounds but to the precursor of pulcherrimin pigment that depletes iron present in the medium, making it not available to the other yeasts. Moreover, cell-to-cell contact and quorum sensing have been investigated as mechanisms involved in non-*Saccharomyces*-mixed fermentation (Oro et la. 2014). Quorum sensing was recently examined in *H. uvarum*, *Torulaspora pretoriensis*, *Zygosaccharomyces bailii*, *C. zemplinina (S. bacillaris*), and *B. bruxellensis*. Results indicated species-specific kinetics for the production of 2-phenylethanol, tryptophol, and tyrosol, considered the main molecules involved in the quorum sensing mechanism (Zupan et al. 2013; Avbelj et al. 2016).

5 Conclusions

Grape must is a complex matrix where grapes, microbes and technological process determine the final composition of wine. Yeasts associated with grapes and winery environment may influence both the analytical composition and sensorial profile of final wine. Indeed, the ecological distribution of microbial community along the whole of the wine production chain plays an important role in the composition of the final wine. For these reasons investigations on yeast microflora of the different wine regions, interactions among them and with other biotic and abiotic factors are of significant importance in wine production. The use of new metagenomic techniques such as new generation sequencing (NGS) strategies will allow to acquire additional knowledge to have a more complete picture on wine yeast ecology. Investigations on physiology features of selected wine yeasts (*Saccharomyces* and non-*Saccharomyces* yeasts) may positively contribute to achieve some goals as: enhanced aroma profile and complexity, ethanol reduction and biocontrol strategy. In this way, applied studies on fermentative yeasts could provide new opportunities in the oenological field, such as the introduction on the market of products with distinctive analytical and sensory characteristics due to recognized yeast strains.

References

Aarnio, T., Suikho, M., & Kauppinen, V. (1991). Isolation of acetic acid tolerant baker's yeast variants in a turbidostat. *Applied Biochemistry and Biotechnology, 27*, 55–63.

Abdelfattah, A., Wisniewski, M., Droby, S., & Schena, L. (2016). Spatial and compositional variation in the fungal communities of organic and conventionally grown apple fruit at the consumer point-of-purchase. *Horticulture Research, 3*, 16047.

Adamo, G. M., Brocca, S., Passolunghi, S., Salvato, B., & Lotti, M. (2012). Laboratory evolution of copper tolerant yeast strains. *Microbial Cell Factories, 11*(1), 1.

Albergaria, H., Francisco, D., Gori, K., Arneborg, N., & Gírio, F. (2010). *Saccharomyces cerevisiae* CCMI 885 secretes peptides that inhibit the growth of some non-*Saccharomyces* wine-related strains. *Applied Microbiology and Biotechnology, 86*, 965–972.

Alston, J. M., Fuller, K. B., Lapsley, J. T., & Soleas, G. (2011). Too much of a good thing? Causes and consequences of increases in sugar content of California wine grapes. *Journal of Wine Economics, 6*, 135–159.

Arneborg, N., Siegumfeldt, H., Andersen, G. H., Nissen, P., Daria, V. R., Rodrigo, P. G., & Glückstad, J. (2005). Interactive optical trapping shows that confinement is a determinant of growth in a mixed yeast culture. *FEMS Microbiology Letters, 245*, 155–159.

Avbelj, M., Zupan, J., Kranjc, L., & Raspor, P. (2016). Quorum-sensing kinetics in *Saccharomyces cerevisiae*: A symphony of ARO genes and aromatic alcohols. *Journal of Agricultural and Food Chemistry, 63*, 8544–8550.

Bacon, C. W., & White, J. F. (2016). Functions, mechanisms and regulation of endophytic and epiphytic microbial communities of plants. *Symbiosis, 68*, 87–98.

Barata, A., González, S., Malfeito-Ferreira, M., Querol, A., & Loureiro, V. (2008). Sour rot-damaged grapes are sources of wine spoilage yeasts. *FEMS Yeast Research, 8*, 1008–1017.

Barata, A., Malfeito-Ferreira, M., & Loureiro, V. (2012). The microbial ecology of wine grape berries. *International Journal of Food Microbiology, 153*, 243–259.

Belda, I., Navascués, E., Marquina, D., Santos, A., Calderon, F., & Benito, S. (2015). Dynamic analysis of physiological properties of *Torulaspora delbrueckii* in wine fermentations and its incidence on wine quality. *Applied Microbiology and Biotechnology, 9*, 1911–1922.

Belda, I., Ruiz, J., Esteban-Fernández, A., Navascués, E., Marquina, D., Santos, A., & Moreno-Arribas, M. (2017a). Microbial contribution to wine aroma and its intended use for wine quality improvement. *Molecules, 22*, 189.

Belda, I., Zarraonaindia, I., Perisin, M., Palacios, A., & Acedo, A. (2017b). From vineyard soil to wine fermentation: Microbiome approximations to explain the "terroir" concept. *Frontiers in Microbiology, 8*, 821.

Bellon, J. R., Schmid, F., Capone, D. L., Dunn, B. L., & Chambers, P. J. (2013). Introducing a new breed of wine yeast: Interspecific hybridisation between a commercial *Saccharomyces cerevisiae* wine yeast and *Saccharomyces mikatae*. *PLoS One, 8*, e62053.

Beltran, G., Torija, M. J., Novo, M., Ferrer, N., Poblet, M., Guillamón, J. M., & Mas, A. (2002). Analysis of yeast populations during alcoholic fermentation: A six year follow-up study. *Systematic and Applied Microbiology, 25*, 287–293.

Benito, S., Palomero, F., Calderón, F., Palmero, D., & Suárez-Lepe, J. A. (2014). Selection of appropriate *Schizosaccharomyces* strains for winemaking. *Food Microbiology, 42*, 218–224.

Bisson, L. F. (1999). Stuck and sluggish fermentations. *American Journal of Enology and Viticulture, 50*, 107–119.

Bisson, L. F., Joseph, C. L., & Domizio, P. (2017). Yeasts. In *Biology of microorganisms on grapes, in must and in wine* (pp. 65–101). Cham: Springer.

Blanco, P., Ramilo, A., Cerdeira, M., & Orriols, I. (2006). Genetic diversity of wine *Saccharomyces cerevisiae* strains in an experimental winery from Galicia (NW Spain). *Antonie Van Leeuwenhoek, 89*, 351–357.

Blanco, P., Orriols, I., & Losada, A. (2011). Survival of commercial yeasts in the winery environment and their prevalence during spontaneous fermentations. *Journal of Industrial Microbiology & Biotechnology, 38*, 235–239.

Bokulich, N. A., & Mills, D. A. (2012). Next-generation approaches to the microbial ecology of food fermentations. *BMB Reports, 45*, 377–389.

Bokulich, N. A., Ohta, M., Richardson, P. M., & Mills, D. A. (2013). Monitoring seasonal changes in winery-resident microbiota. *PLoS One, 8*, e66437.

Bokulich, N. A., Thorngate, J. H., Richardson, P. M., & Mills, D. A. (2014). Microbial biogeography of wine grapes is conditioned by cultivar, vintage, and climate. *Proceedings of the National Academy of Sciences of the United States of America, 111*, E139–E148.

Bokulich, N. A., Collins, T. S., Masarweh, C., Allen, G., Heymann, H., Ebeler, S. E., & Mills, D. A. (2016). Associations among wine grape microbiome, metabolome, and fermentation behavior suggest microbial contribution to regional wine characteristics. *MBio, 7*, e00631–e00616.

Bonnefoy, C., Quénol, H., Bonnardot, V., Barbeau, G., Madelin, M., Planchon, O., & Neethling, E. (2013). Temporal and spatial analyses of temperature in a French wine-producing area: The Loire Valley. *International Journal of Climatology, 33*, 1849–1862.

Borneman, A. R., Desany, B. A., Riches, D., Affourtit, J. P., Forgan, A. H., Pretorius, I. S., Egholm, M., & Chambers, P. J. (2012). The genome sequence of the wine yeast VIN7 reveals an allotriploid hybrid genome with *Saccharomyces cerevisiae* and *Saccharomyces kudriavzevii* origins. *FEMS Yeast Research, 12*, 88–96.

Borneman, A. R., Forgan, A. H., Kolouchova, R., Fraser, J. A., & Schmidt, S. A. (2016). Whole genome comparison reveals high levels of inbreeding and strain redundancy across the spectrum of commercial wine strains of *Saccharomyces cerevisiae*. *G3: Genes, Genomes, Genetics, 6*(4), 115.

Brown, S. W., & Oliver, S. G. (1982). Isolation of ethanol-tolerant mutants of yeast by continuous selection. *European Journal of Applied Microbiology and Biotechnology, 16*, 119–122.
Brysch-Herzberg, M., & Seidel, M. (2015). Yeast diversity on grapes in two German wine growing regions. *International Journal of Food Microbiology, 214*, 137–144.
Čadež, N., Zupan, J., & Raspor, P. (2010). The effect of fungicides on yeast communities associated with grape berries. *FEMS Yeast Research, 10*, 619–630.
Cafarchia, C., Camarda, A., Romito, D., Campolo, M., Quaglia, N. C., Tullio, D., & Otranto, D. (2006). Occurrence of yeasts in cloacae of migratory birds. *Mycopathologia, 16*, 229–234.
Capozzi, V., Garofalo, C., Chiriatti, M. A., Grieco, F., & Spano, G. (2015). Microbial terroir and food innovation: The case of yeast biodiversity in wine. *Microbiological Research, 181*, 75–83.
Capozzi, V., Di Toro, M. R., Grieco, F., Michelotti, V., Salma, M., Lamontanara, A., et al. (2016). Viable But Not Culturable (VBNC) state of *Brettanomyces bruxellensis* in wine: New insights on molecular basis of VBNC behaviour using a transcriptomic approach. *Food Microbiology, 59*, 196–204.
Chatonnet, P., Dubourdie, D., Boidron, J. N., & Pons, M. (1992). The origin of ethylphenols in wines. *Journal of the Science of Food and Agriculture, 60*, 165–178.
Chong, K. P., Rossall, S., & Atong, M. (2009). In vitro antimicrobial activity and fungitoxicity of syringic acid, caffeic acid and 4-hydroxybenzoic acid against *Ganoderma boninense*. *Journal of Agricultural Science, 1*, 15.
Ciani, M. (1995). Continuous deacidification of wine by immobilized *Schizosaccharomyces pombe* cells: Evaluation of malic acid degradation rate and analytical profiles. *Journal of Applied Microbiology, 79*, 631–634.
Ciani, M., & Comitini, F. (2015). Yeast interactions in multi-starter wine fermentation. *Current Opinion in Food Science, 1*, 1–6.
Ciani, M., & Ferraro, L. (1996). Enhanced glycerol content in wines made with immobilized *Candida stellata* cells. *Applied and Environmental Microbiology, 62*, 128–132.
Ciani, M., & Ferraro, L. (1998). Combined use of immobilized *Candida stellata* cells and *Saccharomyces cerevisiae* to improve the quality of wines. *Journal of Applied Microbiology, 85*, 247–254.
Ciani, M., & Maccarelli, F. (1998). Oenological properties of non-*Saccharomyces* yeasts associated with wine-making. *World Journal of Microbiology and Biotechnology, 14*, 199–203.
Ciani M., & Rosini G. (1986). Sulla microflora blastomicetica del Sagrantino D.O.C.: Microflora dei vini e dei locali di vinificazione. Annals Fac. Agr. Univ. Perugia 40: 103–112.
Ciani, M., & Rosini, G. (1990). Selection of strains *Saccharomyces cerevisiae* of "Sagrantino D.O.C." for their wine – making properties: Preliminary results. *Italian Journal of Food Science, 3*, 151–158.
Ciani, M., Mannazzu, I., Marinangeli, P., Clementi, F., & Martini, A. (2004). Contribution of winery-resident *Saccharomyces cerevisiae* strains to spontaneous grape must fermentations. *Antonie Van Leeuwenhoek, 85*, 159–164.
Ciani, M., Beco, L., & Comitini, F. (2006). Fermentation behaviour and metabolic interactions of multistarter wine yeast fermentations. *International Journal of Food Microbiology, 108*, 239–245.
Ciani, M., Comitini, F., Mannazzu, I., & Domizio, P. (2010). Controlled mixed culture fermentation: A new perspective on the use of non-*Saccharomyces* yeasts in winemaking. *FEMS Yeast Research, 10*, 123–133.
Ciani, M., Morales, P., Comitini, F., Tronchoni, J., Canonico, L., Curiel, J. A., & Gonzalez, R. (2016). Non-conventional yeast species for lowering ethanol content of wines. *Frontiers in Microbiology, 7*, 642.
Clavijo, A., Calder6n, I. L., & Paneque, P. (2010). Diversity of *Saccharomyces* and non-*Saccharomyces* yeasts in three red grape varieties cultured in the Serrania de Ronda (Spain) vine-growing region. *International Journal of Food Microbiology, 143*, 241–245.
Cocolin, L., Rantsiou, K., Iacumin, L., Urso, R., Cantoni, C., & Comi, G. (2004). Study of the ecology of fresh sausages and characterization of populations of lactic acid bacteria by molecular methods. *Applied and Environmental Microbiology, 70*, 1883–1894.

Cocolin, L., Alessandria, V., Dolci, P., Gorra, R., & Rantsiou, K. (2013). Culture independent methods to assess the diversity and dynamics of microbiota during food fermentation. *International Journal of Food Microbiology, 167*, 29–43.

Combina, M., Elía, A., Mercado, L., Catania, C., Ganga, A., & Martinez, C. (2005). Dynamics of indigenous yeast populations during spontaneous fermentation of wines from Mendoza, Argentina. *International Journal of Food Microbiology, 99*, 237–243.

Comitini, F., & Ciani, M. (2006). Survival of inoculated *Saccharomyces cerevisiae* strain on wine grapes during two vintages. *Letters in Applied Microbiology, 42*, 248–253.

Comitini, F., Ingeniis De, J., Pepe, L., Mannazzu, I., & Ciani, M. (2004). *Pichia anomala* and *Kluyveromyces wickerhamii* killer toxins as new tools against *Dekkera/Brettanomyces* spoilage yeasts. *FEMS Microbiology Letters, 238*, 235–240.

Comitini, F., Ferretti, R., Clementi, F., Mannazzu, I., & Ciani, M. (2005). Interactions between *Saccharomyces cerevisiae* and malolactic bacteria: preliminary characterization of a yeast proteinaceous compound (s) active against *Oenococcus oeni*. *Journal of Applied Microbiology, 99*, 105–111.

Comitini, F., Gobbi, M., Domizio, P., Romani, C., Lencioni, L., Mannazzu, I., & Ciani, M. (2011). Selected non-*Saccharomyces* wine yeasts in controlled multistarter fermentations with *Saccharomyces cerevisiae*. *Food Microbiology, 28*, 873–882.

Comitini, F., Oro, L., Canonico, L., Marinelli, V., & Ciani, M. (2019). Occurrence of *Brettanomyces bruxellensis* on grape berries and in related winemaking cellar. *Frontiers in Microbiology*, 10, 415.

Connell, L., Stender, H., & Edwards, C. G. (2002). Rapid detection and identification of Brettanomyces from winery air samples based on peptide nucleic acid analysis. *American Journal of Enology and Viticulture, 53*, 322–324.

Contreras, A., Hidalgo, C., Henschke, P. A., Chambers, P. J., Curtin, C., & Varela, C. (2014). Evaluation of non-*Saccharomyces* yeasts for the reduction of alcohol content in wine. *Applied and Environmental Microbiology, 80*, 1670–1678.

Contreras, A., Curtin, C., & Varela, C. (2015). Yeast population dynamics reveal a potential 'collaboration' between *Metschnikowia pulcherrima* and *Saccharomyces uvarum* for the production of reduced alcohol wines during Shiraz fermentation. *Applied Microbiology and Biotechnology, 99*, 1885–1895.

Coombe, P. E., Srinivasan, M. V., & Guy, R. G. (1989). Are the large monopolar cells of the insect lamina on the optomotor pathway. *Journal of Comparative Physiology. A, Neuroethology, Sensory, Neural, and Behavioral Physiology, 166*, 23–35.

Cordero-Bueso, G. A. (2013). Aplicación del Análisis Sensorial de los Alimentos en la Cocina y en la Industria Alimentaria. *Sede de Carmona de la Universidad Pablo de Olavide, XI*, 13–96.

Cordero-Bueso, G., Arroyo, T., Serrano, A., Tello, J., Aporta, I., Vélez, M. D., & Valero, E. (2011). Influence of the farming system and vine variety on yeast communities associated with grape berries. *International Journal of Food Microbiology, 145*, 132–139.

Costantì, M., Poblet, M., Arola, L., Mas, A., & Guillarmon, J. M. (1997). Analysis of yeast population during alcoholic fermentation in a newly established winery. *American Journal of Enology and Viticulture, 48*, 339–344.

Cray, J. A., Bel, l., Bhaganna, P., Mswaka, A. Y., Timson, D. J., & Hallsworth, J. (2013). The biology of habitat dominance; can microbes behave as weeds? *Microbial Biotechnology, 6*, 453–492.

Čuš, F., & Jenko, M. (2013). The influence of yeast strains on the composition and sensory quality of Gewürztraminer wine. *Food Technology and Biotechnology, 51*, 547–553.

De Barros, L. M., Bellon, J. R., Shirley, N. J., & Ganter, P. F. (2002). Evidence for multiple interspecific hybridization in *Saccharomyces* sensu stricto species. *FEMS Yeast Research, 1*, 323–331.

De Filippis, F., La Storia, A., Villani, F., & Ercolini, D. (2013). Exploring the sources of bacterial spoilers in beefsteaks by culture-independent high-throughput sequencing. *PLoS One, 8*, e70222.

del Carmen Portillo, M., & Mas, A. (2016). Analysis of microbial diversity and dynamics during wine fermentation of Grenache grape variety by high-throughput barcoding sequencing. *LWT-Food Science and Technology, 72*, 317–321.
del Monaco, S. M., Barda, N., Rubio, N., & Caballero, A. (2014). Selection and characterization of a Patagonian *Pichia kudriavzevii* for wine deacidification. *Journal of Applied Microbiology, 117*, 451–464.
Domizio, P., Romani, C., Lencioni, L., Comitini, F., Gobbi, M., Mannazzu, I., & Ciani, M. (2011). Outlining a future for non-*Saccharomyces* yeasts: Selection of putative spoilage wine strains to be used in association with *Saccharomyces cerevisiae* for grape juice fermentation. *International Journal of Food Microbiology, 147*, 170–180.
Domizio, P., Liu, Y., Bisson, L. F., & Barile, D. (2017). Cell wall polysaccharides released during the alcoholic fermentation by *Schizosaccharomyces pombe* and *Schizosaccharomyces japonicus*: Quantification and characterization. *Food Microbiology, 61*, 136–149.
Druvefors, U. A., Passoth, V., & Schnürer, J. (2005). Nutrient effects on biocontrol of *Penicillium roqueforti* by *Pichia anomala* J121 during airtight storage of wheat. *Applied and Environmental Microbiology, 71*, 1865–1869.
Duarte, F. L., Pimentel, N. H., Teixeira, A., & Fonseca, A. (2012). Saccharomyces bacillaris is not a synonym of *Candida stellata*: Reinstatement as *Starmerella bacillaris* comb. nov. *Antonie Van Leeuwenhoek, 102*(4), 653–658.
Ekberg, J., Rautio, J., Mattinen, L., Vidgren, V., Londesborough, J., & Gibson, B. R. (2013). Adaptive evolution of the lager brewing yeast *Saccharomyces pastorianus* for improved growth under hyperosmotic conditions and its influence on fermentation performance. *FEMS Yeast Research, 13*, 335–349.
Erten, H. (2002). Relations between elevated temperatures and fermentation behaviour of *Kloeckera apiculata* and *Saccharomyces cerevisiae* associated with winemaking in mixed cultures. *World Journal of Microbiology and Biotechnology, 18*, 373–378.
Escott, C., Del Fresno, J. M., Loira, I., Morata, A., Tesfaye, W., del Carmen González, M., et al. (2018). Formation of polymeric pigments in red wines through sequential fermentation of flavanol-enriched musts with non-*Saccharomyces* yeasts. *Food Chemistry, 239*, 975–983.
Fleet, G. H. (2003). Yeast interactions and wine flavour. *International Journal of Food Microbiology, 86*, 11–22.
Fleet, G. H., & Heard, G. M. (1993). Yeasts-growth during fermentation. In G. H. Fleet (Ed.), *Wine microbiology and biotechnology* (pp. 27–54). Chur: Harwood Academic Publishers.
Fleet, G. H., Lafon-Lafourcade, S., & Ribéreau-Gayon, P. (1984). Evolution of yeasts and lactic acid bacteria during fermentation and storage of Bordeaux wines. *Applied and Environmental Microbiology, 48*, 1034–1038.
Fleet, G. H., Prakitchaiwattana, C., Beh, A. L., & Heard, G. (2002). The yeast ecology of wine grapes. In *Biodiversity and biotechnology of wine yeasts* (Vol. 95, pp. 1–17). Kerala: Research Signpost.
Francesca, N., Canale, D. E., Settanni, L., & Moschetti, G. (2012). Dissemination of wine-related yeasts by migratory birds. *Environmental Microbiology Reports, 4*, 105–112.
Frezier, V., & Dubourdieu, D. (1992). Ecology of yeast strain *Saccharomyces cerevisiae* during spontaneous fermentation in a Bordeaux winery. *American Journal of Enology and Viticulture, 53*, 375–380.
Fugelsang, K. C., & Zoecklein, B. W. (2003). Population dynamics and effects of *Brettanomyces bruxellensis* strains on Pinot noir (*Vitis vinifera* L.) wines. *American Journal of Enology and Viticulture, 54*, 294–300.
Ganga, M. A., & Martinez, C. (2004). Effect of wine yeast monoculture practice on the biodiversity of non-*Saccharomyces* yeasts. *Journal of Applied Microbiology, 96*, 76–83.
Ganucci, D., Guerrini, S., Mangani, S., Vincenzini, M., & Granchi, L. (2018). Quantifying the effects of ethanol and temperature on the fitness advantage of predominant *Saccharomyces cerevisiae* strains occurring in spontaneous wine fermentations. *Frontiers in Microbiology, 9*.
Gao, C., & Fleet, G. H. (1988). The effects of temperature and pH on the ethanol tolerance of the wine yeasts *Saccharomyces cerevisiae*, *Candida stellata* and *Kloeckera apiculata*. *The Journal of Applied Bacteriology, 65*, 405–410.

Garofalo, C., Tristezza, M., Grieco, F., Spano, G., & Capozzi, V. (2016). From grape berries to wine: Population dynamics of cultivable yeasts associated to "Nero di Troia" autochthonous grape cultivar. *World Journal of Microbiology and Biotechnology, 32*, 59.

Gilbert, J. A., Jansson, J. K., & Knight, R. (2014a). The earth microbiome project: Successes and aspirations. *BMC Biology, 12*, 69.

Gilbert, J. A., van der Lelie, D., & Zarraonaindia, I. (2014b). Microbial terroir for wine grapes. *Proceedings of the National Academy of Sciences of the United States of America, 111*, 5–6. https://doi.org/10.1073/pnas.1320471110.

Giorello, F., Valera, M. J., Martin, V., Parada, A., Salzman, V., Camesasca, L., & Berna, L. (2018). Genomic and phenomic analysis of *Hanseniaspora vineae* provides insights for understanding yeast fermentation flavours that contribute to wine quality. *Applied and Environmental Microbiology, 85*(1), e01959–e01918.

Gobbi, M., Comitini, F., Domizio, P., Romani, C., Lencioni, L., Mannazzu, I., & Ciani, M. (2013). *Lachamcea thermotolerans* and *Saccharomyces cerevisiae* in simultaneous and sequential co-fermentation: A strategy to enhance acidity and improve the overall quality of wine. *Food Microbiology, 33*, 271–281.

Gobbi, M., De Vero, L., Solieri, L., Comitini, F., Oro, L., Giudici, P., & Ciani, M. (2014). Fermentative aptitude of non-*Saccharomyces* wine yeast for reduction in the ethanol content in wine. *European Food Research and Technology, 239*, 41–48.

Goddard, M. R., Anfang, N., Tang, R., Gardner, R. C., & Jun, C. (2010). A distinct population of *Saccharomyces cerevisiae* in New Zealand: Evidence for local dispersal by insects and human-aided global dispersal in oak barrels. *Environmental Microbiology, 12*(1), 63–73.

Godden, P., Wilkes, E., & Johnson, D. (2015). Trends in the composition of Australian wine 1984–2014. *Australian Journal of Grape and Wine Research, 21*, 741–753.

Gonzalez, R., Quirós, M., & Morales, P. (2013). Yeast respiration of sugars by non-Saccharomyces yeast species: A promising and barely explored approach to lowering alcohol content of wines. *Trends in Food Science and Technology, 29*, 55–61.

González-Royo, E., Pascual, O., Kontoudakis, N., Esteruelas, M., Esteve-Zarzoso, B., Mas, A., & Zamora, F. (2015). Oenological consequences of sequential inoculation with non-*Saccharomyces* yeasts (*Torulaspora delbrueckii* or *Metschnikowia pulcherrima*) and *Saccharomyces cerevisiae* in base wine for sparkling wine production. *European Food Research and Technology, 240*, 999–1012.

Grangeteau, C., Gerhards, D., von Wallbrunn, C., Alexandre, H., Rousseaux, S., & Guilloux-Benatier, M. (2016). Persistence of two non-*Saccharomyces* yeasts (*Hanseniaspora* and *Starmerella*) in the cellar. *Frontiers in Microbiology, 7*, 268.

Groth, G., Hansen, J., & Piškur, J. (1999). A natural chimeric yeast containing genetic material from three species. *International Journal of Systematic Bacteriology, 49*, 1933–1938.

Guerrero, R. F., & Cantos-Villar, E. (2015). Demonstrating the efficiency of sulphur dioxide replacements in wine: A parameter review. *Trends in Food Science and Technology, 42*, 27–43.

Guillamón, J. M., Barrio, E., & Querol, A. (1996). Characterization of wine yeast strains of the *Saccharomyces* genus on the basis of molecular markers: Relationships between genetic distance and geographic or ecological origin. *Systematic and Applied Microbiology, 19*, 122–132.

Hansen, E. H., Nissen, P., Sommer, P., Nielsen, J. C., & Arneborg, N. (2001). The effect of oxygen on the survival of non-*Saccharomyces* yeasts during mixed culture fermentations of grape juice with *Saccharomyces cerevisiae*. *Journal of Applied Microbiology, 91*, 541–547.

Hart, R. S., Jolly, N. P., Mohamed, G., Booyse, M., & Ndimba, B. K. (2016). Characterisation of *Saccharomyces cerevisiae* hybrids selected for low volatile acidity formation and the production of aromatic Sauvignon blanc wine. *African Journal of Biotechnology, 15*, 2068–2081.

Heard, G. M., & Fleet, G. H. (1985). Growth of natural yeast flora during the fermentation of inoculated wines. *Applied and Environmental Microbiology, 50*, 727–728.

Heard, G. M., & Fleet, G. H. (1988). The effects of temperature and pH on the growth of yeast species during the fermentation of grape juice. *The Journal of Applied Bacteriology, 65*, 23–28.

Henick-Kling, T., Edinger, W., Daniel, P., & Monk, P. (1998). Selective effects of sulfur dioxide and yeast starter culture addition on indigenous yeast populations and sensory characteristics of wine. *Journal of Applied Microbiology, 84*, 865–876.

Hernández, A., Pérez-Nevado, F., Ruiz-Moyano, S., Serradilla, M. J., Villalobos, M. C., Martín, A., & Córdoba, M. G. (2018). Spoilage yeasts: What are the sources of contamination of foods and beverages? *International Journal of Food Microbiology, 286*, 98–110.

Hierro, J. L., Maron, J. L., & Callaway, R. M. (2005). A biogeographical approach to plant invasions: The importance of studying exotics in their introduced and native range. *Journal of Ecology, 93*, 5–15.

Holland, T., & Smit, B. (2014). Recent climate change in the Prince Edward County winegrowing region, Ontario, Canada: Implications for adaptation in a fledgling wine industry. *Regional Environmental Change, 14*, 1109–1121.

Huglin, P. (1978). Nouveau mode d'évaluation des possibilités héliothermiques d'un milieu viticole. *Comptes Rendus de l'Académie de l'Agriculture de France, 64*, 1117–1126.

Jolly, N. P., Augustyn, O. P. H., & Pretorius, I. S. (2003). The occurrence of non-*Saccharomyces cerevisiae* yeast species over three vintages in four vineyards and grape musts from four production regions of the Western Cape, South Africa. *South African Journal of Enology and Viticulture, 24*, 35–42.

Jolly, N. P., Varela, C., & Pretorius, I. S. (2014). Not your ordinary yeast: Non-*Saccharomyces* yeasts in wine production uncovered. *FEMS Yeast Research, 14*, 215–237.

Jones-Vaid, M., Prasad, R., Singh, T., Jones, V., & Katiyar, S. K. (2012). Grape seed proanthocyanidins reactivate silenced tumor suppressor genes in human skin cancer cells by targeting epigenetic regulators. *Toxicology and Applied Pharmacology, 263*, 122–130.

Kapsopoulou, K., Mourtzini, A., Anthoulas, M., & Nerantzis, E. (2007). Biological acidification during grape must fermentation using mixed cultures of *Kluyveromyces thermotolerans* and *Saccharomyces cerevisiae. World Journal of Microbiology and Biotechnology, 23*, 735–739.

Katalinić, V., Možina, S. S., Skroza, D., Generalić, I., Abramovič, H., Miloš, M., & Boban, M. (2010). Polyphenolic profile, antioxidant properties and antimicrobial activity of grape skin extracts of 14 Vitis vinifera varieties grown in Dalmatia (Croatia). *Food Chemistry, 119*, 715–723.

Knight, S., Klaere, S., Fedrizzi, B., & Goddard, M. R. (2015). Regional microbial signatures positively correlate with differential wine phenotypes: Evidence for a microbial aspect to terroir. *Scientific Reports, 5*, 14233.

König, H., & Claus, H. (2018). A future place for *Saccharomyces* mixtures and hybrids in winemaking. *Fermentation, 4*, 67.

Kurita, O. (2008). Increase of acetate ester-hydrolysing esterase activity in mixed cultures of *Saccharomyces cerevisiae* and *Pichia anomala. Journal of Applied Microbiology, 104*, 1051–1058.

Kurtzman, C. P., & Droby, S. (2001). Metschnikowia fructicola, a new ascosporic yeast with potential for biocontrol of postharvest fruit rots. *Systematic and Applied Microbiology, 24*, 395–399.

Kurtzman, C., Fell, J. W., & Boekhout, T. (Eds.). (2011). The yeasts: a taxonomic study. Elsevier, Amsterdam.

Kutyna, D. R., Varela, C., Stanley, G. A., Borneman, A. R., Henschke, P. A., & Chambers, P. J. (2012). Adaptive evolution of *Saccharomyces cerevisiae* to generate strains with enhanced glycerol production. *Applied Microbiology and Biotechnology, 93*, 1175–1184.

Lachance, M. A. (2003). The Phaff school of yeast ecology. *International Microbiology, 63*, 163–167.

Lencioni, L., Taccari, M., Ciani, M., & Domizio, P. (2018). *Zygotorulaspora florentina* and *Starmerella bacillaris* in multistarter fermentation with *Saccharomyces cerevisiae* to reduce volatile acidity of high sugar musts. *Australian Journal of Grape and Wine Research, 24*, 368.

Lereboullet, A. L., Beltrando, G., Bardsley, D. K., & Rouvellac, E. (2014). The viticultural system and climate change: coping with long-term trends in temperature and rainfall in Roussillon, France. *Regional Environmental Change, 14*, 1951–1966.

Liu, X., Hoque, M., Larochelle, M., Lemay, J. F., Yurko, N., Manley, J. L., & Tian, B. (2017). Comparative analysis of alternative polyadenylation in *S. cerevisiae* and *S. pombe*. *Genome Research, 27*(10), 1685–1695.

Magyar, I., & Panyik, I. (1989). Biological deacidification of wine with *Schizosaccharomyces pombe* entrapped in Ca-alginate gel. *American Journal of Enology and Viticulture, 40*, 233–240.

Magyar, I., & Tóth, T. (2011). Comparative evaluation of some oenological properties in wine strains of *Candida stellata, Candida zemplinina, Saccharomyces uvarum* and *Saccharomyces cerevisiae*. *Food Microbiology, 28*, 94–100.

Martini, A. (1993). Origin and domestication of the wine yeast *Saccharomyces cerevisiae*. *Journal of Wine Research, 4*, 165–176.

Martins, G., Lauga, B., Miot-Sertier, C., Mercier, A., Lonvaud, A., Soulas, M. L., Soulas, G., & Masneuf-Pomarède, I. (2013). Characterization of epiphytic bacterial communities from grapes, leaves, bark and soil of grapevine plants grown, and their relations. *PLoS One, 8*, e73013.

Masneuf, I., Hansen, J., Groth, C., Piškur, J., & Dubourdieu, D. (1998). New hybrids between *Saccharomyces* sensu stricto yeast species found among wine and cider production strains. *Applied and Environmental Microbiology, 64*, 3887–3892.

Mehlomakulu, N. N., Setati, B., & Divol, G. (2014). Characterization of novel killer toxins secreted by wine- related non-Saccharomyces yeasts and their action on *Brettanomyces* spp. *International Journal of Food Microbiology, 188*, 83–91.

Mendoza, L. M., Manca de Nadra, M. C., & Farías, M. E. (2007). Kinetics and metabolic behaviour of a composite culture of *Kloeckera apiculata* and *Saccharomyces cerevisiae* wine related strains. *Biotechnology Letters, 29*, 1057–1063.

Mezzasalma, V., Sandionigi, A., Bruni, I., Bruno, A., Lovicu, G., Casiraghi, M., & Labra, M. (2017). Grape microbiome as a reliable and persistent signature of field origin and environmental conditions in Cannonau wine production. *PLoS One, 12*, e0184615.

Milanovic, V., Comitini, F., & Ciani, M. (2013). Grape berry yeast communities: Influence of fungicide treatments. *International Journal of Food Microbiology, 161*, 240–246.

Mills, D. A., Johannsen, E. A., & Cocolin, L. (2002). Yeast diversity and persistence in botrytis-affected wine fermentations. *Applied and Environmental Microbiology, 68*, 4884–4893.

Mira de Orduña, R., Liu, S.-Q., Patchett, M. L., & Pilone, G. J. (2000). Kinetics of the arginine metabolism of malolactic wine lactic acid bacteria *Lactobacillus buchneri* CUC-3 and *Oenococcus oeni* Lo111. *Journal of Applied Microbiology, 89*, 547–552.

Moisselin, J. M., & Dubuisson, B. (2006). Évolution des valeurs extrêmes de température et de précipitations au cours du XXe siècle en France.

Molnar, O., & Prillinger, H. (2005). Analysis of yeast isolates related to *Metschnikowia pulcherrima* using the partial sequences of the large subunit rDNA and the actin gene; description of *Metschnikowia andauensis* sp. Nov. *Systematic and Applied Microbiology, 28*, 717–726.

Moreira, N., Mendes, F., de Pinho, P. G., Hogg, T., & Vasconcelos, I. (2008). Heavy sulphur compounds, higher alcohols and esters production profile of *Hanseniaspora uvarum* and *Hanseniaspora guilliermondii* grown as pure and mixed cultures in grape must. *International Journal of Food Microbiology, 124*, 231–238.

Morrison-Whittle, P., & Goddard, M. R. (2015). Quantifying the relative roles of selective and neutral processes in defining eukaryotic microbial communities. *ISME Journal, 9*, 2003–2011.

Morrison-Whittle, P., & Goddard, M. R. (2018). From vineyard to winery: A source map of microbial diversity driving wine fermentation. *Environmental Microbiology, 20*, 75–84.

Morrissey, J. P., Dow, J. M., Mark, G. L., & O'Gara, F. (2004). Are microbes at the root of a solution to world food production?: Rational exploitation of interactions between microbes and plants can help to transform agriculture. *EMBO Reports, 5*, 922–926.

Muccilli, S., & Restuccia, C. (2015). Bioprotective role of yeasts. *Microorganisms, 3*, 588–611.

Naumova, E. S., Naumov, G. I., Masneuf-Pomarède, I., Aigle, M., & Dubourdieu, D. (2005). Molecular genetic study of introgression between Saccharomyces bayanus and *S. cerevisiae*. *Yeast, 22*, 1099–1115.

Neumann, P. A., & Matzarakis, A. (2014). Potential climate change impacts on winegrape must density and titratable acidity in southwest Germany. *Climate Research, 59*, 161–172.

Nguyen, H. V., & Gaillardin, C. (2005). Evolutionary relationships between the former species *Saccharomyces uvarum* and the hybrids *Saccharomyces bayanus* and *Saccharomyces pastorianus*; reinstatement of *Saccharomyces uvarum* (Beijerinck) as a distinct species. *FEMS Yeast Research, 5*, 471–483.

Nguyen, H. V., Lepingle, A., & Gaillardin, C. (2000). Molecular typing demonstrates homogeneity of *Saccharomyces uvarum* strains and reveals the existence of hybrids between *S. uvarum* and *S. cerevisiae* including the *S. bayanus* type strain CBS 380. *Systematic and Applied Microbiology, 23*, 71–85.

Nissen, P., & Arneborg, N. (2003). Characterization of early deaths of non-Saccharomyces yeasts in mixed cultures with *Saccharomyces cerevisiae*. *Archives of Microbiology, 180*, 257–263.

Nissen, P., Nielsen, D., & Arneborg, N. (2003). Viable *Saccharomyces cerevisiae* cells at high concentrations cause early growth arrest of non-*Saccharomyces* yeasts in mixed cultures by a cell-cell-contact-mediated mechanism. *Yeast, 20*, 331–341.

Ocón, E., Gutierrez, A. R., Garijo, P., Lope, Z. R., & Santamaria, P. (2010). Presence of non-*Saccharomyces* yeasts in cellar equipment and grape juice during harvest time. *Food Microbiology, 27*, 1023–1027.

Ocón, E., Garijo, P., Sanz, S., Olarte, C., Lo'pez, R., Santamaria, P., & Gutierrez, A. R. (2013). Analysis of airborne yeast in one winery over a period of one year. *Food Control, 30*, 585–589.

Orlandini, S., Di Stefano, V., Lucchesini, P., Puglisi, A., & Bartolini, G. (2009). Current trends of agroclimatic indices applied to grapevine in Tuscany (Central Italy). *Idojaras, 113*, 69–78.

Oro, L., Ciani, M., & Comitini, F. (2014). Antimicrobial activity of *Metschnikowia pulcherrima* on wine yeasts. *Journal of Applied Microbiology, 11*, 1209–1217.

Osborne, J. P., & Edwards, C. G. (2007). Inhibition of malolactic fermentation by a peptide produced by *Saccharomyces cerevisiae* during alcoholic fermentation. *International Journal of Food Microbiology, 118*, 27–34.

Padilla, B., Gil, J. V., & Manzanares, P. (2016). Past and future of non-*Saccharomyces* yeasts: From spoilage microorganisms to biotechnological tools for improving wine aroma complexity. *Frontiers in Microbiology, 7*, 411.

Pardo, J. M., & Serrano, R. (1989). Structure of a plasma membrane H+-ATPase gene from the plant *Arabidopsis thaliana*. *The Journal of Biological Chemistry, 264*, 8557–8562.

Peña, R., & Ganga, M. A. (2018). Novel antimicrobial peptides produced by *Candida intermedia* LAMAP1790 active against the wine-spoilage yeast *Brettanomyces bruxellensis*. *Antonie Van Leeuwenhoek, 9*, 1–8.

Pérez-Nevado, F., Albergaria, H., Hogg, T., & Girio, F. (2006). Cellular death of two non-*Saccharomyces* wine-related yeasts during mixed fermentations with *Saccharomyces cerevisiae*. *International Journal of Food Microbiology, 108*, 336–345.

Pérez-Torrado, R., Barrio, E., & Querol, A. (2018). Alternative yeasts for winemaking: *Saccharomyces* non-*cerevisiae* and its hybrids. *Critical Reviews in Food Science and Nutrition, 58*, 1780–1790.

Peter, G., Tornai-Lehoczki, J., Suzuki, M., & Dlauchy, D. (2005). *Metschnikowia viticola* sp. nov., a new yeast species from grape. *Antonie Van Leeuwenhoek, 87*, 155–160.

Petrie, P. R., & Sadras, V. O. (2008). Advancement of grapevine maturity in Australia between 1993 and 2006: putative causes, magnitude of trends and viticultural consequences. *Australian Journal of Grape and Wine Research, 14*, 33–45.

Peynaud, E., & Domercq, S. (1959). Possibilité de provoquer la fermentation malolactique à l'aide de bactéries cultivées. *Comptes Rendus de l'Academie d'Agriculture de France, 45*, 355–358.

Piao, H., Hawley, E., Kopf, S., DeScenzo, R., Sealock, S., Henick-Kling, T., & Hess, M. (2015). Insights into the bacterial community and its temporal succession during the fermentation of wine grapes. *Frontiers in Microbiology, 6*, 809.

Pielke, S. R., Stohlgren, T., Schell, L., Parton, W., Doesken, N., Redmond, K., & Kittel, T. G. F. (2002). Problems in evaluating regional and local trends in temperature: An example from eastern Colorado, USA. *International Journal of Climatology, 22*, 421–434.

Pinto, C., Pinho, D., Cardoso, R., Custódio, V., Fernandes, J., & Susana. (2015). Wine fermentation microbiome: A landscape from different Portuguese wine appellations. *Frontiers in Microbiology, 6*, 905.
Pretorius, I. S. (2000). Tailoring wine yeast for the new millennium: Novel approaches to the ancient art of winemaking. *Yeast, 16*, 675–729.
Pretorius, I. S., Van der Westhuizen, T. J., & Augustyn, O. P. H. (1999). Yeast biodiversity in vineyards and wineries and its importance to the South African wine industry. A review. *South African Journal of Enology and Viticulture, 20*, 61–70.
Ramos, M. C., Jones, G. V., & Martínez-Casasnovas, J. A. (2008). Structure and trends in climate parameters affecting winegrape production in Northeast Spain. *Climate Research, 38*(1), 1–15.
Rankine, B. C., & Pilone, D. A. (1973). *Saccharomyces bailii*, a resistant yeast causing serious spoilage of bottled table wine. *American Journal of Enology and Viticulture, 24*, 55–58.
Rantsiou, K., Dolci, P., Giacosa, S., Torchio, F., Tofalo, R., Torriani, S., Suzzi, G., Rolle, L., & Cocolin, L. (2012). *Candida zemplinina* can reduce acetic acid produced by *Saccharomyces cerevisiae* in sweet wine fermentations. *Applied and Environmental Microbiology, 78*(6), 1987–1994.
Raspor, P., Milek, D. M., Polanc, J., Možina, S. S., & Čadež, N. (2006). Yeasts isolated from three varieties of grapes cultivated in different locations of the Dolenjska vine-growing region, Slovenia. *International Journal of Food Microbiology, 109*, 97–102.
Regueiro, L. A., Costas, C. L., & Rubio, J. E. L. (1993). Influence of viticultural and enological practices on the development of yeast populations during winemaking. *American Journal of Enology and Viticulture, 44*, 405–408.
Renault, P., Coulon, J., de Revel, G., Barbe, J. C., & Bely, M. (2015). Increase of fruity aroma during mixed *T. delbrueckii/S. cerevisiae* wine fermentation is linked to specific esters enhancement. *International Journal of Food Microbiology, 207*, 40–48.
Renouf, V., & Lonvaud-Funel, A. (2007). Development of an enrichment medium to detect *Dekkera/Brettanomyces bruxellensis*, a spoilage wine yeast, on the surface of grape berries. *Microbiological Research, 162*, 154–167.
Riffell, S., Burton, T., & Murphy, M. (2006). Birds in depressional forested wetlands: Area and habitat requirements and model uncertainty. *Wetlands, 26*, 107–118.
Rodríguez, M. E., Lopes, C. A., Valles, S., Giraudo, M. R., & Caballero, A. (2007). Selection and preliminary characterization of β-glycosidases producer Patagonian wild yeasts. *Enzyme and Microbial Technology, 41*, 812–820.
Rodríguez-Gómez, F., Arroyo-López, F. N., López-López, A., Bautista-Gallego, J., & Garrido-Fernández, A. (2010). Lipolytic activity of the yeast species associated with the fermentation/storage phase of ripe olive processing. *Food Microbiology, 27*, 604–612.
Rojas, V., Gil, J. V., Piñaga, F., & Manzanares, P. (2003). Acetate ester formation in wine by mixed cultures in laboratory fermentations. *International Journal of Food Microbiology, 86*, 181–188.
Romano, P., Fiore, C., & Paraggio, M. (2003). Function of yeast species and strains in wine flavour. *International Journal of Food Microbiology, 86*, 169–180.
Rosini, G. (1984). Assessment of dominance of added yeast in wine fermentation and origin of *Saccharomyces cererevisiae* in wine-making. *The Journal of General and Applied Microbiology, 30*, 249–256.
Rosini, G., Federici, F., & Martini, A. (1982). Yeast flora of grape berries during ripening. *Microbial Ecology, 8*, 83–89.
Sabate, J., Cano, J., Esteve-Zarzoso, B., & Guillamon, J. M. (2002). Isolation and identification of yeasts associated with vineyard and winery by RFLP analysis of ribosomal genes and mitochondrial DNA. *Microbiological Research, 157*, 267–274.
Santamarıa, P., Lopez, R., Lopez, E., Garijo, P., & Gutierrez, A. R. (2008). Permanence of yeast inocula in the winery ecosystem and presence in spontaneous fermentations. *European Food Research and Technology, 227*, 1563–1567.
Santos, C. N. S., & Stephanopoulos, G. (2008). Combinatorial engineering of microbes for optimizing cellular phenotype. *Current Opinion in Chemical Biology, 12*, 168–176.

Santos, C., Lima, N., Sampaio, P., & Pais, C. (2011). Matrix-assisted laser desorption/ionization time-of-flight intact cell mass spectrometry to detect emerging pathogenic *Candida* species. *Diagnostic Microbiology and Infectious Disease, 71*, 304–308.

Schuller, D., Cardoso, F., Sousa, S., Gomes, P., Gomes, A. C., & Santos, M. A. S. (2012). Genetic diversity and population structure of *Saccharomyces cerevisiae* strains isolated from different grape varieties and winemaking regions. *PLoS One, 7*, e32507.

Shen, Y., Chen, X., Peng, B., Chen, L., Hou, J., & Bao, X. (2012). An efficient xylose-fermenting recombinant *Saccharomyces cerevisiae* strain obtained through adapted evolution and its global transcription profile. *Applied Microbiology and Biotechnology, 96*, 1079–1091.

Sipiczki, M., Pfliegler, W. P., & Holb, I. J. (2013). Species share a pool of diverse rRNA genes differing in regions that determine hairpin-loop structures and evolve by reticulation. *PLoS One, 8*, e67384.

Sokolowsky, M., & Fischer, U. (2012). Evaluation of bitterness in white wine applying descriptive analysis, time-intensity analysis, and temporal dominance of sensations analysis. *Analytica Chimica Acta, 732*, 46–52.

Steensels, J., Snoek, T., Meersman, E., Nicolino, M. P., Voordeckers, K., & Verstrepen, K. J. (2014). Improving industrial yeast strains: Exploiting natural and artificial diversity. *FEMS Microbiology Reviews, 38*, 947–995.

Stefanini, I., & Cavalieri, D. (2018). Metagenomic approaches to investigate the contribution of the vineyard environment to the quality of wine fermentation: Potentials and difficulties. *Frontiers in Microbiology, 9*.

Stefanini, I., Dapporto, L., Legras, J. L., Calabretta, A., Di Paola, M., & De Filippo, C. (2012). Role of social wasps in *Saccharomyces cerevisiae* ecology and evolution. *Proceedings of the National Academy of Sciences, 109*, 13398–13403.

Suh, S. O., Gibson, C. M., & Blackwell, M. (2004). *Metschnikowia chrysoperlae* sp. nov., *Candida picachoensis* sp. nov. and *Candida pimensis* sp. nov., isolated from the green lacewings *Chrysoperla comanche* and *Chrysoperla carnea* (*Neuroptera*: *Chrysopidae*). *International Journal of Systematic and Evolutionary Microbiology, 54*, 1883–1890.

Swiegers, J. H., Bartowsky, E. J., Henschke, P. A., & Pretorius, I. (2005). Yeast and bacterial modulation of wine aroma and flavour. *Australian Journal of Grape and Wine Research, 11*, 139–173.

Taylor, M. W., Tsai, P., Anfang, N., Ross, H. A., & Goddard, M. R. (2014). Pyrosequencing reveals regional differences in fruit-associated fungal communities. *Environmental Microbiology, 16*, 2848–2858. https://doi.org/10.1111/1462-2920.12456.

Teslić, N., Zinzani, G., Parpinello, G. P., & Versari, A. (2018). Climate change trends, grape production, and potential alcohol concentration in wine from the "Romagna Sangiovese" appellation area (Italy). *Theoretical and Applied Climatology, 131*, 793–803.

Tonietto, J. (1999). Les macroclimats viticoles mondiaux et l'influence du mesoclimat sur la typicite de la Syrah et du Muscat de Hamboug dans le sud de la France: Methodologie de caracterisation. Embrapa Uva e Vinho-Outras publicações científicas. *(ALICE)*.

Ugliano, M., & Henschke, P. A. (2009). Yeasts and wine flavour. In *Wine chemistry and biochemistry* (pp. 313–392). New York: Springer.

Van der Walt, J. P., & van Kerken, A. E. (1961). The wine yeast sf the cape. *Antonie Van Leeuwenhoek, 27*, 81–90.

Van der Westhuizen, T. J., Augustyn, O. P. H., Khan, W., & Pretorius, I. S. (2000). Seasonal variation of indigenous *Saccharomyces cerevisiae* strains isolated from vineyards of the Western Cape in South Africa. *South African Journal of Enology & Viticulture, 21*(1), 10–16.

Van Leeuwen, C., Schultz, H. R., de Cortazar-Atauri, I. G., Duchêne, E., Ollat, N., Pieri, P., & Malheiro, A. C. (2013). Why climate change will not dramatically decrease viticultural suitability in main wine-producing areas by 2050. *Proceedings of the National Academy of Sciences of the United States of America, 110*, E3051–E3052.

Varela, C., Siebert, T., Cozzolino, D., Rose, L., McLean, H., & Henschke, P. A. (2009). Discovering a chemical basis for differentiating wines made by fermentation with 'wild' indigenous and inoculated yeasts: Role of yeast volatile compounds. *Australian Journal of Grape and Wine Research, 15*, 238–248.

Verginer, M., Leitner, E., & Berg, G. (2010). Production of volatile metabolites by grape associated microorganisms. *Journal of Agricultural and Food Chemistry, 58*, 8344–8350.

Viana, M., Kuhlbusch, T. A. J., Querol, X., Alastuey, A., Harrison, R. M., Hopke, P. K., & Hueglin, C. (2008). Source apportionment of particulate matter in Europe: A review of methods and results. *Journal of Aerosol Science, 39*, 827–849.

Williams, T. C., Averesch, N. J. H., Winter, G., Plan, M. R., Vickers, C. E., Nielsen, L. K., & Krömer, J. O. (2015). Quorum-sensing linked RNA interference for dynamic metabolic pathway control in *Saccharomyces cerevisiae*. *Metabolic Engineering, 29*, 124–134.

Winkler, A. J. (1962). *General viticulture*. Berkeley: University of California Press.

Zarraonaindia, I., Owens, S. M., Weisenhorn, P., West, K., Hampton-Marcell, J., Lax, S., & van der Lelie, D. (2015). The soil microbiome influences grapevine-associated microbiota. *MBio, 6*, e02527–e02514.

Zott, K., Miot-Sertier, C., Claisse, O., Lonvaud-Funel, A., & Masneuf-Pomarede, I. (2008). Dynamics and diversity of non-Saccharomyces yeasts during the early stages in winemaking. *International Journal of Food Microbiology, 125*, 197–203.

Zupan, J., Avbelj, M., Butinar, B., Kosel, J., Šergan, M., & Raspor, P. (2013). Monitoring of quorum-sensing molecules during minifermentation studies in wine yeast. *Journal of Agricultural and Food Chemistry, 61*, 2496–2505.

Chapter 2
Yeasts and Their Metabolic Impact on Wine Flavour

Angela Capece and Patrizia Romano

Contents

A. Capece (✉) · P. Romano
University of Basilicata, School of Agriculture, Forest, Food and Environmental Sciences, Potenza, Italy
e-mail: angela.capece@unibas.it; patrizia.romano@unibas.it

P. Romano et al. (eds.), *Yeasts in the Production of Wine*,
https://doi.org/10.1007/978-1-4939-9782-4_2

1 Introduction

The aroma and flavour of wine are one of the main characteristics that define the differences among the vast array of products and wine styles produced throughout the world. The flavour of wine is a sensory perception that varies with the individual, the context of the consumer experience and the chemical composition of the product. The final response is the outcome of complex chemosensory interactions that are difficult to predict because of the influences of many variables. The chemical composition of wine is the foundation of the sensory response and is determined by many factors, determining innumerable possible variations in wine production, both in viticulture and in winemaking. These include grape variety, geographical and viticultural conditions of grape cultivation, microbial ecology of the grape and fermentation processes, winemaking practices and, in the inoculated fermentation, the individual characteristics of the starter culture used.

Microorganisms have a prominent role in determining the chemical composition of wine. They affect the quality of the grape prior to harvest and recent results (Bokulich et al. 2016) suggest that the microbial profile of grapes can influence composition and abundance of secondary compounds affecting wine aroma and many microbial species, both fermentative and dominant grape epiphytes, that can have potential incidence in wine flavour (Belda et al. 2016), although the specific role of microbial species present on the grapes on wine flavour is still poorly understood. In general, the largest percentage of the total aroma composition of wine is represented by fermentation-derived volatile compounds. Yeasts, bacteria and filamentous fungi all contribute to the microbial ecology of wine production and the chemical composition of wine, although yeasts have the dominating influence because of their role in conducting the alcoholic fermentation. Yeasts do not only convert sugars to ethanol and carbon dioxide; they also produce a range of minor but sensorially important metabolites that gives wine its vinous character (Lambrechts and Pretorius 2000). Yeast impact upon wine flavour is largely determined by the array of volatile substances (e.g. higher alcohols, acids, esters, carbonyls, thiols) produced by the metabolism of grape juice components. These reactions vary with the yeast species and strains contributing to the fermentation (Lema et al. 1996; Henick-Kling et al. 1998; Garcia et al. 2002).

2 Origin of Wine Aroma

Many biosynthetic pathways in wine yeasts are involved in the formation of wine aroma and are affected by various factors, such as the composition and pH of the grape must and the nature and prevailing temperature of the fermentation. In addition, viticultural factors influencing the quality of the grapes and, consequently, the wine include the cultivar, soil quality, water management, vine canopy management and the ripeness of the grapes. Technological aspects and vinification practices, like the method of grape crushing, must treatment and skin contact time also significantly influence the final aroma of wine (Lambrechts and Pretorius 2000; Ribéreau-Gayon et al. 2000a, b).

The contribution of yeasts to wine flavour is carried out by several mechanisms: (a) producing enzymes that transform neutral grape compounds into flavour active compounds (pre-fermentative aroma), (b) producing many hundreds of flavour active, secondary metabolites (e.g. acids, alcohols, esters, polyols, aldehydes, ketones, volatile sulphur compounds, volatile phenols) (fermentative and post-fermentative aroma), (c) producing ethanol and other solvents that help to extract flavour components from grape solids, and (d) autolytic degradation of dead yeast cells (Lambrechts and Pretorius 2000).

3 Primary Aroma: Associated with Grape

Although some volatile aroma compounds arise directly from chemical components of the grapes, yeast metabolic activities release or modify grape-derived compounds, determining the formation of a further substantial portion of wine flavour substances (Fig. 2.1).

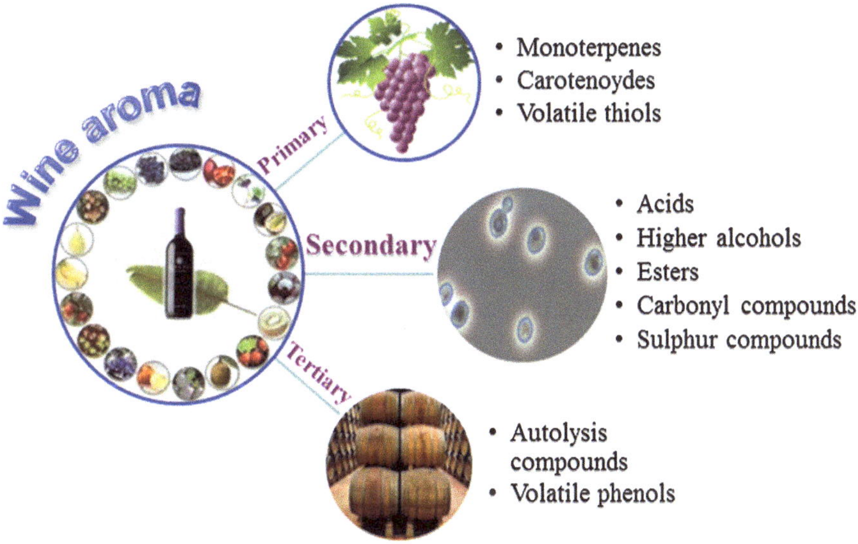

Fig. 2.1 Yeast metabolic activities in determining flavour substances in wine

3.1 Precursors of Primary Aroma

Primary wine aroma is largely due to the presence of **monoterpenes,** which are present as free as well as glycosylated flavourless conjugates amongst the secondary metabolites of certain varieties of *Vitis vinifera*. In general, bound glycosides are prevalent on free terpenoids, and the ratios of bound to free terpenoids can also vary amongst different grape cultivars. The majority of grape terpenes appears as **glycosides**, the sugar moiety of which is represented by glucose, rhamnose, arabinose, and apiose, as well as disaccharides.

Carotenoids, which are structurally related to terpenes, accumulate in ripening grape berries and their oxidation produces the so-called C13-norisoprenoids, which are strongly odoriferous. These compounds include β-ionone (aroma of viola) and damascenone (aroma of exotic fruits), the threshold concentrations of which are equal to 7 and 9 ng/l, as well as β-damascone (aroma of rose and fruits), β-ionol (aroma of fruit and flowers), 3-oxo-β-ionone (tobacco smell), and others. These compounds are present in grape in microgram amounts; for example, the total content of C13-norisoprenoids in muscat wines may be as high as 280 μg/kg.

Terpenes, as well as other volatiles compounds such as straight-chain alcohols, norisoprenoids and benzenoids, contribute to improve wine aroma. These odorous compounds may occur as free forms in grapes, or bounded to sugar molecules (glucose, disaccharides) to form odourless non-volatile glycosidic complexes. The free aroma compounds commonly are dominated by linalool, geraniol and nerol, together with additional monoterpenes, i.e. citronellol, α-terpineol, hotrienol, nerol oxide, plus several other oxides, aldehydes and hydrocarbons. These monoterpenes, as well as their oxides, furan derivatives, and pyran derivatives account for the aroma of muscat wines, where their content ranges from 5 to 30 mg/l. Low aroma grape cultivars contain up to 1 mg/l terpenoids. In wines, several monoterpene ethyl ethers and acetate esters have also been found among the free aroma compounds. Other categories are the polyhydroxylated forms of the monoterpenes, or free odourless polyols. A most significant feature of the polyols is that, although these compounds make no direct contribution to the aroma, some of them are reactive and can break down with great ease to give pleasant and potent volatiles.

Reports indicate that not all glycosides are present in all grape varieties, and that concentrations vary according to variety, ranging from 500 to 1700 μg/l of must (Gunata et al. 1985). Major precursors include structures such as β-D glucopyranoside, 6-*O*-α- L-arabinosil-β-D-glucopyranoside, 6-*O*-α-L-ramnopyranosyl-β-D-glucopyranoside and 6-*O*-β-D-apiofuranosyl-β-D-glucopyranoside apiosylglycosides.

Aglycon chemical structure (volatile when free) may vary, taking the form of a terpenol (linalol, geraniol, nerol, citronelol, ho-trienol or α-terpienol), linalol oxide, linear or cyclic alcohol (hexanol, phenylethanol, benzyl alcohol), C13 norisoprenoid, phenolic acid and/or volatile phenol. It is the mixture of these compounds, rather than the influence of any individual compound, that defines the varietal characteristics of a wine; the olfactory threshold of the mixture is lower than that of any individual component. These compounds are released through the action of β-glucosidase enzymes, which break the terpene–sugar bond, contributing to increased intensity of wine flavour. Mainly, enzymatic hydrolysis of glycosides is carried out with various enzymes which act sequentially according to two steps: firstly, α-L-rhamnosidase, α-L-arabinosidase or β-D-apiosidase make the cleavage of the terminal sugar and rhamnose, arabinose or apiose and the corresponding β-D-glucosides are released; subsequently liberation of monoterpenol takes place after action of a β-D-glucosidase. Hydrolysis during winemaking is caused by enzymes from the grapes themselves or from the microorganisms taking part in the process (Delcroix et al. 1994).

Another set of primary aroma compounds released from odorless bound precursors are volatile thiols, i.e., 4-methyl-4-mercaptopentan-2-one and 3-mercapto-1-hexanol, which impart tropical fruity' characters. These compounds are present in grape must as odorless, non-volatile, cysteine-bound conjugates. During alcoholic fermentation, wine yeasts cleave the cysteinated precursors with a carbon–sulphur lyase enzyme, releasing volatile thiols (Swiegers et al. 2009; Swiegers and Pretorius 2007).

3.2 *Factors Affecting Primary Aroma*

Yeasts developing during grape must fermentation can release glucosidases and other enzymes useful for hydrolysis of the glycosidic bonds of the odorless bound forms of monoterpenes, releasing more odor-contributing compounds to the wine.

Wine yeasts have been studied for the presence of useful hydrolytic activities, mainly β-glucosidase as well as other glycosidases. While *S. cerevisiae*, the main wine yeast, is not believed to be a significant producer of extracellular enzymes, in fact low α-rhamnosidase, α-arabinosidase or β-apiosidase activities were detected in this species (Delcroix et al. 1994), non-*Saccharomyces* wine yeasts are described as potential sources of glycosidases (Mendes-Ferreira et al. 2001; Mateo et al. 2011). Non-*Saccharomyces* yeasts, such as *Brettanomyces/Dekkera, Candida, Debaryomyces, Hanseniaspora* and *Pichia* have been screened for β-glucosidases with properties desirable for winemaking, that are: (a) high affinity for grape-derived terpenoid aglycones; (b) optimal activity at wine pH (pH 2.5–3.8); (c) resistance to glucose inhibition; and (d) high tolerance to ethanol. The β-glucosidase from *Debaryomyces pseudopolymorphus* resulted to be suitable for use under wine conditions (Cordero-Otero et al. 2003) as it was resistant against wine-associated inhibitory compounds, such as glucose, ethanol and sulphur dioxide, its optimum pH lies within the wine range (2.5–3.8) and it has high substrate affinity. Among grape yeasts, some strains belonging to *Candida* spp. and *Debaryomyces* spp. are producers of extracellular β-glucosidase, while in *Hanseniaspora/Kloeckera* spp. the enzyme has been localized essentially within the cell.

Also among wine-spoilage yeasts, some strains belonging to *Brettanomyces* spp., *Dekkera* spp., *Pichia* spp., and *Hansenula* spp. have been reported to produce β-glucosidase, both intra- and extracellularly (Manzanares et al. 2000). In these yeasts, the repression of the enzyme biosynthesis was observed when glucose was used as carbon source.

Data on β-glucosidase activity on *Saccharomyces* are contradictory. First results showed that these yeasts had a very low activity (Gunata et al. 1990), but Delcroix et al. (1994) found three enological strains showing high β-glucosidase activity. However, generally the β-glucosidase activity of *S. cerevisiae* towards glycoside precursors seems to be very low.

Nevertheless, it was found that the formation of some aromas associated with varietal character can be related to yeast metabolism and not to a simple hydrolytic process, as previously thought. In fact, it was found de novo synthesis of monoterpenes by the principal wine yeast species, *Saccharomyces cerevisiae,* demonstrating the biosynthesis of these compounds in the absence of grape derived precursors (Carrau et al. 2005). In this study, several strains of *S. cerevisiae* tested in parallel model fermentations with one *Hanseniaspora uvarum* strain in a simple chemically defined medium, lacking of grape juice, terpenes or their glycoconjugates, exhibited an intraspecific variability to accumulate terpenes in this chemically artificial medium. Among monoterpene alcohols, linalool and α-terpineol were produced in greatest abundance by all the *S. cerevisiae* wine strains tested. It was demonstrated that the contribution of *S. cerevisiae* to the monoterpene composition of a wine could be increased by some fermentation conditions, such as the use of musts containing high concentrations of assimilable nitrogen, like the ammonium ion, in combination with microaerobic fermentation. *H. uvarum* also produced similar concentrations of linalool and α-terpineol, but little citronellol was formed despite a relatively high amount of geraniol being present. This occurrence could be explained by the inability of this *H. uvarum* strain to reduce geraniol to citronellol, as it has been described for *Torulaspora delbrueckii* species when compared to *Saccharomyces* strains during biotransformation experiments. Other studies demonstrated that *S. cerevisiae* is capable of modifying the terpenic profile of the wine; thus, it can produce citronellol from geraniol and nerol; the intensity of this transformation depends on the yeast strain used (Hernandez et al. 2003). Other authors propose a more complex scheme: geraniol was transformed by these yeasts into geranyl acetate, citronellyl acetate and citronellol, while nerol was transformed into neryl acetate; in addition, geraniol was transformed into linalool and nerol was cyclized to α-terpineol at the grape must pH.

Also, the production of a particular terpene alcohol was not only related to the hydrolysis of the corresponding glycoside, but to other reactions involving these terpenic aglycons, such as chemical isomerization, hydration or reduction conducted by metabolic activity of wine yeasts.

4 Secondary Aroma: Must Fermentation

The wine aroma appears mainly during the alcoholic fermentation which represents the secondary aroma (Fig. 2.1). In general, the fermentation-derived volatile compounds compose the largest percentage of the total aroma composition of wine. The primary role of wine yeast is to catalyse the rapid, complete and efficient conversion of grape sugars to ethanol, carbon dioxide and other minor, but sensorially important metabolites without the development of off-flavours (Pretorius 2000). Most of the fermentative aroma compounds have high sensory thresholds and therefore do not contribute, in a significant way, individually to the distinctive aroma of wines, although their combination constitute the basic matrix of wine aroma. Otherwise, most of the aroma impact compounds are present at low concentrations, but, in consequence of their very low sensory thresholds, they have a higher influence on the overall wine aroma.

Metabolites that are direct products and by-products of glycolysis, represented by ethanol, glycerol, and acetic acid, are quantitatively the prevalent compounds, although usually these metabolites present low odor activity value (OAV, also known as flavour activity), which is calculated by dividing the concentration of an aroma component by its detection threshold level.

Ethanol is the main component, which determines the viscosity (body) of wine and acts as a fixer of aroma. Recently, wines with increased ethanol concentration are producing as a consequence of increased grape sugar concentrations due to worldwide climatic changes. The high sugar concentrations affect not only yeast cells, by increasing the osmotic stress that the yeast has to tolerate during the first steps of the fermentation, but also the content of fermentation metabolites, primarily ethanol, as well as other by-products, glycerol, and acetic acid (de Orduña 2010).

The high level of ethanol affects negatively the perception of flavour and aroma in the wine, i.e. higher ethanol levels change the perception of a wine from fruity to herbaceous (Goldner et al. 2009) and can also determine an increase in the perception of astringency of the tannins and the bitterness, roughness, and hotness of wine (Obreque-Slíer et al. 2010). Furthermore, high ethanol content can affect different aspects of yeast cell metabolism, such as by inducing various stress responses, influencing global gene expression, and modifying the structure of the cell membrane.

After ethanol, glycerol is the most abundant by-product of fermentation. Traditionally, this compound is considered affecting the overall mouthfeel of wine, whereas little attention has been given to the interaction of glycerol and flavour compounds and the role that this compound plays in the formation of the aroma profile. Earlier studies on sensorial analysis showed that this compound does not affect wine aroma, whereas further studies demonstrated the existence of a relationship statistically significant between concentration of glycerol and the perceived quality of all styles of white wine (Styger et al. 2011).

Acetic acid is the most important organic acids present in wine. Although the quantity of organic acids in the wine is small, they are sufficiently volatile to contribute to its aroma. Other organic acids, such as propanoic, butanoic and lactic acid are usually below the perception threshold.

Other metabolites produced during fermentation are esters, which are present in small amounts in grapes, but their formation is parallel to ethanol formation. Ethyl and acetate esters, together other flavour compounds, such as fusel alcohols, carbonyls, and volatile fatty acids, form the so-called "yeast bouquet", which are secondary metabolites synthesized by a wide range of yeast species.

4.1 Acid Composition

The acidity of grape juice and wine has a direct impact on sensory quality and physical, biochemical and microbial stability of the final product. Acids can have both positive and negative impacts on aroma and flavour, depending on concentration and the type and style of wine. This acidity, particularly pH, influences the survival and growth of all microorganisms; the effectiveness of anti-oxidants, antimicrobial compounds and enzyme additions; the solubility of proteins and tartrate salts; the effectiveness of bentonite treatment; the polymerisation of the colour pigments; the oxidative and browning reactions; and the freshness of some wine styles. Wine contains a large number of organic and inorganic acids and the wine acid content is sub-divisible in volatile and non-volatile acidity.

4.1.1 Volatile Acidity

Volatile acidity (VA) includes a group of volatile organic acids of short carbon chain-length. The volatile acid content of wine is usually between 500 and 1000 mg/l (10–15% of the total acid content) and of this, acetic acid usually constitutes about 90% of the volatile acids. The rest of the volatile acids, principally propionic and hexanoic acids, are produced as the result of fatty acid metabolism by yeast and bacteria.

Volatile acidity, principally acetic acid, can play a significant role in wine aroma and an excessive concentration of this alcoholic fermentation by-product is highly detrimental to wine quality because at elevated concentrations it imparts a vinegar-like character to wine. The amount of volatile acidity produced is usually low (0.25–0.50 g/l), but may be higher under certain fermentation conditions. Acetic acid becomes objectionable at concentrations of 0.7–1.1 g/l, depending on the style of wine; the optimal concentration is 0.2–0.7 g/l. In particular, during fermentation of high gravity musts, such as botrytized musts, the volatile acidity content may be 1.8 g/l or even higher, which is over the EEC legal limit of 1.5 g/l expressed in acetic acid. The OIV (2010) states that the maximum acceptable limit for volatile acidity in most wines is 1.2 g/l of acetic acid.

Although several enzymatic reactions have been suggested to contribute to acetic acid formation by yeast during fermentation, it seems that under anaerobiosis, acetate is produced by yeast as an intermediate of the pyruvate dehydrogenase (PDH) bypass, a pathway responsible for the conversion of pyruvate into acetyl-CoA through a series of reactions catalysed by pyruvate decarboxylase (PDC), acetaldehyde dehydrogenase (ALD) and acetyl-CoA synthase. The PDH bypass supplies the cell with cytosolic acetyl-CoA, which is needed for anabolic processes, such as lipid biosynthesis. The reaction catalysed by acetaldehyde dehydrogenase (ALD) also generates reducing equivalents, which are needed in many synthetic pathways and for redox reactions (involving NAD(P)H). Acetaldehyde dehydrogenase forms acetate by oxidising the acetaldehyde produced from pyruvate during the fermentation. Five ALD isoforms have now been identified in *S. cerevisiae*. Three of them are cytosolic (encoded by ALD6, ALD2, and ALD3), and the other two are mitochondrial (encoded by ALD4 and ALD5). The major isoforms involved in winemaking are Ald6p, Ald5p and Ald4p.

4.1.2 Non-Volatile Acidity

The predominant non-volatile organic acids are tartaric and malic acids, accounting for 90% of the titratable acidity (TA) of grape juice. Citric and lactic acids also contribute to the acidity of grape juice; succinic and keto acids are present only in trace amounts in grapes, but concentrations are higher in wines as a result of fermentation. Succinic acid produced by yeast is considered responsible for the largest part of the increase in TA. Its taste has been reported to be a mixture of acid, salt and bitterness. Occasionally malic acid has been produced concurrently with succinic acid, however, at a much lower concentration.

Succinic acid is a non-volatile, dicarboxylic organic acid produced during fermentation and can derive from either sugar or amino acid catabolism by yeast, depending on growth conditions and available nitrogen sources. Its direct formation is dependent on the reactions of the tricarboxylic acid (TCA) cycle, also known as the citric acid cycle or the Krebs cycle, of which it is an intermediate. In contrast to other wine non-volatile organic acids, such as tartaric, malic and citric acids, succinic acid is resistant to microbiological attack (particularly by lactic acid bacteria), both anaerobically and aerobically. While succinic acid is only present in trace quantities (<0.1 mg/kg) in the ripe berries of *Vitis vinifera* cultivars, it is a normal by-product of alcoholic fermentations and has been reported to be the main non-volatile carboxylic acid produced by yeasts during wine fermentations, where it typically accumulates to 2 g/l.

4.1.3 Factors Affecting Volatile Acidity

Saccharomyces cerevisiae produces acetic acid as a by-product of alcoholic fermentation; it was demonstrated that, under winemaking conditions, this yeast produces volatile acidity mainly at the beginning of alcoholic fermentation. Wine yeasts also produce acetic acid to equilibrate the redox balance in response to the hyperosmotic stress caused by high sugar concentrations, such as grape must with high °Brix (>35 °Brix) and wines from botrytized grapes.

Although *Saccharomyces* can produce acetic acid, excessive concentrations in wine are largely the result of the metabolism of ethanol by aerobic acetic acid bacteria and by activity of some non-*Saccharomyces* yeasts, such as *Kloeckera/Hanseniaspora* and *Zygosaccharomyces*, which have been traditionally described as producers of excessive amounts of acetic acid (du Toit and Pretorius 2000; Romano et al. 2003; Mendoza et al. 2007) and, in consequence of this, they have been considered for long time as spoilage yeasts. High level of acetic acid is commonly associated also to *Schizosaccharomyces pombe* species. However, results on acetic acid production among non-*Saccharomyces* yeasts are highly variable, probably as this compound is produced with a considerably strain variability. For *H. uvarum* strains, levels of acetic acid ranging from about 0.6 g/l to more than 3.4 g/l were found (Romano et al. 2003) and some strains exhibited a similar behaviour to *S. cerevisiae* in this regard (Romano et al. 1992; Ciani and Maccarelli 1998; Capece et al. 2005). Similar results were found for *Schiz. pombe*; in fact, a screening of *Schiz. pombe* strains allowed to find strains producing less than 0.4 g/l of acetic acid (Benito et al. 2014a).

Different studies on *T. delbrueckii* strains showed that this species is characterized by low production of volatile acidity when compared to *S. cerevisiae* (Ciani and Maccarelli 1998; Comitini et al. 2011). This characteristic is also present in *Lachancea thermotolerans* (previously known as *Kluyveromyces thermotolerans*), together with the high production of L-lactic acid. *Candida stellata/C.zemplinina*, species presenting a strong fructophilic character, is able to produce low amounts of ethanol and acetic acid and high amounts of glycerol (Englezos et al. 2015).

Under usual winemaking conditions, with initial sugar concentrations around 200 g/l, the production of volatile acidity by *S cerevisiae* was correlated to numerous factors. Some authors (Millan et al. 1991) report the impact of physiological conditions and quantity of yeast inoculum on volatile acidity concentration. This production is affected by yeast strain, medium composition, vitamins, initial sugar concentration and fermentation conditions, such as variation in temperature. Other studies have demonstrated the stimulating effect of fermentation by insoluble materials, which reduce the production of volatile acidity by providing saturated and unsaturated fatty acids to yeasts.

Nitrogen has an important impact on volatile acidity production by *S. cerevisiae*, which is especially high in conditions of high sugar fermentation. This production can be reduced by controlled nitrogen addition. Bely et al. (2003) found that a reduction of 40% was obtained during addition up to 210 mg/l of nitrogen at the beginning of fermentation in high sugar must with low assimilable nitrogen content (92 mg /l). In order to limit the production of volatile acidity by *S. cerevisiae* the optimal nitrogen concentration found in the must was 190 mg/l. These authors found that the best moment for nitrogen addition was at the beginning of fermentation, whereas the addition at the end of growth phase had less effect on volatile acidity reduction. They suggested that by stimulating cell growth, nitrogen addition provides NADH in the redox-equilibrating process, which in turn reduces volatile acidity formation.

4.2 Higher Alcohols

The term 'higher alcohol' refers to alcohols that possess more than two carbon atoms and have a higher molecular weight and boiling point than ethanol. Higher alcohols, also known as fusel alcohols, include aliphatic and aromatic alcohols, and are quantitatively the largest group of aroma compounds in many alcoholic beverages. The main aliphatic alcohols include propanol, isoamyl alcohol, isobutanol and active amyl alcohol, whereas tyrosol and 2-phenylethanol are the main aromatic alcohols. During alcoholic fermentation, *S. cerevisiae* produces amounts of higher alcohols that can have a significant effect on the sensorial quality and character of wine (Pretorius and Høj 2005; Swiegers and Pretorius 2005). They can have both positive and negative impacts on the aroma and flavour of wine: excessive concentrations of higher alcohols can result in a strong, pungent smell and taste, whereas optimal levels impart fruity characters. It has been reported that concentrations below 300 mg/l add a desirable level of complexity to wine, whereas concentrations that exceed 400 mg/l can have a detrimental effect. The sulphur-containing alcohols, for example methionol, might also have a strong influence on taste and flavour (Lambrechts and Pretorius 2000). Yeasts produce higher alcohols mainly during fermentation from α-keto acids, involving degradation of an amino acids, by the so-called Ehrlich pathway (reviewed by Hazelwood et al. 2008). Isoamyl alcohol, active amyl alcohol and isobutanol are also known as branched-chain alcohols, being the degradation products of the branched-chain amino acids, leucine, isoleucine and valine, respectively.

Amino acid uptake is performed by different amino acid transporters located in the yeast cell membrane; these transporters have mainly broad substrate specificity, although only few are very specific transporting only one amino acid. The assimilation of the branched chain amino acid valine, leucine and isoleucine is mediated by the branched chain amino acid permease of *S. cerevisiae*, encoded by the *BAP2* gene, which has broad substrate specificity.

The Erlich pathway is composed by three steps:

- an initial transamination, in which the amino group from the amino acid is transferred to α-ketoglutarate to form an α-keto acid (leucine to α-ketoisocaproic acid, valine to α-ketoisovaleric acid, and isoleucine to α-keto-β-methylvaleric acid) and glutamate. Alternatively, these α-keto acids can be generated through the de novo synthesis pathway from glucose via pyruvate;
- decarboxylation of the α-keto acid to form a "fusel aldehyde";
- aldehyde reduction to generate the "fusel alcohol".

In *S. cerevisiae*, it has been shown that the transamination reaction for the branched-chain amino acids is catalyzed by mitochondrial and cytosolic branched-chain amino acid aminotransferases (BCAATases), encoded by *BAT1* and *BAT2* genes respectively. During transaminase reaction of the aromatic amino acids, tryptophan, tyrosine, and phenylalanine, the Aro9p enzyme is involved. It seems that regulation of Ehrlich pathway is dependent on the growth phase, temperature and amino acid content, among others. For example, *BAT1* is preferentially expressed during the exponential growth phase, whereas *BAT2* is overexpressed during the stationary growth phase.

In the further step of Ehrlich pathway, decarboxylation of α-keto into an aldehyde, the pyruvate decarboxylases Pdc1p, Pdcp5, and Pdc6p, are involved, although these proteins are apparently not essential. Other proteins that could possibly be involved in this reaction are the phenylpyruvate decarboxylase Aro10p, and the carboxylase Thi3p (Styger et al. 2011), which plays a role as a regulatory protein of the enzymes regulating thiamine biosynthesis.

The final step of Ehrlich pathway reaches a fork: the aldehyde can be reduced (via a NADH-dependent reaction) or it can be oxidized (via a NAD^+-dependent reaction) to form its respective higher alcohol or volatile carboxylic acid, respectively, in function of redox status of the yeast cell. It was suggested that the reductive reaction is catalyzed by an alcohol dehydrogenase (Adh1p to Adh7p), whereas an aldehyde dehydrogenase is involved in oxidation reaction. Other oxidoreductases involved in the formation of the fusel alcohols are the formaldehyde dehydrogenase Sfa1p, the 3-methylbutanal reductase Gre2p, and the NADPH-dependent aldo-keto reductase Ypr1p, and at least one of the putative aryl-alcohol dehydrogenases (AAD6).

The higher alcohols are transported outside the cell probably by simple passive diffusion across the lipid layer as no membrane transporter in *S. cerevisiae* is known until now.

Enzymes involved in Ehrlich pathway, and those responsible for later ester synthesis, are also present in non-*Saccharomyces* yeasts. Furthermore, as some enzymes of the Ehrlich pathway (such as Aro10p) are specific for broad-substrate, different fusel alcohols can be formed also if the medium contains only one amino acid as nitrogen source. However, different yeast species in similar fermentative conditions produce different higher alcohols and in different amounts, indicating that in non-conventional yeasts the mechanisms involved in Ehrlich pathway are different compared to *Saccharomyces* species.

The physiological function of higher alcohol production by yeast is unclear, although many hypotheses have been postulated. It has been suggested that, physiologically, oxidative deamination provides the yeast with a mechanism for obtaining nitrogen when its pool has become depleted. A second hypothesis proposes that higher-alcohol production contributes to the maintenance of the redox balance in the cell because the final reduction step in higher-alcohol production involves the reoxidation of $NADH+H^+$ to NAD^+. However, it has also been stated that there appears to be enough acetaldehyde to maintain the redox balance and that the formation of higher alcohols is not considered to be an important means for the reoxidation of NADH. Finally, higher alcohol production might act as a detoxification process for the intracellular medium of α-keto acids and aldehydes, or as a means of regulating the metabolism of amino acids (Ribéreau-Gayon et al. 2000a).

4.2.1 Factors Affecting Higher Alcohols

It is widely recognized that the yeast strain involved in wine fermentation influences considerably the higher alcohol profiles and concentrations in wine.

It was found that the total production of higher alcohols by pure cultures of *Hanseniaspora* species is lower than production level exhibited by *S. cerevisiae* (Moreira et al. 2008; Viana et al. 2008). Similar results were found in *Zygosaccharomyces* strains isolated from grape musts. By contrast, wines obtained by fermentation with *C. zemplinina* contained high levels of fusel alcohols, with concentrations exceeding 400 mg/l (Andorrà et al. 2010). High production of 2-phenylethyl alcohol, a compound associated with pleasant aromas, has been described in the non-*Saccharomyces* species *M. pulcherrima* (Clemente-Jimenez et al. 2004), *L. thermotolerans* (Beckner Whitener et al. 2015), and *C. zemplinina* (Andorrà et al. 2010).

The concentration of amino acids (the precursors for higher alcohols) in the must also influences higher alcohol production, which increases as concentrations of the corresponding amino acids increase. The catabolism of amino acids is controlled by NCR (nitrogen catabolite repression), a complex regulation system that enables yeasts to select nitrogen source able to provide the best growth. NCR is mainly mediated by four transcription factors (GATA factors) and by the regulatory protein Ure2p. When in the medium the preferred nitrogen source become limited, genes involved in the use of non-preferred nitrogen source are gradually derepressed and NCR is removed.

Furthermore, ethanol concentration, fermentation temperature, pH and composition of grape must, aeration, level of solids, grape variety, maturity and skin contact time also affect the concentration of higher alcohols in the final product (Fleet and Heard 1993).

Excessive levels of higher alcohols in wines are typically associated with yeast assimilable nitrogen (YAN)-deficient grape must, as reported for amyl alcohols production (Bohlscheid et al. 2007). Increased concentrations of higher alcohols can result from nitrogen catabolism of valine, leucine, isoleucine and phenylalanine, or by overproduction of a-ketoacids because of a loss of feedback inhibition Hernandez-Orte et al. 2002). Other studies, however, suggest that the anabolic pathway of higher alcohol formation appears to predominate over the catabolism of amino acids in low YAN media, but the results of Bohlscheid et al. (2007) indicate a dependence on yeast strain and higher alcohols. Overproduction of higher alcohols can be a result of very low or very high YAN, but has been rarely reported as a result of biotin deficiency.

4.3 Esters

Esters make the greatest contribution to the desirable fermentation bouquet of wine (Nykänen 1986) as they are the most abundant compounds found in wine, with around 160 identified to date. During the primary alcohol fermentation of grape juice, a number of odorous esters are formed. Although various esters can be formed during fermentation, acetate esters and fatty acid ethyl esters are the two main categories of flavour-active esters in fermented beverages. Acetate esters are more important for wine aroma than fatty acids ethyl esters, which derive from the conjugation of an alcohol with an acid. In particular, the acyl group is derived from acetate (as acetyl-CoA), and the alcohol group is ethanol or a complex alcohol derived from amino acid metabolism. The most important acetate esters are ethyl acetate ("fruity", "solvent-like" aromas), isoamyl acetate ("banana" aroma), and 2-phenylethyl acetate ("honey", "roses", "flowery" aromas). In the formation of fatty acid ethyl esters, the alcohol group is ethanol, and the acyl group is derived from activated medium-chain fatty acids. Ethyl hexanoate ("apple-like" aroma), and ethyl octanoate ("apple" aroma) are included in this group.

During wine fermentation, maximum concentration of esters is obtained when yeasts are in the stationary growth phase. Esters can be considered as metabolic by-products of yeasts for three reasons:

1. these compounds usually are less toxic than their correspondent alcohol or acidic precursors, as a consequence, ester formation represents a detoxification mechanism;
2. they are insect attractants, representing a mechanism for yeast spread;
3. their synthesis represents a mechanism for the regeneration of free Coenzyme A from its conjugates.

Volatile esters have a higher impact in wine aroma than higher alcohols, although esters are present in small amounts. However, if these compounds are present in too high amounts, they can mask varietal aromas, decreasing wine complexity. Ethyl acetate is the most abundant ester in wine and it can have favourable effects on wine aroma at concentrations below 80 mg/l. Wines containing more than 90 mg/l of ethyl acetate or 200 mg/l of total esters are considered defective; in particular, high values of ethyl acetate are largely responsible for the altered sensory properties typical of acescency. Isoamyl acetate is one of the esters most markedly contributing to the aroma profile of white wines.

The rate of ester formation during fermentation is function of two factors: (1) the concentration of the co-substrates, the acyl-CoA and the alcohol; (2) the activity of enzymes involved in their synthesis and hydrolysis (acyltransferases and esterases). The best characterized enzymes involved in the formation of esters are alcohol acetyl transferases I and II (AATase I and II), which are encoded by the genes *ATF1* and *ATF2*, respectively. As the most common acyl-CoA molecule found in yeast is acetyl-CoA, the most common esters are acetate esters. These acetyltransferases are sulfhydryl enzymes which react with acetyl coenzyme A (acetyl-CoA) and, depending on the degree of affinity, with various higher alcohols to produce esters. It has also been shown that these enzymatic activities are strongly repressed under aerobic conditions and by the addition of unsaturated fatty acids to a culture. The *ATF1*-encoded AAT activity is the best studied acetyltransferase activity in *S. cerevisiae*. Atflp and Atf2p are partially responsible of isoamyl acetate and ethyl acetate production. It was reported that the differences found among three *Saccharomyces* species (*S. cerevisiae*, *S. kudriavzevii* and *S. uvarum*) for the formation of aroma-active acetate ester are due, to some extent, to the distinct properties of Atf enzymes (Stribny et al. 2016).

The formation of fatty acid ethyl esters is catalysed by Eht1 and Eeb1 acyl-coenzymeA/ethanol O-acyltransferases (AEATases), encoded by the genes *EHT1* and *EEB1*.

The net accumulation of esters in wine is determined by the balance between the activities of enzymes able to synthetize esters and esterase produced by yeasts, responsible for cleavage and, in some cases, in formation of ester bonds (Swiegers and Pretorius 2005).

Extracellular esterases are known to be present in *S. cerevisiae*, although the relevance attributed to the ester synthase (EST, a reverse esterase) as an ester-synthesizing activity is rather limited: two esters, ethyl caprylate and ethyl acetate have been reported as been produced, respectively, by breadmaking and beer yeast strains of *S. cerevisiae* from ethanol and the respective acids (Rojas et al. 2002). As regards extracellular esterases, the situation for non-*Saccharomyces* needs further investigation.

4.3.1 Factor Affecting Esters

Esters are compounds ubiquitous to all wines, but the level formed varies significantly. Apart from being dependent on numerous factors, such as grape cultivar and rootstocks, as well as grape maturity, fermentation temperature, insoluble material in the grape must, vinification methods, skin contact, must pH, sulphur dioxide amount, amino acids present in the must, the concentration of esters produced during fermentation is significantly and sometimes considerably dependent on the yeast strain(s) performing the process (Mateo et al. 1992). Ester production during alcoholic fermentation is closely related to the particular yeast specie/strain involved and with respect to acetate esters is widely believed to be dependent on the balance of ester synthesis by alcohol acetyltransferases and ester hydrolysis by ester-hydrolases.

Traditionally, non-*Saccharomyces* wine yeasts are considered good producers of esters and are associated with negative effects due to the high production of ethyl acetate, whereas these yeasts generally produced much lower amounts of ethyl esters than *S. cerevisiae* (Rojas et al. 2001, 2003).

Species belonging to the genera *Hanseniaspora*, *Candida*, *Hansenula,* and *Pichia* were described as having a greater capacity to produce ethyl acetate than *S. cerevisiae* wine strains.

Hanseniaspora and *Pichia* produced similar levels of ethyl acetate, but *Hanseniaspora* is able to promote the esterification of various alcohols, such as ethanol, geraniol, isoamyl alcohol and 2-phenylethanol, thus increasing concentrations of esters with a fruity aroma, such as fruity acetate esters, i.e. 2-phenylethyl acetate and isoamyl acetate (Rojas et al. 2001; Viana et al. 2008). Among *Hanseniaspora* species, *H. uvarum* is reported to be a good producer of esters in general, whereas *H. guilliermondii* and *H. osmophila* are strong producers of 2- phenylethyl acetate. *Pichia* and *Rhodotorula* are considered producers of remarkable amounts of isoamyl acetate. *Torulaspora delbrueckii* is reported as a producer of small amounts of ethyl acetate, whereas the production of ethyl caprylate seems to be a characteristic of this species (Viana et al. 2008).

The new yeast species *Kazachstania gamospora* produced more esters than the *S. cerevisiae* control strain, in particular phenylethyl propionate, a desirable ester in wine in consequence of its floral aroma (Beckner Whitener et al. 2015).

C. pulcherrima is also known to be a high producer of esters (Clemente-Jimenez et al. 2004), especially the pear-associated ester ethyl caprylate (Lambrechts and Pretorius 2000; Clemente-Jimenez et al. 2004).

4.4 Carbonyl Compounds

A very important volatile aroma fraction of alcoholic beverages is composed of carbonyl compounds, which comprise aldehydes and ketones. They are key compounds in the biochemical reaction when the yeast produces fusel alcohols from amino acids and sugars and are formed in the yeast cells and then transferred to the medium.

4.4.1 Acetaldehyde

Acetaldehyde is one of the most common and important sensory carbonyl compounds formed during vinification and constitutes more than 90% of the total aldehyde content in wine (Nykanen 1986). Various levels of acetaldehyde are found in wine, with average values of about 80 mg/l for white wine, 30 mg/L for red wine and 300 mg/l for sherries. While high levels of acetaldehyde are generally undesirable in table wines, high concentrations of this volatile compound are considered a unique feature of sherry-type wines. Acetaldehyde at low levels gives a pleasant fruity aroma, but at high concentrations it possesses a pungent irritating odour. Indeed, an excess of acetaldehyde produces a green, grassy or apple-like off-flavour in wine.

The sugar is the primary substrate of acetaldehyde formation, but metabolism of amino acids such as alanine also contributes to the formation of this compound. During alcoholic fermentation, acetaldehyde arises from yeast metabolism of sugars via the action of pyruvate decarboxylase (PDC) and alcohol dehydrogenase (ADH).

In addition, acetaldehyde may also be produced after alcoholic fermentation through oxidation of ethanol, which can be chemical (when wine is exposed to air) or biological (by the activity of film yeasts). Acetaldehyde is also extremely reactive and readily binds to proteins or individual amino acids to generate a wide range of flavour and odour compounds.

Acetaldehyde is also a precursor metabolite for acetate, acetoin, and ethanol synthesis. Generally, acetaldehyde reaches a peak value during the early fermentation phases, and is then partly reutilized by yeast. Furthermore, the acetaldehyde levels reach a maximum when the rate of fermentation is at its fastest, then decreases towards the end of fermentation, only to slowly increase again thereafter.

Acetaldehyde is also very polar and may cause water stress in yeasts. Thus, its accumulation (both intracellularly and extracellularly) is one of the central mechanisms of ethanol inhibition of cell yeast growth in ethanol fermentations. Recent evidence shows that acetaldehyde accumulates in fermenting cells of *S. cerevisiae* to concentrations greatly exceeding extracellular levels. In contrast to the inhibitory effect, some evidence indicates that low levels of acetaldehyde stimulate yeast growth under certain conditions. The reason(s) for acetaldehyde stimulation is not fully understood, but may be ascribable to its role in NAD^+ regeneration and energy generation via glycolysis. Acetaldehyde inhibition and stimulation of yeast growth have implications for alcoholic fermentations in winemaking. While low levels of acetaldehyde may be stimulatory to yeast growth, high concentrations of this compound (both intracellular and extracellular) may retard or even inhibit yeast ethanol fermentations, resulting in sluggish or stuck alcoholic fermentations.

In the presence of acetaldehyde, reaction of rapid polymerization between anthocyanins and catechin or tannins occurs with increased colour intensity and stability, but further reaction with polymerized catechin and tannins leads to instability, precipitation and decreased colour. Enhanced colour stability is due presumably to the new compounds formed being partly or wholly resistant to bleaching by sulphur dioxide.

4.4.2 Diacetyl

An important odorant formed from acetaldehyde is the diketone 2,3-butanedione or diacetyl, which is better known for being the compound responsible for the characteristic aroma and flavour of butter. In wine, the concentration of diacetyl is generally low relative to its flavour threshold (8 μg/l) and appears to be important to determining wine style. This diketone is a major flavour metabolite produced by lactic acid bacteria (LAB); yeasts are also able to synthesise diacetyl during alcoholic fermentation, but the majority of this diacetyl is further metabolised to acetoin and 2,3 butanediol. Diacetyl at low concentrations, and in combinations with other wine aroma compounds, imparts yeasty, nutty, toasty aromas, whereas at high concentrations, it has a characteristic buttery aroma that is associated with a lactic character. When present at a high concentration (exceeding 5–7 mg/l), diacetyl is regarded as undesirable in wine, or a spoilage character.

This compound is highly reactive and has been found to react with cysteine, forming sulphur compounds that can influence wine aroma.

4.4.3 Factors Affecting Carbonyl Compounds

There are considerable species and strain differences in acetaldehyde production by yeasts. *Saccharomyces cerevisiae* strains usually produce higher acetaldehyde levels (about 5–280 mg/l) than non-*Saccharomyces* species (up to 40 mg/l), such as *Kloeckera apiculata, Candida krusei, C. stellata, H. anomala*, and *M. pulcherrima* (Fleet and Heard 1993). A mean acetaldehyde concentration of around 25 mg/l was described for *H. uvarum* strains, although significant differences in production among strains were observed (Romano et al. 2003), with a range of 9.5–66 mg/l.

It must be considered that factors, such as temperature, oxygen and sulphur dioxide (SO_2), affect the production of acetaldehyde by yeasts. Fermentation temperature can affect the final total acetaldehyde content in wine, although conflicting results are reported. Some authors found that high fermentation temperatures determined high level of acetaldehyde in wine; i.e. Romano et al. (1994) demonstrated that fermentations at 30 °C produced more acetaldehyde than those at 12 °C, 18 °C or 24 °C and others (Torija et al. 2003) observed that fermentations at 20 °C led to an acetaldehyde content of 90 mg/l in the final wine, while concentrations of 50 and 20 mg/l were obtained at 15 °C and 35 °C, respectively. However, these results are inconsistent with those obtained by other authors. Jackowetz et al. (2011) demonstrated that cool fermentation temperatures (12 °C versus 20 °C) led to higher acetaldehyde residues. Among the enological parameters, the use of SO_2 affects the acetaldehyde concentrations during alcoholic fermentation; in particular, SO_2 induces acetaldehyde formation by yeasts and wines fermented with SO_2 have considerably higher acetaldehyde levels than wines made without SO_2 added. Sulphur dioxide is characterized by the binding capability to acetaldehyde and other compounds, such as pyruvic acid and α-keto-glutaric acid (Romano and Suzzi 1993). The bond SO_2-acetaldehyde is very strong ($Kd = 2.06 \times 10^{-6}$) and hence, SO_2 scavenges the terminal electron acceptor of the alcoholic fermentation. It was demonstrated (Park and Hwang 2008) that SO_2 induced transcription of enzymes involved in carbohydrate metabolism, particularly PDC, and the authors highlighted the role of acetaldehyde in detoxifying SO_2, which can negatively affect energy metabolism in *S. cerevisiae*. As a consequence, the SO_2-induced production of acetaldehyde appears to be related to SO_2 resistance in yeasts.

The total SO_2 consists of bound and free forms. At wine pH of 3–4, free SO_2 consists mainly of bisulphite anion (HSO_3^{-1}) and a small proportion of molecular SO_2 ($SO_2.H_2O$) and sulphite anion (SO_3^{-2}). A number of carbonyl compounds (mainly acetaldehyde, pyruvic acid and α-keto-glutaric acid) can bind with free SO_2 (especially the bisulphite ion) to form a complex compound (bound SO_2), which has only weak antimicrobial properties. One of the properties of added SO_2 is to limit acetaldehyde formation and to bind acetaldehyde formed, so that a wine's taste and aroma are protected or improved. Anaerobiosis, low pH and/or high sugar content apparently promote acetaldehye production by yeasts.

4.5 Sulphur Compounds (hydrogen sulphide, volatile thiols)

Among the volatile metabolites released by yeasts and involved in wine aroma, the sulphur-containing compounds strongly affect wine organoleptic properties in consequence of their very low detection thresholds.

Sulphur compounds comprise a structurally diverse class of molecules that provides a whole range of characteristic aromatic notes. These compounds in wine can be a "double-edged sword". On the one hand, certain sulphur-containing volatile compounds such as hydrogen sulphide, imparting a rotten egg-like aroma, can have a negative impact on the perceived wine quality, and on the other hand, some sulphur compounds such as 3-mercaptohexanol, imparting fruitiness, can have a positive impact on wine flavour and aroma. Furthermore, these compounds can become less or more attractive or repulsive depending on their absolute and relative concentrations.

Generally, the aromatic contributions of these compounds are considered detrimental to wine quality; however, new developments in wine research allowed the differentiation of a family of sulphur compounds responsible for a varietal aroma of wines.

The main volatile sulphur compounds detected in wine (Table 2.1) are: (1) hydrogen sulphide; (2) methanethiol (methylmercaptan); (3) dimethylsulphide, dimethyldisulphide, and dimethyltrisulphide; (4) methylthioesters (S-methyl

Table 2.1 Sensory impact of some volatile sulfur compounds in wine

Compound	Sensory impact
Hydrogen sulfide	Rotten egg aroma
Mercaptans (also known as thiols)	This is a large group of very smelly sulfur compounds. Terms such as cabbagey, rubbery, struck flint or burnt rubber are used as descriptors
Ethyl mercaptan	Burnt match, sulfidy, earthy
Methyl mercaptan (methanethiol)	Rotten cabbage, cooked cabbage, burnt rubber, stagnant water
Dimethyl sulfide	Vegetal, cabbage, onion-like at high levels. Cooked vegetables, cooked corn, canned tomato at high levels; blackcurrant drink concentrate at lower levels. Quince, truffle
Diethyl sulfide	Rubbery
Carbon disulfide	Sweet, ethereal, slightly green, sulfidy
Dimethyl disulfide	Vegetal, cabbage, onion-like at high levels
Diethyl disulfide	Garlic, burnt rubber
4-Mercapto-4-methylpentan-2-one (4MMP), 3-mercaptohexan-1-ol (3MH), 3-mercaptohexyl acetate (3MHA)	Tropical fruit/passion fruit at low levels; cat's urine at higher levels
Benzenemethanethiol	Smoky/gunflint aromas
Methylthioesters (S-methyl thioacetate, S-methyl thiopropanoate, and S-methyl thiobutanoate)	Cooked cauliflower, cheesy, and chives aromas

thioacetate, S-methyl thiopropanoate, and S-methyl thiobutanoate; (5) the "fruity" polyfunctional thiols 3-mercaptohexan-1-ol (3MH), 4-mercapto-4-methyl-pentan-2-one (4MMP), and 3-mercaptohexyl acetate (3MHA).

The mechanism of production of many sulphur compounds is still not well-known. Beside enzymatic formation resulting from yeast metabolism, other non-enzymatic reactions may occur during winemaking and storage through the influence of temperature, light and chemical transformation. Moreover, sulphur-containing pesticides or other breakdown products can be used as precursors in biochemical reactions arising from grape juice enzymes or yeast metabolism. Yeasts form sulphur compounds by three main mechanisms: (1) degradation of amino acids containing sulphur; (2) degradation of sulphur-containing pesticides; (3) release and/or metabolism precursors containing sulphur deriving from grapes.

4.5.1 Hydrogen Sulphide

The best known sulphur compound in wine is probably hydrogen sulphide, a highly volatile thiol, conferring 'rotten egg' aroma. In consequence of the common frequency and very low odour threshold (50–80 μg/l) of this compound, its production during wine fermentation is a frequently encountered problem in winemaking and it's necessary to limit H_2S content in wine in order to avoid a loss in quality and rejection by consumers. In *S. cerevisiae*, H_2S is the product of the sulphate reduction sequence (SRS) pathway. The wine yeasts metabolism forms H_2S from inorganic sulphur compounds (sulphate and sulphite) or organic compounds, such as cysteine and glutathione. Generally, grape must is deficient in organic sulphur compounds, such as the sulphur containing amino acids methionine and cysteine. These amino acids are essential for the growth of *S. cerevisiae* and, if they are not present, or exhausted in the growth medium, sulphur compounds have to be assimilated from inorganic sources, such as extracellular sulphate, usually abundant in grape must.

In the first step of the SRS pathway, sulphate is transported into the cell by two specific permeases; a first step of reduction produces sulphite which is further reduced to sulphide. In the presence of a suitable nitrogen supply, the sulphide produced is combined with a nitrogenous precursor, O-acetyl serine or O-acetyl homoserine, to form cysteine and methionine. When nitrogen sources are insufficient or unsuitable, free H_2S can accumulate in the cell and diffuses from the yeast cell into the wine.

4.5.2 Other Sulphides

Hydrogen sulphide is a highly reactive species, which can take part in a range of reactions to generate compounds that impact on wine flavour. For example, mercaptans, such as ethanethiol can be formed by the reaction of hydrogen sulphide with ethanol or acetaldehyde.

Dimethyl sulphide (DMS), which usually elicits odours described as 'asparagus', 'corn' and 'molasses', is considered a beneficial compound in low concentrations, contributing to the aroma of bottle age. The concentration of DMS found in wine is well above the sensory threshold of 25 μg/l (white wine) and 60 μg/l (red wine). The formation of DMS is not clear. It could be formed in a similar way to other mercaptans. During fermentation, DMS is synthesized by yeast from cysteine, cystine and glutathione or it can be produced from dimethyl sulphoxide by yeast reductase (Ribéreau-Gayon et al. 2000b). During wine maturation, DMS is formed through a yeast mechanism by cleavage of S-methyl-L-methionine to homoserine and dimethyl sulphide.

In wine, the formation of the polysulphides dimethyl disulphide, dimethyl trisulphide, and dimethyl tetrasulphide is thought to occur through the oxidation of mercaptans, e.g. oxidation of methyl mercaptan to form dimethyl disulphide. Yeast can reduce disulphides to mercaptans.

4.5.3 Volatile Thiols

Volatile thiols, in particular 4-mercapto-4-methylpentan-2-one (4MMP), 3-mercaptohexan-1-ol (3MH) and 3 mercaptohexyl acetate (3MHA), are some of the most potent aroma compounds found in wine. At optimal concentrations in wine, these compounds impart flavours of passionfruit, grapefruit, gooseberry, blackcurrant, lychee, guava and box hedge. Of these, 4MMP has the lowest sensory detection threshold of 3 ng/l in wine, followed by 3MH and 3MHA, having reported detection thresholds of 60 ng/l and 4 ng/l, respectively (Tominaga et al. 2000; Murat et al. 2001). These compounds are of particular importance for the varietal character of Sauvignon Blanc wines and were highly appreciated by consumers in some styles of these wines. These volatile thiols are virtually non-existent in grape juice and only develop during fermentation. Although the wine yeast *Saccharomyces cerevisiae* is responsible for the formation of volatile thiols during wine fermentation, it does not synthesize these types of volatile thiols de novo, but yeast is involved in the production of these aromatic compounds during wine fermentation cleaving of the thiols from precursors found in grape juice. For the formation of 4MMP and 3MH, cysteinylated (cys) and glutathionylated (glut) conjugates are taken up by yeast and converted to their respective thiols. No cysteine or glutathione precursor of 3MHA has been identified, and this compound is formed during fermentation and through esterification of 3MH by the alcohol acetyltransferase *ATF1* to yield 3MHA (Dubourdieu et al. 2006; Coetzee and du Toit 2012). Carbon-sulphur lyase enzymes are necessary for the cleaving of cysteine-glutathione conjugated precursor with release of the correspondent volatile thiols. In grape must, the cystenilated forms are generally more abundant than the gluathionylated forms. However, the contribution to total thiols deriving from cys and glut precursors has come under increasing exploration as the yield from the pathway is low, as the concentrations of cys and glut conjugates are high, whereas conversion rates of precursor added to synthetic medium is very low. Moreover, the concentration of 3MH precursors in juice does not correlate with the concentration of thiols in the final wine. It seems that additional pathways contribute to formation of varietal thiols; i.e. an alternative pathway proposed for thiols formation (Schneider et al. 2006) suggests that 3MH is derived from E-2-hexenal, a 'green leaf volatile' found in grape juice, which could react directly with H_2S to form 3MH or react with cysteine and/or glutathione to provide precursors which yeast then biochemically converts to thiols.

The yeast genes involved in the pathway from cys and glut precursors to thiols are not yet clearly established. The uptake of the precursors is mediated by general amino acid transporters, such as *GAP1* and *OPT1* transporters, which are responsible for the uptake of the major part of the cys and glut precursors, respectively. For the cleaving of cysteinylated precursors inside the cell, a carbon-sulphur β-lyase enzyme is involved and four yeast β-lyase genes (*BNA3*, *CYS3*, *GLO1* and *IRC7*) influence the release of volatile thiol 4MMP, with Irc7-p the main enzyme responsible for its production. *STR3* β-lyase is also responsible for thiol release, with high incidence in 3MH but with a low specific activity. The glut precursors, which enter the cell through Opt1p, are degraded to the cys form as an intermediate in a multi-step pathway, in which the genes *DUG1*, *DUG2*, *DUG3*, *CPC*, *CPY* and *ECM38* are involved. Also the gene *CIS2* gene, encoding γ-glutamyl transpeptidase, is also required for the conversion of glutathione precursors to volatile thiols (Belda et al. 2017).

S. cerevisiae is able to release as volatile thiols about 10% of the precursor originally present in the must and the low efficiency is probably due to NCR, one of the most important regulation controls of thiol releasing pathways. NCR affects transport genes (such as *GAP1*) and genes involved in precursors cleavage (such as *IRC7*).

4.5.4 Factors Affecting Sulphur Compounds

It is well established that *S. cerevisiae* is responsible for H_2S off-flavour in wine and that the production is strain dependent, although not all wine strains produce H_2S, as it was found that about 1% of naturally occurring wine strains are unable to produce this compound. The strain ability to produce H_2S in variable amounts might be related to activity of sulphite reductase, one of the main enzymatic activities responsible for the production of this compound. It was found that sulphite reductase activity is a rare feature among the majority of non-*Saccharomyces* species (Belda et al. 2016), since among the 15 non-*Saccharomyces* species tested, only species belonging to *Hanseniaspora* genus (mainly *H. osmophila* and *H. opuntiae*) had quite high sulphite reductase activity. Furthermore, some *T. delbrueckii* strains, apart from *S. cerevisiae*, exhibited a certain H_2S production ability. However, as found in *S. cerevisiae*, a behaviour highly variable in function of strains was found in other wine related yeast species, such as *Dekkera*, *Lachancea*, *Hanseniaspora*, and *Metschnikowia*.

Other environmental and nutritional factors that can affect H_2S production include: (i) high residual levels of elemental sulphur used in the vineyard for plant protection; (ii) presence of sulphur dioxide; (iii) presence of sulphur-containing organic compounds; (iv) vitamin deficiency; (v) high concentration of amino acids like threonine, methionine or cysteine, and nitrogen limitation.

One of the main factors affecting content of sulphur compounds in wine is the yeast metabolism of nitrogen compounds in the media. A starvation for nitrogen could be expected to increase flux of sulphur through the pathway involving the reduction of inorganic sulphur to H_2S by SRS. The activity of the SRS pathway is tightly regulated to match the metabolic demand for methionine and cysteine, which in grape juices usually are not sufficient to meet the metabolic needs of growing cells. A starvation for nitrogen could be expected to deplete the cell of these regulatory end products, resulting in a de-repression of the structural genes of the SRS enzymes and hence an increased flux of sulphur through the pathway. A similar methionine shortage and hence overproduction of H_2S can develop from deficiencies of vitamins which act as cofactors to SRS enzymes, but the common use of vitamin supplements prevents this route of H_2S overproduction from being significant. Furthermore, in some strains the sulphite present in the fermentation medium readily diffuses in the cell, determining a consistent production of H_2S. Therefore, the presence of sulphite, in conditions of nitrogen starvation, gives rise to high and continuous production of H_2S.

Cells which undergo autolysis after fermentation can also release H_2S, probably in consequence of degradation of sulphur-containing amino acids.

Fermentation temperature influences the amount of volatile thiols in wines. Masneuf-Pomarède et al. (2006) found that the final levels of 4MMP and 3MH in wines were higher when the alcoholic fermentation was conducted at 20 °C than at 13 °C. The 3MHA, which was correlated with the amount of 3MH determined in wines, was also higher when the alcoholic fermentation was conducted at 20 °C. Usually, low fermentation temperature enhanced the aromatic characteristics of wines, possibly because of greater synthesis and a greater retention of volatile flavours (Ribéreau-Gayon et al. 2000a), whereas results concerning levels of volatile thiols in wines show opposite conclusions. Probably, the change of membrane fatty acids depending on fermentation temperature can be one hypothesis to explore the negative impact of low temperature on the level of volatile thiols.

Other factors, such as addition of nutrients in rehydration media to active dry yeast (Winter et al. 2011), pre-fermentation operations, such as skin contact (Peyrot Des Gachons et al. 2002), as well as oxygen, phenol, and sulphur dioxide content, affect the production of polyfunctional thiols (Blanchard et al. 2004).

Other than the factors previously cited, the genetic and physiological nature of the yeast strain used to conduct the fermentation is one of the most important factors that affect thiol releasing (Cordente et al. 2012) and therefore selection of yeast strain is highly important to modulate the concentration of these compounds in wine

Several studies (King et al. 2010; Zott et al. 2011) have highlighted natural yeast variation in capacity to release and esterify polyfunctional thiols and the ability of some non-*Saccharomyces* yeasts to contribute positively to release of volatile thios from their cysteinylated precursors, but generally with a higher incidence in 3-MH. It was reported (Anfang et al. 2009) a significant enhancement of 3-MHA production by *Pichia kluyveri* in co-fermentation with *S. cerevisiae*; also Renault et al. (2016) found that an industrial *T. delbrueckii* (Zymaflore® Alpha$^{TDn.sacch}$) strain was able to release 3-MH in co-culture fermentation with *S. cerevisiae*, but not 4-MMP. Belda et al. (2016) evidenced a remarkable β-lyase activity from cysteinylated precursors in *T. delbrueckii*, *Kluyveromyces marxianus* and *Meyerozyma guilliermondii*. It also reported that *Saccharomyces* interspecies hybrid yeasts produce relatively high concentrations of polyfunctional thiols (Swiegers et al. 2009).

5 Tertiary Aroma: Post-Fermentative Aroma

The wine is a dynamic product that is submitted to ageing or maturation (performed in oak barrels or bottle), an important step to improve sensorial characteristics (Fig. 2.1). This step has been traditionally associated with red wines but, nowadays with more frequency, it is used also for white and rosé wines. During aging, the wine acquires a more complex aroma and better taste due to the loss of sensations of astringency and bitterness. In this process, the wine lost characteristic aromas linked to the grape varietal and fermentation, but new compounds appear from oak wood and from the evolution of the primary and secondary aromas. In particular, it was found that the concentrations of ethyl esters of branched-chain fatty acids changed during ageing. If the maturation is performed on lees, deriving from yeast cells autolysis, the concentrations of volatile compounds imparting a fruity aroma decrease, whereas contents of long-chain alcohols and volatile fatty acids increase.

During the process of autolysis, the cells release various cellular components into the wine, such as compounds containing nitrogen, amino acids, peptides, and proteins, mannoproteins and lipids. Lipids released during autolysis liberate fatty acids, that can give rise to volatile components, such as esters, aldehydes, and ketones, affecting aroma and flavour of wine.

Among compounds influencing tertiary aroma, those affected by yeast activity are, principally, the volatile phenols. These compounds greatly influence wine aroma in consequence of their relatively low detection threshold and, therefore, they are easily detected. The most important molecules in this class are 4-ethylphenol and 4-ethylguaiacol, among others (4-ethylcatechol, 4-vinylguaiacol, 4-vinylphenol and 4-vinylcatechol). Although volatile phenols can contribute positively to the aroma of some wines, they are better known for their contribution to off-flavours known as "Brett character". These phenolic off-flavours are described as horse sweat, humid leather, smoky, plastic, phenolic, medical, band-aid and poultry yard. When the sum of the two volatile phenols exceeds 620 μg/l, the 'Brett' character becomes too pronounced. The sensorial threshold of 4-ethylphenol is 230 μg/l, and therefore small quantities of 4-ethylphenol are easily appreciable in wine. 4-ethylguaicol affects wine aromas to a lesser extent, but it is also related to the 'Brett character' of adulterated wines and have been associated with descriptive expressions, such as "bacon" or "smoked". It has been reported that wines with high, medium and no 'Brett' character have average 4-ethylphenol concentrations of 3.0, 1.74 and 0.68 mg/l, respectively.

The origin of volatile phenols involves the sequential action of two enzymes on a hydroxycinnamic acid, which is highly widespread in plants and in grapes primarily consist of ferulic, *p*-coumaric or caffeic acid. Hydroxycinnamate decarboxylase first turns *p*-coumaric and ferulic acids into hydroxystyrenes (4-vinylphenol and 4-vinylguaiacol respectively). Following this, vinylphenol reductase converts 4-vinylphenol and 4-vinylguaiacol into 4-ethylphenol and 4-ethylguaiacol. (Dias et al. 2003). The precursors, *p*-coumaric acid and ferulic acids, are naturally present in must. The enzyme that facilitates the decarboxylation is present in a large number of bacteria, fungi, and yeasts, but the reduction step is only performed by the species *Dekkera bruxellensis*, *D. anomala*, *Pichia guilliermondii*, *Candida versatilis*, *C. halophila* and *C. mannitofaciens*. Initially, the presence of ethylphenols in wine was attributed to lactic acid bacteria. Indeed, these are capable of producing significant quantities of vinylphenols, but under oenological conditions they only produce small amounts. Other yeasts present in wines, such as *S. cerevisiae*, *Pichia* spp., *Torulaspora* spp. and *Zygosaccharomyces* spp. can produce 4-vinylphenol but do not reduce it to 4-ethylphenol (Dias et al. 2003). In particular, *Brettanomyces* yeasts or the members of the sporulating genus *Dekkera*, i.e. *D. bruxellensis* shows hydroxycinnamate decarboxylase and vinyl reductase activities under oenological conditions to the extent that this species is considered an undesirable yeast capable of producing high concentrations of 4-ethylphenol and 4-ethylguaiacol.

5.1 Factor Affecting Tertiary Aroma

The amounts of volatile phenols detected in wine is proportional to the the *Brettanomyces/Dekkera* population (Gerbeaux et al. 2000). However, the capacity to produce volatile phenols is a strain character, different strains of *B. bruxellensis* vary in the production of these compounds (Gerbeaux et al. 2000). Other factors influence the presence of volatile phenols in wine, such as alcohol content, which is inversely related to the concentration of these compounds. Also the temperature affects volatile phenol production: higher levels are produced at 18 °C than at 13 °C. Low influence is attributed to the pH value of wine or the presence of residual sugars. The intensity and temperature of maceration and the use of pectolytic enzymes may be other possible factors conditioning the formation of volatile phenols by *Brettanomyces* and *Dekkera* from hydroxycinnamic acids released from grape skins (Gerbeaux et al. 2002). It also been shown that the volatile phenols can be removed by wine ageing on the lees due to its biosorbent activities (Mazauric and Salmon 2005). Wide and deepened information about spoilage yeasts are dealt in the Chap. 12.

The Fig. 2.2 reports the list of the main yeast genera/species involved in primary, secondary and terziary aroma.

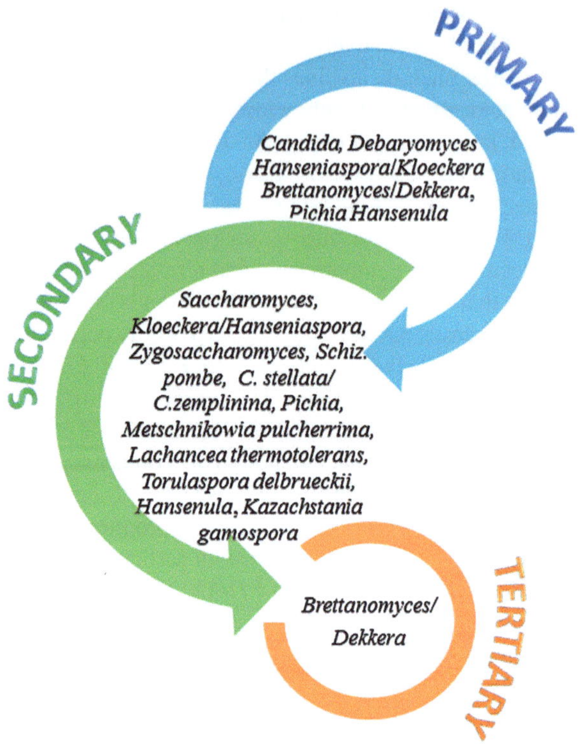

Fig. 2.2 Main yeast genera/species involved in primary, secondary and tertiary aroma

6 Influence of Starter Cultures on Wine Flavour

Taking into account that grape must is a natural substrate, not sterilized or filtered, it can be freely metabolized by the present microorganisms, whose dominance and activity vary in function of different internal and external factors occurrence. For this reason, it is not always possible to obtain the same product from spontaneous wine fermentation, even if the grapes are of the same variety and derive from the same vineyard. In spontaneous fermentation, several different strains of *S. cerevisiae* compete with each other and also with non-*Saccharomyces* yeasts, determining non-homogenous final products in the organoleptic quality.

6.1 Single Starter Cultures: Saccharomyces

The problem of uncontrolled aromatic composition in wines produced by spontaneous fermentation can be solved by the use of starter cultures, which are inoculated into the grape juice in order to establish a high population and accomplish a

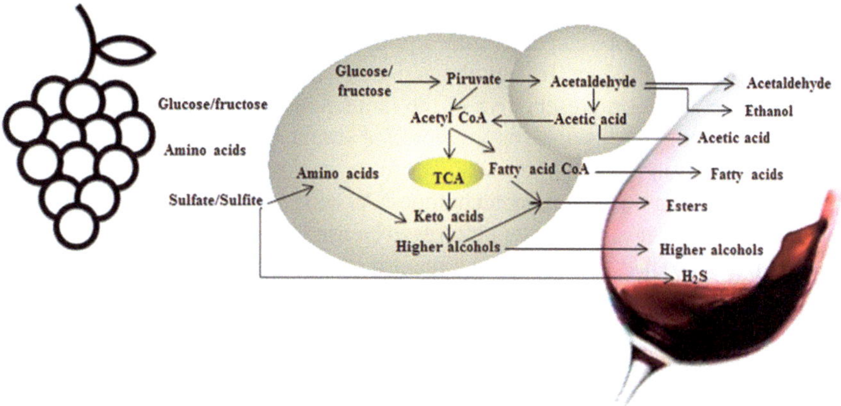

Fig. 2.3 Metabolic scheme of major aroma compounds produced by *Saccharomyces cerevisiae* during fermentation

well-controlled must fermentation. The yeasts used as starter cultures are selected strains, mainly of the species *S. cerevisiae,* Metabolic scheme of major aroma compounds produced by *S. cerevisiae* during fermentation is reported in Fig. 2.3.

Strains of this species have distinct physiological properties rendering them suitable for wine fermentation. This had led to the widespread use of commercially available dried yeast strains of *S. cerevisiae*.

As regards wine aroma, numerous reports confirm that different strains from the same yeast species produce the same fermentation metabolites, but in different concentrations, which affect wine flavour. Fermentations inoculated with different *S. cerevisiae* strains reveal the impact of selected strains on wine quality. It's widely recognized that the metabolic characteristics of a particular yeast strain may lead to the formation of metabolites and the transformation of grape molecules that may sensorially affect the wine. An extensive literature is available reporting the considerable strain variability within the species *S. cerevisiae* in the production of secondary compounds, such as acetaldehyde, fusel alcohols, esters, fatty acids (Estévez et al. 2004; Nikolaou et al. 2006). The wines have significantly different volatile characteristics, which determine large sensory differences in the final product. Production of wines with different sensory characteristics from the same grape variety may be of commercial advantage in order to satisfy the different pleasure of the consumers. In addition, it was reported (Regodón Mateos et al. 2006) that wines elaborated with different grape varieties and under different fermentation conditions exhibit more homogeneous properties when the must fermentations are conducted with the same yeast strain.

As the importance of the role of *S. cerevisiae* in winemaking has been firmly established, the number of wine yeast strains available in the world market for use as winemaking starters has increased in the last years. These commercial strains have been isolated by prestigious wine research Institutes and characterized for their technological traits, but their use for the production of wine of different variety and origin could be inappropriate and detrimental for wine's autochthonous character. Their individual activities can determine a potential uniformity of aromatic characteristics in the final product. In fact, the active dry oenological yeasts available on the market are not yet representative of the indigenous yeasts of most of the wine regions. Today, the need is also for *S. cerevisiae* strains better adapted to the different wine production regions in the world with their respective grape varieties, viticultural practices and winemaking techniques. As wine quality is directly dependent on wine yeast attributes, the starter cultures to be used have to be chosen in function of the wines to produce. Thus, yeast strains for inoculated fermentation can be differentiated in "universal" and "specific" starter cultures. The "universal" starter cultures can be considered strains to be used in inoculated fermentation to produce wide-diffused wines with stable and uniform organoleptic quality. Not all universal strains enhance the organoleptic quality of all wines, but they can guarantee more standardized products, which satisfy a wide-ranging commercial market. On the other hand, specific strains starter cultures can be used in inoculated fermentation for specific variety/grape must, producing niche wines with optimized varietal quality.

The common practice to use active dry "universal" yeasts can determine a progressive substitution of natural local microflora and the consequent reduction of some typical organoleptic properties. In order to avoid a loss of wine typicality it is necessary to highlight the importance of selecting indigenous wine strains from each area with the aim to preserve the typical sensory properties of wine produced in that region. In fact, some researchers believe that each microclimate, such as the vineyard, is characterised by a specific *S. cerevisiae* yeast flora, where some strains can remain for many years and become representatives of an ecological area.

The existence of specific *S. cerevisiae* strains in different wine regions indicates that this species exhibits at least some degree of geographic population structure, perhaps reflecting an adaptation to specific microenvironments (Knight et al. 2015; Capece et al. 2016). It is demonstrated that yeast strains are fully adapted to a certain specific climatic environment and substrate and some oenologists admit that good results can be obtained only with selected yeast starters originating from the microarea where wines are produced (Capece et al. 2010).

This result supports the idea that some strains are better adapted to certain must conditions and, therefore, they should be recommended for those cultivars if the final wine quality is improved. This does not mean that only native local strains can perform their grape must fermentation, but it means that the strain selection for winemaking must consider the individual characteristics of each grape must.

On the other hand, strain metabolic behaviour seems to be correlated also to isolation origin. Some authors reported a common metabolic pattern among strains isolated from the same regions, which differed from that of other wine strains isolated from different regions. This result can be a consequence of a better adaptation of the strains to the chemical and microbiological characteristics of the specific grape must (Lema et al. 1996; Mauriello et al. 2009; Capece et al. 2012).

These findings emphasize the beneficial to select specific strains for specific fermentations as a function of the vine variety characteristics in order to take the major advantage from the combination grape must/*S. cerevisiae* strain.

6.2 Mixed Starter Cultures

The contribution by the numerous grape-must-associated non-*Saccharomyces* yeasts to wine fermentation has been debated extensively. These yeasts, naturally present in all wine fermentations, are metabolically active and their metabolites can impact on wine quality. Although often seen as a source of microbial spoilage, in the past three decades there is substantial contrary evidence pointing to a positive contribution by these yeasts (Jolly et al. 2006). Numerous studies has underlined the important role of non-*Saccharomyces* yeasts to improve wine complexity and specificity (Comitini et al. 2017).

As non-*Saccharomyces* yeasts are in general poor fermenters, the current trend in the wine industry is the design of mixed starters composed by non-*Saccharomyces* with optimized biotechnological characteristics and *S. cerevisiae* to ensure a complete fermentation. The use of mixed cultures in wine fermentation processes, combined with vinification technology, may lead to the production of wines with different characteristics, allowing to winemakers tailoring wines to the changing demands of consumers. Different mixed starters have been designed in order to enhance wine quality (reviewed by Padilla et al. 2016); some of them were formulated with aim of modifying specific targets, such as terpenic profile or concentration of final esters, whereas others to affect the overall complexity of wine aroma. For this reason, several authors have studied fermentation with mixtures of different yeast species, either applied simultaneously or in sequential cultures. In the first case (co-inoculation) the selected non *Saccharomyces* yeasts are inoculated at high viable cell concentration together with *S. cerevisiae*, whereas in the second case (sequential inoculation) the selected non-*Saccharomyces* yeasts are first inoculated at high level, allowing to ferment for a given time, before inoculation of *S. cerevisiae* to complete the fermentation. The use of both practice are feasible, the choice of more appropriate inoculation strategy is based on the potential interaction between yeasts (Ciani et al. 2016).

The use of mixed starter can affect both primary and secondary aroma, by the production of enzymes and metabolites, respectively.

As regards primary aroma, *T. delbrueckii*, *M. pulcherrima*, *D. hansenii*, and *D. pseudopolymorphus*, able to secrete β-D-glucosidase enzyme, were used in combination with *S. cerevisiae* to enhance the terpene content in wine.

The mixed starter composed by *T. delbrueckii*/*S. cerevisiae* increases the content of α-terpineol and linalool in Gewürztraminer wine, enhancing the overall quality of this wine.

The use of *Metschnikowia pulcherrima*, in either simultaneous or sequential inoculation, yielded wines with concentration of nerol and geraniol significantly lower than those observed in grape must, probably in consequence of the *S. cerevisiae* ability to transform nerol and geraniol into α-terpineol at must pH, demonstrating the fundamental role played by yeast interaction in mixed starter fermentation.

Among *Debaryomyces* species, a *D. vanriji* strain was found to influence the concentrations of several volatiles, such as terpene, in consequence of the production of pectinase, amylase, and xylanase activities along the fermentation and a *D. pseudopolymorphus* strain increased the concentration of citronellol, nerol, and geraniol during the fermentation of Chardonnay juice (Cordero-Otero et al. 2003).

Strains of *C. zemplinina and P. kluyvery* produced wines with increased concentration of the volatile thiols 3MH and 3MHA compared with the *S. cerevisiae* single fermentation, although the effect could not be explained as simple additive assumptions, but as the result of interaction between the co-fermenting partners. Furthermore, it was demonstrated that this interaction may not be generalized to the species level, but it is dependent from specific *S. cerevisiae* strain used together the non-*Saccharomyces* yeast. Nowadays, for the improvement of wine primary aroma selected non-*Saccharomyces* strains are available on the market, such as a *M. pulcherrima* strain (recommended for Riesling and Sauvignon Blanc wines) and a selected strain of *P. kluyveri*, recommended for its ability to assure a more efficient conversion of flavour precursors into volatile thiols.

As regards the influence of mixed starter on secondary aroma, one of most investigated use of these starters is correlated to the regulation of wine acidity, in order to reduce the volatile acidity, for acidifying wines or for biological deacidification of must and/or wine. *Schizosaccharomyces pombe* and *Issatchenkia orientalis* (alternative name *Pichia kudriavzevii*) were tested to reduce malic acid in grape juice and/or wine. The combination *S. pombe/S. cerevisiae* was successful in biological deacidification of white and red wines (Benito et al. 2014b), such as wines co-fermented by *I. orientalis* and *S. cerevisiae* showed decrease in malic acid concentrations and high score in sensory evaluation (Kim et al. 2008; Del Mónaco et al. 2014).

Recently, the combined use of selected *S. pombe* and *Lachancea thermotolerans* strains has been reported as an alternative approach to malolactic fermentation as malic acid is totally consumed by *S. pombe*, whereas *L. thermotolerans* produces lactic acid, maintaining or increasing the acidity of wines produced from musts with low acidity. Furthermore, the fruity character of wine was increased, while content of acetic acid or biogenic amines was lower compared to traditional malolactic fermentation controls (Benito et al. 2015). A commercial yeast strain of *S. pombe* is now available in immobilized form to reduce the malic acid content in wine (ProMalics; Proenol, http://www.proenol.pt/files/products/ProMalic_09_2008.pdf).

L. thermotolerans, characterized by high production of fixed acidity and low production of volatile acidity, is a potential acidifying microorganism, very useful to compensate the insufficient acidity of specific grape varieties, an aspect of increasing interest, as global climate change and variations in viticulture and oenology practices determined a reduction in total acidity of wines. Other than enhancement in the total acidity and reduction in the volatile acidity, compared to single

S. cerevisiae cultures, the use of this species determined an increase in glycerol and main esters and sensory analysis tests showed significant increases in the spicy notes (Comitini et al. 2011; Gobbi et al. 2013).

The use of non-*Saccharomyces* yeasts was proposed as a tool to reduce high levels of acetic acid in wine. Different authors (Bely et al. 2008; Comitini et al. 2011) observed significant reductions in volatile acidity by using *T. delbrueckii*, described as a low acetic acid producer both under standard conditions and in high-sugar media, in mixed fermentation with *S. cerevisiae*. Similar results were obtained by inoculating mixed starters composed by *C. stellata*/*C. zemplinina* and *S. cerevisiae*; furthermore, *C. zemplinina* co-inoculated with *S. cerevisiae* reduced the content of acetic acid, maintaining high levels of glycerol and ethanol (Rantsiou et al. 2012). Also the use of *Pichia fermentans*, in different combinations with *S. cerevisiae*, produced less acetic acid than single cultures of *S. cerevisiae*, other than an increase in content of some aromatic compounds, such as acetaldehyde, ethylacetate, 1-propanol, n-butanol, 1-hexanol, ethylcaprilate, 2,3-butanediol and glycerol (Clemente-Jiménez et al. 2005).

Numerous studies reported the use of apiculate yeasts belonging to *Hanseniaspora* genus, the non- *Saccharomyces* yeasts found in the highest numbers in grape must, to make a contribution to wine quality, mainly for the ability to increase the content of fruity acetate esters, such as 2-phenylethyl acetate and isoamyl acetate (Rojas et al. 2001; Moreira et al. 2005; Viana et al. 2008). As regards *Hanseniaspora* genus, *H. uvarum* is reported as a good producer of esters in general, whereas *H. guilliermondii* and *H. osmophila* are considered strong producers of 2-phenylethyl acetate (Rojas et al. 2001, 2003; Viana et al. 2008). In general, non-*Saccharomyces* wine yeasts are considered as good producers of esters and traditionally have been associated with negative influence on aroma in consequence of high production of ethyl acetate. *Candida*, *Hansenula*, and *Pichia* species were reported as producers of higher amounts of ethyl acetate than wine strains of *S. cerevisiae*. As regards ethyl esters, *T. delbrueckii* is described as producer of ethyl caprylate (Viana et al. 2008), whereas the new discovered species *Kazachstania gamospora* produced more esters than *S. cerevisiae*, in particular phenylethyl propionate, an ester conferring to the wine the desirable floral aroma (Beckner Whitener et al. 2015).

The presence of *T. delbrueckii* in mixed starters with *S. cerevisiae* affected the content of other volatile compounds, such as 2-phenylethyl alcohol, isoamyl acetate, fatty acid esters, C4–C10 fatty acids and vinyl phenols. In particular, the presence of *T. delbrueckii* in mixed starters has been associated with increases in the production of 2- phenylethyl alcohol in different wine styles. Other mixed starters, including *M. pulcherrima*, *L. thermotolerans*, *Kazachstania gamospora*, determined an increased production of 2-phenylethyl alcohol (Comitini et al. 2011; Dashko et al. 2015).

In this context, actually blends of active dried yeasts have become commercially available, such as blends of *S. cerevisiae*/*K. thermotolerans*/*T. delbrueckii*, denominated Vinfloras Harmony.nsac (Christian Hansen) and a blend of *K. thermotolerans* and *S. cerevisiae* (Viniflora® SYMPHONY.nsac). This last blend has been developed for the improvement of aroma and flavour in white and red grape varieties as its use can lead to the enhancement of floral and tropical fruit aromas, and more complex and rounded flavours in white and red wine, respectively.

At this aim, it was suggested the use of *C. stellata*, which was reclassified as *Starmerella bombicola* This species frequently predominates in the early fermentation phase together with apiculate yeasts and it is known as a high glycerol producer. It was reported (Ciani and Picciotti 1995) that wines obtained by mixed fermentation at lab scale with *C. stellata* and *S. cerevisiae* showed increased concentrations of glycerol and succinic acid and reduced concentrations of acetic acid and higher alcohols, whereas other by-products were similar to those found in the *S. cerevisiae* control fermentation. These finding were then confirmed at pilot-scale production of wine (Ferraro et al. 2000).

7 Novel Methodologies to Select Wine Yeasts in Function of Their Influence on Wine Aroma

The wine industry requires rapid, comprehensive methods and techniques to answer the new challenges driven by the market demands.

New markets and consumer demands are pushing the need to produce different and novel wine styles with particular characteristics. To do this, understanding how changes in the yeast genome influence potential flavour metabolites is essential. Recent advances in technology have brought about a revolution in the manner in which biological systems are visualized and analysed. The measurements of all small molecules (metabolites) present in the organism, which represent the interaction of the genome, transcriptome and proteome with the environment, is called metabolomics (Dunn and Ellis 2005; Nielsen and Oliver 2005). The development of metabolomics has depended on advances in a diverse range of instrumental techniques, such as liquid chromatography (LC), electrospray ionization mass spectrometry (ESI–MS), capillary electrophoresis (CE) and microchip arrays, among others (Nielsen and Oliver 2005). Each of these methods provides unique capabilities to separate different chemical classes of metabolites. At the same time, developments in mathematics have provided algorithms capable of unravelling the complexity of the datasets generated. Large-scale metabolome analysis is traditionally based on the use of gas chromatography– mass spectroscopy (GC–MS), liquid chromatography–MS (LC–MS), high performance liquid chromatography (HPLC), ESI–MS, mid-infrared (MIR) and high resolution mass spectrometry. It has been generally accepted that a single analytical technique will not provide sufficient visualization of the metabolome, therefore holistic techniques are needed for comprehensive analysis. Near-infrared (NIR) spectroscopy has gained wide acceptance as a rapid analytical tool and it is mainly used in the wine industry to measure the alcohol content of wines. Spectroscopy in the NIR wavelength range offers the advantage of rapid, non-destructive analysis and routine operation can be obtained. One of the advantages of NIR spectroscopy is its ability to assess chemical structures through the analysis of the molecular bonds in the NIR spectrum (O-H, N-H, C-H). Additionally, the spectrum of a given sample is considered to behave as a 'fingerprint' of the sample. The power of spectroscopy techniques as compared with other conventional approaches (i.e. enzymatic, GC–MS, ESI–MS) is that it measures changes in a complex metabolite environment. The use of NIR spectroscopy has been proposed for a rapid screening of yeasts (Cozzolino et al. 2006), allowing the initial clustering of yeast strains with similar extra cellular metabolomes, followed by a more precise method, such as GC–MS, MS or HPLC.

The spectral differences can be used to classify strains on the basis of their metabolome. As the selection of potentially useful commercial wine strains requires the evaluation of their fermentation profiles and of their impact on wine quality by expensive analysis, developing a rapid metabolomic method will be valuable in facilitating the selection of beneficial strains.

References

Andorrà, I., Berradre, M., Rozès, N., Mas, A., Guillamón, J. M., & Esteve-Zarzoso, B. (2010). Effect of pure and mixed cultures of the main wine yeast species on grape must fermentations. *European Food Research and Technology, 231*, 215–224.

Anfang, N., Brajkovich, M., & Goddard, M. R. (2009). Co-fermentation with *Pichia kluyveri* increases varietal thiol concentrations in Sauvignon blanc. *Australian Journal of Grape and Wine Research, 15*, 1–8.

Beckner Whitener, M. E., Carlin, S., Jacobson, D., Weighill, D., Divol, B., Conterno, L., Du Toit, M., & Vrhovsek, U. (2015). Early fermentation volatile metabolite profile of non-*Saccharomyces* yeasts in red and white grape must: A targeted approach. *LWT – Food Science and Technology, 64*, 412–422.

Belda, I., Ruiz, J., Alastruey-Izquierdo, A., Navascués, E., Marquina, D., & Santos, A. (2016). Unraveling the enzymatic basis of wine "flavorome": A phylo-functional study of wine related yeast species. *Frontiers in Microbiology, 7*, 12.

Belda, I., Ruiz, J., Esteban-Fernández, A., Navascués, E., Marquina, D., Santos, A., & Moreno-Arribas, M. V. (2017). Microbial contribution to wine aroma and its intended use for wine quality improvement. *Molecules, 22*, 189.

Bely, M., Rinaldi, A., & Dubourdieu, D. (2003). Influence of assimilable nitrogen on volatile acidity production by *Saccharomyces cerevisiae* during high sugar fermentations. *Journal of Bioscience and Bioengineering, 96*(6), 507–512.

Bely, M., Stoeckle, P., Masneuf-Pomarède, I., & Dubourdieu, D. (2008). Impact of mixed *Torulaspora delbrueckii-Saccharomyces cerevisiae* culture on high-sugar fermentation. *International Journal of Food Microbiology, 122*, 312–320.

Benito, S., Palomero, F., Calderón, F., Palmero, D., & Suárez-Lepe, J. A. (2014a). Selection of appropriate *Schizosaccharomyces* strains for winemaking. *Food Microbiology, 42*, 218–224.

Benito, S., Palomero, F., Gálvez, L., Morata, A., Calderón, F., Palmero, D., & Suárez-Lepe, J. A. (2014b). Quality and composition of red wine fermented with *Schizosaccharomyces pombe* as sole fermentative yeast, and in mixed and sequential fermentations with *Saccharomyces cerevisiae*. *Food Technology and Biotechnology, 52*, 376–382.

Benito, A., Calderón, F., Palomero, F. E., & Benito, S. (2015). Combine use of selected *Schizosaccharomyces pombe* and *Lachancea thermotolerans* yeast strains as an alternative to the traditional malolactic fermentation in red wine production. *Molecules, 20*, 9510–9523.

Blanchard, L., Darriet, P., & Dubourdieu, D. (2004). Reactivity of 3-mercaptohexanol in red wine: Impact of oxygen, phenolic fractions, and sulfur dioxide. *American Journal of Enology and Viticulture, 55*, 115–120.

Bohlscheid, J. C., Fellman, J. K., Wang, X. D., Ansen, D., & Edwards, C. G. (2007). The influence of nitrogen and biotin interactions on the performance of *Saccharomyces* in alcoholic fermentations. *Journal of Applied Microbiology, 102*, 390–400.

Bokulich, N. A., Collins, T. S., Masarweh, C., Allen, G., Heymann, H., Ebeler, S. E., & Mills, D. A. (2016). Associations among wine grape microbiome, metabolome, and fermentation behavior suggest microbial contribution to regional wine characteristics. *MBio, 7*, e00631–e00616.

Capece, A., Fiore, C., Maraz, A., & Romano, P. (2005). Molecular and technological approaches to evaluate strain biodiversity in *Hanseniaspora uvarum* of wine origin. *Journal of Applied Microbiology, 98*, 136–144.

Capece, A., Romaniello, R., Siesto, G., Pietrafesa, R., Massari, C., Poeta, C., & Romano, P. (2010). Selection of indigenous *Saccharomyces cerevisiae* strains for Nero d'Avola wine and evaluation of selected starter implantation in pilot fermentation. *International Journal of Food Microbiology, 144*, 187–192.

Capece, A., Romaniello, R., Siesto, G., & Romano, P. (2012). Diversity of *Saccharomyces cerevisiae* yeasts associated to spontaneously fermenting grapes from an Italian "heroic vine-growing area". *Food Microbiology, 31*, 159–166.

Capece, A., Granchi, L., Guerrini, S., Mangani, S., Romaniello, R., Vincenzini, M., & Romano, P. (2016). Diversity of *Saccharomyces cerevisiae* strains Iiolated from two Italian wine-producing regions. *Frontiers in Microbiology, 7*, 1018.

Carrau, F. M., Medina, K., Boido, E., Farina, L., Gaggero, C., Dellacassa, E., Versini, G., & Henschke, P. A. (2005). De novo synthesis of monoterpenes by *Saccharomyces cerevisiae* wine yeasts. *FEMS Microbiology Letters, 243*, 107–115.

Ciani, M., & Maccarelli, F. (1998). Oenological properties of non-*Saccharomyces* yeasts associated with wine-making. *World Journal of Microbiology and Biotechnology, 14*, 199–203.

Ciani, M., & Picciotti, G. (1995). The growth kinetics and fermentation behaviour of some non-*Saccharomyces* yeasts associated with wine-making. *Biotechnology Letters, 17*, 1247–1250.

Ciani, M., Capece, A., Comitini, F., Canonico, L., Siesto, G., & Romano, P. (2016). Yeast interactions in inoculated wine fermentation. *Frontiers in Microbiology, 7*, 555.

Clemente-Jimenez, J. M., Mingorance-Cazorla, L., Martínez-Rodríguez, S., Las Heras-Vázquez, F. J., & Rodríguez-Vico, F. (2004). Molecular characterization and oenological properties of wine yeasts isolated during spontaneous fermentation of six varieties of grape must. *Food Microbiology, 21*, 149–155.

Clemente-Jiménez, J. M., Mingorance-Cazorla, L., Martínez-Rodríguez, S., Las Heras-Vázquez, F. J., & Rodríguez-Vico, F. (2005). Influence of sequential yeast mixtures on wine fermentation. *International Journal of Food Microbiology, 98*, 301–308.

Coetzee, C., & du Toit, W. J. (2012). A comprehensive review on Sauvignon blanc aroma with a focus on certain positive volatile thiols. *Food Research International, 45*, 287–298.

Comitini, F., Gobbi, M., Domizio, P., Romani, C., Lencioni, L., Mannazzu, I., & Ciani, M. (2011). Selected non-*Saccharomyces* wine yeasts in controlled multistarter fermentations with *Saccharomyces cerevisiae*. *Food Microbiology, 28*, 873–882.

Comitini, F., Capece, A., Ciani, M., & Romano, P. (2017). New insights on the use of wine yeasts. *Current Opinion in Food Science, 13*, 44–49.

Cordente, A. G., Curtin, C. D., Varela, C., & Pretorius, I. S. (2012). Flavour-active wine yeasts. *Applied Microbiology and Biotechnology, 96*, 601–618.

Cordero-Otero, R. R., Úbeda-Iranzo, J. F., Briones-Pérez, A. I., Potgieter, N. M., Arevalo, M. E., Pretorius, I. S., & van Rensburg, P. (2003). Characterization of the β-glucosidase activity produced by enological strains of non-*Saccharomyces* yeasts. *Journal of Food Science, 68*, 2564–2569.

Cozzolino, D., Flood, L., Bellon, J., Gishen, M., & De Barros Lopes, M. (2006). Combining near infrared spectroscopy and multivariate analysis as a tool to differentiate different strains of *Saccharomyces cerevisiae*: A metabolomic study. *Yeast, 23*, 1089–1096.

Dashko, S., Zhou, N., Tinta, T., Sivilotti, P., Sternad Lemut, M., Trost, K., Gamero Lluna, A., Boekhout, T., Butinar, L., Vrhovsek, U., & Piskur, J. (2015). Use of non-conventional yeast improves the wine aroma profile of Ribolla Gialla. *Journal of Industrial Microbiology & Biotechnology, 42*, 997–1010.

de Orduña, R. M. (2010). Climate change associated effects on grape and wine quality and production. *Food Research International, 43*, 1844–1855.

Del Mónaco, S. M., Barda, N. B., Rubio, N. C., & Caballero, A. C. (2014). Selection and characterization of a Patagonian *Pichia kudriavzevii* for wine deacidification. *Journal of Applied Microbiology, 117*, 451–464.

Delcroix, A., Gunata, Z., Sapis, J. C., Salmon, J. M., & Bayonove, C. (1994). Glycosidase activities of three enological yeast strains during wine making. Effect on the terpenol content of Muscat wine. *American Journal of Enology and Viticulture, 45*, 291–296.

Dias, L., Pereira-da-Silva, S., Tavares, M., Malfeito-Ferreira, M., & Loureiro, V. (2003). Factors affecting the production of 4-ethylphenol by the yeast *Dekkera bruxellensis* in enological conditions. *Food Microbiology, 20*, 377–384.

du Toit, M., & Pretorius, I. S. (2000). Microbial spoilage and preservations of wine: Using weapons from nature's own arsenal-a review. *South African Journal of Enology and Viticulture, 21*, 74–96.

Dubourdieu, D., Tominaga, T., Masneuf, I., Peyrot des Gachons, C., & Murat, M. L. (2006). The role of yeasts in grape flavor development during fermentation: The example of Sauvignon blanc. *American Journal of Enology and Viticulture, 57*, 81–88.

Dunn, W. B., & Ellis, D. I. (2005). Metabolomics: Current analytical platforms and methodologies. *Trends in Analytical Chemistry, 24*, 285–294.

Englezos, V., Rantsiou, K., Torchio, F., Rolle, L., Gerbi, V., & Cocolin, L. (2015). Exploitation of the non-*Saccharomyces* yeast *Starmerella bacillaris* (synonym *Candida zemplinina*) in wine fermentation: Physiological and molecular characterizations. *International Journal of Food Microbiology, 199*, 33–40.

Estévez, P., Gil Ma, L., & Falqué, E. (2004). Effects of seven yeast strains on the volatile composition of Palomino wines. *International Journal of Food Science and Technology, 39*, 61–69.

Ferraro, L., Fatichenti, F., & Ciani, M. (2000). Pilot scale vinification process using immobilized *Candida stellata* cells and *Saccharomyces cerevisiae*. *Process Biochemistry, 35*, 1125–1129.

Fleet, G. H., & Heard, G. M. (1993). Yeasts – growth during fermentation. In G. H. Fleet (Ed.), *Wine microbiology and biotechnology* (pp. 27–54). Chur: Harwood Academic Publishers.

Garcia, A., Carcel, C., Dulau, L., Samson, A., Aguera, E., Agosin, E., & Gunata, Z. (2002). Influence of a mixed culture with *Debaryomyces vanriji* and *Saccharomyces cerevisiae* on the volatiles of a Muscat wine. *Journal of Food Science, 67*, 1138–1143.

Gerbeaux, V., Jeudy, S., & Monamy, C. (2000). Study of phenol volatiles in Pinot noir wines in Burgundy. *Bulletin de l'OIV, 73*, 581–599.

Gerbeaux, V., Vincent, B., & Bertrand, A. (2002). Influence of maceration temperature and enzymes on the content of volatile phenols in Pinot Noir wines. *American Journal of Enology and Viticulture, 53*, 131–137.

Gobbi, M., Comitini, F., Domizio, P., Romani, C., Lencioni, L., Mannazzu, I., & Ciani, M. (2013). *Lachancea thermotolerans* and *Saccharomyces cerevisiae* in simultaneous and sequential co-fermentation: A strategy to enhance acidity and improve the overall quality of wine. *Food Microbiology, 33*, 271 281.

Goldner, M. C., Zamora, M., Di Leo Lira, P., Gianninoto, H., & Bandoni, A. (2009). Effect of ethanol level in the perception of aroma attributes and the detection of volatile compounds in red wine. *Journal of Sensory Studies, 24*, 243–257.

Gunata, Z., Bayonove, C., Baumes, R., & Cordonnier, R. (1985). The aroma of grapes. Extraction and determination of free and glycosidically bound fractions of some grape aroma components. *Journal of Chromatography, 331*, 83–90.

Gunata, Z., Dugelay, I., Sapis, J. C., Baumes, R., & Bayonove, C. (1990). Action des glycosidases exogènes au cours de la vinification: Liberation de l'arôme à partir des précurseurs glycosidiques. *Journal International des Sciences de la Vigne et du Vin, 24*, 133–144.

Hazelwood, L. A., Daran, J. M., van Maris, A. J. A., Pronk, J. T., & Dickinson, J. R. (2008). The Ehrlich pathway for fusel alcohol production: A century of research on *Saccharomyces cerevisiae* metabolism. *Applied and Environmental Microbiology, 74*, 2259–2266.
Henick-Kling, T., Edinger, W., Daniel, P., & Monk, P. (1998). Selective effects of sulphur dioxide and yeast starter culture addition on indigenous yeast populations and sensory characteristics of wine. *Journal of Applied Microbiology, 84*, 865–876.
Hernandez, L. F., Espinosa, J. C., Fernandez, M., & Briones, A. (2003). β-Glucosidase activity in a *Saccharomyces cerevisiae* wine strain. *International Journal of Food Microbiology, 80*, 171–176.
Hernandez-Orte, P., Cacho, J. F., & Ferreira, V. (2002). Relationship between varietal amino acid profile of grapes and wine aromatic composition. Experiments with model solutions and chemometric study. *Journal of Agricultural and Food Chemistry, 50*, 2891–2899.
Jackowetz, J. N., Dierschke, S., & Mira de Orduña, R. (2011). Multifactorial analysis of acetaldehyde kinetics during alcoholic fermentation by *Saccharomyces cerevisiae. Food Research International, 44*, 310–316.
Jolly, N. P., Augustyn, O. P. H., & Pretorius, I. S. (2006). The role and use of non-Saccharomyces yeasts in wine production. *South African Journal of Enology and Viticulture, 27*(1), 15–39.
Kim, D. H., Hong, Y. A., & Park, H. D. (2008). Co-fermentation of grape must by *Issatchenkia orientalis* and *Saccharomyces cerevisiae* reduces the malic acid content in wine. *Biotechnology Letters, 30*, 1633–1638.
King, E. S., Kievit, R. L., Curtin, C., Swiegers, J. H., Pretorius, I. S., Bastian, S. E. P., & Francis, I. L. (2010). The effect of multiple yeasts co-inoculations on Sauvignon Blanc wine aroma composition, sensory properties and consumer preference. *Food Chemistry, 122*, 618–626.
Knight, S., Klaere, S., Fedrizzi, B., & Goddard, M. R. (2015). Regional microbial signatures positively correlate with differential wine phenotypes: Evidence for a microbial aspect to terroir. *Scientific Reports, 5*, 14233.
Lambrechts, M. G., & Pretorius, I. S. (2000). Yeast and its importance to wine aroma – a review. *South African Journal of Enology and Viticulture, 21*, 97–129.
Lema, C., Garcia-Jares, C., Orriols, I., & Angulo, L. (1996). Contribution of *Saccharomyces* and non-*Saccharomyces* populations to the production of some components of Albariño wine aroma. *American Journal of Enology and Viticulture, 47*, 206–216.
Manzanares, P., Rojas, V., Genovés, S., & Vallés, S. (2000). A preliminary search for anthocyanin-β-D-glucosidase activity in non-*Saccharomyces* wine yeasts. *International Journal of Food Science and Technology, 35*, 95–103.
Masneuf-Pomarède, I., Mansour, C., Marie-Laure, M., Tominaga, T., & Dubourdieu, D. (2006). Influence of fermentation temperature on volatile thiols concentrations in Sauvignon blanc wines. *International Journal of Food Microbiology, 108*, 385–390.
Mateo, J., Jimenez, M., Herta, T., & Pastor, A. (1992). Comparison of volatiles produced by four *Saccharomyces cerevisiae* strains isolated from Monastrell musts. *American Journal of Enology and Viticulture, 43*, 206–209.
Mateo, J. J., Peris, L., Ibáñez, C., & Maicas, S. (2011). Characterization of glycolytic activities from non-*Saccharomyces* yeasts isolated from Bobal musts. *Journal of Industrial Microbiology & Biotechnology, 38*, 347–354.
Mauriello, G., Capece, A., D'Auria, M., Garde-Cerdán, T., & Romano, P. (2009). SPME-GC method as a tool to differentiate VOC profiles in *Saccharomyces cerevisiae* wine yeasts. *Food Microbiology, 26*, 246–252.
Mazauric, J.-P., & Salmon, J.-P. (2005). Interactions between yeast lees and wine polyphenols during simulation of wine aging: I. Analysis of remnant polyphenolic compounds in the resulting wines. *Journal of Agricultural and Food Chemistry, 53*, 5647–5653.
Mendes-Ferreira, A., Clímaco, M. C., & Mendes Faia, A. (2001). The role of non-*Saccharomyces* species in realising glycosidic fraction of grape aroma components—a preliminary study. *Journal of Applied Microbiology, 91*, 67–71.
Mendoza, L. M., Manca de Nadra, M. C., & Farías, M. E. (2007). Kinetics and metabolic behavior of a composite culture of *Kloeckera apiculata* and *Saccharomyces cerevisiae* wine related strains. *Biotechnology Letters, 29*, 1057–1063.

Millan, M. C., Moreno, J., Medina, M., & Ortega, J. M. (1991). Influence of the physiological state of the inoculum on fermentation of musts from Pedro Ximénez grapes by *Saccharomyces cerevisiae*. *Microbios, 65*, 87–95.

Moreira, N., Mendes, F., Hogg, T., & Vasconcelos, I. (2005). Alcohols, esters and heavy sulphur compounds production by pure and mixed cultures of apiculate wine yeasts. *International Journal of Food Microbiology, 103*, 285–294.

Moreira, N., Mendes, F., Guedes de Pinho, P., Hogg, T., & Vasconcelos, I. (2008). Heavy sulphur compounds, higher alcohols and esters production profile of *Hanseniaspora uvarum* and *Hanseniaspora guilliermondii* grown as pure and mixed cultures in grape must. *International Journal of Food Microbiology, 124*, 231–238.

Murat, M. L., Masneuf, I., Darriet, P., et al. (2001). Effect of *Saccharomyces cerevisiae* yeast strains on the liberation of volatile thiols in Sauvignon blanc wine. *American Journal of Enology and Viticulture, 52*, 136–139.

Nielsen, J., & Oliver, S. (2005). The next wave in metabolome analysis. *Trends in Biotechnology, 23*, 544–546.

Nikolaou, E., Soufleros, E. H., Bouloumpasi, E., & Tzanetakis, N. (2006). Selection of indigenous *Saccharomyces cerevisiae* strains according to their oenological characteristics and vinification results. *Food Microbiology, 23*, 205–211.

Nykanen, L. (1986). Formation and occurrence of flavour compounds in wine and distilled alcoholic beverages. *American Journal of Enology and Viticulture, 37*, 84–96.

Obreque-Slíer, E., Peñ-Neira, A., & Lopez-Solis, R. (2010). Enhancement of both salivary protein-enological tannin interactions and astringency perception by ethanol. *Journal of Agricultural and Food Chemistry, 58*, 3729–3735.

Office Internationale de la Vigne et du Vin. (2010). *International code of oenological practices* (p. 274). Paris: OIV.

Padilla, B., Gil, J. V., & Manzanares, P. (2016). Past and future of non-*Saccharomyces* yeasts: From spoilage microorganisms to biotechnological tools for improving wine aroma complexity. *Frontiers in Microbiology, 7*, 411.

Park, H., & Hwang, Y. S. (2008). Genome-wide transcriptional responses to sulfite in *Saccharomyces cerevisiae*. *Journal of Microbiology, 46*, 542–548.

Peyrot Des Gachons, C., Tominaga, T., & Dubourdieu, D. (2002). Sulfur aroma precursor present in S-glutathione conjugate form: Identification of S-3-(hexan-1-ol)-glutathione in must from *Vitis vinifera* L. cv. Sauvignon blanc. *Journal of Agricultural and Food Chemistry, 50*, 4076–4079.

Pretorius, I. S. (2000). Tailoring wine yeast for the new millennium: Novel approaches to the ancient art of winemaking. *Yeast, 16*, 675–729.

Pretorius, I. S., & Høj, P. B. (2005). Grape and wine biotechnology: Challenges, opportunities and potential benefits. *Australian Journal of Grape and Wine Research, 11*, 83–108.

Rantsiou, K., Dolci, P., Giacosa, S., Torchio, F., Tofalo, R., Torriani, S., Suzzi, G., Rolle, L., & Cocolin, L. (2012). *Candida zemplinina* can reduce acetic acid produced by *Saccharomyces cerevisiae* in sweet wine fermentations. *Applied and Environmental Microbiology, 78*, 1987–1994.

Regodón Mateos, J. A., Pérez-Nevado, F., & Ramírez Fernández, M. (2006). Influence of *Saccharomyces cerevisiae* yeast strain on the major volatile compounds of wine. *Enzyme and Microbial Technology, 40*, 151 157.

Renault, P., Coulon, J., Moine, V., Thibon, C., & Bely, M. (2016). Enhanced 3-sulfanylhexan-1-ol production in sequential mixed fermentation with *Torulaspora delbrueckii/Saccharomyces cerevisiae* reveals a situation of synergistic interaction between two industrial strains. *Frontiers in Microbiology, 7*, 293.

Ribéreau-Gayon, P., Dubourdieu, D., Donéche, B., & Lonvaud, A. (2000a). Biochemistry of alcoholic fermentation and metabolic pathways of wine yeasts. Handbook of enology. In *The microbiology of wine and vinifications* (Vol. 1, pp. 51–74). New York: Wiley.

Ribéreau-Gayon, P., Glories, Y., Maujean, A., & Dubourdieu, D. (2000b). Handbook of enology. In *The chemistry of wine stabilisation and treatments* (Vol. 2). Chichester: Wiley.

Rojas, V., Gil, J. V., Pinaga, F., & Manzanares, P. (2001). Studies on acetate ester production by non-*Saccharomyces* wine yeast. *International Journal of Food Microbiology, 70*, 283–289.
Rojas, V., Gil, J. V., Manzanares, P., Gavara, R., Piñaga, F., & Flors, A. (2002). Measurement of alcohol acetyltransferase and ester hydrolase activities in yeast extracts. *Enzyme and Microbial Technology, 30*, 224–230.
Rojas, V., Gil, J., Piñaga, F., & Manzanares, P. (2003). Acetate ester formation in wine by mixed cultures in laboratory fermentations. *International Journal of Food Microbiology, 86*, 181–188.
Romano, P., & Suzzi, G. (1993). Sulfur dioxide and wine microorganisms. In G. H. Fleet (Ed.), *Wine microbiology and biotechnology* (pp. 373–393). Chur: Harwood Academic Publishers.
Romano, P., Suzzi, G., Comi, G., & Zironi, R. (1992). Higher alcohol and acetic acid production by apiculate wine yeasts. *The Journal of Applied Bacteriology, 73*, 126–130.
Romano, P., Suzzi, G., Turbanti, L., & Polsinelli, M. (1994). Acetaldehyde production in *Saccharomyces cerevisiae* wine yeasts. *FEMS Microbiology Letters, 118*, 213–218.
Romano, P., Fiore, C., Paraggio, M., Caruso, M., & Capece, A. (2003). Function of yeast species and strains in wine flavour. *International Journal of Food Microbiology, 86*, 169–180.
Schneider, R., Charrier, F., Razungles, A., & Baumes, R. (2006). Evidence for an alternative biogenetic pathway leading to 3-mercaptohexanol and 4-mercapto-4-methylpentan-2-one in wines. *Analytica Chimica Acta, 563*, 58–64.
Stribny, J., Querol, A., & Pérez-Torrado, R. (2016). Differences in enzymatic properties of the *Saccharomyces kudriavzevii* and *Saccharomyces uvarum* alcohol acetyltransferases and their impact on aroma-active compounds production. *Frontiers in Microbiology, 7*, 897.
Styger, G., Prior, B., & Bauer, F. F. (2011). Wine flavor and aroma. *Journal of Industrial Microbiology & Biotechnology, 38*, 1145–1159.
Swiegers, J. H., & Pretorius, I. S. (2005). Yeast modulation of wine flavour. *Advances in Applied Microbiology, 57*, 131–175.
Swiegers, J., & Pretorius, I. S. (2007). Modulation of volatile sulfur compounds by wine yeast. *Applied and Environmental Microbiology, 74*, 954–960.
Swiegers, J., Kievit, R., Siebert, T., Lattey, K., Bramley, B., Francis, I., King, E., & Pretorius, I. S. (2009). The influence of yeast on the aroma of Sauvignon Blanc wine. *Food Microbiology, 26*, 204–211.
Tominaga, T., Baltenweck-Guyot, R., Peyrot de Gachons, C., & Dubourdieu, D. (2000). Contribution of volatile thiols to the aromas of white wines made from several *Vitis vinifera* grape varieties. *American Journal of Enology and Viticulture, 51*, 178–181.
Torija, M. J., Rozes, N., Poblet, M., Guillamon, J. M., & Mas, A. (2003). Effects of fermentation temperature on the strain population of *Saccharomyces cerevisiae. International Journal of Food Microbiology, 80*, 47–53.
Viana, F., Gil, J. V., Genovés, S., Vallés, S., & Manzanares, P. (2008). Rational selection of non-*Saccharomyces* wine yeasts for mixed starters based on ester formation and enological traits. *Food Microbiology, 25*, 778–785.
Winter, G., Henschke, P. A., Higgins, V. J., Ugliano, M., & Curtin, C. D. (2011). Effects of rehydration nutrients on H_2S metabolism and formation of volatile sulfur compounds by the wine yeast VL3. *AMB Express, 1*, 36.
Zott, K., Thibon, C., Bely, M., Lonvaud-Funel, A., Dubourdieu, D., & Masneuf-Pomarede, I. (2011). The grape must non-*Saccharomyces* microbial community: Impact on volatile thiol 1. release. *International Journal of Food Microbiology, 151*, 210–215.

Chapter 3
Detection, Quantification, and Identification of Yeast in Winemaking

Severino Zara and Ilaria Mannazzu

Contents

1 Introduction

Winemaking is the process by which grapes are transformed into wine through the fermentative activity of yeast and bacteria. Different factors have key roles in the making of a good wine, including grape quality, winemaking technology, and the properties and performances of the microorganisms that participate in the grape must fermentation. It can be debated which of these factors is the most critical to wine production, and probably all of them are equally important. Indeed, in addition

S. Zara · I. Mannazzu (✉)
Department of Agriculture, University of Sassari, Sassari, Italy
e-mail: szara@uniss.it; imannazzu@uniss.it

P. Romano et al. (eds.), *Yeasts in the Production of Wine*,
https://doi.org/10.1007/978-1-4939-9782-4_3

to sound and healthy grapes and skillful winemaking, yeast choice is essential for successful fermentation.

Investigations of vineyard, grape, and wine mycobiota have revealed great yeast biodiversity within wine-related environments. According to Bisson et al. (2017), grapes and wine mycobiota include over 40 genera and 100 different species, although this composition varies depending on regional and climatic factors, vineyard practices, grape varieties, and the sanitary state of the grapes. In spite of these variations, *Saccharomyces cerevisiae* is always the main actor in grape-must transformation. However, the so-called non-*Saccharomyces* yeast also participate in the winemaking process, although with different effects depending on the quantitative and qualitative variability of the species found on the grapes and in the must in the early stages of fermentation. Due to these variations, spontaneous fermentations are relatively unpredictable, with positive or negative outcomes on the composition, taste, flavor, and appearance of the final product. On the contrary, inoculation of pure or mixed yeast starter cultures allows the fermentation in the winemaking process to be controlled and managed consistently. However, it is well known that in spite of the use of yeast starters, the natural grape microbiota that comprises *S. cerevisiae* and non-*Saccharomyces* yeast can persist during must fermentation, and this can result in poor quality wine, or even wine spoilage.

Thus, to enumerate and identify the wine yeast that operate contemporarily or in succession during grape must fermentation, to detect spoilage yeast, and to evaluate the dominance and persistence of the starter cultures at different stages in the winemaking process, a plethora of different methods have been developed. As reported in the present chapter, these methods have different degrees of complexity and vary in terms of cost, rapidity, sensitivity, and reliability.

For a long time, the study of wine microflora was based on the use of culture-dependent methods. Now microbiologists can take advantage of different culture-independent methods. Depending on the approach followed, these can be used to implement a quality control system based on real-time detection and quantification of specific targets, such as the inoculated starter(s) or the spoilage yeast, or to provide further insights into the composition of the microbial communities involved in the grape must transformation. Indeed, their application to the study of the microbial ecology of wine-related environments now contributes to the determination of the influence of a number of different parameters on wine microbial diversity, such as climate, soil composition, water management, vineyard management, host genotype, and others.

This chapter presents an overview of the methods available for the detection and identification of yeast in winemaking.

2 Monitoring Yeast Cell Populations With Culture-Dependent Methods

Culture-dependent methods for detection, quantification, and identification of wine-related yeast are carried out in two steps, for their cultivation and their identification (Fig. 3.1). For yeast cultivation, Yeast Extract, Peptone, Dextrose agar and Malt

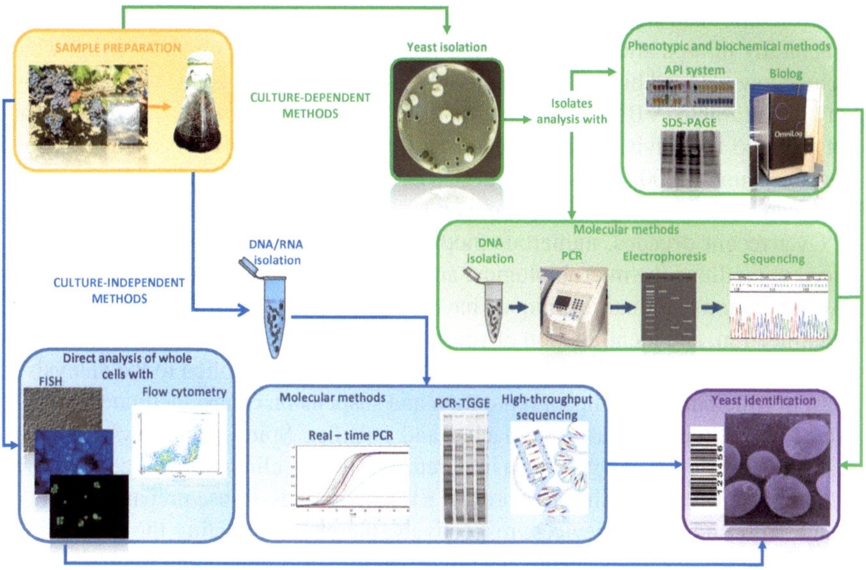

Fig. 3.1 Schematic overview of culture-dependent (green lines and boxes) and culture-independent (blue line and boxes) methods

Extract Agar are commonly used, with antibiotics added (e.g., chloramphenicol, oxytetracycline; 50–100 μg L^{-1}) to inhibit bacterial growth. Molds can be controlled with the addition of Rose Bengal (30 mg L^{-1}) or dichloran (2 mg L^{-1}), which are also available in a commercial form (Dichloran Rose Bengal Chloramphenicol agar). The differential medium known as WL (Wallerstein Laboratory) nutrient agar contains 0.1% bromocresol green, which allows some general discrimination between yeast species based on colony morphology and color (Pallmann et al. 2001). WL medium can be supplemented with 100 mg L^{-1} cycloheximide, 10 mg L^{-1} p-coumaric acid, and 50 mg L^{-1} chloramphenicol for detection of *Dekkera/Brettanomyces* yeast (Couto et al. 2005; Morneau et al. 2011). Malt extract agar with Rose Bengal (30 mg L^{-1}), chloramphenicol (50 μg L^{-1}), erythromycin (70 μg L^{-1}), and dichloran (2 mg L^{-1}) can also be used for some general discrimination between different yeast, again based on colony morphology and color (Pallmann et al. 2001). Lysine agar (Yeast carbon base supplemented with L-lysine-HCl) is used for selective detection of non-*Saccharomyces* yeast, because unlike *Saccharomyces* species, they can use lysine as sole carbon source (Beuchat et al. 1998; Loureiro et al. 2004; Domizio et al. 2011; Wang et al. 2016). Cycloheximide (0.1%) can be added to any general-purpose medium, as this eliminates *Saccharomyces* yeast and allows growth of many non-*Saccharomyces* species. *Saccharomyces* species are generally more tolerant to ethanol and sulfur dioxide than other wine yeast, and can be selectively quantified on an ethanol-sulfite agar medium (Sabate et al. 1998; Li et al. 2010). For the analysis of winery surfaces and equipment, Rodac plates that contain an appropriate agar medium can be used.

Difficulties can arise in the detection of yeast cells that might be metabolically or structurally injured after exposure to the stressful conditions of winemaking, which include low pH, and high ethanol and SO_2 contents. Such cells might be in a viable but nonculturable (VBNC) state, in that they show some metabolic activity but cannot undergo cell division and growth, particularly in selective media (Millet and Lonvaud-Funel 2000; Mills et al. 2002; Divol and Lonvaud-Funel 2005). VBNC yeast can be detected by culture-independent molecular techniques.

Over recent decades, numerous modified cultivation methods have been introduced to facilitate the microbiological analysis of foods and beverages, and various novel, non-conventional techniques have also been developed for rapid detection, quantification, and identification of yeast, often using automated instruments (Deak 2003, 2008). These refinements and developments can be applied to the analysis of yeast in winemaking. Sample preparation and suspension can be facilitated with the use of a gravimetric diluter, stomacher and pulsifier. Spiral plating systems automate the inoculation of agar media into Petri plates, and eliminate the need for serial dilutions. Petrifilm and Simplate present a range of ready-to-use prepared media on membranes and plastic devices, respectively, thereby eliminating the need for the preparation of agar plate media. The hydrophobic grid membrane filter technique has been commercialized as the Iso-Grid/ Neo-Grid system, and this facilitates the estimation of cell populations, including yeast, by the most probable number method. Comparative studies have shown no significant differences between these novel cultivation methods and conventional plate counting (Entis and Lerner 1996; Spangenberg and Ingham 2000; Taniwaki et al. 2001).

Identification of yeast to genus and species levels has traditionally relied on tests for phenotypic characteristics (i.e., morphology, as well as physiological and biochemical tests) (Barnett et al. 2000a). Nowadays, identification using molecular techniques is proving to be faster and more reliable. Nevertheless, some basic phenotypic information is necessary to understand how yeast survive and grow in their habitat.

2.1 Phenotypic Identification

Methods for conducting traditional phenotypic tests and the keys for applying them to genus and species identification are given in the books by Kurtzman and Robnett (1998, 2003) and Barnett et al. (2000a). Generally, about 100 tests need to be performed, which is laborious and time consuming, and requires laboratory experience for accurate application and evaluation. Deak (2008) developed a simplified identification method that usually requires 15–20 physiological tests and a diagnostic key for selected food and beverage yeast. The latest version applies 30 tests and includes 120 yeast species.

Another line to develop rapid and simple techniques of identification led to the miniaturization of tests. Manual methods using serological microtiter plates with small amounts of substrates allow faster developing of the results. This showed the way for the development of commercial ready-to-use systems in various microwell formats, and automated identification systems (Fung 2002). Efforts have been made to make this identification of yeast easier by using commercially available identification kits. Some of these come with automated and computerized processing of data. The performance of these kits has been evaluated in numerous studies, and most compare relatively favorably with standard identification systems. The main limitations include the range of tests covered and the range of species included in their databases, which often have a clinical emphasis (Paugham et al. 1999).

The API 20C system has been widely used, and has often been considered as a reference method for evaluation of other systems (Deak and Beuchat 1993). These systems can be automated, semi-automated or manual, and have been developed based on probabilistic data matrices (Barnett et al. 2000b). Automated systems still require time for preparation and incubation of identification panels, but subjectivity in the reading and evaluation of the results is eliminated. The Biolog system is a semi-automated computer-linked system that is based on 94 tests arranged in a microtiter tray. However, less than one-third of substrates are usually used for identification of the yeast (Stadlwieser et al. 2006).

Barnett and co-workers were the first to construct computerized identification keys for yeast identification (Barnett et al. 2000a), and the program has been commercialized (Barnett et al. 2000b). Another system known as YeastIdent-Food/ProleFood has been developed for the identification of yeast from foods (Velázquez et al. 2001). Automated systems usually provide an on-line database and identification matrix (e.g., API-BioMerieux, Microlog YT for Biolog, ID-YST for Vitek 2). Robert (2003) provided an extensive review of the computerized identification systems for yeast, and also introduced the polyphasic identification systems known as BioloMICS (Robert 2000). BioloMICS was also revised in 2011 (Robert et al. 2011).

2.2 *Identification Based on Biochemical Methods*

To overcome the inherent slowness of traditional phenotypic identification methods or to refine yeast identification, instrumental methodologies based on the analysis of proteins, isoenzymes, and fatty acids have been developed.

Separation of soluble proteins using polyacrylamide gel electrophoresis, with or without sodium dodecyl sulfate, is a well-established technique that has been used to distinguish strains within a single species or to compare strains of closely related species. This method allows good species discrimination, although the preparation of the extracts is time consuming. Electrophoretic protein patterns have been used to differentiate wine and brewing yeast (Dowhanick et al. 1990). The results can be somewhat contradictory though, as the results of protein electrophoresis are dependent on the growth conditions, and therefore the reproducibility of the protein patterns requires rigidly standardized methods.

Another technique for yeast identification and characterization is Fourier transform infra-red spectroscopy (FTIR). This technique is based on irradiation of the yeast with infra-red light at different wavelengths. The different components of the cell (i.e., proteins, fatty acids, polysaccharides, among others) absorb the light and produce specific absorption profiles. Comparison of the profiles obtained with an appropriate database allows identification to the genus, species and even strain level (Grangeteau et al. 2015). Although this technique has attracted interest in many fields, it has rarely been used for any in-depth wine ecological studies (Grangeteau et al. 2016). Some studies have used it to discriminate between *S. cerevisiae*, *Saccharomyces bayanus* (Adt et al. 2010), and *Brettanomyces bruxellensis* strains (Oelofse et al. 2010). Others have applied it to differentiation of *Hanseniaspora guilliermondii* and *Hanseniaspora uvarum* (Grangeteau et al. 2015), and *Starmerella bacillaris* (synonym *C. zemplinina*) at the strain level (Grangeteau et al. 2016).

Recently, matrix-assisted laser desorption/ ionization time-of-flight mass spectrometry (MALDI-TOF MS) has emerged as a reliable, fast, low cost and high-throughput technique for rapid identification and classification of microorganisms. This technique is based on the identification of specific protein patterns (mainly, ribosomal proteins). Basically, cell extracts or whole cells are embedded in a chemical matrix and ionized by a laser. The ionized microbial molecules migrate toward a detector, through a charged field in a vacuum tube and generate a mass spectrum. This is a fingerprint, that is distinctive for each microorganism and that can be compared to a database of known species to arrive at a rapid identification at the genus and species level (Marcos and Pincus 2013). In this regard, MALDI-TOF MS represents a strong challenge to microscopy and molecular biology methods (Chalupová et al. 2014). Du Plessis et al. (2017) used MALDI-TOF MS analyses to identify *H. uvarum*, *M. pulcherrima*, *S. cerevisiae*, *S. pombe*, and *T. delbrueckii* at the species level. More recently, a database of yeast isolated from vineyards and wineries has been developed that is based on MALDI-TOF MS spectra analysis. This database is based on an extensible open-source platform for MALDI data processing and analysis with statistical techniques (Gutiérrez et al. 2017). Recently, kits for the identification of microorganisms through mass spectroscopy have been developed (Vitek MS MALDI-TOF).

2.3 Identification Based on Molecular Methods

Accurate identification is obviously crucial for the study of microbial communities. It is accepted that classical identification techniques can provide incorrect data. Moreover, the reproducibility of classical identification techniques is somewhat questionable, as they can depend on the physiological state of the cells. Molecular techniques circumvent these difficulties by allowing direct analysis of the genome, irrespective of the physiological state of the cell, which provides more precise identification (Barata et al. 2012) (Fig. 3.1).

Since the 1990s, molecular techniques have become the most important tools for such investigations, and they have revolutionized all fields of microbiology and wine microbiology. It is no longer possible to comprehensively review the vast literature on the progress made in recent years. Therefore, selected examples will be referred here to provide an overview and to update the use and applications of molecular techniques for the detection, identification, and typing of wine-related yeast. For reviews, see Deak (2003); Querol and Ramón (1996); van der Vossen et al. (2003); Schuller et al. (2004); Beh et al. (2006); Fernández-Espinar et al. (2006); Ivey and Phister (2011); Fernàndez-Espinar et al. (2011); Barata et al. (2012); and Fröhlich et al. (2017).

Techniques for detection of differences at the nucleic-acid level can be broadly categorized into four groups: (i) direct detection using gel electrophoresis; (ii) detection based on hybridization; (iii) detection based on PCR amplification and gel electrophoresis; and (iv) sequencing of rDNA. Before any investigation, DNA or (sometimes) RNA needs to be extracted from yeast cells that are isolated from the grape or wine samples.

2.3.1 Direct Detection Using Gel Electrophoresis

Separation of DNA molecules and fragments can be achieved by embedding them in agarose or polyacrylamide gels and subjecting these to an electric field. Two methods are used widely: one for separation of smaller DNA fragments obtained by cutting the DNA with restriction enzymes; and the other for separation of large DNA molecules (whole chromosomal DNA) through the use of an alternating (pulsing) electric field.

Mitochondrial DNA-RFLP

Due to distinct polymorphisms of wine *Saccharomyces* strains, analysis of mitochondrial (mt)DNA by restriction fragment length polymorphism (RFLP) has proven useful to explore yeast biodiversity (Sabate et al. 1998; Valero et al. 2005, 2007; Agnolucci et al. 2007; Martinez et al. 2007; Mercado et al. 2007), and to monitor population dynamics during wine fermentation (Querol et al. 1994; Lopes et al. 2002). The efficiency of this technique depends on the choice of the restriction enzymes applied; e.g., *Hin*fI reveals a high level of RFLP for strains belonging to species of *Saccharomyces sensu stricto* (Guillamón et al. 1994; Fernández-Espinar et al. 2001), whereas *Rsa*I cuts more frequently and is used to support the hybrid status of *S. bayanus* and its similarity to *S. uvarum* (Nguyen et al. 2000). Lòpez et al. (2003) simplified this method with a modified protocol that reduced the time required from 77 h to 25 h. This protocol allows analysis of greater numbers of strains in shorter times, and is ideal for industrial applications (Fernàndez-Espinar et al. 2011).

Pulsed-Field Gel Electrophoresis

Pulsed field gel electrophoresis allows separation of large (chromosome sized) DNA molecules. The banding pattern is characteristic of the species (i.e., number and size of chromosomal DNA bands, known as the karyotype), although extensive intraspecific chromosomal length polymorphisms has been reported (Versavaud and Hallet 1995; Pataro et al. 2000; Carro and Piña 2001). These are due to gross chromosomal rearrangements, such as translocations, inversions, duplications, and deletions of large chromosomal regions, which can be associated with physiological characteristics of industrial importance. This technique was used to provide identification and typing, but due to extensive sample preparation, lengthy electrophoretic separation, and expensive equipment, it is not frequently used anymore. Moreover, although pulsed field gel electrophoresis is very reproducible, the high frequency of karyotypic changes in yeast makes it necessary to apply specific hybridizing probes if the method is to be used for species identification. The development of this technique led to the description of electrophoretic karyotypes for a number of species in *Saccharomyces* and other yeast genera that are relevant in winemaking, such as *Dekkera/ Brettanomyces*, and *Hanseniaspora/ Kloeckera* (Vaughan-Martini et al. 1996, 2000; Mitrakul et al. 1999; Povhe Jemec et al. 2001 Guillamón and Barrio 2017). Characteristic banding patterns can distinguish between *Saccharomyces sensu stricto* species and allow recognition of natural hybrids (Le Jeune et al. 2007; González et al. 2006; Nguyen and Gaillardin 2005; Antunovics et al. 2005b). Karyotyping has been extensively applied to the differentiation of wine yeast, and has been used to monitor the diversity and development of species and strains through fermentation (Guillamón and Barrio 2017). Nowadays, its use is mainly limited to karyotype studies.

2.3.2 Detection Based on Hybridization

DNA hybridization is an indirect method to identify microorganisms, and it can include simple probes or microarrays to examine similarities between whole genomes (Ivey and Phister 2011). Hybridization serves to support a number of detection, identification, and typing techniques, and in particular those connected to PCR-based methods. Moreover, under certain conditions, DNA hybridization can be used independently. Basically, single-stranded nucleic-acid molecules are treated with a short, labeled oligonucleotide sequence (i.e., the DNA probe) and allowed to anneal and form a hybrid. Detection of hybridization can be seen directly using fluorescent labels, or indirectly with enzyme reporters. In general, the target nucleic-acid molecules are immobilized on a membrane, either after separation on an electrophoretic gel (i.e., Southern blotting) or from solutions (dot blots).

Hybridization using nuclear DNA sequences (DNA-DNA reassociation) has become a standard procedure in yeast taxonomy and identification (Kurtzman and Robnett 1998; Cardinali et al. 2000; Vaughan-Martini et al. 2000), although in recent years appears to have become less important (see below). Analysis of genomic

DNA is rarely used without probe hybridization. DNA probes provide specific identification and subtyping that depend on the specificity of the oligonucleotide sequence used. Specific DNA probes can be based on conserved coding sequences, such as single copy and polymeric metabolic genes (Querol et al. 1992; Naumov et al. 1994). Variable sequences, such as insertion elements, retrotransposons, repetitive microsatellite or minisatellite sequences, or even synthetic oligonucleotides, are also useful for typing (Baleiras Couto et al. 1994; Nguyen et al. 2000; Casaregola et al. 2001). Species-specific probes can be developed from characterized sequences identified by PCR amplification (Manavathu et al. 1996; Corredor et al. 2000).

2.3.3 Detection Based on PCR Amplification and Gel Electrophoresis

Sequences within mtDNA and ribosomal (r)DNA genes (PCR ribotyping), repetitive regions of genomic DNA, and nuclear genes are most commonly targeted for PCR amplification (Giesendorf et al. 1996). These sequences can contain species-specific sites or universal sequences that are characteristic of all yeast or fungi. The PCR products can be analyzed after separation by agarose gel electrophoresis, analyzed by RFLP (Esteve-Zarzoso et al. 1999), or sequenced to confirm species identification as reported in paragraph 2.3.4.

PCR of Repeated Genomic Elements

Repetitive regions of genomic DNA are good targets for molecular identification of yeast strains. These regions are known as microsatellites and minisatellites, and they vary considerably in length and are present as tandem repeats that are distributed randomly throughout the genome (Fernàndez-Espinar et al. 2011). PCR primers can be devised for amplification of tandem repeated genomic elements of known sequences dispersed throughout the genome of yeast. These are highly variable, and the banding patterns obtained by PCR provide a 'fingerprint' that is useful to discriminate organisms at the strain level. Various simple repeats, such as microsatellites and minisatellites, δ-elements associated with the Ty transposon, intron splice sites, and even synthetic repeats can be amplified by PCR in a rapid and reproducible way, to offer a powerful method for yeast typing (Baleiras Couto et al. 1994; Gallego et al. 1998; Marinangeli et al. 2004; Schuller et al. 2004).

The δ-elements were among the first to be used for yeast identification (Ness et al. 1993; Lavallée et al. 1994). The method has been subsequently improved and is still frequently used (Legras and Karst 2003; Renouf et al. 2006a). Amplification of intron splice sites is based on the use of oligonucleotides that are complementary to the intron splice sites in yeast (de Barros Lopes et al. 1996, 1998), and this has been shown to provide comparable results to other fingerprinting methods (Hierro et al. 2004, 2006). Microsatellites and minisatellites are short and longer sequences that are repeated throughout the yeast genome, of which several can be selected for differentiation between wine strains of *S. cerevisiae* (Gallego et al. 1998; Hennequin et al. 2001; Legras et al. 2005; Pérez et al. 2001b; Marinangeli et al. 2004). Using three primer pairs in a multiplex PCR reaction provides a rapid and powerful fingerprinting method (Legras et al. 2005; Vaudano and Garcia-Moruno 2008). Synthetic repeats such as $(GAC)_5$ or $(GTG)_5$ can also be used to characterize yeast strains (Fernández-Espinar et al. 2001; Caruso et al. 2002; Capece et al. 2003; Senses-Ergul et al. 2006; Ramìrez-Castrillòn et al. 2014; Barbosa et al. 2018).

Random Amplified Polymorphic DNA

Random amplified polymorphic DNA (RAPD) analysis, or arbitrary primed PCR, uses short, synthetic primers with no prior sequence information of the template DNA (Williams et al. 1990; Welsh and McClelland 1990). A variety of 10-base (decamer) oligonucleotide primers are available commercially. RAPD is a very popular method because of its simplicity (i.e., no need for prior sequence data). RAPD methods are particularly useful to determine relationships at strain level, but they can also help to discriminate between species (Baleiras Couto et al. 1994; Di Maro et al. 2007; Urso et al. 2008). RAPD-PCR allowed differentiation of *Saccharomyces* strains (Grando et al. 1994; Martinez et al. 2007), and species and strain discrimination within the genus *Dekkera/ Brettanomyces* (Mitrakul et al. 1999). Pérez et al. (2001a) digested the amplicons obtained by RAPD for the genetic characterization of *S. cerevisiae* isolated from must and wine. This method, termed cleaved amplified polymorphic sequence, detected a greater degree of polymorphism and strain-specific bands.

Based on the principle of RAPD-PCR, nested specifically amplified polymorphic DNA-PCR (nSAPD-PCR) was developed as a useful method for identification and discrimination of strains and genotypes from various yeast, like *S. cerevisiae*, *Dekkera bruxellensis*, and *Candida sp*., and from fungi (Fröhlich et al. 2017). nSAPD-PCR uses specific primers, including the *NotI* recognition site, and additional nucleotides. The whole primer set consists of 20 primers (first PCR, four primers; nested PCR, 16 primers). In contrast to RAPD-PCR, the nSAPD-PCR primers are not restricted to a small group of species, so nSAPD-PCR improves the strain discrimination power of RAPD-PCR in combination with high reproducibility. Pfannebecker et al. (2016) used this method for identification of natural isolates of food spoiling osmophilic species of the genera *Zygosaccharomyces*, *Torulaspora*, *Schizosaccharomyces*, *Candida*, and *Wickerhamomyces*. Christ et al. (2015) studied the restart of stuck fermentations of spontaneously fermented wines and used nSAPD-PCR for discrimination of their species and strains.

PCR of Ribosomal DNA

Ribosomal genes (e.g., 5.8S, 18S, 26S) are grouped in tandem to form transcription units that are repeated 100–200 times throughout the genome. Each transcription unit contains another two regions, the internal transcribed spacer (ITS) and the external transcribed spacer (ETS), both of which are transcribed but not processed. The coding regions are separated by intergenic spacers, which are also known as nontranscribed spacers (NTSs) (Fernàndez-Espinar et al. 2011).

Ribosomal DNA sequences are most frequently chosen for identification, and for taxonomic and phylogenetic studies. Often, PCR-amplified rDNA is further subjected to RFLP analysis or sequencing. This technique is also known as amplified rDNA restriction analysis. Using various primers, part or all of the rRNA genes or the intergenic regions can be amplified (i.e., ITS, ETS, NTS) (White et al. 1990). By way of conservative as well as variable regions of rDNA, the differentiation of organisms at various taxonomic levels is possible (Esteve-Zarzoso et al. 1999; Dlauchy et al. 1999; Cadez et al. 2002; Caruso et al. 2002).

Restriction fragment length polymorphism of the entire ITS1-5.8S rDNA-ITS2 fragment is one of the most frequently used methods for yeast identification and typing (Guillamón et al. 1998; Esteve-Zarzoso et al. 1999; Fernández-Espinar et al. 2000; Renouf et al. 2006a). Using three restriction enzymes (i.e., *HaeIII, HinfI, CfoI*), this has been applied to a large number of yeast species, and a database is available online (www.yeast-id.com). Differentiation is based on the size and number of fragments produced by each enzyme. A similar system based on the region of 18S rDNA and ITS1 was developed by Smole Mozina et al. (1997) and described in full by Dlauchy et al. (1999). This has been applied to studies on the biodiversity of *Saccharomyces* and non-*Saccharomyces* yeast in winery ecosystems (Redzepovic et al. 2002; Raspor et al. 2006; Valero et al. 2007).

Although the variable ITS sequences are widely used in yeast species identification and typing, other parts of the rDNA gene complex have also been chosen to design specific primers in PCR ribotyping. Currently, the variable D1/D2 region of the 26S rRNA gene is the one most used for identification and classification of yeast (Nardi et al. 2006). Although the size of PCR amplicons of this region is species specific, sequencing of the fragments provides more reliable information (see below), in particular, since large databases have been developed for ascomycetous (Kurtzman and Robnett 1998) and basidiomycetous (Fell et al. 2000) yeast species. Moreover, other databases are available, such as NCBI (https://www.ncbi.nlm.nih.gov) and MycoBank (http://www.mycobank.org). A recent study by Arbefeville et al. (2017) showed that D1/D2 and ITS are equivalent for identification of fungi to the genus level, where the percentage of identification depended strongly on the database used.

Other PCR-Based Techniques

In addition to the rRNA genes, primers are known to amplify other nuclear genes and mtDNA sequences. Many specific probes can also be used as primers for amplification in PCR. Among the molecular targets analyzed by PCR elongation factor (EF)3, *ACT*1, and the *MET*2 genes are examples of nuclear genes (Daniel and Meyer 2003; Antunovics et al. 2005a; González et al. 2006), and *COX2* and *ATP8* are examples of mitochondrial genes, with the mitochondrial *COX1* gene used for detecting *Lachancea termotolerans* (Zara et al. 2013). Transposable elements (Ty) and plasmids (2 μm) can also be used as probes (Naumov et al. 1998; Pearson and McKee 1992). Primers have also been developed for species-specific PCR detection, e.g., for *Dekkera* species (Cocolin et al. 2004; De Souza Liberal et al. 2007), *C. krusei* (Manavathu et al. 1996), and *C. albicans* and other *Candida* species of clinical significance (Mannarelli and Kurtzman 1998; Soll 2000).

The sensitivity and specificity of PCR-based analyses has been improved by modification of the standard protocols. One example is given by nested PCR, in which the product of the first PCR amplification is subjected to a second round of PCR using primers internal to the sequence of the first product. Sensitivity is increased because in the second round, the already amplified first-round products are further multiplied. Specificity is increased when broad-specificity primer pairs used in the first round are followed by species-specific primers in the second round. Nested PCR methods have been used for the detection of *Dekkera/ Brettanomyces* strains in sherry (Ibeas et al. 1996).

In multiplex PCR, more than one pair of primers is applied simultaneously under the same PCR conditions (Fujita et al. 2001). Egli and Henick-Kling (2001) used one universal and four species-specific primers in multiplex PCR for differentiation of *Brettanomyces* species. Six primers for introns in the *COX1* (cytochrome oxidase subunit 1) gene of mtDNA were used in multiplex PCR to monitor wine fermentation (López et al. 2003). Nardi et al. (2006) applied specific primers selected from regions of 18S and 26S rDNA for rapid identification of *Saccharomyces sensu stricto* species. Microsatellite repeats have been applied in multiplex PCR (Pérez et al. 2001b; Schuller et al. 2004; Howell et al. 2004); these primer pairs can be used with differential labeling by the HEX (yellow), FAM (blue), and TET (green) fluorescent dyes (Gallego et al. 2005). In both nested and multiplex PCR, care needs to be taken to prevent interference between primers, and also the occurrence of cross-contamination, which can cause false positive results.

Amplified fragment length polymorphism (AFLP) analysis is a specific modification of PCR-RFLP, as a technique that amplifies DNA fragments that are randomly chosen from a restriction digest. The amplified products are separated by denaturing polyacrylamide gel electrophoresis to produce a specific fingerprint of high resolution (de Barros Lopes et al. 1999). Although this method requires additional steps to be performed, it allows more discrimination among strains, and has been applied to the clustering of *Saccharomyces sensu stricto* strains (Azumi and Goto-Yamamoto 2001; Lopandic et al. 2007), and identification of brewery yeast strains (et al. 2001), as well as *D. bruxellensis* in wineries (Curtin et al. 2007). Recently, this technique was used also to characterize genetic variability within the *H. uvarum* species (Albertin et al. 2015). A modified form of AFLP is known as Sau-PCR, which is based on digestion of genomic DNA with the restriction endonuclease *Sau*3AI, and the subsequent amplification with primers where the core sequence is based on the *Sau*3AI recognition site (Corich et al. 2005). This technique has been used recently for the characterization of 36 isolates of *S. bacillaris* (formerly *C. zemplinina*) (Fernandes Lemos Junior et al. 2016).

Another DNA amplification technique that is known as loop-mediated isothermal amplification (LAMP) has been applied for detection and identification of *Dekkera/ Brettanomyces* species from wine, beer, and soft drinks (Notomi et al. 2000; Hayashi et al. 2007), and for differentiation of species belonging to *Saccharomyces* sensu stricto (Hayashi et al. 2009).

Recently, a new rapid PCR protocol based on high-resolution melting analysis was used to identify *Saccharomyces cerevisiae* and other species within this genus. This new approach is based on real-time PCR followed by high-resolution melting analysis. Through the use of this two-step protocol it was possible to differentiate *S. cerevisiae*, *S. uvarum*, and *S. paradoxus*, first at the genus level and then at the species level (Nadai et al. 2018). High-resolution melting analysis provides a rapid, simple, high-throughput, cost-effective, and alternative single-tube approach to direct DNA sequencing for the detection of DNA polymorphisms (Gori et al. 2012).

2.3.4 Sequencing of rDNA

Sequencing is the final and most accurate way of post-PCR analysis. With the emergence of PCR-based direct DNA sequencing technologies that use automated instruments (e.g., ABI Prism capillary electrophoresis sequencer) and gene-bank databases for comparisons of sequences, this technique has become a common, routine tool in molecular studies (Piskur and Langkjaer 2004). In two decades, sequencing has become the most reliable aid to the identification of yeast species. Complete and partial sequences of rRNA genes are most widely used in taxonomic studies (Valente et al. 1999) and to establish phylogenetic relationships (Fell et al. 2000; Kurtzman and Robnett 1998, 2003). Currently, classification of yeast is based on the analysis of rDNA sequences, and in particular those of the ITS1-ITS2 and D1/D2 domains of 26S rDNA regions. Although commercial systems are available, direct sequencing *in house* is still too expensive and laborious for routine use in an industrial setting. Generally, PCR products from yeast isolates are sent off to a specialized gene-sequencing facility for identification within 1-2 days, at reasonable cost.

3 Monitoring Yeast Cell Populations Using Culture-Independent Methods

Culture-independent methods are widely applied in food and beverage microbiology, to either detect targeted groups of microorganisms, or to determine the composition of the microbial populations associated with the transformation of raw materials. In winemaking, their use is aimed at real-time assessing of the dominance of inoculated starters and the occurrence of spoilage yeast and bacteria. They provide rapid quantification and identification of the microbial species associated with grapevines, and they can be used to monitor the evolution of the microflora involved in must fermentation (Andorrà et al. 2010; 2011; Branco et al. 2012; Cocolin and

Mills 2003; Hierro et al. 2006; Röder et al. 2007). To reach these goals culture-independent methods bypass cultivation of the microorganisms, thus avoiding any bias introduced by their isolation (Fig. 3.1). By doing so, culture-independent methods can generally provide more accurate descriptions of the sizes and diversities of wine microbial communities, as compared to culture-dependent methods. In particular, as well as highlighting the dominant species within a population, they can detect VBNC and injured cells, and taxa that do not benefit from the cultivation conditions used for isolation, or that show low abundance in the sample of interest. Indeed, the uncovering of cell populations undetected by culture-dependent methods might be really important, to better understand the impact of all of the components of the wine microbiota on the quality of the final product, also in view of the implementation of quality control systems. For example, culture-independent methods allow the detection of wine yeast that can enter into the VBNC state upon SO_2 exposure (Divol and Lonvaud-Funel 2005; Salma et al. 2013; Capozzi et al. 2016) and that, under these conditions, might alter the final wine, as observed for *B. bruxellensis* (Agnolucci et al. 2013; Serpaggi et al. 2012). Culture-independent methods rely on several approaches that differ in complexity, depth of classification (genus, species, strain), reliability, reproducibility, cost, laboratory skills and equipment required. Thus, the choice of the technique to be used depends on all of these factors and on the question posed by the investigations.

3.1 Direct Microscopy, DEFT and FISH

Direct microscopic observation and hemocytometry can be used to estimate total yeast cell populations, provided they are in excess of 10^6 cell mL^{-1}. Yeast can be stained with methylene blue to differentiate between living and dead cells (Strehiano et al. 1999). This rapid staining can be used in the wine industry for estimating starter yeast vitality, although this stain overestimates yeast viability compared to colony counting. Hence, the use of alternative stains has been suggested, such as methylene violet, trypan blue, rhodamine B, and fluorescent dyes (Smart et al. 1999; Oh and Matsuoka 2002; Agnolucci et al. 2010).

A valid alternative to direct microscopic observation is the Direct Epifluorescent Filter Technique (DEFT). This is based on the use of fluorescent dyes that can highlight total or viable cells. 4,6-Diamidino-2-phenylindole (DAPI) stains DNA and can be used for total cell evaluation, and Propidium Iodide (PI) stains DNA of cells with damaged membranes, and thus has proven useful to quantify yeast and evaluate their viability during fermentation and in the finished wine (Kopke et al. 2000). The commercially available kits called LIVE/DEAD BacLight (Molecular Probes, USA) have been proposed to determine active dry yeast viability after rehydration. These kits include SYTO9, which stains all cells regardless of their viability, and propidium iodide, which stains damaged cells and masks the SYTO9 stain fluorescence, thus providing total and viable cell counts (Rodriguez-Porrata et al. 2009).

Wine yeast can also be detected by fluorescence *in-situ* hybridization (FISH). FISH combines the sensitivity and precision of molecular techniques with the direct visualization of microorganisms within their natural habitat (Amann and Ludwig 2000; Bottari et al. 2006). FISH is based on the use of specific nucleic-acid probes that contain about 15–30 nucleotides and are labeled with different fluorescent dyes, such as fluorescein, Texas red, tetramethyl-rhodamine, carbocyanine, and others (Bottari et al. 2006). Among these, carbocyanine 3 is one of the most commonly used, due to its reduced bleaching, pH insensitivity, quantum yield, and absorption coefficient. FISH probes are designed to hybridize rRNA, due to the abundance and stability of this target in living cells. Moreover, rRNA contains both highly conserved and variable sequence domains (Amman et al. 2001). Thus, depending on the region of rRNA targeted by the probe, it is possible to widen or reduce the range of microbial taxa detected. According to Bottari et al. (2006), special care must be taken in designing FISH probes, which should easily access the target sequence. FISH probes might also consist of peptide nucleic acid (PNA), a pseudopeptide in which the sugar-phosphate backbone of the DNA is replaced by a polyamide chain to which the nucleotide bases are bound with the same spacing as in DNA (Stender et al. 2002). Due to its uncharged backbone, PNA shows more favorable hybridization characteristics than traditional DNA probes, such as higher specificity and more rapid hybridization, and similar to traditional DNA probes, PNA probes can be labeled with fluorescent dyes. For wines, FISH-PNA was successfully used for identification of *Dekkera/ Brettanomyces* wine spoilage yeast (Stender et al. 2001). In general, this technique is well tailored to rapid detection of spoilage or specific groups of microorganisms, but not for the study of microbial communities. Moreover, when FISH is coupled to DEFT and cell quantification is carried out by hemocytometry, the detection limit of this technique is relatively low (about 10^4 cells mL^{-1}) (Andorrà et al. 2011), and visual counting is time consuming and subject to operator variability.

3.2 Flow Cytometry

Flow cytometry has far more potential than microscopic techniques, and can be used for automated detection of fluorescent signals (Longin et al. 2017). Flow cytometry is widely used for the analysis of heterogeneous microbial populations at the single-cell level, it provides morphological and physiological data, it is highly automated, it shows high resolution, and it can be used for rapid detection and quantification of wine microorganisms, to assess their viability and vitality, and to monitor fermentation activity (Attfield et al. 2000; Malacrino et al. 2001; Rodriguez-Porrata et al. 2009).

Flow cytometry relies on fluorescent dyes, but it has advantages over the use of FISH probes or fluorescent antibodies. Numerous different fluorescent dyes are currently commercially available that can be used for the analysis of cell populations. Propidium iodide was used as a marker of membrane permeability to evaluate cell membrane integrity under unfavorable conditions of fermentation (Mannazzu et al. 2008; Landolfo et al. 2010). Fluorescein di-acetate detects esterase activity in metabolically active cells, and this has been applied to the evaluation of yeast vitality in grape must and in red and white wines (Malacrino et al. 2001; Gerbaux and Thomas 2009). FUN-1 differently stains dead (yellow-green) and metabolically active (red) cells, and it can be used to provide information on cell numbers and on the physiological state of a yeast population. Flow cytometry coupled to FISH probes has been used for automated detection and quantification of *Saccharomyces*, non-*Saccharomyces*, and spoilage yeast in finished wines (Andorrà et al. 2010; Serpaggi et al. 2010; Wang et al. 2014). The use of polyclonal fluorescently labeled antibodies has demonstrated high efficiency for flow cytometric discrimination of *S. cerevisiae* from other yeast genera in natural must fermentations (Rodriguez and Thornton 2008), and for detection and quantification of *Brettanomyces* in wine in less than 2 h (Longin et al. 2017). However, antibodies are expensive and their instability limits their use in the wine industry.

3.3 Culture-Independent PCR-Based Methods

Among culture-independent methods, those based on PCR are widely used to detect and identify microorganisms in their ecosystem, through the analysis of their nucleic acids. The great advantage of these techniques is that isolation of total microbial DNA (or RNA) directly from samples without cultivation and isolation eliminates any bias caused by enrichment or selective culture media. While DNA analyses are generally aimed at quantification of microbial cell populations and evaluation of microbial diversity, the analysis of RNA is aimed at highlighting metabolically active populations. Amplification of nucleic acids might be achieved through universal or species-specific primers. Universal primers amplify virtually all microbial DNA/RNA present in the ecosystem analyzed, thus widening the number of microbial targets that can be recognized. On the contrary, species-specific primers are designed to highlight specific targets and their use limits the number of species that can be detected within a single sample. The amplification of nucleic acids is generally coupled to downstream techniques that are aimed at detection/ analysis/ quantification of the amplicons generated. These might work after completion of amplification (i.e., post-amplification techniques), or during amplification, as in the case of quantitative PCR. Indeed, PCR-based techniques represent a true advance in targeted analyses and for microbial community profiling, although all of them require DNA/RNA extraction and PCR primer amplification, and amplicon separation bias might lead to limited understanding of the composition of the microbial community.

3.3.1 PCR–D/TGGE

Denaturing (DGGE) or temperature (TGGE) gradient gel electrophoresis are among the post-amplification techniques utilized to resolve complex mixtures of amplicons that are produced by universal PCR primers. Briefly, after extraction of total DNA/

RNA from the sample, and PCR amplification of target sequences, the PCR products are separated using a polyacrylamide gel containing either a gradient of chemical denaturants (e.g., urea, formamide) or a denaturing temperature gradient. As the DNA molecules reach the appropriate denaturing condition, the double strand partially separates, which causes a change in migration rate. To avoid complete strand separation, a sequence rich in guanine (G) and cytosine (C) is added to the 5'-end of one of the PCR primers. DGGE and TGGE allow separation of DNA molecules that differ by a single base. This results in a band pattern from the mixture of DNA molecules from the different microbial species in the habitat. Using primers specific for relatively conserved regions of the genome, the diversity of complex microbial communities in ecosystems can be analyzed. Accordingly, PCR-D/TGGE were first applied to the study of microbial communities in natural habitats (e.g., soil, sea water, phyllosphere), and subsequently used to investigate microbial diversity in different food systems (Ercolini 2004; Giraffa 2004), which included wine (Cocolin et al. 2000; Manzano et al. 2006; Renouf et al. 2006b; Nisiotou et al. 2007; Di Maro et al. 2007). These methods have greatly contributed to the elucidation of the dynamics and ecology of wine yeast during must fermentation (for a review, see Cocolin et al. 2011). Moreover, detection of RNA through reverse transcript (RT) PCR-DGGE contributed to confirmation of the occurrence of VBNC throughout the transformation of grape must. Indeed, these techniques show a detection limit of about 10^3 CFU mL^{-1} (Cocolin et al. 2000). Thus, although still widely used for the study of wine microbial ecology (Cameron et al. 2013; Wang et al. 2015), these are not suitable for the detection of species that occur below their threshold of sensitivity. In addition, they can lead to underestimation of microbial biodiversity if similar sequences co-migrate or show common melting properties on denaturing gels. On the other hand, biodiversity might be overestimated in the formation of heteroduplexes and chimeric fragments (Bokulich et al. 2012a).

3.3.2 Other Culture-Independent PCR-Based Methods

Other PCR-based methods that have been applied to the study of wine yeast microflora include single-strand conformational polymorphisms (SSCP) (Martins et al. 2014), terminal (T)-RFLP (Martins et al. 2012), amplified fragment length polymorphism (AFLP) (Balselga et al. 2017), and automated ribosomal intergenic spacer analysis (ARISA) (Bagheri et al. 2017). SSCP is based on the evidence that subtle differences in DNA sequences result in the formation of different secondary structures and therefore in measurable differences in electrophoretic migration of single-stranded DNA. Thus, SSCP analysis of PCR-amplified 26S rDNA was carried out to study the composition of the grape must microflora (Duarte and Baleiras-Couto 2012; Martins et al. 2014).

(T)-RFLP relies on PCR amplification of heterogeneous DNA samples with one or more fluorescently labeled primers that targets the rRNA genes. PCR products are digested separately with one or more restriction enzyme(s), and terminal restriction fragments are separated by capillary electrophoresis. Exclusively, fluorescent-labeled restriction fragments are detected, and each of these is meant to represent a single operational taxonomic units or rybotype. Thus, analysis of the size, numbers, and peak heights of the resulting terminal restriction fragments allows an estimation of community diversity. This technique has been applied to the study of the wine yeast community (Bokulich et al. 2012b), although it is characterized by poor discrimination between yeast species (Sun and Liu 2014).

For AFLP, total DNA extracted from a sample of interest is restricted and ligated with enzyme-specific adaptors, and then PCR amplified. A second round of PCR follows, with selective primers labeled with fluorescent dyes, and the PCR fluorescent products are resolved by capillary electrophoresis. Balselga et al. (2017) designed seven pairs of primers that were labeled with fluorescent dyes, to generate a database of AFLP profiles for the differentiation of *Saccharomyces* and non-*Saccharomyces* yeast, and used AFLP for the identification of the yeast species and strains in grape must and wine. They also reported that AFLP is relatively laborious, and requires laboratory skills and use of complex bioinformatic software for data analysis, which renders this as relatively difficult for application to the wine industry.

ARISA is based on amplification of the ITS1-5.8S rRNA-ITS2 gene, a region that is characterized by great length and sequence polymorphism. Using fluorescent primers for PCR and capillary electrophoresis for amplicon analysis, separation of the amplified DNA has improved and this technique has been used in several ecological studies in the wine environment (Ghosh et al. 2015; Bagheri et al. 2017). Indeed, ARISA does not differentiate between strains within a species. Moreover, its sensitivity is comparable to that of PCR-DGGE.

In general, all of these techniques show interesting potential. However, as reported by Morgan et al. (2017), each of these methods has a number of drawbacks that pose a limit on their efficacy for describing the compositions and/or evolution of the yeast communities associated with winemaking.

3.3.3 Quantitative PCR

In contrast to conventional PCR-based techniques, where the amplicon is detected after completion of the amplification (i.e., end-point detection), quantitative (q)-PCR visualizes the synthesis of amplicons during each cycle of amplification through the use of fluorescent reporter molecules like SYBR Green or fluorescent probes. SYBR Green intercalates into dsDNA as this is formed during each PCR cycle, and the fluorescence increases as the target sequence is amplified (Morrison et al. 1998; Rasmussen et al. 1998). Fluorescent probes can be categorized as hydrolysis probes, hairpin probes, and hybridization probes. Among the hydrolysis probes, Taq-man is the most widespread. This is labeled at the 5'-end with a reporter dye and at the 3'-end with a quencher dye. During PCR, when the probe binds to the target sequence, the reporter dye is released from the quencher and activated due to the 5' exonuclease activity of Taq polymerase. This provides an exponential increase in fluorescence that is proportional to the amount of DNA that is amplified

(Brinkman et al. 2003; Hanna et al. 2005). Hairpin probes (Molecular Beacons, Scorpions) have inverted tandem repeats at their 5'-end and 3'-end, and a fluorophore and a quencher linked at the 5'-end and 3'-end, respectively. In the absence of the target sequence, the probe is hairpin shaped, the fluorophore and the quencher are in contact, and fluorescence emission is inhibited. Fluorescence is restored when the probe binds to the target sequence. Hybridization probes contain two oligonucleotide probes that bind the target region and are labeled, one with a donor fluorochrome, and the other with an acceptor dye. The probes are designed to hybridize to the target sequence with a head-to-tail orientation. Thus, the donor fluorochrome of the first probe emits energy that excites the acceptor dye of the second probe, which in turn emits fluorescence at a longer wavelength. On the contrary, if the hybridization probes do not anneal to the target sequence, no fluorescence is generated. In general, although easy to use and relatively cheap, SYBR Green can lead to an overestimation of the concentration of the target sequence, by binding to primer dimers and nonspecific products. On the contrary, the use of fluorescent probes guarantees much higher specificity. However, probe use is more expensive, and their design can be labor intensive. In any case, the final result of q-PCR is that during the amplification of the target sequence there is an increase in the concentration of the fluorescent dye/ probe. Thus, q-PCR enables not only detection, but also quantification of the target molecule, based on the number of cycles at which the amplicons reach a threshold fluorescence level (C_t) above the background fluorescence. On this basis, q-PCR is broadly used to evaluate abundance and/or expression of taxonomic and functional markers (Bustin et al. 2005; Smith and Osborn 2009). The amplified product can be further characterized by the analysis of its melting curves. Based on melting peak (Tm) analysis of the amplified ITS-5.8S rDNA region, Casey and Dobson (2004) differentiated between *Zygosaccharomyce bailii*, *Z. rouxii*, *Candida krusei*, *Rhodotorula glutinis*, and *S. cerevisiae*. A number of different species-specific primers have been designed based on rDNA for q-PCR–based quantification of wine microorganisms. Phister and Mills (2003) detected and quantified *D. bruxellensis* in wine using specific primers of the D1/D2 domains of 26S rDNA in a real-time PCR assay. Martorell et al. (2005) developed a specific primer that differentiated *S. cerevisiae* from its sibling *sensu stricto* species, and used this in real-time quantitative PCR to detect and quantify *S. cerevisiae* directly from wine, without pre-enrichment or culturing. The detection limit was as low as ~5 cells mL^{-1}, and the results were obtained rapidly (within 5 h). Rawsthorne and Phister (2006) and Phister et al. (2007) used primers designed on the D1/D2 domain of the 26S rDNA and were able to detect *Zygosaccharomyces bailii* and *Hanseniaspora* species, respectively, after extraction of DNA directly from fruit juice and wine. The detection limit was only 6–22 cells mL^{-1} even in the presence of 10^7 *S. cerevisiae* cells. Hierro et al. (2007) monitored *Saccharomyces* and *Hanseniaspora* populations during wine fermentation.

Also, primers that target other genes have been used in wine microbiology. For the detection of *D. bruxellensis*, Willenburg and Divol (2012) tested primers designed on the *ACT*1 and on *RAD*4 genes, as well as those designed by Phister and Mills (2003) on the D1/D2 domain of the 26S rDNA, and they tested these primers on either DNA or mRNA. Indeed, if DNA is targeted by PCR, real-time assays correlate with total cell counts that include dead and living cells. Therefore, detection of mRNA by reverse transcription real-time PCR (RT-q-PCR) is the technique of choice to detect and quantify cell viability. Accordingly, Cocolin and Mills (2003) demonstrated that in SO_2-treated wine, while no viable counts were detected by plating, both RT-PCR and PCR-DGGE showed molecular signs for a persistent population of non-*Saccharomyces* yeast. Hierro et al. (2006) targeted 26S rRNA for detection and quantification of viable yeast cells using RT-q-PCR, and showed that rRNA is more stable than mRNA, but much less stable than rDNA.

According to Andorrà et al. (2010), to quantify exclusively viable yeast cells, q-PCR can be coupled to cell staining with ethidium monoazide or propidium monoazide. These are DNA-intercalating dyes that enter cells with damaged cell walls and membranes, and following exposure to light, they covalently bind to DNA and inhibit PCR amplification. Moreover, they reduce DNA levels that are extracted from dead cells (Nocker and Camper 2006). It follows that only the DNA of intact viable cells can be amplified by q-PCR, and so this technique might represent a valid alternative to RT-q-PCR.

Indeed, q-PCR popularity is increasing in the wine industry due to its rapidity and sensitivity. Accordingly, this technique is widely used for the detection of wine microorganisms during fermentation, including VBNC (Rawsthorne and Phister 2006; Andorrà et al. 2008; 2010; Tofalo et al. 2012; García et al. 2017). Due to the use of species-specific primers, q-PCR is suitable for targeted analyses of the samples of interest and cannot be applied to the characterization of any unknown yeast communities. As with other techniques, q-PCR can also underestimate yeast levels due to a number of wine compounds that can affect PCR reactions. For this reason, different DNA extraction methods have been compared, and Tessonnière et al. (2009) indicated that the use of polyvinyl polypyrrolidone enables the elimination of the majority of these PCR inhibitors.

3.4 *High-Throughput Sequencing*

High-throughput sequencing (HTS) approaches are culture-independent techniques that are commonly applied to investigate the microbial ecology of a number of different environments. They are gaining increasing interest also for the study of the microbial ecology of grapevines and wine fermentation. According to HTS approaches, the microbial biodiversity of a matrix of interest can be characterized through PCR-based ITS phylotyping (meta-barcoding) or shotgun metagenomics. Meta-barcoding is based on the use of universal primers that are targeted to rDNA, and that amplify short DNA sequences with high taxonomic resolution (bar-codes). PCR is followed by HTS of the amplicons using next-generation sequencing platforms. Shotgun metagenomics differs from meta-barcoding as it is based on the untargeted sequencing of all microbial genomes present in a sample. In more detail, DNA extracted from all of the components of the microbial community is sheared into small fragments, which are then sequenced. By sequencing the complete genome of all of the microorganisms present, the shotgun metagenomics approach gathers information on the whole gene set of all the members of the community. Thus, it can be used to unveil both the taxonomic composition and functional potential of microbial communities, and to obtain whole-genome sequences (Quince et al. 2017). The main next-generation sequencing platforms differ in terms of the sequencing chemistry, sequence (read) lengths, number of sequences per run, and cost.

The result of HTS is a massive amount of reads, where their multivariate statistical analysis leads to a comprehensive description of the microbial biodiversity (Bokulich and Mills 2013). In particular, the data obtained can be subject to taxonomy-independent or taxonomy-dependent analyses. In the first case, genetic diversity is ordered through grouping of the sequences with those of related species. These groups of sequences are meant to correspond to taxonomic clades, and are defined operational taxonomic units (Santamaria et al. 2012). In the second case, reads are identified according to their similarities to reference sequences in taxonomically annotated databases (Ribeca and Valiente 2011; Santamaria et al. 2012). The quantification of taxonomic diversity within a population reveals which microorganisms are present (i.e., the richness), and at what abundance. This is very useful for the comparison of two or more communities, as their similarity can be evaluated based on the number of shared taxa, and to gather information on the biological function of the community, which will depend on the characteristics of the taxa that occur at higher frequency. Gene prediction and functional annotation of metagenomes is carried out to characterize the biological functions associated with the community and to identify novel genes. Also in this case, the results can be used to evaluate similarities between microbial communities and their variations, to identify new taxa and genes, and to highlight the metabolic pathways within the community (Sharpton 2014).

The application of HTS to the study of wine-related environments is fairly recent. Different studies have shown that the microbial diversity of grapes is shaped by a number of factors, including region, grape variety, and climatic and environmental conditions, and that the wine microbiota depends on the grape microbiota and on the winemaking technology (Bokulich et al. 2014; David et al. 2014; Setati et al. 2015; De Filippis et al. 2017a). Other studies have shown that the soil microbiota is influenced by the viticultural farming practices, and is highly correlated to the grape and wine microbiota (Belda et al. 2017; Grangeteau et al. 2017; Zarraonaindia et al. 2015). The application of HTS enabled Bokulich et al. (2016) to demonstrate that the grape and wine microbiota have regional patterns that affect the fermentative behavior and wine characteristics, and that the microbial composition of grapes serves as biomarker for predicting the final quality of wine.

All of these studies continue to increase our knowledge of microbial diversity during fermentation, and to help in the identification of sources of microbial contamination (Belda et al. 2017). However, further insights into the metabolic functions of wine-related microbial communities might be derived from the application of metatranscriptomics (De Filippis et al. 2017b).

High-throughput sequencing approaches are much more sensitive and accurate in the quantification of target sequences with respect to other culture-independent methods, such as PCR-DGGE. However, they can suffer from the limitations that characterize other culture-independent methods. For example, DNA extraction can be a limiting step, and different DNA extraction methods (Keisam et al. 2016) and DNA contamination might impact on the final results. For meta-barcoding, it is well known that the DNA primer sequence, amplification of the target sequences, and variations in the target sequence repeat numbers can be misleading in the evaluation of the final abundance of the components of the microbial community (Bokulich and Mills 2013). Also, different HTS technologies can generate artifacts. For all of these reasons, Sternes et al. (2017) combined meta-barcoding and shotgun metagenomics analyses to define the dynamics of the yeast microflora during wine fermentation. In doing so, they observed that the meta-barcoding approach generated a significant overestimation of yeast ascribed to *Metschnikowia* spp. Moreover, although the cost is diminishing, this approach is still too expensive for regular application to the wine industry.

4 Conclusions

Yeast have a fundamental role in winemaking. They carry out alcoholic fermentation and they contribute to the quality of the wine, although they can also cause spoilage during production and in the final wine. To control yeast activities, effective methods for yeast detection, quantification and identification need to be used during winemaking.

Not surprisingly, despite the limitations of slowness, inadequacy for detection of VBNC, and poorly represented taxa, culture-dependent methods are still used to define the composition of the microflora associated with wine-related environments. Indeed, as well as allowing yeast identification at the strain level, they provide a means for *ex-situ* preservation of wine yeast biodiversity.

Culture-independent methods generally have several advantages over culture-dependent methods, in terms of rapidity and depth of analysis. Thus, they are better suited for quality assurance and real-time fermentation monitoring and management. In addition, their application to wine-related environments is gathering a wealth of information on the composition and dynamics of microbial communities of grapes and wines.

Indeed, there remain great differences in the complexity, speed, resolution, and cost of various culture-independent techniques, and some of these factors represent obstacles to their use in wineries. Many challenges remain and further advances are required in the analytical techniques to provide an integrated quality assurance system. However, the manifold possibilities offered by the novel molecular methods is opening up new perspectives and providing the basis for future innovations in the wine industry.

5 Dedication

This chapter is dedicated to the memory of Prof. Tibor Deak (Corvinus University, Budapest, Hungary) who was meant to be one of the co-authors of this book and first proposed the topic treated.

References

Adt, I., Kohler, A., Gognies, S., Budin, J., Sandt, C., Belarbi, A., Manfait, M., & Sockalingum, G. D. (2010). FTIR spectroscopic discrimination of *Saccharomyces cerevisiae* and Saccharomyces bayanus strains. *Canadian Journal of Microbiology, 56*, 793–801.

Agnolucci, M., Scarano, S., Santoro, S., Sassano, C., Toffanin, A., & Nuti, M. (2007). Genetic and phenotypic diversity of autochthonous *Saccharomyces* spp. strains associated to natural fermentation of 'Malvasia delle Lipari'. *Letters in Applied Microbiology, 45*, 657–662.

Agnolucci, M., Rea, F., Sbrana, C., Cristiani, C., Fracassetti, D., Tirelli, A., & Nuti, M. (2010). Sulphur dioxide affects culturability and volatile phenol production by Brettanomyces/ Dekkera bruxellensis. *International Journal of Food Microbiology, 143*, 76–80.

Agnolucci, M., Cristani, C., Maggini, S., Rea, F., Cossu, A., Tirelli, A., & Nuti, M. (2013). Impact of sulphur dioxide on the viability, culturability, and volatile phenol production of *Dekkera bruxellensis* in wine. *Annales de Microbiologie, 64*, 653–659.

Albertin, W., Setati, M. E., Miot-Sertier, C., Mostert, T. T., Colonna-Ceccaldi, B., Coulon, J., et al. (2015). *Hanseniaspora uvarum* from winemaking environments show spatial and temporal genetic clustering. *Frontiers in Microbiology, 6*, 1569.

Amann, R., & Ludwig, W. (2000). Ribosomal RNA-target nucleic acid probes for studies in microbial ecology. *FEMS Microbiology Reviews, 24*, 555–565.

Amann, R., Fuchs, B. M., & Behrens, S. (2001). The identification of microorganisms by fluorescence in situ hybridization. *Current Opinion in Biotechnology, 12*(3), 231–236.

Andorrà, I., Landi, S., Mas, A., Guillamón, J. M., & Esteve-Zarzoso, B. (2008). Effect of oenological practices on microbial populations using culture independent techniques. *Food Microbiology, 25*, 849–856.

Andorrà, I., Esteve-Zarzoso, B., Guillamon, J. M., & Mas, A. (2010). Determination of viable wine yeast using DNA binding dyes and quantitative PCR. *International Journal of Food Microbiology, 144*, 257–262.

Andorrà, I., Monteiro, M., Esteve-Zarzoso, B., Albergaria, H., & Mas, A. (2011). Analysis and direct quantification of *Saccharomyces cerevisiae* and *Hanseniaspora guilliermondii* populations during alcoholic fermentation by fluorescence in situ hybridization, flow cytometry and quantitative PCR. *Food Microbiology, 28*, 1483–1149.

Antunovics, Z., Irinyi, L., & Sipiczki, M. (2005a). Combined application of methods to taxonomic identification of *Saccharomyces* strains in fermenting botrytized grape must. *Journal of Applied Microbiology, 98*, 971–979.

Antunovics, Z., Nguyen, H.-V., Gaillardin, C., & Sipiczki, M. (2005b). Gradual genome stabilization by progressive reduction of the *Saccharomyces uvarum* genome in an interspecific hybrid with *Saccharomyces cerevisiae*. *FEMS Yeast Research, 5*, 1141–1150.

Arbefeville, S., Harris, A., & Ferrieri, P. (2017). Comparison of sequencing the D2 region of the large subunit ribosomal RNA gene (MicroSEQ®) versus the internal transcribed spacer (ITS) regions using two public databases for identification of common and uncommon clinically relevant fungal species. *Journal of Microbiological Methods, 140*, 40–46.

Attfield, P. V., Kletas, S., Veal, D. A., van Rooijen, R., & Bell, P. J. L. (2000). Use of flow cytometry to monitor cell damage and predict fermentation activity of dried yeasts. *Journal of Applied Microbiology, 89*, 207–214.

Azumi, M., & Goto-Yamamoto, N. (2001). AFLP analysis of type strains and laboratory and industrial strains of *Saccharomyces* sensu stricto and its application to phenetic clustering. *Yeast, 18*, 1145–1154.

Bagheri, B., Bauer, F. F., & Setati, M. E. (2017). The impact of *Saccharomyces cerevisiae* on a wine yeast consortium in natural and inoculated fermentations. *Frontiers in Microbiology, 8*, 1988.

Baleiras Couto, M. M., van der Vossen, J. M. B. M., Hofstra, H., & Huis in't Veld, J. H. J. (1994). RAPD analysis: A rapid technique for differentiation of spoilage yeasts. *International Journal of Food Microbiology, 24*, 249–260.

Balselga, I., Zafra, O., Pérez-Lago, E., Francisco-Álvarez, R., Rodriguez-Tarduchy, G., & Santos, C. (2017). An AFLP based method for the detection and identification of indigenous yeast in complex must samples without a microbiological culture. *International Journal of Food Microbiology, 241*, 89–97.

Barata, A., Malfeito-Ferreira, M., & Loureiro, V. (2012). The microbial ecology of wine grape berries. *International Journal of Food Microbiology, 153*, 243–259.

Barbosa, C., Lage, P., Esteves, M., Chambel, L., Mendes-Faia, A., & Mendes-Ferreira, A. (2018). Molecular and phenotypic characterization of *Metschnikowia pulcherrima* strains from Douro wine region. *Fermentation, 4*, 8.

Barnett, J. A., Payne, R. W., & Yarrow, D. (2000a). Yeast. In *Characteristics and identification* (3rd ed.). England: Cambridge University Press.

Barnett, J. A., Payne, R. W., & Yarrow, D. (2000b). Yeast. In *Identification PC program* (2nd ed.). England: Cambridge Micro Software, Cambridge University Press.

Beh, A. L., Fleet, G. H., Prakichaiwattana, C., & Heard, G. M. (2006). Evaluation of molecular methods for the analyses of yeasts in foods and beverages. In A. D. Hocking, J. T. Pitt, R. A. Samson, & U. Thrane (Eds.), *Advances in food mycology* (pp. 69–106). Springer.

Belda, I., Zarraonaindia, I., Perisin, M., Palacios, A., & Acedo, A. (2017). From vineyard soil to wine fermentation: Microbiome approximations to explain the "terroir" concept. *Frontiers in Microbiology, 8*, 821.

Beuchat, L. R., Copeland, F., Curiale, M. S., Danisavich, T., Gangar, V., King, B. W., Lawlis, T. C., Likin, R. O., Okwusoa, J., Smith, C. F., & Townsend, D. E. (1998). Comparison of the SimPlate, total plate count method with Petrifilm, Redigel, and conventional pour-plate methods for enumerating aerobic microorganisms in foods. *Journal of Food Protection, 61*, 14–18.

Bisson, L. F., Joseph, C. M. L., & Domizio, P. (2017). Yeasts. In H. König, G. Unden, & J. Fröhlich (Eds.), *Biology of microorganisms on grapes, in must and in wine* (pp. 65–101). Cham: Springer.

Bokulich, N. A., & Mills, D. A. (2013). Improved selection of internal transcribed spacer-specific primers enables quantitative, ultrahigh-throughput profiling of fungal communities. *Applied and Environmental Microbiology, 79*, 2519–2526.

Bokulich, N. A., Bamforth, C. W., & Mills, D. A. (2012a). A review of molecular methods for microbial community profiling of beer and wine. *Journal of the American Society of Brewing Chemists, 70*, 150–162.

Bokulich, N. A., Hwang, C. F., Liu, S., Boundy-Mills, K., & Mills, D. A. (2012b). Profiling the yeast communities of wine using terminal restriction fragment length polymorphism. *American Journal of Enology and Viticulture, 63*, 177–184.

Bokulich, N. A., Thorngate, J. H., Richardson, P. M., & Mills, D. A. (2014). Microbial biogeography of wine grapes is conditioned by cultivar, vintage, and climate. *Proceedings of the National Academy of Sciences of the United States of America, 111*, E139.

Bokulich, N. A., Collins, T. S., Masarweh, C., Allen, G., Heymann, H., Ebeler, S. E., et al. (2016). Associations among wine grape microbiome, metabolome, and fermentation behavior suggest microbial contribution to regional wine characteristics. *mBio, 7*, e00631–e00616.

Bottari, B., Ercolini, D., Gatti, M., & Neviani, E. (2006). Application of FISH technology for microbiological analysis: Current state and prospects. *Applied Microbiology and Biotechnology, 73*, 485–494.

Branco, P., Monteiro, M., Moura, P., & Albergaria, H. (2012). Survival rate of wine-related yeasts during alcoholic fermentation assessed by direct live/dead staining combined with fluorescence in situ hybridization. *International Journal of Food Microbiology, 158*, 49–57.

Brinkman, N. E., Haugland, R. A., Wymer, L. J., Byappanahalli, M., Whitman, R. L., & Vesper, L. J. (2003). Evaluation of a rapid, quantitative real-time PCR method for enumeration of pathogenic *Candida* cells in water. *Applied and Environmental Microbiology, 69*, 1775–1782.

Bustin, S. A., Benes, V., Nolan, T., & Pfaffl, M. W. (2005). Quantitative real-time RT-PCR- a perspective. *Journal of Molecular Endocrinology, 34*(3), 597–601.

Cadez, N., Raspor, P., de Cock, A. W. A. M., Boekhout, T., & Smith, M. T. (2002). Molecular identification and genetic diversity within species of the genera *Hanseniaspora* and *Kloeckera*. *FEMS Yeast Research, 1*, 279–289.

Cameron, M., Siebrits, L., Du Toit, M., & Witthuhn, C. (2013). PCR-based DGGE fingerprinting and identification of the microbial population in South African red grape must and wine. *OENO One, 47*, 47–54.

Capece, A., Salzano, G., & Romano, P. (2003). Molecular typing techniques as a tool to differentiate non-*Saccharomyces* wine species. *International Journal of Food Microbiology, 84*, 33–39.

Capozzi, V., Di Toro, M. R., Grieco, F., Michelotti, V., Salma, M., Lamontanara, A., Russo, P., Orrù, L., Alexandre, H., & Spano, G. (2016). Viable But Not Culturable (VBNC) state of *Brettanomyces bruxellensis* in wine: New insights of molecular basis of VBNC behavior using a transcriptomic approach. *Food Microbiology, 59*, 196–204.

Cardinali, G., Liti, G., & Martini, A. (2000). Non-radioactive dot-blot DNA reassociation for unequivocal yeast identification. *International Journal of Systematic and Evolutionary Microbiology, 50*, 931–936.

Carro, D., & Piña, B. (2001). Genetic analysis of the karyotype instability in natural wine yeast strains. *Yeast, 18*, 1457–1470.

Caruso, M., Capece, A., Salzano, G., & Romano, P. (2002). Typing of *Saccharomyces cerevisiae* and *Kloeckera apiculata* strains from Aglianico wine. *Letters in Applied Microbiology, 34*, 323–328.

Casaregola, S., Nguyen, H., Lapathitis, G., Kotyk, A., & Gaillardin, C. (2001). Analysis of the beer yeast genome by PCR, sequencing and subtelomeric sequence hybridization. *International Journal of Systematic and Evolutionary Microbiology, 51*, 1607–1618.

Casey, G. D., & Dobson, A. D. W. (2004). Potential of using real-time PCR-based detection of spoilage yeast in fruit juice – a preliminary study. *International Journal of Food Microbiology, 91*, 327–335.

Chalupová, J., Raus, M., Sedlářová, M., & Šebela, M. (2014). Identification of fungal microorganisms by MALDI-TOF mass spectrometry. *Biotechnology Advances, 32*, 230–241.

Christ, E., Kowalczyk, M., Zuchowska, M., Claus, H., Löwenstein, R., Szopinska-Morawska, A., Renaut, J., & Kònig, H. (2015). An exemplary model study for overcoming stuck fermentation during spontaneous fermentation with the aid of a *Saccharomyces* triple hybrid. *Journal of Agricultural Science, 7*, 18–34.

Cocolin, L., & Mills, D. A. (2003). Wine yeast inhibition by sulfur dioxide: A comparison of culture–dependent and independent methods. *American Journal of Enology and Viticulture, 54*, 125–130.

Cocolin, L., Bisson, L. F., & Mills, D. (2000). Direct profiling of the dynamics in wine fermentation. *FEMS Microbiology Letters, 189*, 81–87.

Cocolin, L., Rantsiou, K., Iacumin, L., Zironi, R., & Comi, G. (2004). Molecular detection and identification of *Brettanomyces/Dekkera bruxellensis* and *Brettanomyces/Dekkera anomalus* in spoiled wines. *Applied and Environmental Microbiology, 70*, 1347–1355.

Cocolin, L., Campolongo, S., Alessandria, A., Dolci, P., & Rantsiou, K. (2011). Culture independent analyses and wine fermentation: An overview of achievements 10 years after first application. *Annales de Microbiologie, 61*, 17–23.

Corich, V., Mattiazzi, A., Soldati, E., Carraro, A., & Giacomini, A. (2005). Sau-PCR, a novel amplification technique for genetic fingerprinting of microorganisms. *Applied and Environmental Microbiology, 71*, 6401–6406.

Corredor, M., Davila, A.-M., Gaillardin, C., & Casaregola, S. (2000). DNA probes specific for the yeast species *Debaryomyces hansenii*: Useful tools for rapid identification. *FEMS Microbiology Letters, 193*, 171–177.
Couto, J. A., Barbosa, A., & Hogg, T. (2005). A simple cultural method for the presumptive detection of the yeasts *Brettanomyces/Dekkera* in wines. *Letters in Applied Microbiology, 41*, 505–510.
Curtin, C. D., Bellon, J. R., Henschke, P. A., Godden, P. W., & de Barros Lopes, M. A. (2007). Genetic diversity of *Dekkera bruxellensis* yeasts isolated from Australian wineries. *FEMS Yeast Research, 7*, 471–481.
Daniel, H. M., & Meyer, W. (2003). Evaluation of ribosomal RNA and actin gene sequences for identification of ascomycetous yeasts. *International Journal of Food Microbiology, 86*, 71–78.
David, V., Terrat, S., Herzine, K., Claisse, O., Rousseaux, S., Tourdot-Maréchal, R., et al. (2014). High-throughput sequencing of amplicons for monitoring yeast biodiversity in must and during alcoholic fermentation. *Journal of Industrial Microbiology & Biotechnology, 41*, 811–821.
de Barros Lopes, M., Soden, A., Henschke, P. A., & Langridge, P. (1996). PCR differentiation of commercial yeast strains using intron splice primers. *Applied and Environmental Microbiology, 62*, 4514–4520.
de Barros Lopes, M., Soden, A., Henschke, P. A., & Langridge, P. (1998). Differentiation and species identification of yeasts using PCR. *International Journal of Systematic Bacteriology, 48*, 279–286.
de Barros Lopes, M., Raineri, S., Henschke, P. A., & Langridge, P. (1999). AFLP fingerprinting for analysis of yeast genetic variation. *International Journal of Systematic Bacteriology, 49*, 915–924.
De Filippis, F., La Storia, A., & Blaiotta, G. (2017a). Monitoring the micobiota during Greco di Tufo and Anglianico wine fermentation by 18S rRNA gene sequencing. *Food Microbiology, 63*, 117–122.
De Filippis, F., Parente, E., & Ercolini, D. (2017b). Metagenomics insights into food fermentations. *Microbial Biotechnology, 10*, 91–102.
De Souza Liberal, A. T., Basilio, A. C. M., do Monte Resende, A., Brasileiro, B. T. V., da Silva-Filho, E. A., de Morais, J. O. F., Simoes, D. A., & de Morais, M. A., Jr. (2007). Identification of *Dekkera bruxellensis* as a major contaminant yeast in continuous fuel ethanol fermentation. *Journal of Applied Microbiology, 102*, 538–547.
Deak, T. (2003). Detection, enumeration and isolation of yeasts. In T. Boekhout & V. Robert (Eds.), *Yeasts in food, beneficial and detrimental aspects* (pp. 39–64). Hamburg: Behr's Verlag.
Deak, T. (2008). *Handbook of food spoilage yeasts* (2nd ed., p. 325). Boca Raton: CRC Press.
Deak, T., & Beuchat, L. R. (1993). Comparison of the SIM, API20C, and ID32C systems for the identification of yeasts isolated from fruit juice concentrates and beverages. *Journal of Food Protection, 56*, 585–592.
Di Maro, E., Ercolini, D., & Coppola, S. (2007). Yeast dynamics during spontaneous wine fermentation of the Catalanesca grape. *International Journal of Food Microbiology, 117*, 201–210.
Divol, B., & Lonvaud-Funel, A. (2005). Evidence for viable but non-culturable yeasts in botrytis-affected wine. *Journal of Applied Microbiology, 99*, 85–93.
Dlauchy, D., Tornai-Lehoczki, J., & Péter, G. (1999). Restriction enzyme analysis of PCR amplified rRNA as a taxonomic tool in yeast identification. *Systematic and Applied Microbiology, 22*, 445–453.
Domizio, P., Romani, C., Lencioni, L., Comitini, F., Gobbi, M., Mannazzu, I., & Ciani, M. (2011). Outlining a future for non-*Saccharomyces* yeasts: Selection of putative spoilage wine strains to be used in association with Saccharomyces cerevisiae for grape juice fermentation. *International Journal of Food Microbiology, 147*, 170–180.
Dowhanick, T., Sobczak, J., Russel, I., & Stewart, G. (1990). The rapid identification by protein fingerprinting of yeast and brewery contaminants. *Journal of the American Society of Brewing Chemists, 48*, 75–79.

Du Plessis, H. W., du Toit, M., Hoff, J. W., Hart, R. S., Ndimba, B. K., & Jolly, N. P. (2017). Characterization of non-*Saccharomyces* yeasts using different methodologies and evaluation of their compatibility with malolactic fermentation. *South African Journal of Enology and Viticulture, 38*, 46.63.

Duarte, F. L., & Baleiras-Couto, M. (2012). Evaluation of fermenting grape must yeast dynamics by SSCP profiles. *Ciencia e Tecnica Vitivinicola, 27*, 95–101.

Egli, C. M., & Henick-Kling, T. (2001). Identification of *Brettanomyces/Dekkera* species based on polymorphism in the rRNA internal transcribed spacer region. *American Journal of Enology and Viticulture, 52*, 241–247.

Entis, P., & Lerner, I. (1996). Two-day yeast and mold enumeration using the Iso-Grid® membrane filtration system in conjunction with YM-11 agar. *Journal of Food Protection, 59*, 416–419.

Ercolini, D. (2004). PCR-DGGE fingerprinting: novel strategies for detection of microbes in food. *Journal of Microbiological Methods, 56*, 297–314.

Esteve-Zarzoso, B., Belloch, C., Uruburu, F., & Querol, A. (1999). Identification of yeasts by RFLP analysis of the 5.8S rRNA gene and the two ribosomal internal transcribed spacers. *International Journal of Systematic Bacteriology, 49*, 329–337.

Fell, J. W., Boekhout, T., Fonseca, A., Scorzetti, G., & Statzell-Tallman, A. (2000). Biodiversity and systematics of basidiomycetous yeasts as determined by large subunit rD1/D2 domain sequence analysis. *International Journal of Systematic and Environmental Microbiology, 50*, 1351–1371.

Fernandes Lemos Junior, W. J., Bovo, B., Nadai, C., Crosato, G., Carlot, M., Favaron, F., Giacomini, A., & Corich, V. (2016). Biocontrol ability and action mechanism of *Starmerella bacillaris* (Synonym *Candida zemplinina*) isolated from wine musts against gray mold disease agent *Botrytis cinerea* on grape and their effects on alcoholic fermentation. *Food Microbiology, 7*, 1249.

Fernández-Espinar, M. T., Esteve-Zarzoso, B., Querol, A., & Barrio, E. (2000). RFLP analysis of the ribosomal internal transcribed spacers and the 5.8S rRNA gene region of the genus *Saccharomyces*: a fast method for species identification of flor yeast. *Antonie van Leeuwenhoek, 78*, 87–97.

Fernández-Espinar, M. T., López, V., Ramón, D., Bartra, E., & Querol, A. (2001). Study of the authenticity of commercial wine yeast strains by molecular techniques. *International Journal of Food Microbiology, 70*, 1–10.

Fernández-Espinar, M. T., Martorell, P., de Llanos, R., & Querol, A. (2006). Molecular methods to identify and characterise yeasts in foods and beverages. In A. Querol & G. H. Fleet (Eds.), *Yeasts in food and beverages* (pp. 55–82). Berlin: Springer.

Fernández-Espinar, M. T., Llopis, S., Querol, A., & Barrio, E. (2011). Chapter 5 – Molecular identification and characterization of wine yeasts. In A. V. Carrascosa, R. Munoz, & R. Gonzalez (Eds.), *Molecular wine microbiology* (pp. 111–141). San Diego: Academic.

Fröhlich, J., König, H., & Claus, H. (2017). Molecular methods for identification of wine microorganisms and yeast development. In H. König, G. Unden, & J. Fröhlich (Eds.), *Biology of microorganisms on grapes, in must and in wine* (pp. 517–548). Cham: Springer.

Fujita, S. H., Senda, Y., Nakaguchi, S., & Hashimoto, T. (2001). Multiplex PCR using internal transcribed spacer 1 and 2 regions for rapid detection and identification of yeast strains. *Journal of Clinical Microbiology, 39*, 3617–3622.

Fung, D. (2002). Rapid methods and automation in microbiology. *Comprehensive Reviews in Food Science and Food Safety, 1*, 3–22.

Gallego, F. J., Perez, M. A., Martinez, I., & Hidalgo, P. (1998). Microsatellites obtained from database sequences are useful to characterize *Saccharomyces cerevisiae* strains. *American Journal of Enology and Viticulture, 49*, 350–351.

Gallego, F. J., Perez, M. A., Nunez, Y., & Hidalgo, P. (2005). Comparison of RAPDs, AFLPs and SSR markers for the genetic analysis of yeast strains of *Saccharomyces cerevisiae*. *Food Microbiology, 22*, 561–568.

García, M., Esteve-Zarzoso, B., Crespo, J., Cabellos, J. M., & Arroyo, T. (2017). Yeast monitoring of wine mixed or sequential fermentations made by native strains from D.O. "Vinos de Madrid" using Real-Time Quantitative PCR. *Frontiers in Microbiology, 8*, 2520.

Gerbaux, V., & Thomas, J. (2009). Utilisations pratiques de la cytometrie de flux pour le suivi des levures en oenologie. *Revue Francaise d'Oenologie*, 8–13.

Ghosh, S., Bagheri, B., Morgan, H. H., Divol, B., & Setati, M. E. (2015). Assessment of wine microbial diversity using ARISA and cultivation based methods. *Annales de Microbiologie, 65*, 4.

Giesendorf, B. A. J., Quint, W. G. V., Vandamme, P., & van Belkum, A. (1996). Generation of DNA probes for detection of microorganisms by polymerase chain reaction fingerprinting. *Zentbl Bakteriol Parasit Infekt Hyg Abt Orig, 283*, 417–430.

Giraffa, G. (2004). Studying the dynamics of microbial populations during food fermentation: Table 1. *FEMS Microbiology Reviews, 28*(2), 251–260.

González, S. S., Barrio, E., Gafner, J., & Querol, A. (2006). Natural hybrids from *Saccharomyces cereviseae, Saccharomyces bayanus* and *Saccharomyces kudriavzevii* in wine fermentations. *FEMS Yeast Research, 6*(8), 1221–1234.

Gori, A., Cerboneschi, M., & Tegli, S. (2012). High-resolution melting analysis as a powerful tool to discriminate and genotype *Pseudomonas savastanoi* pathovars and strains. *PLoS ONE, 7*, e30199.

Grando, M. S., Ubeda, J., & Briones, A. I. (1994). RAPD analysis of wine *Saccharomyces cerevisiae* strains differentiated by pulsed field gel electrophoresis. *Biotechnology Techniques, 8*, 557–560.

Grangeteau, C., Gerhards, D., Rousseaux, S., von Wallbrunn, C., Alexandre, H., & Guilloux-Benatier, M. (2015). Diversity of yeast strains of the genus *Hanseniaspora* in the winery environment: What is their involvement in grape must fermentation? *Food Microbiology, 50*, 70–77.

Grangeteau, C., Gerhards, D., Terrat, S., Dequiedt, S., Alexandre, H., Guilloux-Benatier, M., von Wallbrunn, C., & Rousseaux, S. (2016). FT-IR spectroscopy: A powerful tool for studying the inter- and intraspecific biodiversity of cultivable non-*Saccharomyces* yeasts isolated from grape must. *Journal of Microbiological Methods, 121*, 50–58.

Grangeteau, C., Roullier-Gall, C., Rousseaux, S., Guogeon, R. D., Schmitt-Kopplin, P., Alexandre, H., et al. (2017). Wine microbiology is driven by vineyard and winery anthropogenic factors. *Microbial Biotechnology, 10*, 354–370.

Guillamón, J. M., & Barrio, E. (2017). Genetic polymorphism in wine yeasts: Mechanisms and methods for its detection. *Frontiers in Microbiology, 8*, 806.

Guillamón, J. M., Barrio, E., Huerta, T., & Querol, A. (1994). Rapid characterization of four species of the *Saccharomyces sensu stricto* complex according to mitochondrial DNA patterns. *International Journal of Systematic Bacteriology, 44*, 708–714.

Guillamón, J. M., Sabaté, J., Barrio, E., Cano, J., & Querol, A. (1998). Rapid identification of wine species based on RFLP analysis of the ribosomal ITS region. *Archives of Microbiology, 169*, 387–392.

Gutiérrez, C., Gómez-Flechoso, M. A., Belda, I., Ruiz, J., Kayali, N., Polo, L., & Santos, A. (2017). Wine yeasts identification by MALDI-TOF MS: Optimization of the preanalytical steps and development of an extensible open-source platform for processing and analysis of an in-house MS database. *International Journal of Food Microbiology, 254*, 1–10.

Hanna, S. E., Christopher, J. C., & Wang, H. H. (2005). Real-time polymerase chain reaction for the food microbiologist: Technologies, applications, and limitations. *Journal of Food Science, 70*(3), R49–R53.

Hayashi, N., Arai, R., Tada, S., Taguchi, H., & Ogawa, Y. (2007). Detection and identification of *Brettanomyces / Dekkera* sp. yeasts with a loop-mediated isothermal amplification method. *Food Microbiology, 24*, 778–785.

Hayashi, N., Minato, T., Kanai, K., Ikushima, S., Yoshida, S., Tada, S., Taguchi, H., & Ogawa, Y. (2009). Differentiation of species belonging to *Saccharomyces sensu stricto* using a loop-mediated isothermal amplification method. *Journal of the American Society of Brewing Chemists, 67*, 118–126.

Hennequin, C., Thierry, A., Richard, G. F., Lecointre, G., Nguyen, H. V., Gaillardin, C., & Dujon, B. (2001). Microsatellite typing as a new tool for identification of *Saccharomyces cerevisiae* strains. *Journal of Clinical Microbiology, 39*, 551–559.

Hierro, N., González, Á., Mas, A., & Guillamón, J. M. (2004). New PCR-based methods for yeast identification. *Journal of Applied Microbiology, 97*, 792–801.

Hierro, N., González, Á., Mas, A., & Guillamón, J. M. (2006). Diversity and evolution of non-*Saccharomyces* yeast populations during wine fermentation: effect of grape ripeness and cold maceration. *FEMS Yeast Research, 6*, 102–111.

Hierro, N., Esteve-Zarzoso, B., Mas, A., & Guillamón, J. M. (2007). Monitoring of *Saccharomyces* and *Hanseniaspora* populations during alcoholic fermentation by real-time quantitative PCR. *FEMS Yeast Research, 7*, 1340–1349.

Howell, K. S., Bartowsky, E. J., Fleet, G. H., & Henschke, P. A. (2004). Microsatellite PCR profiling of *Saccharomyces cerevisiae* strains during wine fermentation. *Letters in Applied Microbiology, 38*, 315–320.

Ibeas, J., Lozano, J., Perdigones, F., & Jimenez, J. (1996). Detection of *Dekkera/Brettanomyces* strains in sherry by a nested PCR method. *Applied and Environmental Microbiology, 62*, 998–1003.

Ivey, M. L., & Phister, T. G. (2011). Detection and identification of microorganisms in wine: A review of molecular techniques. *Journal of Industrial Microbiology and Biotechnology, 38*(10), 1619–1634.

Keisam, S., Romi, W., Ahmed, G., & Jeyaram, K. (2016). Quantifying the biases in metagenome mining for realistic assessment of microbial ecology of naturally fermented foods. *Scientific Reports, 6*, 34155.

Kopke, C., Cristovao, A., Prata, A. M., Silva Pereira, C., Figueiredo Marques, J. J., & San Romao, M. V. (2000). Microbiological control of wine. The application of epifluorescence microscopy method as a rapid technique. *Food Microbiology, 17*, 257–260.

Kurtzman, C. P., & Robnett, C. J. (1998). Identification and phylogeny of ascomycetous yeasts from analysis of nuclear large subunit (26S) ribosomal DNA partial sequences. *Antonie van Leeuwenhoek, 73*, 331–371.

Kurtzman, C. P., & Robnett, C. J. (2003). Phylogenetic relationships among yeasts of the '*Saccharomyces* complex' determined from multigene sequence analyses. *FEMS Yeast Research, 3*, 417–432.

Landolfo, S., Zara, G., Zara, S., Budroni, M., Ciani, M., & Mannazzu, I. (2010). Oleic acid and ergosterol supplementation mitigates oxidative stress in wine strains of *Saccharomyces cerevisiae*. *International Journal of Food Microbiology, 141*, 229–235.

Lavallée, F., Salvas, Y., Lamy, S., Thomas, D. Y., Degree, R., & Dulau, L. (1994). PCR and DNA fingerprinting used as quality control in the production of wine yeast strains. *American Journal of Enology and Viticulture, 45*, 86–91.

Le Jeune, C., Lollier, M., Demuyter, C., Erny, C., Legras, J.-L., Aigle, M., & Masneuf-Pomarede, I. (2007). Characterization of natural hybrids of *Saccharomyces cerevisiae* and *Saccharomyces bayanus* var. *uvarum*. *FEMS Yeast Research, 7*, 540–549.

Legras, J. L., & Karst, F. (2003). Optimisation of interdelta analysis for *Saccharomyces cerevisiae* strain characterisation. *FEMS Microbiology Letters, 221*, 249–255.

Legras, J. L., Ruh, O., Merdinoglu, D., & Karst, F. (2005). Selection of hypervariable microsatellite loci for the characterization of *Saccharomyces cerevisiae* strains. *International Journal of Food Microbiology, 102*, 73–83.

Li, S. S., Cheng, C., Li, Z., Chen, J.-Y., Yan, B., Han, B.-Z., & Reeves, M. (2010). Yeast species associated with wine grapes in China. *International Journal of Food Microbiology, 138*, 85–90.

Longin, C., Petitgonnet, C., Guilloux-Benatier, M., Rousseaux, S., & Alexandre, H. (2017). Application of flow cytometry to wine microorganisms. *Food Microbiology, 62*, 221–231.

Lopandic, K., Gangl, H., Wallner, E., Tscheik, G., Leitner, G., Querol, A., Borth, N., Breitenbach, M., Prillinger, H., & Tiefenbrunner, W. (2007). Genetically different wine yeasts isolated from Austrian vine-growing regions influence wine aroma differently and contain putative hybrids between *Saccharomyces cerevisiae* and *Saccharomyces kudriavzevii*. *FEMS Yeast Research, 7*, 953–965.

Lopes, C. A., van Broock, M., Querol, A., & Caballero, A. C. (2002). *Saccharomyces cerevisiae* wine yeast populations in a cold region in Argentiniean Patagonia. A study at different fermentation scales. *Journal of Applied Microbiology, 93*, 608–615.

López, V., Fernández-Espinar, M. T., Barrio, E., Ramón, D., & Querol, A. (2003). A new PCR-based method for monitoring inoculated wine fermentation. *International Journal of Food Microbiology, 81*, 63–71.

Loureiro, V., Malfeito-Ferreira, M., & Carreira, A. (2004). Detecting spoilage yeasts. In R. R. Steele (Ed.), *Understanding and measuring the shelf-life of food* (pp. 233–288). Cambridge: Woodhead.

Malacrino, P., Zapparoli, G., Torriani, S., & Dellaglio, F. (2001). Rapid detection of viable yeasts and bacteria in wine by flow cytometry. *Journal of Microbiological Methods, 45*, 127–134.

Manavathu, E. K., Vakulenko, S. B., Obedeanu, N., & Lerner, S. A. (1996). Isolation and characterization of a species-specific DNA probe for the detetion of *Candida krusei*. *Current Microbiology, 33*, 147–151.

Mannarelli, B. M., & Kurtzman, C. P. (1998). Rapid identification of *Candida albicans* and other human pathogenic yeasts by using short oligonucleotides in a PCR. *Journal of Clinical Microbiology, 36*, 1634–1641.

Mannazzu, I., Agelozzi, D., Belviso, S., Budroni, M., Farris, G. A., Goffrini, P., Lodi, T., Marzona, M., & Bardi, L. (2008). Behaviour of *Saccharomyces cerevisiae* wine strains during adaptation to unfavourable conditions of fermentation in synthetic medium: Cell lipid composition, membrane integrity, viability and fermentative activity. *International Journal of Food Microbiology, 121*, 84–91.

Manzano, M., Medrala, D., Giusto, C., Bartolomeoli, I., Urso, R., & Comi, G. (2006). Classical and molecular analyses to characterize commercial dry yeasts used in wine fermentations. *Journal of Applied Microbiology, 100*, 599–607.

Marcos, J. Y., & Pincus, D. H. (2013). Fungal diagnostics: Review of commercially available methods. In L. O'Connor & B. Glynn (Eds.), *Fungal diagnostics. Methods in molecular biology (Methods and protocols)* (Vol. 968). Totowa: Humana Press.

Marinangeli, P., Angelozzi, D., Ciani, M., Clementi, F., & Mannazzu, I. (2004). Minisatellites in *Saccharomyces cerevisiae* genes encoding cell wall proteins: A new way towards wine strain characterisation. *FEMS Yeast Research, 4*, 427–435.

Martinez, C., Cosgaya, P., Vásquez, C., Gac, S., & Ganga, A. (2007). High degree of correlation between molecular polymorphism and geographic origin of wine yeast strains. *Journal of Applied Microbiology, 103*, 2185–2195.

Martins, G., Miot-Sertier, C., Lauga, B., Claisse, O., Lonvaud-Funel, A., Soulas, G., et al. (2012). Grape berry bacterial microbiota: Impact of the ripening process and the farming system. *International Journal of Food Microbiology, 158*, 93–100.

Martins, G., Vallance, J., Mercier, A., Albertin, W., Stamatopoulos, P., Rey, P., et al. (2014). Influence of the farming system on the epiphytic yeasts and yeast like fungi colonizing grape berries during the ripening process. *International Journal of Food Microbiology, 177*, 21–28.

Martorell, P., Querol, A., & Fernandez-Espinar, M. T. (2005). Rapid identification and enumeration of *Saccharomyces cerevisiae* cells in wine by real-time PCR. *Applied and Environmental Microbiology, 71*, 6823–6830.

Mercado, L., Dalcero, A., Masuelli, R., & Combina, M. (2007). Diversity of *Saccharomyces* strains on grapes and winery surfaces: Analysis of their contribution to fermentative flora of malbec wine from Mendoza (Argentina) during two consecutive years. *Food Microbiology, 24*, 403–412.

Millet, V., & Lonvaud-Funel, A. (2000). The viable but non-culturable state of wine micro-organisms during storage. *Letters in Applied Microbiology, 30*, 136–141.

Mills, D. A., Johannsen, E. A., & Cocolin, L. (2002). Yeast diversity and persistence in botrytis-affected wine fermentation. *Applied and Environmental Microbiology, 68*, 4884–4893.

Mitrakul, C. M., Henick-Kling, T., & Egli, C. M. (1999). Discrimination of *Brettanomyces / Dekkera* yeast isolates from wine by using various DNA finger-printing methods. *Food Microbiology, 16*, 3–14.

Morgan, H. H., du Toit, M., & Setati, M. E. (2017). The grapevine and wine microbiome: Insights from high-throughput amplicon sequencing. *Food Microbiology, 8*, 820.

Morrison, T. B., Weiss, J. J., & Witter, C. T. (1998). Quantification of low-copy transcript by continuous SYBR Green monitoring during amplification. *BioTechniques, 24*, 954–962.

Morneau, A. D., Zuehlke, J. M., & Edwards, C. G. (2011). Comparison of media formulations used to selectively cultivate *Dekkera/Brettanomyces*. *Letters in Applied Microbiology, 53*(4), 460–465.

Nadai, C., Bovo, B., Giacomini, A., & Corich, V. (2018). New rapid PCR protocol based on high-resolution melting analysis to identify *Saccharomyces cerevisiae* and other species within its genus. *Journal of Applied Microbiology, 124*, 1232–1242.

Nardi, T., Carlot, M., De Bortoli, E., Corich, V., & Giacomini, A. (2006). A rapid method for differentiating *Saccharomyces sensu stricto* strains from other yeast species in an enological environment. *FEMS Microbiology Letters, 264*, 168–173.

Naumov, G., Naumova, E., Gaillardin, C., Turakainen, H., & Korhola, M. (1994). Identification of new chromosomes of *Saccharomyces bayanus* using gene probes from *S. cerevisiae*. *Hereditas, 120*, 121–126.

Naumov, G. I., Naumova, E. S., & Sniegowski, P. D. (1998). *Saccharomyces paradoxus* and *Saccharomyces cerevisiae* are associated with exudates of North American oaks. *Canadian Journal of Microbiology, 44*, 1045–1050.

Ness, F., Lavallée, F., Dubourdieu, D., Aigle, M., & Dulau, L. (1993). Identification of yeast strains using the polymerase chain reaction. *Journal of the Science of Food and Agriculture, 62*, 89–94.

Nguyen, H. V., & Gaillardin, C. (2005). Evolutionary relationships between the former species *Saccharomyces uvarum* and the hybrids *Saccharomyces baya*nus and *Saccharomyces pastorianus*; reinstatement of *Saccharomyces uvarum* (Beijerinck) as a distinct species. *FEMS Yeast Research, 5*, 471–483.

Nguyen, H. V., Lepingle, A., & Gaillardin, C. (2000). Molecular typing demonstrates homogeneity of *Saccharomyces uvarum* strains and reveals the existence of hybrids between *S. uvarum* and *S. cerevisiae*, including the *S. bayanus* Type strain CBS 380. *Systematic and Applied Microbiology, 23*, 71–85.

Nisiotou, A. A., Spiropoulos, A. E., & Nychas, G.-J. E. (2007). Yeast community structures and dynamics in healthy and *Botrytis*-affected grape must fermentations. *Applied and Environmental Microbiology, 73*, 6705–6713.

Nocker, A., & Camper, A. K. (2006). Selective removal of DNA from dead cells of mixed bacterial communities by use of ethidium monoazide. *Applied and Environmental Microbiology, 72*, 1997–2004.

Notomi, T., Okayama, H., Masubuchi, H., Yonekawa, T., Watanabe, K., Amino, N., & Hase, T. (2000). Loop-mediated isothermal amplification of DNA. *Nucleic Acids Research, 28*, 1–7.

Oelofse, A., Malherbe, S., Pretorius, I. S., & Du Toit, M. (2010). Preliminary evaluation of infrared spectroscopy for the differentiation of Brettanomyces bruxellensis strains isolated from red wines. *International Journal of Food Microbiology, 143*, 136–142.

Oh, K. B., & Matsuoka, H. (2002). Rapid viability assessment of yeast cells using vital staining with2-NBDG, a fluorescent derivative of glucose. *International Journal of Food Microbiology, 76*, 47–53.

Pallmann, C., Brown, J. A., Olineka, T. L., Cocolin, L., Mills, D., & Bisson, L. (2001). Use of WL medium to profile native flora fermentations. *American Journal of Enology and Viticulture, 52*, 198–203.

Pataro, C., Guerra, J. B., Petrillo-Peixoto, M. L., Mendonca-Hagler, L. C., Linardi, V. R., & Rosa, C. A. (2000). Yeast communities and genetic polymorphism of *Saccharomyces cerevisiae* strains associated with artisanal fermentation in Brazil. *Journal of Applied Microbiology, 89*, 24–31.

Paugham, A., Benchewtrit, M., Fiarce, A., Tourte-Schafer, C., & Dupouy-Camet, J. (1999). Comparison of four commercialized biochemical systems for clinical yeast identification by colour-producing reactions. *Medical Mycology, 37*, 11–17.

Pearson, B. M., & McKee, R. A. (1992). Rapid identification of *Saccharomyces cerevisiae, Zygosaccharomyces bailii* and *Zygosaccharomyces rouxii. International Journal of Food Microbiology, 16*, 63–67.

Pérez, M. A:., Gallego, F. J., & Hidalgo, P. (2001a). Evaluation of molecular techniques for the genetic characterization of *Saccharomyces cerevisiae* strains. *FEMS Microbiology Letters, 205*, 375–378.

Pérez, M. A., Gallego, F. J., Martinez, I., & Hidalgo, P. (2001b). Detection, distribution and selection of microsatellites (SSRs) in the genome of the yeast *Saccharomyces cerevisiae* as molecular markers. *Letters in Applied Microbiology, 33*, 461–466.

Pfannebecker, J., Schiffer-Hetz, C., Frohlich, J., & Becker, B. (2016). Culture medium optimization for osmotolerant yeasts by use of a parallel fermenter system and rapid microbiological testing. *Journal of Microbiological Methods, 130*, 14–22.

Phister, T., & Mills, D. A. (2003). Real-time PCR assay for the detection and enumeration of Dekkera bruxellensis in wine. *Applied and Environmental Microbiology, 69*, 7430–7434.

Phister, G. T., Rawsthorne, H., Joseph, C. M. L., & Mills, D. A. (2007). Real-time PCR assay for detection and enumeration of *Hanseniaspora* species from wine and juice. *American Journal of Enology and Viticulture, 58*, 229–233.

Piskur, J., & Langkjaer, R. B. (2004). Yeast genome sequencing: The power of comparative genomics. *Molecular Microbiology, 53*, 381–389.

Povhe Jemec, K., Cadez, N., Zagorc, T., Bubic, V., Zupec, A., & Raspor, P. (2001). Yeast population dynamics in five spontaneous fermentations of Malvasia must. *Food Microbiology, 18*, 247–259.

Querol, A., & Ramón, D. (1996). The application of molecular techniques in wine microbiology. *Trends in Food Science and Technology, 7*, 73–78.

Querol, A., Barrio, E., & Ramon, D. (1992). A comparative study of different methods of yeast strain characterization. *Systematic and Applied Microbiology, 15*, 439–446.

Querol, A., Barrio, E., & Ramon, D. (1994). Population dynamics of natural *Saccharomyces* strains during wine fermenmtation. *International Journal of Food Microbiology, 21*, 315–323.

Quince, C., Walker, A. W., Simpson, J. T., Loman, N. J., & Segata, N. (2017). Shotgun metagenomics, from sampling to analysis. *Nature Biotechnology, 35*(9), 833–844.

Ramìrez-Castrillòn, M., Mendes, S. D. C., Inostroza-Ponta, M., & Valente, P. (2014). (GTG)5 MSP-PCR fingerprinting as a technique for discrimination of wine associated yeasts? *PLoS ONE, 9*, e105870.

Rasmussen, R., Morrison, T., Herrmann, M., & Witter, C. (1998). Quantitative PCR by continuous fluorescence monitoring of double strand DNA specific binding dye. *Biochemica, 2*, 8–11.

Raspor, P., Milek, D. M., Polanc, J., Smole Mozina, S., & Cadez, N. (2006). Yeasts isolated from three varieties of grapes cultivated in different locations of the Dolenjska vine-growing region, Slovenia. *International Journal of Food Microbiology, 109*, 97–102.

Rawsthorne, H., & Phister, T. G. (2006). A real-time PCR assay for the enumeration and detection of *Zygosaccharomyces bailii* from wine and fruit juices. *International Journal of Food Microbiology, 112*, 1–7.

Redzepovic, S., Orlic, S., Sikora, S., Majdak, A., & Pretorius, I. S. (2002). Identification and characterization of *Saccharomyces cerevisiae* and *Saccharomyces paradoxus* strains isolated from Croatioan vineyards. *Letters in Applied Microbiology, 35*, 305–310.

Renouf, V., Falcou, M., Miot-Serier, C., Perello, M. C., De Revel, G., & Lonvaud-Funel, A. (2006a). Interactions between *Brettanomyces bruxellensis* and other yeast species during the initial stages of winemaking. *Journal of Applied Microbiology, 100*, 1208–1219.

Renouf, V., Perello, M. C., Strehaiano, P., & Lonvaud-Funel, A. (2006b). Global survey of the microbial ecosystem during alcoholic fermentation in winemaking. *Journal International des Sciences de la Vigne et du Vin, 40*, 101–116.

Ribeca, P., & Valiente, G. (2011). Computational challenges of sequence classification in microbiomic data. *Briefings in Bioinformatics, 12*(6), 614–625.
Robert, V. (2000). BioloMICS a general system for identification, classification and archiving of strains and species data. www.bio-aware.com.
Robert, V. (2003). Data processing. In T. Boekhout & V. Robert (Eds.), *Yeast in food. Beneficial and detrimental aspects* (pp. 139–169). Hamburg: Behr's Verlag.
Robert, V., Szoke, S., Jabas, B., Vu, D., Chouchen, O., Blom, E., & Cardinali, G. (2011). BioloMICS software: Biological data management, identification, classification and statistics. *Open Applied Informatics Journal, 5*, 87–98.
Röder, C., König, H., & Fröhlich, J. (2007). Species-specific identification of *Dekkera / Brettanomyces* yeasts by fluorescently labeled DNA probes targeting the 26S rRNA. *FEMS Yeast Research, 7*, 1013–1026.
Rodriguez, S. B., & Thornton, R. J. (2008). Use of flow cytometry with fluorescent antibodies in real-time monitoring of simultaneously inoculated alcoholic-malolactic fermentation of Chardonnay. *Letters in Applied Microbiology, 46*, 38–42.
Rodriguez-Porrata, B., Novo, M., Guillamón, J., Rozés, N., Mas, A., & Cordero Otero, R. (2009). Vitality enhancement of the rehydrated active dry wine yeast. *International Journal of Food Microbiology, 126*, 116–122.
Sabate, J., Cano, J., Querol, A., & Guillamón, J. M. (1998). Diversity of *Saccharomyces* strains in wine fermentations: Analysis for two consecutive years. *Letters in Applied Microbiology, 26*, 452–455.
Salma, M., Rousseaux, S., Sequeira-Le Grand, A., Divol, B., & Alexandre, H. (2013). Characterization of the viable but nonculturable (VBNC) state in Saccharomyces cerevisiae. *PLOS ONE, 8*, e77600.
Santamaria, M., Fosso, B., Consiglio, A., De Caro, G., Licciulli, F., Liuni, S., Marzano, M., Alonso-Alemany, D., Valiente, G., & Pesole, G. (2012). Reference databases for taxonomic assignment in metagenomics. *Briefings in Bioinformatics, 13*, 682–695.
Schuller, D., Valero, E., Dequin, S., & Casal, M. (2004). Survey of molecular methods for the typing of wine yeast strains. *FEMS Microbiology Letters, 231*, 19–26.
Senses-Ergul, S., Ágoston, R., Belák, Á., & Deak, T. (2006). Characterization of some yeasts isolated from foods by traditional and molecular tests. *International Journal of Food Microbiology, 108*, 120–124.
Serpaggi, V., Remize, F., Suqueira-Le Grand, A., & Alexandre, H. (2010). Specific identification and quantification of the spoilage microorganism *Brettanomyces* in wine by flow cytometry: A useful tool for winemakers. *Citometry Part A, 77A*, 490–499.
Serpaggi, V., Remize, F., Recorbet, G., Gaudot-Dumas, E., Suqueira-Le Grand, A., & Alexandre, H. (2012). Characterization of the "viable but nonculturable" (VBNC) state in the wine spoilage yeast *Brettanomyces*. *Food Microbiology, 30*, 438–447.
Setati, M. E., Jacobson, D., & Bauer, F. F. (2015). Sequence-based analysis of the Vitis vinifera L. cv Cabernet Sauvignon grape must mycobiome in three South African vineyards employing distinct agronomic systems. *Frontiers in Microbiology, 6*, 1358.
Sharpton, T. J. (2014). An introduction to the analysis of shotgun metagenomics data. *Frontiers in Plant Science, 5*, 209.
Smart, K. A., Chambers, K. M., Lambert, I., Jenkins, C., & Smart, C. A. (1999). The use of methylene violet staining procedures to determine yeast viability and vitality. *Journal of the American Society of Brewing Chemists, 57*, 18–23.
Smith, C. J., & Osborn, A. M. (2009). Advantages and limitations of quantitative PCR (QPCR)-based approaches in microbial ecology. *FEMS Microbiology Ecology, 67*(1), 6–20.
Smole Možina, S., Dlauchy, D., Deak, T., & Raspor, P. (1997). Identification of *Saccharomyces sensu stricto* and *Torulaspora* yeasts by PCR ribotyping. *Letters in Applied Microbiology, 24*, 311–315.Soll, D. R. (2000). The ins and outs of DNA fingerprinting the infectious fungi. *Clinical Microbiology Reviews, 13*, 332–370.
Spangenberg, D. S., & Ingham, S. C. (2000). Comparison of methods for enumeration of yeasts and molds in shredded low-moisture part-skim mozzarella cheese. *Journal of Food Protection, 63*, 529–533.

Stadlwieser, P., Domig, K. J., Kögler, B., Kneifel, W., Silhavy, K., & Mandl, K. (2006). Biochemische und molekularbiologische Characterisierung von Reinzuchthefen für die Weinbereitung. *Mitt Klosterneuburg, 56*, 213–223.

Stender, H., Kurtzman, C., Hyldig-Nielsen, J. J., Sørensen, D., Broomer, A., Oliveira, K., Perry-O'Keefe, H., Sage, A., Young, B., & Coull, J. (2001). Identification of *Dekkera bruxellensis (Brettanomyces)* from wine by fluorescence in situ hybridization using peptide nucleic acid probes. *Applied and Environmental Microbiology, 67*, 938–941.

Stender, H., Fiandaca, M., Hyldig-Nielsen, J. J., & Coull, J. (2002). PNA for rapid microbiology. *Journal of Microbiological Methods, 48*, 1–17.

Sternes, P. R., Lee, D., Kutyna, D. R., & Borneman, A. R. (2017). A combined meta-barcoding and shotgun metagenomic analysis of spontaneous wine fermentation. *GigaScience, 6*, 1–10.

Sun, Y., & Liu, Y. (2014). Investigating of yeast species in wine fermentation using terminal restriction fragment length polymorphism method. *Food Microbiology, 38*, 201–207.

Taniwaki, M. H., da Silva, N., Banhe, A. A., & Iamanaka, B. T. (2001). Comparison of culture media, Simplate and Petrifilm for enumeration of yeasts and molds in food. *Journal of Food Protection, 64*, 1592–1596.

Tessonnière, H., Vidal, S., Barnavon, L., Alexandre, H., & Remize, F. (2009). Design and performance testing of a real-time PCR assay for sensitive and reliable direct quantification of *Brettanomyces* in wine. *International Journal of Food Microbiology, 129*, 237–243.

Tofalo, R., Schirone, M., Corsetti, A., & Suzzi, G. (2012). Detection of *Brettanomyces* spp. in red wines using real-time PCR. *Journal of Food Science, 77*, 545–549.

Urso, R., Rantsiou, K., Dolci, P., Rolle, L., Comi, G., & Cocolin, L. (2008). Yeast biodiversity and dynamics during sweet wine production as determined by molecular methods. *FEMS Yeast Research, 8*, 1053–1062.

Valente, P., Ramos, J. P., & Leoncini, O. (1999). Sequencing as a tool in yeast molecular taxonomy. *Canadian Journal of Microbiology, 45*, 946–958.

Valero, E., Schuller, D., Cambon, B., Casal, M., & Dequin, S. (2005). Dissemination and survival of commercial wine yeast in the vineyard: A large-scale, three-years study. *FEMS Yeast Research, 5*, 959–969.

Valero, E., Cambon, B., Schuller, D., Casal, M., & Dequin, S. (2007). Biodiversity of *Saccharomyces* yeast strains from grape berries of wine-producing areas using starter commercial yeasts. *FEMS Yeast Research, 7*, 317–329.

Van der Vossen, J. M. B. M., Rahaoui, H., de Nijs, M. W. C. M., & Hartog, B. J. (2003). PCR methods for tracing and detection of yeasts in the food chain. In T. Boekhout & V. Robert (Eds.), *Yeasts in food, beneficial and detrimental aspects* (pp. 123–138). Hamburg: Behr's Verlag.

Vaudano, E., & Garcia-Moruno, E. (2008). Discrimination of *Saccharomyces cerevisiae* wine strains using microsatellite multiplex PCR and band pattern analysis. *Food Microbiology, 25*, 56–64.

Vaughan-Martini, A., Cardinali, G., & Martini, A. (1996). Differential killer sensitivity as a tool for fingerprintong wine-yeast strains of *Saccharomyces cerevisiae*. *Journal of Industrial Microbiology, 17*, 124–127.

Vaughan-Martini, A., Angelini, P., & Cardinali, G. (2000). Use of conventional taxonomy, electrophoretic karyotyping and DNA-DNA hybridization for the classification fermentative apiculate yeasts. *International Journal of Systematic and Evolutionary Microbiology, 50*, 1665–1672.

Velázquez, E., cruz-Sánchez, J. M., Rivas-Palá, T., Zurdo-Pineiro, J. L., Mateos, P. F., Monte, E., Martinez-Molina, E., & Chordi, A. (2001). YeastIdent-Food/ProleFood, a new system for the identification of food yeasts based on physiological and biochemical tests. *Food Microbiology, 18*, 637–646.

Versavaud, A., & Hallet, J. N. (1995). Pulsed-field gel electrophoresis combined with rare-cutting endonucleases for strain identification of *Candida famata, Kloeckera apiculata* and *Schizosaccharomyces pombe* with chromosome number and size estimation of the two former. *Systematic and Applied Microbiology, 18*, 303–309.

Wang, C., Esteve Zarzoso, B., & Mas, A. (2014). Monitoring of *Saccharomyces cerevisiae*, *Hanseniaspora uvarum*, and *Starmerella bacillaris* (synonym *Candida zemplinina*) populations during alcoholic fermentation by fluorescence in situ hybridization. *International Journal of Food Microbiology, 191*, 1–9.

Wang, C., Esteve Zarzoso, B., Cocolin, L., Mas, A., & Rantsiou, K. (2015). Viable and culturable populations of *Saccharomyces cerevisiae*, *Hanseniaspora uvarum* and *Starmerella bacillaris* (synonym *Candida zemplinina*) during Barbera must fermentation. *Food Research International, 78*, 195–200.

Wang, C., Mas, A., & Esteve-Zarzoso, B. (2016). The interaction between *Saccharomyces cerevisiae* and non-*Saccharomyces* yeast during alcoholic fermentation is species and strain specific. *Frontiers in Microbiology, 7*, 502. https://doi.org/10.3389/fmicb.2016.00502.

Welsh, J., & McClelland, M. (1990). Fingerprint genomes using PCR with arbitrary primers. *Nucleic Acids Research, 18*, 7213–7218.

White, T. J., Bruns, T., Lee, S., & Taylor, J. (1990). Amplification and direct sequencing of fungal ribosomal RNA genes for phylogenetics. In M. A. Innis, D. H. Gelfand, J. I. Sninsky, & T. J. White (Eds.), *PCR protocols: A guide to methods and applications* (pp. 315–322). San Diego: Academic.

Willenburg, E., & Divol, B. (2012). Quantitative PCR: An appropriate tool to detect viable but not culturable *Brettanomyces bruxellensis* in wine. *International Journal of Food Microbiology, 160*, 131–136.

Williams, J. G. K., Kubelik, A. R., Livak, K. J., Rafalski, J. A., & Tingey, S. V. (1990). DNA polymorphisms amplified by arbitrary primers are useful as genetic markers. *Nucleic Acids Research, 18*, 1631–1635.

Zara, G., Ciani, M., Domizio, P., Zara, S., Budroni, M., Carboni, A., & Mannazzu, I. (2013). A culture-independent PCR-based method for the detection of *Lachancea thermotolerans* in wine. *Annales de Microbiologie, 64*, 403–406.

Zarraonaindia, I., Owens, S. M., Weisenhorn, P., West, K., Hampton-Marcell, J., Lax, S., et al. (2015). The soil microbiome influences grapevine-associated microbiota. *MBio, 6*, e02527–e02514.

Chapter 4
New Insights Into Wine Yeast Diversities

Warren Albertin, Isabelle Masneuf-Pomarede, Virginie Galeote, and Jean-Luc Legras

Contents

W. Albertin · I. Masneuf-Pomarede
Institut des Sciences de la Vigne et du Vin, Unité de recherche Oenologie, Villenave d'Ornon Cedex, France
e-mail: warren.albertin@u-bordeaux.fr; isabelle.masneuf@agro-bordeaux.fr

V. Galeote · J.-L. Legras (✉)
Sciences pour l'œnologie (SPO), University of Montpellier, INRA, Montpellier SupAgro, Montpellier Cedex 2, France
e-mail: virginie.galeote@inra.fr; jean-luc.legras@inra.fr

P. Romano et al. (eds.), *Yeasts in the Production of Wine*,
https://doi.org/10.1007/978-1-4939-9782-4_4

1 Introduction

The consumption of fermented fruits is very likely one of the most ancient habits of primates as the specific mutation that enable *ADH4* to oxidize ethanol appeared in our last common ancestor with the chimpanzee and gorilla, more than seven million years ago (Carrigan et al. 2014), far before the advent of fermented beverages. In contrast, archeological evidence suggests that the production of fermented foods appeared simultaneously with the development of agriculture at the Neolithic period (Gibbons and Rinker 2015), and for wine, the most ancient indication of fermented grape products on ceramics excavated in Georgia dates back to 6000 BC (McGovern et al. 2017). The following millennia have witnessed the progressive expansion of grape culture over the world, and the evolution of wine making technologies, leading to actual controlled fermentations with tailored yeast starters (Pretorius 2000).

Multiple studies have revealed the richness of the fungal microbial community of grape must and wine (Fleet et al. 2002; Bokulich et al. 2013), which are under the influence of environmental and anthropogenic factors (Grangeteau et al. 2017). While we have genetic evidence of the origins and evolution of wine grape varieties (Arroyo-García et al. 2006; This et al. 2006), until recently we had no evidence of how these historical processes may have impacted the diversity of the yeasts found on grapes and developing during wine alcoholic fermentation (AF) (Fleet 1993). The different generations of molecular methods have provided progressively more and more information for *S. cerevisiae*, and recently for other yeast species of the grape must microbial communities, revealing progressively how the development of wine technology has shaped the diversities of wine yeasts. This chapter focuses on these recent advances starting from *S. cerevisiae* for which population genomics has offered the most comprehensive view of its evolution including population structure genomic specificities, and demographic history. We then expand to other non-*Saccharomyces* yeast species members of the wine community that also contributes to the wine sensory perception (Fleet 2003).

2 Evolution of Molecular Methods Used for the Characterization of *S. cerevisiae* Diversity

The differences between *S. cerevisiae* strains have long been a matter of debates. The development of the restriction profile of mitochondrial DNA (Aigle et al. 1984) and of pulsed field gel electrophoresis (PFGE) (Schwartz and Cantor 1984) have open the way to a wealth of studies when applied to wine isolates (Dubourdieu et al. 1987; Vezinhet et al. 1990) (Fig. 4.1).

Starting from these pioneering methods, different generations of typing methods relying on DNA polymorphism have been developed. In the case of mtDNA RFLP, after separation from nuclear DNA, the digestion of mitochondrial DNA with appropriate enzymes revealed the polymorphism of *S. cerevisiae* strains and

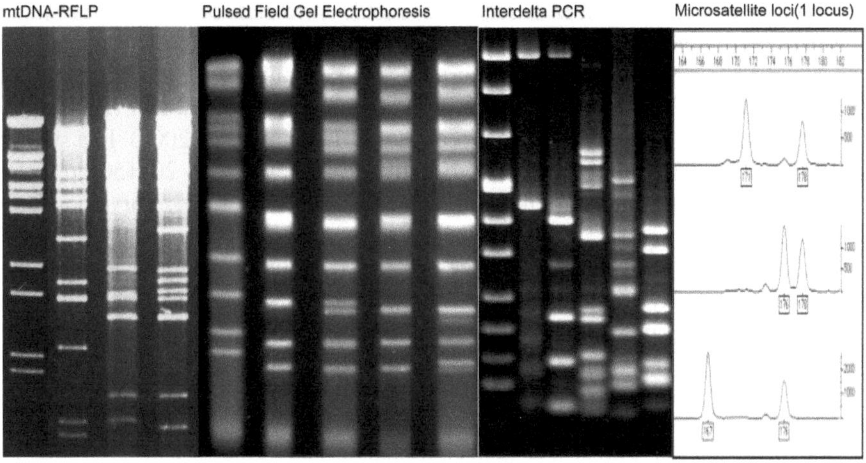

Fig. 4.1 Electrophoretic patterns obtained for different molecular methods for strain characterization. From left to right mtDNA-RFLP, Pulsed Field Gel electrophoresis, interdelta PCR, and fluorescence pattern measured with capillary electrophoresis for a microsatellite loci for 3 strains. A peak is equivalent to a band on one of the gels shown on the left

especially of wine strains (Querol et al. 1992b). This protocol has been simplified with the use of restriction enzymes digesting specifically nuclear DNA (López et al. 2001) and has become the most widely used method (278 citations). mtDNA RFLP and PFGE enabled the first ecological studies of wine yeast such as the study of the implantation of yeast starters in a vat (Querol et al. 1992a), the evaluation of the genetic diversity of yeast strains in a cellar (Frezier and Dubourdieu 1992; Beltran et al. 2002) or in the vineyard (Versavaud et al. 1995), or the estimation of the dissemination of yeast starters in a vineyard (Valero et al. 2005).

The discovery of DNA amplification by polymerase chain reaction (Saiki et al. 1988) has allowed the development of a second generation of methods targeting polymorphic regions. The polymorphism of the distances between insertion sites of delta element dispersed in the genome of strains has been used as the ground for a fingerprint technique (Ness et al. 1993; Legras and Karst 2003), that has become very popular too. Despite the similar resolution obtained by mtDNA RFLP, PFGE, and interdelta typing (Schuller et al. 2004), these three methods present different weaknesses: low resolution of PFGE, the exclusive focus on mitochondrial DNA for mtDNA RFLP, and the small number of bands and possible homoplasy for interdelta typing event for the improved version proposed by Legras et al. 2003. Following the example of plant or animal geneticists, several authors proposed new methods in order to increase discrimination power, AFLP and microsatellite typing have been used for yeast characterization. AFLP has been used in few studies (de Barros Lopes et al. 1999; Azumi and Goto-Yamamoto 2001). In contrast, the analysis of the polymorphism of microsatellite loci has progressively expanded because of their ease to use (González Techera et al. 2001; Hennequin et al. 2001; Pérez et al. 2001; Howell et al. 2004; Legras et al. 2003). Microsatellite loci are regions containing a variable number of repeats of motifs of two, three, or four bases, and provide co-dominant data: in case of a diploid heterozygote the two alleles are observed and taken into account for further genetic analysis. The differentiation of yeast strains can be easily achieved with a combination of 6–12 loci (Hennequin et al. 2001; Pérez et al. 2001; Legras et al. 2003; Bradbury et al. 2006). In a similar manner, minisatellite that are repeated motif of larger size (i.e. 20–150 bp) have also been used for strains differentiation (Marinangeli et al. 2004; Carvalho-Netto et al. 2013).

Multi locus sequence typing has been proposed slightly later for yeast (Fay and Benavides 2005; Ayoub et al. 2006; Eeom et al. 2018), but the first schemes were less efficient for strains differentiation (Ayoub et al. 2006). However the decrease of sequencing costs has progressively made feasible the sequencing of large fraction of the genome or of the whole genome of tenths or hundreds of wine strains (Cromie et al. 2013; Borneman et al. 2016; Legras et al. 2018; Peter et al. 2018). These last population genomic studies present the ultimate vision of *Saccharomyces cerevisiae* diversity

3 Genetic Diversity of *Saccharomyces cerevisiae*

The development of these molecular methods has progressively permitted to untangle the complexity of *Saccharomyces cerevisiae* diversity. Starting in 1996, the first evidence of some regional differentiation by region or by grape varieties were obtained with mtDNA RFLP (Guillamón et al. 1996). However the shift to multilocus sequence, microsatellite typing, which are more efficient and reliable markers for population genetic analysis has offered a more detailed vision of *S. cerevisiae* diversity.

3.1 Population Structure and Demographic History in S. cerevisiae

A first indication of yeast population structure was given by AFLP data (Azumi and Goto-Yamamoto 2001) for Asian strains, but the first MLST study (Fay and Benavides 2005) and a large multilocus microsatellite study (Legras et al. 2007) revealed the high impact of human activities on yeast diversity. Indeed strains isolated from most fermented food are found in specific clusters in the different phylogenetic trees indicating that the denomination of yeasts as "bread", "beer", "wine", "sake", or "palm wine" yeasts is highly relevant.

Population surveys relying on RADSeq (Cromie et al. 2013) or on genome sequencing confirmed these features (Liti et al. 2009; Almeida et al. 2015; Borneman et al. 2016; Gallone et al. 2016; Legras et al. 2018; Peter et al. 2018). Today, the genome of almost 2000 strains has been sequenced, and the most comprehensive view of *S. cerevisiae* diversity has been obtained from the exploration of the genome of 1011 strains from multiple sources (Fig. 4.2). This study revealed that the *S. cerevisiae* species has very likely an Asian origin (Duan et al. 2018; Peter et al. 2018). Wine and flor *S. cerevisiae* strains appear as two closely related monophyletic group with low genetic diversity (π=0.001, π=0.0003 respectively) (Liti et al. 2009; Legras et al. 2014; Coi et al. 2017; Legras et al. 2018; Peter et al. 2018). The wine population is closely related to a population of isolates obtained from oak forests around the Mediterranean sea (Almeida et al. 2015), that represent very likely its ancestral genepool. An historical demographic model inferred for these two populations (wine and Mediterranean oak) indicated a split, accompanied by a bottleneck for the wine population and followed by a demographic expansion for the two groups with bilateral gene flow between them. Dating this historical split between wine and Mediterranean oak strains is difficult given the lack of knowledge of the accurate mutation rate and of the number of generation of wine yeast per year, but the estimates attempts are compatible with the recent development of agriculture. When considering their life style, wine yeast have a mainly clonal lifestyle, with infrequent sex, and an effective outcrossing rate about 12% leading to approximately one outcrossing event every 20000–25000 mitoses (Magwene et al. 2011; Legras et al. 2018), which is clearly more frequent than that of natural oak populations (one outcrossing event every 100 000–1 000 000 mitoses).

3.2 *Genomics Specificities of Wine Yeast*

The most comprehensive analysis of *S. cerevisiae* diversity based on 1011 strains revealed a pangenome containing a total of 7796 open reading frame s (ORFs), composed of 4940 core ORFs and 2856 disposable ORFs (Peter et al. 2018). Among these strains of different ecological niches, wine strains present a specific genomic makeup with tailored gene content, specific copy number variation of several genes, and different genes potentially under selection (Legras et al. 2018).

3.2.1 Horizontally Transferred and Introgressed Genes

Horizontal gene transfers (HGT) and introgressions are the main mechanisms that permitted an expansion of the gene repertoire in *S. cerevisiae* (Liti et al. 2006; Novo et al. 2009; Dunn et al. 2012; Legras et al. 2018; Peter et al. 2018). (Dunn et al. 2012) have analyzed introgressions in wine *S. cerevisiae* yeast strains. Using Array-Comparative Genomic hybridization (aCGH) on 69 commercial *S. cerevisiae* wine strains and 14 strains collected from various environments, they detected five introgressed regions from *S. paradoxus* and one from *S. mikatae*. The five *S. paradoxus* regions were found in 15 among 83 strains and 12 isolates out of 15 were wine strains. Among these regions, some genes could have an interest in sugar-rich

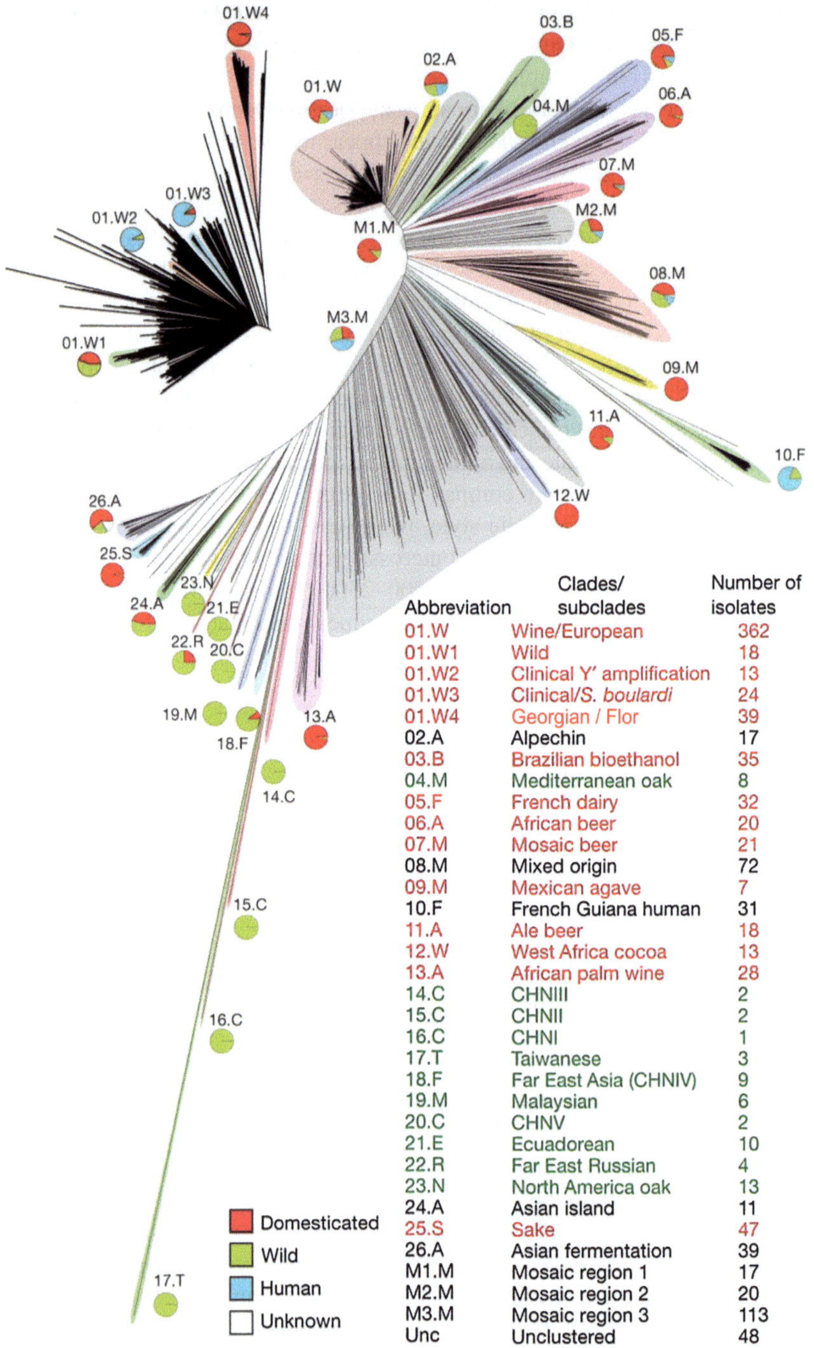

Fig. 4.2 Neighbour-joining tree of 1011 *S. cerevisiae* strains built using the biallelic SNPs identified from genome sequencing. (from Peter et al. 2018)

environments such as grape must or sugar cane juice: *SUC2* gene encoding a sucrose-hydrolyzing invertase, *HPF1* gene encoding a glucan alpha-1,4-glucosidase that when overexpressed reduces protein haze formation in white wines (Brown et al. 2007) and *AWA1*, encoding a putative GPI-anchored protein localized to the cell wall that could be responsible for the foam formation in sake mash (Miyashita et al. 2004).

Besides introgressions, eukaryote-to-eukaryote gene transfers have been found in *S. cerevisiae* strains (Novo et al. 2009; Borneman et al. 2011; Strope et al. 2015; Legras et al. 2018; Peter et al. 2018). The sequencing of 82 *S. cerevisiae* strains from different fermentation environments have highlighted 42 potentially transferred regions (Fig. 4.3) ranging in length from 1.5 to 42kb and predicted to contain 155 hypothetical ORFs encoding proteins longer than 145 aa (Legras et al. 2018). In addition to the 6 first regions that had already been found in the genome of the wine strain EC1118 (A; B; C1 to C5) and originate from a non-*Saccharomyces* species (Novo et al. 2009; Marsit et al. 2015), we identified 12 regions likely to be the result of HGT according to phylogenic tests (Legras et al. 2018). These 6 regions originally called A, B and C (120 kb in total) are prevalent in wine strains. The yeast species *Zygosaccharomyces bailii* and *Torulaspora microellipsoides*, also found in wine fermentations, were identified as the donors of regions B and C, respectively (Novo et al. 2009; Galeote et al. 2011; Marsit et al. 2015), while the origin of the region A was associated to the genus *Torulaspora* (Legras et al. 2018). Assessing the beginning of wine making 9000 ago, the transfer of region B and C should have occurred less than 3000 and 2000 years ago (Galeote et al. 2011; Marsit et al. 2015) and are thus extremely recent. Evidence for the amplification of the region B (17 kb) in the genome of *S. cerevisiae* wine strains was provided (Borneman et al. 2011; Galeote et al. 2011). The organization of this region differs considerably between strains and the identification of an autonomously replicating sequence functional in *S. cerevisiae* strongly suggests an expansion mechanism in yeast genomes involving an extrachromosomal circular DNA molecule. Recently, Peter et al. (2018) proposed that this region results from a size reduction of an ancestral event transfer of 117 kb. The identification of additional genes of region C in some strains suggests a similar transfer event of the whole *T. microellipsoides* chromosome followed by several rearrangements, including gene loss and gene conversion (Marsit et al. 2015; Legras et al. 2018; Peter et al. 2018).

The 3 regions A, B and C comprise 39 genes potentially encoding important metabolic functions in winemaking, such as metabolism and transport of nitrogen and sugar, suggesting a role of HGT in adaptation to the wine environment (Novo et al. 2009). To assess the adaptive value of these events, the functions of several genes of the *T. microellipsoides* were characterized. The first example concerns the *FSY1* gene encoding a high affinity fructose/H^+ symporter (Galeote et al. 2010). This gene that is mainly conserved in flor yeast might be advantageous when traces of fructose are present in wine at the end of the fermentation or in the velum process (Marsit et al. 2015; Coi et al. 2017; Legras et al. 2018). Another gene, *XDH1*, encodes a putative xylitol dehydrogenase involved in xylose metabolism (Wenger et al. 2010). Two other tandem duplicated genes *FOT1-2* encode oligopeptide

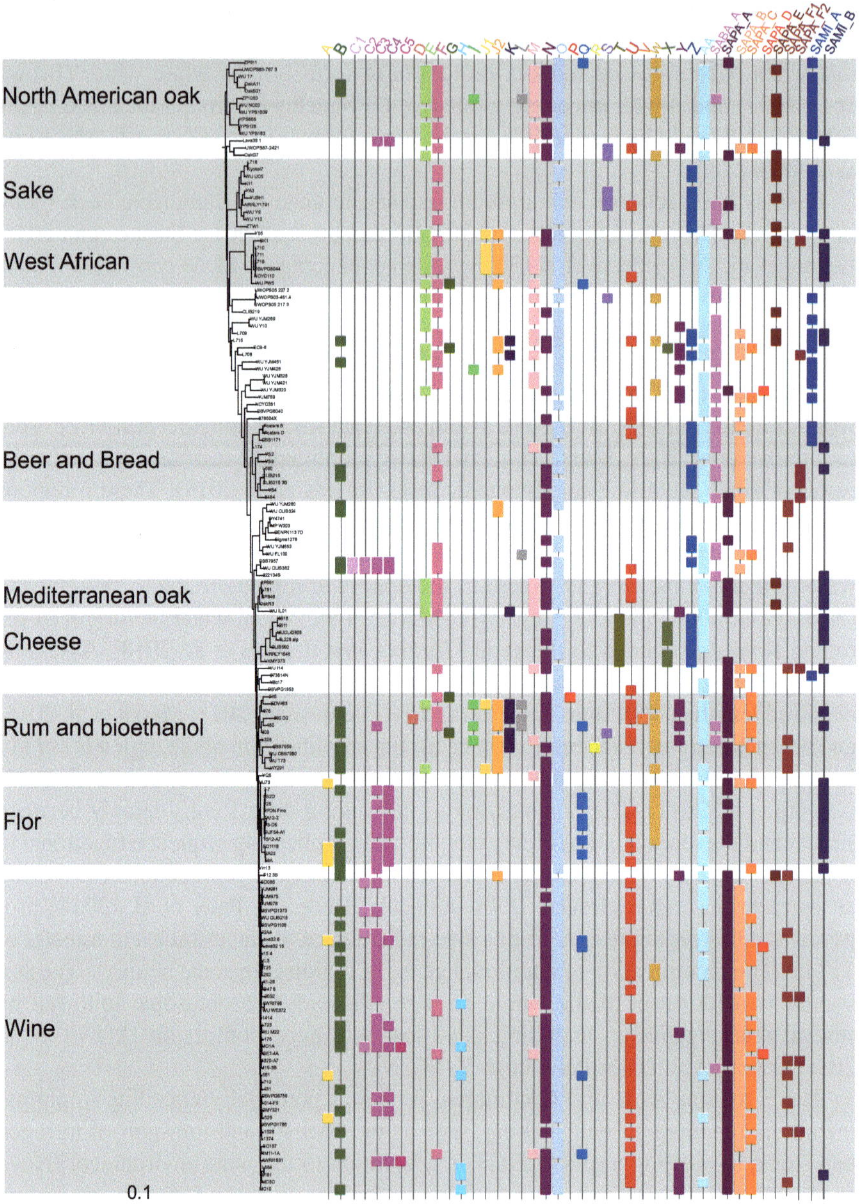

Fig. 4.3 Repartition of different not present in the genome of S288C, putatively transferred horizontally (A to AA) or introgressed (SABA_A to SAMI_B) among a set of 159 strains of different origins

transporters, which considerably increase the range of oligopeptides typically transported by the carrier proteins Ptr2p and Dal5p in *S. cerevisiae* (Damon et al. 2011), especially towards oligopeptides containing glutamate or glutamine which are the most abundant amino acids in oligopetides fraction of the grape must (Marsit et al. 2015). These two *FOT* genes are found in half of wine strains, even when the region C is incomplete. In oenological conditions, these genes improve biomass formation, fermentation efficiency and cell viability that might result from a remodeling of central carbon and nitrogen metabolic pathways. A major modification was observed at the level of the glutamate node and the NADPH/$NADP^+$ balance resulting in decreased acetic acid production and increased ester formation, which might improve the organoleptic balance of wines (Marsit et al. 2016). Furthermore, Marsit et al. (2015) demonstrated by using competition experiments the strong competitive advantage of yeast strains containing these *FOT* genes on grape must. Thus the coexistence of *Z. bailii* and *T. microellipsoides* with *S. cerevisiae* in fermentative environments suggests that these gene transfers could be favored by common ecological niches. In addition, Peter et al. (2018) have found an enrichment for *Torulaspora* and *Zygosaccharomyces* species for 85 potential HGT transfer, among the 183 detected, present in *S. cerevisiae* strains from Wine/European clades which also supports this hypothesis.

3.2.2 Copy Number Variations and Translocations

Copy number variation is a mechanism known for favoring adaptation to a changing environment (Dunham et al. 2002; Pavelka et al. 2010; Kondrashov 2012). Gene losses participate to the variation in the gene content per population, and genome-scans for the variations in the number of genes copies revealed a higher frequency of gene loss in some population such as wine, beer or cheese strains, in a population-dependent manner (Gallone et al. 2016; Legras et al. 2018). In contrast to gene loss, genes amplification was less frequent. The amplification of only two genes *CUP1* and *ADH7* was reported for wine strains with up to 18 or 22 copies of *CUP1* in the genome of some strains (Treu et al. 2014; Strope et al. 2015) and two for *ADH7* (Legras et al. 2018). *CUP1* has a prominent role for yeast in the vine/wine ecosystem as it confers resistance to copper which has been used massively as a fungicide for pest management. The gene *ADH7* codes for an alcohol dehydrogenase which is involved in vanillin detoxification (Nguyen et al. 2015) and may be important for the detoxification of phenolic compounds of the grape must during AF.

Chromosomal rearrangements participate as well to the specificities of the genomic make-up of wine strains. Two translocations giving a positive advantage in the wine environment in presence of sulfite have been described. A translocation between chromosome VIII and XVI is widely spread in wine yeast (Pérez-Ortín et al. 2002). It generates a highly induced allele of the sulfite pump *SSU1* that confers resistance in presence of sulfite. The strength of selection at that loci can be inferred by the finding of a second translocation between chromosomes XV and XVI, asides the *SSU1* gene, that has been detected in a QTL program aimed at reducing lag phase in grape must fermentation and in a second study relating expression levels of different strains to their genomic diversity (Treu et al. 2014; Zimmer et al. 2014).

3.2.3 Indications of Positive Selection

Indication of adaptation can also be obtained from the analysis of the spectrum of genome wide nucleotide diversity. The comparison of wine and Mediterranean oak *S. cerevisiae* strains has revealed a higher proportion of fixed non -synonymous SNPs than synonymous SNPs attesting non neutral evolution at several sites along the genome and the specific differentiation of 9 genomic regions (Almeida et al. 2017). A similar evidence of adaptive divergence can be seen from the comparison of flor and wine yeast genomes, two closely related groups with contrasted lifestyles (Coi et al. 2017). The presence of several highly differentiated regions between the two groups was observed, some of which could be related to the differences in ecological niches. Several statistical tests using population comparison or allelic frequency spectrum analysis of wine strains revealed multiple other sites potentially under selection (Legras et al. 2018). It is noteworthy that these different studies pointed to genes involved in nitrogen assimilation or transport (Almeida et al. 2017; Legras et al. 2018) or to sterol uptake and transport which are the two main limiting steps for AF.

3.3 *Factors Explaining Diversity of Vineyard Associated* S. cerevisiae

In addition to the better knowledge of the genomic specificities of wine yeast, new insights into yeast ecology has been obtained during the last decade. Because of the importance given to the concept of “terroir”, many efforts have been made into the exploration of factors explaining the diversity of *S. cerevisiae* associated to geographic areas.

The first indication of regional differentiation was mentioned in the pioneering study of Guillamón et al. (1996) based of mtDNA RFLP profiles. Further population differentiation was then observed between vineyards of the same region (Schuller and Casal 2007) and between populations of different regions (Legras et al. 2007). The relationships between populations of different regions suggest a striking relationship with migration routes, and are in agreement with the hypothesis of a middle east origin associated with the expansion of viticulture (Legras et al. 2007). Several other studies confirmed population differentiation between different regions or appellations (Anfang et al. 2009; Gayevskiy and Goddard 2012; Schuller et al. 2012; Knight and Goddard 2015; Viel et al. 2017). A recent exploration of *S. cerevisiae* diversity performed in Açores in 32 locations from 6 islands revealed a global pattern of differentiation between islands (with some exceptions) that may result from their geographic separation and also some differentiation between populations isolated from *Vitis labrusca* and *Vitis vinifera* grape varieties (Drumonde-Neves et al. 2018). Analyses led in New Zealand between 5 distant vineyards located in the northern island (Hawke's Bay and Martinborough) and the southern island (Nelson, Marlborough and Central Otago), all planted with Sauvignon blanc grape revealed more complex relationships between metapopulations of the different regions, resulting from variable gene flow (Gayevskiy and Goddard 2012; Knight and Goddard 2015) that could not be solely explained by distance. Lastly, for continental vineyards in Portugal and Italy, differentiation between one appellation and the two others (Schuller et al. 2012; Viel et al. 2017), but this differentiation did not fit strictly with appellation. In addition to site-to-site variation, vintage can also contribute significantly to population differentiation in the vineyard and in the cellar (Schuller and Casal 2007; Börlin et al. 2016).

Besides geographic differentiation, the question of the relationships between cellars and vineyards is another long lasting and debating question. Indeed, given the high diversity of strains isolated from the vineyards and from the cellars, it is difficult to estimate the flow of strains entering or leaving the cellar from the identity of genotypes. The entrance of *S. cerevisiae* strains into a winery could be seen from the entrance of vineyard strains into a newly established winery (Constantí et al. 1997). The first attempt to answer the issue of the release of cellar strains into the vineyard was performed on several wineries in Portugal and France (Valero et al. 2005), and relied on a diversity analysis obtained with pulsed field gel electrophoresis and mtDNA RFLP. This first analysis revealed that 7.8% of the *S. cerevisiae* isolates corresponded to yeast starters used in the nearby cellars. As these strains were found mainly after harvest, it suggests that yeast starters do not remain in the vineyard (Valero et al. 2005). However, recent studies performed in Canada, Italy and France describe a different picture in which yeast starters used in the cellar were isolated on grapes before harvest in higher significant proportions in the population collected in the vineyard (Martiniuk et al. 2016; Viel et al. 2017; Börlin et al. 2018). These three studies, took into account genotypes closely related to the yeast starters used in the nearby cellars, leading to a higher proportion among the population of vineyard isolates: between 25 and 51% in these three studies. This implies that the exchange between these two compartments, cellar and vineyard, are much higher than described before.

The use of co-dominant biallelic loci such as microsatellite or SNPs permits as well to infer the demographic history that has led to actual diversity pattern. Different methods exist such as Approximate Bayesian Computation (ABC) frameworks (Cornuet et al. 2014) that has permitted to infer historical scenario for invasive species (Guillemaud et al. 2009). Interestingly, little attempts have been made in order to infer historical demography between *S. cerevisiae* populations of different vineyards, and compare to known historical data. The sole attempt has been performed in New Zealand, in order to explore the origin of local *S. cerevisiae* population from genomic data: (Gayevskiy et al. 2016) inferred that nowadays

S. cerevisiae population derives from at least 10 founder lineages that have progressively colonized New Zealand vineyards. The development of microsatellite or genomic data should open the way to such inferences.

4 Genetic Diversity of *Saccharomyces* non-*cerevisiae* Species and Their Related Interspecies Hybrids

In the actual yeast classification, the *Saccharomyces* genus comprises eight species: *S. arboricolus*, *S. cerevisiae*, *S. eubayanus*, *S. jurei*, *S. kudriavzevii*, *S. mikatae*, *S. paradoxus* and *S. uvarum* (Libkind et al. 2011; Boynton and Greig, 2014; Naseeb et al. 2017). Apart from *S. cerevisiae,* only two species are reported to be associated to a large extend with the wine environment: *S. uvarum* and *S. kudriavzevii*. The latter has been reported in winemaking production only as *S. cerevisiae* × *S. kudriavzevii* hybrids. As a paradox, *S. kudriavzevii* has however never been isolated from wine-related environment. The only available data concerning the genetic diversity of wine-related *S. kudriavzevii* were reported by Legras et al. (2014). By applying a set of *S. kudriavzevii* microsatellite markers on wine hybrid *S. cerevisiae* × *S. kudriavzevii* isolates, the authors showed a clear population structure for *S. kudriavzevii*. Some clusters were clearly assigned to their source of isolation despite close geographical proximity (Switzerland and Alsace). Due to the lack of data in literature concerning wine *S. kudriavzevii* isolates, only the genetic diversity of *S. uvarum* will be described as non-*S. cerevisiae* species in this chapter.

4.1 S. uvarum, *the Sister Species of* S. cerevisiae

S. uvarum (discovered by Beijerinck in 1894) was initially considered a synonym of *S. bayanus* and later a variety of the latter (Vaughan Martini and Kurtzman 1985; Naumov et al. 2000). However the *S. bayanus* and *S. uvarum* taxonomic position has been confusing for decades in both the literature and databases. Two independent studies, presenting congruent results, proposed to reinstate *S. uvarum* as a biologically meaningful species (Libkind et al. 2011; Nguyen et al. 2011). The species is closely related to *S. eubayanus* but distantly to *S. cerevisiae* from which the separation was estimated to be 20 million years of evolution. Strains of *S. uvarum* have been isolated world-wide from both natural habitats and anthropic environments. The species was associated with plant and trees exudates, barks, insects, in Europe (Naumov et al. 2011), North America (Sampaio and Gonçalves 2008; Almeida et al. 2014), South America and Australasia (Libkind et al. 2011; Almeida et al. 2014; Rodriguez et al. 2014), and coexists in these environments with *S. eubayanus*.

Like its sister species *S. cerevisiae*, *S. uvarum* is associated with different fermented beverages among which cider and apple chicha from Europe (Coton et al.

2006; Suárez Valles et al. 2007) and South America (Rodríguez et al. 2017). Many studies reported the association of *S. uvarum* with wine fermentations, generally in mixed cultures with *S. cerevisiae* (Naumov et al. 2000; Sipiczki 2002; Demuyter et al. 2004; Zhang et al. 2015). Usually described as cryotolerant, *S.uvarum* contributes to AF of must obtained from grapes cultivated in northern European vineyards (Burgundy, Alsace, Champagne and Val de Loire) (Massoutier et al. 1998; Naumov et al. 2000; 2001; Demuyter et al. 2004) or fermentations conducted at low temperatures. Cold climate in the vineyard during harvest period in October and/or low temperature management during the vinification process may constitute favorable conditions for the development of cryotolerant yeasts on the grapes and in the must. *S. uvarum* has also been frequently isolated from natural fermentations of botrytized grape must (Naumov et al. 2000; Sipiczki et al. 2001; Antunovics et al. 2005) and of Recioto and Amarone wines (Torriani et al. 1999; Tosi et al. 2009). *S. uvarum* is less tolerant to ethanol compared to *S. cerevisiae* and produces lower levels of acetic acid, high levels of phenyethanol and its acetate (Castellari et al. 1994; Antonelli et al. 1999; Masneuf-Pomarede et al. 2010).

In a first diversity study, (Naumova et al. 2011) used mitochondrial DNA polymorphism and showed that the mtDNA of *S. uvarum* is much less polymorphic than that of *S. cerevisiae*. This observation is in good agreement with results obtained by electrophoretic karyotyping. Unlike wine *S. cerevisiae*, strains of *S. uvarum* display a low level of chromosome length polymorphism. Later, thanks to the availability of the genome sequence data for *S. uvarum* (Bon et al. 2000; Cliften et al. 2001; Kellis et al. 2003; Scannell et al. 2011), microsatellites markers were proposed and applied to molecular typing of isolates originated from different environments (Wine, ciders and nature) (Masneuf-Pomarède et al. 2007; Zhang et al. 2015; Isabelle Masneuf-Pomarede et al. 2016b). In contrast to *S. cerevisiae* genetic diversity, wild and human origin isolates were intertwined. A total of 75% of strains were proven to be homozygous and estimated heterozygosity suggests a selfing rate above 0.95 for the different population tested. From this point of view, the *S. uvarum* life cycle appears to be more closely related to *S. paradoxus* or *S. cerevisiae* of natural resources than *S. cerevisiae* wine isolates. Population structure could not be correlated to distinct geographic or technological origins suggesting lower differentiation that may result from a large exchange between human and natural populations mediated by insects or human activities. A broad population genomics study carried out on *S. uvarum* strains obtained from diverse geographic origins and habitats (Almeida et al. 2014) revealed the global diversity of the wine and cider yeast *S. uvarum*. Three main clades were clearly distinguished: one clade contained Holarctic strains and a few strains from South America, the second clade phylogenetically related to the first clade and composed of strains from South America and a distant clade corresponding to the Australasian population. The genetic diversity of the South American isolates was higher than that found in the Northern Hemisphere, which contrast with the vast geographical area colonized ranging from temperate North America to Europe and Asia. Moreover, coalescence analyses suggest that a Patagonian subpopulation gave rise to the Holarctic population through a recent bottleneck. As for *S. cerevisiae*, this study highlighted the presence of multiple introgressions, mainly of *S. eubayanus* genomic DNA, into the genomes of European wine or cider strains of *S. uvarum*. These introgressions are absent in the large majority of wild strains and gene ontology analyses indicate that several gene categories relevant for wine fermentation are overrepresented. The analysis of a larger subset of Holarctic isolates confirmed that introgressions are widely disseminated in the holartic population and are more frequently associated with strains isolated from fermentations (Albertin et al. 2018). Some loci were found to be overrepresented thus suggesting their positive selection by human activity (Albertin et al. 2018). Altogether, such findings constitute a first indication of domestication in *S. uvarum*.

4.2 Interspecies Hybridization: A Mechanism Allowing Genome Interaction, Recombination and Evolution in Saccharomyces

Interspecies hybrids are very good models to explore fundamental questions concerning biology, evolution and adaptation. Like in plant kingdom, the production of hybrids in the *Saccharomyces* genus is quite frequent. Indeed, *Saccharomyces* yeasts possess a low prezygotic barrier and through mating in specific conditions, could generate viable interspecies hybrids (Sipiczki 2008; Morales and Dujon 2012). In the wine-related environment, hybrids *S. cerevisiae* × *S. uvarum* and *S. cerevisiae* × *S. kudriavzevii* and even triple hybrids between the three species are the most frequently reported in natural fermentation or as industrial starters (Table 4.1) (Masneuf et al. 1998; Bradbury et al. 2006; González et al. 2006; Lopandić et al. 2007; Masneuf-Pomarède et al. 2007; González et al. 2008; Erny et al. 2012). However, the occurrence of interspecies hybrids in spontaneous fermentation is difficult to assess unless using specific molecular methods to detect the presence of different parental genomes. For that reason, the detection frequency of wine interspecies hybrids may have been underestimated till now.

Several studies suggested a significant role of birds and insects in yeast dissemination and hybrid formation. Indeed, the digestive apparatus of a number of invertebrate species could hydrolyze the ascus wall without affecting spore viability and then, once ejected with the feces, spores may conjugate and form hybrids under advantageous conditions (Pulvirenti et al. 2002; Stefanini et al. 2016). In the case of *S. cerevisiae* × *S. uvarum* hybrids, the two species share the same niche and this situation may lead to the emergence of natural interspecies hybrids. Co-migration with grape varieties has also been suggested as an important means of the dissemination of hybrid yeasts (Sicard and Legras 2011). However, in winery environments, the conditions that favor hybrids formation are not fully understood.

Hybridization of different species produces an alloploid genome consisting of copies of the parental genomes. Hybrids of the first generation possess both parental mtDNA (heteroplasmy) but this status is transient and after few generations, homoplasmic cells with only one parental mtDNA are recovered, with rare cases of

Table 4.1 List of natural and commercial wine interspecies hybrids

Strains	Parental donors	Geographic origin	DNA ploidy	mtDNA	References
S6U	*S. cerevisiae* × *S. uvarum*	Italy	2 or 4	Sc	Ciolfi (1994); Masneuf et al. (1998); Naumov et al. (2000); González et al. (2006); Pérez-Torrado et al. (2015)
RC1-1, RC1-11, RC2-12, RC2-19, RC4-87, RP1-4, RP2-5, RP2-6, RP2-17	*S. cerevisiae* × *S. uvarum*	Alsace, France	2	nd	Le Jeune et al. (2007)
CECT 1885	*S. cerevisiae* × *S. uvarum*	Valadolid, Spain	nd	Su	González et al. (2006)
1T1a, Eg8,Eg2, Vin7, UHA5, UCD505, Eg12, Eg24, Eg 17, Eg6, UHA6, UHA4, UHA3, UHA2, Eg1,	*S. cerevisiae* × *S. kudriavzevii*	Alsace, France	2.7-3.1	nd	Bradbury et al. (2006); Erny et al. (2012); Peris et al. (2012b)
EL1D4	*S. cerevisiae* × *S. kudriavzevii*	Alsace, France	4	nd	Erny et al. (2012)
SOY3	*S. cerevisiae* × *S. kudriavzevii*	Daruvar, Croacia	2.9	nd	
H10418, H10423	*S. cerevisiae* × *S. kudriavzevii*	Alsace, France	3	nd	Erny et al. (2012)
HWD441, HWD77, HWD278, HWD78, HWD205, HWD216, HWD319,	*S. cerevisiae* × *S. kudriavzevii*	Switzerland	3	nd	Erny et al. (2012)
HWD231	*S. cerevisiae* × *S. kudriavzevii*	Switzerland	1.7	nd	Erny et al. (2012)
AWRI 1116	*S. cerevisiae* × *S. kudriavzevii*	Epernay, France	nd	nd	Heinrich (2006)
Uvaferm CEG	*S. cerevisiae* × *S. kudriavzevii*	Epernay, France	1.9	nd	Erny et al. (2012)
HA 1835, HA1844, HA 1842	*S. cerevisiae* × *S. kudriavzevii*	Perchtoldsdorf, Austria	3	nd	Lopandić et al. (2007); Peris et al. (2012b)

(continued)

Table 4.1 (continued)

Strains	Parental donors	Geographic origin	DNA ploidy	mtDNA	References
AMH	*S. cerevisiae* × *S. kudriavzevii*	Germany	nd	nd	Ortiz-Tovar et al. (2018)
CBS 2834	*S. cerevisiae* × *S. kudriavzevii* × *S. uvarum, (unknown species)*	Wadenswill, Switzerland	nd	Sk	González et al. (2006)

nd: not determined

recombination between parental mtDNA (Albertin and Marullo 2012). Molecular analysis of natural and industrial wine hybrids revealed high genetic and chromosomal diversity with extensive variation in the genome organization and structure: number of mosaic genomes, gene copy number variation, ploidy (allodiploids, allotriploids and allotetraploids) and sequence polymorphism (Dunn et al. 2008; Peris et al. 2012b, 2014). Indeed, hybridization event and resulting polyploidy are frequently followed by genomic modifications and genome stabilization such as gross chromosomal rearrangement (Piotrowski et al. 2012; Dunn et al. 2013), loss of heterozygosity (Peris et al. 2012a), formation of particular mitotypes (de Barros Lopes et al. 2002), aneuploidies (Peris et al. 2012a) and introgressions (Almeida et al. 2014). *S. cerevisiae* × *S. kudriavzevii* hybrids genomes diversity have been well studied. They exhibit complex mosaic chromosomal structures (González et al. 2006; Belloch et al. 2009; Navarro 2012) and three types of mitochondrial genome (mtDNA), *S. cerevisiae*-like, *S. kudriavzevii*-like and recombinant (Peris et al. 2012a). The majority of the wine *S. cerevisiae* × *S. kudriavzevii* hybrids are allotriploids, containing a diploid set of *S. cerevisiae* and a haploid set of *S. kudriavzevii* chromosomes (Erny et al. 2012; Peris et al. 2012b). The origin of the *S. kudriavzevii* subgenome donor was questioned and recent genetic analysis of *S. kudriavzevii* isolated from oak trees in Portugal and Spain have suggested that European *S. kudriavzevii* population rather than Japan *S. kudriavzevii* one is more closely related to *S. kudriavzevii* of natural hybrids and could be the true donor (Sampaio and Gonçalves 2008; Lopes et al. 2010; Lopandic 2018). By studying the genotypic profile (microsatellites markers) and ploidy levels of 32 *S. cerevisiae* × *S. kudriavzevii* hybrids isolated from wine and beer, Erny et al. (2012) showed that different hybridization events occurred and some hybrids were widely dispersed, suggesting specific adaptation to winemaking in northern European vineyards.

Environmental stress can contribute to genome instability, thus generating different evolved hybrids. Some *S. cerevisiae* × *S. kudriavzevii* hybrids isolates demonstrated a significant reduction of the *S. kudriavzevii* subgenome (Belloch et al. 2009; Erny et al. 2012; Peris et al. 2012a). One hypothesis to explain this non-symetric genome reduction is a better adaptation of *S. cerevisiae* genome to stressfull fermentation conditions. Only the *S. kudriavzevii* genome portion with adaptive importance (e.g. low temperature tolerance) was maintained (Peris et al. 2012b). Various selective pressures could influence the genome evolution of hybrids; in the case of *S. cerevisiae* × *S. uvarum* hybrids, Piotrowski et al. (2012) showed that increased temperature resulted in loss of the *S. uvarum* genome, whilst elevated ethanol concentration influences neither the *S. cerevisiae* nor the *S. uvarum* genomes. Phenotypic traits are affected by hybridization but also their stability over environmental changes referred as homeostasis (Da Silva et al. 2015). By studying laboratory interspecies *S. cerevisiae* × *S. uvarum* hybrids, the authors showed that interspecies hybridization could generate multi-trait phenotypes with improved oenological performances compared to the parental backgrounds and better homeostasis with respect to temperature. Interspecies hybridization can be considered as an evolutionary mechanism that helps hybrid yeast to overcome the fluctuating conditions of fermentation environment and to colonize a new ecological niches (Lopandic 2018).

Hybrid yeasts usually possess technological traits of the parental strains in new combination and were demonstrated to have higher fermentation abilities and improving qualities of the final product (González et al. 2007; Belloch et al. 2008; Gangl et al. 2009; Pérez-Torrado et al. 2015). The construction of artificial yeast hybrids by crossing different *Saccharomyces* species is a promising approach to select yeast of biotechnological interest, through a non-GMO strategy (Bellon et al. 2011; Bizaj et al. 2012; Bellon et al. 2015; Da Silva et al. 2015).

5 Genetic Diversity of the Main Species of the Wine Microbial Community

In contrast to *S. cerevisiae* and *S. uvarum* that can complete AF in a grape must and are thus able to consume all fermentable sugars, little was known until recently about other yeast species of the grape must microbial community, that can remain throughout alcoholic fermentation. The total cell counts of non-*Saccharomyces* yeasts in grape must usually range from 1.10e3 to 1.10e6 cells/ml (Jolly et al. 2003; Zott et al. 2010). The abundancy of non-*Saccharomyces* yeast usually drops as *S. cerevisiae* grows and AF unfurls, due to alcohol production, temperature increase and negative yeast-yeast interactions or other phenomena (Nissen et al. 2003; Goddard 2008; Salvadó et al. 2011; Varela et al. 2012). In grape must and during the early stages of fermentation, the predominant yeast species include yeast abundantly present at the grape surface like *Hanseniaspora* (syn. *Kloeckera*) spp. or *Starmerella* (syn. *Candida*) spp. Other frequently isolated species include *Debaryomyces* spp., *Rhodotorula* spp., *Kluyveromyces* spp., *Zygosaccharomyces* spp. or *Aureobasidum pullulans* (Esteve-Zarzoso et al. 1998; Jolly et al. 2003; Alessandria et al. 2013; Lederer et al. 2013; David et al. 2014; Nemcová et al. 2015; Wang et al. 2015; Mas et al. 2016). Less abundant species comprise *Metschnikowia* spp., *Torulaspora* spp., *Pichia* (syn. *Hansenula*) spp., *Issatchenkia*, *Lachancea* spp. (Zott et al. 2010; David et al. 2014; Vigentini et al. 2015).

For a long time, these yeast species were considered as undesirable (Padilla et al. 2016) for several reasons. They usually result in stuck or sluggish fermentations when used alone, and some of them produce high levels of acetic acid and other negative compounds (Jolly et al. 2014). However, this last decade, several authors showed that they contributed to the first stages of fermentation and they could modify the organoleptic properties of final wine (Fleet 2008; Jolly et al. 2014; Masneuf-Pomarede et al. 2015a, b). The use of mixed-inoculation, including a *S. cerevisiae* partner to secure AF completion and other yeast species, has partly overcome the problem of stuck/sluggish fermentation and has gained interest for winemakers (Contreras et al. 2014). Indeed, mixed inoculations were reported to have several oenological interests, including positive impact on wine's aroma, or acidity modulation, fructophily (that may improve fermentation completion, particularly in harsh conditions), or the lowering of ethanol content of wines (Ciani and Maccarelli 1998; Domizio et al. 2014; Capozzi et al. 2015; Ciani et al. 2016). Nowadays, several non-*Saccharomyces* (NS) species are commercialized as Active Dried Yeast (ADY), but NS remains a recent niche market for yeast producers: the first commercial release of non-*Saccharomyces* species as ADY for oenology dates back 2003 and involved a blend of 3 species (*S. cerevisiae*, *T. delbrueckii*, *L. thermotolerans* as Melody™ from Chr. Hansen). It was followed by the commercialization of *T. delbrueckii* as pure culture, in 2009 (Azzolini et al. 2015; Benito 2018). Indeed, few yeast species are commercially available as starter culture, among which *Torulaspora delbrueckii*, *Metschnikowia pulcherrima*, *Pichia kluyveri* or *Lachancea thermotolerans* for example. In a context of low/zero SO_2 wine making process, NS are also now proposed as bioprotection agent as an alternative to sulfites (Simonin et al. 2018; Roudil, et al. 2019).

In oenology, these yeast species are collectively referred as 'non-*Saccharomyces*'. The vagueness of the term allows different meanings: for winemakers, 'non-*Saccharomyces*' usually refer to non-*Saccharomyces* species available as ADY and their use is associated with an expected positive oenological benefit. Subsequently, for winemakers this name cannot refer to spoilage yeast. However, the non-*Saccharomyces* term also extends to wine species whose oenological interest is controversial, like *Hanseniaspora uvarum* or *Starmerella bacillaris* (I Masneuf-Pomarede et al. 2016a, b; Chasseriaud et al. 2018). Moreover, at odds with winemakers, many scientific authors design wine spoilage species (like *Brettanomyces bruxellensis*) as 'non-*Saccharomyces*' (Esteve-Zarzoso et al. 1998; Shinohara et al. 2000). To finish with terminology, another expression is sometimes used: non-conventional yeasts usually design non-*Saccharomyces* as well as non-*S. cerevisiae* species used in the wine industry (Mas et al. 2016).

The exploration of the oenological interest of non-*Saccharomyces* species compelled the consideration of their genetic diversity, and the development of dedicated tools. In order to commercialize reliable starter cultures of non-*Saccharomyces*, the first step is usually to compare several strains for the phenotype(s) of interest in order to select the best one(s). Thus, the assembly of a collection as far as possible representative of the genetic diversity of the species is mandatory. The recent advances in next-generation sequencing (NGS) have allowed the *de novo* assembly of several genome sequences of non-*Saccharomyces* (I Masneuf-Pomarede et al. 2016a, b). In its wake, genetic tools were developed, of which AFLP and microsatellite markers, which can be successfully used to address population diversity and structuration.

5.1 Torulaspora delbrueckii, *a Declassified* Saccharomyces

Historically, *Torulaspora delbrueckii* has been used for the production of specific Italian wines with low initial sugar content (Castelli 1955). It is able to produce relatively high ethanol concentrations (up to 12% vol.) for a non-*Saccharomyces* (Renault and Bely 2011), which may explain why *T. delbrueckii* has long been classified within the *Saccharomyces* genus (as *S. rosei* or *S. roseus*). The use of mixed fermentation including *T. delbrueckii/S. cerevisiae* allows a lower production of acetic acid than *S. cerevisiae* alone, and is particularly useful for grape must with high sugar content (Bely et al. 2008). It is reputed as a 'high-purity' fermenter (Renault et al. 2009) because of the low production of undesirable compounds, and it can increase wine complexity and fruitiness (Renault et al. 2015).

Besides its use in winemaking, *T. delbrueckii* has several other biotechnological interests: it is used for bakery applications (Pacheco et al. 2012), it is naturally associated with food and beverage fermentation including dairy product (Albertin et al. 2014a). *T. delbrueckii* also has large natural reservoirs and is frequently isolated from soils, plants, insects.

The genetic diversity of the species was initially addressed using RFLP markers (Pacheco et al. 2012), restriction endonuclease analysis associated with pulse-field gel electrophoresis (REA-PFGE, Renault et al. 2009) or minisatellites (Canonico et al. 2015). However, although discriminant, those markers are not suitable for population genetics studies. In 2014, microsatellites genotyping was performed on a collection of 110 strains, roughly one third from natural environment, one third from grapes and wine and the last third from other bioprocesses (Albertin et al. 2014a). Both dendrogram and population structure (Pritchard et al. 2000) analyses showed a clear clustering of the strains depending on the substrate of origin (Fig. 4.4): two clusters contained most natural strains from America and Old World respectively, while one group contained winemaking isolates. Finally, two clusters were associated with other bioprocesses, one more specifically with dairy products. Analysis of molecular variance (AMOVA) confirmed those results (Table 4.2): the substrate origin explained a significant part (12.3%) of the total variation of the microsatellite dataset, while it was only 7.6% for the geographical origin. Furthermore, when considering only bioprocess and grape/wine isolates, the geographical factor was no longer significant, while it explained up to 17% when considering nature isolates. This confirmed that human activities significantly shaped the genetic variability of the species, with nature isolates still being differentiated on the basis of geographical localization, as expected. The domestication of *T. delbrueckii* could date back to the Neolithic era for bioprocesses (around 4000 years ago).

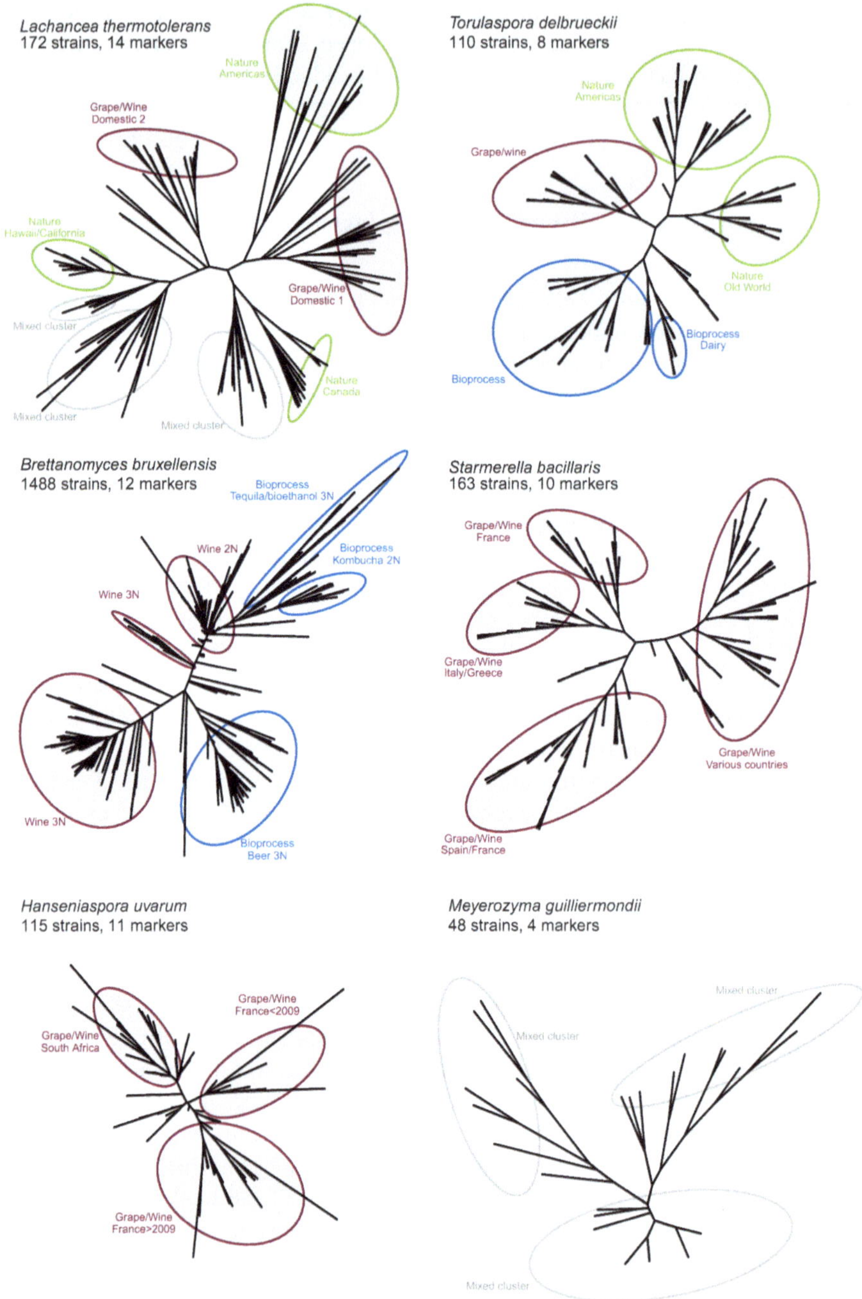

Fig. 4.4 Population genetics of non – *Saccharomyces* wine yeast using microsatellite markers. The original datasets were extracted from the following publications: *T. delbrueckii* (Albertin et al. 2014a), *L. thermotolerans* (Hranilovic et al. 2017), *M.guilliermondii* (Wrent et al. 2015), *Starm. bacillaris* (Masneuf-Pomarede et al. 2015a, b), *H. uvarum* (Albertin et al. 2016) and *B.bruxellensis*

Table 4.2 Analysis of molecular variance (AMOVA) for different factors impacting population genetic diversity

Species	AMOVA, factors tested (p-value): % variance	References
T. delbrueckii	Continent (p-value <0.001): 7.61%	Albertin et al. (2015)
	Continent, wild isolates (p-value <0.001): 17.19%	
	Continent, anthropized isolates: non-significant	
	Substrate (p-value<0.00001): 12.29%	
L. thermotolerans	Continent (p-value <0.001): 20.85%	Hranilovic et al. (2017)
	Substrate (p-value <0.001): 13.58%	
H. uvarum	Country (p-value = 0.0009): 8.54%	Albertin et al. (2016)
	Year of isolation (p-value << 10e-6): 20.62%	
	Substrate: not determined	
Starm. bacillaris	Country (p-value << 10e-6): 12.17%	Masneuf-Pomarede et al. (2015a, b)
	Wine region (p-value << 10e-6): 15.86%	
	Substrate: non-significant	
B. bruxellensis	Country (p-value<0.0001): 4.89%	Albertin et al. (2015)
	Country, non-wine isolates (p-value<0.0001):54.8%	
	Substrate (p-value<0.0001): 5.93%	
	Ploidy (p-value<0.0001): 46.9%	

For winemaking, the estimated domestication period is more recent (around 1900 years ago), and coincides with the expansion of *Vitis vinifera* across Europe and beyond, suggesting that, alike *S. cerevisiae*, *T. delbrueckii* domestication may reflect vine migrations. It has to be noted that, although microsatellite analysis showed strong evidence for wine and bioprocess domestication, substrate clustering was not perfect, suggesting frequent exchanges between wild and domesticated subpopulations. These exchanges could be mediated by insects or human activities (Albertin et al. 2014a). The intra-specific diversity in winemaking conditions was also evaluated: the analysis of strains isolated from the same grape must or wine samples allowed the calculation of diversity index like Shannon's equitability index (Table 4.3). Within oenological environments, the diversity of *T. delbrueckii* varied greatly from one sample to another, ranging from 0.39 (mildly clonal) to 1 (fully diversified population). Fully diversified populations can be interpreted as the absence of competition between genetically different individuals, while more clonal population implicates that a few strains are more fit than others and outcompeted

Fig 4.4 (continued) (Avramova et al. 2018a). All dendrograms were drawn using the same genetic distance (Bruvo's distance) and Neighbour Joining (NJ) clustering (Saitou and Nei 1987; Bruvo et al. 2004) using various R packages including poppr (Kamvar et al. 2014) and ape (Paradis et al. 2004). The substrate origin of the main clusters is represented by different colors: purple for grape/wine, blue for bioprocess, green for nature and grey for mixed anthropic/wild groups. 2N and 3N stand for diploid and triploid respectively

Table 4.3 Genetic diversity in winemaking environments for several wine yeast species. The Shannon's equitability index (EI) was calculated and ranged from 0 to 1 (fully clonal population to fully diversified population)

Species	EI	References
S. cerevisiae	0.39–0.65	Granchi et al. (2003); Albertin et al. (2014b)
T. delbrueckii	0.35–1	Albertin et al. (2015)
H. uvarum	0.86–1	Albertin et al. (2016)
Starm. bacillaris	0.9–0.97	Masneuf-Pomarede et al. (2015a, b)
B. bruxellensis	0–0.31	Albertin et al. (2015)

totally/partly the ecological niche. Here, the large range observed in *T. delbrueckii* species suggests that intraspecific competition may arise, depending on the matrix, and that phenotypic variability exist and could be used for subsequent selection program for winemaking purpose.

5.2 Lachancea thermotolerans, *the Acidifier*

Lachancea thermotolerans (formerly classified as *Kluyveromyces thermotolerans*), was, along with *T. delbrueckii*, the first non-*Saccharomyces* to be proposed as ADY for winemakers. In oenological conditions, it can produce up to 13% v/v ethanol, depending on the strain and grape must (Ciani et al. 2006). *L. thermotolerans* is able to produce relatively high levels of lactic acid (up to 16g/l) during AF, which is quite unique amongst yeast (Banilas et al. 2016). The resulting acidification improves wine microbial stability and organoleptic balance, while limiting the use of external inputs like tartaric acid (Gobbi et al. 2013). This is particularly interesting for grape must from warmer climates/vintages, or in a context of global warming. Another drawback of global warming is an increased sugar content in grape must, leading to high ethanol concentrations in wines while health prevention policies and consumers promote wines with moderate ethanol level (Dequin et al. 2017). *L. thermotolerans* has a lower alcohol yield than *S. cerevisiae*, and several works reported that mixed cultures of *L. thermotolerans*/*S. cerevisiae* showed significantly lower ethanol content (0.2-0.9% v/v) than with *S. cerevisiae* alone (Gobbi et al. 2013; Benito et al. 2015; 2016). Finally, *L. thermotolerans* shows high glycerol production, low acetate release, partial degradation of malate and improvement of wine aroma and flavor (Gobbi et al. 2013; Jolly et al. 2014; Su et al. 2014; Benito et al. 2016), explaining its growing use in winemaking.

Regarding its ecological niches, *Lachancea thermotolerans* occupies a large range of natural habitats, including insects, plants, and soil, as well as anthropized environments (mostly from grape/wine fermentations). A population genomic analysis, using mitochondrial genomes, showed no specific structuration regarding habitats for a subset of 50 strains (Freel et al. 2014). A first set of five microsatellite markers was then developed and applied to 47 grape/wine isolates from Greece

(Banilas et al. 2016). The isolates appeared to be clustered depending on the region of isolation, but no wild isolates were included, limiting the conclusion at the species level. More recently, a second set of 14 microsatellite markers was applied to 172 isolates from various niches including oenological and wild environments (Hranilovic et al. 2017). The resulting dendrogram (Fig. 4.4) showed that the evolution of *L. thermotolerans* has been driven by the geography and the substrate of isolation. Two distinct clusters contained mostly isolates from grape/wine environments, and were called domestic 1 and domestic 2 groups. Three groups contained mostly nature isolates, furthermore differentiated on the basis of geographical localization (Americas, Canada, Hawaii/California), as expected for a wild population (Slatkin 1987). Finally, three clusters were mixed and composed of strains from both wild and anthropic environments, indicating imperfect genetic differentiation depending on the substrate. AMOVA results confirmed the significant impact of both geographical localization and substrate origin that explained respectively 20.8 and 13.6% of the genetic diversity of the species (Table 4.2). Altogether, these results suggested the occurrence of at least two independent domestication events for grape/wine environments, although the actual separation between wild and anthropized group seems less plain than for *S. cerevisiae* or *T. delbrueckii*.

*5.3 Other Non-*Saccharomyces *Commercially Available for Winemakers*

Besides *T. delbrueckii* and *L. thermotolerans*, a few other species of the grape microflora are available for winemaking as starter culture. This is the case of *Metschnikowia pulcherrima* that has been selected because of its positive impact on wine aromas when used in mixed culture with *S. cerevisiae* (Parapouli et al. 2010; Jolly et al. 2017). *M. pulcherrima* also presents an important proteolytic activity, a feature that could help reducing haze formation and release mannoproteins (Chasseriaud et al. 2015). A few records also reported a negative impact, suggesting that the combination of mixed culture and grape variety could impact the oenological outcome (Jolly et al. 2014). *M. pulcherrima* occurs naturally in the fermentation process for wine and cider-making. In addition, a high number of researches considered its potential as biocontrol agent against postharvest decay of various fruits (Spadaro et al. 2008). The genetic diversity of *M. pulcherrima* species remains poorly explored. In fact, much confusion exists within the *Metschnikowia* genus, compelling the need for a deep taxonomic reassessment. However, due to the complexity and intricacy of the genus, its taxonomy can only be resolved using comparative genomics (Kurtzman et al. 2018). AFLP markers were developed and applied to 26 strains of *M. pulcherrima* from Mediterranean regions (Spadaro et al. 2008). A clear distinction between grape/wine isolates and apple ones was observed, suggesting bioprocess specialization. Tandem repeat-tRNA (TRtRNA) was developed and applied to isolates from a unique origin (grape/wine) (Barquet et al. 2012), and was not able to confirm/infirm the possibility of domestication event within the species. The recent genome announcement of strain UCD127 (Venkatesh et al. 2018) will help resolving the taxonomy of the genus and developing future tools for insightful genetic diversity studies.

Schizosaccharomyces pombe is another non-*Saccharomyces* used by winemakers. In fact, it was the first NS to be commercialized in 2003, not as ADY but as immobilized yeast (Rosa et al. 2003). The main use in oenology is the total/partial deacidification of grape must. A comparative genomic approach was applied to 161 isolates from various geographical localisation and substrate. The main result suggested a recent European origin for *S. pombe* in the Americas, coincident with the European colonization of the continent (Jeffares et al. 2015). However, this analysis provided very few input regarding potential domestication of the species or diversity in winemaking conditions, underlying the need for subsequent research program.

Finally, *Pichia kluyveri* has been only recently commercially released, with claims of higher levels of extraction of flavor precursors. Anfang et al. showed that mixed fermentation with *P. kluyveri*/*S. cerevisiae* allowed higher concentrations of varietal thiols, key aromatic compounds for some Sauvignon blanc wines (Anfang et al. 2009). From a population genetics viewpoint, very few data are available: the relationships between 46 strains were assessed using a combination of AFLP and RAPD, and showed a strong influence of geography on the population structure of *P. kluyveri* (Ganter and de Barros Lopes 2000). However, no individuals from grape/wine environments were included, so that no data are available regarding the population structure for winemaking conditions.

5.4 Hanseniaspora *Species, the Profuse non-*Saccharomyces *Species*

Hanseniaspora uvarum (syn. *Kloeckera apiculata*) is probably one of the NS yeasts universally present in grape must, usually at relatively high population level (Jolly et al. 2014; Capozzi et al. 2015). Although some works address the potential use of *H. uvarum*/*S. cerevisiae* mixed cultures for winemaking (Romano et al. 2003; Moreira et al. 2008), *H. uvarum* is viewed by most authors as neutral at best, and as detrimental in many case: it can produce high acetic acid and acetaldehyde contents and other compounds impacting negatively wine quality (Ciani and Maccarelli 1998; Ciani et al. 2006; Moreira et al. 2008; Chasseriaud et al. 2018). Other *Hanseniaspora* species frequently found in grape must are *H. guilliermondii*, *H. vineae* or *H. osmophila*. Alike *H. uvarum*, some works tested the use of mixed cultures of *S. cerevisiae*/*Hanseniaspora* spp. In particular, their use was associated with increased amounts of 2-phenyl-ethyl acetate and other compounds, particularly esters (Rojas et al. 2003; Viana et al. 2009; 2011). However, some of these experiments also showed increased concentrations of heavy sulphur compounds, with negative sensory impact (Moreira et al. 2008). Thus, *Hanseniaspora* species, as a whole, are considered by most as undesirable species to be inhibited (Chasseriaud et al. 2018).

Regarding their genetic diversity, the most studied species is *H. uvarum*. Several approaches have been developed along the years, including RAPD (Random Amplification of Polymorphic DNA) fingerprinting (Capece et al. 2005), or FTIR (Fourier transform infrared spectroscopy) (Grangeteau et al. 2015). In 2015, the genetic relationship between isolates was evaluated using 11 microsatellite markers, in a collection of 115 strains, mostly from grape/wine origin and from France and South Africa (Albertin et al. 2016). A few strains (6) from nature or other bioprocesses were also included: indeed, *H. uvarum* is frequently isolated from natural reservoirs, or from food and beverage fermentations either as part of the fermentation microbiome (cider, tequila, coffee, cocoa, etc.) or as an occasional spoiler (orange juice, yogurt, beer, etc.). The resulting dendrogram (Fig. 4.4) showed three main genetic clusters, one associated with South African isolates, the two other with French strains isolated after or before 2009. These results evidenced for the first time the occurrence of both spatial and temporal genetic differentiation. The isolates from wild or other bioprocesses environments were dispatched within these groups. Although their small number prevents concluding definitively about the structure at the species level, it is noteworthy that no hint of domestication was evidenced so far in *H. uvarum*. Geographical localization explained 8.5% of the total genetic variation (Table 4.2), while the year of isolation explained up to 20.6%. A striking result of the study is the high intraspecific variability at the sample level in oenological conditions. Shannon's equitability index ranged from 0.86 to 1 (Table 4.3), indicating highly diversified populations and suggesting that no individuals were able to outcompete the other in the grape/wine niche. Thus, it can be hypothesized that *H. uvarum* is a species associated with fruit flora, without evident specialization and showing relatively high genetic turnover within a time-scale of a few years. Further analysis of a larger number of isolates will help describing the extent of such results in winemaking.

5.5 **Starmerella bacillaris,** *the Low-Ethanol Producer*

Starmerella bacillaris shows various interesting oenological features: it is highly fructophilic, meaning it is able to consume most fructose contained in grape must, sometimes at the expense of glucose (Englezos et al. 2015). Fructophilic character has gained interest to prevent stuck/sluggish fermentation, particularly in grape must with high sugar content (Magyar and Toth 2011). Another feature lies in its ability to produce low quantities of ethanol and acetic acid, but high amounts of glycerol (Magyar and Toth 2011; Rantsiou et al. 2012; Englezos et al. 2015, 2016; Rantsiou et al. 2017). Thus, *Starm. bacillaris* emerged as a perfect candidate of so-called 'low-alcohol' yeasts (Bely et al. 2013; Dequin et al. 2017; Englezos et al. 2017). Mixed fermentations including both *Starm. bacillaris*/*S. cerevisiae* were shown to increase the production of terpenes and lactone and the degradation of malic acid (Sadoudi et al. 2012; Rantsiou et al. 2017). However, *Starm. bacillaris* may have a negative impact on wine perception due to the production of sulphur compounds (Masneuf-Pomarede et al. 2015a, b) and is thus commonly designed as an undesirable species (Chasseriaud et al. 2018).

Starm. bacillaris is almost exclusively isolated from grape must (red or white), or from winery environments. It is particularly abundant in high sugar grape must, particularly during the fermentation of sweet or botrytized wine (Sipiczki 2003; Urso et al. 2008; Tofalo et al. 2009). *Starm. bacillaris* is rarely found on other substrates, and very few wild isolates are indisputably described. This suggests that *Starm. bacillaris* primary ecological niche is grape/wine environment, with occasional dispersion to other favorable substrates (Masneuf-Pomarede et al. 2015a, b). Different molecular tools were developed to assess the intraspecific diversity of the species, including RAPD-PCR fingerprinting (Tofalo et al. 2012; Csoma et al. 2018), SAU-PCR (Englezos et al. 2015) and tandem repeat-tRNA (TRtRNA) PCR (Barquet et al. 2012). However, these different approaches are not completely congruent (Csoma et al. 2018). Thus, 10 microsatellites were developed and applied to a collection of 163 isolates, 157 from various winemaking regions, and 6 from nature (Masneuf-Pomarede et al. 2015a, b). Four main clusters were identified (Fig. 4.4), associated with specific geographical localization: one cluster contained mostly isolates from France, the second from France/Spain, the third from Italy/Greece and the last one from various countries. The 6 wild strains were distributed within the dendrogram, showing no evidence for a genetic separation between wild and anthropized isolates. STRUCTURE analysis and AMOVA confirms that the genetic diversity of *Starm. bacillaris* was shaped by geographical localization with country or wine region explaining 12–15% of the genetic variability, while the substrate factor (wild versus anthropized) was not significant (Table 4.2). The intraspecific diversity at the sample level was also studied (Table 4.3), and Shannon's equitability index was very high (0.9–0.97), indicating that *Starm. bacillaris* is not under selective pressure in winemaking environments.

5.6 Brettanomyces bruxellensis, *Once a Spoiler, Always a Spoiler*

Brettanomyces bruxellensis (syn. *Dekkera bruxellensis*) is a major cause of wine spoilage worldwide. It affects up to 25% of red wine (Gerbaux et al. 2000) and provokes rejection by consumers and thus important economic losses for winemakers (Wedral et al. 2010). In wine, *B. bruxellensis* produces aromatic molecules (called volatile phenols) associated with unpleasant aromas described as barnyard, horse sweat, burnt plastic, and usually designed as 'Brett' taint (Chatonnet et al. 1992). The main treatment used by winemakers to limit *B. bruxellensis* spoilage is sulphur dioxide addition. However, it was recently demonstrated that 35–40% of

B. bruxellensis strains are tolerant to sulfite treatments at concentrations recommended for winemaking (Avramova et al. 2018b).

Conversely, *B. bruxellensis* is known for its positive, even indispensable, contribution in the elaboration of some beer specialties such as Belgium lambic and Gueuze beers, American coolship ales, and so-called 'Brett beers' (Bokulich et al. 2012; Schifferdecker et al. 2014; Steensels et al. 2015). Likewise, it is considered as a beneficial microorganism in the production of kombucha (a traditional fermented sweetened tea). *B. bruxellensis* is also found in other food processes such as cider, dairy products, sourdough, olives fermentation, soft drinks (Steensels et al. 2015) where its impact (beneficial or detrimental) is often unclear and debated. In addition, *B. bruxellensis* is frequently isolated from bioethanol production where it was initially considered as a contaminant (De Souza Liberal et al. 2007). More recent work demonstrated that it can be exploited for high productivity and quality bioethanol production (Passoth et al. 2007; Reis et al. 2014). So far, all *B. bruxellensis* isolates were sampled exclusively from man-made ecological niches, and no wild isolates were described. *B. bruxellensis* thus forms a remarkable example of anthropized species.

Several molecular tools were developed to explore the genetic diversity of *B. bruxellensis*, including RAPD (Agnolucci et al. 2009), AFLP (Curtin et al. 2007), REA-PFGE (Miot-Sertier and Lonvaud-Funel 2007), SAU-PCR (Di Toro et al. 2015), infrared spectroscopy (Oelofse et al. 2009), mtDNA analysis (Martorell et al. 2006) or genome sequencing (Curtin et al. 2012; Borneman et al. 2014). All these studies highlighted a high intraspecific diversity, with the existence of diploid and allotriploid individuals (Borneman et al. 2014). Recently, microsatellite genotyping was applied to a large collection of 1488 isolates of *B. bruxellensis* from 9 different substrates and 29 countries (Avramova et al. 2018a). Six groups were identified (Fig. 4.4): one group containing diploid strain from wine environments, two distinct groups of triploid strains from wine (that probably arise through independent allotriploidisation events), a diploid group mostly associated with kombucha process, and the last group of triploid strains associated with tequila/bioethanol production. AMOVA confirmed that polyploidy explained 46.9% of the genetic variation of the species. The substrate explained significantly 5.9% of the variation, and the geographical localization shaped drastically non-wine isolates (58.4%).

Noteworthy, the two triploid groups associated with wine were shown to be tolerant/resistant to sulfite treatments, while all other groups were sulfite-sensitive (Avramova et al. 2018b, a), suggesting that allotriploidisation could play a major role on the adaptation of *B. bruxellensis* species to highly selective anthropized niche.

Microsatellites analysis also revealed interesting features regarding clonal persistence: clonal populations were isolated over a long period of time (>20 years) in the same winery (Albertin et al. 2014b; Cibrario and Dols-Lafargue 2017). Consistently, isolates from a given wine sample usually shows relatively high clonality, with Shannon's equitability index ranging from 0-0.31 (Table 4.3). Population genetics using microsatellite genotyping opened new avenues, but more works are needed to understand the different evolutionary forces shaping *B.bruxellensis* genetic diversity. In particular, comparative genomics may help unravel more precisely the selective pressure associated with *B.bruxellensis* specialization to different anthropic environments.

5.7 The Genetic Diversity of Other Wine Spoilage Yeasts

Some other spoilage yeasts are occasionally reported in winemaking. This is the case of *Meyerozyma guilliermondii* (syn. *Pichia guilliermondii*), usually considered as a wine spoiler (Lopes et al. 2009; Wrent et al. 2015) even if some works showed potential application of sequential fermentations regarding color stability (Benito et al. 2011). *M. guilliermondii* is also associated with other fermented food and beverages where it can cause spoilage. However, it has several biotechnological applications like vitamin production, xylose bioconversion, the biocontrol of post-harvest spoilage or more recently dough fermentation (Wrent et al. 2015). In a recent paper, Wrent et al. applied 4 microsatellite markers to the genotyping of 48 strains from various substrates and geographical origins (Wrent et al. 2015). The resulting dendrogram (Fig. 4.4) shows three clusters containing strains from mixed substrate and geographical localization. This may indicate that no specific specialization event is ongoing in this species, although more markers should be tested on a larger collection to confirm/infirm this preliminary result.

Zygosaccharomyces is another spoilage genus associated with wine, but also many other food and beverages product such as fruit juice, soft drinks, cider, honey, vinegar, etc. (Martorell et al. 2007; Dakal et al. 2018). The taxonomy of the *Zygosaccharomyces* genus is in constant evolution: 5 new species have been proposed since 2010, and the existence of hybridization and polyploidization events drastically complicates the precise delineation of the different species (Gordon and Wolfe 2008). Recently, Dakal et al. evaluated the potential of three typing methods, including fingerprinting and RFLP, to 76 strains from various substrates. Their results suggest that the genetic diversity within *Z. sapae* and *Z. mellis* could be shaped by isolation source (Dakal et al. 2018). Further work will help refine the evolutionary forces driving the evolution of this anthropized genus.

6 Conclusion

In the past, wine microbiology has mostly focused on *S. cerevisiae*, for which a vast knowledge has been accumulated on its ecology, phylogeny and population structure, and its specific genomic content that has been strikingly shaped by the grape must/wine environment. Despite its essential role in the completion of alcoholic fermentation, this model species is not representative of all the yeasts of the wine community that also contributes to the complexity of wine perception. In order to explore and exploit pertinently the phenotypic diversity of the non-*Saccharomyces* yeast, a great effort has been done for two decades to investigate their genetic variability, and the use of codominant markers providing reproducible data has changed the resolution that can be attained. Interestingly, the comparison of the population diversity of distinct wine yeast is quite revealing: first, while *S. cerevisiae* stood alone in the podium of domesticated yeast for a few years, it was quickly caught up by its two runners-up *T. delbrueckii* and *L. thermotolerans* that show intermediary degree of specialization for grape/wine environments. The spoilage yeast *B. bruxellensis* also appears as highly specialized and adapted to anthropized environments, indicating that winemaking has significantly shaped the diversity of this species too. Conversely, some highly abundant species in grape must (*H. uvarum*, *Starm. bacillaris*) show no evidence of domestication. This could suggest that their occurrence in grape must and wine is an inevitable and inadvertent outcome of their presence on the grape berry. Altogether, this shows that the development of winemaking by Man has created a new ecological niche colonized by different species that evolved according to different trajectories. Wine fermentation has favored simultaneously the specialization of the desired *S. cerevisiae*, of other passengers of the grape must microflora such as *T. delbrueckii* and of the undesired *B. bruxellensis* species.

However, the field of population genetics applied to non-*Saccharomyces* is still in its infancy, and future works will unravel the extent of wine yeast adaptation and eventually domestication, as well as the underlying selective pressure(s). In particular, there is a great need for genomic approaches applied to non-*Saccharomyces* species to gain thorough insights into their genomic content and into the genetic basis of adaptation and domestication to winemaking environments.

References

Agnolucci, M., Vigentini, I., Capurso, G., Merico, A., Tirelli, A., Compagno, C., Foschino R, & Nuti M. (2009). Genetic diversity and physiological traits of *Brettanomyces bruxellensis* strains isolated from Tuscan Sangiovese wines. *International Journal of Food Microbiology*. [Internet] *130*: 238–244. Available from: http://www.ncbi.nlm.nih.gov/pubmed/19237217.

Aigle, M., Erbs, D., & Moll, M. (1984). Some molecular structures in the genome of lager brewing yeasts. *Journal of the American Society of Brewing Chemists, 42*, 1–7.

Albertin, W., & Marullo, P. (2012). Polyploidy in fungi: Evolution after whole-genome duplication. *Proceedings of the Biological Sciences*. [Internet] *279*: 2497–2509. Available from: http://www.pubmedcentral.nih.gov/articlerender.fcgi?artid=3350714&tool=pmcentrez&rendertype=abstract.

Albertin, W., Chasseriaud, L., Comte, G., Panfili, A., Delcamp, A., Salin, F., Marullo, P., & Bely, M. (2014a). Winemaking and bioprocesses strongly shaped the genetic diversity of the ubiquitous yeast *Torulaspora delbrueckii*. *PLoS One* [Internet] *9*: e94246. Available from: http://www.pubmedcentral.nih.gov/articlerender.fcgi?artid=3981792&tool=pmcentrez&rendertype=abstract.

Albertin, W., Panfili, A., Miot-Sertier, C., Goulielmakis, A., Delcamp, A., Salin, F., Lonvaud-Funel, A., Curtin, C., & Masneuf-Pomarede, I. (2014b). Development of microsatellite markers for the rapid and reliable genotyping of *Brettanomyces bruxellensis* at strain level. *Food Microbiology, 42*, 188–195.

Albertin, W., Marullo, P., Peltier, E., Windholtz, S., Bely, M., Masneuf-Pomarede, I., 2015. Biodiversity of wine yeasts: new insights from population genetics, Oeno2015 proceedings, Bordeaux. https://www.oeno2015.u-bordeaux.fr/en/proceedings-of-the-symposium.

Albertin, W., Setati, M. E., Miot-Sertier, C., Mostert, T. T., Colonna-Ceccaldi, B., Coulon, J., Girard, P., Moine, V., Pillet, M., & Salin, F. (2016). *Hanseniaspora uvarum* from winemaking environments show spatial and temporal genetic clustering. *Food Microbiology, 6*, 1569.

Albertin, W., Chernova, M., Durrens, P., Guichoux, E., Sherman, D. J., Masneuf-Pomarede, I., & Marullo, P. (2018). Many interspecific chromosomal introgressions are highly prevalent in Holarctic *Saccharomyces uvarum* strains found in human-related fermentations. *Yeast* [Internet] *35*: 141–156. Available from: http://doi.wiley.com/10.1002/yea.3248.

Alessandria, V., Giacosa, S., Campolongo, S., Rolle, L., Rantsiou, K., & Cocolin, L. (2013). Yeast population diversity on grapes during on-vine withering and their dynamics in natural and inoculated fermentations in the production of icewines. *Food Research International.* [Internet] *54*: 139–147. Available from: http://www.sciencedirect.com/science/article/pii/S0963996913003451.

Almeida, P., Gonçalves, C., Teixeira, S., Libkind, D., Bontrager, M., Masneuf-Pomarède, I., Albertin, W., Durrens, P., Sherman, D. J., Marullo, P. et al. (2014). A Gondwanan imprint on global diversity and domestication of wine and cider yeast *Saccharomyces uvarum. Nature Communications.* [Internet] *5*: 4044. Available from: http://www.ncbi.nlm.nih.gov/pubmed/24887054.

Almeida, P., Barbosa, R., Zalar, P., Imanishi, Y., Shimizu, K., Turchetti, B., Legras, J.-L., Serra, M., Dequin, S., Couloux, A., et al. (2015). A population genomics insight into the Mediterranean origins of wine yeast domestication. *Molecular Ecology, 24*, 5412–5427.

Almeida, P., Barbosa, R., Bensasson, D., Gonçalves, P., & Sampaio, J. P. (2017). Adaptive divergence in wine yeasts and their wild relatives suggests a prominent role for introgressions and rapid evolution at noncoding sites. *Molecular Ecology*. [Internet] *26*: 2167–2182. Available from: http://doi.wiley.com/10.1111/mec.14071.

Anfang, N., Brajkovich, M., & Goddard, M. R. (2009). Co-fermentation with *Pichia kluyveri* increases varietal thiol concentrations in Sauvignon Blanc. *Australian Journal of Grape and Wine Research.* [Internet] *15*: 1–8. Available from: http://blackwell-synergy.com/doi/abs/10.1111/j.1755-0238.2008.00031.x.

Antonelli, A., Castellari, L., Zambonelli, C., Carnacini, A. 1999. Yeast influence on volatile composition of wines. J. Agric. Food Chem. 47, 1139–1144.

Antunovics, Z., Irinyi, L., & Sipiczki, M. (2005). Combined application of methods to taxonomic identification of *Saccharomyces* strains in fermenting botrytized grape must. *Journal of Applied Microbiology, 98*, 971–979.

Arroyo-García, R., Ruiz-García, L., Bolling, L., Ocete, R., López, M. A., Arnold, C., Ergul, A., Söylemezoğlu, G., Uzun, H. I., Cabello, F., et al. (2006). Multiple origins of cultivated grapevine (*Vitis vinifera* L. ssp. *sativa*) based on chloroplast DNA polymorphisms. *Molecular Ecology.* [Internet] *15*: 3707–3714. Available from: http://www.ncbi.nlm.nih.gov/pubmed/17032268.

Avramova, M., Cibrario, A., Peltier, E., Coton, M., Coton, E., Schacherer, J., Spano, G., Capozzi, V., Blaiotta, G., Salin, F., et al. (2018a). *Brettanomyces bruxellensis* population survey reveals a diploid-triploid complex structured according to substrate of isolation and geographical distribution. *Scientific Reports, 8*, 1–13.

Avramova, M., Vallet-Courbin, A., Maupeu, J., Masneuf-Pomarède, I., & Albertin, W. (2018b). Molecular diagnosis of *Brettanomyces bruxellensis* sulfur dioxide sensitivity through genotype specific method. *Frontiers in Microbiology*. [Internet] *9*. Available from: https://www.frontiersin.org/article/10.3389/fmicb.2018.01260.

Ayoub, M. -J., Legras, J. -L., Saliba, R., & Gaillardin, C. (2006). Application of multi locus sequence typing to the analysis of the biodiversity of indigenous *Saccharomyces cerevisiae* wine yeasts from Lebanon. *Journal of Applied Microbiology*. [Internet] *100*: 699–711. Available from: http://www.ncbi.nlm.nih.gov/pubmed/16553725.

Azumi, M., & Goto-Yamamoto, N. (2001). AFLP analysis of type strains and laboratory and industrial strains of *Saccharomyces sensu stricto* and its application to phenetic clustering. *Yeast* [Internet] *18*: 1145–1154. Available from: http://www.ncbi.nlm.nih.gov/pubmed/11536336.

Azzolini, M., Tosi, E., Lorenzini, M., Finato, F., & Zapparoli, G. (2015). Contribution to the aroma of white wines by controlled *Torulaspora delbrueckii* cultures in association with Saccharomyces cerevisiae. *World Journal of Microbiology and Biotechnology, 31*, 277–293.

Banilas, G., Sgouros, G., & Nisiotou, A. (2016). Development of microsatellite markers for *Lachancea thermotolerans* typing and population structure of wine-associated isolates. *Microbiological Research.* [Internet] *193*: 1–10. Available from: http://www.sciencedirect.com/science/article/pii/S0944501316301938.

Barquet, M., Martin, V., Medina, K., Perez, G., Carrau, F., & Gaggero, C. (2012). Tandem repeat-tRNA (TRtRNA) PCR method for the molecular typing of non-*Saccharomyces* subspecies. *Applied Microbiology Biotechnology* [Internet] *93*: 807–814. Available from: http://www.ncbi.nlm.nih.gov/pubmed/22113560.

Belloch, C., Orlic, S., Barrio, E., & Querol, A. (2008). Fermentative stress adaptation of hybrids within the *Saccharomyces sensu stricto* complex. *International Journal of Food Microbiology.* [Internet] *122*: 188–195. Available from: http://www.ncbi.nlm.nih.gov/pubmed/18222562.

Belloch, C., Pérez-Torrado, R., González, S. S., Pérez-Ortín, J. E., García-Martínez, J., Querol, A., & Barrio, E. (2009). Chimeric genomes of natural hybrids of *Saccharomyces cerevisiae* and *Saccharomyces kudriavzevii. Applied and Environmental Microbiology.* [Internet] *75*: 2534–2544. Available from: http://www.ncbi.nlm.nih.gov/pubmed/19251887.

Bellon, J. R., Eglinton, J. M., Siebert, T. E., Pollnitz, A. P., Rose, L., de Barros Lopes, M., & Chambers, P. J. (2011). Newly generated interspecific wine yeast hybrids introduce flavour and aroma diversity to wines. *Applied Microbiology and Biotechnology.* [Internet] *91*: 603–612. Available from: http://www.ncbi.nlm.nih.gov/pubmed/21538112.

Bellon, J. R., Yang, F., Day, M. P., Inglis, D. L., & Chambers, P. J. (2015). Designing and creating Saccharomyces interspecific hybrids for improved, industry relevant, phenotypes. *Applied Microbiology and Biotechnology* [Internet]: 8597–8609. Available from: http://link.springer.com/10.1007/s00253-015-6737-4.

Beltran, G., Torija, M.J., Novo, M., Ferrer, N., Poblet, M., Guillamón, J. M., Rozès, N., & Mas, A. (2002). Analysis of yeast populations during alcoholic fermentation: A six year follow-up study. *Systematic and Applied Microbiology.* [Internet] *25*: 287–293. Available from: http://www.ncbi.nlm.nih.gov/pubmed/12353885.

Bely, M., Stoeckle, P., Masneuf-Pomarede, I., & Dubourdieu, D. (2008). Impact of mixed *Torulaspora delbrueckii-Saccharomyces cerevisiae* culture on high-sugar fermentation. *International Journal of Food Microbiology* [Internet] *122*: 312–320. Available from: http://www.ncbi.nlm.nih.gov/entrez/query.fcgi?cmd=Retrieve&db=PubMed&dopt=Citation&list_uids=18262301.

Bely, M., Renault, P., da Silva, T., Masneuf-pomarède, I., Albertin, W., Moine, V., Coulon, J., Sicard, D., De Vienne, D., & Marullo, P. (2013). Non-conventional yeasts and alcohol level reduction. In *Alcohol level reduction in wine* (pp. 33–37). Bordeaux: Ed. Vigne et Vin Publications Internationales.

Benito, S. (2018). The impact of Torulaspora delbrueckii yeast in winemaking. *Applied Microbiology and Biotechnology.* [Internet] *102*: 3081–3094. Available from: https://doi.org/10.1007/s00253-018-8849-0.

Benito, S., Morata, A., Palomero, F., González, M. C., & Suárez-Lepe, J. A. (2011). Formation of vinylphenolic pyranoantho;cyanins by *Saccharomyces cerevisiae* and *Pichia guillermondii* in red wines produced following different fermentation strategies. *Food Chemistry.* [Internet] *124*: 15–23. Available from: http://linkinghub.elsevier.com/retrieve/pii/S0308814610006709.

Benito, S., Hofmann, T., Laier, M., Lochbühler, B., Schüttler, A., Ebert, K., Fritsch, S., Röcker, J., & Rauhut, D. (2015). Effect on quality and composition of Riesling wines fermented by sequential inoculation with non-*Saccharomyces* and *Saccharomyces cerevisiae. European Food Research and Technology, 241*, 707–717.

Benito, A., Calderon, F., Palomero, F., & Benito, S. (2016). Quality and composition of airen wines fermented by sequential inoculation of *Lachancea thermotolerans* and *Saccharomyces cerevisiae. Food Technology and Biotechnology* [Internet] *54*: 135–144. Available from: http://www.ncbi.nlm.nih.gov/pubmed/27904403.

Bizaj, E., Cordente, A. G., Bellon, J. R., Raspor, P., Curtin, C. D., & Pretorius, I. S. (2012). A breeding strategy to harness flavor diversity of *Saccharomyces* interspecific hybrids and minimize hydrogen sulfide production. *FEMS Yeast Research*. [Internet] *12*: 456–465. Available from: http://www.ncbi.nlm.nih.gov/pubmed/22385988.

Bokulich, N. A., Bamforth, C. W., & Mills, D. A. (2012). Brewhouse-resident microbiota are responsible for multi-stage fermentation of American coolship ale. *PLoS One* [Internet] *7*: e35507. Available from: http://www.pubmedcentral.nih.gov/articlerender.fcgi?artid=3329477&tool=pmcentrez&rendertype=abstract.

Bokulich, N. A., Thorngate, J. H., Richardson, P. M., & Mills, D. A. (2013). Microbial biogeography of wine grapes is conditioned by cultivar, vintage, and climate. *Proceedings of the National Academy of Sciences*. [Internet]: 1–10. Available from: http://www.pnas.org/cgi/doi/10.1073/pnas.1317377110.

Boynton, P.J., Greig, D., 2014. The ecology and evolution of non-domesticated Saccharomyces species, Yeast 31, 449–62.

Bon, E., Neuvéglise, C., Casaregola, S., Artiguenave, F., Wincker, P., Aigle, M., & Durrens, P. (2000). Genomic Exploration of the Hemiascomycetous Yeasts: 5. *Saccharomyces bayanus* var. *uvarum*. *FEBS Letters, 487*, 37–41.

Börlin, M., Venet, P., Claisse, O., Salin, F., Legras, J-L., & Masneuf-Pomarède, I. (2016). Cellar-associated *Saccharomyces cerevisiae* population structure revealed high diversity and perennial persistence in Sauternes wine estates. *Applied and Environmental Microbiology*. [Internet]: AEM.03627-15. Available from: http://aem.asm.org/lookup/doi/10.1128/AEM.03627-15.

Börlin, M., Claisse, O., Albertin, W., Salin, F., Legras, J. -L., & Masneuf-Pomarede, I. (2018). Quantifying anthropogenic effect on S. cerevisiae vineyard metapopulation diversity. Prep.

Borneman, A.R., Desany, B. A., Riches, D., Affourtit, J.P., Forgan, A.H., Pretorius, I.S., Egholm, M., & Chambers, P. J. (2011). Whole-genome comparison reveals novel genetic elements that characterize the genome of industrial strains of Saccharomyces cerevisiae. Sherlock G, editor. *PLoS Genetics*. [Internet] *7*: e1001287. Available from: http://dx.plos.org/10.1371/journal.pgen.1001287.

Borneman, A. R., Zeppel, R., Chambers, P. J., & Curtin, C. D. (2014). Insights into the *Dekkera bruxellensis* genomic landscape: Comparative genomics reveals variations in ploidy and nutrient utilisation potential amongst wine isolates. *PLoS Genet* [Internet] *10*: e1004161. Available from: 10.1371%2Fjournal.pgen.1004161.

Borneman, A. R., Forgan, A. H., Kolouchova, R., Fraser, J. A., & Schmidt, S. A. (2016). Whole genome comparison reveals high levels of inbreeding and strain redundancy across the spectrum of commercial wine strains of Saccharomyces cerevisiae. *G3 (Bethesda)*. [Internet] *6*: 957–971. Available from: http://www.g3journal.org/content/early/2016/02/10/g3.115.025692%5Cnhttp://www.g3journal.org/content/early/2016/02/10/g3.115.025692.full.pdf%5Cnhttp://www.ncbi.nlm.nih.gov/pubmed/26869621.

Bradbury, J. E., Richards, K. D., Niederer, H. A., Lee, S. A., Rod Dunbar, P., & Gardner, R. C. (2006). A homozygous diploid subset of commercial wine yeast strains. *Antonie Van Leeuwenhoek* [Internet] *89*: 27–37. Available from: http://www.ncbi.nlm.nih.gov/pubmed/16328862.

Brown, S. L., Stockdale, V. J., Pettolino, F., Pocock, K. F., De Barros Lopes, M., Williams, P. J., Bacic, A., Fincher, G. B., Høj, P. B., & Waters, E. J. (2007). Reducing haziness in white wine by overexpression of *Saccharomyces cerevisiae* genes YOL155c and YDR055w. *Applied Microbiology and Biotechnology, 73*, 1363–1376.

Bruvo, R., Michiels, N.K., D'Souza, T.G., & Schulenburg, H. (2004). A simple method for the calculation of microsatellite genotype distances irrespective of ploidy level. *Molecular Ecology*. [Internet] *13*: 2101–2106. Available from: http://www.ncbi.nlm.nih.gov/pubmed/15189230.

Canonico, L., Comitini, F., & Ciani, M. (2015). TdPIR minisatellite fingerprinting as a useful new tool for *Torulaspora delbrueckii* molecular typing. *International Journal of Food Microbiology* [Internet] *200*: 47–51. Available from: http://www.sciencedirect.com/science/article/pii/S0168160515000549.

Capece, A., Fiore, C., Maraz, A., & Romano, P. (2005). Molecular and technological approaches to evaluate strain biodiversity in *Hanseniaspora uvarum* of wine origin. *Journal of Applied*

Microbiology [Internet] *98*: 136–144. Available from: http://www.ncbi.nlm.nih.gov/pubmed/15610426.
Capozzi, V., Garofalo, C., Chiriatti, M.A., Grieco, F., & Spano, G. (2015). Microbial terroir and food innovation: The case of yeast biodiversity in wine. *Microbiological Research.* [Internet] *181*: 75–83. Available from: http://www.sciencedirect.com/science/article/pii/S0944501315300185.
Carrigan, M. A., Uryasev, O., Frye, C. B., Eckman, B. L., Myers, C. R., Hurley, T. D., & Benner, S. A. (2014). Hominids adapted to metabolize ethanol long before human-directed fermentation. *Proceedings of the National Academy of Sciences*, 6–11.
Carvalho-Netto, O. V., Carazzolle, M. F., Rodrigues, A., Bragança, W. O., Costa, G. G. L., Argueso, J. L., & Pereira, G. A. G. (2013). A simple and effective set of PCR-based molecular markers for the monitoring of the *Saccharomyces cerevisiae* cell population during bioethanol fermentation. *Journal of Biotechnology, 168*, 701–709.
Castelli, T. (1955). Yeasts of wine fermentations from various regions of Italy. *American Journal of Enology and Viticulture.* [Internet] *6*: 18–19. Available from: http://www.ajevonline.org/content/6/1/18.abstract.
Castellari, L.;, Ferruzzi, M., Magrini, A., Giudici, P., Passarelli, P., Zambonelli, C. 1994. Unbalanced wine fermentation by cryotolerant vs. non-cryotolerant Saccharomyces strains. Vitis 33, 49-52.
Chasseriaud, L., Miot-Sertier, C., Coulon, J., Iturmendi, N., Moine, V., Albertin, W., & Bely, M. (2015). A new method for monitoring the extracellular proteolytic activity of wine yeasts during alcoholic fermentation of grape must. *Journal of Microbiological Methods* [Internet] *119*: 176–179. Available from: http://www.ncbi.nlm.nih.gov/pubmed/26529648.
Chasseriaud, L., Coulon, J., Marullo, P., Albertin, W., & Bely, M. (2018). New oenological practice to promote non-*Saccharomyces* species of interest: Saturating grape juice with carbon dioxide. *Applied Microbiology and Biotechnology.* [Internet] *102*: 3779–3791. Available from: https://doi.org/10.1007/s00253-018-8861-4.
Chatonnet, P., Dubourdieu, D., Boidron, J., & Pons, M. (1992). The origin of ethylphenols in wines. *Journal of the Science of Food and Agriculture* [Internet] *60*: 165–178. Available from: https://doi.org/10.1002/jsfa.2740600205.
Ciani, M., & Maccarelli, F. (1998). Oenological properties of non-*Saccharomyces* yeasts associated with wine-making. *World Journal of Microbiology and Biotechnology, 14*, 199–203.
Ciani, M., Beco, L., & Comitini, F. (2006). Fermentation behaviour and metabolic interactions of multistarter wine yeast fermentations. *International Journal of Food Microbiology* [Internet] *108*: 239–245. Available from: http://www.sciencedirect.com/science/article/pii/S0168160506000183.
Ciani, M., Morales, P., Comitini, F., Tronchoni, J., Canonico, L., Curiel, J. A. Oro, L., Rodrigues, A. J., & Gonzalez, R. (2016). Non-conventional yeast species for lowering ethanol content of wines. *Frontiers in Microbiology* [Internet] *7*: 642. Available from: http://www.ncbi.nlm.nih.gov/pubmed/27199967.
Cibrario, A., & Dols-Lafargue, M. (2017). Diversité Génétique et Phénotypique de l'Espèce *Brettanomyces bruxellensis*. Influence sur son Potentiel d'Altération des Vins Rouges. PhD thesis.
Ciolfi, G. (1994). Selezione di uno stipite di lievito Saccharomyces della razza fisiologica uvarum e suo impiego enologico allo stato secco. *L'Enotecnico* 71–76.
Ciften, P. F., Hillier, L. W., Fulton, L., Graves, T., Miner, T., Gish, W. R., Waterston, R. H., & Johnston, M. (2001). Surveying *Saccharomyces* genomes to identify functional elements by comparative DNA sequence analysis. *Genome Research, 11*, 1175–1186.
Coi, A. L., Bigey, F., Mallet, S., Marsit, S., Zara, G., Gladieux, P., Galeote, V., Budroni, M., Dequin, S., & Legras, J. -L. (2017). Genomic signatures of adaptation to wine biological ageing conditions in biofilm-forming flor yeasts. *Molecular Ecology.* [Internet] *38*: 42–49. Available from: http://doi.wiley.com/10.1111/ijlh.12426.
Constantí, M., Poblet, M., Arola, L., Mas, A., & Guillamón, J. M. (1997). Analysis of yeast populations during alcoholic fermentation in a newly established winery. *American Journal of*

Enology and Viticulture. [Internet] *48*: 339 LP-339344. Available from: http://www.ajevonline.org/content/48/3/339.abstract.

Contreras, A., Hidalgo, C., Henschke, P. A., Chambers, P.J., Curtin, C., & Varela, C. (2014). Evaluation of non-*Saccharomyces* yeasts for the reduction of alcohol content in wine. *Applied and Environmental Microbiology*. [Internet] 80: 1670–1678. Available from: http://www.ncbi.nlm.nih.gov/pmc/articles/PMC3957604/.

Cornuet, J. M., Pudlo, P., Veyssier, J., Dehne-Garcia, A., Gautier, M., Leblois, R., Marin, J. M., & Estoup, A. (2014). DIYABC v2.0: A software to make approximate Bayesian computation inferences about population history using single nucleotide polymorphism, DNA sequence and microsatellite data. *Bioinformatics, 30*, 1187–1189.

Coton, E., Coton, M., Levert, D., Casaregola, S., & Sohier, D. (2006). Yeast ecology in French cider and black olive natural fermentations. *International Journal of Food Microbiology, 108*, 130–135.

Cromie, G. A., Hyma, K. E., Ludlow, C. L., Garmendia-Torres, C., Gilbert, T. L., May, P., Huang, A. A., Dudley, A. M., & Fay, J. C. (2013). Genomic sequence diversity and population structure of *Saccharomyces cerevisiae* assessed by RAD-seq. *G3 (Bethesda)*. [Internet] *3*: 2163–2171. Available from: http://www.ncbi.nlm.nih.gov/pubmed/24122055.

Csoma, H., Ács-Szabó, L., Papp, L. A., & Sipiczki, M. (2018). Application of different markers and data-analysis tools to the examination of biodiversity can lead to different results: A case study with *Starmerella bacillaris* (synonym *Candida zemplinina*) strains. *FEMS Yeast Research*. [Internet] *18*: foy021–foy021. Available from: https://doi.org/10.1093/femsyr/foy021.

Curtin, C. D., Bellon, J. R., Henschke, P. A., Godden, P. W., & De Barros Lopes, M. A. (2007). Genetic diversity of *Dekkera bruxellensis* yeasts isolated from Australian wineries. *FEMS Yeast Research*. [Internet] *7*: 471–481. Available from: https://doi.org/10.1111/j.1567-1364.2006.00183.x.

Curtin, C., Kennedy, E., & Henschke, P. A. (2012). Genotype-dependent sulphite tolerance of Australian *Dekkera (Brettanomyces) bruxellensis* wine isolates. *Letters in Applied Microbiology*. [Internet] *55*: 56–61. Available from: http://www.ncbi.nlm.nih.gov/pubmed/22537453.

Da Silva, T., Albertin, W., Dillmann, C., Bely, M., la Guerche, S., Giraud, C., Huet, S., Sicard, D., Masneuf-Pomarède, I., De Vienne, D., et al. (2015). Hybridization within *Saccharomyces* genus results in homoeostasis and phenotypic novelty in winemaking conditions. *PLoS One* [Internet] *10*: 1–24. Available from: http://dx.plos.org/10.1371/journal.pone.0123834.

Dakal, T. C., Solieri, L., & Giudici, P. (2018). Evaluation of fingerprinting techniques to assess genotype variation among *Zygosaccharomyces* strains. *Food Microbiology*. [Internet] *72*: 135–145. Available from: http://www.sciencedirect.com/science/article/pii/S0740002017305269.

Damon, C., Vallon, L., Zimmermann, S., Haider, M. Z., Galeote, V., Dequin, S., Luis, P., Fraissinet-Tachet, L., & Marmeisse, R. (2011). A novel fungal family of oligopeptide transporters identified by functional metatranscriptomics of soil eukaryotes. *ISME Journal*. [Internet] *5*: 1871–1880. Available from: http://www.ncbi.nlm.nih.gov/pubmed/21654847.

David, V., Terrat, S., Herzine, K., Claisse, O., Rousseaux, S., Tourdot-Maréchal, R., Masneuf-Pomarede, I., Ranjard, L., & Alexandre, H. (2014). High-throughput sequencing of amplicons to monitor yeast biodiversity in must and wine. *Journal of Industrial Microbiology & Biotechnology, 41*, 811–821.

de Barros Lopes, M., Rainieri, S., Henschke, P.A., & Langridge, P. (1999). AFLP fingerprinting for analysis of yeast genetic variation. *International Journal of Systematic Bacteriology*. [Internet] 49 Pt 2: 915–924. Available from: http://www.ncbi.nlm.nih.gov/pubmed/10319518.

de Barros Lopes, M., Bellon, J.R., Shirley, N.J., & Ganter, P.F. (2002). Evidence for multiple inter-specific hybridization in *Saccharomyces sensu stricto* species. *FEMS Yeast Research*. [Internet] *1*: 323–331. Available from: http://www.ncbi.nlm.nih.gov/pubmed/12702336.

De Souza Liberal, A. T., Basílio, A. C. M., Do Monte Resende, A., Brasileiro, B. T. V., Da Silva-Filho, E. A., De Morais, J. O. F., Simões, D. A., & De Morais, M. A. (2007). Identification of *Dekkera bruxellensis* as a major contaminant yeast in continuous fuel ethanol fermentation. *Journal of Applied Microbiology* [Internet] *102*: 538–547. Available from: https://doi.org/10.1111/j.1365-2672.2006.03082.x.

Demuyter, C., Lollier, M., Legras, J. -L., & Le Jeune, C., (2004). Predominance of *Saccharomyces uvarum* during spontaneous alcoholic fermentation, for three consecutive years, in an Alsatian winery. *Journal of Applied Microbiology*. [Internet] *97*: 1140–1148. Available from: http://www.ncbi.nlm.nih.gov/pubmed/15546404.

Dequin, S., Escudier, J. L., Bely, M., Noble, J., Albertin, W., Masneuf-Pomarede, I., Marullo, P., Salmon, A., & Sablayrolles, J. M. (2017). How to adapt winemaking practices to modified grape composition under climate change conditions. *OENO One, 51*, 205–214.

Di Toro, M. R., Capozzi, V., Beneduce, L., Alexandre, H., Tristezza, M., Durante, M., Tufariello, M., Grieco, F., & Spano, G. (2015). Intraspecific biodiversity and 'spoilage potential' of *Brettanomyces bruxellensis* in Apulian wines. *LWT- Food Science and Technology* [Internet] *60*: 102–108. Available from: http://www.sciencedirect.com/science/article/pii/S0023643814004277.

Domizio, P., Liu, Y., Bisson, L. F., & Barile, D. (2014). Use of non-*Saccharomyces* wine yeasts as novel sources of mannoproteins in wine. *Food Microbiology* [Internet] *43*: 5–15. Available from: http://www.ncbi.nlm.nih.gov/pubmed/24929876.

Drumonde-Neves, J., Franco-Duarte, R., Vieira, E., Mendes, I., Lima, T., Schuller, D., & Pais, C. (2018). Differentiation of *Saccharomyces cerevisiae* populations from vineyards of the Azores Archipelago: Geography vs Ecology. *Food Microbiology*. [Internet] *74*: 151–162. Available from: http://linkinghub.elsevier.com/retrieve/pii/S0740002017308468.

Duan, S. F., Han, P. J., Wang, Q. M., Liu, W. Q., Shi, J. Y., Li, K., Zhang, X. L., & Bai, F. Y. (2018). The origin and adaptive evolution of domesticated populations of yeast from Far East Asia. *Nature Communications*. [Internet] *9*. Available from: https://doi.org/10.1038/s41467-018-05106-7.

Dubourdieu, D., Sokol, A., Zucca, J., Thalouarn, P., Dattee, A., & Aigle, M. (1987). Identification des souches de levures isolées de vin par l'analyse de leur ADN mitochondrial. 267–278.

Dunham, M. J., Badrane, H., Ferea, T. L., Adams, J., Brown, P. O., Rosenzweig, F., & Botstein, D. (2002). Characteristic genome rearrangements in experimental evolution of *Saccharomyces cerevisiae*. *Proceedings of the National Academy of Sciences of the United States of America* [Internet] *99*: 16144–16149. Available from: http://www.ncbi.nlm.nih.gov/pubmed/12446845.

Dunn, B., Sherlock, G., & Franklin, B. (2008). Reconstruction of the genome origins and evolution of the hybrid lager yeast *Saccharomyces pastorianus*. *Genome Research*. [Internet] *18*: 1610–1623. Available from: http://www.ncbi.nlm.nih.gov/pubmed/18787083.

Dunn, B., Richter, C., Kvitek, D. J., Pugh, T., & Sherlock, G. (2012). Analysis of the *Saccharomyces cerevisiae* pan-genome reveals a pool of copy number variants distributed in diverse yeast strains from differing industrial environments. *Genome Research*. [Internet] *22*: 908–924. Available from: http://www.pubmedcentral.nih.gov/articlerender.fcgi?artid=3337436&tool=pmcentrez&rendertype=abstract.

Dunn, B., Paulish, T., Stanbery, A., Piotrowski, J., Koniges, G., Kroll, E., Louis, E. J., Liti, G., Sherlock, G., & Rosenzweig, F. (2013). Recurrent rearrangement during adaptive evolution in an interspecific yeast hybrid suggests a model for rapid introgression. Fay, J. C., editor. PLoS Genetics. [Internet] *9*: e1003366. Available from: http://dx.plos.org/10.1371/journal.pgen.1003366.

Eeom, Y. J., Son, S. Y., Jung, D. H., Hur, M. S., Kim, C. M., Park, S. Y., Shin, W. C., Lee, S. J., Auh, J. H., Kim, G. W., et al. (2018). Diversity analysis of *Saccharomyces cerevisiae* isolated from natural sources by multilocus sequence typing (MLST). *Food Science and Biotechnology, 27*, 1119–1127.

Englezos, V., Rantsiou, K., Torchio, F., Rolle, L., Gerbi, V., & Cocolin, L. (2015). Exploitation of the non-*Saccharomyces* yeast *Starmerella bacillaris* (synonym *Candida zemplinina*) in wine fermentation: Physiological and molecular characterizations. *International Journal of Food Microbiology* [Internet] *199*: 33–40. Available from: http://www.sciencedirect.com/science/article/pii/S0168160515000203.

Englezos, V., Torchio, F., Cravero, F., Marengo, F., Giacosa, S., Gerbi, V., Rantsiou, K., Rolle, L., & Cocolin, L. (2016). Aroma profile and composition of Barbera wines obtained by mixed fermentations of *Starmerella bacillaris* (synonym *Candida zemplinina*) and *Saccharomyces cerevisiae*. *LWT- Food Science and Technology*. [Internet] *73*: 567–575. Available from: http://www.sciencedirect.com/science/article/pii/S0023643816303954.

Englezos, V., Giacosa, S., Rantsiou, K., Rolle, L., & Cocolin, L. (2017). *Starmerella bacillaris* in winemaking: Opportunities and risks. *Current Opinion in Food Science*. [Internet] *17*: 30–35. Available from: http://www.sciencedirect.com/science/article/pii/S2214799317300656.

Erny, C., Raoult, P., Alais, A., Butterlin, G., Delobel, P., Matei-Radoi, F., Casaregola, S., & Legras, J. –L. L. (2012). Ecological success of a group of *Saccharomyces cerevisiae/Saccharomyces kudriavzevii* hybrids in the Northern European wine-making environment. *Applied and Environmental Microbiology*. [Internet] *78*: 3256–3265. Available from: http://aem.asm.org/cgi/doi/10.1128/AEM.06752-11.

Esteve-Zarzoso, B., Manzanares, P., Ramón, D., Querol, A., Ramon, D., & Querol, A. (1998). The role of non-*Saccharomyces* yeasts in industrial winemaking. *International Microbiology*. [Internet] *1*: 143–148. Available from: http://www.ncbi.nlm.nih.gov/pubmed/10943352.

Fay, J. C., & Benavides, J. A. (2005). Evidence for domesticated and wild populations of *Saccharomyces cerevisiae*. *PLoS Genetics*. [Internet] *1*: 66–71. Available from: http://www.ncbi.nlm.nih.gov/pubmed/16103919.

Fleet, G. H. (1993). The microorganisms of winemaking-isolation enumeration and identification. In G. H. Fleet (Ed.), *Wine microbiology and biotechnology* (pp. 1–27). Chur: Harwood Academic Publishers.

Fleet, G. H. (2003). Yeast interactions and wine flavour. *International Journal of Food Microbiology, 86*, 11–22.

Fleet, G. H. (2008). Wine yeasts for the future. *FEMS Yeast Research, 8*, 979–995.

Fleet, G. H., Prakitchaiwattana, C., Beh, A., & Heard, G. M. (2002). The yeast ecology of wine grapes. In M. Ciani (Ed.), *Biodiversity and biotechnology of wine yeast* (pp. 1–17). Kerala: Research Signpost.

Freel, K. C., Friedrich, A., Hou, J., & Schacherer, J. (2014). Population genomic analysis reveals highly conserved mitochondrial genomes in the yeast species *Lachancea thermotolerans*. *Genome Biology and Evolution* [Internet] *6*: 2586–2594. Available from: http://www.ncbi.nlm.nih.gov/pubmed/25212859.

Frezier, V., & Dubourdieu, D. (1992). Ecology of yeast strain *Saccharomyces cerevisiae* during spontaneous fermentation in a Bordeaux winery. *American Journal of Enology and Viticulture, 43*, 375–380.

Galeote, V., Novo, M., Salema-Oom, M., Brion, C., Valério, E., Gonçalves, P., & Dequin, S. (2010). FSY1, a horizontally transferred gene in the *Saccharomyces cerevisiae* EC1118 wine yeast strain, encodes a high-affinity fructose/H+ symporter. *Microbiology* [Internet] *156*: 3754–3761. Available from: http://www.ncbi.nlm.nih.gov/pubmed/20705659.

Galeote, V., Bigey, F., Beyne, E., Novo, M., Legras, J. –L., Casaregola, S., & Dequin, S. (2011). Amplification of a *Zygosaccharomyces bailii* DNA segment in wine yeast genomes by extrachromosomal circular DNA formation. Nikolaidis, N., editor. *PLoS One* [Internet] *6*: e17872. Available from: http://dx.plos.org/10.1371/journal.pone.0017872.

Gallone, B., Steensels, J., Prahl, T., Soriaga, L., Saels, V., Herrera-Malaver, B., Merlevede, A., Roncoroni, M., Voordeckers, K., Miraglia, L., et al. (2016). Domestication and divergence of *Saccharomyces cerevisiae* beer yeasts. *Cell* [Internet] *166*: 1397–1410.e16. Available from: https://doi.org/10.1016/j.cell.2016.08.020.

Gangl, H., Batusic, M., Tscheik, G., Tiefenbrunner, W., Hack, C., & Lopandić, K. (2009). Exceptional fermentation characteristics of natural hybrids from *Saccharomyces cerevisiae* and *S. kudriavzevii*. *N. Biotechnology*. [Internet] *25*: 244–251. Available from: http://www.ncbi.nlm.nih.gov/pubmed/19026772.

Ganter, P. F., & de Barros Lopes, M. (2000). The use of anonymous DNA markers in assessing worldwide relatedness in the yeast species *Pichia kluyveri* Bedford and Kudrjavzev. *Canadian Journal of Microbiology*. [Internet] *46*: 967–980. Available from: http://www.nrcresearchpress.com/doi/abs/10.1139/w00-092.

Gayevskiy, V., & Goddard, M. R. (2012). Geographic delineations of yeast communities and populations associated with vines and wines in New Zealand. *ISME Journal* [Internet] *6*: 1281–1290. Available from: http://www.ncbi.nlm.nih.gov/pubmed/22189497.

Gayevskiy, V., Lee, S., & Goddard, M. R. (2016). European derived *Saccharomyces cerevisiae* colonisation of New Zealand vineyards aided by humans. Pretorius, I., editor. *FEMS Yeast Research.* [Internet] *16*: fow091. Available from: http://www.ncbi.nlm.nih.gov/pubmed/27744274.

Gerbaux, V., Jeudy, S., & Monamy, C. (2000). Étude des phénols volatils dans les vins de Pinot noir en Bourgogne. *Bulletin l'OIV.*

Gibbons, J. G., & Rinker, D. C. (2015). The genomics of microbial domestication in the fermented food environment. *Current Opinion in Genetics & Development.* [Internet] *35*: 1–8. Available from: http://linkinghub.elsevier.com/retrieve/pii/S0959437X15000787.

Gobbi, M., Comitini, F., Domizio, P., Romani, C., Lencioni, L., Mannazzu, I., & Ciani, M. (2013). *Lachancea thermotolerans* and *Saccharomyces cerevisiae* in simultaneous and sequential co-fermentation: A strategy to enhance acidity and improve the overall quality of wine. *Food Microbiology* [Internet] *33*: 271–281. Available from: http://www.ncbi.nlm.nih.gov/pubmed/23200661.

Goddard, M. R. (2008). Quantifying the complexities of *Saccharomyces cerevisiae*'s ecosystem engineering via fermentation. *Ecology* [Internet] *89*: 2077–2082. Available from: http://www.ncbi.nlm.nih.gov/pubmed/18724717.

González Techera, A., Jubany, S., Carrau, F. M., & Gaggero, C. (2001). Differentiation of industrial wine yeast strains using microsatellite markers. *Letters in Applied Microbiology.* [Internet] *33*: 71–75. Available from: http://www.ncbi.nlm.nih.gov/pubmed/11442819.

González, S. S., Barrio, E., Gafner, J., & Querol, A. (2006). Natural hybrids from *Saccharomyces cerevisiae*, *Saccharomyces bayanus* and *Saccharomyces kudriavzevii* in wine fermentations. *FEMS Yeast Research.* [Internet] *6*: 1221–1234. Available from: http://www.ncbi.nlm.nih.gov/pubmed/17156019.

González, S. S., Gallo, L., Climent, M. A. D., Barrio, E., & Querol, A. (2007). Enological characterization of natural hybrids from *Saccharomyces cerevisiae* and *S. kudriavzevii. International Journal of Food Microbiology* [Internet] *116*: 11–18. Available from: http://www.ncbi.nlm.nih.gov/pubmed/17346840.

González, S. S., Barrio, E., & Querol, A. (2008). Molecular characterization of new natural hybrids of *Saccharomyces cerevisiae* and *S. kudriavzevii* in brewing. *Applied and Environmental Microbiology.* [Internet] *74*: 2314–2320. Available from: http://www.ncbi.nlm.nih.gov/pubmed/18296532.

Gordon, J. L., & Wolfe, K. H. (2008). Recent allopolyploid origin of *Zygosaccharomyces rouxii* strain ATCC 42981. *Yeast* [Internet] *25*: 449–456. Available from: http://www.ncbi.nlm.nih.gov/entrez/query.fcgi?cmd=Retrieve&db=PubMed&dopt=Citation&list_uids=18509846.

Granchi, L., Ganucci, D., Viti, C., Giovannetti, L., and Vincenzini, M. (2003). Saccharomyces cerevisiae biodiversity in spontaneous commercial fermentations of grape musts with 'adequate' and 'inadequate' assimilable-nitrogen content. Letters in Applied Microbiology 36(1), 54-58. doi: doi:10.1046/j.1472-765X.2003.01263.x.

Grangeteau, C., Gerhards, D., Rousseaux, S., von Wallbrunn, C., Alexandre, H., & Guilloux-Benatier, M. (2015). Diversity of yeast strains of the genus *Hanseniaspora* in the winery environment: What is their involvement in grape must fermentation? *Food Microbiology* [Internet] 50: 70–77. Available from: http://www.ncbi.nlm.nih.gov/pubmed/25998817.

Grangeteau, C., Roullier-Gall, C., Rousseaux, S., Gougeon, R. D., Schmitt-Kopplin, P., Alexandre, H., & Guilloux-Benatier, M. (2017). Wine microbiology is driven by vineyard and winery anthropogenic factors. *Microbial Biotechnology, 10*, 354–370.

Guillamón, J. M., Barrio, E., & Querol, A. (1996). Characterization of wine yeast strains of the *Saccharomyces* genus on the basis of molecular markers: Relationships between genetic distance and geographic or ecological origin. *Systematic and Applied Microbiology, 19*, 122–132.

Guillemaud, T., Beaumont, M. A., Ciosi, M., Cornuet, J.-M., & Estoup, A. (2009). Inferring introduction routes of invasive species using approximate Bayesian computation on microsatellite data. *Heredity* (Edinb). [Internet]: 1–12. Available from: http://www.ncbi.nlm.nih.gov/pubmed/19654609.

Heinrich, A. J. (2006). Identification of genomic differences between laboratory and commercial strains of *Saccharomyces cerevisiae*. PhD Thesis, The University of Adelaide, Adelaide.

Hennequin, C., Thierry, A., Richard, G. –F., Lecointre, G., Nguyen, H. –V., Gaillardin, C., & Dujon, B. (2001). Microsatellite typing as a new tool for identification of *Saccharomyces cerevisiae* strains. *Journal of Clinical Microbiology*. [Internet] *39*: 551–559. Available from: http://www.ncbi.nlm.nih.gov/pubmed/11158105.

Howell, K. S., Bartowsky, E. J., Fleet, G. H., & Henschke, P. A. (2004). Microsatellite PCR profiling of *Saccharomyces cerevisiae* strains during wine fermentation. *Letters in Applied Microbiology*. [Internet] *38*: 315–320. Available from: http://www.blackwell-synergy.com/links/doi/10.1111%2Fj.1472-765X.2004.01486.x.

Hranilovic, A., Bely, M., Masneuf-Pomarede, I., Jiranek, V., & Albertin, W. (2017). The evolution of *Lachancea thermotolerans* is driven by geographical determination, anthropisation and flux between different ecosystems. *PLoS One* [Internet] *12*: e0184652. Available from: http://dx.plos.org/10.1371/journal.pone.0184652.

Jeffares, D. C., Rallis, C., Rieux, A., Speed, D., Prevorovsky, M., Mourier, T., Marsellach, F. X., Iqbal, Z., Lau, W., Cheng, T. M., et al. (2015). The genomic and phenotypic diversity of *Schizosaccharomyces pombe*. *Nature Genetics* [Internet] *47*: 235–241. Available from: http://www.ncbi.nlm.nih.gov/pubmed/25665008,

Jolly, N. P., Augustyn, O. P. H., & Pretorius, I. S. (2003). The occurrence of Non-Saccharomyces cerevisiae yeast species over three vintages in four vineyards and grape musts from four production regions of the Western Cape. *South African Journal of Enology and Viticulture, 24*, 35–42.

Jolly, N. P., Varela, C., & Pretorius, I. S. (2014). Not your ordinary yeast: Non-*Saccharomyces* yeasts in wine production uncovered. *FEMS Yeast Research, 14*, 215–237.

Jolly, N. P., Augustyn, O. P. R., & Pretorius, I. S. (2017). The effect of non-*Saccharomyces* yeasts on fermentation and wine quality. *South African Journal of Enology and Viticulture* [Internet] *24*: 8. Available from: http://www.journals.ac.za/index.php/sajev/article/view/2638.

Kamvar, Z. N., Tabima, J. F., & Grünwald, N. J. (2014). Poppr: An R package for genetic analysis of populations with clonal, partially clonal, and/or sexual reproduction. *PeerJ, 2*, e281.

Kellis, M., Patterson, N., Endrizzi, M., Birren, B., & Lander, E. S. (2003). Sequencing and comparison of yeast species to identify genes and regulatory elements. *Nature* [Internet] *423*: 241–254. Available from: http://www.ncbi.nlm.nih.gov/pubmed/12748633.

Knight, S., & Goddard, M. R. (2015). Quantifying separation and similarity in a *Saccharomyces cerevisiae* metapopulation. *ISME Journal* [Internet] *9*: 361–370. Available from: http://www.ncbi.nlm.nih.gov/pubmed/25062126.

Kondrashov, F. A. (2012). Gene duplication as a mechanism of genomic adaptation to a changing environment. *Proceedings of the Royal Society B: Biological Sciences*. [Internet] *279*: 5048–5057. Available from: http://www.pubmedcentral.nih.gov/articlerender.fcgi?artid=3497230&tool=pmcentrez&rendertype=abstract%5Cnhttp://rspb.royalsocietypublishing.org/content/early/2012/09/05/rspb.2012.1108.short.

Kurtzman, C. P., Robnett, C. J., Basehoar, E., & Ward, T. J. (2018). Four new species of *Metschnikowia* and the transfer of seven *Candida* species to *Metschnikowia* and *Clavispora* as new combinations. *Antonie Van Leeuwenhoek* [Internet]. Available from: http://www.ncbi.nlm.nih.gov/pubmed/29754318.

Le Jeune, C, et al. (2007). Characterization of natural hybrids of Saccharomyces cerevisiae and *Saccharomyces bayanus var. uvarum*. *FEMS Yeast Research, 7*(4):540–549. PMID:17302940.

Lederer, M. A., Nielsen, D. S., Toldam-Andersen, T. B., Herrmann, J. V., & Arneborg, N. (2013). Yeast species associated with different wine grape varieties in Denmark. *Acta Agriculturae Scandinavica Section B Soil and Plant Science* [Internet] *63*: 89–96. Available from: https://doi.org/10.1080/09064710.2012.723738.

Legras, J.-L., & Karst, F. (2003). Optimisation of interdelta analysis for *Saccharomyces cerevisiae* strain characterisation. *FEMS Microbiology Letters* [Internet] *221*: 249–255. Available from: http://linkinghub.elsevier.com/retrieve/pii/S0378109703002052.

Legras, J. –L., Ruh, O., Merdinoglu, D., & Karst, F. (2003). Selection of hypervariable microsatellite loci for the characterization of strains. *International Journal of Food Microbiology* [Internet] *102*: 73–83. Available from: http://linkinghub.elsevier.com/retrieve/pii/S0168160505000474.

Legras, J. –L., Merdinoglu, D., Cornuet, J. –M., & Karst, F. (2007). Bread, beer and wine: *Saccharomyces cerevisiae* diversity reflects human history. *Molecular Ecology* [Internet] *16*: 2091–2102. Available from: http://www.ncbi.nlm.nih.gov/pubmed/17498234.

Legras, J. –L., Erny, C., & Charpentier, C. (2014). Population structure and comparative genome hybridization of European flor yeast reveal a unique group of *Saccharomyces cerevisiae* strains with few gene duplications in their genome. *PLoS One* [Internet] *9*: e108089. Available from: http://www.ncbi.nlm.nih.gov/pubmed/25272156.

Legras, J.-L, Galeote, V., Bigey, F., Camarasa, C., Marsit, S., Nidelet, T., Sanchez, I., Couloux, A., Guy, J., Franco-Duarte, R., et al. (2018). Adaptation of S. cerevisiae to fermented food environments reveals remarkable genome plasticity and the footprints of domestication. Wittkopp, P., editor. *Molecular Biology and Evolution* [Internet] *35*: 1712–1727. Available from: https://academic.oup.com/mbe/advance-article/doi/10.1093/molbev/msy066/4962507.

Libkind, D., Hittinger, C. T., Valerio, E., Goncalves, C., Dover, J., Johnston, M., Goncalves, P., & Sampaio, J. P. (2011). Microbe domestication and the identification of the wild genetic stock of lager-brewing yeast. *Proceedings of the National Academy of Sciences*. [Internet] *108*: 14539–14544. Available from: http://www.pnas.org/cgi/doi/10.1073/pnas.1105430108.

Liti, G., Barton, D. B. H., & Louis, E. J. (2006). Sequence diversity, reproductive isolation and species concepts in *Saccharomyces*. *Genetics* [Internet] *174*: 839–850. Available from: http://www.ncbi.nlm.nih.gov/pubmed/16951060.

Liti, G., Carter, D. M., Moses, A. M., Warringer, J., Parts, L., James, S. A., Davey, R. P., Roberts, I. N., Burt, A., Koufopanou, V., et al. (2009). Population genomics of domestic and wild yeasts. *Nature* [Internet] *458*: 337–341. Available from: http://www.ncbi.nlm.nih.gov/pubmed/19212322.

Lopandic, K. (2018). Saccharomyces interspecies hybrids as model organisms for studying yeast adaptation to stressful environments. *Yeast* [Internet] *35*: 21–38. Available from: http://doi.wiley.com/10.1002/yea.3294.

Lopandić, K., Gangl, H., Wallner, E., Tscheik, G., Leitner, G., Tiefenbrunner, W., Borth, N., Breitenbach, M., Querol, A., & Prillinger, H. (2007). Genetically different wine yeasts isolated from Austrian vine-growing regions influence wine aroma differently and contain putative hybrids between *Saccharomyces cerevisiae* and *Saccharomyces kudriavzevii*. *FEMS Yeast Research* [Internet] *7*: 953–965. Available from: http://www.ncbi.nlm.nih.gov/pubmed/17484739.

Lopes, C. A., Jofre, V., & Sangorrin, M. P. (2009). Spoilage yeasts in Patagonian winemaking: Molecular and physiological features of *Pichia guilliermondii* indigenous isolates. *Revista Argentina de Microbiología* [Internet] *41*: 177–184. Available from: http://www.ncbi.nlm.nih.gov/pubmed/19831317.

Lopes, C. A., Barrio, E., & Querol, A. (2010). Natural hybrids of *S. cerevisiae*x*S. kudriavzevii* share alleles with European wild populations of *Saccharomyces kudriavzevii*. *FEMS Yeast Research* [Internet] *10*: 412–421. Available from: http://www.ncbi.nlm.nih.gov/pubmed/20337723.

López, V., Querol, A., Ramón, D., & Fernández-Espinar, M. T. (2001). A simplified procedure to analyse mitochondrial DNA from industrial yeasts. *International Journal of Food Microbiology* [Internet] *68*: 75–81. Available from: http://www.ncbi.nlm.nih.gov/pubmed/11545223.

Magwene, P. M., Kayıkçı, Ö., Granek, J. A., Reininga, J. M., Scholl, Z., & Murray, D. (2011). Outcrossing, mitotic recombination, and life-history trade-offs shape genome evolution in *Saccharomyces cerevisiae*. *Proceedings of the National Academy of Sciences of the United States of America* [Internet] *108*: 1987–1992. Available from: http://www.pnas.org/cgi/doi/10.1073/pnas.1012544108.

Magyar, I., & Toth, T. (2011). Comparative evaluation of some oenological properties in wine strains of *Candida stellata*, *Candida zemplinina*, *Saccharomyces uvarum* and *Saccharomyces cerevisiae*. *Food Microbiology* [Internet] *28*: 94–100. Available from: http://www.ncbi.nlm.nih.gov/entrez/query.fcgi?cmd=Retrieve&db=PubMed&dopt=Citation&list_uids=21056780.

Marinangeli, P., Angelozzi, D., Ciani, M., Clementi, F., & Mannazzu, I. (2004). Minisatellites in genes encoding cell wall proteins: A new way towards wine strain characterisation. *FEMS Yeast Research* [Internet] *4*: 427–435. Available from: http://linkinghub.elsevier.com/retrieve/pii/S1567135603001727.

Marsit, S., Mena, A., Bigey, F., Sauvage, F.-X., Couloux, A., Guy, J., Legras, J.-L., Barrio, E., Dequin, S., & Galeote, V. (2015). Evolutionary advantage conferred by an eukaryote-to-eukaryote gene transfer event in wine yeasts. *Molecular Biology and Evolution* [Internet] *32*: 1695–1707. Available from: http://mbe.oxfordjournals.org/cgi/doi/10.1093/molbev/msv057.

Marsit, S., Sanchez, I., Galeote, V., & Dequin, S. (2016). Horizontally acquired oligopeptide transporters favour adaptation of *Saccharomyces cerevisiae* wine yeast to oenological environment. *Environmental Microbiology, 18*, 1148–1161.

Martiniuk, J. T., Pacheco, B., Russell, G., Tong, S., Backstrom, I., & Measday, V. (2016). Impact of commercial strain use on *Saccharomyces cerevisiae* population structure and dynamics in pinot noir vineyards and spontaneous fermentations of a Canadian winery. Schacherer, J., editor. *PLoS One* [Internet] *11*: e0160259. Available from: http://dx.plos.org/10.1371/journal.pone.0160259.

Martorell, P., Barata, A., Malfeito-Ferreira, M., Fernández-Espinar, M. T., Loureiro, V., & Querol, A. (2006). Molecular typing of the yeast species *Dekkera bruxellensis* and *Pichia guilliermondii* recovered from wine related sources. *International Journal of Food Microbiology* [Internet] *106*: 79–84. Available from: http://www.sciencedirect.com/science/article/pii/S0168160505004150.

Martorell, P., Stratford, M., Steels, H., Fernández-Espinar, M. T., & Querol, A. (2007). Physiological characterization of spoilage strains of *Zygosaccharomyces bailii* and *Zygosaccharomyces rouxii* isolated from high sugar environments. *International Journal of Food Microbiology* [Internet] *114*: 234–242. Available from: http://www.sciencedirect.com/science/article/pii/S0168160506005332.

Mas, A., Guillamón, J. M., & Beltran, G. (2016). Non-conventional yeast in the wine industry. *Frontiers in Microbiology, 7*, 1494.

Masneuf, I., Hansen, J., Groth, C., Piškur, J., & Dubourdieu, D. (1998). New hybrids between *Saccharomyces sensu stricto* yeast species found among wine and cider production strains. *Applied and Environmental Microbiology* [Internet] *64*: 3887–3892. Available from: http://www.ncbi.nlm.nih.gov/pubmed/9758815.

Masneuf-Pomarède, I., Le Jeune, C., Durrens, P., Lollier, M., Aigle, M., Dubourdieu, D., & LeJeune, C. (2007). Molecular typing of wine yeast strains *Saccharomyces bayanus* var. *uvarum* using microsatellite markers. *Systematic and Applied Microbiology* [Internet] *30*: 75–82. Available from: http://www.ncbi.nlm.nih.gov/pubmed/16621402.

Masneuf-Pomarède, I., Bely, M., Marullo, P., Lonvaud-Funel, A., Dubourdieu, D., 2010. Reassessment of phenotypic traits for Saccharomyces bayanus var. uvarum wine yeast strains. Int. J. Food Microbiol. 139, 79-86.

Masneuf-Pomarede, I., Juquin, E., Miot-Sertier, C., Renault, P., Laizet, Y., Salin, F., Alexandre, H., Capozzi, V., Cocolin, L., Colonna-Ceccaldi, B., et al. (2015a). The yeast *Starmerella bacillaris* (synonym *Candida zemplinina*) shows high genetic diversity in winemaking environments. *FEMS Yeast Research, 15*, 1–11.

Masneuf-Pomarede, I., Juquin, E., Miot-Sertier, C., Renault, P., Laizet, Y., Salin, F., Alexandre, H., Capozzi, V., Cocolin, L., Colonna-Ceccaldi, B., et al. (2015b). The yeast *Starmerella bacillaris* (synonym *Candida zemplinina*) shows high genetic diversity in winemaking environments. *FEMS Yeast Research* [Internet] *15*: fov045. Available from: http://femsyr.oxfordjournals.org/femsyr/15/5/fov045.full.pdf.

Masneuf-Pomarede, I., Bely, M., Marullo, P., & Albertin, W. (2016a). The genetics of non-conventional yeasts in winemaking: Current knowledge and future challenges. *Frontiers in Microbiology, 6*, 1563.

Masneuf-Pomarede, I., Salin, F., Börlin, M., Coton, E., Coton, M., Jeune, C. Le, & Legras, J. -L. (2016b). Microsatellite analysis of *Saccharomyces uvarum* diversity. Pretorius, I., editor. *FEMS Yeast Research* [Internet] *16*: fow002. Available from: http://femsyr.oxfordjournals.org/lookup/doi/10.1093/femsyr/fow002.

Massoutier, C., Alexandre, H., Feuillat, M., & Charpentier, C. (1998). Isolation and characterization of cryotolerant *Saccharomyces* strains. *Vitis, 37*, 55–59.

McGovern, P., Jalabadze, M., Batiuk, S., Callahan, M. P., Smith, K. E., Hall, G. R., Kvavadze, E., Maghradze, D., Rusishvili, N., Bouby, L., et al. (2017). Early Neolithic wine of Georgia in the South Caucasus. *Proceedings of the National Academy of Sciences* [Internet]: 201714728. Available from: http://www.pnas.org/lookup/doi/10.1073/pnas.1714728114.

Miot-Sertier, C., & Lonvaud-Funel, A. (2007). Development of a molecular method for the typing of *Brettanomyces bruxellensis* (*Dekkera bruxellensis*) at the strain level. *Journal of Applied Microbiology* [Internet] *102*: 555–562. Available from: https://doi.org/10.1111/j.1365-2672.2006.03069.x.

Miyashita, K., Sakamoto, K., Kitagaki, H., Iwashita, K., Ito, K., & Shimoi, H. (2004). Cloning and analysis of the AWA1 gene of a nonfoaming mutant of a sake yeast. *Journal of Bioscience and Bioengineering, 97*, 14–18.

Morales, L., & Dujon, B. (2012). Evolutionary role of interspecies hybridization and genetic exchanges in yeasts. *Microbiology and Molecular Biology Reviews* [Internet] *76*: 721–739. Available from: http://mmbr.asm.org/cgi/doi/10.1128/MMBR.00022-12.

Moreira, N., Mendes, F., Guedes de Pinho, P., Hogg, T., & Vasconcelos, I. (2008). Heavy sulphur compounds, higher alcohols and esters production profile of *Hanseniaspora uvarum* and Hanseniaspora guilliermondii grown as pure and mixed cultures in grape must. *International Journal of Food Microbiology* [Internet] *124*: 231–238. Available from: http://www.sciencedirect.com/science/article/pii/S0168160508001499.

Naseeb, S., James, S.A., Alsammar, H., Michaels, C.J., Gini, B., Nueno-Palop, C., Bond, C.J., McGhie, H., Roberts, I.N., Delneri D., 2017. Saccharomyces jurei sp. nov., isolation and genetic identification of a novel yeast species from Quercus robur. Int J Syst Evol Microbiol. 67, 2046-2052.

Naumov, G. I., Masneuf, I., Naumova, E. S., Aigle, M., & Dubourdieu, D. (2000). Association of *Saccharomyces bayanus* var. *uvarum* with some French wines: Genetic analysis of yeast populations. *Research in Microbiology, 151*, 683–691.

Naumov, G. I., Naumova, E. S., Aigle, M., Masneuf, I., & Belarbi, A. (2001). Genetic reidentification of the pectinolytic yeast strain SCPP as *Saccharomyces bayanus* var. *uvarum. Applied Microbiology and Biotechnology, 55*, 108–111.

Naumov, G. I., Naumova, E. S., Martynenko, N. N., & Masneuf-Pomarède, I. (2011). Taxonomy, ecology, and genetics of the yeast *Saccharomyces bayanus*: A new object for science and practice. *Microbiology* [Internet] *80*: 735–742. Available from: http://link.springer.com/10.1134/S0026261711060154.

Naumova, E. S., Naumov, G. I., Michailova, Y. V., Martynenko, N. N., & Masneuf-Pomarède, I. (2011). Genetic diversity study of the yeast *Saccharomyces bayanus* var. *uvarum* reveals introgressed subtelomeric *Saccharomyces cerevisiae* genes. *Research in Microbiology, 162*, 204–213.

Navarro, D. P. (2012). Genome characterization of natural *Saccharomyces* hybrids of biotechnological interest.

Nemcová, K., Breierová, E., Vadkertiová, R., & Molnárová, J. (2015). The diversity of yeasts associated with grapes and musts of the Strekov winegrowing region, Slovakia. *Folia Microbiologica* (Praha) [Internet] 60: 103–109. Available from: https://doi.org/10.1007/s12223-014-0347-x.

Ness, F., Lavallée, F., Dubourdieu, D., Aigle, M., & Dulau, L. (1993). Identification of yeast strains using the polymerase chain reaction. *Journal of the Science of Food and Agriculture* [Internet] *62*: 89–94. Available from: http://doi.wiley.com/10.1002/jsfa.2740620113.

Nguyen, H. –V., Legras, J. -L., Neuvéglise, C., & Gaillardin, C. (2011). Deciphering the hybridisation history leading to the lager lineage based on the mosaic genomes of *Saccharomyces bayanus* Strains NBRC1948 and CBS380T.Rusche, L. N., editor. *PLoS One* [Internet] *6*: e25821. Available from: http://dx.plos.org/10.1371/journal.pone.0025821.

Nguyen, T. T. M., Iwaki, A., & Izawa, S. (2015). The ADH7 promoter of *Saccharomyces cerevisiae* is vanillin-inducible and enables mRNA translation under severe vanillin stress. *Frontiers in Microbiology, 6*, 1–11.

Nissen, P., Nielsen, D., & Arneborg, N. (2003). Viable *Saccharomyces cerevisiae* cells at high concentrations cause early growth arrest of non-Saccharomyces yeasts in mixed cultures by a cell–cell contact-mediated mechanism. *Yeast* [Internet] *20*: 331–341. Available from: http://www.ncbi.nlm.nih.gov/pubmed/12627399.

Novo, M., Bigey, F., Beyne, E., Galeote, V., Gavory, F., Mallet, S., Cambon, B., Legras, J. –L., Wincker, P., Casaregola, S., et al. (2009). Eukaryote-to-eukaryote gene transfer events revealed by the genome sequence of the wine yeast *Saccharomyces cerevisiae* EC1118. *Proceedings of the National Academy of Sciences*. [Internet] *106*: 16333–16338. Available from: http://www.pnas.org/cgi/doi/10.1073/pnas.0904673106.

Oelofse, A., Lonvaud-Funel, A., & du Toit, M. (2009). Molecular identification of *Brettanomyces bruxellensis* strains isolated from red wines and volatile phenol production. *Food Microbiology*. [Internet] *26*: 377–385. Available from: http://www.ncbi.nlm.nih.gov/pubmed/19376458.

Pacheco, A., Santos, J., Chaves, S., Almeida, J., Leao, C., & Sousa, M. J. (2012). The emerging role of the yeast *Torulaspora delbrueckii* in bread and wine production: Using genetic manipulation to study molecular basis of physiological responses. In A. H. A. Eissa (Ed.), *Structure and function of food engineering*. InTech.

Padilla, B., Gil, J.V., and Manzanares, P., 2016. Past and future of Non-Saccharomyces yeasts: from spoilage microorganisms to biotechnological tools for improving wine aroma complexity. Front Microbiol 7, 411. doi: 10.3389/fmicb.2016.00411.

Paradis, E., Claude, J., & Strimmer, K. (2004). APE: Analyses of phylogenetics and evolution in R language. *Bioinformatics* [Internet] *20*: 289–290. Available from: http://bioinformatics.oxfordjournals.org/content/20/2/289.abstract.

Parapouli, M., Hatziloukas, E., Drainas, C., & Perisynakis, A. (2010). The effect of Debina grapevine indigenous yeast strains of *Metschnikowia* and *Saccharomyces* on wine flavour. *Journal of Industrial Microbiology & Biotechnology*. [Internet] *37*: 85–93. Available from: https://doi.org/10.1007/s10295-009-0651-7.

Passoth, V., Blomqvist, J., & Schnürer, J. (2007). *Dekkera bruxellensis* and *Lactobacillus vini* form a stable ethanol-producing consortium in a commercial alcohol production process. *Applied and Environmental Microbiology* [Internet] *73*: 4354–4356. Available from: http://www.ncbi.nlm.nih.gov/pmc/articles/PMC1932793/.

Pavelka, N., Rancati, G., Zhu, J., Bradford, W. D., Saraf, A., Florens, L., Sanderson, B. W., Hattem, G. L., & Li, R. (2010). Aneuploidy confers quantitative proteome changes and phenotypic variation in budding yeast. *Nature* [Internet] *468*: 321–325. Available from: http://www.ncbi.nlm.nih.gov/pubmed/20962780.

Pérez, M. A., Gallego, F. J., Martínez, I., & Hidalgo, P. (2001). Detection, distribution and selection of microsatellites (SSRs) in the genome of the yeast *Saccharomyces cerevisiae* as molecular markers. *Letters in Applied Microbiology* [Internet] *33*: 461–466. Available from: http://www.ncbi.nlm.nih.gov/pubmed/11737632.

Pérez-Ortín, J.E., Querol, A., Puig, S., & Barrio, E. (2002). Molecular characterization of a chromosomal rearrangement involved in the adaptive evolution of yeast strains. *Genome Research* [Internet] *12*: 1533–1539. Available from: http://www.ncbi.nlm.nih.gov/pubmed/12368245.

Pérez-Torrado, R., González, S. S., Combina, M., Barrio, E., & Querol, A. (2015). Molecular and enological characterization of a natural Saccharomyces uvarum and *Saccharomyces cerevisiae* hybrid. *International Journal of Food Microbiology* [Internet] *204*: 101–110. Available from: http://linkinghub.elsevier.com/retrieve/pii/S0168160515001518.

Peris, D., Lopes, C. A., Arias, A., & Barrio, E. (2012a). Reconstruction of the evolutionary history of *Saccharomyces cerevisiae* × *S. kudriavzevii* hybrids based on multilocus sequence analysis. Moses, A. M., editor. *PLoS One* [Internet] *7*: e45527. Available from: http://dx.plos.org/10.1371/journal.pone.0045527.

Peris, D., Lopes, C. A., Belloch, C., Querol, A., & Barrio, E. (2012b). Comparative genomics among *Saccharomyces cerevisiae* × *Saccharomyces kudriavzevii* natural hybrid strains isolated from wine and beer reveals different origins. *BMC Genomics* [Internet] 13: 407. Available from: http://www.pubmedcentral.nih.gov/articlerender.fcgi?artid=3468397&tool=pmcentrez&rendertype=abstract.

Peris, D., Sylvester, K., Libkind, D., Gonçalves, P., Sampaio, J. P., Alexander, W. G., & Hittinger, C. T. (2014). Population structure and reticulate evolution of *Saccharomyces eubayanus* and its lager-brewing hybrids. *Molecular Ecology* [Internet] *23*: 2031–2045. Available from: http://www.ncbi.nlm.nih.gov/pubmed/24612382.

Peter, J., De Chiara, M., Friedrich, A., Yue, J. –X., Pflieger, D., Bergström, A., Sigwalt, A., Barre, B., Freel, K. C., Llored, A., et al. (2018). Genome evolution across 1,011 *Saccharomyces cerevisiae* isolates. *Nature* [Internet] *556*: 339–344. Available from: http://www.nature.com/articles/s41586-018-0030-5.

Piotrowski, J. S., Nagarajan, S., Kroll, E., Stanbery, A., Chiotti, K. E., Kruckeberg, A. L., Dunn, B., Sherlock, G., & Rosenzweig, F. (2012). Different selective pressures lead to different genomic outcomes as newly-formed hybrid yeasts evolve. *BMC Evolutionary Biology* [Internet] *12*: 46. Available from: http://www.pubmedcentral.nih.gov/articlerender.fcgi?artid=3372441&tool=pmcentrez&rendertype=abstract.

Pretorius, I. S. (2000). Tailoring wine yeast for the new millennium: Novel approaches to the ancient art of winemaking. *Yeast* [Internet] *16*: 675–729. Available from: http://www.ncbi.nlm.nih.gov/pubmed/10861899.

Pritchard, J.K., Stephens, M., Donnelly, P., 2000. Inference of Population Structure Using Multilocus Genotype Data. Genetics 155, 945-959.

Pulvirenti, A., Zambonelli, C., Todaro, A., & Giudici, P. (2002). Interspecific hybridisation by digestive tract of invertebrates as a source of environmental biodiversity within the *Saccharomyces cerevisiae*. *Annales de Microbiologie, 52*, 245–255.

Querol, A., Barrio, E., Huerta, T., & Ramón, D. (1992a). Molecular monitoring of wine fermentations conducted by active dry yeast strains. *Applied and Environmental Microbiology, 58*, 2948–2953.

Querol, A., Barrio, E., & Ramón, D. (1992b). A comparative study of different methods of yeast strain characterization. *Systematic and Applied Microbiology, 15*, 439–446.

Rantsiou, K., Dolci, P., Giacosa, S., Torchio, F., Tofalo, R., Torriani, S., Suzzi, G., Rolle, L., & Cocolin, L. (2012). *Candida zemplinina* can reduce acetic acid produced by *Saccharomyces cerevisiae* in sweet wine fermentations. *Applied and Environmental Microbiology* [Internet] *78*: 1987–1994. Available from: http://www.pubmedcentral.nih.gov/articlerender.fcgi?artid=3298173&tool=pmcentrez&rendertype=abstract.

Rantsiou, K., Englezos, V., Torchio, F., Risse, P. –A., Cravero, F., Gerbi, V., Rolle, L., & Cocolin, L. (2017). Modeling of the fermentation behavior of *Starmerella bacillaris*. *American Journal of Enology and Viticulture* [Internet]. Available from: http://www.ajevonline.org/content/ajev/early/2017/04/04/ajev.2017.16108.full.pdf.

Reis, A. L., de Fatima Rodrigues de Souza, R., Baptista Torres, R. R., Leite, F. C., Paiva, P. M., Vidal, E. E., & de Morais, Jr. M. A. (2014). Oxygen-limited cellobiose fermentation and the characterization of the cellobiase of an industrial *Dekkera/Brettanomyces bruxellensis* strain. *Springerplus* [Internet] *3*: 38. Available from: http://www.ncbi.nlm.nih.gov/pubmed/24498580.

Renault, P., & Bely, M. (2011). No Title. Bordeaux F d'Oenologie de, editor. Oeno.

Renault, P., Miot-Sertier, C., Marullo, P., Hernandez-Orte, P., Lagarrigue, L., Lonvaud-Funel, A., & Bely, M. (2009). Genetic characterization and phenotypic variability in *Torulaspora delbrueckii* species: Potential applications in the wine industry. *International Journal of Food Microbiology* [Internet] *134*: 201–210. Available from: http://www.ncbi.nlm.nih.gov/entrez/query.fcgi?cmd=Retrieve&db=PubMed&dopt=Citation&list_uids=19619911.

Renault, P., Coulon, J., de Revel, G., Barbe, J. C., & Bely, M. (2015). Increase of fruity aroma during mixed *T. delbrueckii/S. cerevisiae* wine fermentation is linked to specific esters enhancement. *International Journal of Food Microbiology* [Internet] *207*: 40–48. Available from: http://www.ncbi.nlm.nih.gov/pubmed/26001522.

Rodriguez, M. E., Perez-Traves, L., Sangorrin, M. P., Barrio, E., & Lopes, C. A. (2014). *Saccharomyces eubayanus* and *Saccharomyces uvarum* associated with the fermentation of Araucaria araucana seeds in Patagonia. *FEMS Yeast Research, 14*, 948–965.

Rodríguez, M., Pérez-Través, L., Sangorrín, M., Barrio, E., Querol, A., & Lopes, C. (2017). *Saccharomyces uvarum* is responsible for the traditional fermentation of apple CHICHA in Patagonia. *FEMS Yeast Research* [Internet] *17*: fow109. Available from: https://academic.oup.com/femsyr/article-lookup/doi/10.1093/femsyr/fow109.

Rojas, V., Gil, J. V., Pinaga, F., & Manzanares, P. (2003). Acetate ester formation in wine by mixed cultures in laboratory fermentations. *International Journal of Food Microbiology* [Internet] *86*: 181–188. Available from: http://www.ncbi.nlm.nih.gov/pubmed/12892933.

Romano, P., Fiore, C., Paraggio, M., Caruso, M., & Capece, A. (2003). Function of yeast species and strains in wine flavour. *International Journal of Food Microbiology* [Internet] *86*: 169–180. Available from: http://www.sciencedirect.com/science/article/pii/S0168160503002903.

Rosa, C. A., Lachance, M. –A., Silva, J. O. C., Teixeira, A. C. P., Marini, M. M., Antonini, Y., & Martins, R. P. (2003). Yeast communities associated with stingless bees. *FEMS Yeast Research* [Internet] *4*: 271–275. Available from: http://linkinghub.elsevier.com/retrieve/pii/S1567135603001739.

Roudil, L., Russo, P., Berbegal, C., Albertin, W., Spano, G., Capozzi, V., 2019. Non-Saccharomyces Commercial Starter Cultures: Scientific Trends, Recent Patents and Innovation in the Wine Sector. Recent Pat Food Nutr Agric. https://doi.org/10.2174/2212798410666190131103713

Sadoudi, M., Tourdot-Maréchal, R., Rousseaux, S., Steyer, D., Gallardo-Chacón, J. –J., Ballester, J., Vichi, S., Guérin-Schneider, R., Caixach, J., & Alexandre, H. (2012). Yeast-yeast interactions revealed by aromatic profile analysis of Sauvignon Blanc wine fermented by single or co-culture of non-*Saccharomyces* and *Saccharomyces* yeasts. *Food Microbiology* [Internet] *32*: 243–253. Available from: http://linkinghub.elsevier.com/retrieve/pii/S074000201200130X.

Saiki, R. K., Gelfand, D. H., Stoffel, S., Scharf, S. J., Higuchi, R., Horn, G. T., Mullis, K. B., & Erlich, H. A. (1988). Primer-directed enzymatic amplification of DNA with a thermostable DNA polymerase. *Science* [Internet] *239*: 487–491. Available from: http://www.ncbi.nlm.nih.gov/pubmed/2448875.

Saitou, N., & Nei, M. (1987). The neighbor-joining method: A new method for reconstructing phylogenetic trees. *Molecular Biology and Evolution* [Internet] *4*: 406–425. Available from: http://www.ncbi.nlm.nih.gov/pubmed/3447015.

Salvadó, Z., Arroyo-López, F. N., Barrio, E., Querol, A., & Guillamón, J. M. (2011). Quantifying the individual effects of ethanol and temperature on the fitness advantage of *Saccharomyces cerevisiae*. *Food Microbiology, 28*, 1155–1161.

Sampaio, J. P., & Gonçalves, P. (2008). Natural populations of *Saccharomyces kudriavzevii* in Portugal are associated with oak bark and are sympatric with *S. cerevisiae* and *S. paradoxus*. *Applied and Environmental Microbiology* [Internet] *74*: 2144–2152. Available from: http://www.ncbi.nlm.nih.gov/pubmed/18281431.

Scannell, D. R., Zill, O. A., Rokas, A., Payen, C., Dunham, M. J., Eisen, M. B., Rine, J., Johnston, M., & Hittinger, C. T. (2011). The awesome power of yeast evolutionary genetics: New genome sequences and strain resources for the *Saccharomyces sensu stricto* genus. *G3 (Bethesda).* [Internet] *1*: 11–25. Available from: http://www.pubmedcentral.nih.gov/articlerender.fcgi?artid=3276118&tool=pmcentrez&rendertype=abstract.

Schifferdecker, A. J., Dashko, S., Ishchuk, O. P., & Piškur, J. (2014). The wine and beer yeast *Dekkera bruxellensis*. *Yeast* [Internet] *31*: 323–332. Available from: https://doi.org/10.1002/yea.3023.

Schuller, D., & Casal, M. (2007). The genetic structure of fermentative vineyard-associated *Saccharomyces cerevisiae* populations revealed by microsatellite analysis. *Antonie Van Leeuwenhoek* [Internet] *91*: 137–150. Available from: http://www.ncbi.nlm.nih.gov/pubmed/17094015.

Schuller, D., Valero, E., Dequin, S., & Casal, M. (2004). Survey of molecular methods for the typing of wine yeast strains. *FEMS Microbiology Letters* [Internet] *231*: 19–26. Available from: http://linkinghub.elsevier.com/retrieve/pii/S0378109703009285.

Schuller, D., Cardoso, F., Sousa, S., Gomes, P., Gomes, A. C., Santos, M. A. S., & Casal, M. (2012). Genetic diversity and population structure of *Saccharomyces cerevisiae* strains isolated from different grape varieties and winemaking regions. *PLoS One* [Internet] *7*: e32507. Available from: http://www.pubmedcentral.nih.gov/articlerender.fcgi?artid=3290581&tool=pmcentrez&rendertype=abstract.

Schwartz, D. C., & Cantor, C. R. (1984). Separation of yeast chromosome-sized DNAs by pulsed-field gradient gel electrophoresis. *Cell, 37*, 67–75.

Shinohara, T., Kubodera, S., & Yanagida, F. (2000). Distribution of phenolic yeasts and production of phenolic off-flavors in wine fermentation. *Journal of Bioscience and Bioengineering* [Internet] *90*: 90–97. Available from: http://www.sciencedirect.com/science/article/pii/S1389172300800407.

Sicard, D., & Legras, J. -L. (2011). Bread, beer and wine: Yeast domestication in the *Saccharomyces sensu stricto* complex. *Comptes Rendus Biologies* [Internet] *334*: 229–236. Available from: http://linkinghub.elsevier.com/retrieve/pii/S1631069110003057.

Simonin, S., Alexandre, H., Nikolantonaki, M., Coelho, C., & Tourdot-Maréchal, R. (2018). Inoculation of *Torulaspora delbrueckii* as a bio-protection agent in winemaking. *Food Research International* [Internet] *107*: 451–461. Available from: https://doi.org/10.1016/j.foodres.2018.02.034.

Sipiczki, M. (2002). Taxonomic and physiological diversity of *Saccharomyces bayanus*. In M. Ciani (Ed.), *Biodiversity and biotechnology of wine yeasts* (pp. 53–69). Kerala: Research Signpost.

Sipiczki, M. (2003). Candida zemplinina sp. nov., an osmotolerant and psychrotolerant yeast that ferments sweet botrytized wines. *International Journal of Systematic and Evolutionary Microbiology* [Internet] *53*: 2079–2083. Available from: http://www.ncbi.nlm.nih.gov/pubmed/14657149.

Sipiczki, M. (2008). Interspecies hybridization and recombination in *Saccharomyces* wine yeasts. *FEMS Yeast Research* [Internet] *8*: 996–1007. Available from: http://www.ncbi.nlm.nih.gov/pubmed/18355270.

Sipiczki, M., Romano, P., Lipani, G., Miklos, I., & Antunovics, Z. (2001). Analysis of yeasts derived from natural fermentation in a Tokaj winery. *Antonie van Leeuwenhoek International Journal of General and Molecular Microbiology, 79*, 97–105.

Slatkin, M. (1987). Gene flow and the geographic structure of natural populations. *Science* (80). [Internet] *236*: 787–792. Available from: http://science.sciencemag.org/content/sci/236/4803/787.full.pdf.

Spadaro, D., Sabetta, W., Acquadro, A., Portis, E., Garibaldi, A., & Gullino, M. L. (2008). Use of AFLP for differentiation of *Metschnikowia pulcherrima* strains for postharvest disease biological control. *Microbiological Research* [Internet] *163*: 523–530. Available from: http://www.sciencedirect.com/science/article/pii/S0944501307000390.

Steensels, J., Daenen, L., Malcorps, P., Derdelinckx, G., Verachtert, H., & Verstrepen, K. J. (2015). *Brettanomyces* yeasts — From spoilage organisms to valuable contributors to industrial fermentations. *International Journal of Food Microbiology* [Internet] *206*: 24–38. Available from: http://www.sciencedirect.com/science/article/pii/S0168160515001865.

Stefanini, I., Dapporto, L., Berná, L., Polsinelli, M., Turillazzi, S., & Cavalieri, D. (2016). Social wasps are a *Saccharomyces* mating nest. *Proceedings of the National Academy of Sciences.* [Internet] *113*: 2247–2251. Available from: http://www.pnas.org/lookup/doi/10.1073/pnas.1516453113.

Strope, P. K., Skelly, D. A., Kozmin, S. G., Mahadevan, G., Stone, E. A., Magwene, P. M., Dietrich, F. S., & McCusker, J. H. (2015). The 100-genomes strains, an *S. cerevisiae* resource that illuminates its natural phenotypic and genotypic variation and emergence as an opportunistic pathogen. *Genome Research* [Internet] *25*: 762–774. Available from: http://genome.cshlp.org/lookup/doi/10.1101/gr.185538.114.

Su, J., Wang, T., Wang, Y., Li, Y. Y., & Li, H. (2014). The use of lactic acid-producing, malic acid-producing, or malic acid-degrading yeast strains for acidity adjustment in the wine industry.

Applied Microbiology and Biotechnology [Internet] *98*: 2395–2413. Available from: http://www.ncbi.nlm.nih.gov/pubmed/24430209.

Suárez Valles, B., Pando Bedriñana, R., González García, A., & Querol Simón, A. (2007). A molecular genetic study of natural strains of *Saccharomyces* isolated from Asturian cider fermentations. *Journal of Applied Microbiology* [Internet] *103*: 778–786. Available from: http://www.ncbi.nlm.nih.gov/pubmed/17897179.

This, P., Lacombe, T., & Thomas, M. R. (2006). Historical origins and genetic diversity of wine grapes. *Trends in Genetics, 22*, 511–519.

Tofalo, R., Chaves-López, C., Di Fabio, F., Schirone, M., Felis, G. E., Torriani, S., Paparella, A., & Suzzi, G. (2009). Molecular identification and osmotolerant profile of wine yeasts that ferment a high sugar grape must. *International Journal of Food Microbiology* [Internet] *130*: 179–187. Available from: http://www.sciencedirect.com/science/article/pii/S0168160509000476.

Tofalo, R., Schirone, M., Torriani, S., Rantsiou, K., Cocolin, L., Perpetuini, G., & Suzzi, G. (2012). Diversity of *Candida zemplinina* strains from grapes and Italian wines. *Food Microbiology* [Internet] *29*: 18–26. Available from: http://www.ncbi.nlm.nih.gov/pubmed/22029914.

Torriani, S., Zapparoli, G., & Suzzi, G. (1999). Genetic and phenotypic diversity of *Saccharomyces sensu stricto* strains isolated from Amarone wine. Diversity of *Saccharomyces* strains from Amarone wine. *Antonie Van Leeuwenhoek* [Internet] *75*: 207–215. Available from: http://www.ncbi.nlm.nih.gov/pubmed/10427409.

Tosi, E., Azzolini, M., Guzzo, F., & Zapparoli, G. (2009). Evidence of different fermentation behaviours of two indigenous strains of *Saccharomyces cerevisiae* and *Saccharomyces uvarum* isolated from Amarone wine. *Journal of Applied Microbiology, 107*, 210–218.

Treu, L., Toniolo, C., Nadai, C., Sardu, A., Giacomini, A., Corich, V., & Campanaro, S. (2014). The impact of genomic variability on gene expression in environmental *Saccharomyces cerevisiae* strains. *Environmental Microbiology* [Internet] *16*: 1378–1397. Available from: http://doi.wiley.com/10.1111/1462-2920.12327.

Urso, R., Rantsiou, K., Dolci, P., Rolle, L., Comi, G., & Cocolin, L. (2008). Yeast biodiversity and dynamics during sweet wine production as determined by molecular methods. *FEMS Yeast Research* [Internet] *8*: 1053–1062. Available from: http://www.ncbi.nlm.nih.gov/pubmed/18341578.

Valero, E., Schuller, D., Cambon, B., Casal, M., & Dequin, S. (2005). Dissemination and survival of commercial wine yeast in the vineyard: A large-scale, three-years study. *FEMS Yeast Research* [Internet] *5*: 959–969. Available from: http://www.ncbi.nlm.nih.gov/pubmed/15975857.

Varela, C., Kutyna, D. R., Solomon, M. R., Black, C. A., Borneman, A., Henschke, P. A., Pretorius, I. S., & Chambers, P. J. (2012). Evaluation of gene modification strategies for the development of low-alcohol-wine yeasts. *Applied and Environmental Microbiology* [Internet] *78*: 6068–6077. Available from: http://aem.asm.org/content/78/17/6068.abstract.

Vaughan Martini, A., & Kurtzman, C. P. (1985). Deoxyribonucleic acid relatedness among species of the genus *Saccharomyces Sensu Stricto*. *International Journal of Systematic Bacteriology* [Internet] *35*: 508–511. Available from: www.microbiologyresearch.org.

Venkatesh, A., Murray, A. L., Boyle, A. B., Quinn Farrington, L., Maher, T. J., Ó'Gaora, P., Wolfe, K. H., O'Brien, C. E., & Butler, G. (2018). Draft genome sequence of a highly heterozygous yeast strain from the <span class="named-content genus-species" id="named-content-1">*Metschnikowia pulcherrima*</span> Subclade, UCD127. *Genome Announcements* [Internet] *6*. Available from: https://mra.asm.org/content/ga/6/25/e00550-18.full.pdf.

Versavaud, A., Courcoux, P., Roulland, C., Dulau, L., & Hallet, J. N. (1995). Genetic diversity and geographical distribution of wild *Saccharomyces cerevisiae* strains from the wine-producing area of Charentes, France. *Applied and Environmental Microbiology* [Internet] *61*: 3521–3529. Available from: http://www.pubmedcentral.nih.gov/articlerender.fcgi?artid=167647&tool=pmcentrez&rendertype=abstract.

Vezinhet, F., Blondin, B., & Hallet, J. -N. (1990). Chromosomal DNA patterns and mitochondrial DNA polymorphism as tools for identification of enological strains of *Saccharomyces cerevisiae*.

Applied and Environmental Microbiology [Internet] *32*: 568–571. Available from: http://link.springer.com/10.1007/BF00173729.

Viana, F., Gil, J. V., Vallés, S., & Manzanares, P. (2009). Increasing the levels of 2-phenylethyl acetate in wine through the use of a mixed culture of *Hanseniaspora osmophila* and *Saccharomyces cerevisiae*. *International Journal of Food Microbiology* [Internet] *135*: 68–74. Available from: http://www.sciencedirect.com/science/article/pii/S0168160509003742.

Viana, F., Belloch, C., Vallés, S., & Manzanares, P. (2011). Monitoring a mixed starter of *Hanseniaspora vineae–Saccharomyces cerevisiae* in natural must: Impact on 2-phenylethyl acetate production. *International Journal of Food Microbiology* [Internet] *151*: 235–240. Available from: http://www.sciencedirect.com/science/article/pii/S016816051100523X.

Viel, A., Legras, J. –L., Nadai, C., Carlot, M., Lombardi, A., Crespan, M., et al. (2017). The geographic distribution of *Saccharomyces cerevisiae* isolates within three Italian neighboring winemaking regions reveals strong differences in yeast abundance, genetic diversity and industrial strain dissemination. *Food Microbiology* [Internet] *8*: 1–15. Available from: http://journal.frontiersin.org/article/10.3389/fmicb.2017.01595/full.

Vigentini, I., De Lorenzis, G., Fabrizio, V., Valdetara, F., Faccincani, M., Panont, C. A., Picozzi, C., Imazio, S., Failla, O., & Foschino, R. (2015). The vintage effect overcomes the terroir effect: A three year survey on the wine yeast biodiversity in Franciacorta and Oltrepò Pavese, two northern Italian vine-growing areas. *Microbiology* [Internet] *161*: 362–373. Available from: http://mic.microbiologyresearch.org/content/journal/micro/10.1099/mic.0.000004.

Wang, C., García-Fernández, D., Mas, A., & Esteve-Zarzoso, B. (2015). Fungal diversity in grape must and wine fermentation assessed by massive sequencing, quantitative PCR and DGGE. *Food Microbiology* [Internet] *6*: 1156. Available from: http://www.ncbi.nlm.nih.gov/pmc/articles/PMC4615962/.

Wedral, D., Shewfelt, R., & Frank, J. (2010). The challenge of *Brettanomyces* in wine. *Food Science and Technology* [Internet] *43*: 1474–1479. Available from: http://www.sciencedirect.com/science/article/pii/S0023643810002288.

Wenger, J. W., Schwartz, K., & Sherlock, G. (2010). Bulk segregant analysis by high-throughput sequencing reveals a novel xylose utilization gene from *Saccharomyces cerevisiae*. *PLoS Genetics* [Internet] *6*: e1000942. Available from: http://www.pubmedcentral.nih.gov/articlerender.fcgi?artid=2869308&tool=pmcentrez&rendertype=abstract.

Wrent, P., Rivas, E. M., Peinado, J. M., & de Siloniz, M. I. (2015). Development of an affordable typing method for *Meyerozyma guilliermondii* using microsatellite markers. *International Journal of Food Microbiology* [Internet] *217*: 1–6. Available from: http://www.ncbi.nlm.nih.gov/pubmed/26476570.

Zhang, H., Richards, K. D., Wilson, S., Lee, S. A., Sheehan, H., Roncoroni, M., & Gardner, R. C. (2015). Genetic characterization of strains of *Saccharomyces uvarum* from New Zealand wineries. *Food Microbiology*. [Internet] *46*: 92–99. Available from: http://linkinghub.elsevier.com/retrieve/pii/S0740002014001920.

Zimmer, A., Durand, C., Loira, N., Durrens, P., Sherman, D. J., & Marullo, P. (2014). QTL dissection of lag phase in wine fermentation reveals a new translocation responsible for *Saccharomyces cerevisiae* adaptation to sulfite. *PLoS One* [Internet] *9*: e86298. Available from: http://www.pubmedcentral.nih.gov/articlerender.fcgi?artid=3904918&tool=pmcentrez&rendertype=abstract.

Zott, K., Claisse, O., Lucas, P. M., Coulon, J., Lonvaud-Funel, A., & Masneuf-Pomarede, I. (2010). Characterization of the yeast ecosystem in grape must and wine using real-time PCR. *Food Microbiology* [Internet] *27*: 559–567. Available from: https://doi.org/10.1016/j.fm.2010.01.006.

Chapter 5
Gene Expression in Yeasts During Wine Fermentation

Linda F. Bisson

Contents

1 Introduction

Wine production is one of the world's oldest biotechnologies. The agent of the alcoholic fermentation, *Saccharomyces cerevisiae*, has been studied extensively as a model eukaryote resulting in a wealth of information on the biological activities of this organism. It was the first eukaryotic organism to have its genome sequenced, thereby enabling systematic analyses of gene expression (Goffeau et al. 1996; Oliver et al. 1998). The vast majority of these studies have been conducted under laboratory conditions using laboratory strains. However, the natural environment of *S. cerevisiae* is quite different from the typical analytical growth conditions that have been investigated.

L. F. Bisson (✉)
Department of Viticulture and Enology, University of California, Davis, CA, USA
e-mail: lfbisson@ucdavis.edu

P. Romano et al. (eds.), *Yeasts in the Production of Wine*,
https://doi.org/10.1007/978-1-4939-9782-4_5

Of the 6604 verified open reading frames (ORFs) in *S. cerevisiae,* 78% have been verified and the rest are uncharacterized or dubious (http://www.yeastgenome.org/). Approximately 12–13% of the verified ORFs are of unknown function despite extensive phenotypic screens of comprehensive mutant databases (Dwight et al. 2002; Winzeler et al. 1999). It is likely that many of these roughly 800 genes function under conditions not being replicated in the laboratory environment (Pena-Castillo and Hughes 2007). Timing of expression has been used to define the potential function of some of the verified ORFs but often this is "function-by-association" rather than a clear demonstration of biological role. The development of quantitative trait loci (QTL) analyses and particularly of expression QTL (eQTL) is enabling analysis of complex phenotypes and the broader assignment of function to verified genes (Liti and Louis 2012). It has been estimated that more than 75% of transcripts map to at least one QTL in analysis of yeast strains (Ehrenreich et al. 2009). Therefore, it is important to analyze both gene function and regulation of expression under conditions more closely mimicking the yeast natural environment.

The use of bulk sequencing technologies in transcript profiling rather than hybridization to sequences based on ORF characterization has revealed the existence of non-coding RNA species or "Stable Untranslated Transcripts" (SUTs) of unknown significance (Sardu et al. 2014). One possible role of SUTs is in the regulation of gene expression as antisense and binding to mRNA and preventing transcription or as regulatory and binding to transcription factors and modifying functionality of those proteins (Sardu et al. 2014).

There are two key considerations in interpretation of gene expression data from wine yeasts operating under winemaking conditions. First, significant genetic diversity exists among the wine strains and many of the genetic differences can lead to large differences in gene expression profiles, suggesting that genetic differences impacting regulatory systems are highly selected in innate populations of wine yeast. Direct comparisons of transcript profiles conducted using different strains will need to appreciate the underlying genetic differences that affect the transcriptional response to specific conditions.

The second important consideration relates to the composition of the yeast's natural environment of grape juice, especially its variable nature, and the presence of concurrent and sequential stressors during fermentation (Bisson et al. 2007; Pizarro et al. 2007). Ethanol accumulates during the fermentation of sugars, attaining concentrations inhibitory to growth and eventually to metabolism (Bisson and Block 2002). In the presence of other stressors, such as high proton concentrations or extremes of temperature, tolerance to ethanol is reduced (Bisson 1999). During fermentation, heat and protons are released to the environment, simultaneously raising the temperature of the medium and lowering the pH, as ethanol is steadily increasing. The combination of these stressors is thought to be the major factor limiting yeast growth and metabolic activities (Marks et al. 2008). Most yeast strains ferment sugars to 11–12% (w/v) ethanol before arrest of fermentation, and there are commercial strains that can easily yield 14–16% (w/v) ethanol (Bisson 1999). Significant changes in gene expression have been observed at ethanol concentrations as low as 2% (Marks et al. 2008). At 6% ethanol, nuclear export of bulk poly $(A)^+$ mRNA is inhibited, and stress-related transcripts are preferentially exported from the nucleus (Izawa et al. 2005). Thus, the accumulation of ethanol has a profound effect not only on gene expression, but also on translation and protein activity. Native yeast strains display significant genetic diversity that, most likely, is a natural consequence of higher rates of spontaneous mutagenesis in combination with the evolutionary catalyst of multiple and simultaneous stressors which vary in nature and severity across grape varieties and seasonal conditions. This review will summarize current information on gene expression in wine yeasts both under laboratory growth conditions and in environments resembling grape juice fermentation. Since comparisons among strains are influenced by the underlying genetic differences of those strains, the review will also discuss wine yeast strain diversity and the impact of that diversity on the interpretation of genomic expression data.

1.1 Origins and Diversity of Saccharomyces

Saccharomyces is considered to be a domesticated organism (Fay and Benavides 2005a: Legras et al. 2007; Liti et al. 2009). It is a common winery contaminant (Boulton et al. 1996). *Saccharomyces* can be found in vineyards; even in those where the practice of placing spent yeast lees in the vineyard as fertilizer has not occurred (Clemente-Jimenez et al. 2004; Martini et al. 1996; Torok et al. 1996; Valero et al. 2005). *Saccharomyces* is commonly isolated from heavily damaged grapes (Mortimer and Polsinelli 1999). Direct inoculation of vineyards with commercial yeasts did not lead to their establishment among the vineyard flora (Comitini and Ciani 2006; Valero et al. 2005), even when damaged berries were inoculated (Comitini and Ciani 2006). Thus, the yeast's natural environment can be considered to be the surface of a grape where it is a minor resident among other yeast genera, but conditions leading to the seepage of berry components enrich for *Saccharomyces*. The creation of fermentative conditions upon crushing of the fruit is more strongly enriching for this microbe. *Saccharomyces,* initially present below detectable levels in spontaneous grape juice fermentations, will often be found as the dominate species at the end of fermentation, even under aseptic grape cluster harvesting conditions (Comitini and Ciani 2006; Valero et al. 2005). These observations have led to the view that *S. cerevisiae* is a minor resident in vineyards with population amplification occurring under high-sugar, low-oxygen environments.

However, alternative views of *Saccharomyces* as a nomadic yeast with no clear single habitat or niche have also been proposed (Goddard and Greig 2015). The observation of recent genetic exchanges between winery and oak strain populations also suggest that a nomadic view may be more consistent with genetic profiling of diversity (Almeida et al. 2015; Marsit and Dequin 2015). The finding that gene transfer, cell fusion and genetic exchange can occur within the gut of insects (Stefanini et al. 2012, 2016) also offers insight into both the origins of yeast diversity and niche variability (Bisson 2016). *Saccharomyces* may be dispersed in the environment and isolatable from diverse niches due to the vectoring role of various insect species. In this case the search for a primary niche may be less important than understanding the relationship between niche and vectors.

Two primary species of *Saccharomyces* are found during the alcoholic fermentation: *S. cerevisiae* and *S. bayanus* (formerly *S. uvarum*), with *S. cerevisiae* being the more prevalent (Sipiczki 2002). Occasionally, *S. pastorianus* can be found as can hybrids of these yeasts (Naumov 1996). Sequence comparisons between *S. cerevisiae* and *S. bayanus* indicate approximately 80% identity of coding sequences and roughly 74% identity of non-coding sequences (Cliften et al. 2003). Significant genetic diversity exists among wine strains of both *S. cerevisiae* (Baleiras Couto et al. 1996; Briones et al. 1996; Gallego et al. 2005; Khan et al. 2000; Lopes et al. 2002; Sabate et al. 1998; Schuller et al. 2005; Schutz and Gafner 1993, 1994; Valero et al. 2006; Van der Westhuizen et al. 2000a, b; Versavaud et al. 1995) and *S. bayanus* (Sipiczki 2002). Analysis of over 1600 isolates of *S. cerevisiae* from 54 spontaneous fermentations demonstrated the existence of 297 unique strains (Schuller et al. 2005). In a more limited study, 13 out of 16 isolates (81%) were determined to be unique strains and the four identical strains were isolated from the same location (Baleiras Couto et al. 1996). Even higher ratios of unique genotypes have been found, 87.5 % (Valero et al. 2006), 81–91% (Gallego et al. 2005) and 91–96% (Schuller et al. 2005), depending on the technique used. The greatest numbers of genotypes in these studies are represented by a single isolate, indicating that the true extent of the diversity present in the wild is still being underestimated. Significant strain diversity exists within the same vineyard environment, suggesting the importance of localized conditions for the selection of genetically modified strains or, alternately, the existence of factors driving genetic change. One such factor may be exposure to ultraviolet light. Metabolites produced either by the plant or by other microbes in the environment, such as mycotoxins, may also serve to accelerate the appearance of genetic differences in the absence of any direct selective pressure.

Not surprisingly, the fitness of strains for specific fermentation niches also has been found to vary. In some cases, only one or a few strains dominate throughout fermentation (Versavaud et al. 1995). In contrast, other researchers have found that different strains dominate at different stages of the fermentation (Sabate et al. 1998) or that several strains of *Saccharomyces* appear to be simultaneously present in equivalently high numbers (Torija et al. 2001; Vezinhet et al. 1992). Presumably, the biodiversity of wine strains in the environment results in these different patterns of dominance in fermentations. Strains that are dominant in one environment may not show the same degree of dominance in another, because the strain attributes conferring dominance may be best suited to the fermentation conditions of a specific winery or vintage. As conditions of production change, different strains may become dominant. Assessment of strain diversity across vintages has shown that different strains are present each year (Gutierrez et al. 1999; Schuller et al. 2005).

The genetic diversity of wine yeasts has also been documented using genomic sequence comparisons and functional genomic analysis of transcript profiles (Borneman et al. 2008; Dunn et al. 2005; Fay et al. 2004; Gresham et al. 2006;

Legras et al. 2007; Liti et al. 2006, 2009; Schacherer et al. 2009; Townsend et al. 2003; Tsai et al. 2008; Winzeler et al. 2003). Strains that are undistinguishable from each other by genomic or mitochondrial DNA profiling may carry mutations leading to changes in important enological phenotypes, particularly if the genetic differences are targeted to high impact genes (such as transcription factors) or genes involved in flavor modification or production. Indeed, analyses of the presence of single nucleotide polymorphisms (SNP) suggest that they exist across populations of *Saccharomyces* with a frequency of approximately 2.8 SNPs per kilobase of DNA (Schacherer et al. 2009). Borneman et al. (2008) found in the sequence comparison of a wine strain AWRI1631 to S288c a SNP frequency of 1 per 150 base pairs or roughly 7 SNPs per kilobase. SNPs occur less frequently in genes located near the centromere and more frequently for genes located in subtelomeric regions (Schacherer et al. 2009). Deletions of genetic material also occur (Schacherer et al. 2009) but are found at a very low frequency in the essential genes. Thus, there is the potential for significant variation in gene expression profiles because of underlying genetic differences across strains, making comparisons of strains grown under different conditions challenging.

Genomic analyses have revealed that many commercial strains have acquired altered signaling properties (Verstrepen et al. 2004) and these signaling differences may be important for differential tolerance to various stressors. It is likely that increased basal levels of expression of genes involved in tolerance to stressful conditions allows more rapid adaptation to those conditions and, therefore, enhance survival (Bisson et al. 2007). However, high basal levels of expression of these genes may result in slower initial growth rates and lack of an ability to dominate fermentations (Bisson et al. 2007). There appears to be a dynamic interplay between expression of genes associated with stress tolerance and those associated with rapid growth (Bisson et al. 2007). Given the existence of multiple stressors in the environment and the feast or famine atmosphere of growth on the surface of fruits, it is not surprising that vast biodiversity of both genetic composition and gene expression profiles is observed in native *Saccharomyces* isolates.

Analysis of the population genomics of commercial, winery and vineyard isolates in comparison to non-wine isolates of *S. cerevisiae* indicates a high degree of relatedness among the wine strains regardless of the analytical methodology used (Legras et al. 2007; Liti et al. 2006, 2009; Schacherer et al. 2007, 2009). These analyses indicate that wine strains appear to have derived from a single common ancestor, and, although some geographically isolated lineages can be observed, there is a strong influence of human migration patterns on the yeast population diversity (Legras et al. 2007; Liti et al. 2009; Schacherer et al. 2009). The first wine strain appears to have emerged somewhere in Mesopotamia/Lebanon with subsequent dispersion consistent with known historical patterns of grape vine migration (Legras et al. 2007; Schacherer et al. 2009). The majority of vineyard and winery isolates appear to be homothallic diploids and the main mode of cellular reproduction appears to be clonal rather than sexual (Legras et al. 2007). Roughly 28% of the over 600 wine and vineyard isolates examined were homozygous suggesting that sporulation and self-diploidization occurs in the wild (Legras et al. 2007).

Commercial and native yeast isolates display greater genomic and genetic instability than laboratory strains (Ambrona et al. 2005), and aberrations in the number of some chromosomes are common (Bakalinsky and Snow 1990). Wild strains are generally homothallic and show low sporulation rates and poor spore viability. They also display high levels of heterozygositiy, chromosomal polymorphisms and rearrangements, and karyotype instability (Carro and Pina 2001; Codon et al. 1998; Hughes et al. 2000; Izquierdo Canas et al. 1997; Johnston et al. 2000; Landry et al. 2006a, b; Longo and Vezinhet 1993; Mortimer 2000; Myers et al. 2004; Oshiro and Winzeler 2000). The dynamic nature of the genome likely poses a distinct advantage in the environment, as evidenced by the extensive diversity observed among native isolates from the same site (Hauser et al. 2001). The biodiversity of wine strains of *Saccharomyces* is likely a consequence of both natural selection and random mutagenesis and accumulation of mutations. Wild yeasts show elevated rates of spontaneous mutagenesis which, if followed by sporulation and diploidization, can lead to the rapid creation of significant diversity across a population. The return to a homozygous state has been termed 'genome renewal' (Ambrona and Ramirez 2007; Mortimer et al. 1994). Some underlying features of gene expression in wine strains will likely be conserved across this rich biodiversity while others may show striking strain dependence. It is important to note that comparisons of gene expression in recently isolated native strains of *S. cerevisiae* versus those that have been cultivated in laboratories, demonstrates clear differences in expression profiles of wild strains and their domesticated derivatives (Kuthan et al. 2003; Palkova 2004). Strains rapidly lose some phenotypes associated with growth in the wild upon laboratory cultivation (Palkova 2004).

1.2 Fermentation Biology and the Impact of Yeast Stressors

Gene expression is strongly influenced by the chemical composition of the environment and the physical conditions of growth. A plethora of integrated and interacting signaling pathways exist in yeast and serve to coordinate biological activities and gene expression (Roosen 2004). Analysis of the gene expression of wine yeast during grape juice fermentation needs to consider multiple variables, each capable of impacting gene regulation. Grape juices are variable in composition, depending upon the varietal, time of harvest, nutritional status of the vine and seasonal influences (Amerine et al. 1980). The juice of ripe grapes is high in sugar content that is an equimolar mixture of glucose and fructose. During fermentation, the sugar is converted to ethanol and carbon dioxide, leading to a dramatic change in osmolarity, and, as ethanol increases, a dramatic decrease of the specific gravity of the environment. High osmolarity is, itself, a stressor and induces a stress-response (Gasch et al. 2000).

Yeast fermentation behavior has been difficult to model kinetically given the number of parameters involved and the varying composition of grape juice (Cramer et al. 2002). Although nitrogen is most often the limiting nutrient, the kinetics of carbon utilization does not correlate well with nitrogen levels, especially toward the end of fermentation (Insa et al. 1995; Maginot et al. 1998). Fermentation rates, likewise, do not correlate well with cell number because the fermentation capacity of individual cells can vary. The challenges in modeling fermentation kinetics likely reflect the significant role of ethanol tolerance as a key driver of metabolic behavior. Ethanol impacts numerous biological functions such as passive proton flux, membrane-protein interactions, and displaces water of hydration, thereby disrupting protein complexes and active sites. Adaptation to steadily increasing ethanol concentrations has been suggested to be the major factor impacting gene regulation and expression during fermentation (Marks et al. 2008). Energy reserves also appear to be a critical factor, with higher levels of glycogen and trehalose being associated with improved survival (Benaroudj et al. 2001; Thomsson et al. 2005).

Most fermentation of grape juice is conducted by non-growing cells. Current understanding of metabolically active but non-proliferative states in yeast is limited. In the case of wine, in addition to the presence of stressors, growth may be limited by the attainment of terminal cell density (Bisson 1999). Cells immediately resume growth with no appreciable lag if the cell number in the non-proliferative condition is reduced. Ethanol-inhibited cells likewise immediately commenced growth as soon as the ethanol content decreases (Marks et al. 2008). Cells under these conditions have not entered a classic resting state but, instead, remain primed to grow immediately upon restoration of permissive conditions.

Grape juice contains variable levels of nitrogen and phosphate and one or both substrates may be limiting, especially if non-*Saccharomyces* microbes are present in high concentration during the early stages of fermentation (Bisson 1999; Fleet et al. 2002). Micronutrients can also be limiting and exacerbated by non-*Saccharomyces* yeasts which have been shown to deplete these components from the medium (Bisson 1999).

Heat is also an end product of yeast fermentative metabolism. For every 100g of sugar consumed, the temperature increases by 1.3 °C (Boulton et al. 1996). Depending upon the type of fermentation vessel, ambient temperature or the use of refrigeration, temperature increases of 12–15°C or higher can be common during batch fermentation. Temperature stratification may develop if mixing is inadequate. The carbon dioxide produced during fermentation allows some mixing and redistribution of yeasts and nutrients throughout the fermentation vessel. Changes in membrane and cell wall composition required for adaptation to high temperature are different from those required for adaptation to ethanol, and the simultaneous presence of both stressors prevents an optimal adaptation to either (Bisson and Block 2002).

The presence of ethanol affects the tolerance of the cell to hydrogen ions (Kudo et al. 1998). Ethanol increases the passive proton flux into the yeast cells and growth inhibition occurs when the influx of protons exceeds the capacity of the plasma membrane ATPase to maintain the cytoplasmic pH (Bisson 1999). The lower the pH, the less tolerant a cell is to high ethanol concentrations, unless other counterbalancing ions are present (Kudo et al. 1998). The pH of grape juice is generally between 3.0–4.0, varying in this range depending upon the metabolic activities of yeast and the other microbes present. As the pH rises above 3.5, a multitude of bacteria that were inhibited at lower pH values can begin to grow, increasing both competition for nutrients and levels of potentially inhibitory end products (Boulton et al. 1996).

Another factor that may be limiting during grape juice fermentation is oxygen. Grape juice fermentations rapidly become anaerobic due to microbial metabolism and the activity of grape-derived polyphenol oxidase, which consumes molecular oxygen (Bisson 1999). The absence of oxygen limits metabolic options for the organism; indeed, nutrient starvation under anaerobic conditions has been shown to be fundamentally different to starvation under aerobiosis (Thomsson et al. 2005). Under aerobic conditions, yeast tolerate limitation of carbon more than of nitrogen, but under anaerobic conditions the opposite is true (Thomsson et al. 2005).

Grape juice also contains a wide array of phenolic compounds such as benzoic and cinnamic acids, flavanols and anthocyanidins, the composition and levels of which vary dramatically with the grape variety, from trace amounts to mg/L concentrations (Amerine et al. 1980). Many of these compounds have been shown to be bioactive in humans, and it is likely that they play roles in yeast biology as well. Some of these compounds are predicted to be able to enter the cell and affect redox status and be capable of interfering with protein function. Phenolic compounds may be stimulatory or inhibitory; their presence has been shown to influence yeast metabolic activities (Cantarelli 1989). The members of the multidrug resistance transporter family, one of the largest in *Saccharomyces* (Goffeau et al. 1997), may play a critical role in the export of toxic phenolic compounds or their derivatives.

Cultivation pre-history may impact the transcriptional response to changing environmental conditions. The basal levels of expression of stress response genes will affect the tolerance to specific stress conditions encountered by the yeast and impact the observations detected at the genomic level (Davidson and Schiestl 2001; Gasch 2003; Ivorra et al. 1999; Zuzuarregui and del Olmo 2004). Strains that show a low-level induction of stress genes often are more tolerant to stress than those showing a high-fold induction, presumably because the difference between basal and maximal expressed level is not as important as the basal level itself or the absolute level of the encoded protein (Siderius and Mager 2003).

Other microbes present in grape juice can impact the nature of stressors present and, therefore, the metabolic activities of *Saccharomyces* (Renouf et al. 2006). Several yeast genera: *Brettanomyces, Candida, Debaryomyces, Hanseniaspora, Kloeckera, Kluyveromyces, Metschnikowia, Pichia, Schizosaccharomyces, Torulaspora*, and *Zygosaccharomyces* have been reported to occur in grape juice (Fleet 1993; Fleet and Heard 1993; Kunkee and Bisson 1993). The levels of the different types of yeasts vary dramatically, depending upon winery practices and the use of antimicrobial agents. Lactic and acetic acid bacteria are also present, the specific genera and species of which largely depend upon grape juice pH, the temperature of fermentation and the sensitivity of the bacteria to the metabolic activities of *Saccharomyces* (Fleet 1993; Fleet and Heard 1993).

To conclude, the inherent genetic diversity of strains, variability of environmental growth parameters, the concurrent occurrence of stressors, and the interaction of these three factors greatly affects the gene expression profile of yeasts. Analysis of a wide array of yeast strains under identical conditions will yield important information on the processes of evolution and selection in this organism.

1.3 Gene Regulation and Transcript Profiling

The transcript profile for a given strain or set of growth conditions reflects the operation of multiple and interactive transcriptional control mechanisms. It can be challenging to define the mode of regulation of specific transcripts in global transcriptional analyses. Transcriptional control has been well studied in *Saccharomyces* and several excellent reviews have appeared (Balaji et al. 2006; Carlson 1997; Hanna-Rose and Hansen 1996; Herschbach and Johnson 1993; Ihmels et al. 2004; Kingston et al. 1996; Orphanides et al. 1996; Schuller 2003; Struhl 1995). The regulation of gene expression is highly hierarchical, with some signals overriding others, and comprised of both general or global regulatory factors and those specific to one or a specialized subset of genes (Schuller 2003). The mode of regulation, induction or repression, caused by a specific transcriptional factor has been shown to be context dependent and likely also a function of whether the gene has been recruited to a sub-nuclear membrane zone of transcriptional activity. Transcription factors may be regulated in multiple ways, by differential location to and within the nucleus, by binding partners and their availability, by protein modification, by competition for DNA binding domains and domain exposure, by conformational change brought about by changes in chromatin structure or position, and by the presence of small molecule effectors (Carlson 1997; Svetlov and Cooper 1995). It is important to note in the context of wine yeast that the impact of ethanol on such binding interactions is expected to be severe, as it readily displaces water of hydration and can interfere in hydrogen bond formation and lead to protein denaturation.

Prior to completion of the sequence of the *S. cerevisiae* genome, analysis of transcription was restricted to one or a handful of genes because of the technical limitations of Northern blot analysis. Availability of the complete genome sequence enabled development of strategies for the simultaneous analysis of global transcript profiles (Table 5.1). In this case, no prior knowledge of the genes expected to be regulated was necessary and, in theory, all genes impacted by a change in conditions could be identified from the change in their transcript levels.

Microarray array analysis is based on use of hybridization of mRNA-derived cDNA to reporter sequences made either from PCR products or from oligonucleotides specific for each gene or ORF in a reference genome. In the case of fragment microarray, the PCR fragments are spotted on a glass grid (DeRisi et al. 1997; Lashkari et al. 1997; Schena et al. 1995), or on a nylon membrane (Alberola et al. 2004). DNA chip microarray analysis uses oligonucleotides specific for each gene also arrayed on a matrix (Draghici et al. 2006). The oligonucleotides may be short (25–30 base pairs) or long (60–70 base pairs), and contact spotted, inkject deposited or synthesized directly on the chip.

Table 5.1 Description of transcriptome methodologies

Technique	Platform	Methodology	Issues and limitations
Nucleic acid hybridization	Fragment microarray	Hybridization to genomic PCR fragments	Cross hybridization
			Reference strain bias
			Intensity and uniformity of spotting
			Dependence upon hybridization kinetics
			Pairwise analysis
	DNAchip microarray	Hybridization to oligonucleotides	Reference strain bias
			Dependence upon hybridization kinetics
Reverse transcription	qPCR	Direct quantification of PCR products	Primer bias
			Kinetics of quantitation primer and target specific
Nucleotide sequencing	SAGE	DNA Sequencing of concatemers	Bias toward highly expressed genes
	RNA-Seq	DNA Sequencing of reverse transcribed mRNA species	Resolution dependent upon availability of target genome sequence
			Inability to use reference genome

In fragment microarray technology, mRNA is purified from reference and experimental samples, labeled with a fluorescent tag directly (biotinylated) or after conversion to cDNA (Cy3 (green) or Cy5 (red)), or during synthesis to cRNA, and the tagged mRNA/cDNA/cRNA is hybridized against the array (Lockhart and Winzeler 2000). Labeled samples may be hybridized to the array singly or after mixture of samples labeled with two different dyes. In the double dye binding method, fluorescence is scanned at two different wavelengths optimized for each dye and used to calculate the dissimilarity in ratios of expression between the reference and experimental sample. Radioactively labeled cDNA preparations have been employed as an alternative to the single dye binding methods (Rep et al. 2000; Zuzuarregui and del Olmo 2004). The advantages of this technique are its relatively inexpensive cost, ability to be utilized in individual laboratories, and relative ease of data manipulation. The disadvantages center on the spotting technology and the inability to achieve uniform spots. If spots are not uniform, scanning of the spots and assessment of the spot signal may be compromised affecting the interpretation of relative transcript levels. A comparative analysis of array platforms found that double dye-binding methodologies were not as reproducible as single signal platforms (Kuo et al. 2006). However, other studies have reached the opposite conclusion, that relative comparisons of expression are more robust than attempts to quantify individual expression patterns (Draghici et al. 2006). Sample replication is vital in these studies as the level of noise in expression is relatively high. Issues with uniformity of spotting and spotting kinetics can increase noise across array grids and it is important to use the same lot of arrays for comparisons. Also, as a consequence of potential heterogeneity of arrays, only pairwise comparisons of the two conditions are possible. Replication of the same sample generally yields acceptable levels of reproducibility (Lee et al. 2000). Greater variation is found if true biological replication is utilized, that is, different cultures of each replicate are processed independently (Quackenbush 2005). Running eight or more replicates of each experiment provides sufficient statistical robustness. Specialized arrays, with a more limited set of genes represented, have also been employed (Rodriguez-Pena et al. 2005).

In DNAchip microarray hybridization to complimentary oligonucleotides is used to identify and quantify the specific cDNA species present (Lockhart et al. 1996; Schadt et al. 2000; Wodicka et al. 1997). Transcripts can be labeled as described above and hybridized to DNA chips. Each transcript is represented by a couple to several oligomers providing independent signals for each gene. As a control for non-specific hybridization, the Affymetrix design also includes a mismatch of each oligonucleotide. The strength of the signal is then estimated both on the absolute signal values as well as on the difference between the perfect match and mismatch signals across the gene. The quality of the mRNA preparation is important in both hybridization-based methods. Rampant degradation or failure to uniformly label the mRNA can lead to misinterpretation of the signal strength and can be difficult to recognize. In Northern based analyses transcripts were separated by size via gel electrophoresis and issues with transcript stability were readily determined. DNA chips have also been used for direct hybridization analysis of DNA to compare DNA sequence diversity across strains and to identify single nucleotide polymorphisms (Gresham et al. 2006; Lashkari et al. 1997; Schacherer et al. 2009; Winzeler et al. 2003).

The limit of detection of mRNA for these technologies is on the order of 1–10 copies of mRNA per cell (Holland 2002; Draghici et al. 2006). Other studies have suggested that most mRNA species in *S. cerevisiae* are at or below this limit (Varela et al. 2005) and are therefore below the sensitivity of current array technologies (Shields 2006). A single mRNA molecule on average can produce approximately 4800 protein molecules, so it is not surprising that the majority of genes would be expressed in this range (Wohlschlegel and Yates 2003). In addition to obstacles imposed by low transcript abundance, accurate quantification of absolute expression levels can be difficult to achieve (Draghici et al. 2006). Highly expressed genes may be inaccurately quantified as signal strength may exceed the dynamic range of the scanner. Cross hybridization can also be a major impediment in transcript quantification (Draghici et al. 2006). Probe selection influences signal strength and it is to take this into account (Wang et al. 2006).

Alternate methods of transcript analysis not relying on hybridization have also been developed. Serial Analysis of Gene Expression or SAGE (Kal et al. 1999; Velculescu et al. 1995) captures and quantifies the poly A ends of the mRNA transcript. The 3' terminal ends of transcripts containing the poly A tail and some flanking sequence are captured, formed into concatemers, and the concatenated molecules sequenced. The relative level of a transcript in the population is directly correlated with the number of times a particular sequence appears in the population. SAGE analysis works well for highly expressed genes but does not accurately detect low abundance species that may be represented only once or twice in the sequenced pool.

Quantitative reverse transcriptase PCR (QRT-PCR) is an alternative technique used for global transcript analysis (Holland 2002). This technology can be scaled to be genome wide, although the major use seems to be in confirming array data generated by other means. QRT-PCR has a broader dynamic range than array technologies but is subject to other types of limitations, such as the nature of the primers used. The comparative analysis by Kuo et al. (2006) indicated good agreement among the array and QRT-PCR methodologies for highly expressed genes.

Advances in sequencing technologies has enabled use of bulk sequencing of RNA transcripts in combination with generation of a partial or nearly complete sequence of the genomes being analyzed. This RNA-seq technology offers advantages over hybridization-based methods and those relying on sequence identity to a single type strain (Powers et al. 2015; Wang et al. 2009). Changes in allele sequence may not alter functionality or level of gene expression but may display altered binding to probes and DNA fragment sequences base on a non-identical strain. Sequence-based methods enable identification of polymorphisms which will have little to no impact on quantitation in contrast to hybridization-base methods. Unbiased sequencing-based methods also allow the detection of novel genes and introgressions not commonly present but that may play an important role in strain phenotype and go undetected if transcript detection and identification are limited to a known reference genome. Transcribed but non-coding RNA species can also be detected (Sardu et al. 2014). Comparison of findings from RNA-seq-based transcriptome analyses to hybridization-based analysis of the same wine strain, EC1118, under the same growth conditions concluded that 92% of the genes displaying changes in expression were identical across both technologies (Treu et al. 2014a). However, a limitation of RNA-Seq technology is the need to have the full genome sequences of all target and reference strains for sufficient resolution.

Analysis of gene expression data can be problematic for reasons independent of the technology platform used. It is a common practice to statistically cluster genes using expression data to identify commonly regulated genes. However, it is difficult, when using these methods, to distinguish between genes that are co-regulated versus those that are co-expressed by different mechanisms (Werner 2003). This is particularly important in view of the discovery of reverse recruitment and the skewed co-expression of adjacent genes. Gene expression data alone are not sufficient to delineate the specific control mechanisms at work in the cell. The integration of transcription factor binding site analysis with array data can better define genes that are truly likely to be co-expressed, especially if this data is confirmed using array data from transcription factors knock-out mutants (Wang et al. 2006; Werner 2003).

It is important to remember that these studies evaluate the mean steady state level of mRNA and that level is influenced by both rates of transcription and rates of degradation of transcripts and governed by the kinetics of these processes that may vary in different cellular states and environments (Pelechano et al. 2010). An alteration in a level of a transcript could be due either to induction or to stabilization of the message which can often not be determined in transcript profiling. Analysis of the kinetics of gene expression suggests that elevation of protein levels requires both an increase in gene expression and a stabilization of the mRNA (Pérez-Ortín et al. 2007). Analysis of transcription rates suggests on average 1000 transcription events per minute or roughly 113,500 events per cell cycle (Pelechano et al. 2010). The mRNA content per cell at any given moment has been estimated to be around 26,000 transcripts (Pelechano et al. 2010) so significant turnover occurs each cell division cycle. Rates of turnover may differ during active cell division and cell cycling versus during non-proliferative phases found in wine production. The process of export of RNA from the nucleus is regulated and can influence transcription and mRNA persistence within the cell (Terry et al. 2007). Changes in nuclear pore complexes in daughter cells impacts transcriptional repressor accumulation such that daughter cell gene expression may vary from that of the mother cell within the same environment (Kumar et al. 2018).

Additionally, analysis of means ignores the noise of transcription and that noise may have important cellular consequences. Analysis of promoters with high transcriptional noise in the commercial wine strain EC1118 found that housekeeping genes tended to have low noise and high noise promoters tended to be associated with stress response genes (Liu et al. 2015). Noise was found to be a function of both *cis* and *trans* acting factors but not necessarily associated with variation in the mean of expression (Liu et al. 2015). A detailed study of gene expression from the MET17 promoter using a cross of the vineyard derivative RM11-1a and the common laboratory strain S288c, identified three loci responsible for noise in the expression (Ansel et al. 2008). Such noise and the dynamic variability of transcript levels across an otherwise isogenic population may play an important role in predisposing some of the population to a more rapid response to stress similar to the situation of variable induction of gene-regulating prions and be manifest as an adaptive prediction of environmental change. The presence of an early stressor in the fermentation environment of wine yeast has been shown to induce expression of genes not needed for resistance to that stress but to other stressors that frequently follow that stress in the environment (Mitchell et al. 2009). In addition, a spike in transcript levels is often initially required to respond quickly and dynamically to changes in the environment. Several gene expression analyses have seen transient increases in expression that are interpreted as decreasing in expression as transcript levels show a drop. However, this is likely not truly a decrease in expression, but the establishment of a new steady state level of expression in real time (Pérez-Ortín et al. 2007). The central role of transcription in adaptive prediction as well as the stress response itself may confound interpretation of transcript profiling data under conditions of stress. Finally, phenotype is correlated with the functionality of the encoded protein not with the transcription of the mRNA molecule. Studies have shown a good correlation of mRNA levels and protein content across strains taken under the identical stage of growth but not across stages of growth (Rossouw et al. 2010). Since protein stability is often unrelated to that of the encoding mRNA it is not surprising that under certain conditions the proteome would not reflect the concurrent mRNA population.

Assuming a mean expression level across all cells in a population may be problematic for other reasons as well. Prions, heritable states of protein structure impacting functionality, have been found in both laboratory and native yeast isolates and can impact transcription directly or indirectly (Garcia and Jarosz 2014; Halfmann et al. 2010, 2012; Holmes et al. 2013). Often these prion states are present in only a proportion of the population allowing the cells to vary in preparation for response to environmental change (Holmes et al. 2013).

If all these caveats are considered, transcript profiling can provide a wealth of information on the physiological status and history of cells and enable identification of the root cause of phenotypic differences across populations of yeast. Gene expression profiling in wine yeast has aimed to better understand cellular physiology and environmental response to enhance our knowledge of the biology of this organism but has more recently been used as an essential tool in the dissection of strain diversity and evolution of complex phenotypes. Both uses of the technology will be discussed.

2 Gene Expression Profiling: Understanding Fermentation Physiology

Analyses of gene expression patterns in wine stains of *S. cerevisiae* have mainly been conducted with the goal of gaining a better understanding of the physiology of the cells during a fermentation and under the spectrum of environmental variation seen during this process. Transcript profiling studies of wine strains of *S. cerevisiae* have largely focused on three areas: the profiling of the time course of a permissive grape juice fermentation in both synthetic media and actual juices; analysis of the impact of normally occurring stress conditions on the wine yeast transcriptome; and, as biodiversity has become more appreciated, the comparison of wine strains to laboratory strains and to each other under various growth and environmental conditions. More recently improvements to and greater ease and availability of genomic DNA sequencing has enable comparison of gene expression profiles in mixed cultures. A permissive fermentation is one in which the strain is able to complete the fermentation under the given nutritional and environmental conditions as evidenced by the complete consumption of available sugar and the concomitant production of ethanol. Although environmental and biotic stressors may be present or arise, the cell is able to adapt to those conditions and maintain metabolic activity. The impact of specific types of stress on wine yeast has also been examined, generally by imposition of the stress followed by analysis of the response to that stress. A summary of the transcript profiling analyses that have been conducted with wine yeast strains is presented in Table 5.2.

Table 5.2 Summary of transcriptome analyses performed on wine yeast strains

Yeast strain	Method of analysis	Medium	Conditions analyzed	Reference
EC1118	Microarray (Affymetrix 2.0)	Synthetic juice, chemostat	Limiting nitrogen response to varying oxygen levels	Aceituno et al. (2012)
UCD2100 (French White)	Microarray	Synthetic juice (MMM)	Differing levels of arginine	Backhus et al. (2001)
UCD522	Microarray (Affymetrix 2.0)	Grape juice	Presence of *Hanseniaspora guilliermondii*	Barbosa et al. (2015a)
Uvaferm CEG, Zimaflore VL1, QA23	Microarray	Synthetic juice (GJM)	Differing nitrogen levels, 67 and 670 mg/L	Barbosa et al. (2015b)
QA23	DNAChip (Biochip)	Muscat grape juice	Differing temperatures: 13 versus 25 °C	Beltran et al. (2006)
M28 (vineyard isolate)	Microarray	YPD	Different spore progeny	Cavalieri et al. (2000)
Enoferm M2 and four M2 hybrids	Microarray (Affymetrix 2.0)	Sauvignon blanc grape juice	Differing temperatures: 12.5 versus 25°C	Deed et al. (2015)
Vin13	DNAChip (Affymetrix YGS98)	Riesling grape juice	High sugar stress	Erasmus et al. (2003)
P5 and P24 (commercial strains)	Microarray	Synthetic juice	Effect of temperature: 4, 8, 12, 15, 22, 28, 33, 37, 40, 42, and 45°C	García-Ríos and López-Malo (2014)
T73	Microarray	YPD	Wine and lab strain comparison	Hauser et al. (2001)
IVC16 (Fermicru Primeur)	Microarray	Synthetic juice (MS300)	Differing nitrogen levels and impact of nitrogen supplementation	Jimenez-Marti et al. (2007)
EC1118	Microarray (Agilent Yeast V2)	Synthetic juice (MS300)	Presence of *Brettanomyces*	Kosel et al. (2017)
BRAIN 97 (native isolate)	Microarray (Hitachi)	GMA (glycerol-yeast extract)	Different colony morphologies	Kuthan et al. (2003)
Vin13	DNAChip (Affymetrix YGS98)	Riesling grape juice	Diammonium phosphate additions	Marks et al. (2003)
Vin13	DNAChip (Affymetrix YGS98)	Riesling grape juice	Fermentation time course	Marks et al. (2008)
PYCC4072	Macroarray	Synthetic grape juice (GJM)	Differing nitrogen conditions	Mendes-Ferreira et al. (2007a)

(continued)

Table 5.2 (continued)

Yeast strain	Method of analysis	Medium	Conditions analyzed	Reference
PYCC4072	Microarray	Synthetic grape juice (GJM)	Differing nitrogen conditions	Mendes-Ferreira et al. (2007b)
QA23	DNAChip (Biochip)	Various synthetic media conditions	Response of ADWY to different conditions of rehydration	Novo et al. (2006)
EC1118	Microarray (Affymetrix 2.0)	Synthetic juice	Oxygen impulse response under limiting nitrogen	Orellana et al. (2013)
EC1118	Microarray(Corning CMT)	Synthetic juice (MS300)	Fermentation time course	Rossignol et al. (2003)
EC1118	Microarray (Eurogenetic)	Synthetic juice (MS300)	Response to rehydration and inoculation	Rossignol et al. (2006)
EC1118	Microarray (Eurogenetic)	Synthetic juice (MS300)	Fermentation time course: comparison to proteome	Rossignol et al. (2009)
Vin13, EC1118, BM45, 285, DV10	Microarray (Affymetrix GeneChip)	Synthetic juice (MS300)	Fermentation time course: comparison to aromatic metabolites	Rossouw et al. (2008)
EC1118, four vineyard isolates (P283, P301, R008, R103), S288c	RNA-seq	Synthetic juice (MS300)	Oxidative stress	Treu et al. (2014a)
FX10	RNA-Seq	Synthetic juice	Presence of *Torulaspora delbrueckii*	Tronchoni et al. (2017)
EC1118	SAGE	Synthetic juice (MS300)	Fermentation time course	Varela et al. (2005)
IVC16, IVC17	Microarray	Synthetic juice (MS300)	Comparison of two strains and of proteome to transcriptome	Zuzuarregui et al. (2006)

2.1 Expression Profiles Under Permissive Conditions

S. cerevisiae is uniquely tailored to dominate natural grape juice fermentations. Analysis of gene expression profiles in this environment should provide a more extensive understanding of the biology of this important model organism. Several investigators have profiled yeast expression patterns using global analyses or assessment of specific genes in natural grape juices or in synthetic juice media under nutrient sufficient conditions to assess the transcriptional changes accompanying normal growth and the transition to non-proliferative fermentation (Aceituno et al. 2012; Backhus et al. 2001; Barbosa et al. 2015a; Marks et al. 2003; Marks et al. 2008; Orellana et al. 2013; Puig and Perez-Ortin 2000; Riou et al. 1997; Rossignol et al. 2003; Varela et al. 2005; Zuzuarregui et al. 2006). Even though different transcript profiling platforms were used and a range of commercial and native isolates analyzed, a common portrait of gene expression during fermentation of synthetic or actual grape juices has consistently emerged. Initial growth arrest in these studies was attributed to eventual nutrient restriction, to attainment of terminal cell density, or to the inhibitory effects of accumulated ethanol. Each of these studies demonstrates a global remodeling of ribosomal composition and translation and mRNA processing upon entry into a non-proliferative state. These responses likely signal exit from active growth and occur regardless of the cause of growth cessation. Initial adaptation to loss of the ability to proliferate is characterized by a gene expression profile which indicates that the cells remain primed to resume growth as soon as the limitation is alleviated. Metabolic activities are maintained while those associated with net growth display decreases in expressed mRNA content. If the stressors are not alleviated, the cell progresses into a highly adapted quiescent state that requires a period of incubation in permissive media for growth to resume. As fermentation continues, ethanol stress increases, activating a stress response. This response appears to be a graded response with a gradual decrease in the expression of genes involved in biosynthesis, and global changes in transport proteins. Changes in global gene expression patterns indicate that the cells undergo a gradual and continual adaptation to the disruptive effects of ethanol (Marks et al. 2008).

There is also an increased expression of genes involved in oxidative stress response. This may appear paradoxical given that these fermentations are largely anaerobic. However, acetaldehyde, an oxidizing agent, is an intermediate in ethanol production and may be responsible for the need to induce these pathways. Further, hydrogen peroxide can be produced from reactions between phenolic compounds and molecular oxygen. Thus, even in the absence of respiration, reactive oxygen species may be present. Osmotic shock has also been shown to generate reactive oxygen species, and there is a relationship between anaerobicity and osmotolerance (Krantz et al. 2004). Increases in expression of some genes known to be involved in ethanol tolerance are observed but, interestingly, expression of other genes that have been shown to lead to ethanol sensitivity when mutated, are not affected (Marks et al. 2008). This observation underscores the challenges associated with using time of expression to define the cellular function and physiological role of the genes expressed. Genes involved in glycogen, trehalose and glycerol metabolism also increase in expression, and these components have been shown to be important in survival of ethanol stress (Benaroudj et al. 2001).

Transcript profiling has revealed many features of the non-proliferative, metabolically active late fermentation stage of *S. cerevisiae*. Upon attainment of maximal cell density or an inhibitory concentration of ethanol, further growth ceases and fermentation rate is maximal (Rossignol et al. 2003). Fermentation rate then gradually decreases as ethanol continues to accumulate in the environment. It is not known if this is due to inhibition by ethanol or to an adaptation of energy generation to match the reduced energy needs of the cell. Genes associated with cell growth and amino acid biosynthesis are also increasingly down-regulated as fermentation progresses, consistent with the disruptive affects of ethanol and concomitant risks associated with attempting cell division under these conditions. The physiological changes needed to resist the disruptive effects of ethanol likely preclude continued growth (Bisson and Block 2002).

The methionine biosynthetic pathway not only continues to be expressed but is often induced under these conditions. Since this pathway is required for the synthesis of factors needed for stress tolerance, particularly S-adenosylmethionine, needed for C1 transfers and the alteration of membrane composition, and cysteine, needed for glutathione required to maintain the redox balance of the cell, it is not surprising that expression of this pathway is maintained (Backhus et al. 2001). Interestingly, the expression of genes required for sterol biosynthesis also gradually decreases, explaining the failure of late oxygen additions to enhance ethanol tolerance. There is also a notable shift in the expression of isoforms of glycolytic enzymes (Marks et al. 2008). The different isoforms may have altered functions or substrate specificity, as is the case in the change of hexokinase P2 for the more fructophilic hexokinase P1 or may reflect the need for a different subcellular localization or complex or, alternately, are simply more resistant to the denaturing effects of ethanol or the oxidative damage from acetaldehyde. Genes involved in vitamin biosynthesis show an increased level of expression, suggesting a role for these compounds in stress tolerance (Backhus et al. 2001; Rossignol et al. 2003). Genes involved in nitrogen recycling increase in expression (Backhus et al. 2001). Certain heat shock proteins are also dramatically induced. Interestingly, in nitrogen limited synthetic juice conditions, a decrease in expression of genes involved in growth was not seen (Backhus et al. 2001). This is again consistent with the observation that nitrogen limitation does not necessarily lead to a quiescent state (Granot and Synder 1993). Analysis of the transcriptional response of a laboratory strain, Σ1278b, to 21 different nitrogen compounds when present as sole nitrogen source concluded that nitrogen sources could be divided into two groups (Godard et al. 2007). One group of nitrogen sources was comprised of alanine, ammonium, arginine, asparagine, aspartate, glutamate, glutamine and serine. These nitrogen sources allowed rapid growth, exhibited varying degrees of nitrogen catabolite repression, and expression patterns suggest a higher protein synthesis rate. These compounds are catabolized to provide carbon molecules that readily enter metabolite pools and are not released as end products. The second group of nitrogen compounds consisted of isoleucine, leucine, methionine, threonine, tryptophan and tyrosine. Growth rates are reduced on these nitrogen sources and nitrogen catabolite repression is absent. The carbon compounds derived from the deamination or transamination of these compounds are released as end products into the medium (Godard et al. 2007). Grape juice contains a mixture of nitrogen sources with the most common nitrogen compounds found being in the first group (Boulton et al. 1996). As these nitrogen sources are consumed and relief from nitrogen catabolite repression occurs, the transcript profile will change consistent with what has been observed in wine strains.Analysis of the impact of nitrogen level on fermentation performance and transcriptome across three strains with differing nitrogen requirements, Uvaferm CEG, Zymaflore VL1 and QA23 showed higher transcriptional variation for all three strains at high nitrogen levels (Barbosa et al. 2015a). A positive correlation between the maximal fermentation rate and the expression of genes associated with response to stress was observed (Barbosa et al. 2015a).

Several of these studies used the same commercial strain of *S. cerevisiae*, EC1118, using different transcript analysis platforms. Transcript profiles for EC1118 during growth and fermentation in a synthetic juice medium, where entry into stationary phase was caused by nitrogen depletion of the medium, has been examined using microarray analysis (Rossignol et al. 2003). Over 2000 genes showed a significant change in expression. However, only 30% of the induced genes corresponded to genes reported to be induced in previous reports of the stationary phase response (Rossignol et al. 2003). Analysis of the response of gene expression to the presence of a variety of stressors identified 367 genes comprising the common stress response (Causton et al. 2001; Gasch et al. 2000; Gasch and Werner-Washburne, 2002). Only 58% of these genes were expressed during fermentation of the synthetic grape juice. In a study using a different commercial strain, Vin13, only 20% of the genes expressed during fermentation corresponded to the environmental stress response genes (Marks et al. 2008). These differences may be due to the simultaneous presence of ethanol stress in addition to arrest of growth. Alternately, the lack of expression of common stress response genes may indicate the differences between stresses that reduce metabolic activity as well as growth, and those only affecting proliferation.

Gene expression in EC1118 under the same general growth conditions but using SAGE analysis for transcript quantification has also been reported (Varela et al. 2005). Since this method relies on sequencing of poly A tail regions, mRNA species not represented on commercial yeast arrays were identified. The authors found expressed sequences from intragenic regions as well as messages that did not match any known sequence in the S288C genome, similar to published reports of other strains (Kumar et al. 2002). Three independent commercial preparations of EC1118 were compared to S288C and, although the presence of additional genomic DNA in EC1118 could not be completely ruled out, it appeared unlikely (Dunn et al. 2005). A descendent of EC1118, AWRI1631, has now been sequenced in its entirety (Borneman et al. 2008) and compared to the sequence of S288C. There were nearly 70,000 instances of genetic variation between the strains, the majority of which were single nucleotide polymorphisms with deletions and insertions occurring at a much lower frequency. AWRI1636 contained approximately 113 kb of unique sequence, sequence not represented in S288C, potentially encoding 37 additional proteins. There were also truncated and extended versions of conserved genes in AWRI1631 (Borneman et al. 2008). Thus, it is likely that a SAGE analysis of transcript expression in EC1118 would indeed identify unique proteins.

For the EC1118 genes that were represented in the S288C genome, the majority (88.6%) were expressed at 10 copies per cell or less (Valero et al. 2005). The relative dynamic ranges of array and SAGE technologies may explain some of the differences in transcript profiles seen. Despite this, many of the same gene families were identified in these two studies. Both groups observed growth arrest coinciding with the depletion of assimilable nitrogen, and an increase in the expression of stress response genes. Similar profiles were also observed for genes involved in carbohydrate metabolism. Many of the transcripts detected as significantly different by microarray were not detected using SAGE analysis. Rossignol et al. (2003) found a decrease in expression of genes involved in protein, nucleotide and amino acid biosynthesis whereas Varela et al. (2005) saw a decrease in transcript levels of biosynthetic pathways as fermentation progressed but were not able to detect most of the transcripts for genes involved in amino acid biosynthesis. Similarly, genes encoding putative cell wall proteins were found to be induced over time in the microarray study but were undetected by SAGE analysis. *MET30*, a key regulator in the methionine biosynthesis pathway, was found to be up regulated during late stationary phase by both the methods.

The expression profiles of some of the hexose transporters showed opposite expression changes in the two studies. The Rossignol group found an increase in expression of *HXT3* and *HXT7* over the time course of the fermentation, whereas Varela et al. (2005) found the transcript levels of *HXT3*, *HXT6* and *HXT7* to decrease over time. Expression of genes involved in the reserve carbohyrdrate biosynthetic pathways also differed in the two studies. Nonetheless, these two methods do provide similar conclusions on the physiology of yeast at different stages of fermentation.

The impact on transcription of the addition of nitrogen in the form of diammonium phosphate at the point of entry into stationary phase has been investigated. In this case, nitrogen was not limiting, and the goal was to determine if the addition of a nutrient above that which is needed would affect the gene expression profile (Marks et al. 2003). Approximately 350 genes changed in expression upon nitrogen addition, with roughly half increasing and half decreasing in expression. Many of the genes increasing in expression were associated with active growth while those that decreased were associated with use of an alternate nitrogen source and the stress response. This result is intriguing because growth was not possible due to the high level of ethanol present at the time of nitrogen addition. The addition of nitrogen appeared to allow the cells to re-enter and maintain a state primed for proliferation. There is a relationship between ethanol tolerance and the nitrogen requirement. Tolerance to higher levels of ethanol requires the presence of higher nitrogen levels (Cramer et al. 2002).

From these studies, several candidate genes have been proposed as markers for normal fermentation progression. Heat shock genes *HSP12*, *HSP26*, *HSP30* and *HSP82* show specific increases in expression at specific times during the fermentation as ethanol increases. *HSP12* and *HSP26* are expressed late in a normal fermentation that is accumulating ethanol (Backhus et al. 2001). Strains with higher basal and induced levels of *HSP12* were found to resist stress more effectively (Ivorra et al. 1999). *HSP12* and *HSP26* expression increased during low temperature stress, in contrast to many other heat shock genes, and may therefore represent generic markers of cellular stress response (Sahara et al. 2002). *HSP30* appears to be expressed to a greater extent in nitrogen limited than in nitrogen sufficient fermentations (Backhus et al. 2001). It may be challenging, however, to develop absolute measures of expression of these genes that would indicate either normal or aberrant stress response was occurring due to differences in basal levels of expression.

Research investigating the relationship between specific gene expression profiles and metabolites important to wine aroma and flavor has also been undertaken (Chidi et al. 2015; Kosel et al. 2017; Rossouw et al. 2008; Schoondermark-Stolk et al. 2006). Five wine strains with different aroma profiles were investigated using a synthetic juice medium (Rossouw et al. 2008). This analysis lead to the identification of five genes, YMR210W, *BAT1*, *AAD10*, *AAD14* and *ACS1* that appeared to play major roles in aroma profiles. When overexpressed in a wine strain, all but one of these genes (YMR210W) impacted the aroma profile as predicted. It was difficult to understand, given what is known about the metabolic functions of their gene products, how these genes were having the impact on volatile compounds that was observed. In addition, genes expected to play a major role in aroma production consistent with the variation seen across the five strains did not show significant differences in expression. Clearly, the integration of transcriptome and metabolomic data is not straightforward, yet this type of research is vital in understanding the true impact of changes in transcript profile. These authors also investigated the relationship between transcript profiles and acid production in this same study (Chidi et al. 2015) and found that trends in acid production were the same across all five strains but that the amount of acids produced varied.

A second study (Schoondermark-Stolk et al. 2006) compared the transcript profile and flavor metabolite production for a laboratory strain grown with either leucine or ammonia as nitrogen source. Volatile components, 3-methyl-1-butanol, 4-methyl-2-oxopentanoate, 3-methyl-2-oxobutanoate and 3-methylbutyrate, were detected when leucine served as sole nitrogen source. A total of 930 genes showed differences in expression between these two conditions showing that nitrogen source exerts a strong influence on transcript profiles. The data obtained were similar to those of Godard et al. (2007) with respect to the major changes seen.

A comparison of the transcriptome to the proteome during a fermentation time course has also been undertaken (Rossignol et al. 2009). Transcript and protein profiles both show dramatic changes, but in general there is only a modest correlation at best between protein and transcript abundance particularly during non-proliferative stages. This suggests that post-transcriptional regulation may be more important in dictating protein levels during stationary phases of growth. A subsequent study (Rossouw et al. 2010) compared the proteome of five industrial yeast strains at three time points during fermentation to the transcriptome and found in contrast that the transcriptome did predict the proteome across the strains at the identical time points but was less well predictive across the time points for each individual strain. These observations suggest that the greater variability seen across the fermentation time course may be due to turnover or half-life differences for the two classes of macromolecules which leads to a temporal disconnect between timing of expression of mRNA and presence of concomitant protein (Rossouw et al. 2010). These authors also used a more accurate method of protein quantitation, based on protein analysis via LC/MS than via 2D SDS PAGE which may also have resulted in better reproducibility across strains.

2.2 The Impact of Biotic and Environmental Stressors on Gene Expression

One of the major areas of research interest in wine strains of *S. cerevisiae* is the analysis of response to stress. The principal stressors encountered by yeast under these conditions are high osmolarity, high ethanol, extremes of temperature, nutrient limitation and presence of inhibitory metabolites as mentioned already (Bisson 1999). Genomic analysis of the response to each of these types of stress has been conducted in laboratory and in wine strains (Alexandre et al. 2001; Aranda and del Olmo 2004; Backhus et al. 2001; Beltran et al. 2006; Celton et al. 2012; Deed et al. 2015; Erasmus et al. 2003; García-Ríos and López-Malo 2014; Jimenez-Marti and del Olmo 2008; Kuhn et al. 2001; Marks et al. 2003; Marks et al. 2008; Mendes-Ferreira et al. 2007a, b, 2010; Novo et al. 2006; Rep et al. 2000; Rossignol et al. 2006; Sahara et al. 2002; Treu et al. 2014a). Several excellent reviews on the yeast stress response have appeared (Gasch 2003; Gasch and Werner-Washburne 2002; Gray et al. 2004; Siderius and Mager 2003).

Existing environmental and growth conditions can profoundly influence the stress response (Siderius and Mager 2003). Plasma membrane composition at the time of stress can impact detection of stress and signal transduction, and the availability of nutrients can be important for the synthesis of stress response factors. Even under permissive growth conditions, the stress response on rich (YPD) varies from that on minimal (YNB) media (Siderius and Mager 2003). Environmental conditions may also mitigate a stress response. Cells were more heat tolerant under anaerobic conditions than under aerobiosis (Davidson and Schiestl 2001). This observation suggests that a primary consequence of heat exposure is the release of reactive oxygen species that accompanies disruption of metabolically active mitochondrial membranes.

Analysis of gene expression during arrest of growth has led to the identification of a set of genes expressed in stationary phase. Most of these genes are regulated by the DNA stress response element (STRE) (Kobayashi and McEntee 1990, 1993; Marchler et al. 1993). Many, but not all, of the genes with STRE elements in their promoter regions are expressed in the later sages of fermentation (Puig and Perez-Ortin 2000). There are clear physiological correlations between arrest of growth due to the sudden imposition of stress and the arrest of growth that accompanies normal batch cultivation and depletion of restricting nutrients. A large group of genes appears to show increases in expression, regardless of the type of factor causing growth arrest, and these have been termed the "environmental stress response" (ESR) (Gasch et al. 2000). Most of the genes showing a decrease in gene expression upon growth arrest (70%) are involved in protein synthesis (Gasch 2003). A much smaller set of roughly 300 genes showed an increase in transcript level (Gasch 2003). In addition, specific patterns of expression unique to each type of stress were observed. Thus, a core set of functions is altered to accommodate arrest of growth, and others are specifically required for a narrower set of biotic and abiotic parameters. Studies of stress responses in other laboratory and wine strains have largely found identical results (Causton et al. 2001; Gasch 2003; Gasch and Werner-Washburne 2002). In general, the imposition of stress results in a transient adaptive phase to the new growth conditions or results in transit of the yeast from growth to a non-proliferative state. When the cells are returned to permissive conditions, the transcript profile returns to the non-stressed state (Gasch and Werner-Washburne 2002). The primary goal of the yeast stress response is to allow adaptation to the new growth conditions to maintain optimal cellular performance or, failing that, to equip the cell for entry into a resting stationary phase. Many of these genes are involved in repair of cellular damage, protein re-folding and denaturation of unsalvageable protein and cellular structures, suggesting that damage control and mitigation is an important response to stress. Other genes that display increased expression are involved in the arrest of *de novo* protein synthesis and redirection of gene expression and the translational apparatus to those proteins required to maintain cell viability. The initial aim of the stress response is to repair damage, restabilize cellular structures, membranes and metabolite gradients and pools, maintain protein function, and to acquire tolerance of the new condition to recommence growth and/or metabolism. Alternately, if the stress is severe, the aim is to attain a resting or hibernating state that protects cellular functionality and viability at the expense of growth. It is not surprising, therefore, that a common group of genes would be expressed regardless of the nature of the stress.

Conditions that lead to enhanced tolerance of stress tend to reduce growth capacity and exit from a non-growing state often requires a lag during which cells re-adapt to a proliferative state. Analysis of a specific mutation that resulted in death in stationary phase demonstrated that there is a distinct difference between arrest of growth and entry into the classically described stationary phase (Drebot et al. 1990).

Changes in gene expression accompanying osmotic stress and accumulation of ethanol have been examined in wine and laboratory strains. Osmotic stress leads to increased expression of the glycolytic and pentose phosphate pathways and a decreased expression of genes involved in biosynthesis (Erasmus et al. 2003; Rep et al. 2000; Zuzuarregui et al. 2005). Osmotolerance is affected by mutations in the genes required for adenine biosynthesis (Ando et al. 2005). However, expression of these genes decreases. Several studies have seen a similar lack of correlation between changes in gene expression and the role of the gene in stress tolerance. A gene required for tolerance of a stressor does not necessarily show regulation by that stressor.

Investigation of the impact of change in growth environments among wine strains on global transcript patterns has also been analyzed. Novo et al. (2006) and Rossignol et al. (2006) explored changes in the transcript profile upon rehydration of commercial strains and after inoculation into grape juice like media. The changes observed were all predictable from studies of laboratory strain adaptations to these two types of growth environments. The rehydrated strain displays transcript profiles consistent with limitation for nitrogen and carbon under aerobic conditions (the conditions of commercial strain preparation) with a shift to a fermentative mode of metabolism upon introduction into grape juice.

The response to increasing ethanol is complex in wine strains because ethanol rarely increases in the absence of other stress factors, such as nutrient limitation, and is accompanied by acetaldehyde production, itself a stress factor (Aranda and del Olmo 2004; Marks et al. 2008). Two types of ethanol tolerance assays have been performed: those in which ethanol accumulates because of metabolism and those in which a high concentration of ethanol is abruptly added to the culture. Short-term ethanol stress has been evaluated in wine yeast (Alexandre et al. 2001). Approximately 3.1% of the transcripts analyzed increased in expression in response to exposure to ethanol. Of these genes, 49.4% were ESR genes and an additional 14.4% of the genes have other known roles in response to stress. Genes involved in energy production, protein localization and ion homeostasis also increased in expression. Genes decreasing in expression were associated with growth and biosynthesis.

To assess the role of increasing ethanol content in a native environment, gene expression during Riesling juice fermentation has been examined (Marks et al. 2008). Physiological analyses indicated that the accumulation of ethanol during fermentation was responsible for limiting cell growth and for the up-regulation of several classes of genes. Of the 2550 genes showing a demonstrable change in expression, 44% showed some level of sustained up-regulation, 6% were transient, and 50% showed decreased expression. No change in expression was observed for 1876 genes. Sixty-two percent of the genes showing an induction of between 4 and 80-fold are not known to be involved in any stress response. This set of genes was termed the "Fermentation Stress Response" (FSR) (Marks et al. 2008). Only 20% of the FSR genes overlap with the other common stress response gene families. An additional 18% have been shown to be induced in response to at least one other type of stress, but most genes increasing in expression have not been previously associated with any stress response.

In general, the changes in gene expression profiles observed in wine strains in response to stress are similar to those observed in laboratory strains, with some differences in the level of change observed in expression. It is important to remember that strains with high basal levels of expression of some stress genes may show a stronger stress tolerance than strains with lower basal levels yet display less of an inductive response. The correlation between stress tolerance and gene expression may, therefore, be indirect in many cases because the absolute level of the protein, not the route to attain that level, is the more critical factor. Messenger and protein turnover rates and efficiencies of translation may also differ between environmental conditions, and changes in expression may occur simply to counter these effects (Kuhn et al. 2001; Sahara et al. 2002; Stahl et al. 2004).

The impact of fermentation temperature on expression profiles has also been explored (Beltran et al. 2006; Deed et al. 2015). Both studies compared low but still permissive of growth and fermentation temperatures of 13 and 12.5°C to the fully permissive 25°C. The latter study included a comparison of a parental strain and four hybrids derived from crosses of that parental strain creating a diversity in tolerance of low temperature. In general, the low temperature fermentations showed a period of adaptation prior to entry into a growth phase, with expression of genes associated with growth trailing and then paralleling that of the 25°C fermentation (Beltran et al. 2006). A subset of the stress response genes was induced at the lower temperature. However, it was not clear if the differences seen were due specifically to temperature or to differences in growth rates at the two temperatures. The lower temperature culture showed enhanced ethanol tolerance which may be due to the early expression of stress response genes or could be due to the changes observed in cell wall lipid composition. In the second study, growth at low temperature reduced the magnitude of the transcriptional transition between exponential and stationary phases of growth (Deed et al. 2015). More changes in gene expression were observed in stationary phase for the samples at different temperatures. Genes enriched in expression at low temperature were associated with nitrogen, sulfur or iron/copper metabolism or with the stress response (Deed et al. 2015). The F1 hybrid strains varied in gene expression at the low temperature with respect to parental strain M2 demonstrating that genetic background influences the expression profile as would be expected. Consistent with these observations, García-Ríos et al. (2014) also comparatively evaluated the impact of low temperature adaptation in two commercial strains and found upregulation of genes involved in sulfur assimilation and glutathione biosynthesis which was correlated with increased protein levels.

Several studies have focused on the impact of nitrogen limitation and addition on transcript expression profiles singly (Backhus et al. 2001; Marks et al. 2003; Jimenez-Marti and del Olmo 2008; Jimenez-Marti et al. 2007; Mendes-Ferreira et al. 2007a, b) or in combination with oxygen levels (Aceituno et al. 2012; Orellana et al. 2013; Treu et al. 2014a). Although different strains and growth conditions were used, some remarkable similarities in response to nitrogen limitation were observed. Cultures grown on ammonia as principal nitrogen source display elevated expression of enzymes involved in amino acid biosynthesis as compared to fermentations on a mixture of amino acids (Jimenez-Marti and del Olmo 2008). In this study, nitrogen catabolite repression appeared stronger with the mixture of amino acids than with ammonia alone. Arginase activity appears to be a good indicator of the status of nitrogen metabolism during fermentation (Jimenez-Marti et al. 2007). Arginase activity increases as ammonia and other preferred nitrogen sources are consumed but is reduced upon supplementation with other nitrogen sources; thus, it would serve as a good indicator of the metabolism of the cells if a baseline of activity is known. Expression levels of the *ACA1* gene also seemed well correlated with nitrogen metabolism (Jimenez-Marti et al. 2007). These authors also observed effects of the nitrogen source and supplementation on the aroma profile of the cultures. Another study also used transcriptome analysis to identify potential markers of nitrogen deficiency (Mendes-Ferreira et al. 2007a, b). Thirty-six genes were identified that seemed to be responsive to the nitrogen status of the medium. However, most of these genes were also known to be regulated in response to other factors such as carbon source limitation or stress and would not be specific to nitrogen limitation. Nitrogen limitation in general seems to lead to increased expression of genes involved in oxidative metabolism and in ribosome remodeling (Backhus et al. 2001; Mendes-Ferreira et al. 2007b) regardless of whether the high nitrogen condition was due to arginine (Backhus et al. 2001) or to ammonia (Mendes-Ferreira et al. 2007b).

Finally, transcriptome analysis of induced stress of increased NADPH oxidation in a wine strain, 59A, has also been evaluated (Celton et al. 2012). The expression of genes involved in meeting the NADPH demand increased as would be predicted. Low concentrations of the stress inducer showed only modest changes in gene expression indicating the cells can buffer low levels of stress without dramatic changes to the transcriptome.

2.3 Impact of Co-Cultivation on Gene Expression During Fermentation

Wine fermentations are never pure culture fermentations. *S. cerevisiae* dominates the environment using a suite of strategies (Ramakrishnan et al. 2016). Gene expression analyses are typically done in pure culture, at least pure culture of *S. cerevisiae*. The presence of native yeast and bacteria early and during the fermentation would be expected to impact gene expression and other adaptive responses. The effect of the presence of yeast residents of grape berry surfaces on the early transcript profile of *Saccharomyces* has been evaluated (Barbosa et al. 2015b; Tronchoni et al. 2017). Co-cultivation with the common grape resident *Hanseniaspora guilliermonii* showed an overall lowered fermentative activity of *S. cerevisiae* compared to strains cultured in isolation with an overall decrease in gene expression of the major families induced early in fermentation (Barbosa et al. 2015b). Genes involved in vitamin biosynthesis were notably upregulated consistent with increased nutrient competition, while genes involved in biosynthesis of amino acids were elevated in the control, consistent with enhanced fermentative activity.

Early gene expression profile has also been evaluated in co-cultivation of S. cerevisiae with another common grape resident, *Torulasopra delbrueckii* (Tronchoni et al. 2017). An increased in gene expression to stimulate metabolic activity was seen in *S. cerevisiae*, particularly of enzymes involved in central metabolism, suggesting an elevated dominance response. Increased production of antimicrobial peptides by S. cerevisiae was also observed. Expression of *HSP2*, a heat shock protein associated with intercellular communication, was increased in both strains (Tronchoni et al. 2017). Hsp2 serves as a stress signal and the cultures could be signaling stress or attempting to send a false signal to the competing yeast present in the environment

The impact of *Brettanomyces* metabolites on the transcriptome of wine strain EC1118 in synthetic media has been assessed (Kosel et al. 2017). The two yeasts were grown in adjacent chambers separated by a membrane. Under these conditions 77 transcripts were increased in *S. cerevisiae*, mostly involved in thiamine biosynthesis, and amino acid and polyamine transport suggesting the presence of the other yeast stimulated medium depletion of nutrients. The aromatic profiles of the two strains in co-cultivation were altered from those of each strain in pure culture presumably either due to cross-consumption or adsorption of these volatile compounds as there was no impact on expression of biosynthetic pathways. These studies demonstrate an impact of the presence of ecosystem members on the expression profile of wine strains of *S. cerevisiae* but the differences in expression are not identical.

2.4 Influence of Strain Background on Gene Expression

Another critical factor in the analysis and interpretation of gene expression data across wine strains is understanding the role of the biodiversity in generating the specific profiles obtained. Several studies have investigated genomic diversity across strains of *S. cerevisiae* from different niches including wine lineages (Bergström et al. 2014; Bornman et al. 2011; Dunn et al. 2012; Hittinger 2013; Salinas et al. 2010; Schacherer et al. 2009; Schaefke et al. 2013; Skelly and Magwene 2016). These analyses identify a breadth of genetic changes of the genome that underpin the rich diversity of phenotype observed across this species. Some of the studies relied on use of tiling on microarrays and others are based on comparisons of DNA sequencing. Polymorphisms include alterations in gene copy number, insertions and deletions (indels), introgressions (insertions of regions of non-self DNA), horizontal gene transfer, single nucleotide polymorphisms (SNPs), aneuploidy (alteration in chromosome number), chromosome structural variation, and variation in location and number of transposable elements (Table 5.3). However, although genome variation is prevalent it is unclear which types of changes most impact phenotypic expression and are drivers of evolution. Evolution of genotype can be surprisingly reproducible with selection or adaptive evolution leading to non-identical mutative changes in the same suite of genes (Hittinger 2013). On the other hand, some genetic changes may enhance fitness in one niche but be detrimental in others so may persist in some populations and be absent in others as a consequence of selective forces (Hittinger 2013).

The most prevalent changes impacting strain phenotype appear to be associated with copy number variation (Dunn et al. 2012) or SNPs leading to *cis* or *trans* changes in gene expression (Fraser et al. 2010; Rossouw et al. 2012; Schaefke et al. 2013). Thus, changes in the regulation of gene expression is a major factor enabling the modification and evolution of novel phenotypes. Consistent with this view, analysis of genomic variability across vineyard strains displaying multiple differing phenotypes concluded that the differences in phenotype were largely due to differences in regulation or protein structure of transcription factors (Treu et al. 2014b). Changes in promoter sequences were also identified. Kvitek et al. (2008) evaluated sensitivity to 14 environmental conditions of stress and gene expression profiles across a phenotypically diverse set of 18 strains and identified both gene copy

Table 5.3 Mechanisms for generation of genomic diversity

Mechanism	Description
Single nucleotide polymorphism (SNP)	Change of individual nucleotide pairs in coding or non-coding regions
Insertions/Deletions (indel)	Small insertions or deletions of sequence in coding or non-coding DNA
Copy number variation	Duplication of coding regions with or without accompanying promoter region; typically found in subtelomeric regions
Transposable Elements	Movement of mobile elements around the genome can lead to changes in coding sequence, loss of coding sequence, change of regulation of coding sequence, affect local chromatin structure
Chromosomal rearrangement	Translocation of chromosomal segments to non-homologous chromosomes
Changes in Ploidy Whole genome	Increase in ploidy from 2N to 3N or 4N
Aneupoidy	Loss or gain of a chromosome leading to the situation of 2N (1N) + 1 or 2N(1N) -1
Lateral/Horizontal Gene Transfer	Transfer of coding information across species, genus and kingdom barriers
Introgression	Transfer of large segments of DNA, coding and non-coding between species or genus
Cross-Species Hybridization	Illicit mating and cell fusion events between members of different species of the same genus; may be resolved by chromosome loss and extensive rearrangement leading to population isolation; alternately may be resolved with extensive loss of one parental genome and appear as introgression

number variation and differences in gene regulation were associated with stress resistance phenotypes.

The influence of strain genetic background on transcript profiles has been compared in domesticated and wild populations of *S. cerevisiae* (Fay and Benavides 2005a, b). The expression profile of nine of these strains grown in the presence of copper sulfate was also investigated (Fay et al. 2004). Copper is commonly used in winemaking to remove sulfides that have formed because of yeast metabolic activity. Over 600 genes showed variation in expression among strains, with only a small subset varying in response to copper addition (Fay et al. 2004). A direct comparison between laboratory and wine yeast identified more than 40 genes that had different expression patterns under the same conditions (Hauser et al. 2001). Upon careful analysis, these differences were attributable to gene copy number and small variations in promoter regions. Comparison of four wine strains with laboratory strain, S288C, in a microarray karyotype analysis found that the four commercial strains displayed common as well as unique differences in expression patterns as compared to S288C (Dunn et al. 2005). There were small but significant differences detected among the wine strains studied. These differences were enough to distinguish the strains from one another, based on a microarray karyotype "signature". Genomic diversity, therefore, leads to changes in expression patterns of gene families, and some stress responses may be restricted by genotype.

Similar expression profiles in standard laboratory media were observed in a comparison of expression levels for genes of Chromosome III between S288C and the commercial strain V5, but, in synthetic grape juice. V5 showed altered expression of several genes, particularly the *PAU* stress response genes, not altered in expression in S288C (Rachidi et al. 2000). Sequence homology between laboratory and wine strains appears to be extensive (Hauser et al. 2001), but the differences observed seem to have profound effects on gene expression. An additional study focused on a comparison of flor strains of *S. cerevisiae*. These strains have adapted to a distinct environment and form a film or "flor" at the air interface of wine during sherry production (Infante et al. 2003). Multiple differences in gene copy number were found, affecting approximately 38% of the genome.

Transcript profiles have been examined in strains displaying differences in fermentative ability (Zuzuarregui and del Olmo 2004). In this study, strains were not nutrient limited, and entered stationary phase presumably because maximal cell density was attained, or ethanol accumulated to inhibitory levels. Both higher and maintained levels of mRNA were found in the strains with more severe fermentation problems as compared to strains that were able to completely consume available sugar, suggesting that these strains were not adapting to the changing environmental conditions. Strains that consumed nitrogen more quickly or more slowly had reduced expression of stress genes. This suggests that specific patterns of gene expression are associated with the ability to utilize nutrients efficiently and to completely consume available sugar. Two strains were not able to adapt to the high osmolarity of the synthetic grape juice media. Another observation of this study was that the appearance of aneuploidy or polyploidy may lead to altered basal or induced levels of gene expression that prevent certain adaptive responses from occurring (Zuzuarregui and del Olmo 2004).

Gene expression profiling of strains displaying differing nitrogen requirements has also been investigated (Brice et al. 2014a). Nitrogen requirements for optimal fermentation were characterized across a setoff 23 strains.

These studies demonstrate significant diversity in the changes in transcript levels as a function of strain genetic background. Analysis of genetic variation in a native vineyard strain of *S. cerevisiae* found just as many differences among spores arising from a single isolate as from different vineyard isolates (Cavalieri et al. 2000). The strain was found to contain multiple heterozygosities, and over 6% of the genome showed a significant change in expression pattern across the different spore types obtained. The major differences occurred in genes encoding proteins involved in protein degradation and amino acid and sulfur metabolism (Townsend et al. 2003).

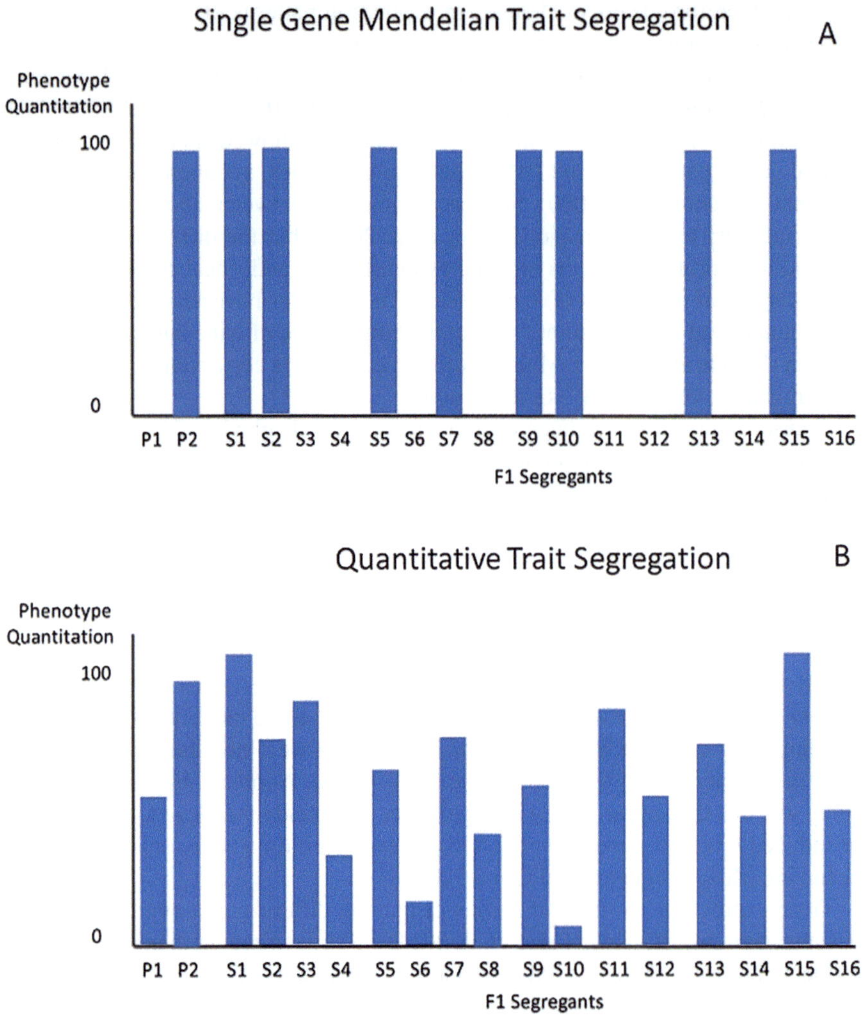

Fig. 5.1 Single gene Mendelian Segregation versus expression of quantitative traits under multi-gene control

3 Gene Expression Profiling: Defining Complex Phenotypes

Some phenotypes are defined by a single gene and mutation of that gene yields a "+" or "-" phenotype. In this case loss of a single gene is sufficient for expression of the phenotype (Fig. 5.1A). This strategy works well for simple phenotypes like auxotrophy/prototrophy where for example loss of a single gene in a biosynthetic pathway leads to a growth requirement for the downstream product of that pathway (auxotrophy). Multiple genes may display the identical phenotype when mutated, as deletion of any gene in a pathway should block the pathway and lead to the same growth requirement. Deletion of any gene in the biosynthetic pathway for histidine leads to a requirement for supplementation of growth media with histidine. However, many phenotypes are complex and not the consequence of loss of a single gene. Growth at low temperature, tolerance to ethanol, fitness for a specific niche, would be expected to be impacted by loss of multiple genes and in a graded rather than a present or absent response (Fig. 5.1B). Phenotypes with graded responses are associated with multiple genetic loci and the loss of any one of these genes may lead to a reduction in phenotype but not loss of phenotype, for example a slower growth rate at low temperature but not loss of growth. These phenotypes are referred to as "quantitative" and the genes involved called quantitative trait loci (QTL).

3.1 *Analysis of Complex Traits*

For wine yeast many of the interesting environmentally related traits are quantitative. The genes responsible for a QTL are typically defined via DNA sequence analysis of a cross between two parental strains that display a quantitative variation in the phenotype (Fig. 5.1B). The heterozygous diploid is then sporulated and the F1 or subsequent generation sequenced, and the sequences obtained compared to the phenotype in order to identify regions correlating with high or low expression of the quantitative trait (Fig. 5.2). Often this technique requires multiple rounds of meiosis from the originating population to generate sufficient density of crossover events and individuals so that the genes involved that will need to be subsequently tested to determine the impact on the phenotype will be a manageable number. In one study

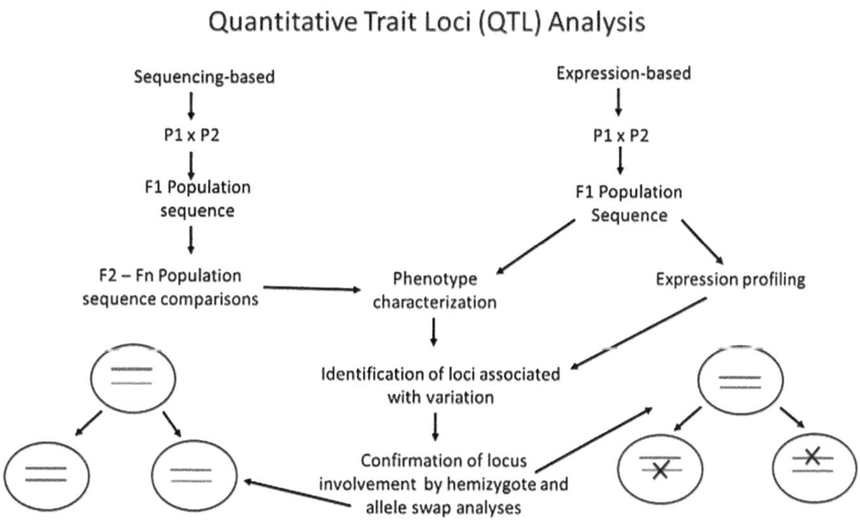

Fig. 5.2 Methods for the dissection of complex multigene traits under the control of quantitative trait loci

with wine yeast the authors had to go to the F13 generation to obtain enough crossover events and rearrangements to identify genes impacting the phenotype (García-Ríos et al. 2017) and another evaluated hundreds of strains from crossed of five yeast strains spanning the known yeast lineages (Cubillos et al. 2011). This methodology can therefore be tedious since all genes identified as co-segregating in the individuals expressing the phenotype will have to be evaluated by both hemizygotic analysis (deletion of each copy of the allele in isolation and assessment of phenotype) and allele swap, swapping one allele for the other and assessment of phenotype (Liti and Louis 2012). QTL analysis has been used in wine yeast strains to identify genes impacting aroma compound production (Eder et al. 2018; Steyer et al. 2012), nitrogen requirements (Brice et al. 2014a, b; Jara et al. 2014), and lag phase duration (Zimmer et al. 2014).

Gene expression profiling of the variants is useful in narrowing down the number of genes to be evaluated within a region identified to contain a QTL (Ehrenreich et al. 2009). In eQTL analysis of transcriptomes is conducted under conditions where the variable trait will be expressed (Fig. 5.2). The assumption is that unexpressed genes will not be drivers of the phenotype, and that differences in the phenotype expressed across a population will be associated with a discrete set of genes differentially expressed under the conditions evaluated. Low resolution is an issue with eQTL and efficient use of this technology requires determination of the complete genome sequences of the strains used (Kita et al. 2017). eQTL enables use of F1 generations eliminating the need for multiple rounds of meiosis and often is successful with fewer individuals, 30–50 versus hundreds. This technique when combined with gene ontology can identify the genes most likely involved as members of the same functional pathway or as downstream targets of the same regulatory circuits. eQTL has been used to evaluate several important wine yeast phenotypes as well as in the comparison of isolates from different environments.

Gene expression variation can be used to define how complex phenotypes evolve (Thompson and Cubillos 2017). Once the QTL gene is identified, subsequent analysis can determine if the QTL is due to mutation of the coding sequence, the upstream non-coding sequence (*cis*-acting) or a regulatory element (*trans*-acting) (Thompson and Cubillos 2017). Comparative analysis of other genes showing similar variation can determine if the *trans*-acting factor is a transcriptional or regulatory element in which case similarly regulated genes will show an impact in expression or due to differences in genes associated with chromatin structure (Thompson and Cubillos 2017).

3.1.1 Gene Expression Profiling and the Dissection of Quantitative Phenotypes in Wine Yeast

Numerous analyses of genetic variability in *Saccharomyces cerevisiae*, as discussed above in detail, have indicated that the two most important drivers of phenotypic variation in wine strains are gene copy number variation and *cis*- and *trans*-mediated changes in gene expression (Fraser et al. 2010; Salinas et al. 2016; Schaefke et al. 2013;

Skelly and Magwene 2016). Critical changes in transcriptional regulators can lead to expression modifications of a suite of downstream genetic and regulatory targets and thereby facilitate genome and phenotype variation from a small number of changes in DNA. Changes in expression may be less severe and therefore less selected against than changes in the coding sequence of functional genes required for fitness (Fraser et al. 2010). Copy number variation can also function to change the level or timing of expression of a gene (Dunn et al. 2012; Steenwyk and Rojas 2017). Changes in basal level of expression of key genes contributing to a quantitative phenotype may be less impactful of fitness across niches than would changes in the functionality of a transcription factor (Skelly and Magwene 2016). Thus, these observations suggest that complex phenotypes can be investigated in wine strains using gene expression profiling techniques and combined with DNA sequencing data of each of the different strains to enable robust eQTL analysis. Comparative analysis of naturally arising and mutationally induced changes in the TDH3 transcription factor showed a congruence between the key nucleotides identified in the saturation mutagenesis study and the nucleotides mutated in the natural population (Duveau et al. 2017).

eQTL analysis has been applied for the dissection of complex traits in wine strains of *S cerevisiae*. Sensitivity to 14 environmental conditions was explored for 52 strains spanning all the major yeast lineages (Kvitek et al. 2008). Associations of fitness phenotypes with both copy number and gene expression variation were observed. Another study similarly used eQTL with segregants of a commercial wine yeast, 59A (a derivative of EC1118), and the common laboratory strain, S288c, to also identify genes important to fitness under stress conditions commonly found during wine production (Brion et al. 2013). From this analysis they were able to identify and map 1465 eQTL genes associated with 5 complex phenotypes: (1) flocculation, (2) clumpiness, (3) metabolite (pyruvic acid, succinic acid and glycerol) production, (4) fermentation rate, and (5) detoxification. This technology will no doubt be used more extensively in the future and is a central component of broader analyses of yeast system biology that include in addition metabolite profiling and whole-cell integration of regulatory and metabolic networks (Österlund et al. 2012).

4 Conclusions

Analysis of gene expression in wine strains of *Saccharomyces* under native growth conditions has provided a dynamic portrait of the changes that accompany the conversion of grape juice to wine. Under these conditions, the yeast must continually adapt to ever increasing levels of stressors such as ethanol that force compromises in cellular activities. The progressive changes seen during fermentation are aimed at maintaining the ability of the cell to resume growth for as long as possible. Eventually, environmental conditions preclude rapid re-entry into active growth and novel non-proliferative metabolically active states are seen. The analysis of transcript profiles during the adaptation to stress indicates that multiple non-proliferative states exist in yeast. The significant biodiversity seen among wild isolates of *Saccharomyces* derives from the natural processes of spontaneous mutagenesis and genome rearrangement followed by selection due to the variable nature of stressors present in the yeast environment. Advances in dissection of complex quantitative traits largely though expression analysis followed by confirmation of the roles of identified genes is and will continue to provide valuable information on the mechanism and driving forces of the modification of existing and development of new phenotypes. The analysis of gene expression across diverse strains with different levels of adaptability to specific growth conditions will lead to the creation of strains with enhanced commercial properties.

References

Aceituno, F. F., Orellana, M., Tores, J., Mendoza, S., Slater, A. W., Melo, F., et al. (2012). Oxygen response of wine yeast *Saccharomyces cerevisiae* EC1118 grown under carbon-sufficient nitrogen-limited enological conditions. *Applied and Environmental Microbiology, 78*, 8340–8352.

Alberola, T. M., Garcia-Martinez, J., Antunez, O., Viladevall, L., Barcelo, A., Arino, J., et al. (2004). A new set of DNA macrochips for the yeast *Saccharomyces cerevisiae*: Features and uses. *International Microbiology, 7*, 199–206.

Alexandre, H., Ansanay-Galeote, V., Dequin, S., & Blondin, B. (2001). Global gene expression during short-term ethanol stress in *Saccharomyces cerevisiae*. *FEBS Letters, 498*, 98–103.

Almeida, P., Barbosa, R., Zalar, P., Imanishi, Y., Shimizu, K., Turchetti, B., et al. (2015). A population genomics insight into Mediterranean origins of yeast domestication. *Molecular Ecology, 24*, 5412–5427.

Ambrona, J., & Ramirez, M. (2007). Analysis of homothallic *Saccharomyces cerevisiae* strain mating during must fermentation. *Applied and Environmental Microbiology, 73*, 2486–2490.

Ambrona, J., Vinagre, A., & Ramirez, M. (2005). Rapid asymmetrical evolution of *Saccharomyces cerevisiae* wine yeasts. *Yeast, 22*, 1299–1306.

Amerine, M. A., Berg, H. W., Kunkee, R. E., Ough, C. S., Singleton, V. L., & Webb, A. D. (1980). *The technology of wine making* (4th ed.). Westport Conn: AVI Publishing Co.

Ando, A., Tanaka, F., Murata, Y., Takagi, H., & Shima, J. (2005). Identification and classification of genes required for tolerance to high-sucrose stress revealed by genome-wide screening of *Saccharomyces cerevisiae*. *FEMS Yeast Research, 6*, 249–267.

Ansel, J., Boltin, H., Rodriquez-Beltran, C., Damon, C., Nagarajan, M., Fehrmann, S., et al. (2008). Cell-to-cell stochastic variation in gene expression is a complex genetic trait. *PloS Genetics, 4*(4), e1000049. https://doi.org/10.1371/journal.pgen.1000049.

Aranda, A., & del Olmo, M. (2004). Exposure of *Saccharomyces cerevisiae* to acetaldehyde induces sulfur amino acid metabolism and polyamine transporter genes, which depend on Met4p and Haa1p transcription factors, respectively. *Applied and Environmental Microbiology, 70*, 1913–1922.

Backhus, L. E., DeRisi, J., & Bisson, L. F. (2001). Functional genomic analysis of a commercial wine strain of *Saccharomyces cerevisiae* under differing nitrogen conditions. *FEMS Yeast Research, 1*, 111–125.

Bakalinsky, A. T., & Snow, R. (1990). The chromosomal constitution of wine strains of *Saccharomyces cerevisiae*. *Yeast, 6*, 367–382.

Balaji, S., Babu, M. M., Iyer, L. M., Luscombe, N. M., & Aravind, L. (2006). Comprehensive analysis of combinatorial regulation using the transcriptional regulatory network of yeast. *Journal of Molecular Biology, 360*, 213–227.

Baleiras Couto, M. M., Eijsma, B., Hofstra, H., & Huis in't Veld JHJ, van der Vossen JMBM. (1996). Evaluation of molecular typing techniques to assign genetic diversity among *Saccharomyces cerevisiae* strains. *Applied and Environmental Microbiology, 62*, 41–46.

Barbosa, C., García-Martínez, J., Pérez-Ortín, J. E., & Mendes-Ferreira, A. (2015a). Comparative transcriptomic analysis reveals similarities and dissimilarities in *Saccharomyces cerevisiae* wine strains response to nitrogen availability. *PloS ONE, 10*(4), e0122709. https://doi.org/10.1371/journal.pone.0122709.

Barbosa, C., Mendes-Faia, A., Lage, P., Mira, N. P., & Mendes-Ferreria, A. (2015b). Genomic expression program of *Saccharomyces cerevisiae* along a mixed-culture wine fermentation with *Hanseniaspora guilliermondii*. *Microbial Cell Factories, 14*, 124. https://doi.org/10.1186/s12934-015-0318-1.

Beltran, G., Novo, M., Leberre, V., Sokol, S., Labourdette, D., Guillamon, J. M., et al. (2006). Integration of transcriptomic and metabolic analysis for understanding the global responses of low-temperature winemaking fermentations. *FEMS Yeast Research, 6*, 1167–1183.

Benaroudj, N., Lee, D. H., & Goldberg, A. L. (2001). Trehalose accumulation during cellular stress protects cells and cellular proteins from damage by oxygen radicals. *The Journal of Biological Chemistry, 276*, 24261–24267.

Bergström, A., Simpson, J. T., Salinas, F., Barré, B., Parts, L., Zia, A., et al. (2014). A high-definition view of functional genetic variation from natural yeast genomes. *Molecular Biology Ecology, 31*(4), 872–888.

Bisson, L. F. (1999). Stuck and sluggish fermentations. *American Journal of Enology and Viticulture, 50*, 107–119.

Bisson, L. F. (2016). Yeast hybrids in winemaking. *Catalyst, 1*, 27–34. https://doi.org/10.5344/catalyst.2016.16001.

Bisson, L. F., & Block, D. E. (2002). Ethanol tolerance in *Saccharomyces*. In M. Ciani (Ed.), *Biodiversity and biotechnology of wine yeasts* (pp. 85–98). Trivandrum: Research Signpost.

Bisson, L. F., Karpel, J. E., Ramakrishnan, V., & Joseph, L. (2007). Functional genomics of wine yeast *Saccharomyces cerevisiae*. *Advances in Food and Nutrition Research, 53*, 65–121.

Borneman, A. R., Forgan, A. H., Pretorius, I. S., & Chambers, P. J. (2008). Comparative genome analysis of a *Saccharomyces cerevisiae* wine strain. *FEMS Yeast Res, 8*, 1185–1195.

Bornman, A. R., Desany, B. A., Riches, D., Affourtit, J. P., Forgan, A. H., Pretorius, I. S., et al. (2011). Whole genome comparison reveals novel genetic elements that characterize the genome of industrial strains of *Saccharomyces cerevisiae*. *PloS Genetics, 7*(2), e1001287. https://doi.org/10.1371/journal.pgen.1001287.

Boulton, R. B., Singleton, V. L., Bisson, L. F., & Kunkee, R. E. (1996). *Principles and practices of winemaking*. New York: Chapman & Hall.

Brice, C., Sanchez, I., Bigey, F., Legras, J.-L., & Blondin, B. (2014a). A genetic approach of wine yeast fermentation capacity in nitrogen-starvation reveals the key role of nitrogen sensing. *BMC Genomics, 15*, 495.

Brice, C., Sanchez, I., Tesnière, C., & Blondin, B. (2014b). Assessing the mechanisms responsible for differences between nitrogen requirements of *Saccharomyces cerevisiae* wine yeasts in alcoholic fermentation. *Applied and Environmental Microbiology, 80*, 1330–1339.

Brion, C., Ambroset, C., Sanchez, I., Legras, J.-L., & Blondin, B. (2013). Differential adaptation to multi-stressed conditions of wine fermentation revealed by variations in yeast regulatory networks. *BMC Genomics, 14*, 681.

Briones, A. I., Ubeda, J., & Grando, M. S. (1996). Differentiation of *Saccharomyces cerevisiae* strains isolated from fermenting musts according to karyotype patterns. *International Journal of Food Microbiology, 28*, 369–377.

Cantarelli, C. (1989). Phenolics and yeast: Remarks concerning fermented beverages. In A. Martini & A. Vaughn-Martini (Eds.), *Proceedings of the seventh international symposium on yeasts* (pp. S51–S52). New York: Wiley.

Carlson, M. (1997). Genetics of transcriptional regulation in yeast: Connections to the RNA polymerase II CTD. *Annual Review of Cell and Developmental Biology, 13*, 1–23.

Carro, D., & Pina, B. (2001). Genetic analysis of the karyotype instability in natural wine yeast strains. *Yeast, 18*, 1457–1470.

Causton, H. C., Ren, B., Koh, S. S., Harbison, C. T., Kanin, E., Jennings, E. G., et al. (2001). Remodeling of yeast genome expression in response to environmental changes. *Molecular Biology of the Cell, 12*, 323–337.

Cavalieri, D., Townsend, J. P., & Hartl, D. L. (2000). Manifold anomalies in gene expression in a vineyard isolate of *Saccharomyces cerevisiae* revealed by DNA Microarray analysis. *Proceedings of the National Academy of Sciences of the United States of America, 97*, 12369–12374.

Celton, M., Sanchez, I., Goelzer, A., Fromion, V., Camarasa, C., & Dequin, S. (2012). A comparative transcriptomic, fluxomic and metabolomic analysis of the response of *Saccharomyces cerevisiae* to increases in NADPH oxidation. *BMC Genomics, 13*, 317.

Chidi, B. S., Rossouw, D., & Bauer, F. F. (2015). Identifying and assessing the impact of wine acid-related genes in yeast. *Current Genetics*. https://doi.org/10.1007/s00294-015-0489-6.

Clemente-Jimenez, J. M., Mingorance-Carzola, L., Martinez-Rodriguez, S., Las Heras-Vazquez, F. J., & Rodriguez-Vico, F. (2004). Molecular characterization and oenological properties of wine yeasts isolated during spontaneous fermentation of six varieties of grape must. *Food Microbiology, 21*, 149–155.

Cliften, P., Sudarsanam, P., Desikan, A., Fulton, L., Fulton, B., Majors, J., et al. (2003). Finding functional features in *Saccharomyces* genomes by phylogenetic footprinting. *Science, 301*, 71–76.

Codon, A. C., Benitez, T., & Korhola, M. (1998). Chromosomal polymorphism and adaptation to specific industrial environments of *Saccharomyces* strains. *Applied Microbiology and Biotechnology, 49*, 154–163.

Comitini, F., & Ciani, M. (2006). Survival of inoculated *Saccharomyces cerevisiae* strain on wine grapes during two vintages. *Letters in Applied Microbiology, 42*, 248–253.

Cramer, A. C., Vlassides, S., & Block, D. E. (2002). Kinetic model for nitrogen-limited wine fermentations. *Biotechnology and Bioengineering, 77*, 49–60.

Cubillos, F. A., Billi, E., Zörgö, E., Parts, L., Fargier, P., Omholt, S., et al. (2011). Addressing the complex architecture of polygenic traits in diverged yeast populations. *Molecular Ecology, 20*, 1401–1413.

Davidson, J. F., & Schiestl, R. H. (2001). Cytotoxic and genotoxic consequences of heat stress are dependent on the presence of oxygen in *Saccharomyces cerevisiae*. *Journal of Bacteriology, 183*, 4580–4587.

Deed, R. C., Deed, N. K., & Gardner, R. C. (2015). Transcriptional response of *Saccharomyces cerevisiae* to low temperature during wine fermentation. *Antonie Van Leeuwenhoek, 107*, 1029–1048.

DeRisi, J. L., Iyer, V. R., & Brown, P. O. (1997). Exploring the metabolic and genetic control of gene expression on a genomic scale. *Science, 278*, 680–686.

Draghici, S., Khatri, P., Eklund, A. C., & Szallasi, Z. (2006). Reliability and reproducibility issues in DNA microarray measurements. *Trends in Genetics, 22*, 101–109.

Drebot, M. A., Barnes, C. A., Singer, R. A., & Johnston, G. (1990). Genetic assessment of stationary phase for cells of the yeast *Saccharomyces cerevisiae*. *Journal of Bacteriology, 172*, 3584–3589.

Dunn, B., Levine, R. P., & Sherlock, G. (2005). Microarray karyotyping of commercial wine yeast strains reveals shared, as well as unique, genomic signatures. *BMC Genomics, 6*, 1–21.

Dunn, B., Richter, C., Kvitek, D., Pugh, T., & Sherlock, G. (2012). Analysis of the *Saccharomyces cerevisiae* pan-genome reveals a pool of copy number variants distributed in diverse yeast strains from differing industrial environments. *Genome Research, 22*, 908–924.

Duveau, F., Yuan, D. C., Metzger, B. P. H., Hodgins-Davis, A., & Wittkopp, P. J. (2017). Effects of mutation and selection on plasticity of a promoter activity in *Saccharomyces cerevisiae*. *Proceedings of the National Academy of Sciences of the United States of America*. https://doi.org/10.1073/pnas.1713960115.

Dwight, S. S., Harris, M. A., Dolinski, K., Bell, C. A., Binkley, G., Christie, K. R., et al. (2002). *Saccharomyces genome* database (SGD) provides secondary gene annotation using the Gene Ontology (GO). *Nucleic Acids Research, 30*, 69–72.
Eder, M., Sanchez, I., Brice, C., Camarasa, C., Legras, J.-L., & Dequin, S. (2018). QTL mapping of volatile compound production in *Saccharomyces cerevisiae* during alcoholic fermentation. *BMC Genomics, 19*, 166. https://doi.org/10.1186/s12864-018-4562-8.
Ehrenreich, I. M., Gerke, J. P., & Kruglyak, L. (2009). Genetic dissection of complex traits in yeast: Insights from studies of gene expression and other phenotypes in the BY × RM cross. *Cold Spring Harbor Symposia on Quantitative Biology, 74*, 145–153.
Erasmus, D. J., van der Merwe, G. K., & van Vuuren, H. J. J. (2003). Genome-wide expression analyses: Metabolic adaptation of *Saccharomyces cerevisiae* to high sugar stress. *FEMS Yeast Research, 3*, 375–399.
Fay, J. C., McCullough, H. L., Sniegowski, P. D., & Eisen, M. B. (2004). Population genetic variation in gene expression is associated with phenotypic variation in *Saccharomyces cerevisiae. Genome Biology, 5*, R26.1–R26.14.
Fay, J. C., & Benavides, J. A. (2005a). Evidence for domesticated and wild populations of *Saccharomyces cerevisiae. PLoS Genetics, 1*, e5.
Fay, J. C., & Benavides, J. A. (2005b). Hypervariable noncoding sequences in *Saccharomyces cerevisiae. Genetics, 170*, 1575–1587.
Fleet, G. H. (1993). The microorganisms of winemaking – Isolation, enumeration and identification. In G. H. Fleet (Ed.), *Wine microbiology and biotechnology* (pp. 1–26). Camberwell: Harwood Academic Publishers.
Fleet, G. H., & Heard, G. M. (1993). Yeasts - Growth during fermentation. In G. H. Fleet (Ed.), *Wine microbiology and biotechnology* (pp. 27–54). Camberwell: Harwood Academic Publishers.
Fleet, G. H., Prakitchaiwattana, C., Beh, A. L., & Heard, G. M. (2002). The yeast ecology of wine grapes. In M. Ciani (Ed.), *Biodiversity and biotechnology of wine yeasts* (pp. 1–17). Kerala: Research Signpost.
Fraser, H. B., Moses, A. M., & Schadt, E. E. (2010). Evidence for widespread adaptive evolution of gene expression in budding yeast. *Proceedings of the National Academy of Sciences of the United States of America, 107*, 2977–2982.
Gallego, F. J., Perez, M. A., Nunez, Y., & Hildago, P. (2005). Comparison of RAPDs, AFLPs and SSR marker for genetic analysis of yeast strains of *Saccharomyces cerevisiae. Food Microbiology, 22*, 561–568.
Garcia, D. M., & Jarosz, D. F. (2014). Rebels with a cause: Molecular features and physiological consequences of yeast prions. *FEMS Yeast Research, 14*, 136–147.
García-Ríos, E., López-Malo, M., & Guillamón, J. M. (2014). Global phenotype and genomic comparison of two Saccharomyces cerevisiae wine strains reveals a novel role of the sulfur assimilation pathway in adaptation at low temperature fermentations. *BMC Genomics, 15*, 1059.
García-Ríos, E., Morard, M., Parts, L., Liti, G., & Guillamón, J. M. (2017). The genetic architecture of low-temperature adaptation in the wine yeast *Saccharomyces cerevisiae. BMC Genomics, 18*, 159. https://doi.org/10.1186/s12864-017-3572-2.
Gasch, A. P. (2003). The environmental stress response: A common yeast response to diverse environmental stresses. *Topics in Current Genetics, 1*, 11–70.
Gasch, A. P., Spellman, P. T., Kao, C. M., Carmel-Harel, O., Eisen, M. B., Stortz, G., et al. (2000). Genomic expression programs in the response of yeast cells to environmental changes. *Molecular Biology of the Cell, 11*, 4241–4257.
Gasch, A. P., & Werner-Washburne, M. (2002). The genomics of yeast responses to environmental stress and starvation. *Functional & Integrative Genomics, 2*, 181–192.
Godard, P., Urrestarazu, A., Vissers, S., Kontos, K., Bontempi, G., van Helden, J., et al. (2007). Effect of 21 different nitrogen sources on global gene expression in the yeast *Saccharomyces cerevisiae. Molecular and Cellular Biology, 27*, 3065–3086.

Goddard, M. R., & Greig, D. (2015). *Saccharomyces cerevisiae*: A nomadic yeast with no niche? *FEMS Yeast Research, 15*, fov009. https://doi.org/10.1093/femsyr/fov009.

Goffeau, A., Barrell, B. G., Bussey, H., Davis, R. W., Dujon, B., Feldmann, H., et al. (1996). Life with 6000 genes. *Science, 274*, 546–567.

Goffeau, A., Park, J., Paulsen, I. T., Jonniaux, J.-L., Dinh, T., Mordant, P., et al. (1997). Multidrug-resistant transport proteins in yeast: Complete inventory and phylogenetic characterization of yeast open reading frames within the major facilitator superfamily. *Yeast, 13*, 43–54.

Granot, D., & Snyder, M. (1993). Carbon source induces growth of stationary phase yeast cells, independent of carbon source metabolism. *Yeast, 9*, 465–479.

Gray, J. V., Petsco, G. A., Johnston, G. C., Ringe, D., Singer, R. A., & Werner-Washburne, M. (2004). "Sleeping beauty": Quiescence in *Saccharomyces cerevisiae*. *Microbiology and Molecular Biology Reviews, 68*, 187–206.

Gresham, D., Ruderfer, D. M., Pratt, S. C., Schacherer, J., Dunham, M. J., Botstein, D., et al. (2006). Genome-wide detection of polymorphisms at nucleotide resolution with a single DNA microarray. *Science, 311*, 1932–1936.

Gutierrez, A. R., Santamaria, P., Epifanio, S., Garijo, P., & Lopez, R. (1999). Ecology of spontaneous fermentation in one winery during 5 consecutive years. *Letters in Applied Microbiology, 29*, 411–415.

Halfmann, R., Alberti, A., & Lindquist, S. (2010). Prions, protein homeostasis, and phenotypic diversity. *Trends in Cell Biology, 20*(3), 125–133.

Halfmann, R., Jarosz, D. F., Jones, S. K., Chang, A., Lancaster, A. K., & Lindquist, S. (2012). Prions are a common mechanism for phenotypic inheritance in wild yeasts. *Nature, 482*((7385) l), 363–368.

Hanna-Rose, W., & Hansen, U. (1996). Active repression mediated by eukaryotic factors: Molecular targets and potential mechanisms. *Trends Genetics, 12*, 229–234.

Hauser, N. C., Fellenberg, K., Gil, R., Bastuck, S., Hoheisel, J. D., & Perez-Ortin, J. E. (2001). Whole genomes of a wine yeast strain. *Comparative and Functional Genomics, 2*, 69–79.

Herschbach, B. M., & Johnson, A. D. (1993). Transcriptional repression in eukaryotes. *Annual Review of Cell Biology, 9*, 479–509.

Hittinger, C. T. (2013). *Saccharomyces* diversity and evolution: A budding model genus. *Trends in Genetics, 29*, 309–317.

Holland, M. J. (2002). Transcript abundance in yeast varies over six orders of magnitude. *The Journal of Biological Chemistry, 277*, 14363–14366.

Holmes, D. L., Lancaster, A. K., Lindquist, S., & Halfmann, R. (2013). Heritable remodeling of yeast multicellularity by an environmentally responsive prion. *Cell*. https://doi.org/10.1016/j.cell.2013.02.026.

Hughes, T. R., Roberts, C. J., Dai, H., Jones, A. R., Meer, M. R., Slade, D., et al. (2000). Widespread aneuploidy revealed by DNA microarray expression profiling. *Nature Genetics, 25*, 333–337.

Infante, J. J., Dombek, K. M., Rebordinos, L., Cantoral, J. M., & Young, E. T. (2003). Genome-wide amplifications caused by chromosomal rearrangements play a major role in the adaptive evolution of natural yeast. *Genetics, 165*, 1745–1759.

Ihmels, J., Levy, R., & Barkai, N. (2004). Principles of transcriptional control in the metabolic network of *Saccharomyces cerevisiae*. *Nature Biotechnology, 22*, 86–92.

Insa, G., Sablayrolles, J.-M., & Douzal, V. (1995). Alcoholic fermentation under oenological conditions. *Bioprocess and Biosystems Engineering, 13*, 171–176.

Ivorra, C., Perez-Ortin, J. E., & del Olmo, M. (1999). An inverse correlation between stress resistance and stuck fermentations in wine yeasts. A molecular study. *Biotechnology and Bioengineering, 64*, 698–708.

Izawa, S., Takemura, R., Miki, T., & Inoue, Y. (2005). Characterization of the export of bulk Poly(A)$^+$ mRNA in *Saccharomyces cerevisiae* during the wine-making process. *Applied and Environmental Microbiology, 71*, 2179–2182.

Izquierdo Canas, P. M., Ubeda Iranzo, J. F., & Briones Perez, A. I. (1997). Study of the karyotype of wine yeasts isolated in the region of Valdepenas in two consecutive vintages. *Food Microbiology, 14*, 221–225.

Jara, M., Cubillos, F. A., García, V., Salinas, F., Aguilera, O., Liti, G., et al. (2014). Mapping genetic variants underlying differences in the central nitrogen metabolism in fermenter yeasts. *PloS One, 9*(1), e86533. https://doi.org/10.1374/journal.pone.0086533.

Jimenez-Marti, E., Aranda, A., Mendes-Ferreira, A., Mendes-Faia, A., & del Olmo, M. (2007). The nature of the nitrogen source added to nitrogen depleted vinifications conducted by a *Saccharomyces cerevisiae* strain in synthetic must affects gene expression and the levels of several volatile compounds. *Antonie Van Leeuwenhoek, 92*, 61–75.

Jimenez-Marti, E., & del Olmo, M. (2008). Addition of ammonia or amino acids to a nitrogen-depleted medium affects gene expression patterns in yeast cells during alcoholic fermentation. *FEMS Yeast Research, 8*, 245–256.

Johnston, J. R., Baccari, C., & Mortimer, R. K. (2000). Genotypic characterization of strains of commercial wine yeasts by tetrad analysis. *Research Microbiology, 151*, 583–590.

Kal, A. J., van Zonneveld, A. J., Benes, V., van den Berg, M., Koerkamp, M. G., Albermann, K., et al. (1999). Dynamics of gene expression revealed by comparison of Serial analysis of gene expression transcript profiles from yeast grown on two different carbon sources. *Molecular Biology of the Cell, 10*, 1859–1872.

Khan, W., Augustyn, O. P. H., Van der Westhuizen, T. J., Lambrechts, M. G., & Pretorius, I. S. (2000). Geographic distribution and evaluation of *Saccharomyces cerevisiae* strains isolated from vineyards in the warmer inland regions of the Western Cape in South Africa. *South African Journal of Enology and Viticulture, 21*, 17–31.

Kingston, R. E., Bunker, C. A., & Imbalzano, A. N. (1996). Repression and activation by multiprotein complexes that alter chromatin structure. *Genes & Development, 10*, 905–920.

Kita, R., Venkataram, S., Zhou, Y., & Fraser, H. B. (2017). High-resolution mapping of cis-regulatory variation in budding yeast. *Proceedings of the National Academy of Sciences of the United States of America*. https://doi.org/10.1073/pnas.1717421114.

Kobayashi, N., & McEntee, K. (1990). Evidence for a heat shock transcription factor –independent mechanism for heat shock induction of transcription in *Saccharomyces cerevisiae*. *Proceedings of the National Academy of Sciences of the United States of America, 87*, 6550–6554.

Kobayashi, N., & McEntee, K. (1993). Identification of cis and trans components of a novel heat shock stress regulatory pathway in *Saccharomyces cerevisiae*. *Molecular and Cellular Biochemistry, 13*, 248–256.

Kosel, J., Cadez, N., Schuller, D., Carreto, L., Franco-Duarté, R., & Raspor, P. (2017). The influence of *Dekkera bruxellensis* on the transcriptome of *Saccharomyces cerevisiae* and on the aromatic profile of synthetic wine must. *FEMS Yeast Research*. https://doi.org/10.1093/femsyr/fox018.

Krantz, M., Nordlander, B., Valadi, H., Johansson, M., Gustafsson, L., & Hohmann, S. (2004). Anerobicity prepares *Saccharomyces cerevisiae* cells for faster adaptation to osmotic shock. *Euc Cell, 3*, 1381–1390.

Kudo, M., Vagnoli, P., & Bisson, L. F. (1998). Imbalance of potassium and hydrogen ion concentrations as a cause of stuck enological fermentations. *American Journal of Enology and Viticulture, 49*, 295–301.

Kuhn, K. M., DeRisi, J. L., Brown, P. O., & Sarnow, P. (2001). Global and specific translational regulation in the genomic response of *Saccharomyces cerevisiae* to a rapid transfer from a fermentable to a nonfermentable carbon source. *Journal of Molecular Cell Biology, 21*, 916–927.

Kumar, A., Harrison, P. M., Cheung, K.-H., Lan, N., Echols, N., Bertone, P., et al. (2002). An integrated approach for finding overlooked genes in yeast. *Nature Biotechnology, 20*, 58–63.

Kumar, A., Sharma, P., Gomar-Alba, M., Shcheprova, Z., Daulny, A., Sanmartín, T., et al. (2018). Daughter-cell-specific modulation of nuclear pore complexes controls cell cycle entry during asymmetric division. *Nature Cell Biology, 20*, 432–442.

Kunkee, R. E., & Bisson, L. F. (1993). Winemaking yeasts. In A. H. Rose & J. S. Harrison (Eds.), *The yeasts: Yeast technology* (pp. 69–126). London: Academic.

Kuo, W. P., Liu, F., Trimarchi, J., Punzo, C., Lombardi, M., Sarang, J., et al. (2006). A sequence-oriented comparison of gene expression measurements across different hybridization-based technologies. *Nature Biotechnology, 24*, 832–840.

Kuthan, M., Devaux, F., Janderova, B., Slaninova, I., Jacq, C., & Palkova, Z. (2003). Domestication of wild *Saccharomyces cerevisiae* is accompanied by changes in gene expression and colony morphology. *Molecular Microbiology, 47*, 745–754.

Kvitek, D. J., Will, J. L., & Gasch, A. (2008). Variations in stress sensitivity and genomic expression in diverse *S. cerevisiae* isolates. *PLoS Genetics, 4*(10), e1000223. https://doi.org/10.1371/journal.pgen.1000223.

Landry, C. R., Oh, J., Hart, D. L., & Cavalieri, D. (2006a). Genome-wide scan reveals that genetic variation for transcriptional plasticity in yeast is biased towards multi-copy and dispensable genes. *Gene, 366*, 343–351.

Landry, C. R., Townsend, J. P., Hartl, D. L., & Cavalieri, D. (2006b). Ecological and evolutionary genomics of *Saccharomyces cerevisiae*. *Molecular Ecology, 15*, 575–591.

Lashkari, D. A., DeRisi, J. L., McCusker, J. H., Namath, A. F., Gentile, C., Hwang, S. Y., et al. (1997). Yeast microarrays for genome wide parallel genetic and gene expression analysis. *Proceedings of the National Academy of Sciences of the United States of America, 94*, 13057–13062.

Lee, M.-L. T., Kuo, F. C., Whitmore, G. A., & Sklar, J. (2000). Importance of replication and microarray gene expression studies: Statistical methods and evidence from repetitive cDNA hybridizations. *Proceedings of the National Academy of Sciences of the United States of America, 97*, 9834–9839.

Legras, J. L., Merdinoglu, D., Cornuet, J. M., & Karst, F. (2007). Bread, beer and wine: *Saccharomyces cerevisiae* diversity reflects human history. *Molecular Ecology, 16*, 2091–2102.

Liti, G., Barton, D. B. H., & Louis, E. J. (2006). Sequence diversity, reproductive isolation and species concepts in *Saccharomyces*. *Genetics, 174*, 839–850.

Liti, G., Carter, D. M., Moses, A. M., Warringer, J., Parts, L., James, S. A., et al. (2009). Population genomics of domestic and wild yeast. *Nature*. preprint. https://doi.org/10.1038/nature07743.

Liti, G., & Louis, E. J. (2012). Advances in quantitative trait analysis in yeast. *PloS Genetics, 8*(8), e1002912. https://doi.org/10.1371/journal.pgen.1002912.

Liu, J., Martin-Yken, H., Bigey, F., Dequin, S., François, J.-M., & Capp, J.-P. (2015). Natural yeast promoter variants reveal epistasis in the generation of transcriptional mediated noise and its potential benefit in stressful conditions. *Genome Biology and Evolution, 7*(4), 969–984.

Lockhart, D. J., Dong, H., Byrne, M. C., Follettie, M. T., Gallow, M. V., Chee, M. S., et al. (1996). Expression monitoring by hybridization to high-density oligonucleotide arrays. *Nature Biotechnology, 14*, 1675–1680.

Lockhart, D. J., & Winzeler, E. A. (2000). Genomics, gene expression and DNA arrays. *Nature, 405*, 827–836.

Longo, E., & Vezinhet, F. (1993). Chromosomal rearrangements during vegetative growth of a wild strain of *Saccharomyces cerevisiae*. *Applied and Environmental Microbiology, 59*, 322–326.

Lopes, C. A., van Broock, M., Querol, A., & Caballero, A. C. (2002). *Saccharomyces cerevisiae* wine yeast populations in a cold region in Argentinean Patagonia. A study at different fermentation scales. *Journal of Applied Microbiology, 93*, 608–615.

Maginot, C., Roustan, J. L., & Sablayrolles, J.-M. (1998). Nitrogen demand of different yeast strains during alcoholic fermentation. Importance of the stationary phase. *Enzyme and Microbial Technology, 23*, 511–517.

Marchler, G., Schuller, C., Adam, G., & Ruis, H. (1993). A *Saccharomyces cerevisiae* UAS element controlled by protein kinase A activates transcription in response to a variety of stress conditions. *The EMBO Journal, 12*, 1997–2003.

Marks, V. D., van der Merwe, G. K., & van Vuuren, H. J. J. (2003). Transcriptional profiling of wine yeast in fermenting grape juice: Regulatory effect of diammonium phosphate. *FEMS Yeast Research, 3*, 269–287.

Marks, V. D., Sui, S. J. H., Erasmus, D., van der Merwe, G. K., Brumm, J., Wasserman, W. W., et al. (2008). Dynamics of the yeast transcriptome during wine fermentation reveals a novel fermentation stress response. *FEMS Yeast Research, 8*, 35–52.

Martini, A., Ciani, M., & Scorzetti, G. (1996). Direct enumeration and isolation of wine yeasts from grape surfaces. *American Journal of Enology and Viticulture, 47*, 435–440.

Marsit, S., & Dequin, S. (2015). Diversity and adaptive evolution of *Saccharomyces* wine yeast: A review. *FEMS Yeast Research, 15*, fov067. https://doi.org/10.1093/femsyr/fov067.

Mendes-Ferreira, A., del Olmo, M., Garcia-Martinez, J., Jimenez-Marti, E., Leao, C., Mendes-Faia, A., et al. (2007a). *Saccharomyces cerevisiae* signature genes for predicting nitrogen deficiency during alcoholic fermentation. *Applied and Environmental Microbiology, 73*, 5363–5369.

Mendes-Ferreira, A., del Olmo, M., Garcia-Martinez, J., Jimenez-Marti, E., Mendes-Faia, A., Perez-Ortin, J. E., et al. (2007b). Transcriptional response of *Saccharomyces cerevisiae* to different nitrogen concentrations during alcoholic fermentation. *Applied and Environmental Microbiology, 73*, 3049–3060.

Mendes-Ferreira, A., Barbosa, C., Jimenez-Marti, E., del Olmo, M & Mendes-Faia, A. (2010) The wine yeast strain-dependent expression of genes implicated in sulfide production in response to nitrogen availability. Journal of Microibology and Biotechnology 20:1314-1321

Mitchell, A., Romano, G. H., Groisman, B., Yona, A., Dekel, E., Kupiec, M., et al. (2009). Adaptive prediction of environmental changes by microorganisms. *Nature*. https://doi.org/10.1038/nature08112.

Mortimer, R. K. (2000). Evolution and variation of the yeast (*Saccharomyces*) genome. *Genome Research, 10*, 403–409.

Mortimer, R., & Polsinelli, M. (1999). On the origin of wine yeast. *Research in Microbiology, 150*, 199–204.

Mortimer, R. K., Romano, P., Suzzi, G., & Polsinelli, M. (1994). Genome renewal: A new phenomenon revealed from a genetic study of 43 strains of *Saccharomyces cerevisiae* derived from natural fermentation of grape musts. *Yeast, 10*, 1543–1552.

Myers, C. L., Dunham, M. J., Kung, S. Y., & Troyanskaya, O. G. (2004). Accurate detection of aneuploidies in array CGH and gene expression microarray data. *Bioinformatics, 20*, 3533–3543.

Naumov, G. (1996). Genetic identification of biological species in the *Saccharomyces* sensu stricto complex. *Journal of Industrial Microbiology, 17*, 295–302.

Novo, M., Beltran, G., Rozes, N., Guillamon, J. M., Sokol, S., Leberre, V., et al. (2006). Early transcriptional response of wine yeast after rehydration: Osmotic shock and metabolic activation FEMS. *Yeast Research, 7*, 304–316.

Oliver, S. G., Winson, M. K., Kell, D. B., & Baganz, F. (1998). Systematic functional analysis of the yeast genome. *Trends Biotechnology, 16*, 373–378.

Orellana, M., Aceituno, F. F., Slater, A. W., Almonacid, L. I., Melo, F., & Agosin, E. (2013). Metabolic and transcriptomic response of the wine yeast *Saccharomyces cerevisiae* strain EC1118 after an oxygen impulse under carbon-sufficient nitrogen-limited fermentation conditions. *FEMS Yeast Research, 14*, 412–424.

Orphanides, G., Lagrange, T., & Reinberg, D. (1996). The general transcription factors of RNA polymerase II. *Genes & Development, 10*, 2657–2683.

Oshiro, G., & Winzeler, E. A. (2000). Aneuploidy- it's more common than you think. *Nature Biotechnology, 18*, 715–716.

Österlund, T., Nookaew, I., & Nielsen, J. (2012). Fifteen years of large scale metabolic modeling of yeast: Development and impacts. *Biotechnology Advances, 30*, 979–988.

Palkova, Z. (2004). Multicellular microorganisms: Laboratory versus nature. *EMBO Reports, 5*, 470–476.

Pelechano, V., Chávez, S., & Pérez-Ortín, J. I. (2010). A complete set of nascent transcription rates for yeast genes. *PloS ONE, 5*(11), e15442. https://doi.org/10.1371/journal.pone.0015442.

Pena-Castillo, L., & Hughes, T. R. (2007). Why are there still over 1000 uncharacterized yeast genes? *Genetics, 176*, 7–14.

Pérez-Ortín, J. E., Alepuz, P. M., & Moreno, J. (2007). Genomics and gene transcription kinetics in yeast. *Trends in Genetics, 23*, 250–257.

Pizarro, F., Vargas, F. A., & Agosin, E. (2007). A systems biology perspective of wine fermentations. *Yeast, 24*, 977–991.

Powers, S., DeJongh, M., Best, A. A., & Tintle, N. L. (2015). Cautions about the reliability of pairwise gene correlations based on expression data. *Frontiers in Microbiology*. https://doi.org/10.3389/fmicb.2015.00650.

Puig, S., & Perez-Ortin, J. E. (2000). Stress response and expression patterns in wine fermentations of yeast genes induced at the diauxic shift. *Yeast, 16*, 139–148.

Quackenbush, J. (2005). Using DNA microarrays to assay gene expression. In A. D. Baxevanis & B. F. F. Ouellette (Eds.), *Bioinformatics: A practical guide to the analysis of genes and proteins* (Third ed., pp. 410–444). New York: Wiley.

Rachidi, N., Barre, P., & Blondin, B. (2000). Examination of the transcriptional specificity of an enological yeast. A pilot experiment on the chromosome-III right arm. *Current Genetics, 37*, 1–11.

Ramakrishnan, V., Walker, G. A., Fan, Q., Ogawa, M., Luo, Y., Leung, P., et al. (2016). Inter-kingdom modification of metabolic behavior: [*GAR*+] prion induction in *Saccharomyces cerevisiae* mediated by wine ecosystem bacteria. *Frontiers in Ecology and Evolution*. https://doi.org/10.3389/fevo.2016.00137.

Renouf, V., Perello, M. C., Strehaiano, P., & Lonvaud-Funel, A. (2006). Global survey of the microbial ecosystem during alcoholic fermentation in winemaking. *Journal International des Sciences de la Vigne et du Vin, 40*, 101–116.

Rep, M., Krantz, M., Thevelein, J. M., & Hohmann, S. (2000). The transcriptional response of *Saccharomyces cerevisiae* to osmotic shock: Hot1p and Msn2p/Msn4p are required for the induction of subsets of high osmolarity glycerol pathway-dependent genes. *The Journal of Biological Chemistry, 275*, 8290–8300.

Riou, C., Nicaud, J.-M., Barre, P., & Gaillardin, C. (1997). Stationary-phase gene expression in *Saccharomyces cerevisiae* during wine fermentation. *Yeast, 13*, 903–915.

Rodriguez-Pena, J. M., Perez-Diaz, R. M., Alvarez, S., Bermejo, C., Garcia, R., Santiago, C., et al. (2005). The 'yeast cell wall chip' – a tool to analyse the regulation of cell wall biogenesis in *Saccharomyces cerevisiae*. *Microbiology, 151*, 2241–2249.

Roosen, J. Oesterhelt, C., Pardons, K., Swinnen, E.& Winderickx, J. (2004). Integration of nutrient signaling pathways in the yeast *Saccharomyces cerevisiae*. In J. Winderickx, P.M. Taylor (Eds) Nutrient-Induced Responses in Eukaryotic Cells*Topics in Current Genetics, 7*, 277–318. Springer-Verlag Berlin Heidelberg

Rossignol, T., Dulau, L., Julien, A., & Blondin, B. (2003). Genome-wide monitoring of wine yeast gene expression during alcoholic fermentation. *Yeast, 20*, 1369–1385.

Rossignol, T., Kobi, D., Jacquet-Gutfreund, L., & Blondin, B. (2009). The proteome of a wine yeast strain during fermentation, correlation with the transcriptome. *Journal of Applied Microbiology*. https://doi.org/10.1111/j.1365-2672.2009.04156:1-10.

Rossignol, T., Postaire, O., Storai, J., & Blondin, B. (2006). Analysis of the genomic response of a wine yeast to rehydration and inoculation. *Applied Microbiology and Biotechnology, 71*, 699–712.

Rossouw, D., Naes, T., & Bauer, F. (2008). Linking gene regulation and the exo-metabolome: A comparative transcriptomics approach to identify genes that impact the production of volatile aroma compounds in yeast. *BMC Genomics*. https://doi.org/10.1186/1471-2164-9-530.

Rossouw, D., van den Dool, A. H., Jacobson, D., & Bauer, F. F. (2010). Comparative transcriptomic and proteomic profiling of industrial wine yeast strains. *Applied and Environmental Microbiology, 76*, 3911–3023.

Rossouw, D., Jacobson, D., & Bauer, F. F. (2012). Transcriptional regulation and the diversification of metabolism in wine yeast. *Genetics, 190*, 251–261.

Sabate, J., Cano, J., Querol, A., & Guillamon, J. M. (1998). Diversity of *Saccharomyces* strains in wine fermentations: Analysis for two consecutive years. *Letters in Applied Microbiology, 26*, 452–455.

Sahara, T., Goda, T., & Ohgiya, S. (2002). Comprehensive expression analysis of time-dependent responses of yeast cells to low temperature. *The Journal of Biological Chemistry, 51*, 50015–50021.

Salinas, F., Mandakovic, D., Urzua, U., Massera, A., Miras, S., Combina, M., et al. (2010). Genomic and phenotypic comparison between similar wine yeast strains of *Saccharomyces cerevisiae* from different geographic origins. *Journal of Applied Microbiology, 108*(5), 1850–1858. https://doi.org/10.1111/j.1365-2672.2010.04689.x.

Salinas, F., de Boer, C. G., Abarca, V., García, V., Cuevas, M., Araos, S., et al. (2016). Natural variation in non-coding regions underlying phenotypic diversity in budding yeast. *Scientific Reports*. https://doi.org/10.1038/srep21849.

Sardu, A., Treu, L., & Campanara, S. (2014). Transcriptome structure variability in *Saccharomyces cerevisiae* strains determined with a newly developed assembly software. *BMC Genomics, 15*, 1045.

Schacherer, J., Ruderfer, D. M., Gresham, D., Dolinski, K., Botstein, D., & Kruglyak, L. (2007). Genome-wide analysis of nucleotide-level variation in commonly used *Saccharomyces cerevisiae* strains. *PLoS One, 3*, e322.

Schacherer, J., Shapiro, J. A., Ruderfer, D. M., & Kruglyak, L. (2009). Comprehensive polymorphism survey elucidates population structure of *Saccharomyces cerevisiae*. Nature preprint https://doi.org/10.1038/nature07670.

Schadt, E. E., Li, C., Su, C., & Wong, W. H. (2000). Analyzing high-density oligonucleotide gene expression array data. *Journal of Cellular Biochemistry, 80*, 192–202.

Schaefke, B., Emerson, J. J., Wang, T.-Y., Lu, M.-Y. J., Hsieh, L.-C., & Li, W.-H. (2013). Inheritance of gene expression level and selective constraints on trans- and cis-regulatory changes in yeast. *Molecular Biology and Evolution, 30*(9), 2122–2133.

Schena, M., Shalon, D., Davis, R. W., & Brown, P. O. (1995). Quantitative monitoring of gene expression patterns with a complementary DNA microarray. *Science, 270*, 467–470.

Schoondermark-Stolk, S. A., Jansen, M., Verkleij, A. J., Verrips, C. T., Euvernink, G. J. W., Dijkhuizen, L., et al. (2006). Genome-wide transcription survey on flavour production in *Saccharomyces cerevisiae*. *World Journal of Microbiology and Biotechnology, 22*, 1347–1356.

Schuller, H.-J. (2003). Transcriptional control of nonfermentative metabolism in the yeast *Saccharomyces cerevisiae*. *Current Genetics, 43*, 139–160.

Schuller, D., Alves, H., Dequin, S., & Casal, M. (2005). Ecological survey of *Saccharomyces cerevisiae* strains from vineyards in the Vinho Verde region of Portugal. *FEMS Microbiology Ecology, 51*, 167–177.

Schutz, M., & Gafner, J. (1993). Analysis of yeast diversity during spontaneous and induced alcoholic fermentations. *Journal of Applied Microbiology, 75*, 551–558.

Schutz, M., & Gafner, J. (1994). Dynamics of the yeast strain population during spontaneous alcoholic fermentation determined by CHEF gel electrophoresis. *Letters in Applied Microbiology, 19*, 253–257.

Shields, R. (2006). MIAME, we have a problem. *Trends Genetics, 22*, 65–66.

Siderius, M., & Mager, W. H. (2003). Conditional response to stress in yeast. *Monatsheft fur Chemie, 134*, 1433–1444.

Sipiczki, M. (2002). Taxonomic and physiological diversity of *Saccharomyces bayanus*. In M. Ciani (Ed.), *Biodiversity and biotechnology of wine yeasts* (pp. 53–69). Kerala: Research Signpost.

Skelly, D. A., & Magwene, P. M. (2016). Population perspectives on functional genomic variation in yeast. *Briefings in Functional Genomics, 15*, 138–146.

Stahl, G., Ben Salem, S. N., Chen, L., Zhao, B., & Farabaugh, P. J. (2004). Translational accuracy during exponential, postdiauxic and stationary growth phases in *Saccharomyces cerevisiae*. *Eukaryotic Cell, 3*, 331–338.

Steenwyk, J., & Rokas, A. (2017). Extensive copy number variation in fermentation-related genes among *Saccharomyces cerevisiae* wine strains. *Genes Genomes Genetics, 7*, 1475–1485.

Stefanini, I., et al. (2012). Role of social wasps in *Saccharomyces cerevisiae* ecology and evolution. *Proceedings of the National Academy of Sciences of the United States of America, 109*, 13398–13403.

Stefanini, I., Dapporto, L., Berna, L., Polsinelli, M., Turillazzi, S., & Cavalieri, D. (2016). Social wasps are a *Saccharomyces* mating nest. *Proceedings of the National Academy of Sciences of the United States of America, 113*, 2247–2251.

Steyer, D., Ambroset, C., Brion, C., Claudel, P., Delobel, P., Sanchez, I., et al. (2012). QTL mapping of the production of wine aroma compounds by yeast. *BMC Genomics, 13*, 573.

Struhl, K. (1995). Yeast transcriptional regulatory mechanisms. *Annual Review of Genetics, 29*, 651–674.

Svetlov, V., & Cooper, T. G. (1995). Review: Compilation and characteristics of dedicated transcription factors in *Saccharomyces cerevisiae. Yeast, 11*, 1439–1484.

Terry, L. J., Shows, E. B., & Vente, S. R. (2007). Crossing the nuclear envelope: Hierarchical regulation of nucleocytoplasmic transport. *Science, 318*, 1412–1416.

Thompson, D. A., & Cubillos, F. A. (2017). Natural gene expression variation studies in yeast. *Yeast, 34*, 3–17.

Thomsson, E., Gustafsson, L., & Larsson, C. (2005). Starvation response of *Saccharomyces cerevisiae* grown in anaerobic nitrogen- or carbon- limited chemostat cultures. *Applied and Environmental Microbiology, 71*, 3007–3013.

Torija, M. J., Rozes, N., Poblet, M., Guillamon, J. M., & Mas, A. (2001). Yeast population dynamics in spontaneous fermentations: comparison between two different wine-producing areas over a period of three years. *Antonie Van Leeuwenhoek, 79*, 345–352.

Torok, T., Mortimer, R. K., Romano, P., Suzzi, G., & Polsinelli, M. (1996). Quest for wine yeasts – an old story revisited. *Journal of Industrial Microbiology, 17*, 303–313.

Townsend, J. P., Cavalieri, D., & Hartl, D. L. (2003). Population genetic variation in genome-wide gene expression. *Molecular Biology and Evolution, 20*, 955–963.

Treu, L., Campanaro, S., Nadai, C., Toniolo, C., Nardi, T., Giacomini, A., et al. (2014a). Oxidative stress response and nitrogen utilization are strongly variable in *Saccharomyces cerevisiae* wine strains with different fermentation performances. *Applied Microbiology and Biotechnology, 98*, 4119–4135.

Treu, L., Toniolo, C., Nadai, C., Sardu, A., Giacomini, A., Corich, V., et al. (2014b). The impact of genomic variability on gene expression in environmental *Saccharomyces* cerevisiae strains. *Environmental Microbiology, 16*, 1378–1379.

Tronchoni, J., Curiel, J. A., Morales, P., Torres-Pérez, R., & Gonzalez, R. (2017). Early transcriptional response to biotic stress in mixed fermentations involving *Saccharomyces cerevisiae* and *Torulaspora delbrueckii. International Journal of Food Microbiology, 241*, 60–68.

Tsai, I. J., Bensasson, D., Burt, A., & Koufopanou, V. (2008). Population genomics of the wild yeast *Saccharomyces paradoxus:* Quantifying the life cycle. *Proceedings of the National Academy of Sciences of the United States of America, 105*, 4957–4962.

Valero, E., Cambon, B., Schuller, D., Casal, M., & Dequin, S. (2006). Biodiversity of *Saccharomyces* yeast strains from grape berries of wine producing areas using starter commercial yeasts. *FEMS Yeast Research, 7*, 317–329.

Valero, E., Schuller, D., Cambon, B., Casal, M., & Dequin, S. (2005). Dissemination and survival of commercial wine yeast in the vineyard: A large-scale, three-years study. *FEMS Yeast Research, 5*, 959–969.

Van der Westhuizen, T. J., Augustyn, O. H. P., & Pretorius, I. S. (2000a). Geographical distribution of indigenous *Saccharomyces cerevisiae* strains isolated from vineyards in the costal regions of the Western Cape in South Africa. *South African Journal of Enology and Viticulture, 21*, 3–9.

Van der Westhuizen, T. J., Augustyn, O. H. P., Kahn, W., & Pretorius, I. S. (2000b). Seasonal variation of indigenous *Saccharomyces cerevisiae* strains isolated from vineyards of the Western Cape in South Africa. *South African Journal of Enology and Viticulture, 21*, 10–16.

Varela, C., Cardenas, J., Melo, F., & Agosin, E. (2005). Quantitative analysis of wine yeast gene expression profiles under winemaking conditions. *Yeast, 22*, 369–383.

Velculescu, V. E., Zhang, L., Vogelstein, B., & Kinzler, K. W. (1995). Serial analysis of gene expression. *Science, 270*, 484–487.

Versavaud, A., Courcoux, P., Roulland, C., Dulau, L., & Hallet, J.-N. (1995). Genetic diversity and geographical distribution of wild *Saccharomyces cerevisiae* strains from the wine-producing area of Charentes, France. *Applied and Environmental Microbiology, 61*, 3521–3529.

Verstrepen, K. J., Iserentant, D., Malcorps, P., Derdelinckx, G., Van Dijck, P., Winderickx, J., et al. (2004). Glucose and sucrose: Hazardous fast-food for industrial yeast? *Trends Biotechnology, 22*, 531–537.

Vezinhet, F., Hallet, J.-N., Valade, M., & Poulard, A. (1992). Ecological survey of wine yeast strains by molecular methods of identification. *American Journal of Enology and Viticulture, 43*, 83–86.

Wang, Y., Barbacioru, C., Hyland, F., Xiao, W., Hunkapiller, K. L., Blake, J., et al. (2006). Large scale real-time PCR validation on gene expression measurements from two commercial long-oligonucleotide microarrays. *BMC Genomics, 7*, 1–16.

Wang, Z., Gerstein, M., & Synder, M. (2009). RNA-Seq: A revolutionary tool for transcriptomics. *Nature Reviews/Genetics, 10*, 57–63.

Werner, T. (2003). Yeast expression-array analysis goes molecular. *Trends in Genetics, 19*, 467–469.

Winzeler, E. A., Castillo-Davis, C. I., Oshiro, G., Liang, D., Richards, D. R., Zhou, Y., et al. (2003). Genetic diversity in yeast assessed with whole-genome oligonucleotide arrays. *Genetics, 163*, 79–89.

Winzeler, E. A., Shoemaker, D. D., Astromoff, A., et al. (1999). Functional characterization of the *S cerevisiae* genome by gene deletion and parallel analysis. *Science, 285*, 901–906.

Wodicka, L., Dong, H., Mittmann, M., Ho, M. H., & Lockhart, D. J. (1997). Genome-wide expression monitoring in *Saccharomyces cerevisiae. Nature Biotechnology, 15*, 1539–1367.

Wohlschlegel, J. A., & Yates, J. R. (2003). Where's Waldo in yeast? *Nature, 425*, 671–672.

Zimmer, A., Durand, D., Loira, N., Durrens, P., Sherman, D. J., & Marullo, P. (2014). QTL dissection of lag phase in wine fermentation reveals a new translocation responsible for *Saccharomyces cerevisiae* adaptation to sulfite. *PloS ONE, 9*(1), e86298. https://doi.org/10.1371/journal.pone.0086298.

Zuzuarregui, A., & del Olmo, M. (2004). Expression of stress response genes in wine strains with different fermentative behavior. *FEMS Yeast Research, 4*, 699–710.

Zuzuarregui, A., Carrasco, P., Palacios, A., Julien, A., & del Olmo, M. (2005). Analysis of the expression of some stress induced genes in several commercial wine yeast strains at the beginning of vinification. *Journal of Applied Microbiology, 98*, 299–307.

Zuzuarregui, A., Monteolivia, L., Gil, C., & del Olmo, M. (2006). Transcriptomic and proteomic approach for understanding the molecular basis of adaptation in *Saccharomyces cerevisiae* to wine fermentation. *Applied and Environmental Microbiology, 72*, 836–847.

Chapter 6
Yeast 2.0 – Synthetic Genome Engineering Pioneers New Possibilities for Wine Yeast Research

Isak S. Pretorius

Contents

1 Solving Yeast Jigsaw Puzzles Over a Glass of Wine

More than 7000 years ago, when grapes and yeast joined forces for the first time to lift the spirit of humankind, the ancients imbibed with joy, unknowingly celebrating the beginnings of one of the world's oldest biotechnological processes. The first 'magical' drops enjoyed by the ancients came from spontaneously fermented grapes cultivated in the Zagros Mountains of Ancient Persia and the Caucasus Mountain Range between the Black and Caspian Seas. The 'mystical art' through which sugary, bland-tasting grape juice is turned into a flavoursome beverage with hedonic and preservative properties, was quickly embraced by others. The knowledge of how to make wine spilled over into neighbouring regions of Mesopotamia, Anatolia, Egypt, Phoenicia, Greece and the Mediterranean Basin.

Following colonisation by the Greeks, Phoenicians and Romans, the 'secret' of winemaking spread throughout Europe and became embedded in the diet and cultural activities of both the aristocracy and proletariat. Roman potters developed large earthenware pots for storage and transport of wine. The Gauls taught the Romans how to fashion barrels from wood. Oak barrels became the vessel of choice for yeast cells to ferment grape must into wine – a skill that survived the crumbling of the Roman Empire and the Dark Ages of economic and cultural decay.

I. S. Pretorius (✉)
Chancellery, Macquarie University, Sydney, NSW, Australia
e-mail: Sakkie.Pretorius@mq.edu.au

P. Romano et al. (eds.), *Yeasts in the Production of Wine*,
https://doi.org/10.1007/978-1-4939-9782-4_6

With the dawn of the Age of Enlightenment, geographic exploration was accompanied by a search for scientific knowledge. As European seafarers set sail to discover and explore far-off continents, their ships carried barrels filled with the 'fermenting knowledge' of winemaking and the 'geminating seeds' of yeast biotechnology. A century before the French biochemist, Louis Pasteur, zoomed in to the bubbling and frothing content of oak barrels and discovered that microscopically-tiny living yeast cells were responsible for the fermentation of grape juice into wine, the Berkshire cartographer, John Spilsbury, zoomed out to reveal the 'Big Picture' of the British Empire in 1767 through the invention of a 'dissected map'. Spilsbury pasted maps onto wood, cut them into small pieces and reconstituted a jigsaw puzzle of the world. Ever since, the jigsaw puzzle evolved into a problem-solving recreational pastime and educational toy.

Oddly, Pasteur and Spilsbury have much in common. In their indomitable quest to solve challenging problems, both were interested in uncovering every bit of detail of the 'problem' and in piecing the 'Big Picture' solution together. Their approach of seeking to understand the 'fundamentals' in the context of potential application to 'grand challenges' served us well through many centuries of scientific endeavour and remains the most powerful dynamo of technological and societal progress today. This statement also applies to the emerging science of synthetic biology.

There is no doubt that harnessing synthetic biology technologies will be crucial to help solve the puzzling challenges of a world with dwindling finite resources and a rapidly-growing and ageing global population. Maximising the bioeconomy – that is, the economic activity derived from scientific advances and innovations in biotechnology and, in particular, the engineering of biology and biomanufacturing – will be one key strategy. However, assessing current global megatrends in today's bioeconomy and finding solutions for the grand challenges of a future world is like trying to solve a complex jigsaw puzzle without all the pieces in the box (Fig. 6.1). The best approach to solving such an incomplete futuristic puzzle is to frame human futures (improved quality of life) in a planetary context (a sustainable environment). Put differently, start by separating the puzzle's edges – well-being, security and sustainability–from the middle pieces, i.e. health, food, water, energy, employment and economy (Kelly et al. 2014; Pretorius 2017a).

Once the frame of the puzzle is pieced together, the middle pieces can be sorted by colour and a more complete picture emerges. These pieces include (i) prediction and prevention of diseases; (ii) tailored, novel and affordable healthcare; (iii) adequate access to clean water and safe, nutritious foods; (iv) energy-rich molecules for renewable biofuels and novel bioenergy resources; (v) compounds for bioremediation of polluted environments and improved land-use; (vi) biodegradable pesticides and sustainable, environmentally-friendly industrial chemicals; and (vii) continuous workforce training in biodesign and biomanufacturing for productive employment in the new bioeconomy (Kelly et al. 2014; Pretorius 2017a, 2018). Bio-based designing, bioengineering and advanced biomanufacturing relevant to global needs are dependent on the development of our biological understanding and the advancement of smart data-intensive technologies (Fig. 6.2).

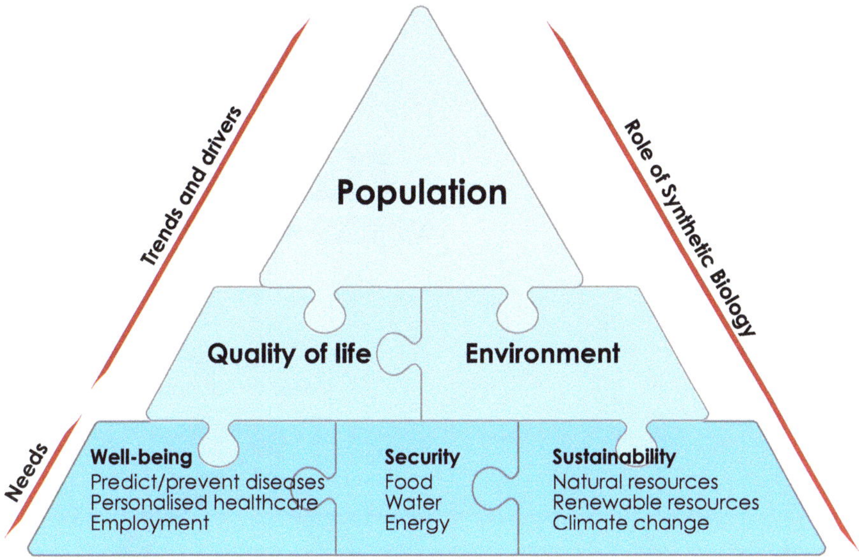

Fig. 6.1 The needs and trends associated with the growing global population require solutions to a wide variety of grand challenges linked to human well-being, security and the sustainability of the environment. This Figure is reproduced from Pretorius (2017b)

The emerging discipline of biodesign – synthetic biology – builds on the rich legacies of several branches of biology (including genetics, molecular biology and systems biology), biomolecular platforms, chemical and physical sciences, mathematical and computational sciences, data science and bioinformatics, as well as engineering and information technology (Pretorius 2017a). The capability to engineer biology resulted in the development of high-throughput analytical technologies and rapid DNA sequencing, synthesis and editing technologies fast-tracked by automated platforms in genome foundries (Fig. 6.3). These advances are making precision genome engineering faster, cheaper and more accurate by the day.

In this context, the well-studied, food-grade yeast, *Saccharomyces cerevisiae*, has become a legacy eukaryotic 'chassis' for synthetic biology (Fig. 6.4). In synthetic biology, the engineering term *chassis* refers to the organism that serves as a framework to physically accommodate new biological parts (genes), devices (gene networks) and modules (biosynthetic pathways) when biological systems (cells and organisms) are being (re)designed (Adams 2016; Pretorius 2017a, 2018). *S. cerevisiae* has had a long history as a model organism for fundamental academic research as well as being a workhorse for a wide range of industrial applications. Based on this track record, it is now the preferred 'cell factory' of semi-synthetic products, such as artemisinic acid (a precursor of the potent anti-malarial compound called artemisinin) as well as food ingredients, including vanillin, resveratrol, saffron, stevia and nootkatone (Brochado et al. 2010; Brochado and Patil 2013; Paddon et al. 2013; Li et al. 2015; Strucko et al. 2015). This means that the successful use of

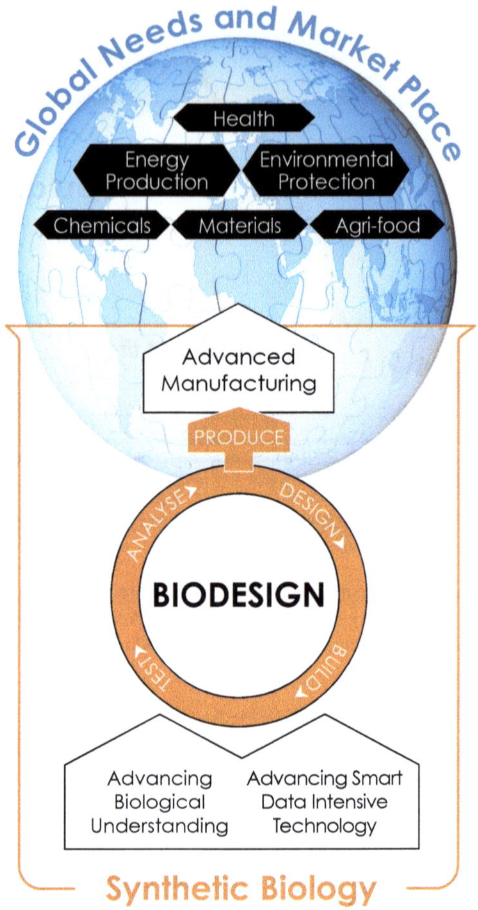

Fig. 6.2 Biodesign and biomanufacturing underpinned by synthetic biology technologies are key success factors in the new bioeconomy. This Figure is reproduced from Pretorius (2017b)

S. cerevisiae to produce these commercial products has moved synthetic biology from the 'laboratory' to the 'field', thereby changing the term 'genetically-modified organism' (GMO) to 'semi-synthetic organism' (SSO) (Jagtap et al. 2017).

2 Building the Ultimate Yeast 2.0 Genome Puzzle on Nature's Design

Recent commercial advances with semi-synthetic yeast cell factories have tossed a fresh pile of jigsaw pieces of a highly complex puzzle on the discussion tables of scientists, industry practitioners, policymakers, regulators, governments, consumers and society at large. One approach to help solve this puzzle is to create a fully man-made genome for *S. cerevisiae* so that we can better understand the biological intricacies of eukaryotic SSOs and be able to more accurately predict and control the

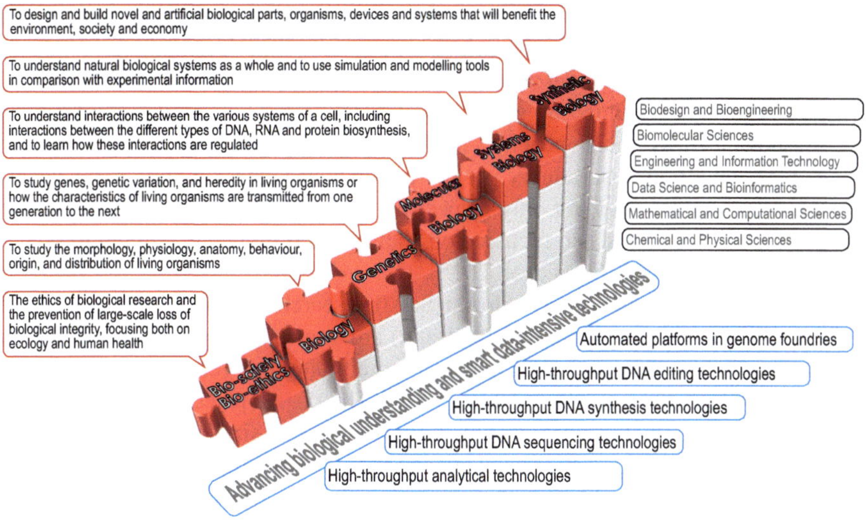

Fig. 6.3 The evolution of biology, genetics, molecular biology and systems biology into synthetic biology

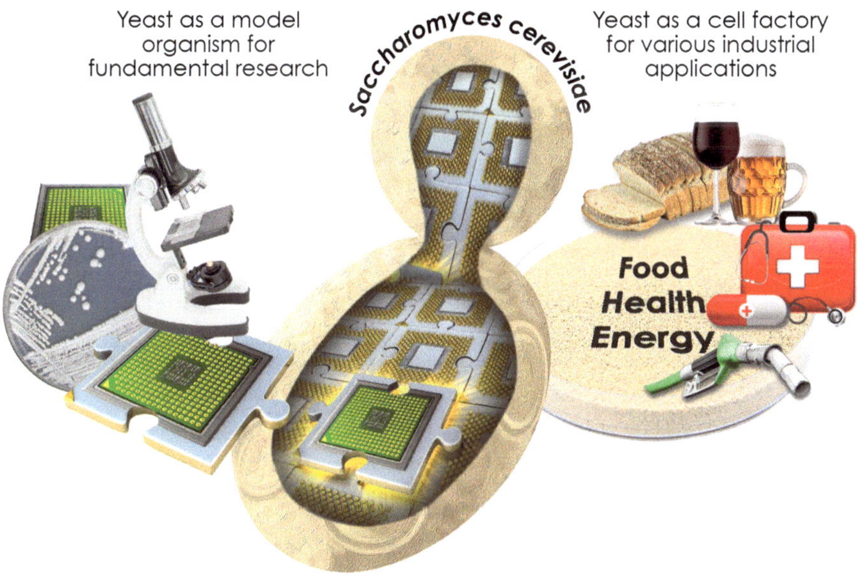

Fig. 6.4 The multi-purpose yeast, *Saccharomyces cerevisiae*, is the best-studied eukaryotic model organism and the most used microbe in the fermentation industry. This Figure is reproduced from Pretorius (2017b)

practical outcomes of *genome engineering* (as opposed to individual gene-based *genetic engineering*). This is the primary purpose of the international *Synthetic Yeast Genome Project*, known as the *Yeast 2.0* or *Sc2.0* project. This ambitious collaborative project is guided by an agreed and legally-binding policy statement on key issues, such as social benefits, intellectual property, safety and governance (Fig. 6.5). This multidisciplinary project is conducted under a sound self-governance structure that does not only respect the scientific achievements and lessons from the past, but that is also focussed on leading the present and helping to secure a brighter future for all (Sliva et al. 2015).

The laboratory benches of a dozen Sc2.0 research groups around the world (USA, UK, China, Singapore and Australia) are strewn with pieces of a complex 6000-piece (i.e. 6000-gene) jigsaw puzzle comprising the genetic make-up of *S. cerevisiae* (Goffeau et al. 1996; Oliver 1996). The puzzle pieces have been sorted into 16 piles (i.e. 16 chromosomes) by colour, shape and size and divided between the puzzle-masters (Fig. 6.6). The challenge is to recreate the guide picture on the front of the box – a round-to-ovoid single-celled fungus, 50–10 μm in diameter and compartmentalised like most other eukaryotic cells, including an encapsulated nucleus. While remaining true to this 'blueprint', each of the collaborating laboratories must meet the challenge of designing, building and interlocking the pieces of the *S. cerevisiae* genome puzzle in order to reveal, for the first time, the completely rebuilt picture of a eukaryote's genome. Each piece of the puzzle is essential if a

Fig. 6.5 The Yeast 2.0 project balances scientific freedom and social responsibility with a legally-binding agreement. This project is guided by a statement on safety and ethics (Sliva et al. 2015). This approach of ongoing oversight, self-regulation and self-governance provides the Yeast 2.0 consortium with an effective and dynamic framework for maximising the benefits of this large-scale project and minimising the risk for harm or damage

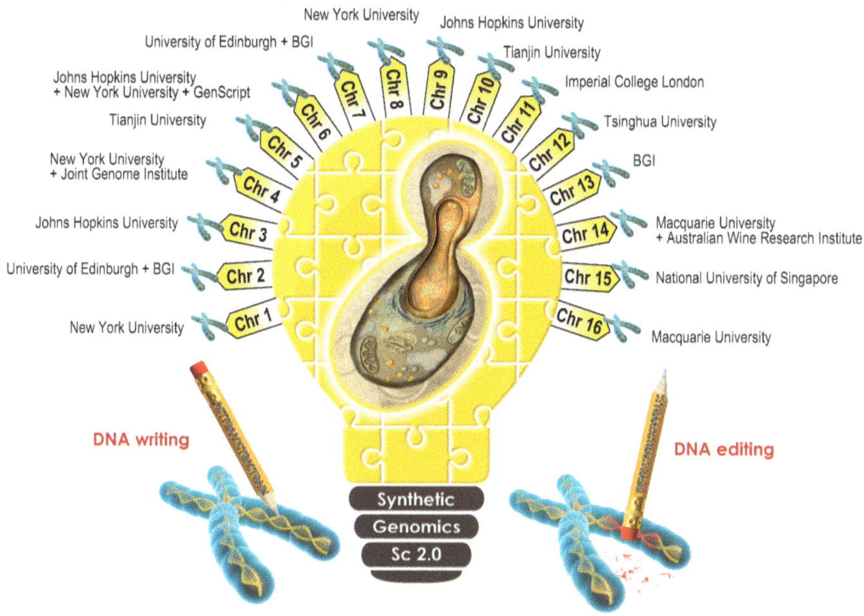

Fig. 6.6 A dozen research laboratories in five countries – USA, UK, China, Singapore and Australia – aim to design and build the world's first synthetic yeast genome (consisting of 16 chromosomes). This Figure is reproduced from Pretorius (2017b)

complete picture is to be produced and the Sc2.0 team is working to have every piece in place by 2018.

The design of the Sc2.0 genome draws on the data from the genome sequence first announced in 1996 for a haploid laboratory strain (S288c) of *S. cerevisiae*. The ~12 Mb (non-redundant) to ~14 Mb (total) genome sequence carries approximately 6000 genes of which about 5000 are individually non-essential. The 6000 genes are distributed along 16 linear chromosomes of varying length (200–2000 kb). The first step towards designing and building *S. cerevisiae*'s genome was taken in 2011 with the successful construction of a manually-designed circular synthetic version of the right arm of chromosome 9 (chr9R), which was labelled syn9R (Dymond et al. 2011; Dymond and Boeke 2012). This opened the way for the synthesis of Chromosome 3 – the third smallest *S. cerevisiae* chromosome – in full in 2014 (Dymond and Boeke 2012; Annaluru et al. 2014; Gibson and Venter 2014). In 2017, the synthetic versions of five more *S. cerevisiae* chromosomes (i.e. Chromosomes 2, 5, 6, 10 and 12) have been published (Fig. 6.7) (Mercy et al. 2017; Mitchell et al. 2017; Richardson et al. 2017; Shen et al. 2017; Wu et al. 2017; Xie et al. 2017; Zhang et al. 2017). It is expected that all 16 chromosomes will be synthesised by the end of 2018. The Sc2.0 project is thus on track to consolidate the 16 chemically-synthesised chromosomes currently being built by the Consortium teams around the world into a single cell of *S. cerevisiae* by early 2019 (Fig. 6.8).

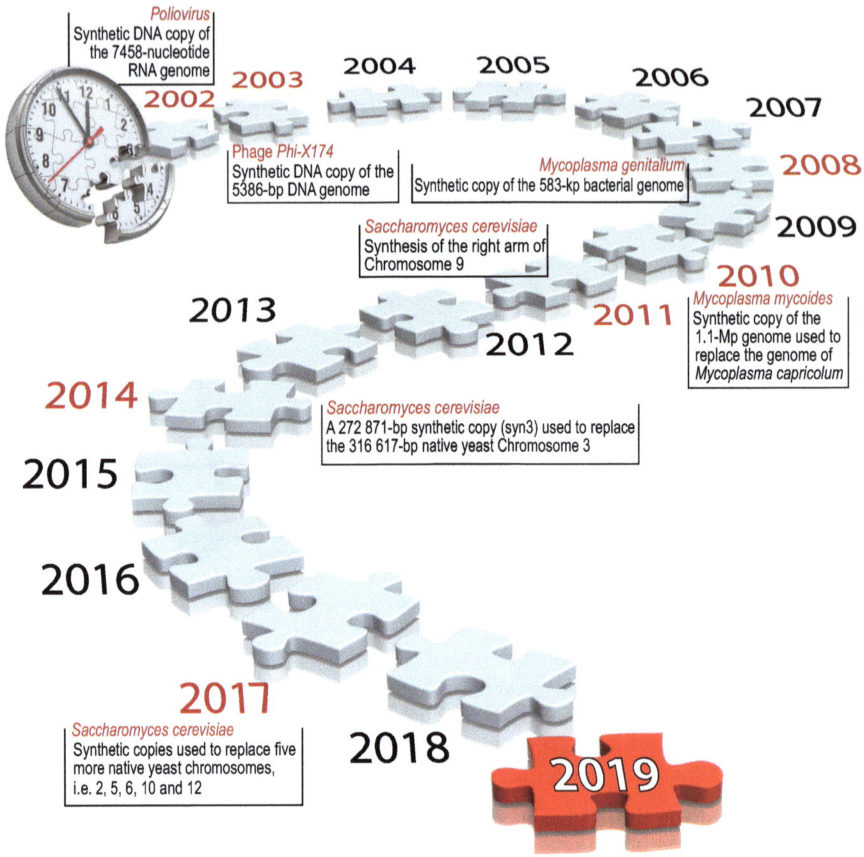

Fig. 6.7 Key milestones in terms of the synthesis of viral and bacterial genomes inspired the idea to chemically synthesise the 16 chromosomes of the yeast *Saccharomyces cerevisiae* and replace the native chromosomes with the synthetic chromosomes. This Figure is adapted from Pretorius (2017b)

This consolidated Sc2.0 genome was designed from the start to be fully customisable so that researchers will be able to ask otherwise intractable questions about the fundamentals of chromosome structure, organisation, function and evolution, as well as gene content, function of RNA splicing and the extent to which small RNAs play a role in yeast biology (Richardson et al. 2017). The guiding design principles for the Sc2.0 genome aspired to balance a desire to preserve the phenotype of the wild-type yeast strain while incorporating inducible flexibility for further manipulation and minimising instability resulting from the repetitive nature of yeast's native genomic DNA. These principles for the design, construction, analysis, fitness testing and curation are most likely to be scalable to future synthetic work on the larger genomes of plants, animals and humans (Richardson et al. 2017).

The final Sc2.0 genome is therefore designed, curated, streamlined and reorganised to encode a slightly modified genetic code (Richardson et al. 2017). To facilitate

Fig. 6.8 The collaborative *Yeast 2.0* (*Sc2.0*) project, which commenced in 2011, has made significant progress with 6 of the 16 chromosomes synthesised – well on track to have all the native chromosomes of the yeast *Saccharomyces cerevisiae* replaced by early 2019. This Figure is reproduced from Pretorius (2017b)

the assembly of the synthetic chromosomes, specific base substitutions within some of the open reading frames (ORFs) are included in the redesigned genome to introduce necessary enzyme recognition sites or remove inconvenient enzyme recognition sites. In addition, recognisable PCRTags – short recoded sequences within certain ORFs facilitating a polymerase chain reaction (PCR)-based assay – are also included in the design of the Sc2.0 genome so that the synthetic DNA can be distinguished from native DNA (Richardson et al. 2017). Other important modifications include the addition of many *lox*Psym sites for future genome scrambling purposes; all TAG stop codons are recoded to TAA; all repetitive and dispensable sequences, such as *Ty* transposons, introns, subtelomeric regions and silent mating-type loci (*HML* and *HMR* located on Chromosome 3) are omitted from the design; and all tRNA genes are relocated to a novel neochromosome in this redesigned yeast genome (Richardson et al. 2017). The expectation is that these designer changes would not cause any fitness defects but would allow a whole gamut of future genome manipulations and research opportunities. To date, about 75% of the DNA synthesis is complete and built into discrete strains by the various Sc2.0 teams; thus this game-changing synthetic biology project is on schedule to deliver the world's first synthetic eukaryotic genome within 12–18 months.

As the Sc2.0 project is progressing, genome engineering technologies are being advanced at a rapid pace while important fundamental biological intricacies of yeast cells are being figured out. By the end of this project, it would be known, for example, if the removal of all introns and transposable elements will affect cell fitness, and whether the relocation of all tRNA genes to a 17th mini-neochromosome will disadvantage the genetic processes and protein synthesis machinery of the redesigned haploid S288c laboratory strain of *S. cerevisiae*. These are just a few examples of the puzzling questions that are being answered as the Sc2.0 picture emerges.

3 Customising Wine Yeast Puzzles Under Challenging Conditions

From a wine scientist's viewpoint, another perplexing question is whether some of these fundamental learnings from the Sc2.0 laboratory-bred yeast strain carrying a man-made genome can be extrapolated to robust industrial wine yeast strains (Pretorius 2016, 2017a, 2018; Goold et al. 2017). Wine yeast researchers expect to gain much knowledge by borrowing some of the Sc2.0 puzzle pieces to help build full pictures of their favourite wine yeast strains. Since yeast fermentation is a centre piece in the process of winemaking, there is much to be gained to unlock the genetic secrets that make different wine yeast strains perform differently. By understanding the fundamentals, the realms of possibility in strain improvement programmes could be stretched by redesigning some the wine yeast strains' natural jigsaw puzzle pieces or inventing totally new ones. The objective of such strain development programmes would be to provide winemakers with a diverse array of wine yeast strains. Each strain would be specifically tailored to produce particular wine styles identified for various markets and market segments the world over.

Not all yeast strains are equally able to catalyse rapid, complete and efficient conversion of grape sugars to ethanol, carbon dioxide and other minor, but important metabolites (e.g., acids, alcohols, carbonyls, esters, terpenes and thiols) without the development of off-flavours (e.g., hydrogen sulphide, volatile acids and volatile phenols) (Pretorius 2000; Pretorius and Bauer 2002; Chambers and Pretorius 2010; Pretorius et al. 2012). Wine yeasts can differ widely in terms of their robustness, fermentation efficiencies and sensory properties, and performance depends on the specific composition of a particular grape juice and specific fermentation conditions and techniques used by the winemaker. During the past three decades or so, a wide variety of strain improvement techniques have been harnessed to optimise fermentation performance, robustness, spoilage-control, processing efficiency, product wholesomeness and sensory quality (Pretorius 2000; Pretorius and Bauer 2002; Chambers and Pretorius 2010; Pretorius et al. 2012).

Non-genetic modification (non-GM) techniques include *hybridisation* (mating or cross-breeding), *mutagenesis* (induction of mutations by exposure to mutagenic chemicals or ultraviolet radiation) and *adaptive evolution* (crossing and back-crossing of selected mutants) (Pretorius 2000). Several hybrid and mutant strains generated by these non-GM techniques have been used successfully in global commercial winemaking. Consumers had no hesitation embracing the many award-winning wines produced with rapid-fermenting and aroma-enhancing hybrid strains originating from mating and cross-breeding, or the many fault-free wines produced with mutants that no longer produced off-flavours, such as hydrogen sulphide (H_2S), volatile acidity (VA) and volatile phenols (Van der Westhuizen and Pretorius 1992; 2000; Pretorius and Bauer 2002; Cordente et al. 2009; Chambers and Pretorius 2010; Pretorius et al. 2012; Cordente et al. 2013; Jagtap et al. 2017; Pretorius 2017a).

Such broad-based acceptance by producers and consumers is, however, not the case for strains generated by genetic engineering. More than 10 years ago, the first two GM wine yeast strains, ML01 and ECMo01, which met all the regulatory requirements, were commercialised in the USA, Canada and Moldova (Volschenk et al. 2004; Coulon et al. 2006; Husnik et al. 2006). Despite the proven success in winemaking trials and the clear benefits to both producers and consumers of the ML01 malolactic strain and the ECMo01 low-ethylcarbamate strain, there is yet to be widespread uptake of these two genetically-engineered wine yeasts in commercial winemaking. ML01 and ECMo01 are not the only GM wine yeasts twiddling their budding thumbs at the entrances of wineries and cellar doors (Fig. 6.9). Several robust and flavour-active strains have been developed to mitigate stuck fermentations during problematic hot vintages and to create market-driven wines with desired alcohol levels (Nevoigt and Stahl 1996; Tilloy et al. 1996; Michnick et al. 1997; Remize et al. 1999; De Barros Lopes et al. 2000; Eglinton et al. 2002; Cambon et al. 2006; Varela et al. 2012; Goold et al. 2017) and aroma profiles (Lilly et al. 2006a, b; Swiegers et al. 2007; Cordente et al. 2009; Roncoroni et al. 2011). So far, the well-orchestrated anti-GMO campaigns, the furore over the labelling of GM food products and associated market sensitivities have deterred winemakers to take full advantage of science and the opportunities afforded by *genetic engineering*, and now more recently, by *genome engineering*.

While the wine industry is caught up in the scientific and cultural maelstrom of the ‘wonders and terrors’ of GMOs and SSOs, researchers continue to mine DNA sequencing data for the responsible design, synthesis and/or editing of wine yeast genomes with huge potential benefits for producers and consumers alike. On one hand, anti-GMO/SSO campaigners and uninformed traditionalists might dismiss such research as an ‘unwanted reality’ that will eventually vanish into thin air. However, future-focussed innovators, on the other hand, are highly supportive of these synthetic genomic research efforts because they recognize that researchers are gaining invaluable insights into the molecular intricacies of wine yeast cells. Akin to what research into the Higgs boson elementary particle is revealing about the ‘Standard Model’ of particle physics, synthetic genomic experimentation is illuminating the biomolecular mysteries of wine yeast cells. Factually-correct information and knowledge gained from such fundamental research and evidence-based data are the only way to counteract ideologically-driven doomsday prophecies, exaggerated fantasies, empty promises and guesswork about the future of SSOs.

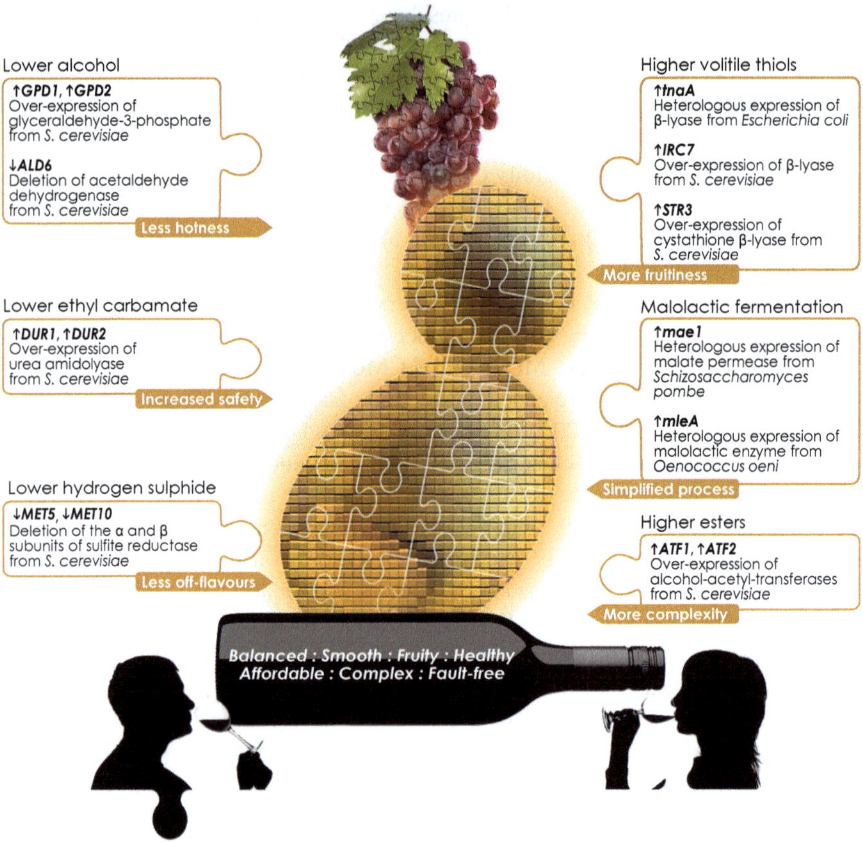

Fig. 6.9 Examples of wine yeasts bioengineered for improved robustness, fermentation efficiency and sensory attributes. This Figure is reproduced from Pretorius (2017b)

4 Uncorking a Raspberry-Tasting Chardonnay Produced by a Semi-Synthetic Wine Yeast

History was made with the recent successful development of the world's first semi-synthetic wine yeast as a 'demonstrator' project. With this ground-breaking work, a cassette of four synthetic genes encoding the production of a highly-desirable fragrant raspberry ketone – 4-[4-hydroxyphenyl]-butan-2-one – was embedded into the genetic blueprint of a wine yeast strain (Fig. 6.10) (Lee et al. 2016). This phenylpropanoid is the primary aroma compound found in several fruits, vegetables and berries (including raspberries, blackberries, grapes and rhubarb) but, due to the low concentrations present in these plants, it is not economical to extract this flavoursome compound from its natural sources. However, thanks to market preferences, chemically-manufactured derivatives of this flavouring agent fetch much lower

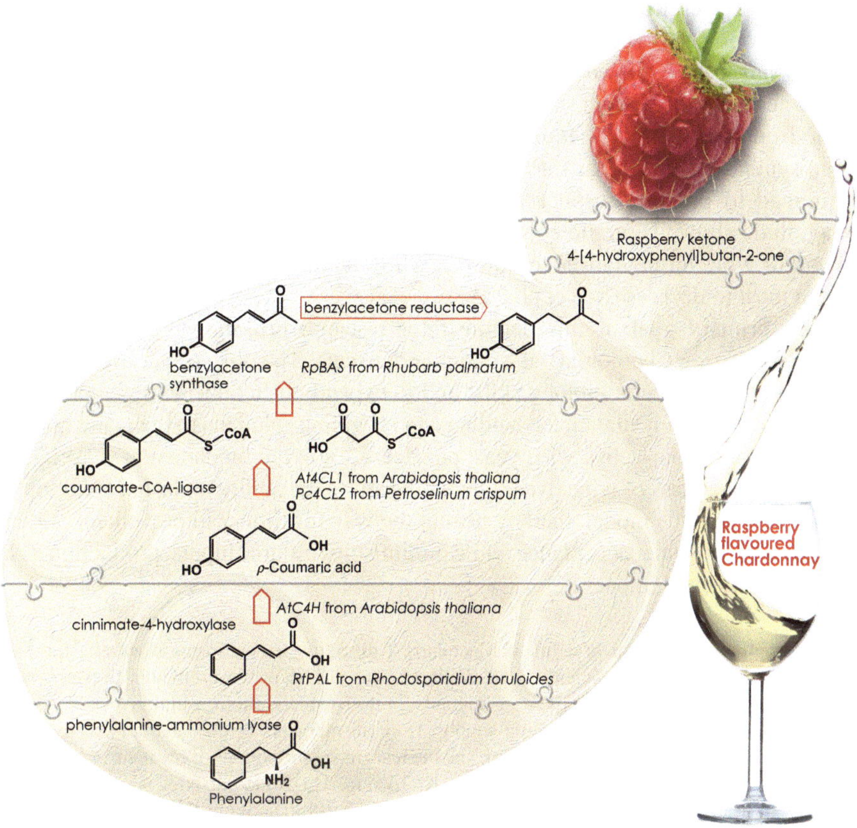

Fig. 6.10 The first semi-synthetic wine yeast capable of producing Chardonnay wine with a raspberry aroma. This figure is reproduced from Pretorius (2017b)

prices than the naturally-derived form. This led to early attempts to produce raspberry ketone from *p*-coumaric acid in heterologous bacterial and yeast strains. However, the high cost of *p*-coumaric acid as a substrate and the trace amounts of raspberry ketone obtained in these GM strains, prevented commercial production of this phenylpropanoid as a food-grade flavouring agent. The missing puzzle pieces in this work are the ability to eliminate the requirement for supplementing the culture medium with expensive *p*-coumaric acid and to increase the yield of 4-[4-hydroxyphenyl]-butan-2-one (Lee et al. 2016).

Solving such a scientific puzzle starts with the unravelling of the phenylpropanoid biosynthetic pathway. This pathway commences with the conversion of phenylalanine to *p*-coumaric acid via cinnamate or directly from tyrosine to *p*-coumaric acid. Conversion of *p*-coumaric acid to raspberry ketone requires three additional enzymatic steps including a condensation reaction between coumaroyl-CoA and malonyl-CoA. To design a biosynthetic pathway for the de novo production of 4-[4-hydroxyphenyl]-butan-2-one in a wine yeast, the following codon-optimised genes were chemically synthesised and integrated into *HO* locus of a wine yeast strain (AWRI1631): the phenylalanine ammonia lyase from an oleaginous yeast, *Rhodosporidium toruloides*; the cinnamate-4-hydroxylase from the well-characterized model plant, *Arabidopsis thaliana*; and the coumarate CoA ligase 2 gene from parsley, *Petroselinum crispum*, fused by a rigid linker to the benzalacetone synthase from rhubarb, *Rheum palmatum*. This semi-synthetic organism was equipped to produce raspberry ketone at concentrations almost two orders of magnitude above its predicted sensory threshold in Chardonnay grape juice under standard wine fermentation conditions, while retaining the ability to ferment the grape must to dryness (Lee et al. 2016).

The primary goal of this demonstrator research project was not to produce raspberry-tasting Chardonnay at a commercial-scale. The objective was to hone our technological synthetic biology skills and to expand our toolkit with which we can advance our fundamental understanding and providing solutions to the many riddling questions of flavour-active wine yeast puzzles. By solving fundamental yeast jigsaw puzzles over a glass of wine, we might well acquire the ability to design the ultimate wine yeast genome model, thereby paving the way for further improvement of wine quality and consumer acceptance while minimising resource inputs, production costs and environmental impact.

Acknowledgements I am grateful to Macquarie University, Bioplatforms Australia, the New South Wales (NSW) Chief Scientist and Engineer, and the NSW Government's Department of Primary Industries for providing the start-up funds for the Synthetic Biology initiative at Macquarie University. I acknowledge the ongoing support from all members of the international Synthetic Yeast Genome Project (Sc2.0) consortium, and the research contributions of my team at Macquarie University: Drs Thomas Williams, Heinrich Kroukamp, Hugh Goold, Niël van Wyk, Natalie Curach and Ian Paulsen, and my collaborators at The Australian Wine Research Institute: Drs Anthony Borneman, Dariusz Kutyna and Daniel Johnson. Rae Blair and Tori Hocking are gratefully acknowledged for their editorial assistance, and Bronte Turner for the artwork. This chapter is a slightly re-worked version of an article published in 2017 in *EMBO Reports* 18 (11):1875-1884. Information is reproduced here with kind permission of the publisher, EMBO Press.

Conflict of Interest The author declares that he has no conflict of interest.

References

Adams, B. L. (2016). The next generation of synthetic biology chassis: Moving synthetic biology from the laboratory to the field. *ACS Synthetic Biology, 5*, 1328–1330.

Annaluru, N., Muller, H., Mitchell, L. A., Ramalingam, S., Stracquadanio, G., Richardson, S. M., et al. (2014). Total synthesis of a functional designer eukaryotic chromosome. *Science, 344*, 55–58.

Brochado, A. R., & Patil, K. R. (2013). Overexpression of *O*-methyltransferase leads to improved vanillin production in baker's yeast only when complemented with model-guided network engineering. *Biotechnology and Bioengineering, 110*, 656–659.

Brochado, A. R., Matos, C., Møller, B. L., Hansen, J., Mortensen, U. H., & Patil, K. R. (2010). Improved vanillin production in baker's yeast through *in silico* design. *Microbial Cell Factories, 9*, 1–15.

Cambon, B., Monteil, V., Remize, F., Camarasa, C., & Dequin, S. (2006). Effects of *GPD1* overexpression in *Saccharomyces cerevisiae* commercial wine yeast strains lacking *ALD6* genes. *Applied and Environmental Microbiology, 72*, 4688–4694.
Chambers, P. J., & Pretorius, I. S. (2010). Fermenting knowledge: The history of winemaking, science and yeast research. *EMBO Reports, 11*, 1–7.
Cordente, A. G., Heinrich, A. J., Pretorius, I. S., & Swiegers, J. H. (2009). Isolation of sulfite reductase variants of a commercial wine yeast with significantly reduced hydrogen sulfide production. *FEMS Yeast Research, 9*, 446–459.
Cordente, A. G., Cordero-Bueso, G., Pretorius, I. S., & Curtin, C. D. (2013). Novel wine yeast with mutations in *YAP1* that produce less acetate during fermentation. *FEMS Yeast Research, 13*, 62–73.
Coulon, J., Husnik, J. I., Inglis, D. L., Van der Merwe, G. K., Lonvaud, A., Erasmus, D. J., et al. (2006). Metabolic engineering of *Saccharomyces cerevisiae* to minimize the production of ethyl carbamate in wine. *American Journal of Enology and Viticulture, 57*, 113–124.
De Barros Lopes, M., Rehman, A., Gockowiak, H., Heinrich, A. J., Langridge, P., & Henschke, P. A. (2000). Fermentation properties of a wine yeast over-expressing the *Saccharomyces cerevisiae* glycerol-3-phosphate dehydrogenase gene (*GPD2*). *Australian Journal of Grape and Wine Research, 6*, 208–215.
Dymond, J., & Boeke, J. (2012). The *Saccharomyces cerevisiae* SCRaMbLE system and genome minimization. *Bioengineered, 3*, 170–173.
Dymond, J. S., Richardson, S. M., Coombes, C. E., Babatz, T., Muller, H., Annaluru, N., et al. (2011). Synthetic chromosome arms function in yeast and generate phenotypic diversity by design. *Nature, 477*, 471–476.
Eglinton, J. M., Heinrich, A. J., Pollnitz, A. P., Langridge, P., Henschke, P. A., & de Barros Lopes, M. A. (2002). Decreasing acetic acid accumulation by a glycerol overproducing strain of *Saccharomyces cerevisiae* by deleting the *ALD6* aldehyde dehrydrogenase gene. *Yeast, 19*, 295–301.
Gibson, D. G., & Venter, J. G. (2014). Construction of a yeast chromosome. *Nature, 509*, 168–169.
Goffeau, A., Barrell, B. G., & Bussey, H. (1996). Life with 6000 genes. *Science, 274*, 546–567.
Goold, H. D., Kroukamp, H., Williams, T. C., Paulsen, I. T., Varela, C., & Pretorius, I. S. (2017). Yeast's balancing act between ethanol and glycerol production in low-alcohol wines. *Microbial Biotechnology, 10*, 264–278.
Husnik, J. I., Volschenk, H., Bauer, J., Colavizza, D., Luo, Z., & Van Vuuren, H. J. J. (2006). Metabolic engineering of malolactic wine yeast. *Metabolic Engineering, 8*, 315–323.
Jagtap, U. B., Jadhav, J. P., Bapat, V. A., & Pretorius, I. S. (2017). Synthetic biology stretching the realms of possibility in wine yeast research. *International Journal of Food Microbiology, 252*, 24–34.
Kelly, N. J., Whelan, D. J., Kerr, E., Apel, A., Beliveau, R., & Scanlon, R. (2014). Engineering biology to address global problems: Synthetic biology markets, needs, and applications. *Industrial Biotechnology, 10*, 140–149.
Lee, D., Lloyd, N., Pretorius, I. S., & Borneman, A. R. (2016). Heterologous production of raspberry ketone in the wine yeast *Saccharomyces cerevisiae* via pathway engineering and synthetic enzyme fusion. *Microbial Cell Factories, 15*, 49–55.
Li, M., Kildegaard, K. R., Chen, Y., Rodriguez, A., Borodina, I., & Nielsen, J. (2015). *De novo* production of resveratrol from glucose or ethanol by engineered *Saccharomyces cerevisiae*. *Metabolic Engineering, 32*, 1–11.
Lilly, M., Bauer, F. F., Lambrechts, M. G., Swiegers, J. H., Cozzolino, D., & Pretorius, I. S. (2006a). The effect of increased alcohol acetyl transferase and esterase activity on flavour profiles of wine and distillates. *Yeast, 23*, 641–659.
Lilly, M., Styger, G., Bauer, F. F., Lambrechts, M. G., & Pretorius, I. S. (2006b). The effect of increased yeast branched-chain amino acid transaminase activity and the production of higher alcohols on the flavour profiles of wine and distillates. *FEMS Yeast Research, 6*, 726–743.
Mercy, G., Mozziconacci, J., Scolari, V. F., Yang, K., Zhao, G., Thierry, A., et al. (2017). 3D organization of synthetic and scrambled chromosomes. *Science, 355*, 1050–eaaf4597.

Michnick, S., Roustan, J. L., Remize, F., Barre, P., & Dequin, S. (1997). Modulation of glycerol and ethanol yields during alcoholic fermentation in *Saccharomyces cerevisiae* strains overexpressed or disrupted for *GPD1* encoding glycerol 3-phosphate dehydrogenase. *Yeast, 13*, 783–793.

Mitchell, L. A., Wang, A., Stracquadanio, G., Kuang, Z., Wang, X., Yang, K., et al. (2017). Synthesis, debugging, and effects of synthetic chromosome consolidation: SynVI and beyond. *Science, 355*, 1045–eaaf4831.

Nevoigt, E., & Stahl, U. (1996). Reduced pyruvate decarboxylase and increased glycerol-3-phosphate dehydrogenase [NAD^+] levels enhance glycerol production in *Saccharomyces cerevisiae*. *Yeast, 12*, 1331–1337.

Oliver, S. G. (1996). From DNA sequence to biological function. *Nature, 379*, 597–600.

Paddon, C. J., Westfall, P. J., Pitera, D. J., Benjamin, K., Fisher, K., McPhee, D., et al. (2013). High-level semisynthetic production of the potent antimalarial artemisinin. *Nature, 496*, 528–532.

Pretorius, I. S. (2000). Tailoring wine yeast for the new millennium: Novel approaches to the ancient art of winemaking. *Yeast, 16*, 675–729.

Pretorius, I. S. (2016). Conducting wine symphonics with the aid of yeast genomics. *Beverages, 2*. https://doi.org/10.3390/beverages2040036.

Pretorius, I. S. (2017a). Synthetic genome engineering forging new frontiers for wine yeast. *Critical Reviews in Biotechnology, 37*, 112–136.

Pretorius, I. S. (2017b). Solving yeast jigsaw puzzles over a glass of wine: Synthetic genome engineering pioneers new possibilities for wine yeast research. *EMBO Reports, 18*, 1875–1884.

Pretorius, I. S. (2018). Yeast 2.0 – connecting the dots in the construction of the world's first functional synthetic eukaryotic genome. *FEMS Yeast Research* (in press), *18*.

Pretorius, I. S., & Bauer, F. F. (2002). Meeting the consumer challenge through genetically customised wine yeast strains. *Trends in Biotechnology, 20*, 426–432.

Pretorius, I. S., Curtin, C. D., & Chambers, P. J. (2012). The winemaker's bug: From ancient wisdom to opening new vistas with frontier yeast science. *Bioengineered, 3*, 149–158.

Remize, F., Roustan, J., Sablayrolles, J., Barre, P., & Dequin, S. (1999). Glycerol overproduction by engineered *Saccharomyces cerevisiae* wine yeast strains leads to substantial changes in by-product formation and to a stimulation of fermentation rate in stationary phase. *Applied and Environmental Microbiology, 65*, 143–149.

Richardson, S. M., Mitchell, L. A., Stracquadanio, G., Yang, K., Dymond, J. S., DiCarlo, J. E., et al. (2017). Design of a synthetic yeast genome. *Science, 355*, 1040–1044.

Roncoroni, M., Santiago, M., Hooks, D. O., Moroney, S., Harsch, M. J., Lee, S. A., et al. (2011). The yeast *IRC7* gene encodes a β-lyase responsible for production of the varietal thiol 4-mercapto-4-methylpentan-2-one in wine. *Food Microbiology, 28*, 926–935.

Shen, Y., Wang, Y., Chen, T., Gao, F., Gong, J., Abramczyk, D., et al. (2017). Deep functional analysis of synII, a 770-kilobase synthetic yeast chromosome. *Science, 355*, 1047–eaaf4791.

Sliva, A., Yang, H., Boeke, J. D., & Mathews, D. J. (2015). Freedom and responsibility in synthetic genomics: The synthetic yeast project. *Genetics, 200*, 1021–1028.

Strucko, T., Magnesko, O., & Mortensen, U. H. (2015). Benchmarking two commonly used *Saccharomyces cerevisiae* strains for heterologous vanillin-β-glucoside production. *Metabolic Engineering Communications, 2*, 99–108.

Swiegers, J. H., Capone, D. L., Pardon, K. H., Elsey, G. M., Sefton, M. A., Francis, I. L., et al. (2007). Engineering volatile thiol release in *Saccharomyces cerevisiae* for improved wine aroma. *Yeast, 24*, 561–574.

Tilloy, V., Cadiere, A., Ehsani, M., & Dequin, S. (1996). Reducing alcohol levels in wines through rational and evolutionary engineering of *Saccharomyces cerevisiae*. *International Journal of Food Microbiology, 213*, 49–58.

Van der Westhuizen, T. J., & Pretorius, I. S. (1992). The value of electrophoretic fingerprinting and karyotyping in wine yeast breeding programmes. *Antonie van Leeuwenhoek, 61*, 249–257.

Varela, C., Kutyna, D. R., Solomon, M. R., Black, C. A., Borneman, A. R., Henschke, P. A., et al. (2012). Evaluation of gene modification strategies to develop low-alcohol wine yeasts. *Applied and Environmental Microbiology, 78*(17), 6068–6077.

Volschenk, H., Viljoen-Bloom, M., Van Staden, J., Husnik, J., & Van Vuuren, H. J. J. (2004). Genetic engineering of an industrial strain of *Saccharomyces cerevisiae* for L-malic acid degradation via an efficient malo-ethanolic pathway. *South African Journal of Enology and Viticulture, 25*, 63–73.

Wu, Y., Li, B. Z., Zhao, M., Mitchell, L. A., Xie, Z. X., Lin, Q. H., et al. (2017). Bug mapping and fitness testing of chemically synthesized chromosome X. *Science, 355*, 1048–eaaf4706.

Xie, Z. X., Li, B. Z., Mitchell, L. A., Wu, Y., Qi, X., Jin, Z., et al. (2017). Perfect designer chromosome V and behavior of a ring derivative. *Science, 355*, 1046–eaaf4704.

Zhang, W., Zhao, G., Luo, Z., Lin, Y., Wang, L., Guo, Y., et al. (2017). Engineering the ribosomal DNA in a megabase synthetic chromosome. *Science, 355*, 1049–eaaf3981.

Chapter 7
Yeasts in Botrytized Wine Making

Matthias Sipiczki

Contents

1 Introduction

The plant pathogenic fungus, *Botrytis cinerea* Pers. ex Fr (teleomorph *Botryotinia fuckeliana* (de Bary) Whetzel) is responsible for the destructive grey rot (or grey mould) on grapes (Elad et al. 2004). It is one of the most serious threats for vine growers, and causes heavy losses of yield in wine grapes in many places around the world. But in certain terroirs characterized by very specific microclimatic

M. Sipiczki (✉)
Department of Genetics and Appled Micobiology, University of Debrecen, Debrecen, Hungary

P. Romano et al. (eds.), *Yeasts in the Production of Wine*,
https://doi.org/10.1007/978-1-4939-9782-4_7

conditions, it can have a magical effect on the grape: it generates the benevolent process called noble rot, described for the first time in Germany 130 years ago (Müller-Thurgau 1888). The grape undergoes drastic desiccation and chemical transformations leading to extremely high sugar concentrations and unique aroma composition. Wines made from nobly rotten grapes, the so-called botrytized sweet wines, belong to the greatest white wines of the world. Although the key element in noble rot is *Botrytis* invasion, many other fungi, yeasts and bacteria can co-colonise the *Botrytis*-infected berries and make beneficial contributions to the quality of the grape juice as well. These pre-harvest colonists commence fermentation in the berries and then become the initial microflora of the must. The botrytized must is particularly rich in non-*Saccharomyces* yeasts. Some of them can persist throughout most of the fermentation-vinification process. *Saccharomyces cerevisiae* is the principal yeast also in botrytized wine fermentation, but two other species, *Saccharomyces uvarum* and *Candida zemplinina*, are nearly as important. This chapter reviews the growth and activity of yeasts during the development of noble rot, in the course of the fermentation of botrytized grape must and after the completion of fermentation.

2 Terminology

2.1 Sweet Wine – Botrytized Wine – Botrytis-Affected Wine

All sweet wines have high sugar concentrations which can be achieved, broadly speaking, in two ways. The first way involves arresting the fermentation before completion with the addition of spirit (fortification), resulting in what is really a blend of wine consisting of partially fermented grape juice and added grape spirit (fortified wines, see Chap. 10). The other method is based on dehydration of grapes; a reduction of the grape water content either before or after the harvest. A common practise is leaving the grapes on the vine longer than usual, allowing them to shrivel (late harvest wine). Many wine regions use this method and produce nice sweet wines. If dehydration is associated with the *Botrytis*-generated "noble rot", the grapes not only loose water but also gain specific properties that make them suitable for making a very specific type of wines referred to as botrytized sweet wines (Doneche 1993).

Regulations in certain countries make distinctions between "botrytized wines" and "*Botrytis*-affected wines" [e.g. Austria: law BGBl. I Nr. 111/2009 (https://www.ris.bka.gv.at/eli/bgbl/I/2009/111); Czech Republic: law 321/2004 (https://web.archive.org/web/20110718190125/http://www.vinarskyfond.cz/legislativa/1-vinarska_legislativa_2004.pdf); Germany: law BGBl. I S. 66 (https://www.gesetze-im-internet.de/weing_1994/index.html); Hungary: law XVIII/2004 (https://net.jogtar.hu/jr/gen/hjegy_doc.cgi?docid=a0400018.tv); Slovakia: law 313/2009 (http://www.epi.sk/zz/2009-313); European Union: Commission Regulation 607/2009 (http://eur-lex.europa.eu/LexUriServ/LexUriServ.do?uri=OJ:L:2009:193:0060:013

9:EN:PDF)]. According to these classifications, the botrytized wines are made from selectively harvested nobly rotten grapes or from heavily *Botrytis*-infected grapes. One essential process when producing a great botrytized wine is to harvest the nobly rotted berries selectively because various stages of healthy, noble-, and grey-rotted grapes frequently occur in the same bunch. The most renowned botrytized wines are select wines, made from noble-rotted berries separated from the rest of the grape crop. *Botrytis*-affected wines are made from mixed grapes in which the proportion of healthy and nobly rotten berries varies from region to region and from year to year. In this chapter no distinction will be made between these categories because most research papers do not specify the "degree" of botrytization of the must or wine investigated. In the literature the term "Botrytis-affected" is not used exclusively for grapes undergoing noble rot but also for grapes affected by the destructive grey rot and it remains sometimes unclear which type of rotting was investigated (e.g. Nisiotou et al. 2007).

2.2 Yeast Taxonomic Names

Over the long history of yeast taxonomy and wine microbiology, the taxonomic names of many yeast species have been changed many times. Old (obsolete) and new (valid) taxonomic names are used inconsistently in the wine-yeast literature so that it is frequently impossible to sort out in older papers which strains belongs to which currently valid taxon names. *Saccharomyces bayanus* and *Candida stellata* are good examples of confusion in the literature around the taxonomic definitions and usage of species names.

Application of molecular taxonomic methods has revealed that *S. bayanus* is a heterogeneous taxon composed of diverse groups of strains (for a review see Sipiczki 2002). Recently two taxonomic modifications have been proposed to make clear distinction between the wine strains that form a homogeneous group, and the rest of *S. bayanus*. Naumov (2000) retained the wine strains in *S. bayanus* but established a variety for them (*S. bayanus* var. *uvarum*), whereas other researchers (Pulvirenti et al. 2000; Nguyen et al. 2000) separated them and proposed to reinstate the old taxonomic name, *S. uvarum,* used for this type of yeast before its merger with *S. bayanus*. As the *S. bayanus* and *S. bayanus* var. *uvarum* wine isolates usually turned out in molecular tests to belong to *S. uvarum*, only the latter name will be used in this chapter. The situation is further complicated by the *Saccharomyces* strains that cannot be unambiguously assigned to any species because of their chimerical genomes consisting of mosaics from the genomes of two or even more *Saccharomyces* species (for more details, see Sect. 8).

Molecular taxonomic analysis also revised *C. stellata* (old synonyms: *S. stellatus, S. bacillaris, Brettanomyces italicus, Torulopsis stellata*). Sequencing of the D1/D2 domains of rDNA repeats transferred most of its wine strains into other species, mainly into *C. zemplinina* (Sipiczki et al. 2005; Csoma and Sipiczki 2008), a species described from botrytized wines (Sipiczki 2003). Other strains turned out to belong to *Starmerella bombicola* (Sipiczki et al. 2005), *Debaromyces hanseni, Pichia anomala, C. bombi, C. lactis-condensi, Torulaspora delbrueckii* (Csoma and Sipiczki 2008). *Candida zemplinina* was later merged with *S. bacillaris (T. bacillaris)* but none of these names were retained after the merger because of the large phylogenetic distance from *Saccharomyces* and because the old genus name *Torulopsis* in no longer legitimate, but mainly because of the new rules of taxonomic naming of fungi laid down in the *International Code of Botanical Nomenclature*, which stipulates the taxonomic reclassification and renaming of the species of *Candida* and other anamorph yeast genera (Daniel et al. 2014). In the literature, both the "pre-merger" name *C. zemplinina* and the "post-merger" name *Starmerella bacillaris* (Duarte et al. 2012) are used.

Other examples of alternatively used names (synonyms): *Candida glabrata* and Torulopsis glabrata, Candida krusei and Issatc*henkia orientalis, Candida vini* and *Kregervanrija fluxuum, Hanseniaspora uvarum* and *Kloeckera apiculata, Hanseniaspora guilliermondii and Kloeckera apis, Lachancea thermotholerans* and *Kluyveromyces thermotolerans, Pichia anomala and Hansenula anomala, Issatchenkia terricola and Pichia terricola, Zygosaccharomyces florentinus* and *Zygotorulaspora florentina, Torulopsis pulcherrima* and *Candida pulcherrima* and *Metschnikowia pulcherrima, Saccharomyces rosei* and *Torulaspora delbrueckii, Pichia angophorae* and *Ambrosiozyma angophorae*, etc. (for currently valid species names and synonyms, see the Mycobank Database: http://www.mycobank.org).

The "pigmented" *Metschnikowia* strains represent a different type of challenge to taxonomic identification. Wine researchers tend to assign the yeasts producing the red-maroon pigment pulcherrimin to one single species, *M. pulcherrima (T. pulcherrima, C. pulcherrima)*, ignoring the fact that 6 species of the genus can produce this complex. Their excuse can be, that these species cannot be reliably differentiated by physiological tests used in conventional yeast taxonomy (Lachance 2011) and the sequences of their PCR-amplified rDNA repeats (the D1/D1 domain and ITS spacer sequences are routinely used in molecular taxonomy) which frequently contain ambiguous nucleotides (Kurtzman and Droby 2001; Sipiczki 2006; Brysch-Herzberg and Seidel 2015). The molecular analysis of the D1/D2 domains of the rDNA repeats revealed that the rDNA of these species, in contrast to those of other yeasts, is not homogenized, which makes them unsuitable for taxonomic identification (Sipiczki et al. 2013). Since most papers on wine yeasts do not present the details of taxonomic identification, it often remains obscure whether the strains assigned by the authors to *M. pulcherrima* belong to this species indeed.

Another difficulty encountered mainly in review articles is the usage of genus names without species epithets. This practice can be misleading because the larger genera, such as *Candida. Pichia, Hanseniaspora, Metschnikowia*, consist of species with very heterogeneous properties. To get an idea of how misleading the omission of species names can be it is worth comparing the very different oenological characteristics of the *Candida* species *C. glabrata, C. vini, C. oleophila* and *C. stellata,* which all have been detected in botrytized grapes and juices (see later). This is also a major problem with metagenomic analyses which frequently identify the organisms only to genus or higher taxonomic level (e.g. Salvetti et al. 2016).

3 Wine Regions Producing Botrytized and/or *Botrytis*-Affected Wines

Botrytized or *Botrytis*-affected wines are produced in numerous localities in the cooler parts of the wine-producing zone of Europe and in certain wine growing regions of other continents, wherever conditions are favourable to noble-rot development. Infection of ripening berries by *B. cinerea* requires moist conditions, and if the weather stays wet, it causes the malevolent "grey rot" ("pourriture grise", "Graufäule"), which destroys the grape. This is the usual form of *Botrytis*-generated bunch rot which occurs all over the world. If humidity fluctuates (e.g. humid nights followed by dry sunny days), the infected grapes develop "noble rot" ("pourriture noble", Edelfäule") (e.g. Vannini and Chilosi 2013; Magyar and Soos 2016).

The best known regions that produce the greatest botrytized white wines are Sauternes-Barsac (France), Rheingau and Mosel-Saar-Ruwer (Germany) and Tokaj (shared by Hungary and Slovakia) (Nelson and Amerine 1957; Jackson 2000), with the longest documented tradition in Tokaj (Greger 1881; Allen 1928; Magyar 2011). Apart from these major, traditional regions, over 30 other wine regions are known to have climatic conditions that allow at least moderate noble rot (Sipiczki et al. 2010) (Table 7.1). Thus, a chapter on the yeast communities associated with botrytization and botrytized wine making can garner attention also outside the "great" regions. Botrytized wines are usually white. *Botrytis* rarely has positive effect on red grapes because it increases the risk of bacterial infections (red grapes typically have a higher pH) and can change the colour (Ky et al. 2012). Recioto Valpolicella (Italy) is an exception (Jackson 2000).

4 Effects of *Botrytis*-Generated Changes in the Noble-Rotting Grape on the Microbiota

4.1 Botrytis Invasion and Noble Rot

Although *B. cinerea* can attack any part of the plant during the vegetation period, the infection that leads to noble rot does not appear before grape maturity (Pucheu-Planté and Mercier 1983) and is critically dependent on the proper fluctuation of humidity. The higher humidity in the night permits fungal infection, whereas the fall of humidity during the day limits fungal growth and modifies the metabolism of the fungus. The fungus invades the berries through microfissures and then its hyphae

Table 7.1 Wine-growing localities in which botrytized or *Botrytis*-affected wines can be produced

Country	Region/community
The "Greats"	
France	Sauternes-Barsac
Germany	Rheingau, Mosel-Saar-Ruwer
Hungary-Slovakia	Tokaj
Australia	Riverina
Austria	Rust, Illmitz
Canada	Niagara Peninsula, Oliver (British Columbia)
Chile	Valle del Maule
Czech Republic	Novosedly
France	Loupiac, Sainte-Croix du Mont near Sauternes, Monbazillac in Dordogne, Bonnezeaux, Quarts de Chaume, and Premier Cru Chaume in Anjou, Vouvray (Touraine), Cote de Beaune (Burgundy), Alsace
Hungary	Mor
Italy	Castello della Sala (Umbria), Valpolicella, Soave, Breganze (Veneto)
Japan	Nagano, Yamanashi
Moldova	Orhei
New Zealand	Te Kauwhata
Portugal	Duoro
Romania	Cotnari, Murfatlar
Slovakia	Strekov
Slovenia	Podravje
Spain	Penedes
Switzerland	Vetroz (Wallis)
South Africa	Paarl, Devon Valley, Breedekloof Valley (Western Cape)
Ukraine	Transcarpathia
Uruguay	Juanico
USA	Napa Valley, Santa Barbara (California)

grow into the skin tissue. Then the thinned grape skin breaks easily and water can evaporate through the ruptures, leading to gradual desiccation of the berry (for a recent review, see Vannini and Chilosi 2013). In drier regions, such as California, it is less common to get natural infection, so inoculation of grapes with *B. cinerea* must be performed (Akau et al. 2004). Induction of noble rotting by *Botrytis* inoculation has been attempted in several other wine-growing regions with climatic conditions unfavourable for the natural process as well (e.g. Preobrazhenskii 1947; Nelson and Amerine 1957; Gangl et al. 2004; Wang et al. 2017). Noble rot can also take place after harvest, either spontaneously or upon inoculation with *B. cinerea*, in wine-making technologies involving post-harvest withering of grapes (e.g. Nelson and Amerine 1956, 1957; Negri et al. 2017; Lorenzini et al. 2013; Tosi et al. 2013). During fungal invasion and grape desiccation a number of important transformations occur that have a large impact on the activity of the fermenting microorganisms and on the organoleptic quality of the finished wine (for reviews see Doneche 1993; Ribéreau-Gayon et al. 2000; Magyar 2011; Blanco-Ulate et al. 2015). Most of these changes have no effect on yeast activity and thus will not be discussed in this chapter.

As Müller showed as early as in 1888, the modifications in the nobly rotten grape mainly concern the sugars, the organic acids and the nitrogen-containing compounds. Dehydration concentrates both sugar and acidity, even though 35–45% of the sugar and much of the organic acids are metabolized by the mould (e.g. Dittrich et al. 1974; Sponholz et al. 1987). High sugar in the must inhibits the activity (growth and fermentation) of yeasts (Müller-Thurgau 1888). The fungus preferably metabolizes glucose, resulting in musts with atypically high fructose/glucose ratios (Sponholz et al. 1987) which could be another factor with adverse effect on yeast activity (Minarik et al. 1978; Gafner and Schütz 1996). Another negative consequence of the *Botrytis* invasion is associated with the consumption of certain nutrients essential for yeast propagation and activity. *B. cinerea* degrades grape proteins, can consume 41% of the total amino acid concentration in the berry, causes as much as 51% reduction in proline (e.g. Rapp and Reuther 1971; Dittrich et al. 1975; Sponholz 1991) and reduces the thiamin and pyridoxine contents (Dietrich and Sponholz 1975). It produces galacturonic and glucuronic acids through degradation of pectic compounds in the grape cell walls (Sponholz and Dittrich 1984). These acids can bind SO_2, so more SO_2 is required to stop yeast growth in botrytized wines than in normal wines. Several agents with negative effects on yeasts and other cocolonising microbes are also produced by the invading *Botrytis* (see Sect. 4.3).

Two more types of important alterations (microbial secondary colonisation and antagonistic interactions) occur during the development of noble rot and will be discussed in the following sections (see Sects 4.2 and 4.3).

4.2 Secondary Colonisation by Moulds and Bacteria

The *Botrytis*-generated ruptures of the grape skin make the berry accessible to other microorganisms. The nobly rotten grapes are usually very rich in microorganisms (e.g. Fleet et al. 1984; Joyeux et al. 1984; Magyar 1996; Duhail et al. 1999; Kalmar et al. 1999; Barbe et al. 2001; Antunovics et al. 2003; Sipiczki 2016) including moulds, yeasts and bacteria. These secondary colonists further reduce the level of certain nutrients and produce additional metabolites that may beneficially or adversely affect the composition of the juice. Some of these metabolites may even be toxic to other microorganisms. Duhail et al. (1999) found a correlation between the number of microorganisms in the berries and the SO_2 binding power of the juice produced.

Penicillium species, *Aspergillus niger, Rhizopus nigricans, Cladosporium herbarum* and *Alternaria tenuis* are the most abundant filamentous fungi detected in nobly rotten grapes (Fugelsang 1997; Kalmar et al. 1999; Bene and Magyar 2002; Sipiczki et al. 2006). They also metabolize important nutrients and some of them have the potential to produce mycotoxins and other metabolites that can retard the growth of yeasts during fermentation. But they may also have beneficial role: certain grape fungi were found to stimulate the alcoholic fermentation (cited in Reed and Nagodawithana 1988).

Acetic acid bacteria and lactic acid bacteria (LAB) are another characteristic group of microorganisms growing in *Botrytis*-infected grapes. In the Sauternes area 10^2–10^4 times more bacteria were found in botrytized grapes than in non-botrytized berries (Joyeux et al. 1984; Barbe et al. 2001). Most acetic acid bacteria belonged to *Glucanobacter*, *Acetobacter aceti* and *Acetobacter pasteurianus*. The occurrence of *Ac. aceti* was ascribable to the production of ethanol by fermentative yeasts which also colonise the decaying grape tissues (see below). The presence of acetic acid bacteria in grapes and during fermentation has several important effects on the wine, either directly by affecting the level of acidity, or indirectly by affecting yeast activities. For example they produce substances that retard yeast growth during fermentation (Drysdale and Fleet 1989). Another example is the production of gluconic acid, 5-oxofructose, and dihydroxyacetone that reduce the effective concentration of SO_2, making these wines more difficult to stabilize against further yeast growth and activity (Barbe et al. 2001). A recent barcoded amplicon sequencing of fermenting botrytised Californian wines (Bokulich et al. 2012b) revealed a broad diversity of low-abundance taxa not traditionally associated with wine, but most OTUs could be resolved only to family-level. Lactic acid bacteria-specific TRFLP (Terminal Restriction Fragment Length Polymorphism) performed with the same samples identified *Lactobacillus kunkeei*, *Lactococcus lactis*, *Lc. raffinolactis*, *Weissella minor*, *Lb. sakei*. *Oenococcus* was entirely absent.

Yeasts and yeast-like fungi represent the third large group of secondary colonists. Because of their particular importance, a separate section will be devoted to them (see Sect. 5).

4.3 Antagonistic Interactions Among the Colonising Microorganisms

Co-colonisation of the nobly rotting grape by various microorganisms results in mixed populations that compete for nutrients and interact by various mechanisms.

One type of interaction is antagonism. Several species of the colonising population have antagonistic effect on other members of the population. *B. cinerea* produces a group of heteropolysaccharides collectively referred to as "botryticine" (reviewed in Doneche 1993). Botryticine is supposed to be an antimicrobial substance, which adversely affects the growth of *S. cerevisiae* (Minárik et al. 1977). Nevertheless, *Botrytis* infection of grape was found to cause fermentative retardation during alcoholic fermentation (Hong et al. 2011). Apart from botryticine, *B. cinerea* also produces compounds with fungistatic activities and various antibiotic substances in grapes: botrydial (Fehlaber et al. 1974), norbotryal acetate (Cuevas and Hanson 1977) and botrylactone (Weimar et al. 1979). These substances can cause fermentation difficulties. Other fungi occurring in nobly rotten grapes are also known to produce compounds with antimicrobial activities but their possible role in botrytized winemaking is an unexplored issue.

In contrast to the presumable adverse effect of *B. cinerea* on yeast activity, Minarik (1983, 1986) found that the addition of *B. cinerea* extracts to grape must accelerated the fermentation activity of *S. cerevisiae* and *S. oviformis* (probably *S. uvarum*) resulting in wines with lower residual sugar and volatile acid content. Moreover, *B. cinerea* has great capacity for detoxification by converting phytoalexins and certain antifungal compounds (for a review see Aleu and Collado 2001), which may impair the protecting mechanisms of the plant tissues and thus facilitate the invasion of the *Botrytis*-infected berry by other microorganisms.

A different type of antagonism was found between pigment(pulcherrimin)-producing *Metschnikowia* strains and the moulds colonising botrytized grapes (Sipiczki 2006, 2016). Most pigmented *Metschnikowia* strains isolated from nobly rotten grapes in Tokaj inhibited the germination of *Botrytis, Penicillium* and *Aspergillus* conidia and induced lysis of their hyphae. The inhibitory effect was attributed to the irreversible binding of free ferric ions by a compound (pulcherriminic acid, a derivative of leucine) secreted by the *Metschnikowia* cells. The immobilization of iron by the secreted compound can also inhibit the bacterium *Oenococcus oeni* and the yeasts *C. stellata* but not *C. zemplinina*, *S. cerevisiae* and *S. uvarum* (Sipiczki 2006). The resistance of the latter two species indicates that iron depletion by pulcherrimin synthesis may not affect alcoholic fermentation. Apart from these pigmented yeasts, *C. oleophila*, *H. osmophila*, *H. vineae*, *L. thermotolerans*, *P. fermentans*, *P. kluyvery*, *S. paradoxus* (but not *S. cerevisiae*), *Z. bailii* and *Z. florentina* strains isolated from botrytized grapes also showed some antagonism against *B. cinerea* (Sipiczki 2016). The yeasts can also interact with each other. *K. dobzhanskii* and *W. anomalus* strains were found to have adverse effects on the growth of many yeasts including strains of *S. cerevisiae* and *S. uvarum* (Sipiczki 2016).

5 Preharvest Yeast Colonisation

Since the surface of the grape is poor in nutrients, usually low numbers of yeasts are found on unripe grapes. Before the unset of ripening, the yeast community residing on the surface of healthy berries is dominated by basidiomycetous genera (reviewed in Bisson and Joseph 2009; Barata et al. 2012). Among the less abundant ascomyteous yeasts the apiculate *Klo. apiculata/H. uvarum,* the pigment producing *Metschnikowia* species and the dimorphic *Aureobasidium pullulans* are the most frequent but 9–15 other species have also been found (e.g. Fleet et al. 2002; Jolly et al. 2006; Brysch-Herzberg and Seidel 2015). Nobly rotten berries are exposed to heavier infection than the healthy berries because the sweet juices leaching out from inner tissues through the *Botrytis*-generated lesions can bind more dust grains and attract more insects. The sugary juice seeping through the wounds favours the proliferation of acetic bacteria and various yeasts. Mortimer and Polsinelli (1999) found that the total numbers of microorganisms in damaged berries were in the range of 10–100 million cells per berry. A quantitative comparison of yeast numbers in a freshly pressed botrytized Sauternes must and in a non-botrytized must detected 10 times more yeasts in the previous one (Fleet et al. 1984). In a study of Tokaj grapes, ten million yeast cells per gram of nobly rotten grape were counted (Magyar 1996).

When compared to healthy berries, nobly rotten berries not only contained more yeasts but also differed in the composition of the yeast population (Table 7.2). In an early study, Le Roux et al. (1973) found that *Botrytis*-infected grapes were particularly rich in *K. apiculata* (*H. uvarum*) and *T. stellata* (probably *C. zemplinina).* Rosini et al. (1982) also reported on the occurrence of the latter species on the surface of botrytized grapes. Fleet et al. (1984) found that *H. uvarum* and *T. stellata* (probably *C. zemplinina)* predominated in a must freshly extracted from Sauternes botrytized grapes. The same authors also detected large populations of *M. pulcherrima* and *Candida krusei.* Investigation of the nobly rotten berries in the Tokaj region (Magyar 1996; Sipiczki and Csoma 2002; Bene and Magyar 2002, 2004; Antunovics et al. 2003; Magyar and Bene 2006; Sipiczki et al. 2006; Csoma and Sipiczki 2007; Sipiczki 2016) identified 15 basidiomycetous and over 25 ascomycetous species (Table 7.2). *C. stellate (probably C. zemplinina)* was more abundant in berries being in more advanced stages of rotting or being stored in the winery. *Saccharomyces* may not belong to the usual microflora; its occurrence seems to be more occasional than regular. Fleet et al. (1984) found high number of *S. cerevisiae* cells in a fresh Sauternes must, Antunovics et al. (2003) detected more *S. uvarum* than *S. cerevisiae* in botrytized Tokaj grapes, Magyar (1996) and Magyar and Bene (2006) found that *Saccharomyces* was present in certain Tokaj samples and missing in other samples. Sipiczki (2016) could isolate *S. cerevisiae*, *S. paradoxus* and *S. uvarum* from samples of only 5 out of 16 Tokaj vineyards. All basidiomycetous species identified in this study were known phylloplane yeasts. In Californian Dolce-type must (the grape was spray-inoculated with *Botrytis*) *S. cerevisiae, H. uvarum, P. kluyveri, M. pulcherrima,* and *C. zemplinina* were detected (Mills et al. 2002). The fermentative yeasts most probably produce some alcohol in the berries by the time of harvest (Joyeux et al. 1984).

It is not clear how the non-phylloplane yeasts show up on the ripening and noble-rotting berries. Their origin and the way of their dispersal on the vineyard are poorly understood. During the long period of time separating the vintage seasons, the vineyard conditions are too harsh for these yeasts for survival. It has been proposed that wine yeasts can tide over the hard times either in inactive (dormant) state in the vineyard (e.g. as spores in the soil) (Cordero-Bueso et al. 2011) or in vector organisms such as insects (e.g. Stefanini et al. 2012; Lam and Howell 2015) and birds (Francesca et al. 2012). These mechanisms can be assumed to be involved in the colonization of noble-rotting berries by yeasts, but a recent comprehensive survey of yeast communities of overwintering berries in the Tokaj region (covering 9 vineyards

Table 7.2 Yeast and yeast-like species detected on botrytized grapes, in botrytized must and wine

Stage	Yeast species[a]	References
Botrytized grape and fresh must[b]	*Aureobasidium pullulans*	Le Roux et al. (1973)
	Aureobasidium subglaciales	Rosini et al. (1982)
	Bulleromyces albus	Magyar (1996)
	Candida catenulate	Fleet et al. (1984), (2002)
	Candida dendrica	Bene and Magyar (2002), (2004)
	Candida glabrata	Sipiczki and Csoma (2002)
	Candida krusei	Antunovics et al. (2003)
	Candida lactis-condensi	Jolly et al. (2006)
	Candida oleophila	Magyar and Bene (2006)
	Candida paludigena	Mills et al. (2002)
	Candida stellata	Sipiczki et al. (2006)
	Candida zemplinina	Csoma and Sipiczki (2003), (2007)
	Cryptococcus albidus	Bokulich et al. (2012a)
	Cryptococcus carnescens	Azzolini et al. (2013)
	Cryptococcus flavescens	Sipiczki (2016)
	Cryptococcus keelungensis	
	Cryptococcus macerans	
	Cryptococcus magnus var. *magnus*	
	Cryptococcus stepposus	
	Cryptococcus victoriae	
	Curvibasidium cygneicollum	
	Curvibasidium pallidicorallinum	
	Filobasidium elegans	
	Hanseniaspora osmophila	
	Hanseniaspora uvarum	
	Hanseniaspora vineae	
	Kabatiella microsticta	
	Kloeckera apiculata	
	Kluyveromyces dobzhanskii	
	Kluyveromyces thermotolerans	
	Kregervanrija fluxuum	
	Lachancea thermotolerans	
	Metschnikowia fructicola	
	Metschnikowia pulcherrima	
	Pichia angophorae	
	Pichia fermentans	
	Pichia kluyveri	
	Pichia membranifaciens	
	Pichia scaptomyzae	
	Rhodotorula graminis	
	Rhodotorula kratochwilovae	
	Rhodotorula nothofagi	
	Saccharomyces bayanus	
	Saccharomyces cerevisiae	
	Saccharomyces paradoxus	
	Saccharomyces uvarum	
	Sporobolomyces coprosmae	
	Sporobolomyces oryzicola	
	Sporidiobolus pararoseus	
	Torulaspra delbrueckii	
	Torulopsis stellata	
	Wickerhamomyces anomalus	
	Zygoascus meyerae	
	Zygosaccharomyces bailii	
	Zygosaccharomyces rouxi	
	Zygotorulaspora florentina	

(continued)

Table 7.2 (continued)

Stage	Yeast species[a]	References
Alcoholic fermentation[b]	*Candida stellata* *Candida zemplinina* *Kluyveromyces thermotolerans* *Pichia angophorae* *Saccharomyces bayanus* *Saccharomyces cerevisiae* *Saccharomyces oviformis* *Saccharomyces paradoxus* *Saccharomyces rosei* *Saccharomyces uvarum* *Saccharomyces vini* *Saccharomyces* with chimerical genomes ("interspecies hybrids") *Torulaspora delbrueckii* *Torulopsis bacillaris* *Zygosaccharomyces rouxii*	Minarik and Laho (1962) Minarik (1965), (1969) Minarik et al. (1978) Usseglio-Tomasset et al. (1980) Frezier and Dubourdieu (1992) Magyar (1996) Torriani et al. (1999) Naumov et al. (2000), (2002), (2011) Sipiczki (2001) Sipiczki et al. (2001) Cocolin et al. (2001) Mills et al. (2002) Antunovics et al. (2003) Antunovics et al. (2005a) Masneuf-Pomarede et al. (2007) Magyar et al. (2008) Miki et al. (2008) Bokulich et al. (2012a)
Post-fermentation stage[b]	*Candida krusei* *Candida lactis-condensi* *Candida mycoderma* *Candida stellata* *Candida zemplinina* *C. zeylanoides* *Pichia membranifaciens* *Rhodotorula mucilaginosa* *Saccharomyces bailii* *Saccharomyces bayanus* *Saccharomyces cerevisiae* *Saccharomyces cerevisiae* race *capensis* *Saccharomyces cerevisiae* race *aceti* *Saccharomyces globosus* *Saccharomyces oviformis* *Saccharomyces pastorianus* *Saccharomyces uvarum* *Saccharomyces vini* *Zygosaccharomyces bailii* *Zygosaccharomyces rouxii*	Soos and Asvany (1950) Minarik and Nagyova (1964) Fleet et al. (1984) Miklos et al. (1994) Divol et al. (2005), (2006) Divol and Lonvaud-Funel (2005) Csoma and Sipiczki (2007), (2008)

[a]Note that certain species names listed in the table are no longer valid and are considered synonyms of currently valid names. Also, keep in mind that before the advent of molecular taxonomic methods, less sensitive methods were applied to taxonomic identification

[b]These lists contains also species detected by culture independent techniques (e.g. TRFLP: Terminal Restriction Fragment Length Polymorphism) which do not differentiate between active and inactive (e.g. dead) cells (e.g. Bokulich et al. 2012a)

in Hungary and 7 vineyards in Slovakia) revealed that non-phylloplane yeasts can survive the harsh winter season also in grapes left behind on the vines at harvest (Sipiczki 2006).

6 Yeasts in the Fermenting Must

6.1 *Population Dynamics*

Due to preharvest yeast growth in nobly rotten grapes, the botrytized must contains much more yeasts than the must made from healthy grapes. Fleet et al. (1984) found that the initial yeast population in a botrytized Sauternes must was 10 times higher than in a non-botrytized must. These yeasts are important for the fermentation because, as in the case of other wines (e.g. Rementeria et al. 2003), the composition of the microflora of the harvested berries determines the composition of the starting microbial population of the fresh must (Table 7.2). *Botrytis* and other filamentous fungi disappear soon from the must (Cocolin et al. 2001; Divol and Lonvaud-Funel 2005), which is probably due to their inability to switch from aerobic to the anaerobic conditions. Although it is logical to assume that the composition of the initial fermenting yeast community is primarily determined by the yeasts of the grape, it can later be enriched by yeasts residing in the vinery environment (Börlin et al. 2016; Magyar et al. 2017).

During fermentation, there is a sequential succession of yeasts. Since the specific environmental conditions in botrytized grape must are limiting and hostile to many types of yeasts, certain species cannot propagate efficiently and drop behind in the competition or even die off. The changes in the yeast population seem to depend on the degree of botrytization (Fleet et al. 1984; Csoma and Sipiczki 2003, 2007). At the start of fermentation of a Sauternes botrytized grape must an initial proliferation of apiculate yeasts (*Kloeckera*, *Hanseniaspora*), *M. pulcherrima* and *C. krusei* was observed but these yeasts died off soon (Fleet et al. 1984). In Essence (Eszencia/Esencia), the Tokaj wine brand made from juice seeping out spontaneously (free-run) from stored nobly rotten berries, the major yeast species were *Z. bailii, Z. rouxii, C. zemplinina, C. lactis-condensi, S. uvarum* and *S. cerevisiae* (Csoma and Sipiczki 2003, 2007). Tokaj Essence has extremely high concentration of sugar (up to 60–70%) and is fermented at low temperature (12 °C or below, determined by the temperature of the cellar). In other botrytized types of Tokaj wines (made from blends of botrytized and normal musts or from mixed grapes), the dynamics of the yeast population is more similar to that of the non-botrytized wines. The yeasts belonging to the genera *Aureobasidium, Metschnikowia, Rhodosporidium* and *Rhodotorula* die within 1 or 2 days. The rest of the non-*Saccharomyces* yeasts, such as *Hanseniaspora*, *Cryptococcus. T. delbrueckii* and certain *Candida* strains live longer but only *C. zemplinina* can survive and persist until the end of fermentation (Minarik et al. 1978; Sipiczki 2001; Antunovics et al. 2003). In the Californian

Dolce-type must the initial population consisting mainly of *C. zemplinina* and less of *M. pulcherrima, K. thermotholerans, H. uvarum, H. osmophila, P. kluyveri, S. cerevisiae* rapidly changed in favour of *S. cerevisiae* (which became the dominating yeast) and *C. zemplinina* and *K. thermotolerans* that decreased in cell number but persisted until the end of the fermentation (Cocolin et al. 2001; Mills et al. 2002; Bokulich et al. 2012b). *P. kluyveri, H. uvarum* and *M. pulcherrima* were present for 6 days, 8 days and 10 days, respectively, when fermentation was carried out at 18 °C and for much shorter periods of time when the temperature was 28 °C. Interestingly, inoculation of the must with a *S. cerevisiae* starter culture did not significantly alter the yeast population dynamics (Mills et al. 2002). The death and autolysis of the non-*Saccharomyces* yeasts can be a significant source of nutrients for the surviving yeasts. When comparing results obtained in various laboratories, one has to bear in mind that the degree of botrytization (or proportion of botrytized grapes) and thus the chemical composition (e.g. sugar concentration) might have been very different in the musts and wines studied.

Long persistence of yeasts referred to as *C. stellata* (or *T. stellata* in older literature) during wine fermentation has been observed in many wine-growing regions (reviewed in Csoma and Sipiczki 2008), but when certain Tokaj isolates were characterized by molecular methods, the results raised doubts about the correctness of their taxonomic classification. The 26S rDNA and ITS1–5S rDNA- ITS2 sequences of the Tokaj isolates differed from the corresponding sequences of the type strain of *C. stellata* to an extent that demanded the establishment of a separate species for them. The new species, closely related both to *C. stellata* and to the osmotolerant *Candida davenportii*, was designated *C. zemplinina* (Sipiczki 2003). A recent study revealed that none of the wine strains deposited under the name *C. stellata* in four official culture collections were conspecific with the *C. stellata* type strain but belonged to different species, most of them to *C. zemplinina* and the rest either to *Starmerella bombicola* or to *Candida lactis-condensi* (Csoma and Sipiczki 2008). The latter species was also found in Essence and Aszú/Výber type of Tokaj wines (Csoma and Sipiczki 2007, 2008).

Concomitant with the decline of the non-*Saccharomyces* populations, rapid proliferation of *S. cerevisiae* and/or *S. uvarum* can be observed and these yeasts then dominate throughout the rest of fermentation. A peculiarity of botrytized wine fermentation is the frequent occurrence of *S. uvarum* (Minarik and Laho 1962; Minarik 1965, 1969; Usseglio-Tomasset et al. 1980; Magyar 1996; Torriani et al. 1999; Sipiczki et al. 2001; Naumov et al. 2000, 2002; Antunovics et al. 2005a; Masneuf-Pomarede et al. 2007). This yeast can even be dominating, particularly when fermentation is carried out at low temperatures (Usseglio-Tomasset et al. 1980; Sipiczki et al. 2001). All of its wine strains subjected to molecular analysis were more similar to the type strain of *S. uvarum* than to that of *S. bayanus* (see Sect. 2.2), therefore many authors refer these yeasts as to *S. uvarum* or *S. bayanus* var. *uvarum*. *S. uvarum* appears to be particularly able to withstand the high su\gar concentration and low nitrogen, thiamine, and sterol conditions found in botrytized juice (for a review see Sipiczki 2002). When Tokaj isolates were compared, the *S. uvarum* strains had lower optimal growth temperature, grew much better than

S. cerevisiae at low temperatures (e.g. 10 °C) and were more osmotolerant (Sipiczki et al. 2001; Antunovics et al. 2003, 2005a). Due to these specific abilities, *S. uvarum* may have a selective advantage over *S. cerevisiae* in fermenting botrytized grape must. Surprisingly, a *S. pastorianus*, a *S. paradoxus*-like strain and an interspecies hybrid were also identified among the Tokaj isolates (Sipiczki et al. 2001; Antunovics et al. 2005a).

It is a general experience that musts derived from grapes affected by noble rot are more difficult to ferment than musts originating from healthy grapes. Botrytized must usually ferments slowly and sometimes with difficulty. Yeast growth is slow and fermentation can become sluggish (e.g. Azzolini et al. 2013) or even ceases prematurely, before the wine contains an appropriate level of alcohol (stuck fermentation). Sluggish fermentation is not a peculiarity of botrytized wines, it also occurs in other sweet wines (for reviews, see Alexandre and Charpentier 1998; Bisson 1999). A number of factors are suspected of causing sluggish fermentation and premature cessation of fermentation in wines, including high sugar concentration (e.g. Lafon-Lafourcade et al. 1979), deficiencies in nitrogen substrates and vitamins (Ingledew and Kunkee 1985; Bataillon et al. 1996), low temperature (e.g. Llaurado et al. 2002), unequal concentrations of fructose and glucose (Gafner et al. 2000), the presence of botryticine and other anti-yeast/antifungal agents (Ribéreau-Gayon et al. 1979), and accumulation of fatty acids and their esters (Lafon-Lafourcade et al. 1984). The addition of nutrients (nitrogen, vitamins) stimulates the initial yeast growth but may be ineffective in the later phases of fermentation (Dittrich et al. 1975; Lafon-Lafourcade et al. 1979; Minarik 1983; Bely et al. 2003). Some stimulation could also be achieved by supplementation with steroids and oxytocin (collectively named "survival factors") (Lafon-Lafourcade et al. 1979). The inhibition of growth by fatty acids can be overcome by addition of yeast cell walls (yeast ghosts, yeast hulls) to the stuck wine (Ribéreau-Gayon et al. 1979). Preparations of dry *Botrytis* mycelium and supplementation of the must with thiamine can have positive impacts on the fermentation as well (Dittrich et al. 1975; Minarik 1986). These stimulators were shown to speed up fermentation onset and improve the overall fermentation performance of *S. oviformis* (*S. uvarum*) in Tokaj wines (Minarik 1986).

As stated above, *Botrytis* and other moulds usually die off very fast in the grape must, but their DNA can be detected almost until the end of fermentation when culture-independent methods are applied to population analysis (Bokulich et al. 2012a). These methods cannot differentiate between the DNA of active and dead cells.

6.2 Roles of Yeasts in Vinification

The principal role of *S. cerevisiae* and *S. uvarum* is the conversion of sugar, both glucose and fructose, into ethanol. Wine strains of *S. cerevisiae* are known to display a preference for glucose (Berthels et al. 2004) and *S. uvarum* strains can be even more glucophilic (Minarik et al. 1978; Gafner and Schütz 1996; Schütz and Gafner 1995). Consequently, glucose is consumed at higher rate than fructose and fructose becomes the main sugar in the later stages of fermentation.

Saccharomyces strains produce a broad spectrum of metabolites that have significant impact on wine quality, but little is known about the production of these compounds in *Saccharomyces* strains isolated from botrytized wines. Sipiczki et al. (2001) reported on the production of six metabolites of a *S. cerevisiae*, a *S. uvarum* and a *S. paradoxus*-like strain derived from natural fermentation in Tokaj. The *S. cerevisiae* strain had the highest level of acetic acid production and the *S. uvarum* strain showed high production of isoamyl alcohol and isobutanol. Experiments with Zymaflore ST, a commercial *S. cerevisiae* strain isolated in Sauternes showed that volatile acidity was inversely correlated with the maximum cell population and the assimilable nitrogen concentration (Bely et al. 2003). The formation of volatile acidity was stimulated when yeast growth was inhibited, and addition of nitrogen at the beginning of fermentation enhanced cell propagation and reduced the production of volatile acidity. Magyar et al. (2008) observed that *S. bayanus* (probably *S. uvarum*) strains produced several times more acetaldehyde and significantly more glycerol than the *S. cerevisiae* and *C. zemplinina* strains in lab-scale fermentation of botrytized must. The highest levels of acetic acid production were detected in the *S. cerevisiae* cultures, whereas the lowest levels were measured in the *C. zemplinina* cultures. A comparative phenotypic analysis of 28 *S. uvarum* strains including those isolated from botrytized musts and wines reinforced previous observations that this species is cryotolerant, poorly ethanol tolerant and diverse in acetic acid productionthat (Masneuf-Pomarede et al. 2010).

As discussed already, botrytized musts are rich in non-*Saccharomyces* yeasts. Since fermentation proceeds slowly, these yeasts are present longer in botrytized musts than in non-botrytized musts. Their prolonged persistence can be attributed to the slower increase of ethanol concentration and to the low fermentation temperature that enhances their ethanol tolerance (e.g. Gao and Fleet 1988). Numerous studies carried out with non-botrytized wines demonstrated that the non-*Saccharomyces* yeasts have considerable impact on the chemical composition and organoleptic properties of the wine. These species can increase the levels of acetic acid, esters and higher alcohols (Shimizu and Watanabe 1981; Sponholz and Dittrich 1974). For example, the presence of apiculate yeasts in the initial phase of wine fermentation contributes to a more complex and better aroma of the wine because of higher production of glycerol, esters and acetoin (Romano 2002). Generally, however, the impact of the prolonged persistence of the non-*Saccharomyces* yeasts during botrytized fermentations is poorly explored.

Next to *S. cerevisiae* and *S. uvarum*, *C. zemplinina* (*T. stellata, T. bacillaris, C. stellata*) is quantitatively the most important yeast in the fermentation of botrytized grape musts. Its success can be attributed to its fructophilic character and resistance to botryticine (Minárik et al. 1977). Wine strains of this species have been the subject of numerous biochemical and technological studies (reviews, see Jolly et al. 2006; Sipiczki et al. 2010; Magyar 2011). However, most of the strains examined were not from botrytized wines and turned out to have been taxonomically misidentified

(Sipiczki et al. 2005; Csoma and Sipiczki 2008). For example most DBVPG *C. stellata* strains proved to belong to *Starmerella bombicola*. Other *C. stellata* strains were found conspecific with *C. zemplinina*. It is difficult to sort out in the older literature which published properties belonged to *C. zemplinina* strains. If we leave the results obtained with the DBVPG *St. bombicola* strains out of consideration, and suppose that most *T. stellata, T. bacillaris* and *C. stellata* wine strains described in the older literature might have been strains of *C. zemplinina* (as suggested by the results of the molecular reclassification by Csoma and Sipiczki 2008), the probable oenological properties of this wine yeast species can be summarised as shown in Table 7.3.

A major characteristic feature of *C. zemplinina* is fructophily, described by many authors (Table 7.3). Benda (1988) found an interesting correlation between

Table 7.3 Certain oenological properties of *C. zemplinina*[a,b]

C. zemplinina	Reference
is fructophilic	Minárik et al. (1977), (1978); Benda (1988); Mills et al. (2002); Magyar and Tóth (2011); Di Maio et al. (2012); Cabral et al. (2015)
tolerates high sugar concentrations	Sipiczki (2003); Tofalo et al. (2012)
is psychrotolerant: has growth rates superior to that of *S. cerevisiae* at low temperatures	Gao and Fleet (1988); Charoenchai et al. (1998); Sipiczki (2004); Zott et al. (2008)
tolerates botryticine	Minárik et al. (1977)
has antagonistic effect against *B. cinerea*	Lemos et al. (2016)
tolerates up to 14% ethanol	Combina et al. (2005); Rantsiou et al. (2012); Englezos et al. (2015)
produces up to 10% ethanol	Gao and Fleet (1988)
produces more glycerol than most *S. cerevisiae* strains	Soden et al. (2000); Jolly et al. (2003); Magyar and Tóth (2011)
produces variable amount of acetic acid, occasionally more than most *S. cerevisiae* strains	Shimizu and Watanabe (1981); Romano et al. (1997); Jolly et al. (2003); Sipiczki (2004); Di Maio et al. (2012)
produces more higher alcohols, isobutyric acid, than *S. cerevisiae*	Andorra et al. (2010)
tolerates free SO_2 concentrations up to 50 mg/litre	Moore et al. (1988); Englezos et al. (2015)
can produce high amount of 2-methyl-1-propanol	Holloway and Subden (1991)
is strong producer of terpens	Sadoudi et al. (2012)
facilitates Vitisin A production in mixed fermentation	Romboli et al. (2015)
can reduce acetic acid and ethyl acetate level in mixed fermentation with *Saccharomyces cerevisiae*	Rantsiou et al. (2012); Suzzi et al. (2012)
can degrade malic acid	du Plessis et al. (2017)

[a]*T. bacillaris, T. stellata* or *C. stellata* in older literature and *St. bacillaris* after its recent transfer to the genus *Starmerella*

[b]Most features listed in the table were studied in laboratory media or in non-botrytized must

morphology and sugar preference of strains thought to belong to *C. stellata*. The "bacillaris" type (long-oval cells) strains were always fructophilic, whereas the "stellata" type (round-oval cells) strains were glucophilic. Since the type strain of *C. stellata* has round cells and the type strain of *C. zemplinina* has elongated cells (Sipiczki 2003), this correlation supports the recent assumption that the fructophilic wine yeasts identified as *C. stellata* in the literature were actually strains of *C. zemplinina* (Csoma and Sipiczki 2008). The *C. zemplinina* strains show some diversity in fructose preference. For instance, strain EJ1 was found to be extremely fructophilic: it fermented no glucose in a grape juice even after fructose was completely exhausted (Mills et al. 2002). Further important features of the *C. zemlinina* strains are osmotolerance, psychrotolerance (Sipiczki 2003), and high ethanol tolerance, which can be further increased at low temperatures (Gao and Fleet 1988; Fleet 1990). These features together with botryticine-tolerance may account for their sustained presence during the fermentation of botrytized wines.

Except for a few reports, the contribution of *C. zemplinina* to wine aroma has been studied in non-botrytized musts and wines. The effect of the high volatile acidity levels (for references, see Table 7.3) can be negative on the sensory character of the non-botrytized wines but it may beneficially contribute to the sugar-acid balance of botrytized wines. A *C. zemplinina* strain was found to produce 5 times more higher alcohols than *S. cerevisiae* (Andorra et al. 2010). There is evidence that some *C. zemplinina* strains are capable of producing 240 mg of 2-methyl-1-propanol per litre of wine (Holloway and Subden 1991), a concentration that has a negative effect on wine quality.

Soden et al. (2000) found that the Chardonnay wine fermented with *C. zemplinina* had more apricot honey and sauerkraut aromas than the *S. cerevisiae*-fermented wine. Other reports could not corroborate all of these features. For example Mills et al. (2002) found no coincidence between the persistence of *C. zemplinina* EJ1 and increased glycerol or acetic acid concentration in botrytized wine. Strain EJ1 is more sensitive than *S. cerevisiae* to SO_2 (Cocolin and Mills 2003).

Although *C. zemplinina* appears much more common than *C. stellata* in wine fermentation, the latter may also be present in certain wines and in certain periods of fermentation (Magyar and Bene 2006). Nevertheless, the *C. stellata* type strain, which was isolated from a must made in Germany from overripe, somewhat shrivelled grapes ("Trockenbeerenauslese") (Kroemer and Krumbholz 1931), is less osmotolerant, less alcoholtolerant and less psychrotolerant than the type strain of *C. zemplinina* isolated from a botrytized Tokaj wine and the *C. zemplinina* strain EJ1 isolated from a botrytized Californian wine (Sipiczki 2003, 2004). These differences possibly make *C. zemplinina* better adapted to the highly stressful conditions during botrytized wine making.

Besides their direct effects on wine composition, the non-*Saccharomyces* yeasts can also indirectly affect wine quality by modulating the growth and metabolism of other microorganisms. The iron-binding agent secreted by *Metschnikowia* strains was found to inhibit the growth of the type strain of *C. stellata* (but not that of *C. zemplinina*) and to have a dual effect on the growth of *O. oeni* (Sipiczki 2006). *K. apiculata* depletes the must of thiamine, leading to a deficient situation for *Saccharomyces* (Bataillon et al. 1996). The fructophilic nature of *C. zemplinina* may positively affect the fermentative activity of glucophilic *Saccharomyces* strains by increasing the glucose/fructose ratio and can be exploited for the remediation of some types of sluggish or stuck fermentation (Gafner et al. 2000). *Zygosaccharomyces bailii* and *Z. rouxii* whose strains were found on botrytized grapes and in botrytized wines are also considered fructophilic (Minarik et al. 1978; Cabral et al. 2015). Early reports described fructophilic "Sauternes yeasts" of unknown taxonomic affiliation and origin (Sobotka and Reiner 1930; Gottschalk 1946; Sols 1956). The dying cells of the non-*Saccharomyces* yeasts release compounds that can be used by the surviving population as valuable nutrients.

One of the main trends in modern wine-making technologies is the enrichment (inoculation) of the natural fermenting yeast communities with yeast strains to render the fermentation process more controllable and improve the quality of the wine (for a review, see Ciani et al. 2010). The must or wine can be inoculated with starters containing pure or mixed populations of carefully selected and improved strains of *S. cerevisiae*, *S. uvarum* and various non-*Saccharomyces* species. Non-*Saccharomyces* yeasts are mainly used to modulate/improve the aromatic complexity of the wine or diminish the unfavourable effects of other yeasts. Many commercially available starters developed for the fermentation of sweet wines and dessert wines have been successfully applied to botrytized wine-making. Watanabe and Shimizu (1980) reported on the fermentation of an artificially botrytized Koshu grape must with pure and mixed yeast cultures. The best quality was obtained when they used *S. cerevisiae* together with *Z. bailii* and *Klo. apiculata*. Obviously, the ideal solution would be the application of region-specific starters. Many of the works reporting on strain isolation and cited above aimed at the selection of strains potentially suitable for inoculated fermentation of botrytized musts (e.g. Dubourdieu 1999; Sipiczki et al. 2001; Bely et al. 2005; Miki et al. 2008; Magyar et al. 2008; Azzolini et al. 2013). An attempt: to improve the quality of Italian passito wine produced from withered grapes infected by *B. cinerea* during post-harvest drying, resulted in the selection of *Botrytis*-*S. cerevisiae* pairs suitable for combined application (Azzolini et al. 2013). Quality improvement by inoculation of fermenting botrytized musts with non-*Saccharomyces* yeasts was also attempted several times. For example *T. delbrueckii* (*S. rosei*) strains were tested in Sauternes (Laffon-Lafourcade et al. 1981; Bely et al. 2008). Inoculation with a mixed *T. delbrueckii*/*S. cerevisiae* culture at a 20:1 ratio produced 53% less in volatile acidity and 60% less acetaldehyde than a pure culture of *S. cerevisiae*.

7 Post-Fermentation Yeast Activities

Microbiological stabilization is difficult with botrytized and *Botrytis*-affected wines. Due to the high levels of residual sugar, these wines can undergo refermentation, particularly when the temperature increases To terminate alcoholic fermentation and prevent refermentation, high concentration of sulphur dioxide is added, after which the wine can be cellared for decades of years. The RNAse inhibitor diethylpyrocarbamate (DEPC) and the beverage preservative dimethyldicarbonate (DMDC) can be more effective alternatives of SO_2 (Minarik and Nagyova 1964; Divol et al. 2005). However, yeasts can persist (or perhaps also re-infect) in botrytized wines even after stabilization. In Tokaj wines with 25–120 g/l residual sugar and 11.5–16% alcohol, Minarik and Nagyova (1964) found 9 yeast species (Table 7.2), with *S. oviformis* being the most abundant (up to 78%). If reactivated, some of these yeasts can reduce the sugar content, change the character of the wine and result in extreme high alcohol content (Minarik and Nagyova 1964). Fleet et al. (1984) found a yeast population of 10^3 cells per ml in botrytized Sauternes wine samples taken 2 months after termination of fermentation by SO_2. These yeasts belonged to *S. bailii* (*Zygosaccharomyces bailii*), *S. globosus* (*S. uvarum*), *P. membranifaciens* and *C. krusei*. A recent study on aging Sauternes wines detected *S. cerevisiae, C. zemplinina, R. mucilaginosa* and *Z. bailii* 26S rRNA sequences in DNA isolated directly from wine samples (Divol and Lonvaud-Funel 2005). The latter two species were only detected after a few months of maturation. It was proposed that these yeasts may survive in a so-called VBNC (viable but non-culturable) state associated with decreased metabolic activity and perhaps enhanced resistance. As for the role of *C. stellata* (perhaps *C. zemplinina*) in refermentation, Divol and Lonvaud-Funel (2005) hypothesized that it may not be direct. Consistent with this, they only identified *S. cerevisiae* strains in refermenting Sauternes wines (Divol et al. 2006). Two isolated strains exhibited high expression level of *SSU1*, a gene coding for a major facilitator superfamily protein required for sulphite efflux. The overexpression of this gene might be important for the adaptation of the yeast cells to the high sulphite level in the wine stabilized with SO_2. Other reports also detected *Z. rouxii* in Sauternes wines after stabilisation with SO_2 (Divol et al. 2005). *Zygosaccharomyces* yeasts are considered to be winery spoilage organisms and their presence in aging wine presents a threat to stability (Loureiro and Malfeito-Ferreira 2003). In *Zygosaccharomyces* strains isolated from Sauternes wines, the putative counterpart of the *S. cerevisiae SSU1* was also very active, suggesting that this yeast may adapt to sulphite stress in a way similar to that of *S. cerevisiae*. In spite of its involvement in wine spoilage, *Zygosaccharomyces* may also contribute positively to wine fermentation (Romano and Suzzi 1993; Sütterlin et al. 2004). Because of its fructophilic nature (Minarik et al. 1978), *Z. bailii* was suggested (together with *C. zemplinina*) for alleviation of sluggish fermentation by improving the inhibitory glucose-fructose unbalance (Sütterlin et al. 2004). *P. membranifaciens* is also regarded as a spoilage yeast that can produce increased levels of acids and esters in wines (Shimizu and Watanabe 1981; Sponholz and Dittrich 1974).

Velum formation is also a sort of microbial instability. From two types of botrytized Tokaj wines osmotolerant *S. cerevisiae* strains were isolated that had low ethanol tolerance and were poor ethanol producers (Miklos et al. 1994). The latter two features and the propensity of one of them to form pseudomycelium indicated that they were from the surface flora. Detailed taxonomic examination identified them as *S. cerevisiae* race *capensis* and *S. cerevisiae* race *aceti* (Miklos et al. 1994). Both races are known as typical surface "flor yeasts" involved in the biological aging of sherry wines (Guijo et al. 1986). In sherry making the flor yeasts form a biofilm (velum) on the surface of wine where they oxidise ethanol, glycerol, organic acids and produce higher alcohols, acetaldehyde and acetoin (e.g. Martinez et al. 1998). Non-*Saccharomyces* film-forming yeasts (*C. mycoderma*, *C. zeylanoides* and *C. krusei*) were also detected in 2–3 year-old Tokaj wines (Minarik and Nagyova 1964). The presence of flor-like yeasts in botrytized wines with extremely high sugar concentration is an intriguing phenomenon that requires investigation.

8 Genetics

8.1 Intraspecies Diversity

All yeast species associated with noble rotting and botrytized wines occur also on healthy gapes and in other types of wines. Attempts have been made to find out if variants of the species are specific for or at least more common in wine regions in which botrytized wines are made. Numerous molecular markers were used for comparing and clustering of *S. cerevisiae*, *S. uvarum* and *C. zemplinina* strains. Masneuf and Dubourdieu (2000) compared the karyotypes of 199 *S. cerevisiae* strains isolated from indigenous fermentation of botrytized must and found high profile diversity, with no dominant ones. A microsatellite analysis of over 600 *S. cerevisiae* strains isolated in 3 Sauternes wineries revealed enormous diversity but could not group the strains in clearly distinct clades corresponding to the wineries (Börlin et al. 2016). 15 strains isolated from *Botrytis*-infected grapes in the passito technology also showed high diversity in microsatellite and interdelta profiles (Azzolini et al. 2013).

S. uvarum is much less polymorphic in the molecular tests than *S. cerevisiae*. A microsatellite-primed analysis of a large number of *S. uvarum* strains including isolates from botrytised wines of several geographically distant regions failed to reveal clear correlation between the substrate (e.g. botrytized vs. non-botrytized) and the geographical location (Naumova et al. 2010). The strains had very similar karyotypes and RFLP patterns. This observation was in agreement with an earlier study which found highly uniform karyotypes in 18 Tokaj isolates of the species (Antunovics et al. 2005a, b). A recent microsatellite analysis of 108 *S. uvarum* strains identified 16 alleles but failed to group the isolates obtained from botrytized wines together (Masneuf-Pomarede et al. 2016).

The *C. zemplinina* strains isolated from botrytized grapes or wines did not form compact clades either and their grouping varied in the dendrogrammes derived from different markers (microsatellites, mitochondrial RFLP) with different clustering methods (Masneuf-Pomerede et al. 2007; Pfliegler et al. 2014; Pfliegler and Sipiczki 2016; Masneuf-Pomarede et al. 2015; Csoma et al. 2018).

8.2 Genome Structure, Heterozygosity and Polygenic Control

Natural *Saccharomyces* wine strains usually show chromosome length polymorphism and are frequently heterozygous. The few genetic data available indicates that the yeasts of botrytized wines are not exceptions. For example considerable chromosome size variability was revealed by electrophoretic karyotyping among Tokaj strains of *S. cerevisiae* (Sipiczki et al. 2001; Naumov et al. 2002; Antunovics et al. 2003, 2005a). Interestingly, the *uvarum* isolates showed almost no polymorphism. The analysis also identified strains with supernumerary chromosomes that might have been (intra- or interspecies) hybrids or alloaneuploids. From three Tokaj strains tetrads of spores were isolated, and the chromosomal patterns and the phenotypes of the spore clones were compared with each other and with those of the parental strains. No segregation was seen in electrophoretic karyotyping but the *S. uvarum* strain was heterozygous for the abilities to utilize maltose (*MAL/mal*) and to selfsporulate (*HO/ho*). The *S. cerevisiae* strains were homozygous for *MAL* and *HO*, but in all three strains the production levels of secondary metabolites segregated. The segregation patterns indicated high degree of heterozygosity and suggested that the production of the metabolites examined might be under polygenic control. None of the strains proved to have favourable features sufficient for a starter culture for fermentation of botrytized Tokaj wines but it was concluded that the combination of their positive traits might lead to a powerful genotype. To this end, the hybridisation of the *S. uvarum* isolates with *S. cerevisiae* isolates was proposed (see next section).

8.3 Interspecies Hybridisation, Genome Chimerisation and Genetic Modification

S. uvarum and *S. cerevisiae* are closely related species separated from each other and from the other members of the genus by post-zygotic sterility barriers: their allodiploid hybrids are viable but sterile because their ascospores (meiospores, equivalents of gametes of higher organisms) are not viable (hybrid sterility) (Naumov 1996). Nevertheless, the viability of the hybrids offers a possibility for bringing the genomes of these species together to create new phenotypes (for a review see Sipiczki 2008). Hybrids could be obtained between a *S. uvarum* strain isolated in the Tokaj region and various *S. cerevisiae* and *S. kudriavzevii* strains (Antunovics et al. 2005b; Pfliegler et al. 2012; Karanyicz et al. 2017). These studies not only demonstrated that the hybrids differ in phenotype from both parents but also revealed that the species are separated by a double sterility barrier ensuring that both the allodiploids and the allotetraploids are sterile. Interestingly, hybrid-like strains were found among "natural" yeasts isolated from botrytized grapes. For example Antunovics 2005a found a strain in Tokaj wine samples which had *S. uvarum*-type karyotype but its ITS-RFLP pattern was heterozygous and its cells grew at 37 °C, a temperature restrictive for *S. uvarum*. Strains heterozygous at the 26S rDNA and the ITS1–5.8S rDNA-ITS2 loci were isolated also from refermenting Sauternes wines (Divol et al. 2006). In a different collection of Tokaj strains of *S. uvarum*-type chromosomal sets, Naumova et al. (2010) found *S. cerevisiae*-type subtelomeric sequences in certain chromosomes. *Saccharomyces* strains containing genes from two or more species were found in other types of wines as well (for reviews see Sipiczki 2008; Morales and Dujon 2012). These strains are frequently referred to as "interspecies hybrids" although this term can be misleading because they rarely have complete partner genomes. Most of them have only mosaic (chimerical) genomes or only a few genes from a different species.

The mechanism underlying genome chimerisation ("gene transfer") is a matter of debate. It can be due to introgressive hybridisation (as hypothesized by many authors; e.g. Naumova et al. 2010; Marsit and Dequin 2015) or GARMe, a recently described mechanism in synthetic hybrids of a *S. uvarum* strain isolated from botrytized wine and strains of *S. cerevisiae* or *S. kudriavzevii* (Karanyicz et al. 2017). The caveat with the previous model is that introgression requires backcrossing of the hybrid with one of the parents which, however, is hampered by the sterility of hybrids (see above). GARMe (Genome AutoReduction in Meiosis) is a process triggered in allotetraploids by the loss of the *MAT*-carrying chromosome of one of the partner genomes which is then followed by gradual genome reduction resulting in various chimerical genomes consisting of mosaics of the partner genomes. Depending on the combinations of the inherited genes, these chimeras can show diverse phenotypes (Lopandic et al. 2016). Chromosomes can be spontaneously lost also in the course of vegetative propagation (e.g. Pérez-Través et al. 2014). Interspecies hybridization and posthybridisation genome reduction accompanied by genome chimerisation offer a "natural" alternative to GMO methodology for combination of genes of the *Saccharomyces* species in breeding of new starters for inoculated fermentation. This strategy is, however, limited to the genus *Saccharomyces* and cannot be used for importing genes from other genera into the genomes of *Saccharomyces* wine strains. Enrichment of their genomes of wine strains with alien genes can be achieved by the application of methods of recombinant DNA which produces GMO yeasts. For example by transforming two bacterial genes and a yeast gene encoding the three major classes of glucanases, a GMO *S. cerevisiae* was constructed which can degrade beta-glucan polysaccharides produced by *B. cinerea* in grape berries (van Rensburg et al. 1997).

Many other genetically modified wine strains containing constructs suitable for improving the fermentation of botrytized grape musts are already available in research laboratories. However, these strains cannot be used in wine making because of the strict *restrictions on GMOs* in foodstuffs within the European Union (where most botrytized wine is produced) and in numerous other countries.

References

Akau, H. L., Miller, K. M., Sabeh, N. C., Allen, R. G., Block, D. E., & Vander Gheynst, J. S. (2004). Production of *Botrytis cinerea* for potential introduction into a vineyard. *Bioresource Technology, 92*, 41–48.

Aleu, J., & Collado, G. I. (2001). Biotransformations by *Botrytis* species. *Journal of Molecular Catalysis B: Enzymatic, 13*, 77–93.

Alexandre, H., & Charpentier, C. (1998). Biochemical aspects of stuck and sluggish fermentation in grape must. *Journal of Industrial Microbiology & Biotechnology, 21*, 20–27.

Allen, H. W. (1928). *The romance of tokay*. London: Berry Bross.

Andorra, I., Berradre, M., Rozes, N., Mas, A., Guillamo'n, J. M., & Esteve-Zarzoso, B. (2010). Effect of pure and mixed cultures of the main wine yeast species on grape must fermentations. *European Food Research and Technology, 231*, 215–224.

Antunovics, Z., Csoma, H., & Sipiczki, M. (2003). Molecular and genetic analysis of the yeast flora of botrytized Tokaj wines. Bulletin de l'OIV (Office International de la Vigne et du Vin Paris) *76*: 380–397.

Antunovics, Z., Irinyi, L., & Sipiczki, M. (2005a). Combined application of methods to taxonomic identification of *Saccharomyces* strains in fermenting botrytized grape must. *Journal of Applied Microbiology, 98*, 971–979.

Antunovics, Z., Nguyen, H.-V., Gaillardin, C., & Sipiczki, M. (2005b). Gradual genome stabilisation by progressive reduction of the *S. uvarum* genome in an interspecific hybrid with *S. cerevisiae*. *FEMS Yeast Research, 5*, 1141–1150.

Azzolini, M., Tosi, E., Faccio, S., Lorenzini, M., Torriani, S., & Zapparoli, G. (2013). Selection of *Botrytis cinerea* and *Saccharomyces cerevisiae* strains for the improvement and valorization of Italian passito style wines. *FEMS Yeast Research, 13*, 540–552.

Barata, A., Malfeito-Ferreira, M., & Loureiro, V. (2012). The microbial ecology of wine grape berries. *International Journal of Food Microbiology, 153*, 243–259.

Barbe, J.-C., de Revel, G., Joyeux, A., Bertrand, A., & Lonvaud-Funel, A. (2001). Role of botrytized grape microorganisms in SO_2 binding phenomena. *Journal of Applied Microbiology, 90*, 34–42.

Bataillon, M., Rico, A., Sablayrolles, J. M., Salmon, J. M., & Barre, P. (1996). Early thiamine assimilation by yeasts under enological conditions: Impact on alcoholic fermentation kinetics. *Journal of Fermentation and Bioengineering, 82*, 145–150.

Bely, M., Rinaldi, A., & Dubourdieu, D. (2003). Influence of asssimilable nitrogen on volatile acidity production by *Saccharomyces cerevisiae* during high sugar fermentation. *Journal of Bioscience and Bioengineering, 96*, 507–512.

Bely, M., Masneuf-Pomerede, I., & Dubourdieu, D. (2005). Influence of physiological state of inoculum on volatile acidity production by *Saccharomyces cerevisiae* during high sugar fermentation. *Journal International des sciences de la vigne et du vin, 39*, 191–197.

Bely, M., Stoeckle, P., Masneuf-Pomarede, I., & Dubourdieu, D. (2008). Impact of mixed *Torulaspora delbrueckii–Saccharomyces cerevisiae* culture on high-sugar fermentation. *International Journal of Food Microbiology, 122*, 312–320.

Benda, I. (1988). Untersuchungen zur Frage der Fruktofilie und Osmotoleranz bei der Hefeart *Candida stellata* (syn. *Torulopsis stellata*). *Mitt Klosterneuburg, 38*, 60–65.

Bene, Z., & Magyar, I. (2002). Study of the yeast and mould biota of the botrytized grapes in Tokaj region in two years. *International Journal of Horticultural Science, 8*, 61–65.

Bene, Z., & Magyar, I. (2004). Characterization of yeast and mould biota of botrytized grapes in Tokaj wine region in the years 2000 and 2001. *Acta Alimentaria, 33*, 259–267.

Berthels, N. J., Cordero Otero, R. R., Bauer, F. F., Thevelein, J. M., & Pretorius, I. S. (2004). Discrepancy in glucose and fructose utilisation during fermentation by *Saccharomyces cerevisiae* wine yeast strains. *FEMS Yeast Research, 4*, 683–689.

Bisson, L. F. (1999). Stuck and sluggish fermentations. *American Journal of Enology and Viticulture, 50*, 107–119.

Bisson, L. F., & Joseph, C. M. L. (2009). Yeasts. In H. König, G. Unden, & J. Fröhlich (Eds.), *Biology of microorganisms on grapes, in must and in wine* (pp. 45–60). Berlin: Springer-Verlag.

Blanco-Ulate, B., Amrine, K. C., Collins, T. S., Rivero, R. M., Vicente, A. R., Morales-Cruz, A., Doyle, C. L., Ye, Z., Allen, G., Heymann, H., Ebeler, S. E., & Cantu, D. (2015). Developmental and metabolic plasticity of white-skinned grape berries in response to *Botrytis cinerea* during noble rot. *Plant Physiology, 169*, 2422–2443.

Bokulich, N. A., Hwang, C. F., Liu, S., Boundy-Mills, K. L., & Mills, D. A. (2012a). Profiling the yeast communities of wine fermentations using terminal restriction fragment length polymorphism analysis. *American Journal of Enology and Viticulture, 63*, 185–194.

Bokulich, N. A., Joseph, C. M., Allen, G., Benson, A. K., & Mills, D. A. (2012b). Next-generation sequencing reveals significant bacterial diversity of botrytized wine. *PLoS One, 7*, e36357.

Börlin, M., Venet, P., Claisse, O., Salin, F., Legras, J. L., & Masneuf-Pomarede, I. (2016). Cellar-associated *Saccharomyces cerevisiae* population structure revealed high-level diversity and perennial persistence at sauternes wine estates. *Applied and Environmental Microbiology, 82*, 2909–2918.

Brysch-Herzberg, M., & Seidel, M. (2015). Yeast diversity on grapes in two German wine growing regions. *International Journal of Food Microbiology, 214*, 137–144.

Cabral, S., Prista, C., Loureiro-Dias, M. C., & Leandro, M. J. (2015). Occurrence of *FFZ* genes in yeasts and correlation with fructophilic behaviour. *Microbiology, 161*, 2008–2018.

Charoenchai, C., Fleet, G. H., & Henschke, P. A. (1998). Effects of temperature, pH, and sugar concentration on the growth rates and cell biomass of wine yeasts. *American Journal of Enology and Viticulture, 49*, 283–288.

Ciani, M., Comitini, F., Mannazzu, I., & Domizio, P. (2010). Controlled mixed culture fermentation: A new perspective on the use of non-*Saccharomyces* yeasts in winemaking. *FEMS Yeast Research, 10*, 123–133.

Cocolin, L., & Mills, D. A. (2003). Wine yeast inhibition by Sulphur dioxide: A comparison of culture-dependent and independent methods. *American Journal of Enology and Viticulture, 54*, 125–130.

Cocolin, L., Heisy, A., & Mills, D. A. (2001). Direct identification of the indigenous yeasts in commercial wine fermentations. *American Journal of Enology and Viticulture, 52*, 49–53.

Combina, M., Elia, A., Mercado, L., Catania, C., Ganga, A., & Martinez, C. (2005). Dynamics of indigenous yeast populations during spontaneous fermentation of wines from Mendoza, Argentina. *International Journal of Food Microbiology, 99*, 237–243.

Cordero-Bueso, G., Arroyo, T., Serrano, A., Tello, J., Aporta, I., Velez, M. D., & Valero, E. (2011). Influence of the farming system and vine variety on yeast communities associated with grape berries. *International Journal of Food Microbiology, 145*, 132–139.

Csoma, H., & Sipiczki, M. (2003). Investigation of the yeast microflora of "Tokaj essence" 1st FEMS Congress of European Microbiologists, Ljubljana, Abstract Book p. 213.

Csoma, H., & Sipiczki, M. (2007). Taxonomic investigation of the yeast biota of botrytized crapes and "Essence" in the Tokaj wine region. 8th International Enology Symposium, Bordeaux, Book of Abstracts, p.: 174.

Csoma, H., & Sipiczki, M. (2008). Taxonomic reclassification of *Candida stellata* strains reveals frequent occurrence of *Candida zemplinina* in wine fermentation. *FEMS Yeast Research, 8*, 328–336.

Csoma, H., Acs-Szabo, L., Papp, L. A., & Sipiczki, M. (2018). Application of different markers and data-analysis tools to the examination of biodiversity can lead to different results: A case study with *Starmerella bacillaris* (synonym *Candida zemplinina*) strains. *FEMS Yeast Research* (in press), *18*.

Cuevas, O., & Hanson, J. R. (1977). Norbotryal acetate, a norsesquiterpenoid aldehyde from *Botrytis cinerea*. *Phytochemistry, 16*, 1061–1062.

Daniel, H. M., Lachance, M. A., & Kurtzman, C. P. (2014). On reclassification of species assigned to *Candida* and other anamorphic ascomycetous yeast genera based on phylogenetic circumscription. *Antonie Van Leeuwenhoek, 106*, 67–84.

Di Maio, S., Genna, G., Gandolfo, V., Amore, G., Ciaccio, M., & Oliva, D. (2012). Presence of *Candida zemplinina* in Sicilian musts and selection of a strain for wine mixed fermentations. *South African Journal of Enology and Viticulture, 33*, 80–87.

Dittrich, H. H., Sponholz, W. R., & Kast, W. (1974). Vergleichende Untersuchungen von Mosten und Weinen aus gesunden und aus *Botrytis*-infizierten Traubenbeeren. I. Säurestoffwechsel, Zuckerstoffwechselprodukte, Leucoanthocyangehalte. *Vitis, 13*, 36–49.

Dittrich, H. H., Sponholz, W. R., & Göbel, H. G. (1975). Vergleichende Untersuchungen von Mosten und Weinen aus gesunden und aus *Botrytis*-infizierten Traubenbeeren. II. Modelversuche zur Veränderungen des Mostes durch *Botrytis*-Infection und ihre Konsequenzen für die Nebenproduktbildung bei der Gärung. *Vitis, 13*, 336–347.

Divol, B., & Lonvaud-Funel, A. (2005). Evidence for viable but nonculturable yeasts in *Botrytis*-affected wine. *Journal of Applied Microbiology, 99*, 85–93.

Divol, B., Strehaiano, P., & Lonvaud-Funel, A. (2005). Effectiveness of dimethyldicarbonate to stop alcoholic fermentation in wine. *Food Microbiology, 22*, 169–178.

Divol, B., Miot-Sertier, C., & Lonvaud-Funel, A. (2006). Genetic characterization of strains of *Saccharomyces cerevisiae* responsible for "refermentation" in *Botrytis*-affected wines. *Journal of Applied Microbiology, 100*, 516–526.

Doneche, B. J. (1993). Botrytized wines. In G. H. Fleet (Ed.), *Wine microbiology and biotechnology* (pp. 327–351). Philadelphia: Harwood Academic Publishers.

Drysdale, G. S., & Fleet, G. H. (1989). The effect of acetic acid bacteria upon the growth and metabolism of yeast during the fermentation of grape juice. *The Journal of Applied Bacteriology, 67*, 471–481.

du Plessis, H. W., du Toit, M., Hoff, J. W., Hart, R. S., Ndimba, B. K., & Jolly, N. P. (2017). Characterisation of non-*Saccharomyces* yeasts using different methodologies and evaluation of their compatibility with malolactic fermentation. *South African Journal of Enology and Viticulture, 38*, 45–63.

Duarte, F. L., Pimentel, N. H., Teixeira, A., & Fonseca, A. (2012). *Saccharomyces bacillaris* is not a synonym of *Candida stellata*: Reinstatement as *Starmerella bacillaris* comb. nov. *Antonie Van Leeuwenhoek, 102*, 653–658.

Dubourdieu, D. (1999). La vinification des vins liquoreux de pourriture noble. *Rev Fr Oenologie, 176*, 32–35.

Duhail, C., Rousseau, S., l'Hyvernay, A., & Doneche, B. (1999). Nouvelles acquisitions concernant l'obtention d'une pourriture de qualité et la vinification de vendanges botrytisées. *Revue Française d'Œnologie, 176*, 28–31.

Elad, Y., Williamson, B., Tudzynski, P., & Delen, N. (2004). *Botrytis: biology, pathology and control*. Bordrecht: Kluyver Academic Publisher.

Englezos, V., Rantsiou, K., Torchio, F., Rolle, L., Gerbi, V., & Cocolin, L. (2015). Exploitation of the non-*Saccharomyces* yeast *Starmerella bacillaris* (*synonym Candida zemplinina*) in wine fermentation: Physiological and molecular characterizations. *International Journal of Food Microbiology, 199*, 33–40.

Fehlaber, H. W., Geipel, R., Mercker, H. J., Tschesche, R., & Welmar, K. (1974). Botrydial, ein Antibiotikum aus der Nährlösung des Pilzes *Botrytis cinerea*. *Chemische Berichte, 107*, 1720–1724.

Fleet, G. H. (1990). Which yeast species really conducts the fermentation? In P. J. Williams, D. M. Davidson, & T. H. Lee (Eds.), *Proc 7th Aust Wine Ind Tech Conf Adelaide* (pp. 153–156).

Fleet, G. H., Lafon-Lafourcade, S., & Ribéreau-Gayon, P. (1984). Evolution of yeasts and lactic acid bacteria during fermentation and storage of Bordeaux wines. *Applied and Environmental Microbiology, 48*, 1034–1038.

Fleet, G. H., Prakitchaiwattana, C., Beh, A. L., & Heard, G. (2002). The yeast ecology of wine grapes. In M. Ciani (Ed.), *Biodiversity and biotechnology of wine yeasts* (pp. 1–17). Kerala: Research Signpost.

Francesca, N., Canale, D. E., Settanni, L., & Moschetti, G. (2012). Dissemination of wine-related yeasts by migratory birds. *Environmental Microbiology Reports, 4*, 105–112.

Frezier, V., & Dubourdieu, D. (1992). Ecology of yeast strains *Saccharomyces cerevisiae* during spontaneous fermentation in Bordeaux winery. *American Journal of Enology and Viticulture, 43*, 375–380.

Fugelsang, K. C. (1997). *Wine microbiology*. New York: Chapman and Hall.

Gafner, J., & Schütz, M. (1996). Impact of glucose-fructose ratio on stuck fermentations: Practical experiences to restart stuck fermentations. *Wine-Wissenschaft, 51*, 214–218.

Gafner, J., Hoffmann-Boller, P., Porret, N. A., & Pulver, D. (2000). *Restarting sluggish and stuck fermentations* (2nd ed.). Cape Town: International Viticulture and Enology Congress.

Gangl, H., Leitner, G., Tiefenbrunner, W., & Redl, H. (2004). Die Induktion der Edelfäule (*Botrytis cinerea* Pers.) mittels einer Sporensuspension im Freiland. *Mitt Klosterneuburg, 54*, 214–222.

Gao, C., & Fleet, G. H. (1988). The effects of temperature and pH on ethanol tolerance of the wine yeasts *Saccharomyces cerevisiae, Candida stellata* and *Kloeckera apiculata*. *The Journal of Applied Bacteriology, 65*, 405–409.

Gottschalk, A. (1946). The mechanism of selective fermentation of D-fructose from invert sugar by sauternes yeast. *The Biochemical Journal, 40*, 621–626.

Greger, M. (1881). *Notes on the pure or natural wines of Hungary, their properties and uses*. London: Max Gregor, Ltd.

Guijo, S., Millan, C., & Ortega, J. M. (1986). Fermentative features of vinification and maturation yeast isolated in the Montilla-Moriles region of Southern Spain. *Food Microbiology, 3*, 133–142.

Holloway, P., Subden, R. E. (1991). Volatile metabolites produced in a Riesling must by wild yeasts. *Canadian Institute of Food Science and Technology Journal, 24*, 57–59.

Hong, Y. S., Cilindre, C., Liger-Belair, G., Jeandet, P., Hertkorn, N., & Schmitt-Kopplin, P. (2011). Metabolic influence of *Botrytis cinerea* infection in champagne base wine. *Journal of Agricultural and Food Chemistry, 59*, 7237–7245.

Ingledew, W. M., & Kunkee, R. E. (1985). Factors influencing sluggish fermentations of grape juice. *American Journal of Enology and Viticulture, 36*, 65–75.

Jackson, R. S. (2000). *Wine science: Principle, practice, perception*. San Diego: Academic.

Jolly, N. P., Augustyn, O. P. H., & Pretorius, I. S. (2003). The effect of non-*Saccharomyces* yeasts on fermentation and wine quality. *South African Journal of Enology and Viticulture, 24*, 55–62.

Jolly, N. P., Augustyn, O. P. H., & Pretorius, I. S. (2006). The role and use of non-Saccharomyces yeasts in wine production. *South African Journal of Enology and Viticulture, 27*, 15–39.

Joyeux, A., Lafon-Lafourcade, S., & Ribereau-Gayon, P. (1984). Evolution of acetic acid bacteria during fermentation and storage of wine. *Applied and Environmental Microbiology, 48*, 153–156.

Kalmar, Z. P., Miklosy, E., Polos, V., & Kerenyi, Y. (1999). Les effets de la qualité des grains et les différents modes de vinification sur la constitution des vins d'aszu de Tokajhegyalja. Œnologie 99, 6e Symposium international d'œnologie. Proceedings, pp. 191–195.

Karanyicz, E., Antunovics, Z., Kallai, Z., & Sipiczki, M. (2017). Non-introgressive genome chimerisation by malsegregation in autodiploidised allotetraploids during meiosis of *Saccharomyces kudriavzevii* x *Saccharomyces uvarum* hybrids. *Applied Microbiology and Biotechnology, 101*, 4617–4633.

Kroemer, K., & Krumbholz, G. (1931). Untersuchungen über osmophile Sprosspilze. *Archiv für Mikrobiologie, 2*, 352–410.

Kurtzman, C. P., & Droby, S. (2001). Metschnikowia fructicola, a new ascosporic yeast with potential for biocontrol of postharvest fruit rots. *Systematic and Applied Microbiology, 24*, 395–399.

Ky, I., Lorrain, B., Jourdes, M., Pasquier, G., & Fermau, M. (2012). Assessment of grey mould (*Botrytis cinerea*) impact on phenolic and sensory quality of Bordeaux grapes, musts and wines for two consecutive vintages. *Australian Journal of Grape and Wine Research, 18*, 215–226.

Lachance, M. A. (2011). *Metschnikowia* Kamienski (1899). In C. P. Kurtzman, J. W. Fell, & T. Boekhout (Eds.), *The yeasts, a taxonomic study* (pp. 575–619). San Diego: Elsevier.

Laffon-Lafourcade, S., Lucmaret, V., Joyeux, A., & Ribéreau-Gayon, P. (1981). Utilisation de levains mixtes dans l'élaboration des vins de pourriture noble, en vue de réduire l'acidité volatile. Comptes Rendus de l'Académie d'Agriculture de France 67: 616–622.

Lafon-Lafourcade, S., Larue, F., & Ribereau-Gayon, P. (1979). Evidence for the existence of "Survival Factors" as an explanation for some peculiarities of yeast growth, especially in grape must of high sugar concentration. *Applied and Environmental Microbiology, 38*, 1069–1073.

Lafon-Lafourcade, S., Geneix, C., & Ribéreau-Gayon, P. (1984). Inhibition of alcoholic fermentation of grape must by fatty acids produced by yeasts and their elimination by yeast ghosts. *Applied and Environmental Microbiology, 47*, 1246–1249.

Lam, S. S., & Howell, K. S. (2015). *Drosophila*-associated yeast species in vineyard ecosystems. *FEMS Microbiology Letters, 362*, fnv170.

Le Roux, G., Eschenbruch, R., & De Bruin, S. I. (1973). The microbiology of South African wine making. Part VIII – The microflora of healthy and *Botrytis cinerea* infected grapes. *Phytophylactica, 5*, 51–54.

Lemos, W. J., Jr., Bovo, B., Nadai, C., Crosato, G., Carlot, M., Favaron, F., Giacomini, A., & Corich, V. (2016). Biocontrol ability and action mechanism of *Starmerella bacillaris* (Synonym *Candida zemplinina*) isolated from wine musts against gray mold disease agent *Botrytis cinerea* on grape and their effects on alcoholic fermentation. *Frontiers in Microbiology, 7*, 1249.

Llaurado, J., Rozes, N., Bobet, R., Mas, A., & Constanti, M. (2002). Low temperature alcoholic fermentations in high sugar concentration grape musts. *Journal of Food Science, 67*, 268–273.

Lopandic, K., Pfliegler, W. P., Tiefenbrunner, W., Gangl, H., Sipiczki, M., & Sterflinger, K. (2016). Genotypic and phenotypic evolution of yeast interspecies hybrids during high-sugar fermentation. *Applied Microbiology and Biotechnology, 100*, 6331–6343.

Lorenzini, M., Azzolini, M., Tosi, E., & Zapparoli, G. (2013). Postharvest grape infection of *Botrytis cinerea* and its interactions with other moulds under withering conditions to produce noble-rotten grapes. *Journal of Applied Microbiology, 114*, 762–770.

Loureiro, V., & Malfeito-Ferreira, M. (2003). Spoilage yeasts in the wine industry (review). *International Journal of Food Microbiology, 86*, 23–50.

Magyar, I. (1996). Study of the yeast flora of Tokaj wine district. 11th International oenological symposium, Sopron, Hungary proceedings, pp 30–40.

Magyar, I. (2011). Botrytized wines. *Advances in Food and Nutrition Research, 63*, 147–206.

Magyar, I., & Bene, Z. (2006). Morphological and taxonomic study on mycobiota of noble rotted grapes in the Tokaj wine district. *Acta Alimentaria, 35*, 237–246.

Magyar, I., & Soos, J. (2016). Botrytized wines – Current perspectives. *International Journal of Wine Research, 8*, 29–39.

Magyar, I., & Tóth, T. (2011). Comparative evaluation of some oenological properties in wine strains of *Candida stellata, Candida zemplinina, Saccharomyces uvarum* and *Saccharomyces cerevisiae*. *Food Microbiology, 28*, 94–100.

Magyar, I., Toth, T., & Pomazi, A. (2008). Oenological characterization of indigenous yeasts involved in fermentation of Tokaji aszu. *Bulletin de l'OIV, 81*, 35–43.

Magyar, D., Kallai, Z., Sipiczki, M., Dobolyi, C., Sebok, F., Beregszaszi, T., Bihari, Z., Kredics, L., & Oros, G. (2017). Survey of viable airborne fungi in wine cellars of Tokaj, Hungary. *Aerobiologia* (in press).

Marsit, S., & Dequin, S. (2015). Diversity and adaptive evolution of *Saccharomyces* wine yeast: A review. *FEMS Yeast Research, 15*, fov067.

Martinez, P., Valcarcel, M., Perez, L., & Benitez, T. (1998). Metabolism of *Saccharomyces cerevisiae* flor yeasts during fermentation and biological aging of fino sherry: By-products and aroma compounds. *American Journal of Enology and Viticulture, 49*, 240–250.

Masneuf, I., & Dubourdieu, D. (2000). Rôle de la souche de levures sur les combinaisons du dioxyde de soufre des vins issus de raisins botrytisés et passerillés. *Journal International des Sciences de la vigne et du vin International Journal of Vine and Wine Sciences, 34*, 27–32.

Masneuf-Pomarede, I., Le Jeune, C., Durrens, P., Lollier, M., Aigle, M., & Dubourdieu, D. (2007). Molecular typing of wine strains *Saccharomyces bayanus* var. *uvarum* using microsatellite markers. *Systematic and Applied Microbiology, 30*, 75–82.

Masneuf-Pomarede, I., Bely, M., Marullo, P., Lonvaud-Funel, A., & Dubourdieu, D. (2010). Reassessment of phenotypic traits for *Saccharomyces bayanus* var. *uvarum* wine yeast strains. *International Journal of Food Microbiology, 139*, 79–86.

Masneuf-Pomarede, I., Juquin, E., Miot-Sertier, C., Renault, P., Laizet, Y., Salin, F., Alexandre, H., Capozzi, V., Cocolin, L., Colonna-Ceccaldi, B., Englezos, V., Girard, P., Gonzalez, B., Lucas, P., Mas, A., Nisiotou, A., Sipiczki, M., Spano, G., Tassou, C., Bely, M., & Albertin, W. (2015). The yeast *Starmerella bacillaris* (synonym *Candida zemplinina*) shows high genetic diversity in winemaking environments. *FEMS Yeast Research, 15*, fov045.

Masneuf-Pomarede, I., Salin, F., Borlin, M., Coton, E., Coton, M., Le Jeure, C., & Legras, J.-L. (2016). Microsatellite analysis of *Saccharomyces uvarum* diversity. *FEMS Yeast Research, 16*, fov002.

Miki, T., Ito, Y., Kuroha, K., Izawa, S., & Shinohara, T. (2008). Potential of yeasts isolated in botrytized grape juice to be new wine yeasts. *Food Science and Technology Research, 14*, 345–350.

Miklos, I., Sipiczki, M., & Benko, Z. (1994). Osmotolerant yeasts isolated from Tokaj wines. *Journal of Basic Microbiology, 34*(6), 379–385.

Mills, D. A., Johannsen, E. A., & Cocolin, L. (2002). Yeast diversity and persistence in *Botrytis*-affected wine fermentations. *Applied and Environmental Microbiology, 68*, 4884–4893.

Minarik, E. (1965). Ecology of natural species of wine yeasts in Czechoslovakia. *Mikrobiologija (Beograd), 20*, 29–37.

Minarik, E. (1969). Zur Ökologie von Hefen und hefeartigen Mikroorganismen sekundärer Standorte im Tokayer Weinbaugebiet. *Mitt Klosterneuburg, 19*, 40–45.

Minarik, E. (1983). Zur Aktivierung der alkoholischen Gärung zuckerreicher Moste. *Wein-Wissenschaft, 38*, 202–209.

Minarik, E. (1986). Zur Aktivierung der alkoholischen Gärung schwer vergärbarer Moste durch Hefezellwände. *Mitt Klosterneuburg, 36*, 194–197.

Minarik, E., & Laho, L. (1962). Die Hefen des Tokayer Weinbaugebietes. *Mitt Klosterneuburg, 12A*, 7–10.

Minarik, E., & Nagyova, M. (1964). Microflora of sweet tokay wines. *Kvasny Prum, 10*, 40–42.

Minárik, E., Emeriaud, M., & Jungova, O. (1977). Importance of preferential glucose and fructose fermentation by wine yeast for natural sweet wines. *Kvasny Prum, 23*, 281–284.

Minarik, E., Jungova, E., & Emeriaud, M. (1978). Fruktophile Hefen und deren Einfluss auf süsse Naturweine. *Wein-Wissenschaft, 33*, 42–47.

Moore, K. J., Johnson, M. G., & Morris, J. R. (1988). Indigenous yeast microflora on Arkansas white Riesling (*Vitis vinifera*) grapes and in model systems. *Journal of Food Science, 53*, 1725–1728.

Morales, L., & Dujon, B. (2012). Evolutionary role of interspecies hybridisation and genetic exchange in yeasts. *Microbiology and Molecular Biology Reviews, 76*, 721–739.

Mortimer, R., & Polsinelli, M. (1999). On the origin of wine yeast. *Research in Microbiology, 150*, 199–204.

Müller-Thurgau, H. (1888). Die Edelfäule der Trauben. *Landw Jbr, 17*, 83–90.

Naumov, G.I. (1996) Genetic identification of biological species in the Saccharomyces sensu stricto complex. Journal of Industrial Microbiology, 17, 295–302.

Naumov, G. I. (2000). *Saccharomyces bayanus* var. *uvarum* comb. nov., a new variety established by genetic analysis. *Microbiologiya, 69*, 410–414.

Naumov, G. I., Masneuf, I., Naumova, E. S., Aigle, M., & Dubourdieu, D. (2000). Association of *Saccharomyces bayanus* var. *uvarum* with some French wines: Genetic analysis of yeast populations. *Research in Microbiology, 151*, 683–691.

Naumov, G. I., Naumova, E. S., Antunovics, Z., & Sipiczki, M. (2002). *Saccharomyces bayanus* var. *uvarum* in Tokaj wine-making of Slovakia and Hungary. *Applied Microbiology and Biotechnology, 59*, 727–730.

Naumov, G. I., Naumova, E. S., Martynenko, N. N., & Masneuf-Pomarede, I. (2011). Taxonomy, ecology, and genetics of the yeast *Saccharomyces bayanus*: A new object for science and practice. *Microbiology, 80*, 735–742.

Naumova, E. S., Naumov, G. I., Barrio, E., & Querol, A. (2010). Mitochondrial DNA polymorphism of the yeast *Saccharomyces bayanus* var. *Uvarum*. *Mikrobiologiya, 79*, 543–550.

Negri, S., Lovato, A., Boscaini, F., Salvetti, E., Torriani, S., Commisso, M., Danzi, R., Ugliano, M., Polverari, A., Tornielli, G. B., & Guzzo, F. (2017). The induction of noble rot (*Botrytis cinerea*) infection during postharvest withering changes the metabolome of grapevine berries (*Vitis vinifera* L., cv. Garganega). *Frontiers in Plant Science, 8*, 1002.

Nelson, K. E., & Amerine, M. A. (1956). Use of *Botrytis cinerea* for the production of sweet table wines. *American Journal of Enology and Viticulture, 7*, 131–136.

Nelson, K. E., & Amerine, M. A. (1957). The use of *Botrytis cinerea* Pers. in the production of sweet table wines. *Hilgardia, 26*, 521–563.

Nguyen, H. V., Lepingle, A., & Gaillardin, C. (2000). Molecular typing demonstrates homogeneity of *Saccharomyces uvarum* strains and reveals the existence of hybrids between *S. uvarum* and *S. cerevisiae*, including the *S. bayanus* type strain CBS380. *Systematic and Applied Microbiology, 23*, 71–85.

Nisiotou, A. A., Spiropoulos, A. E., & Nychas, G. J. E. (2007). Yeast community structures and dynamics in healthy and *Botrytis*-affected grape must fermentations. *Applied and Environmental Microbiology, 73*, 6705–6713.

Pérez-Través, L., Lopes, C. A., Barrio, E., & Querol, A. (2014). Stabilization process in *Saccharomyces* intra and interspecific hybrids in fermentative conditions. *International Microbiology, 17*, 213–224.

Pfliegler, W. P., & Sipiczki, M. (2016). Does fingerprinting truly represent the diversity of wine yeasts? A case study with interdelta genotyping of *Saccharomyces cerevisiae* strains. *Letters in Applied Microbiology, 63*, 406–411.

Pfliegler, W. P., Antunovics, Z., & Sipiczki, M. (2012). Double sterility barrier between *Saccharomyces* species and its breakdown in allopolyploid hybrids by chromosome loss. *FEMS Yeast Research, 12*, 703–718.

Pfliegler, W. P., Horvath, E., Kallai, Z., & Sipiczki, M. (2014). Diversity of *Candida zemplinina* isolates inferred from RAPD, micro/minisatellite and physiological analysis. *Microbiological Research, 169*, 402–410.

Preobrazhenskii, A. A. (1947). Methods of artificial inoculation of grapes with *Botrytis cinerea* for the production of Shato-Ikem wine. *Akademiya Nauk, Biokhemiya Vinodeliya, 1*, 77–97.

Pucheu-Planté, B., & Mercier, M. (1983). Étude ultrastructurale de l'interrelation hôte-parasite entre le raisin et le champignon *Botrytis cinerea*: Example de la pourriture noble en Sauternais. *Canadian Journal of Botany, 61*, 1785–1797.

Pulvirenti, A., Nguyen, H. V., Caggia, C., Guidici, P., Rainieri, S., & Zambonelli, C. (2000). *Saccharomyces uvarum*, a proper species within *Saccharomyces sensu stricto*. *FEMS Microbiology Letters, 192*, 191–196.

Rantsiou, K., Dolci, P., Giacosa, S., Torchio, F., Tofalo, R., Torriani, S., Suzzi, G., Rolle, L., & Cocolin, L. (2012). *Candida zemplinina* can reduce acetic acid produced by *Saccharomyces cerevisiae* in sweet wine fermentations. *Applied and Environmental Microbiology, 78*, 1987–1994.

Rapp, A., & Reuther, K. H. (1971). Der Gehalt an freien Aminosäuren in Traubenmosten von gesunden und edelfaulen Beeren verschiedener Rebesorten. *Vitis, 10*, 51–58.

Reed, G., & Nagodawithana, T. W. (1988). Technology of yeast usage in winemaking. *American Journal of Enology and Viticulture, 39*, 83–90.

Rementeria, A., Rodriguez, J. A., Cadaval, A., Amenabar, R., Muguruza, J. R., Hernendo, F. L., et al. (2003). Yeast associated with spontaneous fermentations of white wines from the "Txakoli de Bizkaia" region (Basque Country, North Spain). *International Journal of Food Microbiology, 86*, 201–207.

Ribéreau-Gayon, P., Lafon-Lafourcade, S., Dubourdieu, D., Lucmaret, V., & Larue, F. (1979). Métabolisme de *Saccharomyces cerevisiae* dans les moûts de raisins parasites par *Botrytis cinerea*. Inhibition de la fermentation: formation d'acide acétique et de glycerol. *Comptes Rendus de l'Académie des Sciences, 289*, 441–444.

Ribéreau-Gayon, P., Dubourdieu, D., Donéche, B., & Lonvoud, A. (2000). Handbook of enology. In *The microbiology of wine and vinifications* (Vol. 1). Chichester: Wiley.

Romano, P. (2002). Role of apiculate yeasts on organoleptic characteristics of wine. In M. Ciani (Ed.), *Biodiversity and biotechnology of wine yeasts* (pp. 99–109). Kerala: Research Signpost.

Romano, P., & Suzzi, G. (1993). Potential use of *Zygosaccharomyces* species in winemaking. *Journal of Wine Research, 4*, 87–94.

Romano, P., Suzzi, G., Domizio, P., & Fatichenti, F. (1997). Secondary products formation as a tool for discriminating non-*Saccharomyces* wine strains. *Antonie Van Leeuwenhoek, 71*, 239–242.

Romboli, Y., Mangani, S., Buscioni, G., Granchi, L., & Vincenzini, M. (2015). Effect of *Saccharomyces cerevisiae* and *Candida zemplinina* on quercetin, vitisin a and hydroxytyrosol contents in Sangiovese wines. *World Journal of Microbiology and Biotechnology, 31*, 1137–1145.

Rosini, G., Federici, F., & Martini, A. (1982). Yeast flora of grape berries during ripening. *Microbial Ecology, 8*, 83–89.

Sadoudi, M., Tourdot-Maréchal, R., Rousseaux, S., Steyer, D., Gallardo-Chacon, J. J., Ballester, J., Vichi, S., Guerin-Schneider, R., Caixach, J., & Alexandre, H. (2012). Yeast-yeast interactions revealed by aromatic profile analysis of Sauvignon Blanc wine fermented by single or co-culture of non-*Saccharomyces* and *Saccharomyces* yeasts. *Food Microbiology, 32*, 243–253.

Salvetti, E., Campanaro, S., Campedelli, I., Fracchetti, F., Gobbi, A., Tornielli, G. B., Torriani, S., & Felis, G. E. (2016). Whole-metagenome-sequencing-based community profiles of *Vitis vinifera* L. cv. Corvina berries withered in two post-harvest conditions. *Frontiers in Microbiology, 7*, 937.

Schütz, M., & Gafner, J. (1995). Lower fructose uptake capacity of genetically characterized strains of *Saccharomyces bayanus* compared to strains of *Saccharomyces cerevisiae*: A likely cause of reduced alcoholic fermentation activity. *American Journal of Enology and Viticulture, 46*, 175–180.

Shimizu, Y., & Watanabe, M. (1981). Effects of yeast strains and environmental conditions on formation of organic acids in must during fermentation. *Journal of Fermentation Technology, 59*, 27–32.

Sipiczki, M. (2001). Characterisation of *Candida stellata* strains isolated from botrytized grapes and wines in Tokaj. 21st international specialized symposium on yeasts. Lviv. Book of Abstracts p. 61.

Sipiczki, M. (2002). Taxonomic and physiological diversity of *Saccharomyces bayanus*. In M. Cian (Ed.), *Biodiversity and biotechnology of wine yeasts* (pp. 53–69). Kerala: Research Signpost.

Sipiczki, M. (2003). *Candida zemplinina* sp. nov., an osmotolerant and psychrotolerant yeast that ferments sweet botrytized wines. *International Journal of Systematic and Evolutionary Microbiology, 53*, 2079–2083.

Sipiczki, M. (2004). Species identification and comparative molecular and physiological analysis of *Candida zemplinina* and *Candida stellata*. *Journal of Basic Microbiology, 44*, 471–479.

Sipiczki, M. (2006). *Metschnikowia* strains isolated from botrytized grapes antagonize fungal and bacterial growth by iron depletion. *Applied and Environmental Microbiology, 72*, 6716–6724.

Sipiczki, M. (2008). Interspecies hybridisation and recombination in *Saccharomyces* wine yeasts. *FEMS Yeast Research, 8*, 996–1007.

Sipiczki, M. (2016). Overwintering of vineyard yeasts: Survival of interacting yeast communities in grapes mummified on vines. *Frontiers in Microbiology, 7*, 212.

Sipiczki, M., & Csoma, H. (2002). An investigation into the yeast flora of botrytized grapes in Tokaj. 22nd international specialized symposium on yeasts. Pilanesberg, Programme and abstracts, p. 106.

Sipiczki, M., Romano, P., Lipani, G., Miklos, I., & Antunovics, Z. (2001). Analysis of yeasts derived from natural fermentation in a Tokaj winery. *Antonie Van Leeuwenhoek, 79*, 97–105.

Sipiczki, M., Ciani, M., & Csoma, H. (2005). Taxonomic reclassification of *Candida stellata* DBVPG 3827. *Folia Microbiologica, 50*, 494–498.

Sipiczki, M., Csoma, H., & Antunovics, Z. (2006). Biodiversity of yeast microbiota of botrytized Tokaj grapes and wines. ECCO XXV. The role of culture collections at the beginning of the XXIst century. Budapest, Proceedings pp. 55–65.

Sipiczki, M., Csoma, H., Antunovics, Z., & Pfliegler, W. (2010). Biodiversity in yeast populations associated with botrytised wine making. *Mitt Klosterneuburg, 60*, 387–394.

Sipiczki, M., Pfliegler, W. P., & Holb, I. J. (2013). Species share a pool of diverse rRNA genes differing in regions that determine hairpin-loop structures and evolve by reticulation. *PLoS One, 8*, e67384.

Sobotka, H., & Reiner, M. (1930). CII. Selective fermentation. I. Alcoholic fermentation of glucose, fructose and mannose mixtures. *The Biochemical Journal, 24*, 926–931.

Soden, A., Francis, I. L., Oakey, H., & Henschke, P. A. (2000). Effects of co-fermentation with *Candida stellata* and *Saccharomyces cerevisiae* on the aroma and composition of Chardonnay wine. *Australian Journal of Grape and Wine Research, 6*, 21–30.

Sols, A. (1956). Selective fermentation and phosphorylation of sugars by sauternes yeast. *Biochimica et Biophysica Acta, 20*, 62–68.

Soos, I., & Asvany, A. (1950). Morphological and physiological investigation of the Hungarian wine yeast collection. *Yearbook of the Research Institute for Ampelology, 10*, 255–290.

Sponholz, W. R. (1991). Nitrogen compounds in grapes, must, and wine. In J. Rantz (Ed.), *International symposium on nitrogen in grapes and wine* (pp. 67–77). Davis: American Society for Enology and Viticulture.

Sponholz, W. R., & Dittrich, H. H. (1974). Die Bildung von SO_2 bindenden Gärungsnebenprodukten, höheren Alkoholen und Estern bei einigen Reinzuchthefestämmen und bei einigen für die Weinbereitung wichtigen "wilden" Hefen. *Wein-Wissenschaft, 29*, 301–314.

Sponholz, W. R., & Dittrich, H. H. (1984). Über das vorkommen von Galacturon- und Glucuronsäure in Weinen, Sherries, Obst- und Dessertweinen. *Vitis, 23*, 214–224.

Sponholz, W. R., Dittrich, H. H., & Linssen, U. (1987). Die Veränderungen von Most-Inhaltsstoffen durch *Botrytis cinerea* in edelfaulen Traubenbeeren definierter Auslese-Stadien. *Wein-Wissenschaft, 42*, 266–284.

Stefanini, I., Dapporto, L., Legras, J. L., Calabretta, A., Di Paola, M., De Filippo, C., Viola, R., Capretti, P., Polsinelly, M., Turillazi, S., & Cavalieri, D. (2012). Role of social wasps in *Saccharomyces cerevisiae* ecology and evolution. *Proceedings of the National Academy of Sciences of the United States of America, 109*, 13398–13403.

Sütterlin, K. A., Hoffman-Boller, P., & Gafner, J. (2004). Kurieren von Gährstockungen mit der fructophilen Wienhefe *Zygosaccharomyces bailii*. 7th International symposium on innovations in enology, Intervitis Interfructa 2004, Stuttgart-Killesberg.

Suzzi, G., Schirone, M., Sergi, M., Marianella, R. M., Fasoli, G., Aguzzi, I., & Tofalo, R. (2012). Multistarter from organic viticulture for red wine Montepulciano d'Abruzzo production. *Frontiers in Microbiology, 3*, 135.

Tofalo, R., Schirone, M., Torriani, S., Rantsiou, K., Cocolin, L., Perpetuini, G., & Suzzi, G. (2012). Diversity of *Candida zemplinina* strains from grapes and Italian wines. *Food Microbiology, 29*, 18–26.

Torriani, S., Zapparoli, G., & Suzzi, G. (1999). Genetic and phenotypic diversity of *Saccharomyces sensu stricto* strains isolated from Amarone wine. *Antonie Van Leeuwenhoek, 75*, 207–215.

Tosi, E., Azzolini, M., Lorenzini, M., Torriani, S., Fedrizzi, B., Finato, F., Cipriani, M., & Zapparoli, G. (2013). Induction of grape botrytization during withering affects volatile composition of Recioto di Soave, a "passito"-style wine. *European Food Research and Technology, 236*, 853–862.

Usseglio-Tomasset, L., Bosia, P. D., Delfini, C., & Ciolfi, G. (1980). I vini Recioto e Amarone della Valpolicella. *Vini d'Italia, 22*, 85–97.

van Rensburg, P., Zyl, W. H., & Pretorius, I. S. (1997). Over-expression of the *Saccharomyces cerevisiae* exo-β-1, 3-glucanase gene together with the *Bacillus subtilis* endo-β-1, 3-1, 4-glucanase gene and the *Butyrivibrio fibrisolvens* endo-β-1, 4-glucanase gene in yeast. *Journal of Biotechnology, 55*, 43–53.

Vannini, A., & Chilosi, G. (2013). *Botrytis* infection: Grey mould and noble rot. In F. Mencarelli & P. Tonutti (Eds.), *Sweet, reinforced and fortified wines: Grape biochemistry, technology and vinification* (pp. 159–169). Oxford: Wiley.

Wang, X. J., Tao, Y. S., Wu, Y., An, R. Y., & Yue, Z. Y. (2017). Aroma compounds and characteristics of noble-rot wines of chardonnay grapes artificially botrytized in the vineyard. *Food Chemistry, 226*, 41–50.

Watanabe, M., & Shimizu, Y. (1980). Effect of yeasts on botrytized wine making: Application of *Botrytis cinerea* in wine making (III). *Journal of Fermentation Technology, 58*, 227–235.

Weimar, K., Tschesche, R., & Breitmaler, E. (1979). Botrylacton, ein neuer Wirkstoff aus der Nährlösung des Pilzes *Botrytis cinerea. Chemische Berichte, 112*, 3598–3602.

Zott, K., Miot-Sertier, C., Claisse, O., Lonvaud-Funel, A., Masneuf-Pomarede, I. (2008). Dynamics and diversity of non-Saccharomyces yeasts during the early stages in winemaking. International Journal of Food Microbiology, 125, 197–203.

Chapter 8
Production and Utilisation of Yeast Biomass for Wine Fermentation

Richard Degré, Anne Ortiz-Julien, Forbes Wardrop, and Zhigen Zhang

Contents

R. Degré
Lallemand Inc., Montréal, Québec, Canada
e-mail: rdegre@lallemand.com

A. Ortiz-Julien
Lallemand S.A.S., Blagnac Cedex, France
e-mail: ajulien@lallemand.com

F. Wardrop
Lallemand Inc., Montréal, Québec, Canada
e-mail: fwardrop@lallemand.com

Z. Zhang (✉)
Lallemand Inc., Montréal, Québec, Canada
e-mail: zzhang@lallemand.com

P. Romano et al. (eds.), *Yeasts in the Production of Wine*,
https://doi.org/10.1007/978-1-4939-9782-4_8

1 Introduction

Today hundreds of wine yeast strains are available worldwide, offering winemakers a significant variety of biological diversity. The yeasts are mostly strains of *Saccharomyces cerevisiae*, but some belong to the other *Saccharomyces* species and a few non-*Saccharomyces* strains are also finding their way into the market. The wine yeasts available in the dry format are produced by baker's yeast producers. Other formats such as cream yeast tend to come from specialty Enology Technical Centers.

Isolation of wine yeast strains has normally been carried out by enologists and scientists working with research institutes, universities, and large wineries. They have been selected after several rounds of microvinification for desirable traits like rate of fermentation, influence on aroma, flavor and even color of wines, to name only a few. Their ability to be propagated and dried is of course not a criterion at that stage; hence the challenge for the yeast producer is to ensure that selected strains will retain their technological properties after propagation and drying. With the large number of strains available nowadays some have proven to be less robust than others and therefore come with corresponding quality attributes and specifications. It is not possible for a yeast producer, for instance, to guarantee for a flocculent strain the same number of viable cells as with a regular non-flocculating strain; it does not mean though that the yeast will not perform properly. Moreover the reality is that it is difficult for a yeast producer to become an expert at producing some of the most difficult strains when they are required only once per year, while baker's yeast is produced several times a day. Some strains will therefore receive a lot of attention from the research and production departments of yeast companies, sometimes with success, sometimes unsuccessfully. Development efforts are usually conducted under an appropriate agreement between the selectors and the producers and necessitate a lot of effort from both parties. Even though these isolates cannot be protected by patents, it is nevertheless important to recognize that considerable investments are made to develop new strains. It is more the required know-how and the fair competition between producers that limit the number of copies. This concept is now well accepted by customers who very often will sign agreements recognizing the ownership, know-how and efforts that have been devoted to the selection and development of new strains.

2 The Market

If 280 million hectoliters of wine are produced every year and if all of it is inoculated with dry wine yeast at 15 g per hectolitre, the total potential wine yeast market is in the range of 4200 tons of dry yeast per year or 14,000 tons fresh yeast equivalent. This remains a relatively small market considering that yeast plants with production capacities of more than 50,000 tons per year are not uncommon. Although most strains are currently being offered as dry yeast, the advances of non-*Saccharomyces* strains as well as the developments of immobilized cultures and other yeast formats such as stabilized liquid yeast might represent other opportunities in the near future.

The wine market remains traditional and more than ever the different players want to promote the image of a natural process and a healthy product. It is, therefore, not surprising that commercialization of genetically modified yeasts have not been very successful although plenty of these genetically modified strains already exist in laboratories around the globe,(Volschenk et al. 2004; Coulon et al. 2006).

3 Production

3.1 Stock Cultures

This is where all yeast propagations begin and thus suitable conservation techniques are required. In yeast industries the control of pure cultures is normally limited to a few well-trained individuals in order to avoid errors. Unlike most other commercial yeast applications where only a very limited number of yeast strains are used, wine yeast culture collections contain a much greater range of diversity not only in *Saccharomyces* species but also in yeasts that would be considered "non-conventional" such as *Torulaspora* species, or *Kluyveromyces* species.

It has been a practice for years at Lallemand to test for yeast genetic stability before depositing it in the culture collection, by running pulse field gel electrophoresis of the chromosomes on at least 10 clones per strain; some have been found unstable from the beginning. Several authors have studied this phenomenon and have reported chromosome size variations (Nadal et al. 1999; Carro and Pina 2001; Schuller et al. 2007). Genetic variation does not necessarily mean that the technological properties of those strains will be affected after industrial propagation but it is possible that the "unique" genetic fingerprint of the dry yeast may vary. It becomes important nevertheless to let the owner of the yeast strain know about this property of their commercialized product, and to run appropriate field trials to establish if there are any changes to the technological behavior.

The yeast isolates that have passed this first step are then characterized for their sugar assimilation and fermentation profiles, their killer properties and their DNA profile, better characterized by other techniques: PCR (RAPD), RFLP and microsatellite PCR. These techniques can be used to establish that the yeast genetic profile is unique in comparison to those DNA profiles already present in the culture collection database. New isolates which happened to already have an existing fingerprint within the database would require greater information on the technological properties of the new isolate before committing this strain to commercialization.

Longer term preservation is normally achieved by storage of the culture under liquid nitrogen or by keeping the yeast frozen at – 80 °C in a 10% glycerol solution in order to preserve their genetic and physiological stability. Lyophilization of yeast cultures is still a common method used by some yeast culture collections, but increased risk of respiratory mutant has meant that it is generally avoided by commercial yeast producers.

After this initial phase of characterization the yeast is ready to be taken to the next step or the production phase itself.

3.2 Commercial Propagation of Yeast

As mentioned before, baker's yeast companies have traditionally produced wine yeast. Dry wine yeast is offered either in the form of small rods or as beads. The process is an adaptation of the classical process used to grow dry baker's yeast and several reviews have described the process (Reed 1982; Chen and Chiger 1985; Reed and Nagodawithana 1991; Degré 1993). The reader is encouraged to consult those articles but it can be briefly described as follows.

The production of yeast for grape must fermentation is, in itself, a multi-step process. Generally, manufacturers of yeast for winemaking have to produce yeast that must be packaged, stored and shipped in large quantities in a manner that guarantees the purity, the viability and the technological properties of the final yeast product.

Wine yeast production starts with a pure culture tube or frozen vial of the appropriate yeast strain. This yeast serves as the inoculum for the pre-pure culture tank, a small pressure vessel where seed is grown in medium under strict sterile conditions. Following growth, the contents of this vessel are transferred to a larger pure culture fermenter where propagation is carried out with some aeration, again under sterile conditions. These early stages are conducted as set-batch fermentations. In set-batch fermentation, all the growth media and nutrients are introduced to the tank prior to inoculation.

From the pure culture vessel, the grown cells are transferred to a series of progressively larger fermenters. These later stages are conducted as fed-batch fermentations. During fed-batch fermentation, molasses, phosphoric acid, ammonia, vitamins and minerals are fed to the yeast at a controlled rate. This rate is designed to feed just enough sugar and nutrients to the yeast to maximize multiplication and minimize the production of alcohol. Extensive cleaning of the equipment, steaming of pipes and tanks and filtering of the air are practiced to ensure as aseptic conditions as possible.

At the end of the seed fermentation, the contents of the vessel are pumped to a series of centrifugal separators that separate the yeast from the spent molasses. The yeast is then washed with cold water and pumped to a seed yeast storage tank where the yeast cream is held at approximately 1–2 C until it is used to inoculate the commercial fermentation tanks. These commercial fermenters are the final step in the fermentation process and are often referred to as the final or trade fermentation.

Trade fermentations are carried out in large fermenters with working volumes typically in excess of 200,000 liters. To start the commercial fermentation, a volume of water, referred to as set water, is pumped into the fermenter. Next, in a process referred to as pitching, seed yeast from the storage tank is transferred into the fermenter. Following addition of the seed yeast, aeration, cooling and nutrient additions are started to begin the 14–20 h fermentation. At the start of the fermentation, the liquid seed yeast and additional water may occupy only about one-third to one-half of the fermenter volume. Constant additions of nutrients during the course of fermentation bring the fermenter to its final volume. The rate of nutrient addition increases throughout the fermentation because more nutrients have to be supplied to support growth of the increasing cell population. The number of yeast cells increases about five- to ten-fold during this fermentation.

The classical way of providing air to the fermenter is through a series of perforated tubes located at the bottom of the vessel. The rate of airflow is about one volume of air per fermenter volume per minute. A large amount of heat is generated during yeast growth and cooling is accomplished by pumping and circulating the fermentation liquid, also known as broth, through an external heat exchanger. The addition of nutrients and regulation of pH, temperature and airflow are carefully monitored and controlled by computer systems during the entire production process. Throughout the fermentation and for the most typical strains, the temperature is controlled around 30–32 C with peaks at 35–37 C towards the end; the pH is generally in the range of 3.5–6.0. Certain isolates and non-*Saccharomyces* may require lower temperatures.

At the end of fermentation, the fermenter broth is separated by nozzle-type centrifuges, washed with water and re-centrifuged to yield a yeast cream with a solids concentration in the 18% range. The yeast cream is cooled to about 1 °C and stored in a separate, refrigerated stainless steel cream tank. The yeast cream can be pumped to a plate and frame filter press or a rotary vacuum filtration system and dewatered to a cake-like consistency containing 30–35% yeast solids that is further processed into dried yeast. Contrary to a common belief wine yeast is not freeze-dried but rather dried in fluid bed dryers that allow very rapid drying, typically in less than 30 min, without exposing yeast to high temperatures, typically at 35–40 C. It allows the most resistant strains to be available on the market in active dry yeast (ADY) format with viabilities approaching 20 billion cells per gram.

At the end of the drying cycle the ADY product is immediately packaged in alum foil packages and put under vacuum; it will protect its integrity and prevent cross contamination from the environment. When kept under vacuum and at controlled temperature (4–15 C) the ADY products made from most yeast strains will remain active for 4 years.

3.3 Process Control: Fuzzy Logic

As previously reported there is a great diversity in the number of strains available and different propagation strategies have to be deployed in order to cope with the physiological traits of some of those strains. Process control is certainly one of the tools available to the yeast producer to resolve some of the problems raised by this greater number of strains. Among the different approaches that have been developed, fuzzy logic is certainly promising.

Metabolism of *Saccharomyces cerevisiae* can be respiratory or/and fermentative depending on culture conditions (pH, temperature, etc.) and availability of the substrates (dissolved oxygen, sugar, etc.). Under oxygen limitation, fermentative pathway dominates in yeast and ethanol is produced. When the sugar concentration is high, *Saccharomyces cerevisiae* also produces ethanol even if the oxygen is sufficient for growth, a phenomenon referred to as the Crabtree effect (De Dekken 1966). Use of effective process control during fermentation is very important to ensure the quality and consistency of wine yeast; fermentation predictability in grape must is certainly the major challenge for a yeast producer.

Due to the complexities and limited understanding of microbial systems, it is difficult to develop mathematical models quantitatively describing these biological systems. As a result, a model-based approach has rarely succeeded in the control and optimization of microbial fermentation systems.

Fuzzy logic is an innovative approach that was developed by professor Lotfi Zadeh of University of California at Berkeley (Zadeh 1965, 1976), but this method had not gained popularity until recently.

Fuzzy logic, which is the logic on which fuzzy control is based, is much closer in spirit to human thinking and natural language than the traditional logical systems. It provides an effective means of capturing the approximate, inexact nature of the real world. When someone is asked how hot it is today, he may tell us that it is "hot", "moderate hot" or "cold", even though he cannot tell us the exact temperature. Fuzzy logic also uses similar linguistic labels such as "very", "moderate", "somewhat", "a little" to express the degree of intensity. For example, a process can be "very fast", "somewhat slow" or "moderate hot". The quantitative meanings of these linguistic labels are hard to define by traditional mathematics. Fuzzy logic is able to deal with these ambiguous linguistic expressions. In a nutshell, fuzzy logic is computing by words.

The ambiguous linguistic nature of these expressions is used in conjunction with experiences from human operators, this would include collective understanding of how the process is expected to work and the rational decision-making that is made during real time operation. This then represents the knowledge and set of rules that allows a fuzzy logic controller to more closely approximate human thought process than the standard control methods, such as proportional, integral and derivative (PID) controllers. Fuzzy logic can provide excellent solutions to some process control systems that are difficult to control; fuzzy logic controllers are expected to be more flexible, effective and robust in the control of microbial systems (Fig. 8.1).

Establishing fuzzy sets for input and output variables is the first step in constructing a fuzzy logic controller. Using typical ethanol analyzers to monitor ethanol produced by yeast and measured either in the vapor phase or in the propagation broth itself, two input variables are usually included in the ethanol fuzzy controller. The two input variables are Ethanol-Error (EtE), which is the difference between the actual ethanol concentration and the ethanol set point, and Ethanol-Rate (EtR),

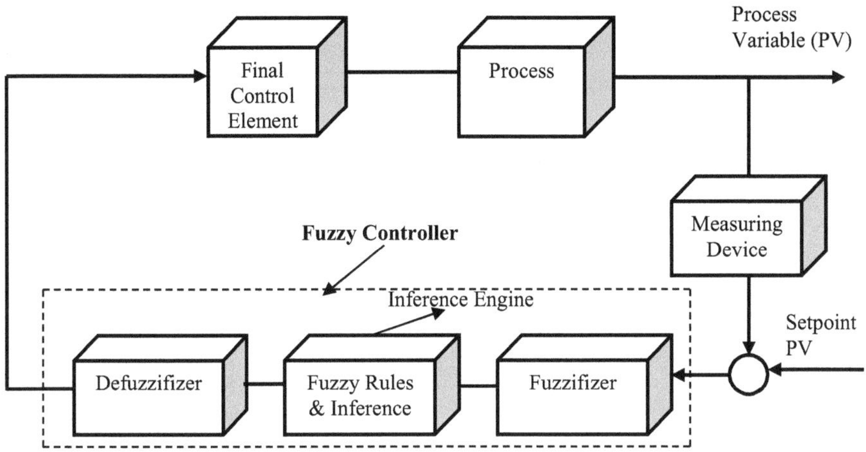

Fig. 8.1 A typical control loop with fuzzy logic controller. A fuzzy controller works similar to a conventional control system: it accepts an input value, performs some calculations according to fuzzy control algorithm, and generates an output value. This process is called the Fuzzy Inference Process and works in three steps: (a) Fuzzification where a crisp input is translated into a fuzzy value, (b) Rule Evaluation and Inference, where the fuzzy output truth values are computed, and (c) Defuzzification where the fuzzy output is translated to a crisp value. The three steps are realized in Fuzzfizer, Fuzzy Inference Engine and Defuzzifier, respectively, as shown in the figure

which is the ethanol concentration changing rate at a given time. Other input variables like fermentation time, fermentation temperature, broth volume, oxygen utilization rate, carbon dioxide production rate, respiration quotient rate, etc., can easily be incorporated into the input fuzzy sets to make a more robust controller. Capability to readily handle multiple input variables is one of the important advantages of the fuzzy logic controller.

The output variable for the fuzzy logic controller is the Molasses Control Factor, which modifies the molasses feeding rate and controls ethanol production. Of course other nutrients such as nitrogen and phosphorus sources can be linked to different degrees to the molasses response in order to make the model even more flexible. With rapid advances in disruptive technologies such as artificial intelligence (AI) and big data technologies, process control and automation would play more important roles in wine yeast production in near future.

3.3.1 Test Trials of the Fuzzy Logic Controller for Ethanol Control

Several 16 h-fed-batch propagations were carried out to test the ethanol control performance of a fuzzy logic controller with difficult to grow strains and with strains not reabsorbing ethanol properly at the end of the batch fermentation, or having a strong tendency to form ethanol even at low growth rates. Under the control of the fuzzy logic controller, a typical profile for the ethanol level in the exhaust gases is

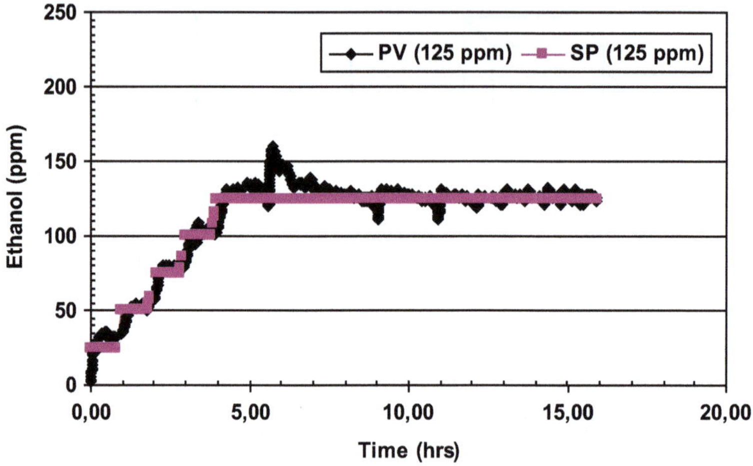

Fig. 8.2 VOC profile for fermentation IRB03227 under the control of the Fuzzy logic controller at lower plateau setpoint (125 ppm)

shown in Fig. 8.2. During the fermentations, the controller generates output based on ethanol level and control algorithm, and applies the output to manipulate the molasses feeding rates of the fermentations, so that the ethanol emission can been controlled at the desired level of 125 ppm. It can be seen that excellent ethanol control performance was achieved with the fuzzy logic controller. This superior control during the active phase of growth makes it easier later in the propagation to follow up with a maturation phase aimed at improving the resistance of the yeast to drying and the fermentation performance in the field.

4 Quality Control and Fermentation Predictability

Basic Quality Control of active dry wine yeast includes residual humidity, number of viable cells, a quick assessment of fermentative activity and other criteria suggested by OIV; serious producers would also include DNA fingerprinting of the finished product against the original strain. The reader is encouraged to consult the reviews cited earlier to learn more about those basic quality criteria (Reed 1982; Chen and Chiger 1985; Reed and Nagodawithana 1991; Degré 1993).

In order to go a step further, Lallemand has adapted from Sablayrolles et al. (1996) a technique to monitor yeast performance: it consists of running alcoholic fermentation in fermenters of 1,1 liter volume and to monitor CO_2 evolution by measurement of weight loss (Fig. 8.3). It allows evaluation of yeast performance in synthetic and real musts at different temperatures. During such fermentations parameters like fermentation kinetics, risk of stuck or sluggish fermentation, lag-phase, nutritional needs, and impact of specific nutrition can be assessed. Other

Fig. 8.3 Fermentation room for yeast phenotyping

fermentation conditions like optimum temperature, ethanol resistance and even foam production can also be monitored. Alcohol, residual sugar, total and volatile acidity, total and free SO_2 are also analyzed, post-fermentation, in order to appreciate the effect on fundamental wine character.

5 Advances in Wine Active Dry Yeast (WADY) Production

Due to the high sensitivity of most biological materials to high temperature and water activity, such as wine yeast, its preservation is a challenge. In order to prolong its shelf life, it is necessary to maintain its activities over a period of time. One typical format for wine yeast is active dry yeast (ADY) produced with thermal drying. During thermal drying, however, the wine yeast may undergo numerous changes such as damage of cell membranes, denaturation of proteins or enzymes, or even death. Therefore, suitable type of dryers under optimum operational conditions are required in order to minimize such adverse effects of thermal drying. Different dryers are available in drying industries; in which fluid bed dryer is widely used for industrial drying of wine yeast due to its many advantages. Fluid bed drying technology has been used in industry already for many years. The main advantages of this technology are linked with large contact surface area between solids (yeast particles) and gas, good degree of solids mixing, and rapid transfer of heat and moisture between solids and gas, which allow to lower drying temperature, to shorten drying time, and to reduce the level of damaging of heat-sensitive yeast cells. The fluid bed technology allow drying wine yeast at low temperature that is similar to yeast fermentation temperature (35–40 C) for a shorter period of time (20–30 min) to produce a wine active dry yeast (WADY) with over 94% solids.

One can see that even though equipment currently used to produce yeast has not significantly changed in the last 25 years or so, progress has been made on the process control front. In addition, our understanding of yeast physiology has tremendously improved, which would be beneficial to improve WADY production. During yeast fermentation and drying, there are a wide variety of stressors for yeast such as temperature, osmotic pressure, ethanol, oxidation, aging, desiccation etc. The evolution of our understanding of yeast signalling pathways and of stress resistance for instance, to name only a few, allow better appreciation of the types of organism *Saccharomyces* really are and how they would respond when facing various stresses. The existence of a general stress response is nowadays well recognized (Kobayashi and McEntee 1993; Mager and De Kruijff 1995; Martinez-Pastor et al. 1996) and this phenomenon can be described as a cross-protection (Treger et al. 1998), since exposure to a mild stress will contribute to protect the cells against a more severe similar stress and to other stresses as well.

It is also interesting to reflect upon the fact that wine yeast is typically produced in conditions that are completely the opposite of what the cells are going to face after inoculation in grape must; (i) fully aerated vessels as opposed to anaerobic or at best micro-aerophilic vessels, and (ii) the presence at any time of a minute amount of sugar as compared to high concentrations; nevertheless dry wine yeast finds the way to perform well under those completely different sets of conditions. Part of this successful adaptation is certainly related to the fact that *Saccharomyces* is Crabtree positive (De Dekken 1966; Wardrop et al. 2004), but also to its reaction to stresses in general. Contrary to the general belief, yeast is indeed exposed to stresses even during propagation under conditions that some people might consider ideal. During industrial propagation yeast can be exposed to a mild heat shock in order to increase its trehalose content. Such heat shock has been shown to also increase protection against osmotic or oxidative stress (Mager and De Kruijff 1995; Ruis and Schuller 1995). Yeast is also exposed to osmotic stress since the better performance of today's fermenters allows to grow more biomass per cubic meter; therefore more molasses is being fed and yeast becomes exposed to a greater osmotic pressure associated with minerals and other molecules accumulating in the fermentation broth. Yeast is of course as well submitted to oxidative stress by the nature of the aerobic propagation (Perez-Torrado et al. 2005). Nutrient limitation cannot be excluded either (De Winde and Thevelein 1997) since it is a common practice under industrial conditions to completely stop nitrogen feeding a few hours before the end of the propagation in order to build up more trehalose. Even successions of mild ethanol stresses exist since the first propagation steps in a plant environment involve a series of batch fermentations beginning in the laboratory during which ethanol is produced and reabsorbed. Although most of these examples are related to the propagation phase of the process, it is important to consider that the drying itself in a fluid bed environment contributes to expose yeast to oxidative stress as well as to osmotic stress since minerals inside the cell tend to become more and more concentrated as water is being evaporated. All these stresses have been shown to trigger different responses involving heat shock proteins and other molecules like zinc finger-containing proteins with a central role for cAMP in nutrient signaling pathways (Bauer and Pretorius 2000). The reader is encouraged to consult publications (Gancedo 1998; Lee et al. 2002; Magasanik and Kaiser 2002) pertaining to signaling pathways since they represent a subject that is itself worthy of a separate chapter.

Another interesting observation has been the discovery of the role of sirtuins on yeast longevity and its relationship with caloric restriction (Bordone and Guarente 2005; Sinclair 2005). Again, according to this hypothesis, caloric restriction acts as mild stress and contributes to stimulate respiration in yeast with NAD activating the sirtuin enzyme which in turn causes the DNA to be more tightly coiled. As a result this increased sirtuin enzyme activity protects yeast cells from forming extra DNA rings that are detrimental and allow the yeast to continue to remain youthful and divide longer. The connection here is easy to make with the fed-batch mode favored during yeast propagation and during which yeast is literally kept starved since sugar is almost not detectable during the whole growth process. We can therefore speculate that the preferred way of growing yeast on a large scale plays in favor of more robust cells that remain healthy and ready to divide when the time comes to face the stresses associated with grape must fermentation. Particularly when the yeast may be faced with reproduction in a medium where there are not sufficient quantities of nutrients to allow proper reproduction.

Technological developments that are not perhaps derived from winemaking or yeast production specifically, but nonetheless have driven the current state of knowledge, have come from the different "-omics" technologies, i.e. transcriptomics, proteomics, metabolomics, etc., which have lead to a greater understanding of the behavior of a variety of different organisms under many different conditions. With the use of *S. cerevisiae* as a model system, much has been obtained from the "whole cell" approach to understanding what is happening in a cell or a population of cells under changing environmental conditions (Gasch et al. 2000; James et al. 2003; Rossignol et al. 2003).

In discussing the general production of dry yeast earlier in this chapter, it was noted how yeast producers were aware of how to adapt the yeast during the production process to the rigors of drying. This was achieved while still maintaining two very simple parameters at acceptable levels, namely viability and activity. These two are closely related and are very important in allowing the dry yeast to function. However for winemaking purposes the activity of the dry yeast strains is somewhat more complicated. Not only does the winemaker require a good population of viable cells that are also significantly active but perhaps there are other characteristics of the yeast in wine fermentations that can be adjusted during propagation.

To answer this it is necessary to utilize the new technologies that allow studies of global gene transcription or protein expression. The first step is to define what characteristics are of interest or are perceived to be less well understood in terms of wine fermentation. Most of the descriptors currently in use are derived from baker's yeast where performance over minutes or hours is all that is required of the yeast.

However, wine yeast performance is a much longer affair and, as is becoming more apparent, is not limited to its living activity but also to the activities associated with its decay towards the end of wine fermentation. Wine yeast activity is also more complex than a baker's yeast activity where gassing power is the main effect. Certainly with *S. cerevisiae* conversion of sugar to ethanol is a major factor, but so too are some of the other metabolic characters of the yeast. Some yeasts are particularly good at production of fruity esters that are attractive in some wine styles. Some strains have enzymes which act on other molecules in the wine must, releasing more aromatic compounds or providing further intermediates which can be acted on by the yeast or by other indigenous microflora to add more flavor complexity to the wine. Using more advanced techniques such as genetic modification of yeast in the lab can lead to a greater understanding of how these processes occur (Sweigers et al. 2007) and how future selection or improvement of yeast strains may be carried out to further enhance the palette of flavors that are available to the winemaker through application of different yeast strains.

Using techniques such as DNA microarrays it is possible to study how the industrial propagation of active dry wine yeast affects different patterns of gene expression during the yeast production. It is then possible to study the yeast behavior and gene expression in subsequent wine fermentations, especially with more easy-to-grasp concepts such as lag phase in winemaking, or perhaps volatile acidity production or propensity to produce H_2S. The great challenge with this type of technique is to bring out the positive characteristics and the gene expression pattern obtained from the yeast under wine fermentation conditions, and to transfer that knowledge to what the yeast is doing in terms of gene expression during its propagation.

Needless to say the expression profile of yeast under commercial propagation conditions is much different from that of yeast in the early stages of a wine fermentation. However, much progress has been made in the field of yeast DNA microarrays and in particular there have been interesting articles published on gene expression of vineyard *Saccharomyces cerevisiae* (Cavalieri et al. 2000; Rossignol et al. 2003), stress response of yeast at the diauxic shift (Puig and Perez-Ortin 2000), and indeed during rehydration (Rossignol et al. 2006). This study of the genome expression profile of an industrial wine yeast strain during rehydration and fermentation and the results they have obtained have also helped to understand more of the complexities of the yeast behavior during vinification.

In some recent research on the behavior of rehydrated wine yeast, the outcome was slightly different than what may have been assumed. One striking difference was the apparent lack of response from genes involved in the stress response when the rehydrated yeast was inoculated into the wine must. Given earlier discussion of how the stress response was employed to prepare yeast for drying and storage, the lack of stress response in rehydrating yeast seems conflicting. Moreover, during rehydration the authors of this work noted that the yeast displayed a response to organic acids (Rossignol et al. 2006), consistent with a stress response to pH, but it is not clear as to why this occurred or indeed if this response by the rehydrating yeast is a good or bad characteristic. More research like this would certainly help elucidate further understanding of active dry yeast and aid in development of production techniques that can enhance or eliminate perceived strengths or weaknesses in the dry yeast.

6 Yeast Utilization

6.1 Wine Yeast Characterization: Oxygen & Nitrogen Requirements

Oxygen and nitrogen are essential yeast nutrients during grape must fermentation. Their deficiencies can lead to stuck or sluggish fermentations. Moreover, the needs of these essential factors can be affected by other winemaking parameters such as temperature and sugar concentration (Coleman et al. 2007).

Many studies have shown that yeast assimilable nitrogen content of must determines yeast population and fermentation activity. Nitrogen is often limiting (Amerine et al. 1980) and a relationship has been found between its initial concentration and maximum fermentation rate (Bely et al. 1991). The addition of nitrogen during the stationary phase is very effective because of the stimulation of protein synthesis and particularly on sugar transport systems (Bely et al. 1994). Some authors have demonstrated that this effect is strain specific (Jiranek et al. 1991; Manginot et al. 1998; Julien et al. 2001).

A study of yeast nitrogen requirements (Julien et al. 2001) showed that consumption during the stationary phase has a high technological impact since most of the sugar is metabolized after growth has stopped (Bely et al. 1994); adding ammonium salts during this phase is at least as effective as adding it in the must (Bely et al. 1991). Another study (Manginot et al. 1997) explained how the maximum fermentation rate can be maintained throughout most of the fermentation by nitrogen addition (Fig. 8.4).

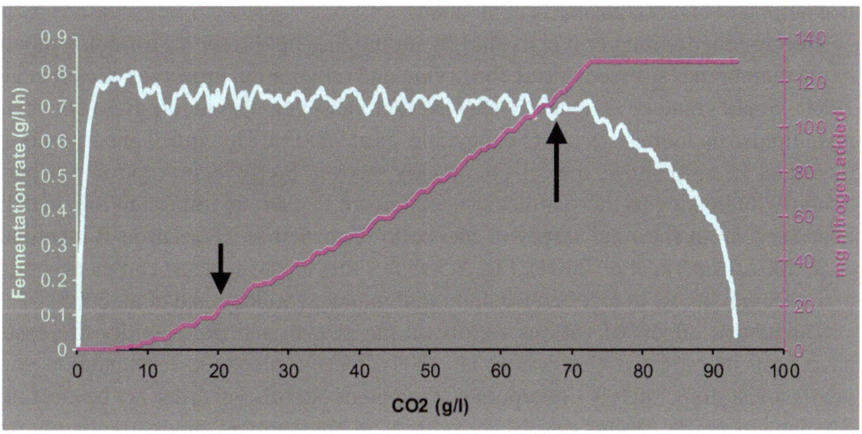

Fig. 8.4 Monitoring of the fermentation speed and the nitrogen added in function of the CO_2 released during alcoholic fermentation by commercial yeast of *Saccharomyces cerevisiae*. To calculate the nitrogen needs in mg of nitrogen added per gram of CO_2 released and per hour, we used the slope in between the arrows

Results of this study indicate that enological yeasts have very different nitrogen demands, some yeasts requiring three times more nitrogen than others to ferment the same must at the same fermentation rate. A good relationship was demonstrated between this criterion (yeast nitrogen demand) and the fermentation duration of deficient nitrogen grapes, showing that available nitrogen is a good indicator of the yeast ability to ferment nitrogen-deficient must.

Oxygen is another important factor for yeast metabolism during winemaking since it is required for synthesis of sterols and unsaturated fatty acids. The importance of controlled oxygen supply for efficient fermentation has been widely described (Ribereau-Gayon and Peynaud 1960; Sablayrolles and Barre 1986); Sablayrolles et al. pointed out the effectiveness of combined additions of oxygen and nitrogen to prevent stuck or sluggish fermentations (Sablayrolles et al. 1996). It was also recently reported (Julien et al. 2001) that yeast oxygen requirements among yeast strains vary much more than what was observed for nitrogen requirements among the same strains.

6.2 *Sterol Protection Role During Active Dry Yeast Rehydration*

Dry wine yeast needs to be rehydrated in lukewarm water before inoculation into the must. This phase of rehydration is necessary to obtain healthy membranes, a condition essential to allow optimum viability and subsequent optimum technological performance. The contribution of specific sterols at this stage improves the structure of the plasma membrane, and in the long term prepares yeast to better ensure alcoholic fermentation, particularly under difficult conditions (Fornairon-Bonnefond et al. 2002; Luparia et al. 2004).

During rehydration yeast cells start by mobilizing lipid reserves for quick repair of the membranes stressed during the drying and rehydration processes (Beker et al. 1984). Soubeyrand et al. recently showed that yeast can also incorporate extracellular lipids, including sterols (Soubeyrand et al. 2005). This quick incorporation (less than 15 min) allows the cells to quickly recover. Sterol incorporation was particularly interesting as these molecules are known to exert an important role in cellular survival at the final stages of alcoholic fermentation (Fornairon-Bonnefond et al. 2002; Luparia et al. 2004). Fig. 8.5 shows that dry wine yeast can incorporate extracellular sterols in an effective and rapid manner (Soubeyrand et al. 2005).

Qualitative effects of various sterols on the growth and the viability of yeasts have been investigated (Luparia et al. 2004). During alcoholic fermentation, the yeasts must imperatively incorporate exogenous sterols in order to be able to develop. In grape musts, sterols are present in the form of phytosterols, whose chemical nature differs from the sterols normally synthesized by yeast under aerobic conditions (Luparia et al. 2004). These phytosterols are primarily localized in the grape skin (Valero et al. 1998). However a recent study showed that such phytosterols were not sufficient to guarantee yeast integrity during the whole fer-

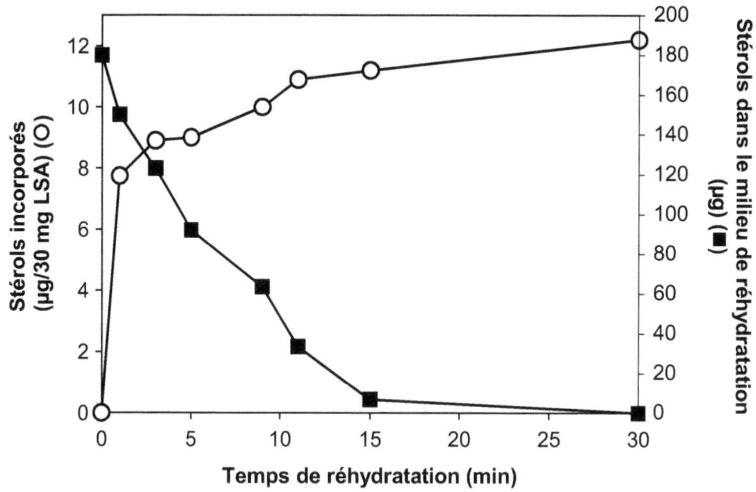

Fig. 8.5 Kinetics of incorporation of sterols measured using [4–14C] cholesterol during the rehydration of strain Lallemand EC-1118 (1 g) in 10 ml of water glucose (0,5 G) in the presence of a preparation of solubilized sterols (25 mg) at 37 °C

mentation because of the difference in chemical structure between the phytosterols and the sterols normally found in the yeast membranes (Luparia et al. 2004).

Many pre-fermentative treatments, and in particular clarification usually practiced in white and *rosé* winemaking, involve flocculation of pectin aggregates on which phytosterols can remain adsorbed; grape must is therefore depleted of assimilable sterols. Hence it is imperative to try as much as possible during the rehydration phase to promote conditions that will favor optimum conformation of the cellular membranes: it can be done by the incorporation of specific yeast sterols. It is possible to do so by using, at the time of yeast rehydration, inactive yeast made from yeast naturally rich in sterols. The impact obtained on the later performance of yeasts is remarkable. Under difficult winemaking conditions (depletion in anaerobic survival factors, highly clarified must, low or high temperatures, high potential alcohol) sterol protection provided by an inactive yeast rich in sterols during the rehydration phase of the active dry yeast allowed promotion of a higher viability (Fig. 8.6) until the end of fermentation and a reduction in volatile acidity.

Figure 8.7 illustrates the impact of rehydration in the presence of yeast sterols on the cell surface, those being rehydrated in the presence of yeast sterols (Fig. 8.7b) showing a smoother and healthier surface than those rehydrated without (Fig. 8.7a)

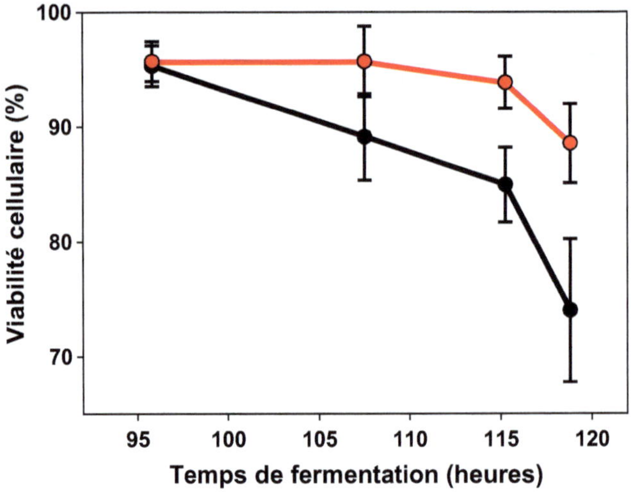

Fig. 8.6 Effect of an addition of solubilised yeast sterols at the time of the phase of rehydration of the stock Lallemand EC-1118. Rehydration without addition (black curve), in the presence of solubilised yeast sterols (24 mg L^{-1}, curves red)

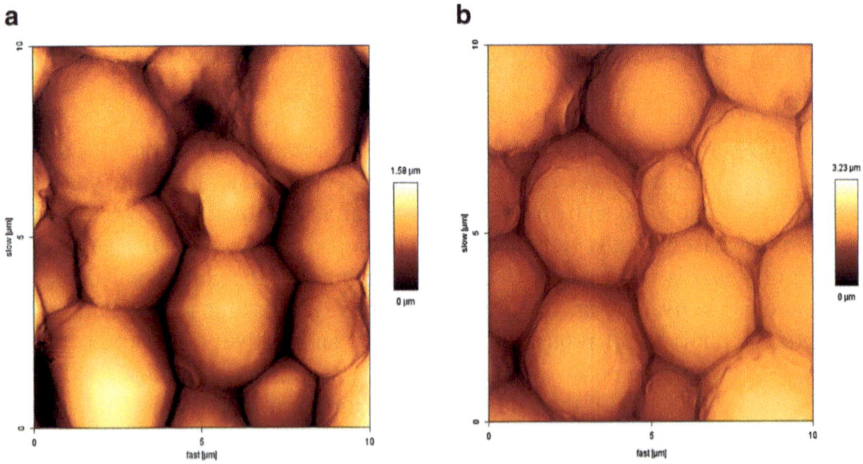

Fig. 8.7 It illustrates the impact of rehydration in the presence of yeast sterols on the cell surface (**a**) represents atomic force microscopy images of yeast cells of ADY n°77 after 20 min of rehydration in water (**b**) represents atomic force microscopy images yeast cells of ADY n°77 after 20 min of rehydration in water in presence of solubilised sterols issued from a high sterol content Inactive Dry Yeast. (Images courtesy of Ashok Adya, University of Abertay Dundee, Dundee, Scotland)

References

Amerine, M. A., Berg, H. W., Kunkee, R. E., Ough, C. S., Singleton, V. L., & Webb, A. D. (1980). The composition of grapes and wines. In *The technology of wine making* (4th ed.). Westport: Avi Publishing Company.

Bauer, F. F., & Pretorius, I. S. (2000). Yeast stress response and fermentation efficiency: How to survive the making of wine – A review. *South African Journal of Enology and Viticulture, 21*, 27–51.

Beker, M. J., Blumbergs, J. E., Ventina, E. J., & Rapoport, A. I. (1984). Characteristics of cellular membranes at rehydration of dehydrated yeast *Saccharomyces cerevisiae*. *Applied Microbiology and Biotechnology, 19*, 347–352.

Bely, M., Sablayrolles, J. M., & Barre, P. (1991). Automatic detection of assimilable nitrogen deficiencies during alcoholic fermentation in enological conditions. *Journal of Fermentation and Bioengineering, 70*, 246–252.

Bely, M., Salmon, J. M., & Barre, P. (1994). Assimilable nitrogen addition and hexose transport system activity during enological fermentations. *Journal of the Institute of Brewing, 100*, 279–282.

Bordone, L., & Guarente, L. (2005). Calorie restriction, SIRT1 and metabolism: Mnderstanding longevity. Nature Reviews. *Molecular Cell Biology, 6*, 298–305.

Carro, D., & Pina, B. (2001). Genetic analysis of the karyotype instability in natural wine yeast strains. *Yeast, 18*, 1457–1470.

Cavalieri, D., Townsend, J. P., & Hartl, D. L. (2000). Manifold anomalies in gene expression in a vineyard isolate of *Saccharomyces cerevisiae* revealed by DNA microarray analysis. *Proceedings of the National Academy of Sciences of the United States of America, 97*, 12369–12374.

Chen, S. L., & Chiger, M. (1985). Production of baker's yeast. In M. Moo-Young (Ed.), *Comprehensive biotechnology*. Oxford: Pergamon Press.

Coleman, M. C., Fish, R., & Block, D. E. (2007). Temperature-dependent kinetic model for nitrogen-limited wine fermentations. *Applied and Environmental Microbiology, 73*, 5875–5884.

Coulon, J., et al. (2006). Metabolic engineering of *Saccharomyces cerevisiae* to minimize the production of ethyl carbamate in wine. *American Journal of Enology and Viticulture, 57*, 113–124.

De Dekken, R. H. (1966). The crabtree effect: A regulatory system in yeast. *Journal of General Microbiology, 44*, 149–156.

De Winde, J. H., & Thevelein, J. M. (Eds.). (1997). *From feast to famine: Adaptation to nutrient depletion in yeast*. Austin: R.G. Landes Company.

Degré, R. (1993). Selection and commercial cultivation of wine yeast and bacteria. In G. H. Fleet (Ed.), *Wine microbiology and biotechnology*. Chur: Harwood Academic Publishers.

Fornairon-Bonnefond, C., Demaretz, V., Rosenfeld, E., & Salmon, J.-M. (2002). Oxygen addition and sterol synthesis in *Saccharomyces cerevisiae* during enological fermentation. *Journal of Bioscience and Bioengineering, 93*, 176–182.

Gancedo, J. M. (1998). Yeast carbon catabolite repression. *Microbiology and Molecular Biology Reviews, 62*, 334–361.

Gasch, A. P., et al. (2000). Genomic expression programs in the response of yeast cells to environmental changes. *Molecular Biology of the Cell, 11*, 4241–4257.

James, T. C., Campbell, S., Donnelly, D., & Bond, U. (2003). Transcription profile of brewery yeast under fermentation conditions. *Journal of Applied Microbiology, 94*, 432–448.

Jiranek, V., Langridge, P., & Henschke, P. A. (1991). Yeast nitrogen demand: Selection criterion for wine yeasts for fermenting low nitrogen musts. In J. Rantz (Ed.), *Proceedings of the International symposium on nitrogen in grapes and wines. The American society for enology and viticulture* (pp. 266–269). Washington: Seattle.

Julien, A., Roustan, J.-L., Dulau, L., & Sablayrolles, J.-M. (2001). Variabilité des besoins en oxygène et en azote assimilable suivant les souches de levures œnologiques. *Revue Française d'œnologie, 189*, 20–22.

Kobayashi, N., & McEntee, K. (1993). Identification of *Cis* and *Trans* components of a novel heat shock stress regulatory pathway in *Saccharomyces cerevisiae*. *Molecular and Cellular Biology, 13*, 248–256.

Lee, T. I., et al. (2002). Transcriptional regulatory networks in *Saccharomyces cerevisiae*. *Science, 298*, 799–804.

Luparia, V., Soubeyrand, V., Berges, T., Julien, A., & Salmon, J.-M. (2004). Assimilation of grape phytosterols by *Saccharomyces cerevisiae* and their impact on enological fermentations. *Applied Microbiology and Biotechnology, 65*, 25–32.

Magasanik, B., & Kaiser, C. A. (2002). Nitrogen regulation in *Saccharomyces cerevisiae*. *Gene, 290*, 1–18.

Mager, W. H., & De Kruijff, A. J. (1995). Stress-induced tanscriptional activation. *Microbiological Reviews, 59*, 506–531.

Manginot, C., Sablayrolles, J. M., Roustan, J. L., & Barre, P. (1997). Use of constant rate alcoholic fermentations to compare the effectiveness of different nitrogen sources added during the stationary phase. *Enzyme and Microbial Technology, 20*, 373–380.

Manginot, C., Roustan, J. L., & Sablayrolles, J. M. (1998). Nitrogen demand of different yeast strains during alcoholic fermentation. Importance of the stationary phase. *Enzyme and Microbial Technology, 23*, 511–517.

Martinez-Pastor, M. T., Marchler, G., Schuller, C., Marchler-Bauer, A., Ruis, H., & Estruch, F. (1996). The *Saccharomyces cerevisiae* zinc finger proteins *Msn2p* and *Msn4p* are required for transcriptional induction through the Stress Response Element (STRE). *The EMBO Journal, 15*, 2227–2235.

Nadal, D., Carro, D., Fernandez-Larrea, J., & Pina, B. (1999). Analysis and dynamics of the chromosomal complements of wild sparkling-wine yeast strains. *Applied and Environmental Microbiology, 65*, 1688–1695.

Perez-Torrado, R., Bruno-Barcena, J. M., & Matallana, E. (2005). Monitoring stress-related genes during the process of biomass propagation of *Saccharomyces cerevisiae* strains used for wine making. *Applied and Environmental Microbiology, 71*, 6831–6837.

Puig, S., & Perez-Ortin, J. E. (2000). Stress response and expression patterns in wine fermentations of yeast genes expressed at the diauxic shift. *Yeast, 16*, 139–148.

Reed, G. (1982). Production of baker's yeast. In G. Reed (Ed.), *Prescott and dunn's industrial microbiology*. Westport: Avi Publishing Company.

Reed, G., & Nagodawithana, T. W. (1991). *Yeast technology*. New York: Van Nostrand Reinhold.

Ribereau-Gayon, J., & Peynaud, E. (1960). Traité d'œnologie. In *Librairie polytechnique* (pp. 325–333 et 455–458). Paris: Béranger.

Rossignol, T., Dulau, L., Julien, A., & Blondin, B. (2003). Genome-wide monitoring of wine yeast gene expression during alcoholic fermentation. *Yeast, 20*, 1369–1385.

Rossignol, T., Postaire, O., Storai, J., & Blondin, B. (2006). Analysis of the genomic response of a wine yeast to rehydration and inoculation. *Applied Microbiology and Biotechnology, 71*, 699–712.

Ruis, H., & Schuller, C. (1995). Stress signaling in yeast. *BioEssays, 17*, 959–965.

Sablayrolles, J. M., & Barre, P. (1986). Évaluation des besoins en oxygène de fermentations alcooliques en conditions œnologiques simulées. *Sciences des Aliments, 6*, 373–383.

Sablayrolles, J. M., Dubois, C., Manginot, C., Roustan, J. L., & Barre, P. (1996). Effectiveness of combined ammoniacal nitrogen and oxygen additions for completion of sluggish and stuck wine fermentations. *Journal of Fermentation and Bioengineering, 82*, 377–381.

Schuller, D., Pereira, L., Alves, H., Cambon, B., Dequin, S., & Casal, M. (2007). Genetic characterization of commercial *Saccharomyces cerevisiae* isolates recovered from vineyard environments. *Yeast, 24*, 625–636.

Sinclair, D. A. (2005). Toward a unified theory of caloric restriction and longevity regulation. *Mechanisms of Ageing and Development, 126*, 987–1002.

Soubeyrand, V., et al. (2005). Formation of micella containing solubilized sterols during rehydration of active dry yeasts improves their fermenting capacity. *Journal of Agricultural and Food Chemistry, 53*, 8025–8032.

Sweigers, J. H., et al. (2007). Engineering volatile thiol release in Saccharomyces cerevisiae for improved wine aroma. *Yeast, 24*, 561–574.

Treger, J. M., Magee, T. R., & McEntee, K. (1998). Functional analysis of the stress response element and its role in the multistress response of *S. cerevisiae*. *Biochemical and Biophysical Research Communications, 243*, 13–19.

Valero, E., Millan, M. C., Mauricio, J. C., & Ortega, J. M. (1998). Effect of grape skin maceration on sterol, phospholipid, and fatty acid contents of *Saccharomyces cerevisiae* during alcoholic fermentation. *American Journal of Enology and Viticulture, 49*, 119–124.

Volschenk, H., Viljoen-Bloom, M., Van Staden, J., Husnik, J., & van Vuuren, H. J. J. (2004). Genetic engineering of an industrial strain of *Saccharomyces bayanus*, for L-malic acid degradation via an efficient malo-ethanolic pathway. *South African Journal of Enology and Viticulture, 25*, 63–73.

Wardrop, F. R., Liti, G., Cardinali, G., & Walker, G. M. (2004). Physiological responses of crabtree positive and crabtree negative yeasts to glucose upshifts in a chemostat. *Annals of Microbiology, 54*, 103–114.

Zadeh, L. A. (1965). Fuzzy sets. *Information and Control, 8*, 339–353.

Zadeh, L. A. (1976). A fuzzy-algorithmic approach to the definition of complex or imprecise concepts. *International Journal of Man-Machine Studies, 8*, 249–291.

Chapter 9
Kinetics and Control of Alcoholic Fermentation During Wine Production

Jean-Marie Sablayrolles

Contents

J.-M. Sablayrolles (✉)
UMR Sciences for Enology, INRA, Montpellier Cedex 2, France
e-mail: jean-marie.sablayrolles@inra.fr

P. Romano et al. (eds.), *Yeasts in the Production of Wine*,
https://doi.org/10.1007/978-1-4939-9782-4_9

1 Introduction

Alcoholic fermentation is a key step in winemaking. During this process, hexoses and other grape must constituents are converted to ethanol, carbon dioxide, and many other secondary by-products that affect the sensorial properties of wine. The species and strains of yeast responsible for fermentation have a major effect on the process, but must composition and fermentation control are also important. Fermentation kinetics, i.e., the rate of sugar consumption and ethanol or CO_2 production, depend on interactions between yeasts, the availability of nutrients in the must and parameters controlling fermentation (e.g., temperature).

Optimizing the control of alcoholic fermentation for winemaking is a difficult challenge. In contrast to other industrial fermentation processes, the aim is not to maximize the concentration or yield of a defined metabolite or the productivity of the process. Rather, in winemaking, the main objective is to optimize product quality, which is very difficult to quantify (Francis and Newton 2005). The control of technological parameters, such as sugar exhaustion, duration of fermentation and the amount of energy required to regulate fermentation temperature, is a general prerequisite for managing the characteristics of wine. Legal constraints are also responsible for certain aspects of fermentation control during winemaking. Certain traditional practices are of particular historical and cultural importance, especially in the context of AOP (*Protected Designation of Origin*), in which regulations have been established to preserve these practices, and only a limited and pre-defined list of operations is possible. However, this list is rapidly changing to allow winemakers to adapt to new wine markets and consumer expectations.

This chapter describes the kinetics of alcoholic fermentation in winemaking conditions, the main parameters that affect these kinetics, and the main factors that can be controlled to improve fermentation and modify wine properties.

2 Description of Fermentation Kinetics

2.1 Kinetics of a Standard Fermentation

In a standard fermentation, i.e., performed by a pure culture of *Saccharomyces cerevisiae* at constant temperature, three main phases are observed: lag, growth and stationary phases, which will be discussed in the following sections.

2.1.1 Lag Phase

The lag phase is usually defined as the phase before 'active fermentation'. The phase corresponds to the progressive saturation of the medium in CO_2 (approximately 1.5 g/l). A cell population of approximately 10^7 cells/ml is reached by the end of this phase, corresponding to 2–3 generations, and the yeasts are depleted of accumulated trehalose and glycogen (Novo et al. 2003). Less than 4 g/l sugar is consumed, and 2 g/l ethanol is produced, and the composition of the medium changes only very slightly, with the exception of a rapid decrease in thiamine concentration. Bataillon et al. (1996) showed that in a must inoculated with 10^6/ml cells, thiamine was exhausted in less than 5 h (24 h following inoculation with 10^4/ml). Many factors affect the length of the stationary phase in a strain-dependent manner, as recently reported by Ferreira et al. (2017). These authors demonstrated that low temperature and osmotic stress substantially affected all strains, promoting considerably extended lag phases. SO_2 addition had a partially temperature-dependent effect, whereas low phytosterol and thiamine concentrations affected the lag phase in a strain-dependent manner. Zhao (2005) noted both the major effect of SO_2 and the differences between yeast strains. These differences, with MET2 and SKP2 playing a key role, were explained by Noble et al. (2015).

2.1.2 Growth Phase

The growth phase, during which 20–40% of the sugar is consumed, lasts from the end of the lag phase until the maximum population is reached. At the beginning of this phase, the yeast population increases exponentially, i.e., at a constant specific growth rate. As described by Bely et al. (1990a), this growth rate decreases during the second half of this phase, which corresponds to the last generation, and the final population ranges between 50 and 250 × 10^6 cells/ml. These authors also used an online monitoring device for measuring the amount of CO_2 evolved – proportional to the amount of ethanol produced and the amount of sugar consumed – to calculate and describe precisely the evolution of the other kinetic parameters. They observed the following:

(i) The maximum specific CO_2 production rate (proportional to the maximum metabolic activity of the yeast) is reached very soon, i.e., when the sugar concentration has been reduced by less than 10 g/l. This finding indicates that the decrease in yeast activity due to limitations and inhibition phenomena occurs at the very beginning of fermentation.
(ii) The maximum CO_2 production rate $(dCO_2/dt)_{max}$ (proportional to the maximum fermentation rate) is reached later but always before the end of cell growth. This maximum has technological importance because it is proportional to the maximum power generated by fermentation and therefore to the maximum power required to cool the tank.

In most cases, growth stops due to the exhaustion of assimilable nitrogen in the must. This exhaustion occurs when (dCO_2/dt) is maximal, as demonstrated by Colombie et al. (2007a), who measured changes in medium conductivity (directly proportional to ammoniacal nitrogen content) online. Crepin et al. 2012 demonstrated that nitrogen compounds could be classified into three groups according to their order of use: prematurely consumed (Lys), early consumed (Asp, Thr, Glu,

Leu, His, Met, Ile, Ser, Gln, and Phe), and late consumed (ammonium, Val, Arg, Ala, Trp, and Tyr). This classification corresponds to a strategy of distribution of metabolic fluxes implemented by yeast as a means of adapting to environments with changing and scarce nitrogen resources (Crepin et al. 2017).

2.1.3 Stationary Phase

The stationary phase starts when the yeast cells have reached the maximum population. Most of the sugar (between 50 and 80%) is fermented during this phase by yeast that are not actually growing. This feature is specific to the winemaking conditions. The fermentative activity of the yeast gradually decreases during this phase due to a decrease in the hexose transport rate (Salmon et al. 1993) and in the expression of the main glycolytic genes (Puig and Perez-Ortin 2000). Various mechanisms that inhibit yeast growth and activity operate during this phase. The mechanisms involve highly different compounds, such as mid-length-chain fatty acids, acetic acid and killer toxins, but the main inhibitor is ethanol, as demonstrated by Ansanay-Galéote et al. (2001), who studied the effect of adding ethanol and observed that the fermentation rate correlated directly with ethanol concentration in the fermenting must. Other changes have been described during the stationary phase, especially in nitrogen metabolism (Noti et al. 2018). It should be noted that dedicated fermentation devices have been developed for physiological analysis in continuous fermentation processes mimicking different phases of the stationary phase (Clement et al. 2011; Vazquez-Lima et al. 2014).

During the stationary phase, yeast cell size increases due to the synthesis of trehalose and glycogen, as indicated in Fig. 9.1. Glycogen is considered to be a storage carbohydrate (Parrou et al. 1999), whereas trehalose plays a more complex role.

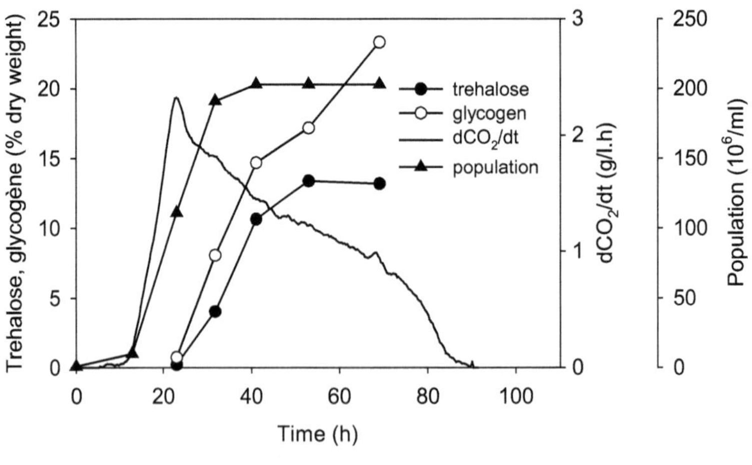

Fig. 9.1 Changes in yeast population, CO_2 production rate, and trehalose and glycogen content of yeast (Roustan and Sablayrolles 2004)

Trehalose has been implicated in response to stresses caused by ethanol, temperature (Vianna et al. 2008) and nitrogen depletion (Varela et al. 2004). However, under winemaking conditions, trehalose undoubtedly replaces glycerol as a 'compatible solute' during the course of fermentation, counterbalancing the osmotic pressure, which may exceed 2000 mOsm/kg (Roustan and Sablayrolles 2004). Trehalose may also be involved in an adaptive mechanism for fine tuning glycolysis and carbon storage management during alcoholic fermentation (Jules et al. 2008; Novo et al. 2005).

2.2 *Effect of Hydrodynamic Conditions*

Most studies published on the kinetic parameters of wine fermentation have been carried out at the laboratory scale. However, the hydrodynamics of reactions within small fermentors differ considerably from those within large tanks (several hundred hectolitres). When trying to extrapolate between scales, the hydrodynamics within tanks of laboratory, pilot and industrial scales must be considered. The different factors affecting this parameter are described in the following sections.

2.2.1 Homogeneity

Homogeneity of the suspension within the tank is essential for effective exchange between the yeast and the medium. In winemaking, mixing is caused solely by the production of CO_2. Garcia et al. (1994) and Muller and Velten (2015) proposed models for calculating mixing power and a heterogeneity factor for fermentation tanks.

In practice, Garcia et al. (1994) considered homogeneity to be achieved in most tanks. Malherbe (2003) found no heterogeneity when measuring the temperature, sugar and ethanol concentrations at different locations within an 11,000-l tank during white wine fermentation. The cell population was found to have decreased by 30% at the end of fermentation, with the population beginning to decrease at a CO_2 production rate of approximately 0.2 g/l.h (i.e., 0.4 g/l.h sugar consumption). However, the liquid phase remained homogeneous, as confirmed by Aguera and Sablayrolles (2005) in a study of hydrodynamics at the pilot scale (100 l).

2.2.2 Effect of Agitation, Turbidity and Tank Size on Fermentation Kinetics

Plouy (2000) and Casalta et al. (2010) studied the effect of stirring and turbidity on fermentation kinetics in 1-l and 100-l fermentors. Stirring considerably decreased the duration of fermentation but had little effect on the maximum fermentation rate, indicating that stirring was most effective at the end of fermentation, i.e., when the fermentation rate was low. This effect was confirmed by the absence of a decrease in the yeast cell population in the stirred fermentors. When solids were added to the musts to increase their turbidity to 200 turbidity units (NTU), stirring had a much weaker effect, confirming (i) the positive effect of the presence of solid particles and (ii) the interactions between must turbidity and agitation (Fig. 9.2). In small fermentors (laboratory scale), stirring has several advantages: (i) it increases kinetic reproducibility, (ii) it abolishes the decrease in cell population and (iii) it decreases the negative effect of excessive must clarification.

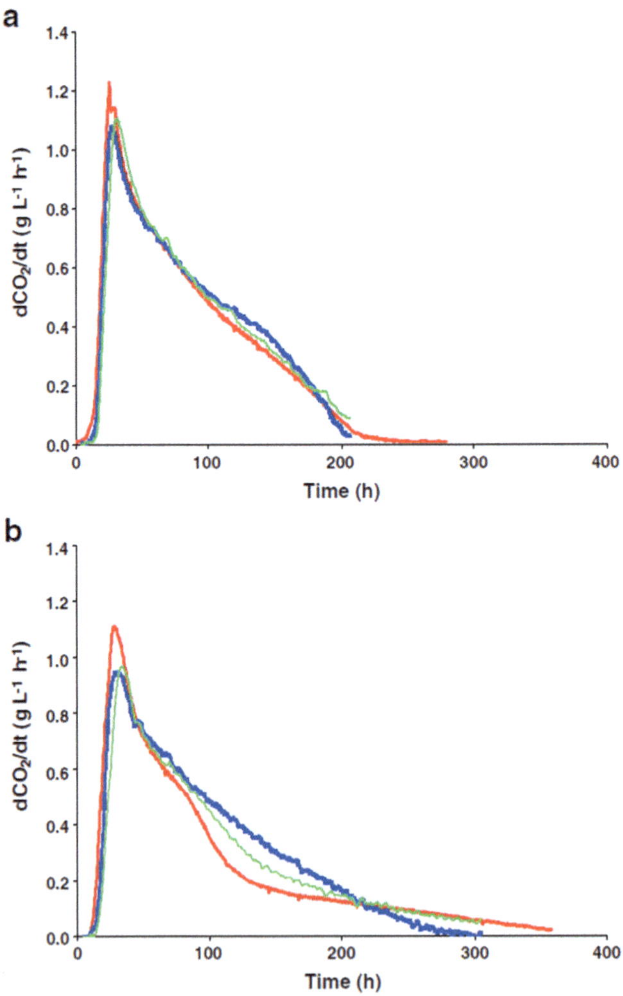

Fig. 9.2 Changes in CO_2 production rate during the fermentation of Grenache must with grape must solid particles (**a**) and without solid particles (**b**) in a 100-L (red), a stirred 1-L (blue), and a static 1-L fermentor (green) (Casalta et al. 2010)

Malherbe (2003) investigated the effect of pilot- and industrial-scale tank sizes. The author compared kinetics of fermentation in 100-l and 11,000-l tanks and found them highly reproducible. However, further experiments indicated slight differences, particularly in sluggish fermentations, with very low fermentation rates at the end.

2.2.3 Red Winemaking

The hydrodynamic conditions in traditional red winemaking fermentation processes are highly specific and cannot be simulated satisfactorily in laboratory fermentors. Therefore, the conditions should be studied at the pilot scale, with tank volumes of at least several tens of litres. During fermentation (Fig. 9.3), the solid particles rise to the surface to form a cap, which becomes increasingly compact. This cap is only partly submerged (approximately half the cap is submerged). The upper part of the cap contains very little or no interstitial liquid and is highly heterogeneous in terms of both temperature (Schmid et al. 2009) and yeast population. In contrast, the must phase may be considered homogeneous throughout most of the fermentation process. Cap punching (or pumping over) is usually carried out to homogenize the contents of the tank. In studies of changes in the rate of production of CO_2 under such fermentation conditions, Aguera and Sablayrolles (2005) observed a highly significant increase in fermentation rates after cap punching (Fig. 9.4). This increase was (i) nearly instantaneous and continued over several hours and (ii) particularly

Fig. 9.3 Cap formation during red wine vinification. (1): Tank filling, (2): after fermentation of 5% of the sugar, and (3): after fermentation of 40% of the sugar. One hundred-litre tank with transparent walls (Aguera and Sablayrolles 2005)

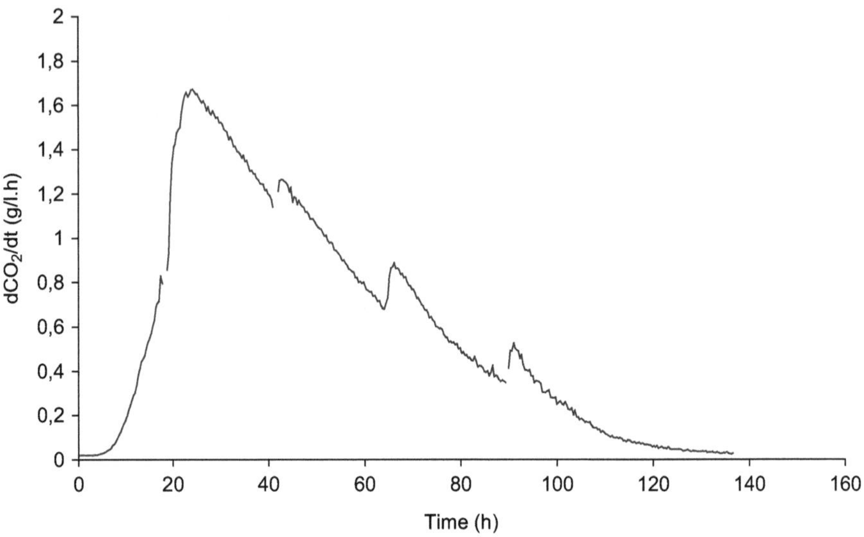

Fig. 9.4 Effect of cap punching (↑) on fermentation kinetics in red winemaking. One hundred-litre tank (Aguera and Sablayrolles 2005)

large when pumping over was carried out during the second part of the fermentation (stationary phase). Cell population measurements indicated that this kinetic acceleration was mostly due to a transfer of yeasts from the cap to the liquid phase, increasing the size of the cell population in the liquid (by more than 50%).

3 Control of Fermentation Kinetics

Despite strong constraints, due in particular to regulations in place, it is possible to control – at least partially – the fermentation kinetics by acting on several parameters.

3.1 Importance of Nutrients

Fermentation kinetics may vary considerably with must composition. Indeed, nutrients are essential for yeast growth and survival.

3.1.1 Assimilable Nitrogen

Casalta et al. (2013) compared different methods for the determination of assimilable nitrogen in musts and confirmed that ammoniacal nitrogen and free amino acids, but also some peptides, can be assimilated by yeasts. Bely et al. (1990a) tested one hundred musts from the main winemaking regions of France under standardized conditions. These authors demonstrated that changes in fermentation rates, expressed as dCO_2/dt, were mostly attributable to differences in the assimilable nitrogen content of the musts (Bely 1990; Bely et al. 1990b). There was a direct relationship between fermentation rate and assimilable nitrogen concentration, which varied from 53 to 444 mg/l. Nicolini et al. (2004) also reported a wide variability in the assimilable nitrogen content of musts. The authors analysed 600 Italian wines and reported assimilable nitrogen concentrations ranging from a few tens of mg/l to nearly 400 mg/l, with an average value of 136 mg/l, whereas Butzke (1998) reported a mean concentration of 213 mg/l in 1500 musts originating from California. Year and maturity (Dubois et al. 1996), as well as vineyard treatments (Garde-Cerdan et al. 2014; Holzapfel et al. 2015; Perez-Alvarez et al. 2015; Verdenal et al. 2016), also affect assimilable nitrogen content and, consequently, fermentation kinetics.

The addition of ammonium salts (diammonium phosphate (DAP) or ammonium sulphate) efficiently increases the fermentation rate and decreases the duration of fermentation (Agenbach 1977), but the timing of this addition is important. If nitrogen is added at the time of inoculation, it is metabolized and used for additional yeast growth. If added during the stationary phase, nitrogen rapidly increases the rate of CO_2 production by reactivating hexose transport (Bely et al. 1994), and it is at least as efficient than when added initially (Bely et al. 1990b; Seguinot et al. 2018). It is notable that additions made after the consumption of 80–100 g/l sugar lead to a lower increase in the CO_2 production rate and therefore to a lower peak of energy demand for temperature regulation than those observed with addition to the must, while the fermentation durations observed in both situations are very similar (Malherbe et al. 2004). The effectiveness is also dependent on the initial concentration in the must. Adding ammoniacal salts consistently affects the kinetics, but additions are usually considered necessary when the assimilable nitrogen concentration in the must is lower than 150–180 mg/l. Adding a mixture of amino acids has a highly similar effect to that of adding ammonium salts on the fermentation kinetics (Fig. 9.5), although the effect on yeast secondary metabolism may be significantly different (cf 3.2).

3.1.2 Oxygen and Lipids

Yeasts require oxygen for their development and the maintenance of viability at the end of fermentation. Oxygen is required for the synthesis of sterols and unsaturated fatty acids, which are essential constituents of the plasma membrane (Andreasen and Stier 1953; Parks 1978). Under typical winemaking conditions, the yeasts used

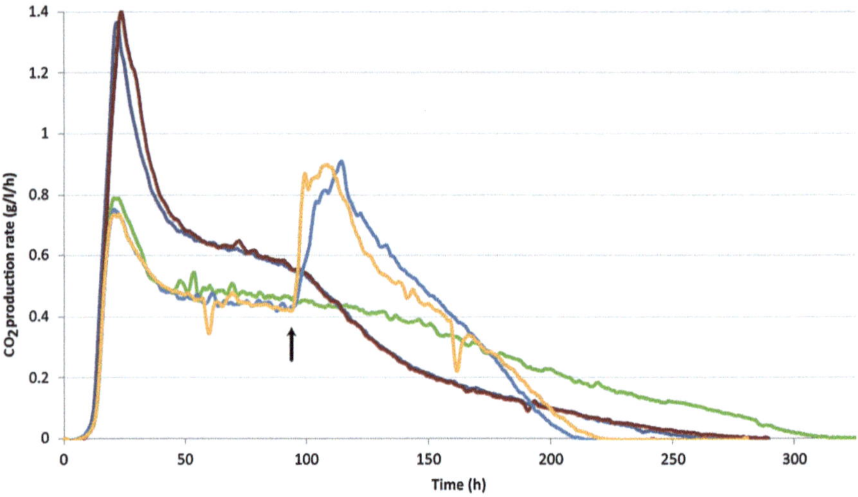

Fig. 9.5 Changes in fermentation kinetics depending on the addition of nitrogen: control without addition (green curve); addition at the beginning of fermentation of DAP (dark blue curve) or amino acids (red curve); and addition during the stationary phase of DAP (light blue curve) or amino acids (orange curve). The arrow indicates the timing of addition during the stationary phase (Seguinot et al. 2018)

for inoculation (in the form of active dry yeasts) contain large amounts of lipids, either incorporated into the membranes or stored as lipid reserves (triglycerides, etc.). It is therefore possible for the yeast to persist several generations without oxygen, but under these conditions, the yeast cells will have low sterol and fatty acid levels at the end of the growth phase and during the stationary phase.

Even if oxygen is usually not limiting for yeast growth and activity, oxygenation may strongly affect yeast survival, particularly if there is a risk of fermentation becoming stuck. Sablayrolles and Barre (1986) estimated oxygen requirements at approximately 10 mg/l. Musts generally contain lipids likely to be directly incorporated by yeast, decreasing oxygen requirements (Luparia et al. 2004). These compounds are found mainly in grape residues in wine (Houtman and du Plessis 1986). The timing of oxygen addition is at least as important as the quantity added. Oxygen is most effective when added at the end of the growth phase (Sablayrolles and Barre 1986; Vivas and Cros 1991), when yeasts have utilized their lipid reserves. Another advantage of adding oxygen at this time point is that it limits the risk of oxygen use by polyphenol oxidases. Indeed, oxygen consumption by oxidases is likely much less effective than oxygen assimilation by yeasts at this time point, whereas this is not the case during inoculation.

It is difficult to control oxygen addition precisely because only very small quantities (a few mg/l) are added and oxygenation is carried out when the fermentation rate is high, i.e., when oxygen is consumed by the yeast as it is added. Moreover, only the oxygen transferred into the must be metabolized by the yeasts. Several authors (Moenne et al. 2014; Cerda-Drago et al. 2016) have proposed methods, including modelling approaches, to quantify this amount in industrial tanks. Blateyron et al. (1998) showed that (i) during oxygen sparging, the quantity of oxygen transferred to the fermenting must by most spargers is at least 50% of the amount added and (ii) when using pumping over, it is necessary to pump approximately twice the volume of the tank to transfer 10 mg/l.

Adding lipids is also a good way to provide major constituents of the yeast plasma membrane and to limit oxygen requirements. Solid particles of the must are the major source of lipids (Alexandre et al. 1994; Casalta et al. 2016), and their management is essential, especially in white winemaking. Indeed, in highly clarified musts, solid particles can dramatically limit the fermentation kinetics, as illustrated in Fig. 9.6. Casalta (2017, personal communication) demonstrated that the effect of solid addition was mostly due to phytosterols. Indeed, adding 1 mg/l of phytosterol of different origins (synthetic or issued from grape solids of different varieties) had the same effect on the kinetics (Fig. 9.7). These results explain the major interest in estimating phystosterol concentrations in musts. Unfortunately, there is no simple method for such estimation, and turbidity measurements provide only rough estimates.

3.1.3 Other Nutrients

Other nutrients may also affect fermentation kinetics (Bisson 1999). Magnesium concentration may directly affect the yeast growth rate and sugar degradation (Birch et al. 2003). Magnesium ions play an important role in protecting cells from stress factors. Magnesium and phospholipids stabilize the cell membrane, protecting yeast cells from the toxic effects of ethanol. The loss of magnesium and excessive

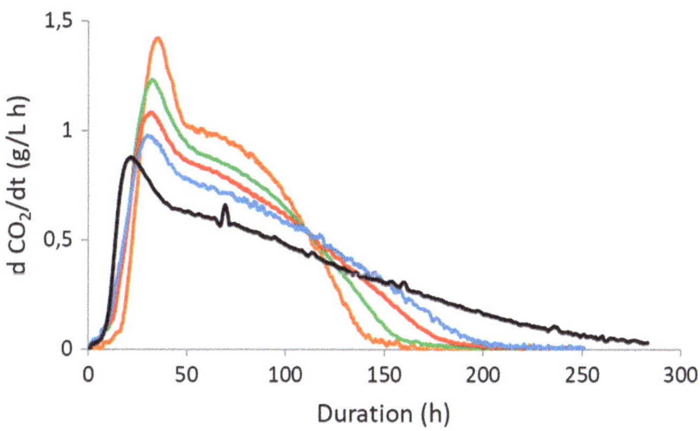

Fig. 9.6 Effect of adding solid particles, in a highly clarified must, on fermentation kinetics (Casalta, personal communication). Control (in black) and additions of 0.5%, 1%, 1.5% and 2%, respectively

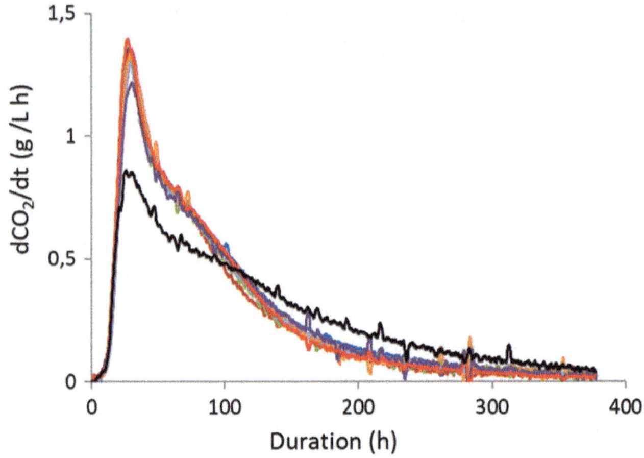

Fig. 9.7 Effect of adding 1 mg/l of β-sitosterol on fermentation kinetics. Comparison of synthetic β-sitosterol (positive control, in red) and β-sitosterol from solid particle samples issued from different musts: Maccabeu, Viognier, Chardonnay, Marselan, Cabernet, Cabernet Sauvignon, Merlot (Casalta et al. 2016)

calcium uptake increases the susceptibility of yeast to alcohol and high temperature stresses.

Vitamin concentrations are not usually limiting in musts: the addition of a mixture of biotin, thiamine, (meso) inositol, pantothenic acid, nicotinic acid and pyridoxin to 17 different musts had very little effect (less than 10% difference) on the duration of fermentation (Sablayrolles and Salmon 2001). However, thiamine was the first nutrient eliminated from a contaminated must (Bataillon et al. 1996). This biological depletion of thiamine led to low kinetics of fermentation, and this phenomenon was amplified in musts with high assimilable nitrogen concentrations. The addition of thiamine may, therefore, have a positive effect under certain winemaking conditions.

3.1.4 Interest of Combined Nutrient Management

All nutrients combine their effects, and an optimal strategy for controlling fermentation requires considering several of them. Lowering the risks of stuck fermentation and managing fermentation of highly clarified musts represent good examples.

Stuck and Sluggish Fermentations Despite improvements in fermentation management, such as using selected strains and controlling temperature, certain fermentation problems persist, mostly because of an increase in the average sugar content of the musts.

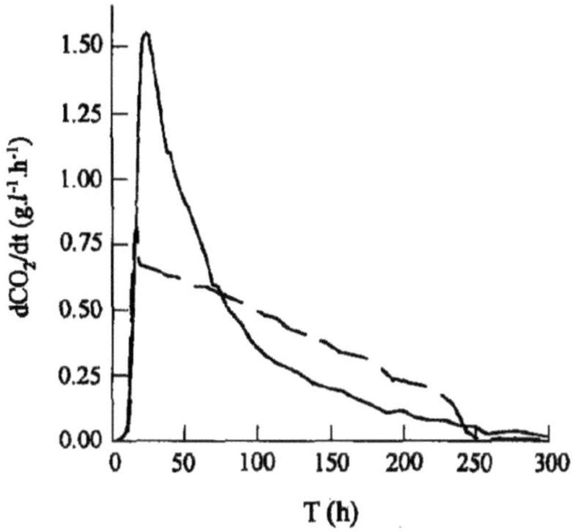

Fig. 9.8 Differences in CO_2 production rate between a 'slow fermentation', characterized by low CO_2 production rates throughout fermentation (dotted line), and a 'sluggish fermentation', with a very low rate of fermentation at the end of the process (continuous line) (Blateyron and Sablayrolles 2001)

It is first necessary to describe and define these problems because they may correspond to different situations (with different solutions) (Fig. 9.8), which are not always distinguished in practice or even in the literature (Alexandre and Charpentier 1998; Bisson 1999; Bell and Henschke 2005). To overcome this difficulty, Blateyron and Sablayrolles (2001) proposed distinguishing (i) fermentations with a low fermentation rate throughout the process, which they called 'slow fermentations', from (ii) fermentations with a very slow fermentation rate only at the end, called 'stuck' (when remaining residual sugar) or 'sluggish' (without residual sugar) fermentations. The authors tested 178 difficult-to-ferment musts and, using statistical analysis, demonstrated that 'slow fermentations' are mainly due to must nitrogen deficiencies. In contrast, sluggish and stuck fermentations are closely related to yeast mortality, which is highly favoured by high ethanol concentrations.
In practice, a precise analysis of the type of fermentation problem is not usually feasible, and the combined addition of ammoniacal nitrogen and oxygen is a good solution because it combines the positive effects of oxygen on yeast survival and those of nitrogen on yeast activity. However, this addition has to be correctly timed, i.e., at the start of the stationary phase, when approximately 5% ethanol has been produced (Blateyron et al. 2000). Blateyron and Sablayrolles (2001) assessed the efficacy of combined additions on 72 musts leading to sluggish or stuck fermentations. In all cases, the addition of 5 mg/l oxygen and 300 mg/l di-ammonium phosphate when the ethanol concentration was approximately 5% led to (i) a dramatic decrease (by 44% on average) in the duration of the fermentation (for sluggish fermentations) or (ii) sugar exhaustion (for stuck fermentations). This effect was independent of the variety and origin of the must and the yeast strain.

Clarified Musts As previously described (2.1.2), the addition of solid particles to highly clarified musts dramatically increases the fermentation rate (Fig. 9.6). This effect is due to greater consumption of assimilable nitrogen (Fig. 9.9) with a switch

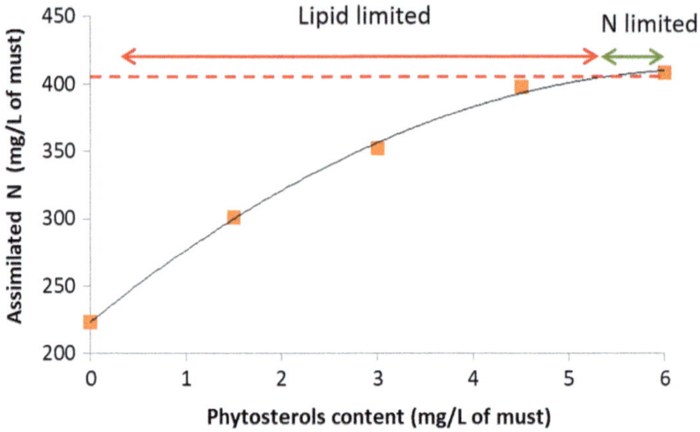

Fig. 9.9 Effect of adding solids on yeast nitrogen assimilation. Switch from a lipid deficiency to a nitrogen deficiency

from a lipid deficiency to a nitrogen deficiency. In this example, the switch occurred when 5 mg/l phytosterols was added, but this value is must- and strain-dependent. Using another must and strain, Ochando et al. (2017) showed that 8 mg/l phytosterol was necessary to exhaust assimilable nitrogen, while only 3 mg/l was necessary when 10 mg/l oxygen was added.

The interest in avoiding residual assimilable nitrogen was recently explained by Duc et al. (2017), who demonstrated that (i) cell death results from yeast's inability to trigger an appropriate stress response under certain conditions characterized by nutrient limitations, particularly a limitation by lipids, and that (ii) limitation by nitrogen is the most favourable situation for inducing such stress responses.

Therefore, it can be considered that, in the case of nitrogen-rich musts, the best strategy is to combine (i) sufficient turbidity, i.e., 50–150 NTU, to increase nitrogen assimilation and therefore increase the fermentation rate and move towards a nitrogen-limited situation with (ii) oxygen addition, which is the best way to improve yeast viability.

To limit the risks of fermentation problems, especially in the case of clarified musts, it is also possible to use commercial products that are combinations of different nutrients, including at least ammoniacal nitrogen, thiamine and inactivated yeasts. Some activators containing inactivated dry yeasts (IDY) are usable during the rehydration phase because they facilitate the rehydration of active dry yeasts (ADY) due to the transfer of sterols from IDY to ADY (Soubeyrand et al. 2005).

3.2 Influence of the Yeast Strain

Saccharomyces cerevisiae is the main yeast used in winemaking due to its high fermentation capacity, and more than 200 different *S. cerevisiae* strains are currently available commercially, with the strains showing highly diverse fermentation properties.

In a comparison of 20 randomly chosen commercial strains (i.e., without taking their fermentative capabilities into account), Colombie et al. (2005) observed differences in fermentation kinetics in a synthetic medium, with the duration of fermentation ranging from 80 to 109 h. Several studies (Julien et al. 2000; Gutierrez et al. 2012) showed that differences between strains in easy-to-ferment musts were mostly due to differences in assimilable nitrogen demands. Brice et al. (2014) compared two groups of strains, one with low nitrogen requirements (LNRs) and the other with high nitrogen requirements (HNRs) and suggested that differences may originate from variations in nitrogen sensing and signalling under conditions of starvation.

In the case of difficult-to-ferment musts, it is essential to choose strains with a good ability to resist high ethanol concentrations. Blateyron et al. (2000) compared 13 randomly chosen strains cultured in a natural must leading to sluggish or stuck fermentations: three strains yielded stuck fermentations, with residual sugar contents of 10–56 g/l, whereas the remaining 10 strains fermented all the sugar, with a fermentation duration of 119–170 h.

Ethanol concentration is also an important parameter, and due to the higher average sugar concentration of modern musts, some winemakers would like to lower the final ethanol concentration by 1 or 2% v/v. Could this be achieved with commercial yeast strains with a lower ethanol yield? In a study of 55 different strains inoculated in a synthetic medium containing 200 g/l sugar, final ethanol concentrations ranged between 11.72% and 12.09%, with a mean concentration of 11.95% and a standard deviation of 0.08%, clearly indicating a low level of variability in ethanol yield, with differences of less than 0.5%. This finding is not surprising because a large amount of sugar must be metabolized to obtain a single percent decrease in alcohol content (16.8 g/litre), and the accumulation of undesirable by-products in the wine must be avoided. Various genetic engineering strategies based on modification of the carbon and redox balance have been developed to lower the ethanol yield. These approaches have often faced problems of indirect accumulation of undesirable products that may affect wine quality, demanding several gene modifications to obtain a fine adjustment of the level of metabolites. To date, the best strategy complying with wine quality and strain performance requirements is the concomitant overproduction of glycerol and 2,3-butanediol at the expense of ethanol obtained by multi-step engineering (Tilloy et al. 2014). In addition to the development and exploitation of new strains of *Saccharomyces* yeasts, the use of non-*Saccharomyces* yeasts in combination with *S. cerevisiae* has attracted interest in recent years for lowering ethanol yield (Ciani et al. 2016a; Contreras et al. 2014, 2015; Englezos et al. 2015).

Table 9.1 Effect of temperature regime on energy requirement for temperature regulation

Temperature (°C)	Duration (days)	Peak of energy demand (kcal/h)	Total energy required (10^3 kcal)
15	23	7960	1411
15–22 (early increase)	12,5	9780	811
15–22 (late increase)	16	7760	935
22	11	13230	1103

3.3 *Effect of Temperature*

Temperature has a major effect on fermentative kinetics. Between 15 and 25 °C, the fermentation rate is doubled when the temperature is increased by approximately 8 C (Bely et al. 1990a). The temperature profile is also important; an increase of several degrees during fermentation – a frequent situation in oenology – greatly changes the fermentation kinetics. Table 9.1 illustrates the effect of temperature on the fermentation time as well as on the maximum fermentation rate. In all cases, the temperature is between 15 °C and 22 °C, a typical range in white winemaking. The values indicated in this table have been calculated using simulation software (Goelzer et al. 2009). Logically, non-isothermal fermentations have intermediate values between the 2 isotherms at 15 °C and 22 °C, with faster fermentations occurring when the temperature increases from the beginning. However, we note the highly significant effect of the increase in temperature during the second part of the fermentation. Indeed, compared with the isotherm at 15 °C, this temperature regime (which corresponds to a common practice) can shorten alcoholic fermentation to 7 days. Moreover, it should be noted that a final temperature of at least 20 °C is usually necessary to perform a subsequent malolactic fermentation.

The temperature regime also has a highly significant effect on the energy requirements for temperature control. Table 9.1 indicates: (i) the peak energy (Pmax), at which the fermentation rate is maximum, and (ii) the total energy (E_{tot}) necessary for the whole fermentation. These parameters are of great technological interest because they are proportional to the power of the refrigeration unit and the energy consumption, respectively. Compared with the isothermal at 15 °C (E_{tot} = 1411 Mcal), increasing the temperature to 22 °C allows for highly significant energy savings of 43% (E_{tot} = 811 Mcal) and 34% (E_{tot} = 935 Mcal). The early increase in temperature allows for a slightly greater saving, but the later increase (at mid-fermentation) has the advantage of leading to a moderate value of Pmax.

4 Modulation of Wine Characteristics

The main objectives of fermentation control have shifted from technological aims to quality-related issues. The effect of yeast strain on wine quality, especially aroma production, has been emphasized by many authors, but fermentation management, especially the addition of nutrients and temperature control, also represents a powerful way to modulate wine characteristics.

4.1 Importance of the Yeast Strain

The use of starter cultures has improved the overall quality of wines. This positive contribution is first due to the reduction of many off flavours (volatile acidity, H_2S, SO_2, phenolic off flavours) that have a negative masking effect on wine quality. The yeast also contributes positively by several mechanisms to aroma synthesis: (i) de novo synthesis of flavour-active metabolites (e.g., esters, higher alcohols, fatty acids), (ii) biotransformation of grape juice constituents into aroma- or flavour-impacting components (e.g., thiols) (Swiegers et al. 2009) and (iii) production of enzymes that transform neutral grape compounds into flavour-active compounds (Styger et al. 2011). Enhancing colour extraction (Fernandez-Gonzalez et al. 2005) and optimizing ageing on lees (del Barrio-Galan et al. 2015) also represent possible contributions of yeasts and criteria of choice.

Some authors have proposed to enhance the technological interest of yeast strains by using different strategie (Dequin 2001; Marullo et al. 2006; Schuller and Casal 2005; Pretorius and Hoj, 2005). On the contrary, others have suggested that active dry yeasts have led to an organoleptic standardization of wines (Vigentini et al. 2016) and could be responsible for the loss of typicality of wines (Di Maro et al. 2007). Consequently, one practice consists in the selection of native *S. cerevisiae* local strains that are believed to be able to enhance the peculiarities of a wine and to show better acclimation to their original environment (Grieco et al. 2011; Comitini et al. 2017).

Several studies have described differences between *Saccharomyces cerevisiae* strains (Takush and Osborne 2012; Blanco et al. 2013), but there is also an increasing interest in non-conventional strains (Egli et al. 1998). Indeed, some strains have several advantages, such as the excretion of enzymes of oenological interest (de Ovalle et al. 2018; Hu et al. 2016), the production of secondary metabolites (Ciani et al. 2016b), and the release of mannoproteins (Padilla et al. 2016).

4.2 Effect of Nutrient Addition

The synthesis of fermentation aromas is affected by the addition of nitrogen, as summarized by Bell and Henschke (2005), with nature (ammoniacal or organic) (Torrea et al. 2011; Barbosa et al. 2012) and the timing of addition (Barbosa et al.

2009) providing important effects. Recently, Seguinot et al. (2018) observed that the effect of timing was stronger than that of nitrogen composition. The strongest effect of nitrogen was observed for acetate esters. In particular, the production of these compounds increased greatly after the addition of nitrogen, especially during the stationary phase. Very low YAN levels favour the production of undesirable sulphur compounds (e.g., hydrogen sulphide), and it is widely reported that H_2S production decreases in response to DAP addition. However, (Ugliano et al. 2011) found a non-linear relationship such that, under the conditions examined, moderate DAP supplementation resulted in a remarkable increase in H_2S formation. High YAN content is also associated with the production of higher concentrations of haze-causing proteins, urea and ethyl carbamate and biogenic amines and higher risks of microbial instability and atypical ageing. Several studies have also shown that residual assimilable nitrogen after alcoholic fermentation may affect subsequent malolactic fermentation (Alexandre et al. 2004).

The addition of oxygen may also affect yeast metabolism and the concentrations of fusel alcohols and esters. According to Blateyron and Sablayrolles (2001), the addition of 5 mg/l oxygen halfway through fermentation had no organoleptic consequences, whereas excessive addition (e.g., 50 mg/l) had a highly negative organoleptic effect on both aroma compounds and wine oxidation.

The nutrients combine their effects on the synthesis of fermentation by-products, as reported by Rollero et al. (2015) and Bloem et al. (2018), with those of nitrogen and lipids. Therefore, complex activators have beneficial effects, including those caused by the addition of nitrogen. Solid particles and lipids tend to lower volatile acidity (Delfini and Costa 1993; Belviso et al. 2004), whereas thiamine limits SO_2 binding and the production of keto acids (Delfini et al. 1980).

4.3 Impact of Temperature

Low temperatures increase the production of volatile compounds (esters, acetates, medium-chain fatty acids) by yeast during alcoholic fermentation (Killian and Ough 1979; Cotrell and MC Lellan 1986; Torija et al. 2003; Beltran et al. 2008a, b) and lower losses caused by evaporation (Ferreira et al. 1996). Such temperatures (10–15 °C) may be used by winemakers to enhance the production of these volatile compounds, improving the aromatic profile of the wine. However, low temperatures may also result in sluggish or stuck fermentations, and the choice of yeast strain and an increase in temperature at the end of fermentation are essential. Other studies have also highlighted the importance of temperature for varietal aromas, focusing mostly on volatile thiol production during the fermentation of Sauvignon musts (Howell et al. 2004; Masneuf-Pomarede et al. 2006). Temperature has been shown to affect the concentrations of 4-MMP, 3-MH, and 3-MHA, with high fermentation temperatures (20 °C) resulting in significantly higher concentrations of all volatile thiols than lower fermentation temperatures (13 °C), regardless of the yeast strain used. Therefore, in white winemaking, temperature may have a direct effect on aromatic characteristics, favouring either fermentative or varietal aromas.

By monitoring the main aroma compounds online and modelling their gas-liquid partitioning during fermentation, Morakul et al. (2011) calculated losses in the exhausted gas throughout fermentation. Negligible amounts of fusel alcohols were lost, regardless of the fermentation temperature. In contrast, 56% of\ ethyl hexanoate and 34% of isoamyl acetate were stripped by CO_2 when the temperature profile simulated red winemaking conditions. Even at a moderate temperature of 20 °C, typical of white wine fermentations, 40% of ethyl hexanoate and 21% of isoamyl acetate were lost. Loss rates were maximal at the end of fermentation, indicating that high final temperatures, although promising in preventing sluggish fermentations, can be very detrimental to aroma losses.

In conventional red winemaking, the aroma compounds produced by yeast fermentation have a much weaker effect. The fermentation temperature is mostly regulated to favour the transfer of polyphenol compounds from the solid to the liquid phase (reviewed by Sacchi et al. 2005). A combination of high temperatures (up to 30°C) and an increase in ethanol concentration is generally considered to favour this extraction.

5 Future Prospects

The oenological sector is highly diverse, as are the prospects for the sector.

One main question is becoming how new technologies, especially during the fermentation process, can assist winemakers in elaborating wines with predefined characteristics. Additionally, process performance must be improved, and environmental issues must be taken into account. Online fermentation monitoring and control represents a promising prospect in this respect.

One very different question is how to increase wine complexity and diversity. As mentioned previously (3.1), using non-conventional strains is an interesting prospect, but optimizing such cultures remains a challenge, and new fermentation strategies have to be elaborated.

5.1 Online Fermentation Monitoring and Control

Control over fermentation kinetics (i) has direct technological advantages, in terms of tank use optimization in the winery and control over the energy expenses for the regulation of temperature, and (ii) is generally a prerequisite for controlling the characteristics of wine. Recently, studies have also directly considered the online monitoring and control of ‘quality markers’.

During fermentation, temperature is usually measured and automatically controlled. For red winemaking, new devices have also been proposed to increase the transfer between the cap and the liquid phase. There are also examples of the automation of aeration, pumping over, cap punching, etc. However, some operations still remain manual, the main one being monitoring density or sugar consumption.

Online fermentation monitoring is clearly a promising method for improving process control because (i) it is much more accurate than manual measurements (Roger et al. 2002) and (ii) it makes new control strategies in which winemaking operating conditions are adapted to actual fermentation behaviour possible. The potential value of online monitoring for controlling wine fermentation was first discussed many years ago (Sablayrolles 1988), but at the time, the necessary technology (computers, sensors) and the proposed changes in fermentation strategies were largely incompatible with the oenological context. Few industrial fermentation processes are currently monitored online, but this situation is subject to change, and commercial products are now available.

5.1.1 Online Monitoring

Many strategies for the online monitoring of fermentation kinetics have been proposed. These include density measurement, usually calculated based on differential pressure (Bruch 2001; Blankenhorn and Neumann 2004); refractometry (Kovacs et al. 2015); the estimation of sugar and ethanol concentrations by electrochemical biosensors (Warriner et al. 2002; Esti et al. 2003), fibre-optic Fourier transform-near-infrared (FT-NIR) spectrometry (Wang and Peng 2017), piezoelectric MEMS resonators (Toledo et al. 2018; Pfusterschmied et al. 2017), vibrational spectroscopy (Cozzolino 2016), optoelectronic sensors (Jimenez-Marquez et al. 2016), chemocapacitor sensor arrays (Oikonomou et al. 2014), and amperometric sensors (Piermarini et al. 2011); and the measurement of CO_2 production (Corrieu et al. 1997; Saur 2004; Leo Kuebler Gmbh 2007; Martinez-Landa et al. 2006). CO_2 measurement has the advantage of involving the use of moderately priced sensor providing online estimations of density, sugar or ethanol concentration by using simple correlations described by Corrieu et al. (1997). For example, the production of 1 g/l CO_2 corresponds to the consumption of 2.17 g/l sugar. CO_2 measurement is, above all, interesting because it allows for the precise calculation of the instantaneous fermentation rate. This rate is of prime importance both technologically (proportional to the amount of energy produced in the fermenting tank) and microbiologically (proportional to the fermentative activity of yeasts).

Different methods for online measurement of quality markers have also been proposed. Several authors have developed sensors to monitor the evolution of colour during maceration in red winemaking (Jimenez-Marquez et al. 2013, 2015; Shrake et al. 2014). Such measurements strongly correlate with measurements performed with a reference UV-vis spectrophotometer. The measurements may offer advantages under standardized and well-known conditions, but because of the complexity of the associated polyphenol chemistry, the main difficulty is finding general relationships between this information and wine quality. Different devices for the online measurement of fermentation aromas have also been studied. Electronic noses have been tested by several authors (Jiang et al. 2015; Buratti and Benedetti 2016). This device is of particular interest for the analysis of complex mixtures, but it is still in an early stage of development, and more research is needed before using it for the online monitoring of wine aromas during fermentation. Morakul et al. (2011) proposed an online GC system for monitoring the synthesis of the main aroma compounds. The system is difficult to implement at the industrial level, but it is very powerful for research (3.1) because it allows for (i) highly precise monitoring of the kinetics with the possibility of calculating the rates of synthesis and (ii) an estimation of total production, taking into account losses in the exhaust gas.

5.1.2 Online Control

With online monitoring of the fermentation rate, winemaking operations can be adapted to actual fermentation behaviour, with the possibility of taking into account the variability of the must composition as a function of the tank used and optimizing the control of individual tanks. For example, a fermentation control device can perform the following:

- Estimate the concentration of assimilable nitrogen from the measurement of the maximum CO_2 production rate and add ammoniacal nitrogen when necessary and at the best moment;
- Control oxygenation by, for example, (i) the addition of 10 mg/l at the end of the growth phase to ensure fermentation and (ii) secondary addition during the stationary phase to act on the metabolism of by-products by yeast, mostly higher alcohols, ester and sulphur compounds productions, with the possibility of adding oxygen very progressively and proportionally to the fermentation progress (for example, 0.2 mg of oxygen per g of ethanol produced);
- Control temperature as a function of fermentation progress or rate.

Online monitoring may be particularly useful in certain oenological situations in which it is necessary to stop fermentation when a predefined alcohol concentration has been reached (example of production of natural sweet wines of the Muscat type; Perret et al. 1997).

In the future, preventive control, based on modelling the fermentation process, may be possible. Several models of fermentation kinetics have been proposed (del Nobile et al. 2003; Sainz et al. 2003; Cramer et al. 2002; Insa et al. 1995). These models differ considerably, some having a physiological basis, and others being black box models. Malherbe et al. (2004) proposed a physiological model integrating the effects on the fermentation kinetics of anisothermal conditions and DAP addition. Colombie et al. (2005) tested this model under highly different conditions and concluded that this model had a sufficiently large domain of validity to be of potential interest for practical use as a simulator. This kinetic model was combined with (i) a thermal model (Colombie et al. 2007b) and (ii) a decision support module based on fuzzy logic to propose new strategies for optimizing tank and energy use in wineries (Goelzer et al. 2009). Commercial software is now available. The next step will be to integrate such models into control systems for optimizing fermentation.

Recently, Mouret et al. (2015) combined kinetic and thermal models with one modelling the partition of volatile compounds between the liquid and gas phases (Morakul et al. 2011) and a first model for the prediction of the production kinetics of the main fermentative aromas in winemaking fermentations. This research represents the first step in developing a multicriteria optimization strategy. In the future, such strategies should permit, for example, optimization of the production of several aroma compounds while minimizing the energy needs for temperature regulation.

5.2 *Management of Mixed and Immobilized Cultures*

Undoubtedly, most fermentations will continue to be run in conventional batches in the near future, but new processes may also be used to face challenges such as increasing aromatic complexity (Tempère et al. 2018) or reducing the ethanol content of wines. The use of non-conventional yeast is a main issue. Indeed, non-*Saccharomyces* (NS) wine yeast species comprise a large number or species, thus encompassing a wider physiological diversity than does *Saccharomyces cerevisiae*.

NS strains are not able to complete wine alcoholic fermentation. However, this limitation can be overcome through the use of mixed inoculations with an *S. cerevisiae* strain. Controlling such cultures remains a difficult challenge. Sequential inoculation (the inoculation of *S. cerevisiae* being delayed to allow for the implantation of the NS strain) represents the best strategy, but it is complex and strain-dependent. Several parameters must be considered, in particular, inoculation concentrations, timing between the first and the second inoculation and sulphite content. The consumption of nitrogen sources and vitamins from grape must by NS yeasts during the first stage of sequential inoculation fermentation is of particular importance (Kemsawasd et al. 2015), and this consumption often requires compensation by suitable yeast nutrients to prevent stuck fermentations after inoculation of *S. cerevisiae* (Medina et al. 2012; Lage et al. 2014). As a general trend, optimizing the management of NS/*S. cerevisiae* mixed fermentations requires better understanding of the interactions between strains (Renault et al. 2013; Ciani et al. 2016b), especially cell-to-cell mechanisms and the role of acetaldehyde.

To reduce ethanol content, several strategies using NS strains, reviewed by Ciani et al. (2016a), have been described. Nevertheless, the oxygen supply amounts required to ensure efficient yeast respiration are far beyond those typically required by yeast strains, and there is a risk that the strong oxygenation levels required for yeast respiration would promote, as a side effect, the oxidation of key components for the sensory quality of wines, namely, phenolics and aroma compounds.

Yeast immobilization is a promising way to improve the management of mixed cultures. For example, co-immobilized non-*Saccharomyces* yeasts in Ca-alginate coupled with a final inoculation of free *S. cerevisiae* cells have been used to reduce ethanol content in wine (Canonico et al. 2016). There are several methods for yeast immobilization (Kourkoutas et al. 2004; Fleet 2008): the use of natural supports (e.g., fruit pieces), organic supports (e.g., alginate), inorganic supports (e.g., porous ceramics), membrane systems, and multifunctional agents. However, such technologies are still arising, and there are not many applications at the industrial level because of their lack of feasibility, their cost or the limited knowledge of winemakers (Moreno-Garcia et al. 2018). To overcome these difficulties, new concepts are emerging, such as the co-immobilization of yeasts and filamentous fungus categorized as GRAS (Peinado et al. 2006; Nyman et al. 2013).

References

Agenbach, W. A. (1977). A study of must nitrogen content in relation to incomplete fermentations, yeast production and fermentation activity. In *Proceedings South African Society of Enology and Viticulture* (pp. 66–87). Cape Town.

Aguera, E., & Sablayrolles, J. M. (2005). Pilot scale vinifications (100l). II Characterisation–advantages. Wine Internet Journal n°7 (www.vinidea.net).

Alexandre, H., & Charpentier, C. (1998). Biochemical aspects of stuck and sluggish fermentation in grape must. *Journal of Industrial Microbiology and Biotechnology, 20*(1), 20–27.

Alexandre, H., Rousseaux, I., & Charpentier, C. (1994). Ethanol adaptation mechanisms in *Saccharomyces cerevisiae. Biotechnology and Applied Biochemistry, 20*(2), 173–183.

Alexandre, H., Costello, P. J., Remize, F., Guzzo, J., & Guilloux-Benatier, M. (2004). *Saccharomyces cerevisiae-Oenococcus oeni* interactions in wine: Current knowledge and perspectives. *International Journal of Food Microbiology, 93*(2), 141–154.

Andreasen, A. A., & Stier, T. J. B. (1953). Anaerobic nutrition of *Saccharomyces cerevisiae*. I: Ergosterol requirement for growth in a defined medium. *Journal of Cellular and Comparative Physiology, 41*, 23–36.

Ansanay-Galéote, V., Blondin, B., Dequin, S., & Sablayrolles, J. M. (2001). Stress effect of ethanol on fermentation kinetics by stationary-phase cells of *Saccharomyces cerevisiae. Biotechnology Letters, 23*, 677–681.

Barbosa, C., Falco, V., Mendes-Faia, A., & Mendes-Ferreira, A. (2009). Nitrogen addition influences formation of aroma compounds, volatile acidity and ethanol in nitrogen deficient media fermented by Saccharomyces cerevisiae wine strains. *Journal of Bioscience and Bioengineering, 108*(2), 99–104.

Barbosa, C., Mendes-Faia, A., & Mendes-Ferreira, A. (2012). The nitrogen source impacts major volatile compounds released by Saccharomyces cerevisiae during alcoholic fermentation. *International Journal of Food Microbiology, 160*(2), 87–93.

Bataillon, M., Rico, A., Sablayrolles, J. M., Salmon, J. M., & Barre, P. (1996). Early thiamine assimilation by yeasts under enological conditions: Impact on fermentation kinetics. *Journal of Fermentation and Bioengineering, 82*(2), 145–150.

Bell, S. J., & Henschke, P. A. (2005). Implications of nitrogen nutrition for grapes, fermentation and wine. *Australian Journal of Grape and Wine Research, 11*(3), 242–295.

Beltran, G., Novo, M., Guillamón, J. M., Mas, A., & Rozès, N. (2008a). Effects of temperature, ammonium and glucose concentrations on yeast growth in a model wine system. *International Journal of Food Science and Technology, 2006*(41), 1152–1157.

Beltran, G., Novo, M., Guillamon, J. M., Mas, A., & Rozes, N. (2008b). Effect of fermentation temperature and culture media on the yeast lipid composition and wine volatile compounds. *International. Journal of Food Microbiology, 121*(2), 169–177.

Belviso, S., Bardi, L., Bartolini, B. A., & Marzona, M. (2004). Lipid nutrition in *Saccharomyces cerevisiae* in winemaking. *Canadian Journal of Microbiology, 50*, 669–674.

Bely, M. (1990). Automatic detection and correction of deficiencies in assimilable nitrogen during alcoholic fermentations in winemaking conditions. PhD Thesis. University of Montpellier II.

Bely, M., Sablayrolles, J. M., & Barre, P. (1990a). Description of alcoholic fermentation kinetics: Its variability and interest. *American Journal of Enology and Viticulture, 41*(4), 319–324.

Bely, M., Sablayrolles, J. M., & Barre, P. (1990b). Automatic control of assimilable nitrogen addition during alcoholic fermentation in enological conditions. *Journal of Fermentation and Bioengineering, 70*, 1–6.

Bely, M., Salmon, J. M., & Barre, P. (1994). Assimilable nitrogen addition and hexose transport system activity during enological fermentations. *Journal of the Institute of Brewing, 100*, 279–282.

Birch, R. M., Ciani, M., & Walker, G. M. (2003). Magnesium, calcium and fermentative metabolism in wine yeasts. *Journal of Wine Research, 14*(1), 3–15.

Bisson, L. F. (1999). Stuck and sluggish fermentations. *American Journal of Enology and Viticulture, 50*(1), 107–119.

Blanco, P., Miras-Avalos, J. M., Pereira, E., & Orriols, I. (2013). Fermentative aroma compounds and sensory profiles of Godello and Albarino wines as influenced by Saccharomyces cerevisiae yeast strains. *Journal of the Science of Food and Agriculture, 93*(11), 2849–2857.

Blankenhorn, D., & Neumann, M. (2004). Fermentation control with density probes. *Der Deutsche Weinbau, 16*(17), 28–30.

Blateyron, L., & Sablayrolles, J. M. (2001). Stuck and slow fermentations in enology: Statistical study of causes and effectiveness of combined additions of oxygen and diammonium phosphate. *Journal of Bioscience and Bioengineering, 91*, 184–189.

Blateyron, L., Aguera, E., Dubois, C., & Sablayrolles, J. M. (1998). Control of oxygen additions during alcoholic fermentations. *Wein-Wissenschaft, 53*(3), 131–135.

Blateyron, L., Julien, A., & Sablayrolles, J. M. (2000). *Stuck fermentations–O_2 and nitrogen requirements - importance of optimizing their addition*. Vienne: Lallemand Research Meeting.

Bloem, A., Rollero, S., Seguinot, P., et al. (2018). Workflow based on the combination of isotopic tracer experiments to investigate microbial metabolism of multiple nutrient sources. *Jove-Journal of Visualized Experiments*, (131).

Brice, C., Sanchez, I., Tesniere, C., & Blondin, B. (2014). Assessing the mechanisms responsible for differences between nitrogen requirements of Saccharomyces cerevisiae wine yeasts in alcoholic fermentation. *Applied and Environmental Microbiology, 80*(4), 1330–1339.

Bruch, G. (2001). Automatic densimeter. French patent n°FR2801676-A1.

Buratti, S., & Benedetti, S. (2016). Alcoholic fermentation using electronic nose and electronic tongue (2016). In M. L. R. Mendez (Ed.), *Electronic noses and tongues in food science* (pp. 291–299). London: Academic Press LTD-Elsevier Science LTD.

Butzke, C. E. (1998). Study of yeast assimilable N status in musts from California, Oregon and Washington. *American Journal of Enology and Viticulture, 49*(2), 220–224.

Canonico, L., Comitini, F., Oro, L., & Ciani, M. (2016). Sequential fermentation with selected immobilized non-Saccharomyces yeast for reduction of ethanol content in wine. *Frontiers in Microbiology, 7*. https://doi.org/10.3389/fmicb.2016.00278.

Casalta, E., Aguera, E., Picou, C., Rodriguez-Bencomo, J. J., Salmon, J. M., & Sablayrolles, J. M. (2010). A comparison of laboratory and pilot-scale fermentations in winemaking conditions. *Applied Microbiology and Biotechnology, 87*(5), 1665–1673.

Casalta, E., Sablayrolles, J. M., & Salmon, J. M. (2013). Comparison of different methods for the determination of assimilable nitrogen in grape musts. *Lwt-Food Science and Technology, 54*(1), 271–277.

Casalta, E., Vernhet, A., Sablayrolles, J. M., Tesniere, C., & Salmon, J. M. (2016). Review: Characterization and role of grape solids during alcoholic fermentation under enological conditions. *American Journal of Enology and Viticulture, 67*(2), 133–138.

Cerda-Drago, T. G., Agosin, E., & Perez-Correa, J. R. (2016). Modelling the oxygen dissolution rate during oenological fermentation. *Biochemical Engineering Journal, 106*, 97–106.

Ciani, M., Capece, A., Comitini, F., Canonico, L., Siesto, G., & Romano, P. (2016a). Yeast interactions in inoculated wine fermentation. *Frontiers in Microbiology, 7*. https://doi.org/10.3389/fmicb.2016.00555.

Ciani, M., Morales, P., Comitini, F., et al. (2016b). Non-conventional yeast species for lowering ethanol content of wines. *Frontiers in Microbiology, 7*. https://doi.org/10.3389/fmicb.2016.00642.

Clement, T., Perez, M., Mouret, J. R., Sablayrolles, J. M., & Camarasa, C. (2011). Use of a continuous multistage bioreactor to mimic winemaking fermentation. *International Journal of Food Microbiology, 150*, 42–49.

Colombié, S., Malherbe, S., & Sablayrolles J. M. (2005). Modeling alcoholic fermentation in enological conditions: feasibility and interest. *American Journal of Enology and Viticulture, 56*, 238–245.

Colombie, S., Latrille, E., & Sablayrolles, J. M. (2007a). On-line estimation of assimilable nitrogen by electrical conductivity measurement during alcoholic fermentation in enological conditions. *Journal of Bioscience and Bioengineering, 103*, 229–235.

Colombié, S., Malherbe, S., & Sablayrolles, J. M. (2007b). Modeling of heat transfer in tanks during wine-making fermentation. *Food Control, 18*, 953–960.

Comitini, F., Capece, A., Ciani, M., & Romano, P. (2017). New insights on the use of wine yeasts. *Current Opinion in Food Science, 13*, 44–49.

Contreras, A., Hidalgo, C., Henschke, P. A., Chambers, P. J., Curtin, C., & Varela, C. (2014). Evaluation of non-Saccharomyces yeasts for the reduction of alcohol content in wine. *Applied and Environmental Microbiology, 80*(5), 1670–1678.

Contreras, A., Hidalgo, C., Schmidt, S., Henschke, P. A., Curtin, C., & Varela, C. (2015). The application of non-Saccharomyces yeast in fermentations with limited aeration as a strategy for the production of wine with reduced alcohol content. *International Journal of Food Microbiology, 205*, 7–15.

Corrieu, G., Perret, B., & Merican, F. (1997). Pressure drop and flow measurement apparatus and its application to control of fermentation. In a brewery, French patent FR2746914-A1.

Cotrell, T. H. E., & Mc Lellan, M. R. (1986). The effect of fermentation temperature on chemical and sensory characteristics of wines from seven white grape cultivars grown in New York state. *American Journal of Enology and Viticulture, 37*(3), 190–194.

Cozzolino, D. (2016). State-of-the-art advantages and drawbacks on the application of vibrational spectroscopy to monitor alcoholic fermentation (beer and wine). *Applied Spectroscopy Reviews, 51*(4), 282–297.

Cramer, A. C., Vlassides, S., & Block, D. E. (2002). Kinetic model for nitrogen-limited wine fermentations. *Biotechnology and Bioengineering, 77*(1), 49–60.

Crepin, L., Nidelet, T., Sanchez, I., Dequin, S., & Camarasa, C. (2012). Sequential use of nitrogen compounds by Saccharomyces cerevisiae during wine fermentation: A model based on kinetic and regulation characteristics of nitrogen permeases. *Applied and Environmental Microbiology, 78*(22), 8102–8111.

Crepin, L., Truong, N. M., Bloem, A., Sanchez, I., Dequin, S., & Camarasa, C. (2017). Management of Multiple Nitrogen Sources during wine fermentation by Saccharomyces cerevisiae. *Applied and Environmental Microbiology, 83*(5). https://doi.org/10.1128/aem.02617-16.

de Ovalle, S., Cavello, I., Brena, B. M., Cavalitto, S., & Gonzalez-Pombo, P. (2018). Production and characterization of a beta-glucosidase from Issatchenkia terricola and its use for hydrolysis of aromatic precursors in cabernet sauvignon wine. *Lwt-Food Science and Technology, 87*, 515–522.

del Barrio-Galan, R., Caceres-Mella, A., Medel-Maraboli, M., & Pena-Neira, A. (2015). Effect of selected Saccharomyces cerevisiae yeast strains and different aging techniques on the polysaccharide and polyphenolic composition and sensorial characteristics of cabernet sauvignon red wines. *Journal of the Science of Food and Agriculture, 95*(10), 2132–2144.

del Nobile, M. A., D'Amato, D., Altieri, C., Corbo, M. R., & Sinigaglia, M. (2003). Modelling the yeast growth-cycle in a model wine system. *Journal of Food Science, 68*(6), 2080–2085.

Delfini, C., & Costa, A. (1993). Effects of the grape must lees and insoluble materials on the alcoholic fermentation rate and the production of acetic acid, pyruvic acid, and acetaldehyde. *American Journal of Enology and Viticulture, 44*(1), 86–92.

Delfini, C., Castino, M., & Ciolfi, G. (1980). Thiamin addition to musts to reduce ketoacid formation and increase the efficacy of SO_2 in wines. *Rivista di Viticoltura e di Enologia, 33*(12), 572–589.

Dequin, S. (2001). The potential of genetic engineering for improving brewing, wine-making and baking yeasts. *Applied Microbiology and Biotechnology, 56*, (5–6), 577–588.

Di Maro, E., Ercolini, D., & Coppola, S. (2007). Yeast dynamics during spontaneous wine fermentation of the Catalanesca grape. *International Journal of Food Microbiology, 117*(2), 201–210.

Dubois, C., Manginot, C., Roustan, J. L., Sablayrolles, J. M., & Barre, P. (1996). Effect of variety, year, and grape maturity on the kinetics of alcoholic fermentation. *American Journal of Enology and Viticulture, 47*(4), 363–368.

Duc, C., Pradal, M., Sanchez, I., Noble, J., Tesniere, C., & Blondin, B. (2017). A set of nutrient limitations trigger yeast cell death in a nitrogen-dependent manner during wine alcoholic fermentation. *PLoS One, 12*(9), e0184838. https://doi.org/10.1371/journal.pone.0184838.

Egli, C. M., Edinger, W. D., Mitrakul, C. M., & Henick-Kling, T. (1998). Dynamics of indigenous and inoculated yeast populations and their effect on the sensory character of Riesling and Chardonnay wines. *Journal of Applied Microbiology, 85*(5), 779–789.

Englezos, V., Rantsiou, K., Torchio, F., Rolle, L., Gerbi, V., & Cocolin, L. (2015). Exploitation of the non-Saccharomyces yeast Starmerella bacillaris (synonym Candida zemplinina) in wine fermentation: Physiological and molecular characterizations. *International Journal of Food Microbiology, 199*, 33–40.

Esti, M., Volpe, G., Compagnone, D., Mariotti, G., Moscone, D., & Palleschi, G. (2003). Monitoring alcoholic fermentation of red wine by electrochemical biosensors. *American Journal of Enology and Viticulture, 54*(1), 39–45.

Fernandez-Gonzalez, M., Ubeda, J. F., Cordero-Oterob, R. R., Gururajan, V. T., & Briones, A. I. (2005). Engineering of an oenological Saccharomyces cerevisiae strain with pectinolytic activity and its effect on wine. *International Journal of Food Microbiology, 102*(2), 173–183.

Ferreira, V., Pena, C., Escudero, A., & Cacho, J. (1996). Losses in volatile compounds during fermentation. *Lebensmittel-Wissenschaft und -Technologie, 29*(3), 251–259.

Ferreira, D., Galeote, V., Sanchez, I., Legras, J. L., Ortiz-Julien, A., & Dequin, S. (2017). Yeast multistress resistance and lag-phase characterisation during wine fermentation. *FEMS Yeast Research, 17*(6). https://doi.org/10.1093/femsyr/fox051.

Fleet, G. H. (2008). Wine yeasts for the future. *FEMS Yeast Research, 8*, 879–995.

Francis, I. L., & Newton, J. L. (2005). Determining wine aroma from compositional data, Australian. *Journal of Grape and Wine Research, 11*(2), 114–126.

Garcia, A., Garcia, L. A., & Diaz, M. (1994). Mechanisms for mixing and homogenisation in beer fermentation. *Bioprocess Engineering, 10*, 179–184.

Garde-Cerdan, T., Portu, R. L. J., Gonzalez-Arenzana, L., Lopez-Alfaro, I., & Santamaria, P. (2014). Study of the effects of proline, phenylalanine, and urea foliar application to Tempranillo vineyards on grape amino acid content. Comparison with commercial nitrogen fertilisers. *Food Chemistry, 163*, 136–141.

Goelzer, A., Charnomordic, B., Colombié, S., Fromion, V., & Sablayrolles, J. M. (2009). Simulation and optimization software for alcoholic fermentation in winemaking conditions. *Food Control, 20*, 635–642.

Grieco, F., Tristezza, M., Vetrano, C., et al. (2011). Exploitation of autochthonous micro-organism potential to enhance the quality of Apulian wines. *Annals of Microbiology, 61*(1), 67–73.

Gutierrez, A., Chiva, R., Sancho, M., Beltran, G., Arroyo-Lopez, F. N., & Guillamon, J. M. (2012). Nitrogen requirements of commercial wine yeast strains during fermentation of a synthetic grape must. *Food Microbiology, 31*(1), 25–32.

Holzapfel, B. P., Watt, J., Smith, J. P., Suklje, K., & Rogiers, S. Y. (2015). Effects of timing of N application and water constraints on N accumulation and juice amino N concentration in 'Chardonnay' grapevines. *Vitis, 54*(4), 203–211.

Houtman, A. C., & Du Plessis, C. S. (1986). Nutritional deficiencies of clarified white grape juices and their correction in relation to fermentation. *South African Journal of Enology and Viticulture, 7*(1), 39–46.

Howell, K. S., Swiegers, J. H., Elsey, G. M., Siebert, T. E., Bartowsky, E. J., Fleet, G. H., Pretorius, I. S., & Barros Lopes, M. A. (2004). Variation in 4-mercapto-4-methyl-pentan-2-one released by *Saccharomyces cerevisiae* commercial wine strains. *FEMS Microbiology Letters, 240*(2), 125–129.

Hu, K., Qin, Y., Tao, Y. S., Zhu, X. L., Peng, C. T., & Ullah, N. (2016). Potential of glycosidase from non-Saccharomyces isolates for enhancement of wine aroma. *Journal of Food Science, 81*(4), M935–M943.

Insa, G., Sablayrolles, J. M., & Douzal, V. (1995). Alcoholic fermentation under enological conditions. Use of a combination of data analysis and neural networks to predict sluggish and stuck fermentations. *Bioprocess Engineering, 13*, 171–176.

Jiang, H., Zhang, H., Chen, Q. S., Mei, C. L., & Liu, G. H. (2015). Recent advances in electronic nose techniques for monitoring of fermentation process. *World Journal of Microbiology & Biotechnology, 31*(12), 1845–1852.

Jimenez-Marquez, F., Vazquez, J., Ubeda, J., & Sanchez-Rojas, J. L. (2013). High-precision optoelectronic sensor device for monitoring fermentation kinetics and maceration of wine. In: Schmid U, Aldavero J, LeesterSchaedel M (eds) Smart sensors, actuators, and Mems Vi. Proceedings of SPIE, vol 8763.

Jimenez-Marquez, F., Vazquez, J., & Sanchez-Rojas, J. L. (2015). Optoelectronic sensor device for monitoring the maceration of red wine: Design issues and validation. *Measurement, 63*, 128–136. https://doi.org/10.1016/j.measurement.2014.12.009.

Jimenez-Marquez, F., Vazquez, J., Ubeda, J., & Sanchez-Rojas, J. L. (2016). Optoelectronic sensor for measuring ethanol content during grape must fermentation using NIR spectroscopy. *Microsystem Technologies-Micro-and Nanosystems-Information Storage and Processing Systems, 22*(7), 1799–1809.

Jules, M., Beltran, G., Francois, J., & Parrou, J. L. (2008). New insights into trehalose metabolism by Saccharomyces cerevisiae: NTH2 encodes a functional cytosolic trehalase, and deletion of TPS1 reveals Ath1p-dependent trehalose mobilization. *Applied and Environmental Microbiology, 74*(3), 605–614.

Julien, A., Roustan, J. L., Dulau, L., & Sablayrolles, J. M. (2000). Comparison of nitrogen and oxygen demands of enological yeasts: Technological consequences. *American Journal of Enology and Viticulture, 51*(3), 215–222.

Kemsawasd, V., Viana, T., Ardo, Y., & Arneborg, N. (2015). Influence of nitrogen sources on growth and fermentation performance of different wine yeast species during alcoholic fermentation. *Applied Microbiology and Biotechnology, 99*(23), 10191–10207.

Killian, E., & Ough, C. S. (1979). Fermentation esters – formation and retention as affected by fermentation temperature. *American Journal of Enology and Viticulture, 30*, 301–305.

Kourkoutas, Y., Bekatorou, A., Banat, I. M., Marchant, R., & Koutinas, A. A. (2004). Immobilization technologies and support materials suitable in alcoholic beverages production: A review. *Food Microbiology, 21*, 377–397.

Kovacs, A., Ivanov, A., Mescheder, U., & Ieee. (2015). Refractometer using photonic crystals for fermentation process characterization 2015 Ieee Sensors. IEEE Sensors, p 1363–1366.

Lage, P., Barbosa, C., Mateus, B., Vasconcelos, I., Mendes-Faia, A., & Mendes-Ferreira, A. (2014). H. guilliermondii impacts growth kinetics and metabolic activity of S. cerevisiae: The role of initial nitrogen concentration. *International Journal of Food Microbiology, 172*, 62–69.

Leo Kuebler GmbH. (2007). Method and apparatus for monitoring of a fermentation process. German Patent n° DE102005015530-A1.

Luparia, V., Soubeyrand, V., Berges, T., Julien, A., & Salmon, J. M. (2004). Assimilation of grape phytosterols by *Saccharomyces cerevisiae* and their impact on enological fermentations. *Applied Microbiology and Biotechnology, 65*(1), 25–32.

Malherbe, S. (2003). Modelling of alcoholic fermentation in winemaking conditions. PhD Thesis. University of Montpellier II.

Malherbe, S., Fromion, V., Hilgert, N., & Sablayrolles, J. M. (2004). Modelling the effects of assimilable nitrogen and temperature on fermentation kinetics in enological conditions. *Biotechnology and Bioengineering, 86*, 261–272.

Martinez-Landa, P., Alba-Elias, F., Perez-Ezcurdia, A., Benito-Amurrio, M., Martinez de P., Ascacibar, F. J., Ordieres-Mere, J., Pernia-Espinosa, A., Vergara-Gonzalez, E., & Castejon-Limas, M. (2006). System for monitoring alcoholic fermentation. Spanish patent n°ES2241457-A1.

Marullo, P., Bely, M., Masneuf-Pomarede, I., Pons, M., Aigle, M., & Dubourdieu, D. (2006). Breeding strategies for combining fermentative qualities and reducing off-flavour production in a wine yeast model. *FEMS Yeast Research, 6*(2), 268–279.

Masneuf-Pomarede, I., Mansour, C., Murat, M. L., Tominaga, T., & Dubourdieu, D. (2006). Influence of fermentation temperature on volatile thiols concentrations in Sauvignon blanc wines. *International Journal of Food Microbiology, 108*(3), 385–390.

Medina, K., Boido, E., Dellacassa, E., & Carrau, F. (2012). Growth of non-Saccharomyces yeasts affects nutrient availability for Saccharomyces cerevisiae during wine fermentation. *International Journal of Food Microbiology, 157*(2), 245–250.

Moenne, M. I., Saa, P., Laurie, V. F., Perez-Correa, J. R., & Agosin, E. (2014). Oxygen incorporation and dissolution during industrial-scale red wine fermentations. *Food and Bioprocess Technology, 7*(9), 2627–2636.

Morakul, S., Mouret, J. R., Nicolle, P., Trelea, I. C., Sablayrolles, J. M., & Athes, V. (2011). Modelling of the gas-liquid partitioning of aroma compounds during wine alcoholic fermentation and prediction of aroma losses. *Process Biochemistry, 46*(5), 1125–1131.

Moreno-Garcia, J., Garcia-Martinez, T., Mauricio, J. C., & Moreno, J. (2018). Yeast immobilization systems for Alcoholic wine fermentations: Actual trends and future perspectives. *Frontiers in Microbiology, 9*. https://doi.org/10.3389/fmicb.2018.00241.

Mouret, J. R., Farines, V., Sablayrolles, J. M., & Trelea C. (2015). Prediction of the production kinetics of the main fermentative aromas in winemaking fermentations. *Biochemical Engineering Journal, 103*, 211–218.

Muller, J., & Velten, K. (2015). Application of computational fluid dynamics for the optimization of homogenization processes in wine tanks. In: JeanMarie A (ed) *38th World Congress of Vine and Wine*. BIO Web of Conferences, vol 5.

Nicolini, G., Larcher, R., & Versini, G. (2004). Status of yeast assimilable nitrogen in Italian grape musts and effects of variety, ripening and vintage. *Vitis, 43*(2), 89–96.

Noble, J., Sanchez, I., & Blondin, B. (2015). Identification of new Saccharomyces cerevisiae variants of the MET2 and SKP2 genes controlling the sulfur assimilation pathway and the production of undesirable sulfur compounds during alcoholic fermentation. *Microbial Cell Factories, 14*, 68. https://doi.org/10.1186/s12934-015-0245-1.

Noti, O., Vaudano, E., Giuffrida, M. G., et al. (2018). Enhanced arginine biosynthesis and lower proteolytic profile as indicators of Saccharomyces cerevisiae stress in stationary phase during fermentation of high sugar grape must: A proteomic evidence. *Food Research International, 105*, 1011–1018.

Novo, M. T., Beltran, G., Torija, M. J., Poblet, M., Rozes, N., Guillamon, J. M., & Mas, A. (2003). Changes in wine yeast storage carbohydrate levels during preadaptation, rehydration and low temperature fermentations. *International Journal of Food Microbiology, 86*(1–2), 153–161.

Novo, M. T., Beltran, G., Rozes, N., Guillamon, J. M., & Mas, A. (2005). Effect of nitrogen limitation and surplus upon trehalose metabolism in wine yeast. *Applied Microbiology and Biotechnology, 66*(5), 560–566.

Nyman, J., Lacintra, M. G., Westman, J. O., et al. (2013). Pellet formation of zygomycetes and immobilization of yeast. *New Biotechnology, 30*(5), 516–522.

Ochando, T., Mouret, J. R., Humbert-Goffard, A., Sablayrolles, J. M., & Farines, V. (2017). Impact of initial lipid content and oxygen supply on alcoholic fermentation in champagne-like musts. *Food Research International, 98*, 87–94.

Oikonomou, P., Raptis, I., & Sanopoulou, M. (2014). Monitoring and evaluation of alcoholic fermentation processes using a Chemocapacitor sensor Array. *Sensors, 14*(9), 16258–16273.

Padilla, B., Gil, J., & Manzanares, P. (2016). Past and future of non-Saccharomyces yeasts: From spoilage microorganisms to biotechnological tools for improving wine aroma complexity. *Frontiers in Microbiology, 7*. https://doi.org/10.3389/fmicb.2016.00411.

Parks, L. W. (1978). Metabolim of sterol in yeast. *CRC Critical Reviews in Microbiology, 6*(4), 301–341.

Parrou, J. L., Enjalbert, B., Plourde, L., Bauche, A., Gonzalez, B., & Francois, J. M. (1999). Dynamic responses of reserve carbohydrate metabolism under carbon and nitrogen limitations in *Saccharomyces cerevisiae*. *Yeast, 15*(3), 191–203.

Peinado, R. A., Moreno, J. J., Villalba, J. M., Gonzalez-Reyes, J. A., Ortega, J. M., & Mauricio, J. C. (2006). Yeast biocapsules: A new immobilization method and their applications. *Enzyme and Microbial Technology, 40*(1), 79–84.

Perez-Alvarez, E. P., Garcia-Escudero, E., & Peregrina, F. (2015). Soil nutrient availability under cover crops: Effects on vines, must, and wine in a Tempranillo Vineyard. *American Journal of Enology and Viticulture, 66*(3), 311–320.

Perret, B., Merican, F., & Corrieu, G. (1997). A new method for following the fermentation of the natural sweet wine of Muscat type by online assessment of the density of the fermenting must. *Revue Française d'Oenologie, 163*, 14–17.

Pfusterschmied, G., Toledo, J., Kucera, M., et al. (2017). Potential of piezoelectric MEMS resonators for grape must fermentation monitoring. *Micromachines, 8*(7). https://doi.org/10.3390/mi8070200.

Piermarini, S., Volpe, G., Esti, M., Simonetti, M., & Palleschi, G. (2011). Real time monitoring of alcoholic fermentation with low-cost amperometric biosensors. *Food Chemistry, 127*(2), 749–754.

Plouy, D. (2000). *Characterisation of winemaking fermentors at the pilot scale*. Germany: Master of Science. Fachhochschule Giessen-Friedberg.

Pretorius, I. S., & Hoj, P. B. (2005). Grape and wine biotechnology: Challenges, opportunities and potential benefits. *Australian Journal of Grape and Wine Research, 11*(2), 83–108.

Puig, S., & Perez-Ortin, J. E. (2000). Expression levels and patterns of glycolytic yeast genes during wine fermentation. *Systematic and Applied Microbiology, 23*(2), 300–303.

Renault, P. E., Albertin, W., & Bely, M. (2013). An innovative tool reveals interaction mechanisms among yeast populations under oenological conditions. *Applied Microbiology and Biotechnology, 97*(9), 4105–4119.

Roger, J. M., Sablayrolles, J. M., Steyer, J. P., & Bellon-Maurel, V. (2002). Pattern analysis techniques to process fermentation curves: Application to discrimination of enological alcoholic fermentations. *Biotechnology and Bioengineering, 79*, 804–815.

Rollero, S., Bloem, A., Camarasa, C., et al. (2015). Combined effects of nutrients and temperature on the production of fermentative aromas by Saccharomyces cerevisiae during wine fermentation. *Applied Microbiology and Biotechnology, 99*(5), 2291–2304.

Roustan, J. L., & Sablayrolles, J. M. (2004). Role of trehalose and glycogen in alcoholic fermentation in wine-making conditions. *Journal of Wine Research, 15*, 189–202.

Sablayrolles, J. M. (1988). Pilotage de la fermentation alcoolique en oenologie. In P. Bidan & J. R. Bonneviale (Eds.), *Applications à l'œnologie des progrès récents en microbiologie et en fermentation* (pp. 235–256). Paris: Office International de la Vigne et du Vin.

Sablayrolles, J. M., & Barre, P. (1986). Evaluation des besoins en oxygène de fermentations alcooliques en conditions oenologiques simulées. *Sciences des Aliments, 6*, 373–383.

Sablayrolles, J. M., & Salmon, J. M. (2001). Vitamins: Importance in alcoholic fermentation. In *Lallemand research meeting*. Perugia.

Sacchi, K. L., Bisson, L. F., & Adams, D. O. (2005). A review of the effect of winemaking techniques on phenolic extraction in red wines. *American Journal of Enology and Viticulture, 56*(3), 197–206.

Sainz, J., Pizarro, F., Perez-Correa, J. R., & Agosin, E. (2003). Modelling of yeast metabolism and process dynamics in batch fermentation. *Biotechnology and Bioengineering, 81*(7), 818–828.

Salmon, J. M., Vincent, O., Mauricio, J. C., Bely, M., & Barre, P. (1993). Sugar transport inhibition and apparent loss of activity in *Saccharomyces cerevisiae* as a major factor of enological fermentations. *American Journal of Enology and Viticulture, 44*(1), 56–64.

Saur, D. (2004). Apparatus for monitoring and control of alcoholic fermentation of grape musts and other fruit juices. German patent n°DE102004011413-A1.
Schmid, F., Schadt, J., Jiranek, V., & Block, D. E. (2009). Formation of temperature gradients in large- and small-scale red wine fermentations during cap management. *Australian Journal of Grape and Wine Research, 15*(3), 249–255.
Schuller, D., & Casal, M. (2005). The use of genetically modified *Saccharomyces cerevisiae* strains in the wine industry. *Applied Microbiology and Biotechnology, 68*(3), 292–304.
Seguinot, P., Rollero, S., Sanchez, I., et al. (2018). Impact of the timing and the nature of nitrogen additions on the production kinetics of fermentative aromas by *Saccharomyces cerevisiae* during winemaking fermentation in synthetic media. *Food Microbiology, 76*, 29–39.
Shrake, N. L., Amirtharajah, R., Brenneman, C., Boulton, R., & Knoesen, A. (2014). In-line measurement of color and Total Phenolics during red wine fermentations using a light-emitting diode sensor. *American Journal of Enology and Viticulture, 65*(4), 463–470.
Soubeyrand, V., Luparia, V., Williams, P., Doco, T., Vernhet, A., Ortiz-Julien, A., & Salmon, J. M. (2005). Formation of micella containing solubilized sterols during rehydration of active dry yeasts improves their fermenting capacity. *Journal of Agricultural and Food Chemistry, 53*(20), 8025–8032.
Styger, G., Prior, B., & Bauer, F. F. (2011). Wine flavor and aroma. *Journal of Industrial Microbiology & Biotechnology, 38*(9), 1145–1159.
Swiegers, J. H., Kievit, R. L., Siebert, T., Lattey, K. A., Bramley, B. R., Francis, I. L., King, E. S., & Pretorius, I. S. (2009). The influence of yeast on the aroma of Sauvignon Blanc wine. *Food Microbiology, 26*(2), 204–211.
Takush, D. G., & Osborne, J. P. (2012). Impact of yeast on the aroma and flavour of Oregon Pinot Noir wine. *Australian Journal of Grape and Wine Research, 18*(2), 131–137.
Tempère, S., Marchal, A., Barbe, J. C., et al. (2018). The complexity of wine: Clarifying the role of microorganisms. *Applied Microbiology and Biotechnology, 102*, 3995–4007.
Tilloy, V., Ortiz-Julien, A., & Dequin, S. (2014). Reduction of ethanol yield and improvement of glycerol formation by adaptive evolution of the wine yeast Saccharomyces cerevisiae under hyperosmotic conditions. *Applied and Environmental Microbiology, 80*(8), 2623–2632.
Toledo, J., Ruiz-Diez, V., Pfusterschmied, G., Schmid, U., & Sanchez-Rojas, J. L. (2018). Flow-through sensor based on piezoelectric MEMS resonator for the in-line monitoring of wine fermentation. *Sensors and Actuators B-Chemical, 254*, 291–298.
Torija, M. J., Beltran, G., Novo, M., Poblet, M., Guillamon, J. M., Mas, A., & Rozes, N. (2003). Effects of fermentation temperature and *Saccharomyces* species on the cell fatty acid composition and presence of volatile compounds in wine. *International Journal of Food Microbiology, 85*(1–2), 127–136.
Torrea, D., Varela, C., Ugliano, M., Ancin-Azpilicueta, C., Francis, I. L., & Henschke, P. A. (2011). Comparison of inorganic and organic nitrogen supplementation of grape juice – effect on volatile composition and aroma profile of a chardonnay wine fermented with Saccharomyces cerevisiae yeast. *Food Chemistry, 127*(3), 1072–1083.
Ugliano, M., Kolouchova, R., & Henschke, P. A. (2011). Occurrence of hydrogen sulfide in wine and in fermentation: Influence of yeast strain and supplementation of yeast available nitrogen. *Journal of Industrial Microbiology & Biotechnology, 38*(3), 423–429.
Varela, C., Pizarro, F., & Agosin, E. (2004). Biomass content governs fermentation rate in nitrogen-deficient wine musts. *Applied and environmentam microbiology, 70*, 3392–3400.
Vazquez-Lima, F., Silva, P., Barreiro, A., et al. (2014). Use of chemostat cultures mimicking different phases of wine fermentations as a tool for quantitative physiological analysis. *Microbial Cell Factories, 13*, 85. https://doi.org/10.1186/1475-2859-13-85.
Verdenal, T., Spangenberg, J. E., Zufferey, V., et al. (2016). Leaf-to leaf ratio affects the impact of foliar-applied nitrogen on N accumulation in the grape must. *Journal International Des Sciences De La Vigne Et Du Vin, 50*(1), 23–33.
Vianna, C. R., Silva, C. L. C., Neves, M. J., & Rosa, C. A. (2008). Saccharomyces cerevisiae strains from traditional fermentations of Brazilian cachaca: Trehalose metabolism, heat and ethanol resistance. *Antonie Van Leeuwenhoek, 93*, 205–217.

Vigentini, I., Maghradze, D., Petrozziello, M., et al. (2016). Indigenous Georgian wine-associated yeasts and grape cultivars to edit the wine quality in a precision oenology perspective. *Frontiers in Microbiology, 7*. https://doi.org/10.3389/fmicb.2016.00352.

Vivas, N., & Cros, M. (1991). Control of anaerobiosis and temperature and its effect on the growth cycle and metabolism of yeasts. Application to tanks with sequential remontages. *Progres Agricole et Viticole, 108*, 163–165.

Wang, B. Q., & Peng, B. Z. (2017). A feasibility study on monitoring residual sugar and alcohol strength in kiwi wine fermentation using a fiber-optic FT-NIR spectrometry and PLS regression. *Journal of Food Science, 82*(2), 358–363.

Warriner, K., Morrissey, A., Alderman, J., King, G., Treloar, P., & Vadgama, P. M. (2002). Modified microelectrode interfaces for in-line electrochemical monitoring of ethanol in fermentation processes. *Sensors and Actuators, 84*(2–3), 200–207.

Zhao, H. (2005). Effect of sulfur dioxide on fermentation kinetics. In *Master of Sciences*. France: University of Montpellier II.

Chapter 10
Genetic Improvement of Wine Yeasts

Ileana Vigentini, Ramon Gonzalez, and Jordi Tronchoni

Contents

1 Introduction

For many years, wine biotechnology was developed in the absence of genetic improvement of the microorganisms responsible for wine fermentation. Nevertheless, unconscious domestication seems to have played a relevant role on shaping wine yeast strains as we currently find them (Pérez-Ortín et al. 2002; Querol et al. 2003; Legras et al. 2007; Warringer et al. 2011; Sicard and Legras 2011). Indeed, compared to other biotechnological industries, including industrial food production like dairy or beer, also the use of starter cultures became generalized relatively. In this context, genetic improvement was not even considered.

By the early 90's, both the use of wine yeast starter cultures and genetic engineering of microorganisms were well stablished late (Gonzalez et al. 2011). Some researchers in the field considered that natural genetic variation found in wild

I. Vigentini (✉)
Department of Food, Environmental and Nutritional Sciences, University of Milan, Milan, Italy
e-mail: ileana.vigentini@unimi.it

R. Gonzalez · J. Tronchoni
Instituto de Ciencias de la Vid y del Vino (ICVV), Logroño, Spain
e-mail: rgonzalez@icvv.es; jordi.tronchoni@icvv.es

P. Romano et al. (eds.), *Yeasts in the Production of Wine*,
https://doi.org/10.1007/978-1-4939-9782-4_10

isolates of wine yeasts, that was the basis for improvement and innovation in wine biotechnology until that moment, could benefit from genetic engineering tools (Pérez-González et al. 1993; Dequin and Barre 1994; Laing and Pretorius 1993; González-Candelas et al. 1995). These tools would allow combining features not ordinarily found in the same strains; or jumping the species barrier, by incorporating mostly extracellular hydrolytic enzymes, as well as malolactic fermentation capabilities, to the genome of wine yeasts. However, only two genetically engineered wine yeasts reached the market (Husnik et al. 2006; Coulon et al. 2006). These strains were positively evaluated by FDA and Health Canada (Cebollero et al. 2007). Given the restricted marketing area, the market share of these strains is probably very low, but no official data are available. Negative public perception, the cost of putting a new genetically modified organism (GMO) in the European market as well as other countries, and the poor support by OIV (Organisation Internationale de la vigne et du vin), are surely among the main reasons for the lack of success of these recombinant wine yeasts in most wine producing countries.

During the XXI century, most researchers in the field have turned to more conventional genetic improvement tools, which had been the basis for microbial biotechnology in other sectors ever since the 50's. In first instance, these involved mostly random mutagenesis (Rous et al. 1983; Gonzalez et al. 2003; Cebollero et al. 2005; McBryde et al. 2006; Cordente et al. 2009; Quirós et al. 2010; González Ramos et al. 2010), as well as intra an interspecific hybridization (Kishimoto 1994; Marullo et al. 2006; Bellon et al. 2011; Thornton 1985; Pérez-Través et al. 2015). One important advantage of strains developed by the use of these techniques is they are free of the prejudices (and regulatory restrictions) surrounding the use of GMOs in the food industry. At the same time, the fact that these tools, especially interspecific hybridization, induce a much higher genetic variability than genetic engineering constitutes both a challenge and an opportunity. Indeed, sometimes, these random methods, relying on phenotypic selection, allow attaining targets almost unreachable by a rational design. Nevertheless, the challenge is sometime designing the right phenotypic test to screen the yeast populations obtained by random mutagenesis or genetic hybridization. Probably one of the most powerful tool for genetic improvement recently incorporated to our panoply is adaptive laboratory evolution (McBryde et al. 2006; Tilloy et al. 2014; Novo et al. 2014; Cadière et al. 2011; Kutyna et al. 2012). It also benefits from a rational design, especially when technological targets cannot be directly selected for. Anyway, the revival of all these genetic tools in wine biotechnology is already releasing strains to the market that are not direct natural isolates.

Nowadays, Systems Biology and NGS technologies are providing new tools to boost the potential of non-GMO genetic improvement of wine yeasts. Metabolic modelling provides new ways to design experimental evolution conditions, targeting features that might not be easily selectable in a direct way. High throughput yeast phenotyping and automatic culture management allow performing quick analyses of the strains derived from in vitro hybridization, as well as running multiple experimental evolution runs in parallel, so increasing the odds of success. Moreover, NGS analyses of these new strains can help redesigning the parameters for new experimental evolution of hybrid selection experiments. Finally, synthetic biology and genome edition based on CRISPR are providing new alternatives for the genetic improvement of wine yeasts (Lee et al. 2016; Pretorius 2017). However, these later techniques will also face difficulties to get the market, at least in European and other wine producing countries, because despite they can be tuned to avoid the introduction of DNA from other species in wine yeasts, they will still fall under the definition of GMOs of the European regulations, according to a recent opinion of the European Commission (Kupferschmidt 2018).

2 Targets for Genetic Improvement of Wine Yeasts

S. cerevisiae has two important roles for the development of human well-being, on the one hand it has been in charge for thousands of years of producing some of our most important foods (wine, bread and partly beer), on the other hand, more recently, it is one of the model organisms for research (Botstein and Fink 2011). Being a model organism and having such an extended biotechnological use, gives it a unique status. This has allowed many of the tools developed as a model organism to reach the biotechnology field to improve different areas of the yeast fermentative process.

One of the areas of improvement has been the fermentation process itself, increasing the ability of *S. cerevisiae* to assimilate sugars or nitrogen sources, its fermentation rate or its performance at low temperatures. Different tools have been applied to improve the fermentation kinetics of the yeast *S. cerevisiae*, random mutagenesis targeting nitrogen assimilation and fermentation performance (Salmon and Barre 1998), experimental evolution more recently for similar purposes (Novo et al. 2014) or genetic hybridization (Romano et al. 1985). Hybrid strains of a *S. cerevisiae* with a second more cryotolerant parental species like *S. kudriavzevii* or *S. uvarum* have also been shown as a natural approximation to improve fermentation performance at lower temperatures. Hybrids of these species can be isolated from fermentation in areas with cold environments (González et al. 2006).

Another main target of improvement in wine is the aroma composition and complexity. In this sense, almost all the different approaches discussed in this chapter have been used for this purpose. Some of these examples are, for the release of aroma precursors, the development of recombinant yeast strains with the ability to express pectin degrading enzymes that increase varietal aroma compounds (González-Candelas et al. 1995; Manzanares et al. 2003). Another target for genetic engineering has been changing the levels of yeast metabolites for aromatic purposes (Swiegers et al. 2007). Hybrid strains have also been reported to improve secondary aromas (Steensels et al. 2014), experimental evolution trials have achieved the same goal (Cadière et al. 2011).

Like aroma compounds, the ethanol content of the wine plays an important role in the final product. The recent increase of this metabolite mainly for global climate warming, but also for consumer trends looking for more mature and fully body wines, is also an objective for improvement. Different experimental approaches have been conducted in order to reduce the final concentration of this metabolite. Usually by changing the yeast metabolic carbon flux towards the production of other metabolites through genetic engineering (Remize et al. 2000) or experimental evolution (Tilloy et al. 2014).

Mannoproteins contribute to the wine quality by the stabilization of the final product or retention of aroma compounds among others (Waters 1994; Núñez et al. 2006). For this reason, its over-production has been achieved by the construction of genetically improved strains (Gonzalez-Ramos et al. 2008, 2009) or by random mutagenesis (Gonzalez-Ramos et al. 2010).

Another interesting trait with biotechnological character is flocculation. Flocculation is a well-studied mechanism of *S. cerevisiae* that is mainly under the control of the FLO family of genes, although it has also been shown to be environmentally dependent (Govender et al. 2010). The up-regulation of certain genes inside the FLO family leads to an increase in cell aggregation (Verstrepen and Klis 2006). This is an interesting feature that helps to avoid problems with the clarification of the wine, once the alcoholic fermentation has finished, removing easily the yeast cells (Pretorius and Bauer 2002; Soares 2011). Several works have succeeded in the construction of yeast strains using genetic engineering that increase their flocculation phenotype (Verstrepen et al. 2001; Wang et al. 2008).

Other traits that have been targets for improvement are the reduction of volatile acidity, accomplish by random mutagenesis (Cordente et al. 2013) and also using the hybridization of wine yeasts as approximation (Bellon et al. 2011, 2013); the process of malolactic fermentation performed by lactic acid bacteria after the alcoholic fermentation has finished, has been genetically engineering in a recombinant yeast strain (Husnik et al. 2007); reduction of foam production by yeast hybridization processes (Eschenbruch et al. 1982); or the increase in SO_2 tolerance by intraspecific breeding (Thornton 1982).

3 Genetic Engineering Based on Standard Homologous Integration

Starting early in the 1990s, pioneers in genetic engineering of wine yeasts took advantage of all the tools already developed and available for laboratory strains of *S. cerevisiae*. In most cases, genetic improvement involved the introduction of new enzymatic activities (Pérez-González et al. 1993; González-Candelas et al. 1995; Sánchez-Torres et al. 1996; Ganga et al. 1999; Volschenk et al. 2001), or enhancement of existing ones (Michnick et al. 1997). Gene disruption was also a target in many cases, (Cambon et al. 2006; Tabera et al. 2006; Gonzalez Ramos et al. 2009) as well as promoter substitution to change the expression pattern (Cardona et al. 2007; Govender et al. 2010), or the expression of dominant defective alleles (Cebollero et al. 2005). The scope of this section is genetic engineering tools actually employed on *S. cerevisiae* wine yeast strains with applied purposes, either the modified yeast strains were finally commercialized or not. It is not intended as a general review of genetic modification on this model species. Indeed, despite almost all techniques developed for laboratory yeast strains have the potential to be useful on wine yeast improvement, there are several genetic and physiological features of wine yeast strains that differentiate them from model strains and become a hurdle for the transfer of some genetic tools. Some of the most relevant features to consider in this context are prototrophy, homothallism, and diploidy or aneuploidy (Bisson 2004; Novo et al. 2009). Technologies used for the construction of recombinant wine yeast strains have evolved over time in response to the genetic tools becoming available for yeast research (in any context). However, the choices were not only driven by technical considerations. Researchers took also decisions based on their perception of public opinion on GMOs in foods, and to what was expected for easy approval by health and food authorities; despite only two strains have gone through an official approval process, and this only in very few countries (Grossmann et al. 2011). A summary of the features of these two strains is shown in Table 10.1.

The main features of the genetic modification systems that must be taken in consideration are the way to make DNA go through the different cell layers up to the nucleus, integrative or replicative nature of the transformation vector, selection markers, and elements required to drive gene expression (i.e. promoter and terminator) and protein secretion. Concerning ways to introduce transforming DNA into wine yeast cells, fortunately, systems already working for laboratory strains use to work reasonably well for most wine yeast strains. From the first trials, lithium acetate transformation was shown to be effective (Pérez-González et al. 1993), but electroporation has also been extensively used (Husnik et al. 2006).

Replicative vectors were used on early times for the construction of some strains. Those were based on the 2 μm replication origin or on yeast chromosomal replication origins (episomal vectors) (Pérez-González et al. 1993; Volschenk et al. 2001). In the late case, CEN sequences (centromeric vectors) are also included in order to

Table 10.1 Main features of the two only wine yeast strains that have ever been commercially available

Commercial name	ML01	ECMo01
Genetic background	S92	UCD522
Expressed genes	Malate permease from *Schizosaccharomyces pombe*	*DUR1,2* from *Saccharomyces cerevisiae*
	Malic enzyme from *Oenococcus oeni*	
Control sequences	*PGK1* promoter and terminator sequences	*PGK1* promoter and terminator sequences
Tranformation marker	Phleomycin resistance (by co-transformation; cured from final strain)	Phleomycin resistance (by co-transformation; cured from final strain)
Integration locus	*URA3*	*URA3*

help mitotic heritability (Cebollero and Gonzalez 2004). However, in order to improve genetic stability of the engineered strains, integration in the genome, usually by site-directed recombination, has usually been the method of choice. Genetic stability is not only a requirement for a true industrial usefulness of the improved yeast strains, but also a key feature for commercial GMOs. Often integration sites are targeted by including in the same plasmid the construction of interest together with sequences homologous to the target site. These plasmids are usually linearized by restriction enzyme digestion in this homologous sequence before yeast transformation (Volschenk et al. 2001; Swiegers et al. 2007). Alternatively, site directed integration can be attained with linear DNA fragments on which the construction is flanked by sequences homologous to those flanking a genomic region to be replaced. This is usually done for total or partial gene deletion (Gonzalez-Ramos et al. 2009; Cuello et al. 2017). These homologous flanking sequences can be part of a genetic construction carried on a plasmid vector or be introduced by PCR amplification as 5′ extensions of the amplification primers, similar to the construction of the whole genome yeast knockout collection (Baudin et al. 1993; Giaever et al. 2002). Homologous flanking sequences can also be used for promoter replacement strategies (Cardona et al. 2007; Jiménez-Martí et al. 2009; Govender et al. 2010).

Since, for any transformation procedure, the number of viable non-transformed cells recovered is highly in excess over transformed ones, the use of selection markers during transformation is unavoidable. For many years, auxotrophic selection markers have been the most popular alternative for laboratory yeast strains. These strains are typically auxotroph for two or three nutrients among a number of amino acids or uridine (Pronk 2002). The defective alleles are also very well known in each case, so that transformants can be easily selected by incorporating a functional copy of the cognate gene in the transforming vector or DNA fragment. However, wine yeast strains are typically phototrophs, precluding the direct use of auxotrophic selection markers. Puig et al. (1998) addressed this issue by engineering a *URA3* auxotrophic wine yeast strain, by using a geneticin resistance selection and marker-rescue strategy. The resulting strain (T73–4), has been used by them, as well as other research groups, for developing multiple recombinant strains with an "industrial" background, using the *URA3* marker, but extrapolating this strategy to other genetic backgrounds would require performing all the construction steps for each of these strains. Alternatively, it was found to be relatively easy to select for induced or spontaneous auxotrophic variants, in the case of genetic markers that can be selected for both positively and negatively, like *URA3*, *LYS2*, or *MET15* among others (Boeke et al. 1984; Ito-Harashima and McCusker 2004; Cost and Boeke 1996; Hashimoto et al. 2005). Notwithstanding, most recombinant wine yeasts have been developed using dominant selection markers based on the resistance to drugs and chemicals. The main advantage of these markers is they do not depend on a previous genetic defect of the host strain, although basal tolerance might be a limitation in some cases. Some examples of dominant markers employed in this field include cycloheximide, sulfometuron, sulphite, p-fluorophenyl-alanine, based on *S. cerevisiae* mutant or overexpressed alleles; or geneticin (G-418), phleomycin, nourseothricin, hygromycin, deriving from bacterial antibiotic resistance genes (Pérez-González et al. 1993; Casey et al. 1988; Goldstein and McCusker 1999; Cebollero and Gonzalez 2004; Coulon et al. 2006; López-Malo et al. 2014; Lilly et al. 2000; Swiegers et al. 2007). An important disadvantage of most of these dominant markers is they rely on the use of antibiotics, which constitutes a handicap for the commercial application of recombinant yeast strains. In order to clear the way to the market, researchers have used a co-transformation strategy (Husnik et al. 2006; Coulon et al. 2006). The approach is based on transforming yeast with of linear, integrative DNA fragment, carrying the construction of interest, together with a replicative plasmid carrying an antibiotic resistance marker. A good percentage of antibiotic resistant cells usually has also integrated the construction. Finally, the plasmid vector cured by culturing the transformed strain under non-selective conditions (Fig. 10.1).

Most recombinant wine yeast strains express heterologous or own genes (usually coding for intra- or extracellular enzymes), under the control of different *S. cerevisiae* promoters. Constitutive promoters are often preferred, but not all promoters considered as constitutive under standard laboratory growth conditions can be taken as such for winemaking conditions (Puig et al. 1996). It must be considered that most of the sugar consumption during wine fermentation usually takes place after the whole of yeast biomass was already produced. For example, the *ACT1* promoter (from the actin encoding gene) was used in some of the original constructs (Pérez-González et al. 1993). Its expression is associated to cell growth, so expression of these constructs is restricted to the beginning of the fermentation process, since most of the sugar is transformed into alcohol after yeast growth arrest. The promoters of

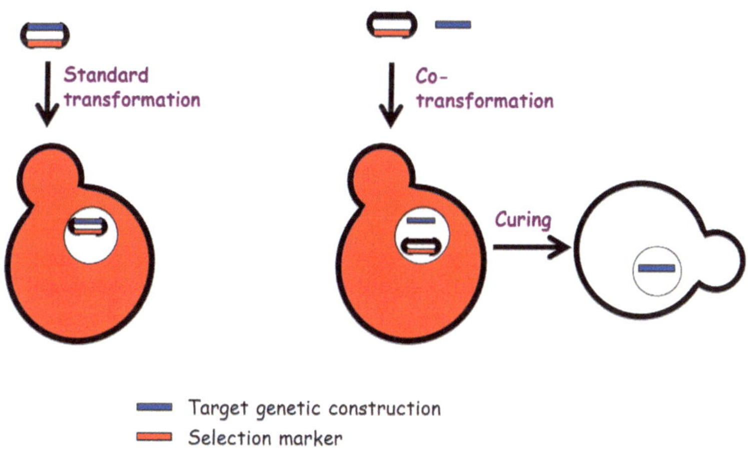

Fig. 10.1 Schematic representation of standard transformation and co-transformation of wine yeasts; the later allowing to easily curing the recombinant strain of auxiliary sequences and phenotypes not required in the final strain (e.g. antibiotic resistance). This is attained by sub-culturing the co-transformed strain in the absence of selective pressure. The target genetic construction is carried in a linear DNA fragment and stably integrated by homologous recombination

TDH3 (for glyceraldehyde-3-P-dehydrogenase), *ADH1* (alcohol dehydrogenase), and *PGK1* (phoshopglycerate-kinase) were among the most popular choices (Malherbe et al. 2003; Vilanova et al. 2000; Ehsani et al. 2009; Cambon et al. 2006. Efficient heterologous expression also requires the use of suitable transcription termination and polyadenylation sequences. Often, but not always, these sequences are derived from the same gene as the promoter used in that construction.

Extracellular enzymes derived from other yeast or fungal species do not usually require an exchange of signal peptides in order to be secreted by wine yeasts (Pérez-González et al. 1993; Sánchez-Torres et al. 1996; Ganga et al. 1999). Otherwise, mating pheromone α-factor signal peptide is considered a good choice to drive protein secretion in case extracellular expression is required (Laing and Pretorius 1993; Malherbe et al. 2003).

In some instances, the genetic improvement relies on the loss-of-function of a given gene, rather than on the acquisition of a new activity (Gonzalez-Ramos et al. 2009). Gene disruption on wine yeast strain might also aim to the characterization of the oenological impact of the genes involved. The fact that most wine yeast strains are at least 2n in terms of cell DNA content, meaning at least two copies of each chromosome are present, constitutes an additional challenge for the development of wine yeast recombinant strains by gene disruption. This has been addressed by several ways, including obtaining haploid derivatives, developing homozygous strains, marker rescue, or use of multiple selection markers. Haploid derivatives can be obtained from strains which are naturally heterologous for the HO *locus* (HO/ho), or by disruption of one copy of HO (Walker et al. 2003). Sporulation of these strains gives raise to both stable haploids and stable diploid strains in equal proportions. Homothallic strains can be sporulated, after a first gene disruption event, so that half of the haploid spores will become double disrupted after spontaneous diploidization. However, since most of these strains are heterozygous to some degree (sometimes highly heterozygous) there is a risk of obtaining strains that phenotypically differ to a significant extent from the parent strain, beyond the intended gene disruption. To avoid this, a homozygous strain can be obtained in the first instance. A confirmation of the phenotypic similarity with the original strain, including behaviour under wine fermentation conditions is advised, both for haploid and homozygous derivatives of industrial wine strains (Mangado et al. 2018). Using these strains homozygous deleted strains, isogenic to the parent strains can be obtained by first generating a hemizygous strain, deleted for one allele, followed by sporulation and spontaneous diploidization of isolated spores (Curiel et al. 2016). One half of the single spore clones obtained are homozygous for the intended deletion.

For gene deletion based genetic improvement, the availability of the whole genome YKO collection constitutes an interesting advantage (Giaever et al. 2002). Each strain from such collection already carries a copy of the cassette, flanked by the homologous upstream and downstream region of one of the *S. cerevisiae* ORFs. This construct can be "transplanted" to wine yeast be using genomic DNA from the appropriate strain as template for PCR amplification with primers upstream and downstream the target gene. This PCR fragment can then be used to transform wine yeast for geneticin resistance (Curiel et al. 2016; Salvadó et al. 2016).

4 Inter and Intraspecific Hybridization

Cross breeding has been a classic methodology to obtain individuals that have the best characteristics of their parents. This technique has been used with great success in the breeding of animals and plants. In yeast, intra-specific hybridization has also been widely used to introduce interesting characteristics from a strain into a second preferred strain. There are numerous examples where this type of approach has been successful in *S. cerevisiae* strains. In the oenology context the approach has been used to improve aroma production (Shinohara et al. 1994; Dufour et al. 2013; Steensels et al. 2014), to enhance fermentation performance coupled with low H_2S production (Romano et al. 1985), to reduce foam production (Romano et al. 1985) or to increase temperature tolerance (Marullo et al. 2009).

Perhaps more interesting than intra-specific hybridization are the phenomenons of inter-species hybridization in industrial applications among the *Saccharomyces* genus (Fig. 10.2). One of the most well studied cases occurs in the brewing industry where the species responsible for most of the fermentation is *Saccharomyces pastorianus*, a hybrid between *S. cerevisiae* and the recently described *Saccharomyces eubayanus* (Libkind et al. 2011; Gibson and Liti 2015). Also in the brewing industry, mostly found as contaminants is easy to find S. bayanus hybrid strains (*S. eubayunas* × *S. uvarum*) (Libkind et al. 2011; Nguyen et al. 2011). On the contrary, in the wine making industry, the major species responsible for the

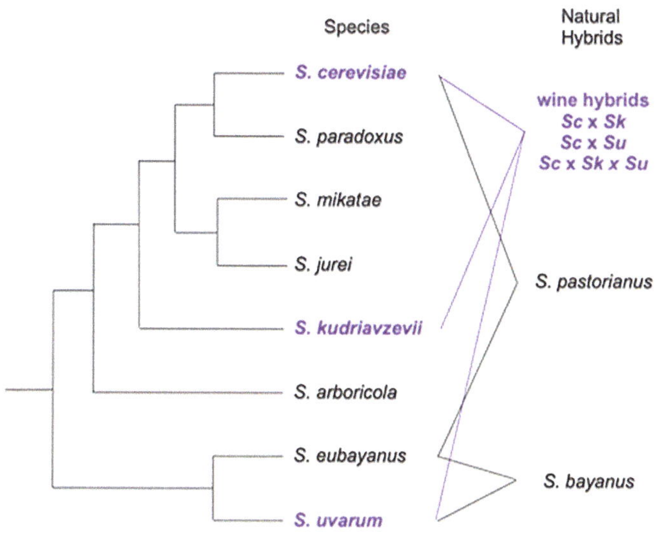

Fig. 10.2 Schematic cladogram of the phylogenetic relationship of the different *Saccharomyces* species, their recognized hybrids and the hybrids that can be isolated from oenological environments. Cladogram topography from Boynton and Greig, (2014; Fig. 1). The recently novel yeast species *Saccharomyces jurei* described by Naseeb et al. 2017 is included

fermentation is not a hybrid, but *Saccharomyces* cerevisiae. This does not mean that we can not find hybrids in an oenological context, the first reported case was a hybrid between *S. cerevisiae* × *S. uvarum* (*S. bayanus* var. *uvarum*) found in Italian wines (Masneuf-Pomarède et al. 1998). Since then hybrid strains have been isolated in wine making environments across Europe, North America, South Africa, Australia and New Zealand, involving *S. cerevisiae*, *S. uvarum* and *S. kudriavzevii* and in one occasion a triple hybrid (Barros Lopes et al. 2002; Belloch et al. 2009; Bradbury et al. 2006; Dunn et al. 2012; Erny et al. 2012; González et al. 2006, 2008; Heinrich 2006; Lopandic et al. 2007; Naumova et al. 2005; Peris et al. 2012a, b, c). These hybrids have been mostly isolated from low-temperature fermentations, providing the first clue for the success of these hybridization processes, both S. kudriavzevii and S. uvarum are better suited to this condition than *S. cerevisiae* (González et al. 2006; Belloch et al. 2008; Salvadó et al. 2011; Sampaio and Gonçalves 2008) (Fig. 10.2).

The ability of these hybrid strains to grow at lower temperatures than *S. cerevisiae* decreases the probabilities of halted or sluggish fermentations, a common problem under these conditions (Bisson 1999). Therefore, they are an interesting choice when conducting this kind of fermentations that have being shown to improve the wine aromatic profile (Boulton et al. 2013). Hybrid strains have also been shown to provide other advantages that the increase in performance at low temperature. They produce wines with higher glycerol content, reduced acetic acid production and lower ethanol concentration (González et al. 2006; Lopandic et al. 2007; Tronchoni et al. 2009; Masneuf-Pomarède et al. 2010; Paget et al. 2014).

These attractive features have promoted the interest of different research groups to create artificial hybrids in an attempt to increase the phenotypic variability of the yeast strains used to different biotechnological applications. *Saccharomyces* species share the same number of chromosomes (16), and have similar morphologic characteristics (Duina et al. 2014). They are defined as species by the biological concept and, therefore being postzygotically isolated. This allows the establishment of hybrid strains that typically will have alloploid genomes (usually being the combination of two diploid genomes). The main mechanisms to create these hybrid strains are: (a) spore to spore mating; (b) protoplast fusion (Curran and Bugeja 1996); (c) mass-mating (Nakazawa et al. 1999) and (d) rare-mating (Spencer and Spencer 1996) and its variant methodology HyPr (hybrid production), which involves the use of a set of plasmids converting the rare-mating into a frequent-mating event (Alexander et al. 2016; Peris et al. 2017). The different investigations that have being conducted in the study of artificial hybrids has also shown that these strains are unstable. Since hybrids have postzygotic barriers, they have to propagate by asexual reproduction. During this increase in number, the hybrid genome will be stabilized, usually by the loss of chromosomes from one of the parental species (Pfliegler et al. 2014; Steensels et al. 2014). An example of a hybrid with genome instability is the industrial wine yeast strain Vin7 (*S. cerevisiae* × *S. kudriavzevii*), one of the first commercial strains to be sequenced (Borneman et al. 2012). This strain can loss its copy of the

S. kudriavzevii chromosome III. The mitochondrial genome can also influence the stability of the hybrid genome. *S. cerevisiae* × *S. kudriavzevii* hybrid strains had a higher *S. kudriavzevii* contribution when possessing a *S. kudriavzevii* mitochondrial genome instead of the *S. cerevisiae* version (Peris et al. 2012c). Therefore, in order to use these hybrid strains, it is important to develop protocols for its genome stabilization, that could allow them to be commercially available. In this sense, it has been shown that adaptive evolution experiments may be a way to achieve this. Under the selection pressure of choice, this methodology stabilizes the genome of these hybrids and improves the key traits that were affected by hybridization (Pérez-Través et al. 2014a, b; Lopandic et al. 2016; Peris et al. 2017; Krogerus et al. 2018).

The combination of these two techniques, artificial hybrids breeding plus adaptive evolution, is one of the most promising tools to increase yeast variability for different industrial processes. Since these two methodologies are defined by the European Union as GM-free organisms, these strains can be commercialized and used by in the different industrial food processes.

5 Genetic Improvement by Random Mutagenesis

Random mutagenesis represents a classical mutagenesis approach useful to increase the rate of appearance of genetic mutations in a large population of cells. The introduction of genetic variability in a cell can be exploited for basic research, for the elucidation of protein structure-function relationships, and industrial sector, for proteins modification to improve or alter their characteristics thus generating improved phenotypes. In the latter context, the resulting mutant cells undergo to a higher general acceptance and possibility of commercialisation in comparison to genetically engineering products. The technique is simple and any prior knowledge about the genetic background of the investigated system is not mandatory, thus allowing for the unbiased discovery of novel or beneficial mutations. However, once mutants are generated, the problem then becomes how to isolate them from the treated population. It is clearly impractical to examine millions of microorganisms individually. Therefore, the isolation of mutants relies on a good screening system, better if direct, from which it is possible to fish out the desired mutant; usually, samples are analysed in cultural medium designed to allow the growth of the desired mutants only.

The induction of random mutations can be caused by agents that damage the DNA, called mutagens, that are of three main types: mutagenic chemicals, radiations and heat. When these agents are deliberately used on living cells, we talk about "*in vivo* mutagenesis"; alternatively, the mutagenic agent can be applied directly on purified DNA that has to be transferred into the living cell before to screen for the desired mutations (in vitro mutagenesis). Common mutagens are toxic compounds that can alter the chemical structure of the nucleotides in DNA. For example, EMS (ethyl methane sulfonate) is widely used to mutagenize growing cells by introducing an ethyl group to bases in DNA; this event changes the shape and base-pairing properties of nucletides. Nitrite is used to mutate purified DNA because it converts amino groups to hydroxyl groups and, thus, converting the base cytosine to uracil. Radiations, such as high frequency electromagnetic radiation, ultraviolet radiation (UV light), X-rays and gamma rays (γ-rays), cause a direct damage on DNA. X-rays and γ-rays can react with water and other molecules to generate ions and free radicals, mainly hydroxyl radicals, or to interact directly with DNA; about 70% of the DNA damage is caused by hydroxyl radicals while the other 30% of the radiation damage is due to direct interaction of X-rays and γ-rays with DNA itself. X-rays tend to produce multiple mutations and often yield rearrangements of the DNA, such as deletions, inversions and translocations (Clark 2005). Ultraviolet radiation works within a wavelength from 100 to 400 nm. It acts directly on the DNA by forming dimers between two neighbouring pyrimidine bases that cross-react with each other. Thymine dimers are particularly frequent and at their level the DNA polymerase leaves a single-stranded region that needs repairing. The repair process can provoke the insertion of incorrect nucleotides in the synthesized strand, resulting in mutations.

Few studies on the use of random mutagenesis in wine yeasts are available in literature. They mainly refer to the genetic improvement of wine quality features that are linked to carbon, nitrogen and sulphur metabolism of wine yeasts, such as the production of acetic acid (Cordente et al. 2009), the nitrogen assimilation under oenological conditions (Salmon and Barre 1998) and the release of hydrogen sulphide (Cordente et al. 2013) in *S. cerevisiae*, and to secondary characteristics of wine like the generation of mutants overproducing mannoproteins (González Ramos et al. 2010; Quirós et al. 2010) or showing accelerated autolysis (Gonzalez et al. 2003; Nunez et al. 2005).

In the case of secondary characteristics an interesting example is the release of mannoprotein that can occur during the alcoholic fermentation of grape must in wine. These structural components of the cell have been recognized to impart beneficial properties of wine in terms of protein and tartaric stability, increasing perceptions of body and roundness, reducing astringency, retaining aromatic molecules and helping in the maintenance of sparkling wines. The increase of the release of mannoproteins represents an example of synergy between classical and genetic engineering approaches for the genetic improvement of wine yeast. Since mannoprotein overproduction cannot be directly selected, an overproducing mutant strain was first obtained by brute-force approach (González Ramos et al. 2010). In parallel, researchers identified *KNR4* inactivation as a potential genetic target for improvement, and found that strains deleted in some other interesting genes were killer-nine-resistant just as *KNR4* (Gonzalez-Ramos et al. 2009). This led to the development of a more efficient mannoprotein overproducing mutant selection scheme (Quirós et al. 2010).

6 Experimental Evolution of Wine Yeasts

Adaptive, directed or experimental evolution are all terms referred to engineering yeast strains in laboratory conditions by using the intrinsic mechanisms of adaptation of these microorganisms. The continuous growth for many cell divisions under a stress condition will select for individuals with improved fitness due to genomic changes (Kawecki et al. 2012). *S. cerevisiae* was the model organism when experimental evolution was proposed by Francis and Hansche (1972, 1973). The experimental setup was firstly though in a chemostat using continues culture but it can also be done in batch cultures keeping the conditions of choice for many generations or cell divisions. The probability of success in an experimental evolution assay varies depending on the starting population (Elena and Lenski 2003). For industrial purposes, usually, this population comes from an isogenic preferred strain to be improved under a given condition but, it can also be a population with genomic diversity. The starting variability will increase the chances of obtaining the desired phenotype. For this reason, experimental evolution is usually coupled to a first technique that increases the genomic variability of the starting population. This can be random mutagenesis, genetic engineering, hybridization events or genome shuffling of a heterogenic population by mass-mating. During the adaptation to the trait of evolution yeast cells can undergo different genomic changes. These changes can be at the genomic level with changes in the ploidy level, at chromosome level with chromosome copy number variation (CCNV), at gen level with changes in the number of copies of specific genes and also small changes at nucleotide level (base insertions, deletions or substitutions) (Dunham et al. 2002; Mangado et al. 2018).

In recent years this methodology has acquired great attention thanks to the last advances in sequencing techniques that allow to better understand which have been the different mechanisms that have taken place during the evolution experiment (Burke et al. 2014). The cheapening of the sequencing techniques allows not only to sequence several isolated clones at the end of the experiment but also to do it in several points for the whole population. This has allowed us to have a better follow-up of the different beneficial mutations that appear during the evolution and how they are fixed in the population. The technique has been improved in resolution, it has been showed that by tagging the whole population, each of the different beneficial mutations that arise in the population can be observed and not only those present in high frequency (Cvijović et al. 2018). Also, it has been improved in efficacy, mass scaled evolution experiments are platforms were millions of yeast clones undergo evolution independently and its performance can be followed by image analysis. This allows to cover a larger spectrum of the genomic landscape of evolution, identifying the mutations that occur in lower frequency but still contribute to increase the performance of the population under a certain stress (Zackrisson et al. 2016).

It must be taken into account that once the conditions of the evolution experiment have finished the evolved population could be unstable and lose its recently acquired fitness for a certain condition. It has been shown that some genomic changes like aneuploidies can be beneficial under certain stress but to reduce fitness under normal conditions (Yona et al. 2012; Chang et al. 2013). Zhu et al. (2018) showed that even when no significant difference in growth can be observed in evolved aneuploid cells compared to the parental strain under regular conditions, changes in the number of chromosomes promote oxidative stress. This will drive the cell to lose again these aneuploidies back to regular ploidy levels.

In an oenological context, experimental evolution has been used to improve sugar consumption rates by growing yeast in wine-like fermentation conditions (McBryde et al. 2006), improved fermentation kinetics by cultivating in the presence of ethanol (Novo et al. 2014). There are also interesting examples where the growing conditions affect indirectly the metabolic target of study. Some of these examples are the reduction of ethanol yields by growing in gluconate as unique carbon source (Cadière et al. 2011) or in a hyperosmotic media (Tilloy et al. 2014). *Oenococcus oeni* has also being subjected to directed evolution in a multi-stressors environment in order to improve malolactic fermentation of the strain with satisfactory results (Jiang et al. 2018). It has been also used to understand the underlying mechanism behind the evolution of wine yeast strains under oenological conditions, showing that genomic changes occur at different genomic levels to accomplish the adaptation to a wine-like environment (Mangado et al. 2018). Although evolution experiments related to the wine industry usually involved one unique strain, Morrison-Whittle et al. (2018) recently showed how the co-evolution of *Candida glabrata* and *Pichia kudriavzevii* had a significant impact of on the production of metabolites that affect the flavour and aroma of experimental wines.

An important aspect of experimental evolution is that the microorganism resulting from this technique can be easily reach experiment the market, as in the case of Cadière et al. (2011), since they are considered non-GM organisms.

7 Genome Editing by the CRISPR/Cas9 Approach

Genome editing refers to a set of recent technologies able to modify DNA in a very precise way using programmable nucleases, including Zinc Finger Nucleases (ZFNs), TALENs (transcription-activator-like effector nucleases) and CRISPR (Clustered Regularly Interspected Short Palindromic Repeats) – Cas9 (CRISPR-associated protein 9) RNA-guided endonucleases (RGENs). These are useful in genome editing as they have specific endonuclease activity to a target sequence. ZFNs are proteins with a modular structure; each module, about 30 amino acids, with ββα structure, recognizes a single codon in an active site, where a zinc atom is present. Currently, there are modules for all 64 existing codons; thus, it is possible to form a protein with the desired modules with codons and selected sequences. Similar are the TALENs, endonucleases extracted from *Xanthomonas* with a modular and assembling structure; a domain of 33–35 amino acids recognizes a single base pair. More versatile for this reason than the ZFN but, however, limited by the necessity that the first recognized nucleotide is a thymine (Gaj et al. 2013). CRISPR/

Cas RGENs are considered more precise and versatile than the first two technologies that target specificity is determined by the modification of their DNA-binding domains. Moreover, being able to recognize the nucleotide sequence to be cut thanks to an association with a guide RNA (gRNA), the system "CRISPR/Cas9" can be customized by replacing the guide RNAs, making the system much more affordable and scalable (Kim 2016).

In general, technologies for genome editing exploit both the action of a nuclease to generate double strand breakings (DSBs) in the DNA at a desired point and the DNA repair mechanisms of the cell to insert a specific mutation at the cutting site. When the DNA is damaged on both strands with a DBS, the cell can rely on two repair methods: (i) non-homologous recombination (NHEJ – non-homologous end-joining), consisting of simple ligation of the two ends broken by the cut; (ii) homologous recombination (HR – homologous recombination) that allows the repair of the damaged genome site on the basis of a homologous fragment at the two ends where the cut occurred. While the NHEJ event can cause point mutations, or the insertion or deletion of some nucleotides, the manipulation of the nucleotide sequence at the cutting site by the HR takes place with the insertion in the cell of a 'donor' sequence, a fragment of double-stranded DNA that can provoke two types of changes (Kim 2016; Gratz et al. 2013; Mahfouz et al. 2014): (i) point mutations or non-sense mutations to induce an amino acid change or to insert a stop codon. In this case, the donor will be almost identical to the sequence in the proximity of the cutting site except for some nucleotides; (ii) insertion of a heterologous gene for its expression in a new organism. The donor will bring the gene sequence flanked by homologous sequences adjacent to the cutting site in order to trigger the recombination.

CRISPR elements were first discovered in *Haloferax mediterranei*, an archaeal microbe with extreme salt tolerance (Mojica et al. 1993; Mojica et al. 2005) even though repeated sequences with similar structure were already described in *Escherichia coli* (Ishino et al. 1987). They are also present in many other eubacteria and their role is to provide resistance against invading exogenous DNA such as that of bacteriophage or conjugative plasmids (Barrangou et al. 2007; Hryhorowicz et al. 2017; Ishino et al. 1987; Lander 2016). Foreign invading genetic material that is incorporated between CRISPR is transcribed and processed into CRISPR RNAs (crRNAs) (including both foreign and CRISPR repeat DNA). The crRNAs hybridize with transactivating CRISPR RNAs (tracrRNAs) and the resulting crRNA/tracrRNA complex acts as a guide for the endonuclease Cas, which cleaves invading nucleic acid sequences (Brouns 2012; DiCarlo et al. 2013).

The main elements of the CRISPR/Cas system are a bacterial CRISPR-associated protein nuclease (Cas) and a short guide of RNA. The type II CRISPR system is the most widely studied system and it exploits the Cas9 nuclease enzyme from *Streptococcus pyogenes*. The Cas9 contains two distinct endonuclease domains, a HNH domain and a RuvC-like domain, that independently cleave both stands at the target site to generate a DSBs. As far the RNA, it has alternately been referred to as a guide RNA (gRNA), a single-guide RNA (sgRNA) or a chimeric RNA (chiRNA). In the simplest form of the type II CRISPR system, Cas9 is guided by the gRNA to a cleavage site; this latter is a specific DNA *locus* composed by 20 nucleotides and a protospacer adjacent motif (PAM). The PAM consists of a NGG sequence located at the 3′ end of the target sequence. The gRNA-Cas9 complex generates DSBs immediately before the PAM site on the target DNA (Ryan and Cate 2014) in the nuclease domains RuvC and HNH (Mahfouz et al. 2014). Finally, the DSBs in the chromosomal DNA are repaired with knockouts/deletions or knock-ins/insertion by NHEJ and HR (Gratz et al. 2013) (Fig. 10.3).

In 2012, the CRISPR/Cas9 system was used for the first time as a 'molecular machine'; this study shows that the engineering of the gRNA complex can guide the Cas9 to cut at a specific DNA sequence provided by an adjacent PAM sequence (Jinek et al. 2012). However, the attribution of the discovery of the CRISPR/Cas approach as a new technique for genome editing remains contested. Following the

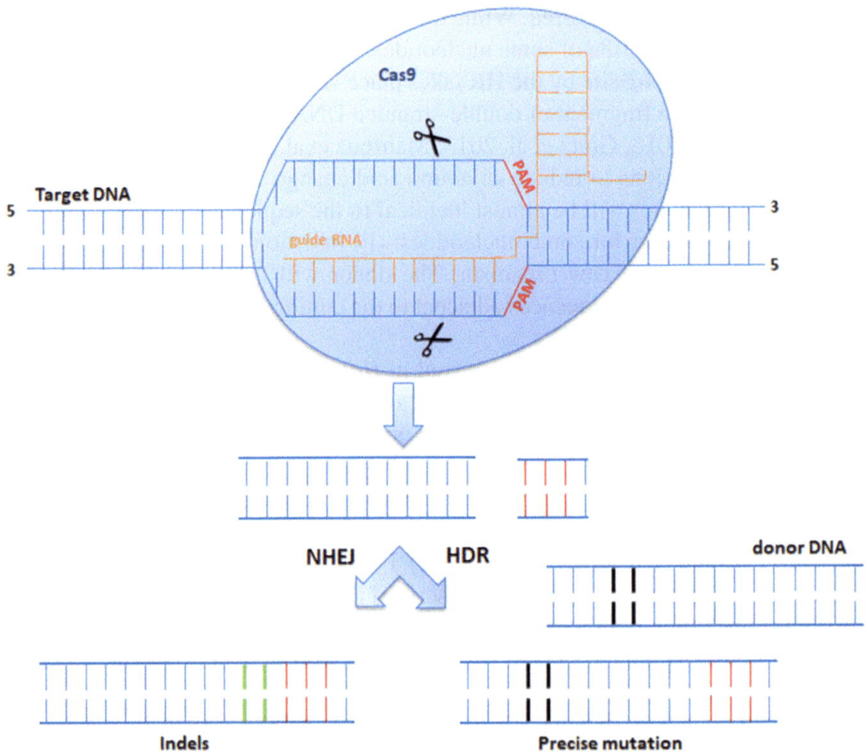

Fig. 10.3 Schematic representation of the CRISPR/Cas9 system. The Cas9 nuclease (blue circle) interacts with a guide RNA (orange lines) and the complex is directed to the target DNA at a specific locus adjacent to a PAM site (red lines). The presence of a PAM site adjacent to the targeted sequence (blue line, 3′–5′ direction) allow the cleavage at the *locus*. Double strand breaks (DSBs) between the third and the fourth bases 5′ to the PAM site on both strands of DNA are formed. After the endonuclease cleaves the DNA can be repaired by non-homologous end joining (NHEJ) or homologous recombination (Homology Directed Repair, HDR). Donor DNA: foreign DNA with flanking homologous regions to the target *locus*; indel: insertion/deletion

first publication, the CRISPR/Cas9 system has found wide use in various fields: the Cas9 has been modified and adapted to various organisms, including eukaryotes, thanks to the optimization of the codon usage. For example, a Cas9 firstly optimized for humans has been applied in the genome editing in yeast (DiCarlo et al. 2013). From 2015 the CRISPR/Cpf1 system is also applied. The Cpf1 nuclease works like the Cas9 but with some differences, offering different advantages: it recognizes a rich PAM (and therefore different from the NGG), is driven only by a crRNA and it cuts the two strands of DNA generating sticky ends (Zetsche et al. 2015; Verwaal et al. 2018). Transcriptional regulation, via the use of a nuclease deficient ("dead") Cas9 (dCas9) has been developed for repression of gene expression of endogenous genes (Gilbert et al. 2013; Jensen et al. 2017; Deaner and Alper 2017).

CRISPR/Cas9 is exploited today in the biotechnologies of every field, from the application on bacteria to that on eukaryotic cells: it is used on food-borne microorganisms, for the production of biofuels and other molecules of industrial interest, on fungi, on plants for genetic improvement of functional characters and to increase resistance to pathogens, on animals both for food and for disease carriers, as well as on some mosquitos, and finally on human cells and on humans, especially in the medical field, for the development of innovative gene therapies (Ledford 2015; Wang et al. 2016; Gorter de Vries et al. 2017; Fraczek et al. 2018; Alexander 2018). However, applications of CRISPR/Cas9 gene editing technology in *S. cerevisiae* are continuing developed (Giersch and Finnigan 2017).

The first application of the CRISPR/Cas9 system in yeast was reported in 2013 (DiCarlo et al. 2013). The Cas9 and the gRNA were expressed in the cell by two different plasmids, each with a different selectable marker. The cell was also co-transformed with a linear donor-DNA bringing the modified sequence for homologous recombination. A couple of year later, a single plasmid was developed, the pCRCT, which included the information for the Cas9, one or more guide RNAs and the corresponding donor-DNA (Bao et al. 2015). This last approach represents a great advantage in metabolic engineering studies; indeed, using a single Cas9 nuclease it is possible to modify multiple genes in the presence of several gRNAs, each carrying a target gene, and of the donors with the modified sequences to be inserted (Wang et al. 2016). There are at least four methods to modify multiple genes simultaneously using the CRISPR/Cas9 system (Stovicek et al. 2017): (1) construction of plasmids containing up to two gRNA expression cassettes with as many selectable markers; each cassette is first individually cloned into a plasmid p426-SNR52p-gRNA.CAN1.Y-SUP4t and then fused thanks to the Gibbson assembly (Mans et al. 2015); (2) HI-CRISPR system, involves the use of a single plasmid, the aforementioned pCRCT, containing the sequence of Cas9, plus the gRNA organized as an array in 'interspaced' crRNA and the respective donors (Bao et al. 2015); (3) co-transformation of cells, already equipped with Cas9, with multiple plasmids each with a different gRNA and a different selectable gene (Horwitz et al. 2015); (4) addition to a plasmid of a USER (Uracil Specific Excision Reaction) sequence which allow to include in itself tandem cassettes of gRNAs (Jakočiūnas et al. 2015).

The wine industry could particularly gain an advantage from this engineering system (Pretorius 2017); indeed, molecular studies should help understanding the contribution of *Saccharomyces* and non-*Saccharomyces* species to several wine features such as those linked to wine quality and safety (i.e. aroma and off-flavours compounds, ethanol and glycerol production, sulphur dioxide resistance, toxic compound formation, etc.). Recently, Vigentini and co-authors successfully established the CRISPR/Cas9 system in two commercial starter yeasts of *S. cerevisiae* (EC1118, AWRI796), modifying the *CAN1* gene encoding for an arginine permease, in order to generate strains with reduced urea production (Vigentini et al. 2017). This can be useful to limit the formation of ethylcarbamate in wine, a carcinogenic compound in a number of mammalian species. In this study, the yeast strains carrying the *can1* mutation failed to produce urea in oenological conditions suggesting that the genetic modification could impaired the arginine metabolism with a consequent potential decrease in ethylcarbamate production. In the same year (de Trindade et al. 2017), a polygenic analysis (pooled-segregant whole-genome sequence Analysis) was combined with CRISPR/Cas9-mediated allele exchange approach in order to identify novel *S. cerevisiae* genes affecting the production of phenylethyl acetate (2-PEAc), a desirable flavor compound of major importance in alcoholic beverages imparting rose- and honey-like aromas. In particular, *FAS2* gene and a mutant allele of *TOR1* gene were found to be responsible for high 2-PEAc.

Thus, winemakers might benefit by the application of this new approach to yeasts and to grapes as well, enabling better understanding of the connections between wine features and wine yeast genetics. Nevertheless, the recent interpretation of genome editing in the EU GMO legislation by the ECJ (European Court of Justice) seems to close the door the CRISPR/Cas approach. Indeed, the court deliberates that genome editing fall under the techniques to obtain OGMs; this means that all genome edited organisms will have to comply with all provisions of the EU GMO legislation. However, the ECJ in its ruling does not consider all the risk-related issues opening the road toward possible criticisms. This decision could negatively affect the innovative research and development involving modern genetic engineering techniques such as CRISPR in Europe. Against the tide seems to be the Swedish Board of Agriculture (SBA) regarding the modified plants; indeed, SBA has recently debated that plants modified using CRISPR/Cas9 where DNA only has been deleted (and no exogenous DNA is inserted), should not be regulated as GMOs. This has opened up the possibility of producing such plants as "normal crops" (i.e. without the supervision of any authority) in Sweden.

8 Conclusion

An extensive repertoire of molecular tools is available for the genetic engineering of *S. cerevisiae* wine strains and some of them are working properly in non-*Saccharomyces* yeasts as well. In wine field, genetically modified yeasts can contribute to improve the winemaking process overall and the final characteristics of wines in terms of quality and complexity. Targets of these modifications are indeed several and they include traits of the primary and secondary metabolism of yeasts.

The success stories reporting the use of modified yeasts in the production of wines are numerous but sometimes the real exploitation of these microorganisms is limited. In fact, two different classes of microorganisms can be obtained with the molecular tools discussed in this chapter, non-GM and GM organisms. From a technical point of view, if the genetic background of the investigated yeasts is known, there are several advantages in using molecular strategies that edit specific genes because the desired modification can be easily achieved. Unfortunately, these approaches normally generate GMOs with all public concerns that can arise from their use in food production. On the other hand, non-GMOs are better accepted by consumers and they can be quickly introduced on the market. Among the promising techniques generating GM-free organisms, artificial hybrids breeding coupled with experimental evolution is of course a tangible chance for wine industry.

Modern viticulture and oenology could benefit by the CRISPR/Cas9 approach. Aside from the molecular advantage of producing quick genome changes by using a unique gene-editing approach, the CRISPR/Cas9 system has the potential to become soon the gold standard technique for the production of novel microorganisms suitable for the food industry. However, the scientific community needs actions to bring the EU regulatory framework in line with our current scientific understanding and with international developments.

References

Alexander, W. G. (2018). A history of genome editing in *Saccharomyces* cerevisiae. *Yeast, 35*, 355–360. https://doi.org/10.1002/yea.3300.

Alexander, W. G., Peris, D., Pfannenstiel, B. T., Opulente, D. A., Kuang, M., & Hittinger, C. T. (2016). Efficient engineering of marker-free synthetic allotetraploids of *Saccharomyces. Fungal Genetics and Biology, 89*, 10–17.

Bao, Z., Xiao, H., Liang, J., Zhang, L., Xiong, X., Sun, N., Si, T., & Zhao, H. (2015). Homology-integrated CRISPR–Cas (HI-CRISPR) system for one-step multigene disruption in *Saccharomyces* cerevisiae. *ACS Synthetic Biology, 4*(5), 585–594. https://doi.org/10.1021/sb500255k.

Barrangou, R., Fremaux, C., Deveau, H., Richards, M., Boyaval, P., Moineau, S., Romero, D. A., & Horvath, P. (2007). CRISPR provides acquired resistance against viruses in prokaryotes. *Science, 315*(5819), 1709–1712.

Baudin, A., Ozier-Kalogeropoulos, O., Denouel, A., Lacroute, F., & Cullin, C. (1993). A simple and efficient method for direct gene deletion in *Saccharomyces* cerevisiae. *Nucleic Acids Research, 21*(14), 3329.

Belloch, C., Orlic, S., Barrio, E., & Querol, A. (2008). Fermentative stress adaptation of hybrids within the *Saccharomyces* sensu stricto complex. *International Journal of Food Microbiology, 122*(1), 188–195.

Belloch, C., Pérez-Torrado, R., González, S. S., Pérez-Ortín, J. E., García-Martínez, J., Querol, A., & Barrio, E. (2009). Chimeric genomes of natural hybrids of *Saccharomyces* cerevisiae and *Saccharomyces* kudriavzevii. *Applied and Environmental Microbiology, 75*(8), 2534–2544.

Bellon, J. R., Eglinton, J. M., Siebert, T. E., Pollnitz, A. P., Rose, L., de Barros Lopes, M., & Chambers, P. J. (2011). Newly generated interspecific wine yeast hybrids introduce flavour and aroma diversity to wines. *Applied Microbiology and Biotechnology, 91*(3), 603–612.

Bellon, J. R., Schmid, F., Capone, D. L., Dunn, B. L., & Chambers, P. J. (2013). Introducing a new breed of wine yeast: Interspecific hybridisation between a commercial *Saccharomyces* cerevisiae wine yeast and *Saccharomyces* mikatae. *PLoS One, 8*(4), e62053.
Bisson, L. F. (1999). Stuck and sluggish fermentations. *American Journal of Enology and Viticulture, 50*(1), 107–119.
Bisson, L. F. (2004). The biotechnology of wine yeast. *Food Biotechnology, 18*(1), 63–96.
Boeke, J. D., La Croute, F., & Fink, G. R. (1984). A positive selection for mutants lacking orotidine-5′-phosphate decarboxylase activity in yeast: 5-fluoro-orotic acid resistance. *Molecular and General Genetics MGG, 197*(2), 345–346.
Borneman, A. R., Desany, B. A., Riches, D., Affourtit, J. P., Forgan, A. H., Pretorius, I. S., Egholm, M., & Chambers, P. J. (2012). The genome sequence of the wine yeast VIN7 reveals an allotriploid hybrid genome with *Saccharomyces* cerevisiae and *Saccharomyces* kudriavzevii origins. *FEMS Yeast Research, 12*(1), 88–96.
Botstein, D., & Fink, G. R. (2011). Yeast: An experimental organism for 21st century biology. *Genetics, 189*(3), 695–704.
Boulton, R. B., Singleton, V. I., Bisson, I. F., & Kunkee, R. E. (2013). *Principles and practices of winemaking*. Berlin: Springer Science & Business Media.
Bradbury, J. E., Richards, K. D., Niederer, H. A., Lee, S. A., Dunbar, P. R., & Gardner, R. C. (2006). A homozygous diploid subset of commercial wine yeast strains. *Antonie Van Leeuwenhoek, 89*(1), 27–37.
Brouns, S. J. (2012). Molecular biology. A swiss army knife of immunity. *Science, 337*, 808–809. https://doi.org/10.1126/science.1227253.
Burke, M. K., Liti, G., & Long, A. D. (2014). Standing genetic variation drives repeatable experimental evolution in outcrossing populations of *Saccharomyces cerevisiae*. *Molecular Biology and Evolution, 31*(12), 3228–3239.
Cadière, A., Ortiz-Julien, A., Camarasa, C., & Dequin, S. (2011). Evolutionary engineered *Saccharomyces* cerevisiae wine yeast strains with increased in vivo flux through the pentose phosphate pathway. *Metabolic Engineering, 13*(3), 263–271.
Cambon, B., Monteil, V., Remize, F., Camarasa, C., & Dequin, S. (2006). Effects of GPD1 overexpression in *Saccharomyces* cerevisiae commercial wine yeast strains lacking ALD6 genes. *Applied and Environmental Microbiology, 72*(7), 4688–4694.
Cardona, F., Carrasco, P., Pérez-Ortín, J. E., lí del Olmo, M., & Aranda, A. (2007). A novel approach for the improvement of stress resistance in wine yeasts. *International Journal of Food Microbiology, 114*(1), 83–91.
Casey, G. P., Xiao, W., & Rank, G. H. (1988). A convenient dominant selection marker for gene transfer in industrial strains of *Saccharomyces* yeast: SMRI encoded resistance to the herbicide sulfometuron methyl. *Journal of the Institute of Brewing, 94*(2), 93–97.
Cebollero, E., & Gonzalez, R. (2004). Comparison of two alternative dominant selectable markers for wine yeast transformation. *Applied and Environmental Microbiology, 70*(12), 7018–7023.
Cebollero, E., Martinez-Rodriguez, A., Carrascosa, A. V., & Gonzalez, R. (2005). Overexpression of csc1-1. A plausible strategy to obtain wine yeast strains undergoing accelerated autolysis. *FEMS Microbiology Letters, 246*(1), 1–9.
Cebollero, E., González-Ramos, D., Tabera, L., & González, R. (2007). Transgenic wine yeast technology comes of age: Is it time for transgenic wine? *Biotechnology Letters, 29*, 191–200.
Chang, S., Wang, Y., Lu, J., Gai, J., Li, J., Chu, P., Guan, R., Zhao, T., & Xu, Y. (2013). The mitochondrial genome of soybean reveals complex genome structures and gene evolution at intercellular and phylogenetic levels. *PLoS One, 8*(2), e56502.
Clark, D. (2005). *Molecular biology*. Elsevier Academic Press. ISBN: 0-12-175551-7.
Cordente, A. G., Heinrich, A., Pretorius, I. S., & Swiegers, J. H. (2009). Isolation of sulfite reductase variants of a commercial wine yeast with significantly reduced hydrogen sulfide production. *FEMS Yeast Research, 9*(3), 446–459.
Cordente, A. G., Cordero-Bueso, G., Pretorius, I. S., & Curtin, C. D. (2013). Novel wine yeast with mutations in YAP1 that produce less acetic acid during fermentation. *FEMS Yeast Research, 13*(1), 62–73.

Cost, G. J., & Boeke, J. D. (1996). A useful colony colour phenotype associated with the yeast selectable/counter-selectable marker MET15. *Yeast, 12*(10), 939–941.

Coulon, J., Husnik, J. I., Inglis, D. L., van der Merwe, G. K., Lonvaud, A., Erasmus, D. J., & van Vuuren, H. J. (2006). Metabolic engineering of *Saccharomyces* cerevisiae to minimize the production of ethyl carbamate in wine. *American Journal of Enology and Viticulture, 57*(2), 113–124.

Cuello, R., Flores Montero, K. J., Mercado, L. A., Combina, M., & Ciklic, F. (2017). Construction of low-ethanol–wine yeasts through partial deletion of the *Saccharomyces cerevisiae PDC2* gene. *AMB Express, 7*(1), 67.

Curiel, J. A., Salvadó, Z., Tronchoni, J., Morales, P., Rodrigues, A. J., Quirós, M., & Gonzalez, R. (2016). Identification of target genes to control acetate yield during aerobic fermentation with *Saccharomyces* cerevisiae. *Microbial Cell Factories, 15*(1), 156.

Curran, B. P., & Bugeja, V. C. (1996). Protoplast fusion in *Saccharomyces* cerevisiae. In *Yeast protocols* (pp. 45–49). New Jersey: Humana Press.

Cvijović, I., Nguyen Ba, A. N., & Desai, M. M. (2018). Experimental studies of evolutionary dynamics in microbes. *Trends in Genetics, 34*(9), 693–703.

de Barros Lopes, M., Bellon, J. R., Shirley, N. J., & Ganter, P. F. (2002). Evidence for multiple interspecific hybridization in *Saccharomyces* sensu stricto species. *FEMS Yeast Research, 1*(4), 323–331.

de Carvalho, B. T., Holt, S., Souffriau, B., Lopes Brandão, R., Foulquié-Moreno, M. R., & Thevelein, J. M. (2017). Identification of novel alleles conferring superior production of rose flavor phenylethyl acetate using polygenic analysis in yeast. *mBio, 8*, e01173–e01117. https://doi.org/10.1128/mBio.01173-17.

Deaner, M., & Alper, H. S. (2017). Systematic testing of enzyme perturbation sensitivities via graded dCas9 modulation in *Saccharomyces* cerevisiae. *Metabolic Engineering, 40*, 14–22.

Dequin, S., & Barre, P. (1994). Mixed lactic acid–alcoholic fermentation by Saccharomyes cerevisiae expressing the Lactobacillus casei L (+)–LDH. *Nature Biotechnology, 12*(2), 173.

DiCarlo, J. E., Norville, J. E., Mali, P., Rios, X., Aach, J., & Church, G. M. (2013). Genome engineering in *Saccharomyces* cerevisiae using CRISPR-Cas systems. *Nucleic Acid Research, 41*(7), 4336–4343.

Dufour, M., Zimmer, A., Thibon, C., & Marullo, P. (2013). Enhancement of volatile thiol release of *Saccharomyces* cerevisiae strains using molecular breeding. *Applied Microbiology and Biotechnology, 97*, 5893–5905.

Duina, A. A., Miller, M. E., & Keeney, J. B. (2014). Budding yeast for budding geneticists: A primer on the *Saccharomyces* cerevisiae model system. *Genetics, 197*(1), 33–48.

Dunham, M. J., Badrane, H., Ferea, T., Adams, J., Brown, P. O., Rosenzweig, F., & Botstein, D. (2002). Characteristic genome rearrangements in experimental evolution of *Saccharomyces cerevisiae. Proceedings of the National Academy of Sciences, 99*(25), 16144–16149.

Dunn, B., Richter, C., Kvitek, D. J., Pugh, T., & Sherlock, G. (2012). Analysis of the *Saccharomyces* cerevisiae pan-genome reveals a pool of copy number variants distributed in diverse yeast strains from differing industrial environments. *Genome Research*. https://doi.org/10.1101/gr.130310.

Ehsani, M., Fernández, M. R., Biosca, J. A., Julien, A., & Dequin, S. (2009). Engineering of 2, 3-butanediol dehydrogenase to reduce acetoin formation by glycerol-overproducing, low-alcohol *Saccharomyces* cerevisiae. *Applied and Environmental Microbiology, 75*(10), 3196–3205.

Elena, S. F., & Lenski, R. E. (2003). Evolution experiments with microorganisms: The dynamics and genetic bases of adaptation. *Nature Reviews Genetics, 4*(6), 457–469.

Erny, C., Raoult, P., Alais, A., Butterlin, G., Delobel, P., Matei-Radoi, F., Casaregola, S., & Legras, J. L. (2012). Ecological success of a group of *Saccharomyces* cerevisiae/*Saccharomyces* kudriavzevii hybrids in the northern European wine making environment. *Applied and Environmental Microbiology*. AEM-06752.

Eschenbruch, R., Cresswell, K. J., Fisher, B. M., & Thornton, R. J. (1982). Selective hybridisation of pure culture wine yeasts. *European Journal of Applied Microbiology and Biotechnology, 14*(3), 155–158.

Fraczek, M. G., Naseeb, S., & Delneri, D. (2018). History of genome editing in yeast. *Yeast, 35*, 361–368. https://doi.org/10.1002/yea.3308.

Francis, J. C., & Hansche, P. E. (1972). Directed evolution of metabolic pathways in microbial populations. I. Modification of the acid phosphatase pH optimum in *S. cerevisiae*. *Genetics, 70*, 59–73.

Francis, J. C., & Hansche, P. E. (1973). Directed evolution of metabolic pathways in microbial populations II. A repeatable adaptation in *Saccharomyces* cerevisiae. *Genetics, 74*, 259–265.

Gaj, T., Gersbach, C. A., & Barbas, C. F. (2013). ZFN, TALEN, and CRISPR/Cas-based methods for genome engineering. *Trends in Biotechnology, 31*, 397–405.

Ganga, M. A., Pinaga, F., Valles, S., Ramón, D., & Querol, A. (1999). Aroma improving in microvinification processes by the use of a recombinant wine yeast strain expressing the Aspergillus nidulans xlnA gene. *International Journal of Food Microbiology, 47*(3), 171–178.

Giaever, G., Chu, A. M., Ni, L., Connelly, C., Riles, L., Veronneau, S., et al. (2002). Functional profiling of the *Saccharomyces* cerevisiae genome. *Nature, 418*(6896), 387.

Gibson, B., & Liti, G. (2015). *Saccharomyces* pastorianus: Genomic insights inspiring innovation for industry. *Yeast, 32*(1), 17–27.

Giersch, R. M., & Finnigan, G. C. (2017). Yeast still a beast: Diverse applications of CRISPR/Cas editing Technology in *S. cerevisiae*. *Yale Journal of Biology and Medicine, 90*, 643–651.

Gilbert, L. A., Larson, M. H., Morsut, L., Liu, Z., Brar, G. A., Torres, S. E., et al. (2013). CRISPR-mediated modular RNA-guided regulation of transcription in eukaryotes. *Cell, 154*(2), 442–451.

Goldstein, A. L., & McCusker, J. H. (1999). Three new dominant drug resistance cassettes for gene disruption in *Saccharomyces* cerevisiae. *Yeast, 15*(14), 1541–1553.

González Ramos, D., Muñoz, A., Ortiz Julien, A., Palacios García, A. T., Heras Manso, J. M., & González García, R. (2010). A *Saccharomyces* cerevisiae wine yeast strain overproducing mannoproteins selected through classical genetic methods. *International Journal of Vine and Wine Sciences, 44*(4), 243–249.

Gonzalez, R., Martínez-Rodríguez, A. J., & Carrascosa, A. V. (2003). Yeast autolytic mutants potentially useful for sparkling wine production. *International Journal of Food Microbiology, 84*(1), 21–26.

González, S. S., Barrio, E., Gafner, J., & Querol, A. (2006). Natural hybrids from *Saccharomyces* cerevisiae, *Saccharomyces* bayanus and *Saccharomyces* kudriavzevii in wine fermentations. *FEMS Yeast Research, 6*(8), 1221–1234.

González, S. S., Barrio, E., & Querol, A. (2008). Molecular characterization of new natural hybrids of *Saccharomyces* cerevisiae and S. kudriavzevii in brewing. *Applied and Environmental Microbiology, 74*(8), 2314–2320.

Gonzalez, R., Munoz, R., & Carrascosa, A. V. (2011). In A. V. Carrascosa, R. Munoz, & R. González (Eds.), *Production of wine starter cultures. Molecular wine microbiology* (pp. 279–302). Amsterdam/Boston: Elsevier.

González-Candelas, L., Cortell, A., & Ramón, D. (1995). Construction of a recombinant wine yeast strain expressing a fungal pectate lyase gene. *FEMS Microbiology Letters, 126*(3), 263–269.

Gonzalez-Ramos, D., Cebollero, E., & Gonzalez, R. (2008). A recombinant Saccharomyces cerevisiae strain overproducing mannoproteins stabilizes wine against protein haze. *Applied and Environmental Microbiology, 74*, 5533–5540.

Gonzalez-Ramos, D., Quiros, M., & Gonzalez, R. (2009). Three different targets for the genetic modification of wine yeast strains resulting in improved effectiveness of bentonite fining. *Journal of Agricultural and Food Chemistry, 57*(18), 8373–8378.

Gorter de Vries, A. R., de Groot, P. A., van den Broek, M., & Daran, J.-M. G. (2017). CRISPR-Cas9 mediated gene deletions in lager yeast *Saccharomyces* pastorianus. *Microbial Cell Factories, 16*, 222. https://doi.org/10.1186/s12934-017-0835-1.

Govender, P., Bester, M., & Bauer, F. F. (2010). FLO gene-dependent phenotypes in industrial wine yeast strains. *Applied Microbiology and Biotechnology, 86*(3), 931–945.

Gratz, S. J., Harrison, M. M., O'Connor-Giles, K. M., & Wildonger, J. (2013). CRISPR/Cas9-mediated genome engineering and the promise of designer flies on demand. *Fly, 7*(4), 249–255.

Grossmann, M., Kießling, F., Singer, J., Schoeman, H., Schröder, M. B., & von Wallbrunn, C. (2011). Genetically modified wine yeasts and risk assessment studies covering different steps within the wine making process. *Annals of Microbiology, 61*(1), 103–115.

Hashimoto, S., Ogura, M., Aritomi, K., Hoshida, H., Nishizawa, Y., & Akada, R. (2005). Isolation of auxotrophic mutants of diploid industrial yeast strains after UV mutagenesis. *Applied and Environmental Microbiology, 71*(1), 312–319.

Heinrich, A. J. (2006). Identification of genomic differences between laboratory and commercial strains of *Saccharomyces* cerevisiae (Doctoral dissertation).

Horwitz, A. A., Walter, J. M., Schubert, M. G., Kung, S. H., Hawkins, K., Platt, D. M., Hernday, A. D., Mahatdejkul-Meadows, T., Szeto, W., Chandran, S. S., & Newman, J. D. (2015). Efficient multiplexed integration of synergistic alleles and metabolic pathways in yeasts via CRISPR-Cas. *Cell Systems, 1*(1), 88–96.

Hryhorowicz, M., Lipinski, D., Zeyland, J., & Stomski, R. (2017). CRISPR/Cas9 immune system as a tool for genome engineering. *Archivum Immunologiae et Therapiae Experimentalis, 65*, 233–240. https://doi.org/10.1007/s00005-016-0427-5.

Husnik, J. I., Volschenk, H., Bauer, J., Colavizza, D., Luo, Z., & van Vuuren, H. J. (2006). Metabolic engineering of malolactic wine yeast. *Metabolic Engineering, 8*(4), 315–323.

Husnik, J. I., Delaquis, P. J., Cliff, M. A., & van Vuuren, H. J. (2007). Functional analyses of the malolactic wine yeast ML01. *American Journal of Enology and Viticulture, 58*(1), 42–52.

Ishino, Y., Shinagawa, H., Makino, K., Amemura, M., & Nakata, A. (1987). Nucleotide sequence of the iap gene, responsible for alkaline phosphatase isozyme conversion in *Escherichia coli*, and identification of the gene product. *Journal of Bacteriology, 169*(12), 5429–5433.

Ito-Harashima, S., & McCusker, J. H. (2004). Positive and negative selection LYS5MX gene replacement cassettes for use in *Saccharomyces* cerevisiae. *Yeast, 21*(1), 53–61.

Jakočiūnas, T., Bonde, I., Herrgård, M., Harrison, S. J., Kristensen, M., Pedersen, L. E., Jensen, M. K., & Keasling, J. D. (2015). Multiplex metabolic pathway engineering using CRISPR/Cas9 in *Saccharomyces* cerevisiae. *Metabolic Engineering, 28*, 213–222.

Jensen, E. D., Ferreira, R., Jakociunas, T., Arsovska, D., Zhang, J., Ding, L., et al. (2017). Transcriptional reprogramming in yeast using dCas9 and combinatorial gRNA strategies. *Microbial Cell Factories, 16*(1), 46.

Jiang, J., Sumby, K. M., Sundstrom, J. F., Grbin, P. R., & Jiranek, V. (2018). Directed evolution of Oenococcus oeni strains for more efficient malolactic fermentation in a multi-stressor wine environment. *Food Microbiology, 73*, 150–159.

Jiménez-Martí, E., Zuzuarregui, A., Ridaura, I., Lozano, N., & Del Olmo, M. (2009). Genetic manipulation of HSP26 and YHR087W stress genes may improve fermentative behaviour in wine yeasts under vinification conditions. *International Journal of Food Microbiology, 130*(2), 122–130.

Jinek, M., Chylinski, K., Fonfara, I., Hauer, M., Doudna, J. A., & Charpentier, E. (2012). A programmable dual-RNA–guided DNA endonuclease in adaptive bacterial immunity. *Science, 337*(6096), 816–821.

Kawecki, T. J., Lenski, R. E., Ebert, D., Hollis, B., Olivieri, I., & Whitlock, M. C. (2012). Experimental evolution. *Trends in Ecology & Evolution, 27*(10), 547–560.

Kim, J. S. (2016). Genome editing comes of age. *Nature Protocols, 11*(9), 1573–1578. https://doi.org/10.1038/nprot.2016.104.

Kishimoto, M. (1994). Fermentation characteristics of hybrids between the cryophilic wine yeast *Saccharomyces* bayanus and the mesophilic wine yeast *Saccharomyces* cerevisiae. *Journal of Fermentation and Bioengineering, 77*(4), 432–435.

Krogerus, K., Holmström, S., & Gibson, B. (2018). Enhanced wort fermentation with de novo lager hybrids adapted to high-ethanol environments. *Applied and Environmental Microbiology, 84*(4), e02302–e02317.

Kupferschmidt, K. (2018). EU verdict on CRISPR crops dismays scientists.
Kutyna, D. R., Varela, C., Stanley, G. A., Borneman, A. R., Henschke, P. A., & Chambers, P. J. (2012). Adaptive evolution of *Saccharomyces* cerevisiae to generate strains with enhanced glycerol production. *Applied Microbiology and Biotechnology, 93*(3), 1175–1184.
Laing, E., & Pretorius, I. S. (1993). Co-expression of an Erwinia chrysanthemi pectate lyase-encoding gene (pelE) and an E. carotovora polygalacturonase-encoding gene (peh1) in *Saccharomyces* cerevisiae. *Applied Microbiology and Biotechnology, 39*(2), 181–188.
Lander, E. S. (2016). The heroes of CRISPR. *Cell, 164*(1–2), 18–28.
Ledford, H. (2015). Crispr, the disruptor. *Nature, 522*, 20.
Lee, D., Lloyd, N. D., Pretorius, I. S., & Borneman, A. R. (2016). Heterologous production of raspberry ketone in the wine yeast *Saccharomyces* cerevisiae via pathway engineering and synthetic enzyme fusion. *Microbial Cell Factories, 15*(1), 49.
Legras, J. L., Merdinoglu, D., Cornuet, J. M., & Karst, F. (2007). Bread, beer and wine: *Saccharomyces* cerevisiae diversity reflects human history. *Molecular Ecology, 16*(10), 2091–2102.
Libkind, D., Hittinger, C. T., Valério, E., Gonçalves, C., Dover, J., Johnston, M., Gonçalves, P., & Sampaio, J. P. (2011). Microbe domestication and the identification of the wild genetic stock of lager-brewing yeast. *Proceedings of the National Academy of Sciences*. https://doi.org/10.1073/pnas.1105430108.
Lilly, M., Lambrechts, M. G., & Pretorius, I. S. (2000). Effect of increased yeast alcohol acetyltransferase activity on flavor profiles of wine and distillates. *Applied and Environmental Microbiology, 66*(2), 744–753.
Lopandic, K., Gangl, H., Wallner, E., Tscheik, G., Leitner, G., Querol, A., Borth, N., Breitenbach, M., Prillinger, H., & Tiefenbrunner, W. (2007). Genetically different wine yeasts isolated from Austrian vine-growing regions influence wine aroma differently and contain putative hybrids between *Saccharomyces* cerevisiae and *Saccharomyces* kudriavzevii. *FEMS Yeast Research, 7*(6), 953–965.
Lopandic, K., Pfliegler, W. P., Tiefenbrunner, W., Gangl, H., Sipiczki, M., & Sterflinger, K. (2016). Genotypic and phenotypic evolution of yeast interspecies hybrids during high-sugar fermentation. *Applied Microbiology and Biotechnology, 100*(14), 6331–6343.
López-Malo, M., García-Rios, E., Chiva, R., Guillamon, J. M., & Martí-Raga, M. (2014). Effect of deletion and overexpression of tryptophan metabolism genes on growth and fermentation capacity at low temperature in wine yeast. *Biotechnology Progress, 30*(4), 776–783.
Mahfouz, M. M., Piatek, A., & Stewart, C. N. (2014). Genome engineering via TALENs and CRISPR/Cas9 systems: Challenges and perspectives. *Plant Biotechnology Journal, 12*, 1006–1014. https://doi.org/10.1111/pbi.12256.
Malherbe, D. F., Du Toit, M., Otero, R. C., Van Rensburg, P., & Pretorius, I. S. (2003). Expression of the Aspergillus Niger glucose oxidase gene in *Saccharomyces* cerevisiae and its potential applications in wine production. *Applied Microbiology and Biotechnology, 61*(5–6), 502–511.
Mangado, A., Morales, P., Gonzalez, R., & Tronchoni, J. (2018). Evolution of a yeast with industrial background under winemaking conditions leads to diploidization and chromosomal copy number variation. *Frontiers in Microbiology, 9*, 1816.
Mans, R., van Rossum, H. M., Wijsman, M., Backx, A., Kuijpers, N. G., van den Broek, M., Daran-Lapujade P., Pronk J. T., van Maris A. J., Daran J. M. (2015). CRISPR/Cas9: A molecular Swiss army knife for simultaneous introduction of multiple genetic modifications in *Saccharomyces*. *FEMS Yeast Research, 15*(2). pii: fov004. https://doi.org/10.1093/femsyr/fov004.
Manzanares, P., Orejas, M., Gil, J. V., De Graaff, L. H., Visser, J., & Ramón, D. (2003). Construction of a genetically modified wine yeast strain expressing the Aspergillus aculeatus rhaA gene, encoding an alpha-L-rhamnosidase of enological interest. *Applied and Environmental Microbiology, 69*, 7558–7562.
Marullo, P., Bely, M., Masneuf-Pomarède, I., Pons, M., Aigle, M., & Dubourdieu, D. (2006). Breeding strategies for combining fermentative qualities and reducing off-flavor production in a wine yeast model. *FEMS Yeast Research, 6*(2), 268–279.

Marullo, P., Mansour, C., Dufour, M., Albertin, W., Sicard, D., Bely, M., & Dubourdieu, D. (2009). Genetic improvement of thermo-tolerance in wine *Saccharomyces* cerevisiae strains by a back-cross approach. *FEMS Yeast Research, 9*, 1148–1160.

Masneuf-Pomarède, I., Hansen, J., Groth, C., Piskur, J., & Dubourdieu, D. (1998). New hybrids between *Saccharomyces* sensu stricto yeast species found among wine and cider production strains. *Applied and Environmental Microbiology, 64*, 3887–3892.

Masneuf-Pomarède, I., Bely, M., Marullo, P., Lonvaud-Funel, A., & Dubourdieu, D. (2010). Reassessment of phenotypic traits for *Saccharomyces* bayanus var. uvarum wine yeast strains. *International Journal of Food Microbiology, 139*(1), 79–86.

McBryde, C., Gardner, J. M., de Barros Lopes, M., & Jiranek, V. (2006). Generation of novel wine yeast strains by adaptive evolution. *American Journal of Enology and Viticulture, 57*, 423–430.

Michnick, S., Roustan, J. L., Remize, F., Barre, P., & Dequin, S. (1997). Modulation of glycerol and ethanol yields during alcoholic fermentation in *Saccharomyces* cerevisiae strains overexpressed or disrupted for GPD1 encoding glycerol 3-phosphate dehydrogenase. *Yeast, 13*(9), 783–793.

Mojica, F. J. M., Juez, G., & Rodriguez-Valera, F. (1993). Transcription at different salinities of Haloferax mediterranei sequences adjacent to partially modified PstI sites. *Molecular Microbiology, 9*(3), 613–621.

Mojica, F. J., García-Martínez, J., & Soria, E. (2005). Intervening sequences of regularly spaced prokaryotic repeats derive from foreign genetic elements. *Journal of Molecular Evolution, 60*(2), 174–182.

Morrison-Whittle, P., Lee, S. A., Fedrizzi, B., & Goddard, M. R. (2018). Co-evolution as tool for diversifying flavour and aroma profiles of wines. *Frontiers in Microbiology, 9*, 910.

Nakazawa, N., Okawa, K., Sato, T., Enei, H., & Harashima, S. (1999). Mass mating method in combination with G418-and aureobasidin A-resistance markers for efficient selection of hybrids from homothallic strains in *Saccharomyces* cerevisiae. *Journal of Bioscience and Bioengineering, 88*(5), 468–471.

Naseeb, S., James, S. A., Alsammar, H., Michaels, C. J., Gini, B., Nueno-Palop, C., Bond, C. J., McGhie, H., Roberts, I. N., & Delneri, D. (2017). *Saccharomyces jurei* sp. nov., isolation and genetic identification of a novel yeast species from Quercus robur. *International Journal of Systematic and Evolutionary Microbiology., 67*, 2046–2052.

Naumova, E. S., Naumov, G. I., Masneuf-Pomarède, I., Aigle, M., & Dubourdieu, D. (2005). Molecular genetic study of introgression between *Saccharomyces* bayanus and *S. cerevisiae*. *Yeast, 22*(14), 1099–1115.

Nguyen, H. V., Legras, J. L., Neuvéglise, C., & Gaillardin, C. (2011). Deciphering the hybridisation history leading to the lager lineage based on the mosaic genomes of *Saccharomyces* bayanus strains NBRC1948 and CBS380T. *PLoS One, 6*(10), e25821.

Novo, M., Bigey, F., Beyne, E., Galeote, V., Gavory, F., Mallet, S., Cambon, B., Legras, J. L., Wincker, P., Casaregola, S., & Dequin, S. (2009). Eukaryote-to-eukaryote gene transfer events revealed by the genome sequence of the wine yeast *Saccharomyces* cerevisiae EC1118. *Proceedings of the National Academy of Sciences*. https://doi.org/10.1073/pnas.0904673106.

Novo, M., Gonzalez, R., Bertran, E., Martínez, M., Yuste, M., & Morales, P. (2014). Improved fermentation kinetics by wine yeast strains evolved under ethanol stress. *LWT-Food Science and Technology, 58*(1), 166–172.

Núñez, Y. P., Carrascosa, A. V., Gonzalez, R., Polo, M. C., & Martínez-Rodriguez, A. (2006). Isolation and characterization of a thermally extracted yeast cell wall fraction potentially useful for improving the foaming properties of sparkling wines. *Journal of Agricultural and Food Chemistry, 54*, 7898–7903.

Nunez, Y. P., Carrascosa, A. V., Gonzalez, R., Polo, M. C., & Martínez-Rodríguez, A. J. (2005). Effect of accelerated autolysis of yeast on the composition and foaming properties of sparkling wines elaborated by a Champenoise method. *Journal of Agricultural and Food Chemistry, 53*, 7232–7237.

Paget, C. M., Schwartz, J. M., & Delneri, D. (2014). Environmental systems biology of cold-tolerant phenotype in *Saccharomyces* species adapted to grow at different temperatures. *Molecular Ecology, 23*(21), 5241–5257.
Pérez-González, J. A., Gonzalez, R., Querol, A., Sendra, J., & Ramón, D. (1993). Construction of a recombinant wine yeast strain expressing beta-(1, 4)-endoglucanase and its use in microvinification processes. *Applied and Environmental Microbiology, 59*(9), 2801–2806.
Pérez-Ortín, J. E., Querol, A., Puig, S., & Barrio, E. (2002). Molecular characterization of a chromosomal rearrangement involved in the adaptive evolution of yeast strains. *Genome Research, 12*(10), 1533–1539.
Pérez-Través, L., Lopes, C. A., Barrio, E., & Querol, A. (2014a). Stabilization process in *Saccharomyces* intra and interspecific hybrids in fermentative conditions. *International Microbiology, 17*, 213–224. https://doi.org/10.2436/20.1501.01.224.
Pérez-Través, L., Lopes, C. A., Querol, A., & Barrio, E. (2014b). On the complexity of the *Saccharomyces* bayanus taxon: Hybridization and potential hybrid speciation. *PLoS One, 9*(4), e93729.
Pérez-Través, L., Lopes, C. A., González, R., Barrio, E., & Querol, A. (2015). Physiological and genomic characterisation of *Saccharomyces* cerevisiae hybrids with improved fermentation performance and mannoprotein release capacity. *International Journal of Food Microbiology, 205*, 30–40.
Peris, D., Belloch, C., Lopandić, K., Álvarez-Pérez, J. M., Querol, A., & Barrio, E. (2012a). The molecular characterization of new types of *Saccharomyces* cerevisiaex S. kudriavzevii hybrid yeasts unveils a high genetic diversity. *Yeast, 29*(2), 81–91.
Peris, D., Lopes, C. A., Arias, A., & Barrio, E. (2012b). Reconstruction of the evolutionary history of *Saccharomyces* cerevisiae x S. kudriavzevii hybrids based on multilocus sequence analysis. *PLoS One, 7*(9), e45527.
Peris, D., Lopes, C. A., Belloch, C., Querol, A., & Barrio, E. (2012c). Comparative genomics among *Saccharomyces* cerevisiae × *Saccharomyces* kudriavzevii natural hybrid strains isolated from wine and beer reveals different origins. *BMC Genomics, 13*(1), 407.
Peris, D., Moriarty, R. V., Alexander, W. G., Baker, E., Sylvester, K., Sardi, M., Leducq, J. B., et al. (2017). Hybridization and adaptive evolution of diverse *Saccharomyces* species for cellulosic biofuel production. *Biotechnology for Biofuels, 10*(1), 78.
Pfliegler, W. P., Atanasova, L., Karanyicz, E., Sipiczki, M., Bond, U., Druzhinina, I. S., Sterflinger, K., & Lopandic, K. (2014). Generation of new genotypic and phenotypic features in artificial and natural yeast hybrids. *Food Technology and Biotechnology, 52*(1), 46–57.
Pretorius, I. S. (2017). Synthetic genome engineering forging new frontiers for wine yeast. *Critical Reviews in Biotechnology, 37*(1), 112–136.
Pretorius, I. S., & Bauer, F. F. (2002). Meeting the consumer challenge through genetically customized wine-yeast strains. *Trends in Biotechnology, 20*, 426–432.
Pronk, J. T. (2002). Auxotrophic yeast strains in fundamental and applied research. *Applied and Environmental Microbiology, 68*(5), 2095–2100.
Puig, S., Querol, A., Ramón, D., & Pérez-Ortín, J. E. (1996). Evaluation of the use of phase-specific gene promoters for the expression of enological enzymes in an industrial wine yeast strain. *Biotechnology Letters, 18*(8), 887–892.
Puig, S., Ramón, D., & Pérez-Ortín, J. E. (1998). Optimized method to obtain stable food-safe recombinant wine yeast strains. *Journal of Agricultural and Food Chemistry, 46*(4), 1689–1693.
Querol, A., Fernández-Espinar, M. T., lı del Olmo, M., & Barrio, E. (2003). Adaptive evolution of wine yeast. *International Journal of Food Microbiology, 86*(1–2), 3–10.
Quirós, M., Gonzalez-Ramos, D., Tabera, L., & Gonzalez, R. (2010). A new methodology to obtain wine yeast strains overproducing mannoproteins. *International Journal of Food Microbiology, 139*(1), 9–14.
Remize, F., Andrieu, E., & Dequin, S. (2000). Engineering of the pyruvate dehydrogenase bypass in *Saccharomyces* cerevisiae: Role of the cytosolic mg(2+) and mitochondrial K(+) acetaldehyde dehydrogenases Ald6p and Ald4p in acetate formation during alcoholic fermentation. *Applied and Environmental Microbiology, 66*, 3151–3159.

Romano, P., Soli, M. G., Suzzi, G., Grazia, L., & Zambonelli, C. (1985). Improvement of a wine *Saccharomyces* cerevisiae strain by a breeding program. *Applied and Environmental Microbiology, 50*, 1064–1067.

Rous, C. V., Snow, R., & Kunkee, R. E. (1983). Reduction of higher alcohols by fermentation with a leucine-auxotrophic mutant of wine yeast. *Journal of the Institute of Brewing, 89*(4), 274–278.

Ryan, O. W., & Cate, J. H. D. (2014). Multiplex engineering of industrial yeast genomes using CRISPRm. *Methods in Enzymology, 546*, 473–489. https://doi.org/10.1016/B978-0-12-801185-0.00023-4.

Salmon, J. M., & Barre, P. (1998). Improvement of nitrogen assimilation and fermentation kinetics under enological conditions by derepression of alternative nitrogen-assimilatory pathways in an industrial *Saccharomyces* cerevisiae strain. *Applied and Environmental Microbiology, 64*(10), 3831–3837.

Salvadó, Z., Arroyo-López, F. N., Guillamón, J. M., Salazar, G., Querol, A., & Barrio, E. (2011). Temperature adaptation markedly determines evolution within the genus *Saccharomyces*. *Applied and Environmental Microbiology, 77*(7), 2292–2302.

Salvadó, Z., Ramos-Alonso, L., Tronchoni, J., Penacho, V., García-Ríos, E., Morales, P., Gonzalez, R., & Guillamón, J. M. (2016). Genome-wide identification of genes involved in growth and fermentation activity at low temperature in *Saccharomyces* cerevisiae. *International Journal of Food Microbiology, 236*, 38–46.

Sampaio, J. P., & Gonçalves, P. (2008). Natural populations of *Saccharomyces* kudriavzevii in Portugal are associated with oak bark and are sympatric with *S. cerevisiae* and S. paradoxus. *Applied and Environmental Microbiology, 74*(7), 2144–2152.

Sánchez-Torres, P., González-Candelas, L., & Ramón, D. (1996). Expression in a wine yeast strain of the Aspergillus Niger abfB gene. *FEMS Microbiology Letters, 145*(2), 189–194.

Shinohara, T., Saito, K., Yanagida, F., & Goto, S. (1994). Selection and hybridization of wine yeasts for improved winemaking properties: Fermentation rate and aroma productivity. *Journal of Fermentation and Bioengineering, 77*, 428–431.

Sicard, D., & Legras, J. L. (2011). Bread, beer and wine: Yeast domestication in the *Saccharomyces* sensu stricto complex. *Comptes Rendus Biologies, 334*(3), 229–236.

Soares, E. V. (2011). Flocculation in *Saccharomyces* cerevisiae: A review. *Journal of Applied Microbiology, 110*, 1–18.

Spencer, J. F., & Spencer, D. M. (1996). Rare-mating and cytoduction in *Saccharomyces* cerevisiae. In *Yeast protocols* (pp. 39–44). New Jersey: Humana Press.

Steensels, J., Meersman, E., Snoek, T., Saels, V., & Verstrepen, K. J. (2014). Large-scale selection and breeding to generate industrial yeasts with superior aroma production. *Applied and Environmental Microbiology, 80*, 6965–6975.

Stovicek, V., Holkenbrink, C., & Borodina, I. (2017). CRISPR/Cas system for yeast genome engineering: Advances and applications. *FEMS Yeast Research, 17*(5). https://doi.org/10.1093/femsyr/fox030.

Swiegers, J. H., Capone, D. L., Pardon, K. H., Elsey, G. M., Sefton, M. A., Francis, I. L., & Pretorius, I. S. (2007). Engineering volatile thiol release in *Saccharomyces* cerevisiae for improved wine aroma. *Yeast, 24*, 561–574.

Tabera, L., Muñoz, R., & Gonzalez, R. (2006). Deletion of BCY1 from the *Saccharomyces* cerevisiae genome is semidominant and induces autolytic phenotypes suitable for improvement of sparkling wines. *Applied and Environmental Microbiology, 72*(4), 2351–2358.

Thornton, R. J. (1982). Selective hybridisation of pure culture wine yeasts. *European Journal of Applied Microbiology and Biotechnology, 14*(3), 159–164.

Thornton, R. J. (1985). The introduction of flocculation into a homothallic wine yeast. A practical example of the modification of winemaking properties by the use of genetic techniques. *American Journal of Enology and Viticulture, 36*(1), 47–49.

Tilloy, V., Ortiz-Julien, A., & Dequin, S. (2014). Reduction of ethanol yield and improvement of glycerol formation by adaptive evolution of the wine yeast *Saccharomyces* cerevisiae under hyperosmotic conditions. *Applied and Environmental Microbiology, 80*(8), 2623–2632.

Tronchoni, J., Gamero, A., Arroyo-López, F. N., Barrio, E., & Querol, A. (2009). Differences in the glucose and fructose consumption profiles in diverse *Saccharomyces* wine species and their hybrids during grape juice fermentation. *International Journal of Food Microbiology, 134*(3), 237–243.

Verstrepen, K. J., & Klis, F. M. (2006). Flocculation, adhesion and biofilm formation in yeasts. *Molecular Microbiology, 60*, 5–15.

Verstrepen, K. J., Derdelnckx, G., Delvaux, F. R., Winderickx, J., Thevelein, J. M., Bauer, F. F., & Pretorius, I. S. (2001). Late fermentation expression of FLO1 in *Saccharomyces* cerevisiae. *Journal of the American Society of Brewing Chemists, 59*, 69–76.

Verwaal, R., Buiting-Wiessenhaan, N., Dalhuijsen, S., & Roubos, J. A. (2018). CRISPR/Cpf1 enables fast and simple genome editing of *Saccharomyces* cerevisiae. *Yeast, 35*, 201–211. https://doi.org/10.1002/yea.3278.

Vigentini, I., Gebbia, M., Belotti, A., Foschino, R., & Roth, F. P. (2017). CRISPR/Cas9 system as a valuable genome editing tool applicable to wine yeasts: The decrease of urea production as a case of study. *Frontiers in Microbiology, 8*(2194), 2017. https://doi.org/10.3389/fmicb.2017.02194.

Vilanova, M., Blanco, P., Cortés, S., Castro, M., Villa, T. G., & Sieiro, C. (2000). Use of a PGU1 recombinant *Saccharomyces* cerevisiae strain in oenological fermentations. *Journal of Applied Microbiology, 89*(5), 876–883.

Volschenk, H., Viljoen-Bloom, M., Subden, R. E., & Van Vuuren, H. J. J. (2001). Malo-ethanolic fermentation in grape must by recombinant strains of *Saccharomyces* cerevisiae. *Yeast, 18*(10), 963–970.

Walker, M. E., Gardner, J. M., Vystavelova, A., McBryde, C., de Barros Lopes, M., & Jiranek, V. (2003). Application of the reuseable, KanMX selectable marker to industrial yeast: Construction and evaluation of heterothallic wine strains of *Saccharomyces* cerevisiae, possessing minimal foreign DNA sequences. *FEMS Yeast Research, 4*(3), 339–347.

Wang, D., Wang, Z., Liu, N., He, X., & Zhang, B. (2008). Genetic modification of industrial yeast strains to obtain controllable NewFlo flocculation property and lower diacetyl production. *Biotechnology Letters, 30*, 2013–2018.

Wang, H., La Russa, M., & Qi, L. S. (2016). CRISPR/Cas9 in genome editing and beyond. *Annual Review of Biochemistry, 85*, 227–264.

Warringer, J., Zörgö, E., Cubillos, F. A., Zia, A., Gjuvsland, A., Simpson, J. T., et al. (2011). Trait variation in yeast is defined by population history. *PLoS Genetics, 7*(6), e1002111.

Waters, E. J. (1994). A *Saccharomyces* mannoprotein that protects wine from protein haze. *Carbohydrate Polymers, 23*, 185.

Yona, H., Manor, Y. S., Herbst, R. H., Romano, G. H., Mitchell, A., Kupiec, M., Pilpel, Y., & Dahan, O. (2012). Chromosomal duplication is a transient evolutionary solution to stress. *Proceedings of the National Academy of Sciences, 109*(51), 21010–21015.

Zackrisson, M., Hallin, J., Ottosson, L.-G., Dahl, P., Fernandez-Parada, E., Ländström, E., Fernandez-Ricaud, L., Kaferle, P., Skyman, A., Stenberg, S., Omholt, S., Petrovič, U., Warringer, J., & Blomberg, A. (2016). Scan-o-matic: High-resolution microbial phenomics at a massive scale. *G3: Genes, Genomes, Genetics, 6*(9), 3003–3014.

Zetsche, B., Gootenberg, J. S., Abudayyeh, O. O., Slaymaker, I. M., Makarova, K. S., Essletzbichler, P., Volz, S. E., Joung, J., van der Oost, J., Regev, A., Koonin, E. V., & Zhang, F. (2015). Cpf1 is a single RNA-guided endonuclease of a class 2 CRISPRCas system. *Cell, 163*(3), 759–771. https://doi.org/10.1016/j.cell.2015.09.038.

Zhu, J., Tsai, H.-J., Gordon, M. R., & Li, R. (2018). Cellular stress associated with aneuploidy. *Developmental Cell, 44*(4), 420–431.

Chapter 11
Wine Yeasts and Consumer Health

Lisa Granchi, Marilena Budroni, Doris Rauhut, and Giacomo Zara

Contents

1 Introduction

During the last 30 years, the relationship between the wine and consumers health has raised a great interest due to the different bioactive compounds occurring in wine. A bioactive compound is, by definition, a natural or chemical compound that

L. Granchi (✉)
Department of Agriculture, Food, Environment and Forestry, University of Florence, Florence, Italy
e mail: lisa.granchi@unifi.it

M. Budroni · G. Zara
Department of Agriculture, University of Sassari, Sassari, Italy
e-mail: mbudroni@uniss.it; gzara@uniss.it

D. Rauhut
Department of Microbiology and Biochemistry, Hochschule Geisenheim University, Geisenheim, Germany
e-mail: Doris.Rauhut@hs-gm.de

P. Romano et al. (eds.), *Yeasts in the Production of Wine*,
https://doi.org/10.1007/978-1-4939-9782-4_11

can display a positive or negative effect on a living organism and, in this case, on wine consumer health (Guaadaoui et al. 2014). Some bioactive compounds, such as polyphenols and sulphur dioxide, can also affect organoleptic properties and the aging potential of the wine. Therefore, it is important, from both a technological and a healthy point of view, to assess the mechanisms affecting bioactive compounds content in wine. In particular, this chapter is focused on the current knowledge on the role of yeasts in the production and/or reduction of the main bioactive compounds in wine.

2 Role of Yeasts in the Production of Health-Promoting Compounds in Wine

Renaud and de Lorgeril published in 1992 on "The Lancet" journal a study revealing that a higher wine consumption in the French population in comparison with other industrialized countries caused a lower incidence of coronary heart disease, despite the intake of high levels of saturated fat associated with the traditional French diet (French paradox). Since then, many epidemiological studies were carried out on this matter. They demonstrated that individuals consuming daily moderate amounts of wine (i.e., 30 g of ethanol for men and 15 g for women), display a reduction of cardiovascular mortality and an improvement of antioxidant parameters, when compared with individuals who abstain or who drink alcohol to excess (Renaud and Gueguen 1998; Poli et al. 2013). These health benefits are principally attributed to different phenolic compounds (such as anthocyanins and flavan-3-ols) and non-flavonoids (such as resveratrol, cinnamates, and gallic acid) and precisely to their antioxidant power capable of counteracting the negative action of free radicals. Actually, red wine is one of the most important dietary sources of these compounds (Monagas et al. 2005; Ferreira et al. 2016) and their health-promoting properties on several human disorders are widely discussed in scientific literature (Fragopoulou et al. 2018). These disorders are cardiovascular and neurodegenerative diseases, some cancers, obesity, diabetes, allergies, and osteoporosis (Fernandes et al. 2017; Wang et al. 2006; Lindberg and Amsterdam 2008; Zell et al. 2007; Pierini et al. 2008; Fenwick et al. 2015; Bassig et al. 2012). Phenolic compounds are naturally present in grape berries and seeds and their amounts are affected by wide range of factors such as grape variety, geopedological characteristics, climate, and agronomical practices. Moreover, the phenolic content in wine can be influenced by the different oenological practices and in details by fermentation temperature, maceration length, use of clarifying agents, oak-wood aging, duration of wine aging and storage procedures (Guilford and Pezzuto 2011; Ribéreau-Gayon et al. 2000; El Darra et al. 2016; Muller and Fugelsang 1997). Finally, native phenols can be modified or new healthy compounds can be formed by metabolic activities of yeasts, so influencing the antioxidant capacity of the wine.

The following sections describe the ability of yeasts to produce antioxidant metabolites from amino acids during alcoholic fermentation and deal with the effect of yeasts on the content of reduced-glutathione, which is another important antioxidant compound occurring in grapes and wine, as well as the influencing on the total antioxidant power of wines.

2.1 Production of Antioxidant Compounds by Amino Acids Metabolism of Yeasts

Bioactive compounds with antioxidant activity in wine can derive not only from grapes but also from the yeast metabolism of aromatic amino acids during alcoholic fermentation (Mas et al. 2014) (Fig. 11.1). They include higher alcohols, such as tryptophol and tyrosol that originate from tyrosine and melatonin, an indoleamine, synthesized from L-tryptophan (Rodriguez-Naranjo et al. 2011, 2012, 2013; Fernández-Mar et al. 2012; Romboli et al. 2015). Despite the interest on the subject, the contribution of the yeasts to produce in wine these health-promoting compounds has been poorly investigated (Mas et al. 2014). Romboli et al. (2015) demonstrated that the concentrations in wine of tyrosol and hydroxytyrosol produced by *Saccharomyces cerevisiae* increased under non-aerated conditions and in case of slow fermentation kinetics due to a lower concentration of amino acids in grape musts. As regards the melatonin synthesis, it has been undoubtedly related to the yeast growth phase (Rodriguez-Naranjo et al. 2012) both of different *S. cerevisiae*

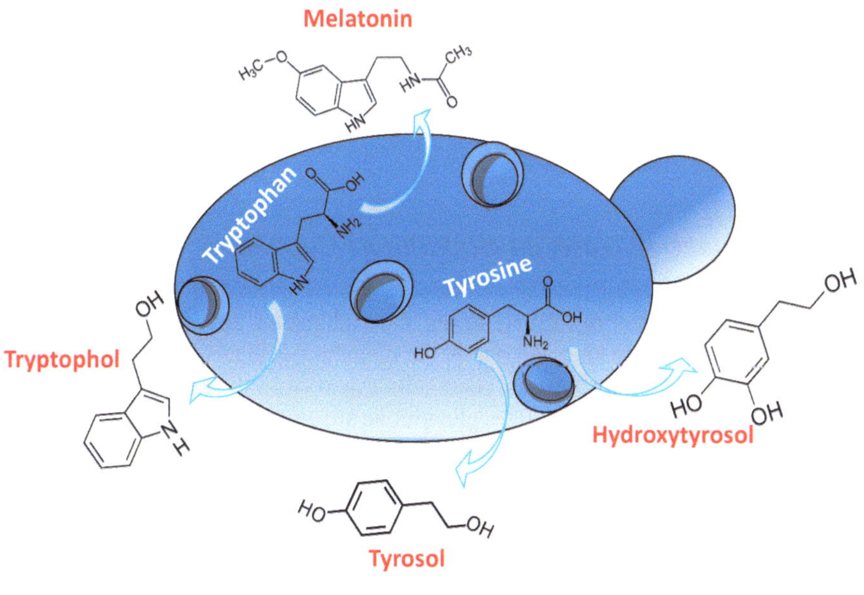

Fig. 11.1 Production of antioxidant compounds by amino acids metabolism of yeasts

strains and two non-*Saccharomyces* species (*Torulaspora delbrueckii* and *Metschnikowia pulcherrima*) (Fernández-Cruz et al. 2017) and it was detected during red and white winemaking steps of five grape varieties (Rodriguez-Naranjo et al. 2013). Moreover, a recent study investigated the protective effect of this compound on *S. cerevisiae* against different stresses (Bisquert et al. 2018). *S. cerevisiae* cells under stress and in normal growth conditions showed a significant shortened lag phase compared to the control cells when treated with melatonin. The gene expression analysis showed that melatonin significantly modulated gene expression in the unstressed cells in the exponential growth phase and during various stress treatments. Therefore, this study supports the role of melatonin as an antioxidant molecule in yeast providing new evidences for its ability to confer yeast cells protection against oxidative stress.

Recently, another study (Guerrini et al. 2018) assessed the ability of four commercial and two indigenous *S. cerevisiae* strains to produce tyrosol, hydroxytyrosol, tryptophol and melatonin during alcoholic fermentation of a synthetic must. The results demonstrated that all the six *S. cerevisiae* strains were able to produce tyrosol, tryptophol, and melatonin during alcoholic fermentation. In particular, the commercial strain Lalvin BM45® (Lallemand, Canada) was able to produce the highest amounts of all the bioactive compounds assayed. Moreover, ex-vivo assays on cultured cells (RAW 264.7 murine macrophages) and immunological assays demonstrated that freeze-dried samples, originating from the experimental wines of all the six *S. cerevisiae* strains, exhibited antioxidant and anti-inflammatory properties, but with different intensity. In other studies, tyrosol has been described as an antioxidant in human cell lines (Giovannini et al. 1999) and as a cardio protective agent (Samuel et al. 2008), while melatonin in humans modulates circadian rhythms and seems to be related to learning and memory processes, ageing, and treatment for Alzheimer's disease, amyotrophic lateral sclerosis, or migraine (Mas et al. 2014).

Therefore, the use of an appropriate *S. cerevisiae* starter strain could represent a suitable tool to enhance the content of some health-promoting compounds in wine, in addition to the well-known phenolic compounds occurring in grapes.

2.2 Influence of Yeasts on Glutathione Content

Glutathione (reduced form, GSH) is the most abundant non-protein thiol (L-γ-glutamyl-L-cystinyl-glycine) and one of the major antioxidants in living organisms (Anderson 1998) and thus, also in yeast cells. Apart from the reduced form, GSH, which occurs normally over 90%, glutathione disulphide (oxidized glutathione, GSSG) is present and can be reduced to GSH by glutathione reductase. In addition, other forms of disulphide are of the mixed type, GSSR (e. g. GS-S-CoA and GS-S-Cys), and proteins can be glutathionylated (GS-S-protein) (Li et al. 2004; Penninckx 2002). Glutathione has received much interest due to its multiple functions, including the control of redox potential in cells, involvement in many detoxification processes, e.g. scavenging toxic reactive oxygen species (ROS), protein folding, transport of organic sulphur, and protecting against heavy metals toxicity (Duncan and Derek 1996; Mezzetti et al. 2014; Penninckx 2002; Wu et al. 2004; Zechmann et al. 2011). Moreover, in wine GSH seems to play an important role in limiting the amount of browning pigments by competitive o-quinones reduction and in protecting volatile thiols that are responsible for the varietal flavour during aging of bottled white wines (Lavigne-Cruège and Dubourdieu 2002; Belda et al. 2017; Parker et al. 2017). Therefore, it is desirable to select yeast strains with a higher formation and release of GSH during fermentation and during storage of wines on lees as a tool to stabilize varietal aromas that are related to specific thiols and to prevent from browning to a certain level.

GSH may account for 0.5–1% of the dry weight in the yeast of the species *Saccharomyces cerevisiae* and represents more than 95% of the low-molecular-mass thiol pool (Elskens et al. 1991; Mehdi and Penninckx 1997; Penninckx 2002). The low redox potential (E'o = – 240 mV for thiol disulphide exchange) and that its reduced form is maintained by NADPH-dependent glutathione reductase offer the tripeptide the properties of a cellular redox buffer (Penninckx 2002).

In *Saccharomyces* and non-*Saccharomyces* yeasts, GSH may be involved to maintain mitochondrial and membrane integrity. It serves as a storage compound, which can be mobilized during nitrogen and sulphur starvation and/or reproduction. About 50% of the GSH was in the yeast cytoplasm and the remaining in the central vacuole during growth on nitrogen-sufficient medium. In case of total sulphur deprivation, GSH stored in the yeast cell is used as an endogenous sulphur source. More than 90% of the cellular GSH was transferred to the central vacuole of the yeast cell (*S. cerevisiae*) at nitrogen starvation (Mehdi and Penninckx 1997). Vacuolar transport of metal derivatives of the tripeptide ensures resistance to metal stress (Penninckx 2000). GSH can be taken up from the medium by the yeast cell via two transport systems. Sulphur flows from GSH to other S-containing metabolites along the sulphur metabolic pathway (Penninckx 2002; Rauhut 2017). GSH is biosynthesized in two ATP-dependent steps in *S. cerevisiae*. First, cysteine is connected with glutamate by γ-glutamylcysteine synthetase to γ-glutamylcysteine. Then glycine is bound by glutathione synthetase to complete the formation of GSH (Li et al. 2004; Mezzetti et al. 2014; Penninckx 2002; Zechmann et al. 2011) (Fig. 11.2).

Zechmann et al. (2011) studied the subcellular distribution of glutathione in *S. cerevisiae* and identified three cell types due to their behaviour to incorporate GSH. The highest amounts of GSH were found in the mitochondria. High and stable concentrations of GSH in mitochondria seem to be important for the protection and survival of cells at oxidative stress.

Glutathione plays also a central role during the winemaking process and is an important ingredient in grapes, must and wine (Kritzinger et al. 2013a).

GSH ranged from 14 to 102 mg/L in grape musts from different grape varieties (Cheynier et al. 1989). Park et al. (2000a, 2000b) detected up to 1.3 mg/L GSH in grape musts and up to 5.1 mg/L GSH in wines. Their investigations also indicated that final wine concentration of GSH was correlated with both total nitrogen and assimilable amino acid concentration. It was also demonstrated that yeast can change the levels of GSH during fermentation. An increase towards the end of

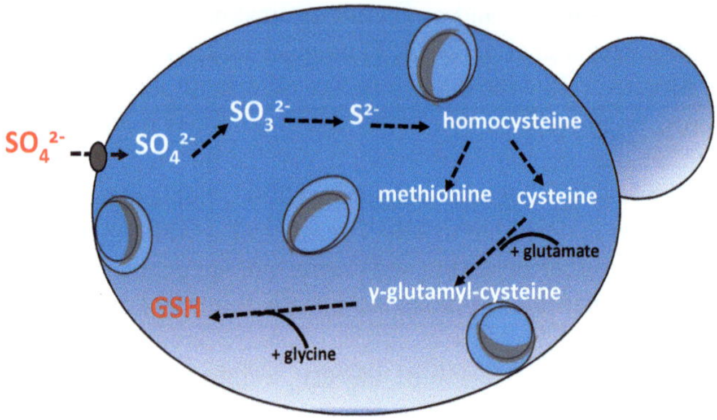

Fig. 11.2 Metabolic pathway to produce reduced glutathione (GSH) in yeast

fermentation was noticed. Concentrations of GSH in musts ranged from 3 to 71 mg/L. Its amount in wines was detected up to about 35 mg/L. The amount of GSH after fermentation varied depending on the yeast strain. The use of specific yeasts led to higher GSH levels in wine than in must. Storing wines on lees assists to maintain GSH, which is important for preventing the oxidation of white wines to a certain extent and for protecting volatile thiols that are responsible for the varietal flavour during aging of bottled white wines. Therefore, it is desirable to select yeast strains with a higher formation and release of GSH during fermentation and during storage of wines on lees as a tool to stabilize varietal aromas that are related to specific thiols and to prevent from browning to a certain level.

Several investigations indicated that *S. cerevisiae* strains differ in their formation and release of glutathione during fermentation (Lavigne et al. 2007; Kritzinger et al. 2013b). Indeed, different strategies were investigated to obtain yeasts with higher GSH formation and release (Mezzetti et al. 2014; Nisamedtinov et al. 2011; De Vero et al. 2017).

The impact of an addition of GSH for the preservation of the varietal character and colour stability was investigated in several studies (Lavigne-Cruège and Dubourdieu 2002; Roussis et al. 2007; Ugliano et al. 2011), demonstrating that the management of the antioxidant metabolome is very important since the beginning of the winemaking process in order to estimate wine ageing potential (Nikolantonaki et al. 2018).

Certain inactive dry yeast (IDY) preparations enriched with glutathione and probably other non-volatile thiols have been offered to enhance flavour intensity and stabilize wine colour (Gabrielli et al. 2017; Pozo-Bayón et al. 2009). There is also evidence that GSH could partially replace the application of SO_2 filling to wines (Badea and Antoce 2015). Webber et al. (2017) showed that the addition of GSH after disgorging of sparkling wine reduces browning and acetaldehyde formation for up to 12 months, but they observed no effect on the concentration of free SO_2 and total phenols.

Finally, the addition of glutathione in concentrations up to a maximum of 20 mg/L to must or wine was adopted by the International Organization of Vine and Wine (OIV) with the Resolutions OIV-OENO 445–2015 and OIV-OENO 446–2015 (OIV 2015).

On the other hand, it has to be indicated that the supplementation of GSH to must can lead to unpleasant volatile sulphur compounds and off-flavours at particular circumstances (e. g. deficiency of assimilable nitrogen; yeast strain) (Rauhut 2003, 2017). Furthermore, research trials with glutathione supplementation and glutathione-enriched IDY-products demonstrated that pleasant flavours (caused by thiols like 3-sulfanylhexanol) could be achieved (Peyrot des Gachons et al. 2002), but also unpleasant volatile sulphur compounds like hydrogen sulphide were detected at certain conditions and specific wine treatments (Wegmann-Herr et al. 2016). Therefore, additional investigations are required on the supplementation of GSH and with GSH and other thiols enriched IDY preparations during the wine-making process and their impact on fermentation development and sensory impression of wines.

Further research on glutathione formation and metabolism might be profitable for oenology and biotechnology to take advantage of the differences among *S. cerevisiae* strains during the winemaking process and storage, in order to obtain strains better adapted to the frequent environmental stresses occurring before, during and after alcoholic fermentation (Penninckx 2002).

Focus should be also related to specific non-*Saccharomyces* yeasts and their function in GSH formation, release and metabolism, because they are more and more discussed as important contributors to certain desirable flavours and characteristics or to improve wine organoleptic properties (Romano et al. 2003; Jolly et al. 2014; Gamero-Sandemetrio et al. 2018). In addition, the application of mixed yeast cultures with sequential and simultaneous inoculation should be studied related to GSH and other non-volatile thiols, their metabolism and impact on wine quality and stability during ageing.

In general, further studies should be carried out in order to investigate yeast metabolic behaviour to produce bioactive compounds during wine fermentation. Indeed, the accumulation of these antioxidant compounds during alcoholic fermentation might enhance the protection of wines from oxidation, thus potentially assuming an important role in sensorial quality and potential longevity of the final products.

2.3 Total Antioxidant Capacity Affected by Yeast Strain

Starter cultures performing the fermentation process can have a significant effect on the total antioxidant capacity (TAC) of wine (Brandolini et al. 2007) To understand the impact of the yeasts on this wine characteristic, Brandolini et al. (2011) used 40 *S. cerevisiae* strains to obtain different experimental wines that were analyzed for the total antioxidant capacity (TAC). The TAC of these 40 wines ranged from 0.625

to 3.3 mM/L ascorbic acid equivalents, and the wines could be placed in three classes as a function of TAC: low, medium, and high. The majority of strains tested produced wines of the medium TAC class, whereas only a few strains exhibited a high TAC. Guerrini et al. (2018) confirmed these results demonstrating that fermented samples obtained with different strains of *S. cerevisiae* showed different antioxidant capabilities. Probably these TAC variations are mainly attributable to yeast action on wine phenolic component. In fact, a recent study demonstrated that yeast ecology influenced the accumulation of phenolic compounds, such as quercetin and vitisin A (Romboli et al. 2015). In particular, the accumulation of quercetin and its glycosides resulted influenced by must aeration and by yeast species occurring in the fermentative process, while vitisin A contents were significantly higher in wines produced in presence of *Starmerella bacillaris* (syn. *Candida zemplinina*). Another recent study compared the capability of *S. cerevisiae* and *Saccharomyces bayanus* to influence the polyphenol content of wine during fermentation (Samoticha et al. 2017). The wines obtained with *S. cerevisiae* were characterized by higher total polyphenol content and lower reduction of antioxidant activity. Finally, Datta et al. (2017) demonstrated that the extracellular fraction of *Saccharomyces cerevisiae* var. *boulardii* NCYC 3264 strain was rich in polyphenolic metabolites: vanillic acid, cinnamic acid, phenyl ethyl alcohol (rose oil), erythromycin, amphetamine and vitamin B6.

3 Role of Yeasts in Formation or Reduction of Toxic Compounds in Wine

Nowadays, it is well known that some compounds possibly influencing negatively wine consumer's health may occur in wine, mainly because of microbial metabolism. Particularly, assimilation, degradation as well as biosynthesis of different nitrogen compounds by some wine microorganisms (yeasts and bacteria) may lead to direct or indirect formation of biogenic amines and ethyl carbamate (also referred as urethane). Biogenic amines can cause a wide variety of undesirable physiological effects to sensitive humans, especially when alcohol and acetaldehyde are present (Alvarez and Moreno-Arribas 2014; EFSA 2011; Lehtonen 1996; Soufleros et al. 1998). These compounds have been studied extensively for 30 years and particularly over the last 10 years, because of the increasing attention to consumer protection (Restuccia et al. 2018). Likewise, ethyl carbamate is now considered probably carcinogenic to humans, as it has been recently reclassified as a Group 2A carcinogen by the World Health Organization's International Agency for Research on Cancer.

Another compound of interest for human health that can occur in wine is Ochratoxin A. Ochratoxin A is the main mycotoxin in wine, and its levels are regulated in different countries due to its toxic effects. As all the mycotoxins, Ochratoxin A is a secondary metabolite that is produced by some mould species and yeasts can detoxify wine from this compound.

Furthermore, during the last decades, the formation of sulphur-containing compounds by wine yeast itself has become a central and crucial point of research (Divol et al. 2012; Guerrero and Cantos-Villar 2015) because of health concerns and an unfavourable public perception of sulphite that have led to demands for restriction of its use in all wine growing countries (Costanigro et al. 2014).

Therefore, the following section not only deals with the role of wine yeasts in producing biogenic amines and ethyl carbamate but also in detoxifying wine from Ochratoxin A and in producing sulphites by yeasts.

3.1 Biogenic Amines Formation

Biogenic amines (BA) are volatile and non-volatile nitrogen low-molecular-weight organic bases of different nature, including aliphatic, aromatic and heterocyclic compounds, that can be found in wines in variable quantities and that are generally known to possess high biological activity. Volatile amines (i.e. methylamine, ethylamine, isoamylamine, dymethylamine, diethylamine etc.) seem to be formed by the amination of non-nitrogen compounds, such as aldehydes and ketones but their origin is not really demonstrated. During the fermentation the concentrations of such amines tend to decrease because of yeast metabolism (Herbert et al. 2005) while they are released during yeast autolysis. However, most of them are related to negative effects on wine taste and flavour rather than on human health (Lehtonen 1996; Torrea and Ancin 2002) although, because of their reaction with the nitrites, some of them can be involved in the formation of nitrosamines that are carcinogenic compounds. Non-volatile amines (histamine, tyramine, putrescine, cadaverine, agmatine) and the two volatile amines phenylethylamine and ethanolamine have their origin in the microbial decarboxylation of their precursor amino acids (Table 11.1). Moreover, the polyamines spermine and spermidine can be produced from putrescine through methylation reactions involving S-adenosyl-methionine. Such biogenic amines can cause different toxicological effects to humans, depending on the specific BA, its concentration and individual sensitivity (Vincenzini et al. 2017).

Table 11.1 Biogenic amines produced by microbial decarboxylation of their precursor amino acids

Aminoacid precursor	Biogenic amines
Histidine	Histamine
Tyrosine	Tyramine
Lysine	Cadaverine
Phenylalanine	Phenylethylamine
Ornithine	Putrescine, spermine, spermidine
Arginine	Agmatine
Serine	Ethanolamine

The heterocyclic (histamine) and aromatic (tyramine, phenylethylamine) amines are the most toxic and are known to cause a wide variety of symptoms including hypotension, headache, heart palpitation, cutaneous and gastrointestinal disorders in the case of histamine and migraines and hypertension crisis in the case of aromatic amines. The aliphatic polyamines (putrescine, agmatine, cadaverine, spermine, spermidine), although not toxic in themselves, potentiate the effects of other amines, especially histamine, inhibiting the monoamino-oxidases (MAO) and diamino-oxidases (DAO) that, catalysing the oxidative deamination of amines, represent the physiological system of detoxification in humans. In this connection, it is worth mentioning that ethanol is known to be among the most active inhibitors of such enzymes increasing the risks of BA effects in wine consumers. Nevertheless, some studies about the toxicological effects of BA occurring in wines are contradictory; some Authors report that the presence of BA in wine could cause intolerance and intoxication symptoms and others that the concentrations of BA in wine are not correlated to the occurrence of symptoms (Marques et al. 2007). Anyway, the toxic doses of BA are difficult to establish as the activity of MAO and DAO enzymes can vary in different individuals. Therefore, the following intervals of BA concentrations in wine that should not be overcome to avoid negative physiological effects are simply recommended: for histamine 8 and 20 mgL^{-1}, for tyramine 25 and 40 mgL^{-1}, for phenylethylamine 3 mgL^{-1} (Soufleros et al. 1998). However, in 2011, the biological hazards (BIOHAZ) panel of the European Food Safety Authority (EFSA 2011) reported that the present knowledge and data on toxicity of BA, individually and in combination(s), are limited and insufficient to conduct quantitative risk assessment; nevertheless, histamine and tyramine are the most toxic and relevant BA for food safety.

Now, the possible role of yeast populations in the formation of BA during alcoholic fermentations appears to be poorly investigated and sometimes results on the various BA produced, are contradictory. However a general statement can be pointed out: yeasts are not the microorganisms most responsible for BA production in wines (Torrea and Ancin 2001; Caruso et al. 2002; Landete et al. 2007) and their BA producing capability has to be accounted in order to minimize the accumulation of these undesirable compounds throughout the winemaking process.

Indeed, some studies performed on different wine yeasts inoculated in sterilized or sulphited musts from various grape varieties under laboratory conditions revealed that they are able to produce BA. In particular, Caruso et al. (2002) assayed different yeast strains of the species most frequently found during alcoholic fermentations (*Saccharomyces cerevisiae*, *Kloeckera apiculata*, *Candida stellata* and *Metschnikowia pulcherrima*) or during wine storage such as *Brettanomyces bruxellensis*. The results demonstrated that the yeast behaviour resulted to be strain dependent within the single species but, on the average, the highest BA concentration was formed by *B. bruxellensis* (15 mg/L), while the other species produced amines at lower levels. Considering the individual amines, independently of the yeast species, histamine, putrescine, cadaverine and methylamine concentrations were lower than 2 mg/L; while agmatine, phenylethylamine and ethanolamine were the most representative, with agmatine produced by all the yeast species tested, phenylethylamine by *K. apiculata*, *M. pulcherrima* and *B. bruxellensis* and ethanolamine by *C. stellata* and *S. cerevisiae*. Particularly, concentrations of phenylethylamine formed by *B. bruxellensis* strains deserve some attention since they reached an average of about 10 mg/L, a considerable value, considering that 3 mg/L of such amine could provoke negative physiological effects. Moreover, some *S. cerevisiae* strains were found to be able to produce significant levels of ethanolamine and agmatine, underlining the importance of using selected strains. Conversely, a low BA producing capability of *S. cerevisiae* was pointed out by other studies, including also commercial active dry yeast strains inoculated in white and red grape musts (Torrea and Ancin 2001, 2002). In these microvinifications, carried out under laboratory conditions, histamine, tyramine, putrescine, phenylethylamine and spermine as well as some volatile amines (i.e. ethylamine and dimethylamine) were found at different extent depending on the yeast strain involved in the fermentation, but they never exceeded the concentrations considered toxic for humans. Anyway, putrescine was always the most abundant amine in the experimental wines likely because its precursor amino acid, the ornithine, can originate from the breakdown of arginine that is one of the major amino acids found in grape juices and mostly metabolised by yeasts during alcoholic fermentation. In this context, it was also reported that, throughout the fermentation, the highest synthesis of putrescine took place after the consumption of the first 25% of sugar according to the fact that arginine is usually considered as a secondary nitrogen source for yeast. Nevertheless, a relationship between the content of non-volatile BA occurring in wine and the utilization of their amino acid precursors by yeasts during fermentation was not always found due to a possible use of BA by yeasts as nitrogen source.

In contrast to above mentioned findings, Landete et al. (2007) reported that no strains of *S. cerevisiae* were able to produce histamine, tyramine, putrescine, cadaverine, and phenylethylamine in both synthetic medium and grape must. On the other hand, analysis of BA content in wines at the end of alcoholic fermentations carried out at industrial level confirmed most of the results obtained under laboratory conditions. Indeed, in several assayed wines, independently of geographical origin and grape variety low amounts of BA were usually detected, the most frequently being phenylethylamine, ethylamine, tyramine and putrescine. Confirming the results of Torrea and Ancin (2001, 2002), a *S. cerevisiae* commercial strain proved capable to produce histamine and tyramine in a barrel Chardonnay fermentation (Medina et al. 2013). On the contrary, no significant level of putrescine, cadaverine and spermidine were found in wines inoculated with *B. bruxellensis* by Vigentini et al. (2008).

All the contradictory data reported above regarding the BA producing capability of wine yeasts could be the consequence of a high strain dependent variability of this metabolic feature (Granchi et al. 2005). It is pointed out that, in a study on the evolution of BA content throughout industrial manufacture of red wines, a decrease of putrescine (usually occurring in fresh grape juices due to the synthesis by vine), consistent with high cell densities of non-*Saccharomyces* yeast populations, was found in both spontaneous and induced alcoholic fermentations (Granchi et al. 2005). Therefore, it could be suggested that some non-*Saccharomyces* yeast species or otherwise different strains of the same species, are able to remove putrescine originating from *Vitis* plant. Concerning this matter, Uemura et al. (2004) found that in *S. cerevisiae* the product of *UGA4* gene, which is a membrane protein catalyzing the transport of 4-aminobutiric acid (GABA), also catalyzes the transport of putrescine. Moreover, they also reported that expression of the *UGA4* gene is enhanced in the presence of GABA and at acidic pH values typical of musts and wines.

In conclusion, although some aspects on the role of yeasts in BA formation in wine deserve to be more elucidated, at the present, it is reasonable to sustain that these microorganisms usually scarcely contribute to BA accumulation in wine. However, since a certain yeast strain dependent variability was demonstrated in BA producing ability, this feature should be included as additional criterion for yeast starter selection in order to contribute to the safeguarding of consumer health. Indeed, even though there are no precise limits, several countries like Canada, Switzerland or South Africa are requiring BAs analysis before placing the wine on the market (Restuccia et al. 2018). Few years ago Switzerland established an official maximum limit of 10 mg/L for histamine in wines, but this limit was removed in 2011 (Restuccia et al. 2018). Today, the recommended upper limits for histamine are 10 mg/L in Australia and Switzerland, 8 mg/L in France, 3.5 mg/L in Netherlands, 6 mg/L in Belgium and 2 mg/L in Germany (Martuscelli et al. 2013; Restuccia et al. 2018). Anyway, in view to safeguard wine consumer's health these rules should be observed. Since, the variability in wine biogenic amine contents, in addition to the microbial metabolic activity, could be due to several factors such as raw material quality, differences in winemaking processes, time and storage conditions, etc., OIV published in 2011 a code of good vitivinicultural practices in order to minimize the presence of BA in wine (OIV 2011). As concern the biogenic amines production by yeasts, recommendations during alcoholic fermentation are the following:

- the addition of ammoniacal nitrogen, inactivated yeasts, yeast cell walls, or yeast auto-lysates should be restricted to minimal concentrations;
- alcoholic fermentation should be carried out using *Saccharomyces* with a low predisposition for the formation of biogenic amines.

Recently, with the aim of reducing the formation of biogenic amines, Benito et al. (2016) proposed the use of *Schizosaccharomyces pombe* to solve the problem of BA formation in wine. Indeed, at present, wine is generally produced using *S. cerevisiae* followed by *Oenococcus oeni*, the lactic acid bacterium often responsible for the formation of biogenic amines in wine, to complete malolactic fermentation. The Authors explored the potential of some *S. pombe* strains as substitute of both microorganisms. These strains demonstrated to be not only as efficient as *S. cerevisiae* winemaking strains, but also able to perform malic acid deacidification without production of biogenic amines.

3.2 Ethyl Carbamate Formation

Besides BA, another health concern in wine is the presence of ethyl carbamate (EC). EC, also referred to as urethane, forms spontaneously in wine because of a chemical reaction of ethanol with a compound containing a carbamyl group. The interest for this compound has attracted due to its animal carcinogen potential and its possible origin from precursors produced by microbial catabolism of amino acids. Hence, wine yeasts are indirectly involved in formation of EC and their contribution, besides to ethanol generation during alcoholic fermentation, consists of producing precursor compounds containing a carbamyl group. These compounds include urea, carbamyl phosphate and citrulline, which can be excreted into the wine at different extent depending on several factors, including yeast strain, amounts of assimilable nitrogen and temperatures occurring during vinification and storage. Consequently, EC levels are usually low or non-detectable in young wines and variable to different extent in aged or stored wines, depending on cellar or storage temperature. For what concern the methods of analysis, Xia et al. (2018) provided a complete description of the traditional detection techniques and newly introduced methods for accurate and convenient determination of EC in fermented food matrices.

The origin and the importance of precursor compounds in EC production in wine is here discussed.

3.2.1 Origin of Urea

Urea is formed by yeast as an intermediate product during the degradation of arginine, purines and pyrimidines. However, due to their relatively low levels in grape juices purine and pyrimidine are not usually utilized by yeasts as nitrogen sources during vinification and thus scarcely contribute to urea amount. Consequently, arginine, being one of the major amino acids of grape must and an important nitrogen source for yeasts, is the most significant source of urea in wine (Ough et al. 1988). Nevertheless, the amount of urea in wine depends strictly on the complex regulation of arginine metabolism, which has been well defined in *Saccharomyces cerevisiae* yeast. According to this pathway arginine is transported into yeast cell through the general amino acid permease (GAPp) or the arginine specific permease (CAN1p) and then it is cleaved by the enzyme arginase, which is encoded by the *CAR*1 gene, into ornithine and urea. Ornithine is metabolized via proline and glutamate pathway (for synthesis of amino acids), while urea can be further catabolized to ammonia and CO_2 in an energy-dependent two-step reaction catalyzed by urea amidolyase (the product of the *DUR*1, 2 gene) or, otherwise, it can be excreted from the yeast cell likely via a facilitated transporter (Dur4p). Moreover, urea may be reabsorbed by yeast cell for additional nitrogen requirement through an active transporter (Dur3p). During wine fermentation, different steps of the arginine degradation pathway are subject to regulation essentially depending on availability of assimilable nitrogen sources. Indeed, in the presence of high concentrations of readily used nitrogen sources, such as ammonium and glutamine, transport and catabolism of arginine are delayed and urea is not degraded because of the repression of genes encoding GAPp, arginase (*CAR*1) and urea amidolyase (*DUR*1,2) (according to the phenomenon called Nitrogen Catabolite Repression). In this case, that is usually during the first stages of alcoholic fermentation, urea-producing yeast strains excrete urea into the wine. On the contrary, when low levels of ammonium or poorly used nitrogen sources are available in fermentation medium, arginine metabolism generally proceeds without or with little urea excretion and possible urea secreted is reabsorbed through the active transporter Dur3p and it is degraded to ammonia, which can be used for biosynthesis of cell nitrogen compounds. However, it has been demonstrated that wine yeast strains differ in their ability to rapidly catabolize arginine and urea during fermentation in relation to different expression of the genes *CAR*1 and *DUR*1,2 (encoding arginase and urea amidolyase, respectively) as they are regulated by distinctive mechanisms. High urea producing yeasts could hence possess a high capacity to degrade arginine to urea but a low urea metabolizing ability owing to a lower level of *DUR*1,2 expression in comparison to the expression of the *CAR*1 gene. Moreover, the timing and degree of aeration seem to affect urea production during alcoholic fermentation likely because of an increased nitrogen demand for anabolic reactions that are stimulated by oxygen. (Henschke and Jiranek 1993).

In conclusion, the amount of urea in wine at the end of the fermentation is affected by urea excretion and urea reabsorption by yeast cells whose metabolic activities differentiate depending on genetic as well as environmental factors. Furthermore, urea may be released in wine during the storage because of yeast autolysis. Since the chemical reaction between urea and ethanol is exponentially accelerated at elevated temperatures in acidic conditions, it is essential that wine be not exposed to elevated temperatures of storage (above 37 °C).

3.2.2 Origin of Carbamyl Phosphate and Citrulline

Carbamyl phosphate as well as citrulline are intermediate products in the synthesis of arginine in yeast cells. Particularly, in *S. cerevisiae* carbamyl phosphate is produced from ATP, CO_2 and ammonium, resulting from glutamine, by a carbamyl-phosphate synthetase in a reaction which is enhanced when ammonia levels are high (Ingledew et al. 1987). If ornithine is available, carbamyl phoshate may be converted into citrulline through the ornithine transcarbamylase activity. Then, citrulline can be excreted or transformed via argininosuccinate to arginine, which is catabolized by arginase pathway. Generally, carbamyl phosphate and citrulline occurr in wine at low concentrations so that they may account only for a small fraction of EC. Moreover, on a molar basis, citrulline has an ethyl carbamate formation rate of approximately one quarter of that of urea.

3.3 Ethyl Carbamate Reduction

Although a legal limit for EC levels in wine is recognized only in Canada, the potential reduction of EC content is of general concern in wine industry. In this context, in 1997, the U.S. Food and Drug Administration (FDA), in collaboration with the Department of Viticulture and Enology at the University of the California (Davis), published an EC preventative action manual containing several advisory recommendations, especially to minimise practices that can affect the urea levels in wine. On the contrary, no restriction or limits now exist in Europe. However, European food and health regulators, such as the European Food Safety Authority in 2006, have launched a study on the risks to human health related to the presence of Ethyl Carbamate in foods and beverages.

Concerning wine yeasts, since as above mentioned they display a certain variability in the production of urea, the inoculation of selected commercial low urea-producing yeast strains or, otherwise, the control of arginine content in spontaneous alcoholic fermentations are recommended in order to reduce EC formation in wine.

More recently, improved urea-degrading *S. cerevisiae* strains developed by recombinant DNA techniques have been proposed. In particular, a metabolically engineered *S. cerevisiae* (the UC Davis 522^{EC} strain) able to continually break down urea produced during grape must fermentation, even in presence of a good nitrogen source, has been obtained owing to the constitutive expression of *DUR*1,2, the gene encoding the enzyme urea amidolyase which catalyze the degradation of urea to ammonia and CO_2 (Coulon et al. 2006). This result was achieved by introducing an additional copy of the *DUR*1,2 gene under the control of *PGK1* promoter and terminator sequences for high expression. Since the *S. cerevisiae* 522^{EC} strain (whose trade name is ECMo01) does not contain any antibiotic resistance marker genes or genetic material from any other organism it is not transgenic. Moreover, it is genetically stable and exhibits oenological properties corresponding to those of the parental strain. Analyses of EC content in commercial trials performed by ECMo01 yeast revealed EC reductions ranging from 76 to 92% in comparison with the control yeast. Therefore, considering also that the FDA proved that there are no safety concerns associated with the use of the ECMo01 strain in the production of fermented alcoholic beverages, metabolically engineered urea-degrading wine yeasts could be a possible alternative to minimizing production of EC in wines.

3.4 Detoxifycation of Ochratoxin A

Mycotoxins are secondary metabolites that are produced by some mould species. Contamination of food and beverages by mycotoxins is a serious and recurring problem worldwide, which can result in economic losses and health concerns. The best-characterized mycotoxins are mainly produced by members of three fungal genera: *Aspergillus* (e.g., aflatoxin, ochratoxin A), *Fusarium* (e.g., T-2 toxin, deoxynivalenol), and *Penicillium* (e.g., ochratoxin A, patulin) (Moss 2002). The fumonisin aflatoxins are found in grapes and wine, and while patulin is seldom detected, *Alternaria* mycotoxins (e.g., alternariol) are also frequently found. Therefore, as well as their individual considerations, the combined effects of mycotoxins in wine also require consideration. Among these mycotoxins, ochratoxin A (OTA) is the main one found in wine. Ochratoxin A was originally detected in a culture of *Aspergillus ochraceus* and then was identified in numerous mycotoxinogenic species of *Aspergillus* and *Penicillium*. The OTA molecule is a pentaketide that consists of a dihydroisocumarine linked to an L-ß-phenylalanine that contains two ionisable functional groups: a carboxylic group (-COOH) in the phenylalanine, and a hydroxyphenolic group (Fig. 11.3). OTA is a white or colourless crystalline solid that has weak acidic properties, is soluble in aqueous sodium hydrogen carbonate, and emits blue fluorescence under ultraviolet light irradiation (maximum emission at 467 nm in 97% ethanol). When recrystallized from benzene and xylene, OTA shows melting points of 90 °C and 171 °C, respectively. The most relevant metabolites related to OTA are the dechloro analog of OTA (ochratoxin B), its ethyl ester (ochratoxin C), and the isocoumaric derivatives of OTA (ochratoxin α and its dechloro analog ochratoxin β). Generally, OTA is the most relevant in vineyards, and its occurrence in grapes and wine has been reviewed (Hocking et al. 2007; Luo et al. 2018).

OTA is considered cytotoxic, nephrotoxic, immunotoxic, myelotoxic, teratogenic, carcinogenic, genotoxic, and mutagenic (Pfohl-Leszkowicz and Manderville 2007). OTA has been detected in a variety of foods and feed, including grapes, grape juice, beer, and wine. In humans, it is frequently cited as a possible causative agent of Balkan endemic nephropathy, a syndrome that is characterized by contracted kidneys with tubular degeneration, interstitial fibrosis, and hyalinization of the glomeruli. This endemic disease was described about 30 years ago in populations in isolated villages in the Balcans (i.e., Bulgaria, Croatia). Moreover, OTA has been implicated in testicular cancer (Schwartz 2002). For these reasons, the International Agency for Research on Cancer has classified OTA as a possible human renal carcinogen (group 2B) (IARC 2002). Children of 4–6 years old represented the highest risk group, as their daily intake was 16-fold the average intake of adults. Indeed, grape juice can contain more OTA than some table wines, and this is a serious problem: fruit juices appear to increase the daily OTA intake of children (Zimmerli and Dick 1996).

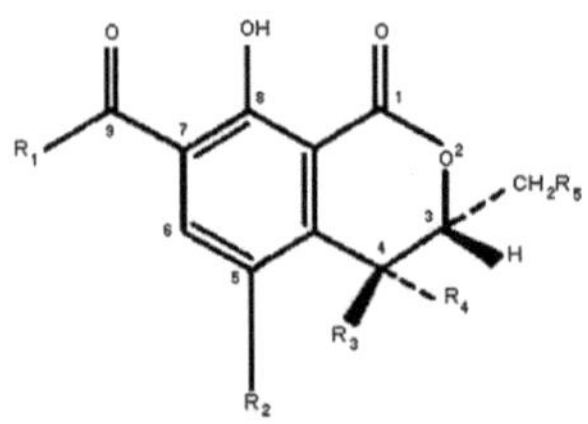

Name	R1	R2	R3	R4	R5
Ochratoxin A	phenylalanine	Cl	H	H	H
Ochratoxin B	phenylalanine	H	H	H	H
Ochratoxin C	phenylalanine ethyl ester	Cl	H	H	H
Ochratoxin A methyl ester	phenylalanine methyl ester	Cl	H	H	H
Ochratoxin B methyl ester	phenylalanine methyl ester	H	H	H	H
Ochratoxin B ethyl ester	phenylalanine ethyl ester	H	H	H	H
Ochratoxin alpha	OH	Cl	H	H	H
Ochratoxin ß	OH	H	H	H	H
4R-Hydroxyochratoxin A	phenylalanine	Cl	H	OH	H
4S-Hydroxyochratoxin A	phenylalanine	Cl	OH	H	H
10-Hydroxyochratoxin A	phenylalanine	Cl	H	H	OH

Fig. 11.3 Chemical structures of Ochratoxin A and related metabolites

Elimination and detoxification of OTA in humans is very slow, where it shows the longest half-life known for living mammals (840 h, or 35 days) (Petzinger and Weidenbach 2002) (Fig. 11.4). A preliminary survey by the Food and Agriculture Organization of the United Nations (FAO) and the World Health Organization (WHO) Commission for the "Codex Alimentarius" that was based on information coming from various countries of the European Union, indicated that red wine is the second highest source of human exposure to OTA, after cereals, and before coffee and beer. For this reason, European regulations (Reg. CE N° 123/2005 of 26 January, 2005) have established the maximum permissible levels of OTA in grapes and grape-derived products: 10.0 μg/kg (ppb) in dried grapes (i.e., Corinto dried grapes, dried grapes in general, as raisins and sultanas); and 2.0 μg/kg (ppb) in wines (red, white, rosé; excluded fortified wines and wine with alcohol levels higher than 15% by volume). The same OTA levels (2.0 μg/kg) must not be exceeded in other drinks made from grape must, and in grape juice, grape must itself, and grape must concentrates destined directly for human consumption. Therefore, the Joint FAO/WHO Expert Committee on Food Additives has established the provisional maximum weekly intake of 120 ng OTA/kg body weight.

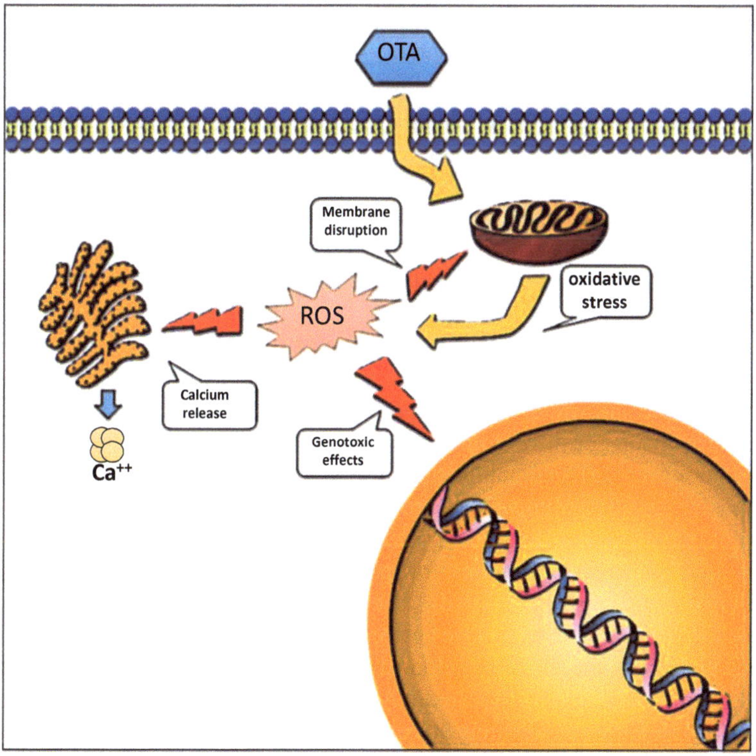

Fig.11. 4 OTA toxicity on humans

Several studies have indicated that OTA levels in grapes and wines are very variable and can be influenced by different factors. OTA levels in wines from southern wine-growing regions of Europe and northern Africa are higher than those in wines from the northern areas of Europe. Climate and high temperatures in particular are the major factors for OTA contamination. Therefore, OTA is found at higher levels in wines from the warmer southern Europe, compared to northern Europe (Otteneder and Majerus 2000). In the Mediterranean basin, the incidence of OTA in wine is very high (>50% of wines analyzed) (El Khoury and Atoui 2010). OTA contamination of grapes and wines also varies according to the latitude of the vineyard and from year to year in the same vineyard, depending on the climate. In addition, water activity influences the physiology of the fungal species that produce OTA. In colder regions, OTA is produced by *Penicillium verrucosus* and *Penicillium nordicum*, while in temperate regions it is mainly produced by *Aspergillus carbonarius* (optimum temperature, 32–35 °C). Finally, in tropical and semitropical regions, OTA is more often produced by *Aspergillus niger* (optimum temperature, 35–37 °C).

The occurrence of OTA-derived products on grapes is due to the frequent presence of *Aspergillus* spp. and *Penicillium* spp. in vineyards, particularly during the maturation phase, from veraison onwards. In particular, black Aspergilli are the causal agents of several plant diseases. They are considered as opportunistic pathogens of grapes, and can cause bunch rot (i.e., sour rot) or berry rot, and raisin mould. Among the black Aspergilli, *A. carbonarius* is the most important, as OTA-producing isolates represent 41–100% of isolates examined (Battilani et al. 2003). Thus, *A. carbonarius* and some members of the *A. niger* aggregate are considered to be the main sources of OTA contamination in grapes and wines.

After its first detection in wine (Zimmerli and Dick 1996), records of OTA in grapes, grape juice, must, and wine were reported with increasingly alarming frequencies. This aroused the awareness that consumption of food contaminated with OTA represents a risk that should not be underestimated, in particular for consumers of red wines, dessert wines, and dried grapes.

The occurrence of OTA depends on the winemaking technology used, and it is higher in red and sweet wines than in rosé and white wines. In particular, the maximum OTA levels detected in wines have been reported as 15.6, 6.32, and 8.86 mg/kg for red, rosé and white wines, respectively. Here, inhibition of OTA-producing fungal growth on sensitive commodities appears to be by far the most reliable method to prevent OTA contamination of food and feed.

The adoption of preventive measures in the vineyard can help to reduce contamination of mycotoxigenic fungi on grapes. However, severe OTA contamination still occurs in certain high-risk regions, for specific vintages, and on must obtained from susceptible grape varieties. Thus, different technological and biological methods to remove OTA in must and wine have been evaluated by the International Organization of Vine and Wine that, in 2005, adopted the "Code of sound vitivinicultural practices to minimise levels of Ochratoxin A in vine-based products" (Resolution Viti-Oeno 1/2005). The basic requirements of these methods should be their effectiveness, simplicity of use, low cost, and absence of negative effects on wine quality.

The use of SO_2, fining agents, and adsorbent materials can control the growth of OTA-producing moulds and the resultant mycotoxin contamination. However, consumer trends are directed toward limited use of chemical products in wines. Hence, there is growing interest to use biological strategies to detoxify OTA, which include inhibition of *Aspergillus* spp. growth, degradation of OTA, and adsorption of OTA onto yeast cell surface (Fig. 11.5).

Cubaiu et al. (2012) showed that *Saccharomyces cerevisiae* wine yeasts can prevent *A. carbonarius* growth, and that yeast culture filtrates inhibit OTA production and polyketide synthase transcription in *A. carbonarius* and *A. ochraceus*. They concluded that the inhibition mechanism observed is likely to involve synergistic mechanisms, which include effects mediated at the level of gene transcription.

Microbiological degradation of OTA can occur through hydrolysis of the amine bond, to release L-ß-phenylalanine and Ochratoxin α, or through hydrolysis of the lactone ring, with an opened lactone as the final degradation product. In addition, Angioni et al. (2007) suggested another possible pathway for degradation of OTA that is different from the previous mechanisms. Indeed, they reported that some yeasts can degrade OTA, but they could not find any products of this hydrolysis. Biological degradation of OTA has also been observed for some bacteria (e.g., *Streptococcus*, *Bifidobacterium*, *Bacillus*) and for living and heat-treated dead conidia of black *Aspergillus* isolates (e.g., *A. carbonarius*, *A. niger*, *Aspergillus japonicus*) (Petruzzi et al. 2014). In addition, it has been reported that microbial-derived enzymes with carboxypeptidase A activity can degrade OTA (Amézqueta et al. 2009). Finally, considering that the opened lactone OTA form that can result from OTA hydrolysis has shown toxicity toward rats, mice, and *Bacillus brevis*, and that undesirable effects on non-targeted microorganisms might result in reduced wine quality, biological degradation of OTA is less desirable than its adsorption.

Ochratoxin A adsorption onto yeast cell surfaces is the consequence of the chemical nature of OTA, which is partially dissociated at the pH of wine, and carries

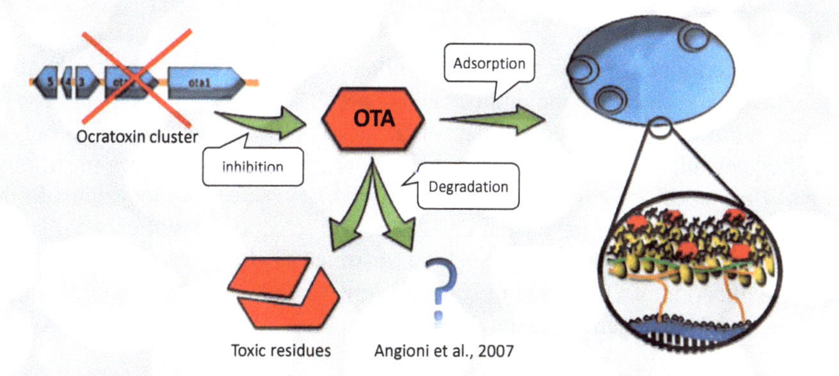

Fig. 11.5 Biocontrol of OTA by yeasts

a positive charge on the amine group. Moreover, OTA can bind to hydrophobic surfaces through the phenol group. At the same time, yeast biomass can be regarded as a good source of adsorbent material. Indeed, the yeast cell wall is made up of beta D-glucans and mannoproteins. Due to the presence of mannosylphosphate, at the pH of wine, mannoproteins carry negative charges and can establish both polar and nonpolar interactions with OTA. Further, mannoproteins and cell wall components are released during spontaneous or induced cell autolysis, which thus increases their adsorption capacity. Heat-treated and acid-treated cells bind higher levels of OTA than viable yeast cells. In addition, yeast 'hulls' have been used to significantly reduce OTA in wine. Piotrowska et al. (2013) used thermally inactivated yeast biomass and achieved 95% OTA reduction. Garcia Moruno et al. (2005) evaluated the reduction of ochratoxin A in red wines by using active dry yeast and yeast lees obtained from alcoholic fermentation. The results pointed out that the effectiveness of the treatment depends upon the quantity of the active dry yeast used, while in the case of yeast lees, the greatest reduction was found in those obtained from white wine. Also Petruzzi et al. (2015) reported that yeast cell walls have higher adsorption capacity than thermally inactivated cells (50% vs. 43% OTA reduction).

Some variables have major roles in the efficiency of OTA adsorption and removal, including the protocol used for production of yeast derivatives, the must composition, and finally, the yeast strain. Heating yeast cell walls leads to protein denaturation and the production of Maillard products. Acid treatments might cause the release of monomers and aldehydes from cell-wall polysaccharides, thus increasing the possible adsorption sites (Piotrowska et al. 2013). Petruzzi and co-workers (2014) showed that the sugar concentration in the must, fermentation temperature, and diammonium phosphate supplementation all have major roles in OTA adsorption. They thus suggested that all of these factors allow better release of the mannoproteins in the must. Further compounds in wine might also compete with OTA for adsorption onto the yeast cell wall. Some of these, such as medium-chain and short-chain fatty acids and sulphur compounds, are considered detrimental to winemaking process and wine quality. Thus, the adsorption and removal of these molecules by yeast cells might be considered of great interest.

On the contrary, the reduction of some other compounds might negatively affect wine quality, such as for the anthocyanins, which are largely responsible for the colour of red wine. However, Petruzzi et al. (2015) showed that adsorption of OTA and anthocyanins occurs through two different phenomena that act in different ways. Thus, specific yeast strains might adsorb OTA but not anthocyanins. In this respect, various studies have been devoted to the selection of starter yeast strains that can adsorb OTA from wine. Such yeast strains might thus have a great future for detoxification of OTA without using harmful chemicals and without any negative effects on the organoleptic quality of wine.

3.5 Effect on Sulphite Content

The application of sulphur dioxide (sulphite or 'SO_2') to treat must and wine is one of the most important oenological additives affecting character and quality and microbial stability of wine. The use of sulphur dioxide has always been regulated and oenological methods have always been developed to lower its concentrations in wines, because its addition can raise health-related objections (Romano and Suzzi 1993; Ribéreau-Gayon et al. 2006; Fleet 2007). Health concerns and an unfavourable public perception of sulphite have led to demands for restriction of its use in all wine growing countries (Costanigro et al. 2014). Furthermore, a reassessment of all aspects of sulphite accumulation in wine is conducted and more and more alternative treatments are observed and tested to diminish the application of SO_2 in recent years, in particular in organic wine production (Rauhut and Micheloni 2010). For this reason, the formation of sulphur-containing compounds by wine yeast itself has become a central and crucial point of research during the last decades (Divol et al. 2012; Guerrero and Cantos-Villar 2015).

The occurrence of sulphur dioxide in food and wine can cause health-related problems. At the beginning of the nineteenth century research on the possible toxicity of sulphur dioxide has been started. An intermediate product of the degradation of S-containing amino acids is sulphite, which is oxidized to sulphate. Adults that react normal in the metabolism of S-containing compounds, daily excrete about 2.400 mg sulphate (25 mmol) with the urine or faeces. Sulphite is produced at a concentration of about 1000 mg per day during the metabolism of S-containing amino acids and inorganic sulphate in humans (Bakalinsky 1996). Consumption of food or wines treated with sulphites is usually not a problem except for a few persons who are deficient in the natural enzyme (sulphite oxidase) to break it down. A number of studies showed reactions by sensitive humans consuming food, wine or other beverages with sulphites. The lethal dose for 50% of individuals (LD50) is between 0.7 and 2.5 g of SO_2 per kilogram of body weight. Studies on chronic toxicity in animals showed that the following three main complications could occur: thiamine insufficiency through its decomposition by sulphur dioxide, histopathological change of the stomach and reduced growth (Til et al. 1972). Research investigations in humans demonstrated that SO_2 could lead to the following intoxication symptoms such as nausea, vomiting and gastric irritation at significantly high-absorbed concentrations (4 g of sodium sulphite in a single concentration). No secondary effects were noticed with an amount of 400 mg of SO_2 during 25 days. Its possible toxicity in humans was often linked to the destruction of thiamine (vitamin B1) by sulphites, but this reaction is very limited at a pH of around 2, which is similar to stomach pH (Ribéreau-Gayon et al. 2006).

It could be shown that SO_2 can cause allergic reactions at the beginning of the seventies. Allergic reactions occur at very low ingested concentrations (about 1 mg) and mainly concern asthmatics (4–10% of the general population). Therefore, the US FDA (Food and Drug Administration) decided to declare the presence of sulphites on wine labels ('contains sulfites') in the United States when the concentration exceeds 10 mg/L, since 2005 it has also to be indicated in the EU. In Australia and New Zealand wine containing SO_2 would be labelled as 'contains preservative 220' or 'sulphur dioxide added' (Ribéreau-Gayon et al. 2006; Guerrero and Cantos-Villar 2015).

The World Health Organization (WHO) set the RDA (recommended daily allowance) at 0.7 mg of SO_2/kg of body weight. The acceptable concentration for an individual is 49 mg per day (at a body weight of 70 kg). Therefore the consumption of half a bottle of wine per day (375 ml) can supply an amount of SO_2 higher than the RDA, if the total SO_2 concentration is at the maximum limit allowed by the EU (e. g. 150 mg/L for red dry wines and 200 mg/L for white dry wines). Although the average amounts of SO_2 in wine is much lower in white and red wines, national and international health authorities recommend and demand for additional decreases in the legal limits of SO_2 (Ribéreau-Gayon et al. 2006).

Often people seem to associate their problems with the occurrence of headaches after the consumption of wines with higher sulphite levels, but more extensive medical studies are still necessary to clarify the cause of these experiences (Bakalinsky 1996; Costanigro et al. 2014).

Sulphur is essential for yeast growth. *S. cerevisiae* can use sulphate, sulphite and elemental sulphur as source of sulphur to synthesize sulphur-containing amino acids. Yeasts are also able to produce and to liberate sulphite. This can lead under usual wine making conditions and with *S. cerevisiae* strains that have a normal ability for its production to levels about 10–20 mg/L 'bound' sulphite in wine and in certain cases 30 mg/L can be exceeded. Therefore, it is nearly impossible to produce wines without any sulphite, even if no sulphite was added.

S. cerevisiae can synthesize sulphur-containing amino acids (methionine and cysteine) from an inorganic sulphur source as sulphate (SO_4^{2-}), SO_2 or elemental sulphur (applied as fungicide in form of wettable sulphur). SO_2 and SO_4^{2-} will be sequentially reduced to sulphide (S^{2-}) through the sulphate reduction sequence (SRS) pathway. The first step of the SRS pathway involves the transport extracellular sulphate from the medium into the yeast cell through two specific sulphate permeases in co-transport with 3H+ (Belda et al. 2017; Rauhut 2017). Then, sulphate is activated by an adenylation and reduced by two successive reactions which require four NADPH + H+ molecules and two ATP molecules. The adenylation of sulphate decreases the electropotential of sulphate that the following reduction into sulphite and sulphide through NADPH + H^+ can take place.

Sulphate activation occurs by two reactions. First, the adenosyl-phosphoryl-moiety of ATP is transferred to sulphate and 5′-adenylylsulphate (APS) is formed, which is catalyzed through ATP sulphurylase. Then APS is phosphorylated and 3′-phospho-5′-adenylylsulphate (PAPS) is produced through APS kinase. In the next step, PAPS is reduced through PAPS reductase to sulphite. The reduction from sulphite to sulphide occurs through the activity of sulphite (S^{2-}) reductase. The reduced sulphur atom can be incorporated into carbon chains to produce methionine and cysteine.

O-acetylhomoserine (OAS) is the amino acid precursor to form methionine with sulphide via homocysteine, whereas cysteine is formed through the trans-sulphuration pathway by the reaction of homocysteine and serine via the further intermediate cystathionine in yeasts (Thomas and Surdin-Kerjan 1997). The antioxidant glutathione is then formed on the basis of cysteine with a reaction of glutamate and glycine (Thomas and Surdin-Kerjan 1997; Rauhut 2017).

Intermediates of the sulphate reduction pathway can be released out of the yeast cell and contribute to flavour and conservation of wine. Increased levels of sulphite can stabilize flavour by forming adducts with acetaldehyde. On the other hand, also a binding to further metabolites, e. g. pyruvate and 2-ketoglutarate can take place, which can increase the bound SO_2-concentrations to inadequate amounts. Sulphide can be used to form the required sulphur-containing amino acids or it can be released out of the cell as hydrogen sulphide (H_2S) if the necessary precursors are limited. H_2S has an odour reminiscent of 'rotton eggs' and a very low odour threshold in wine (about 10 μg/L). H_2S is the basis for various other flavour active compounds that also contribute to reduced sulphur off-flavours in wine (Belda et al. 2017; Rauhut 2003, 2017). Liberation of sulphide is one of the main problems that affect wine fermentation.

The formation of sulphite by *S. cerevisiae* is strain-dependent and is influenced by the grape must composition and fermentation conditions. Sulphite is usually produced by *S. cerevisiae* in concentrations of 10–30 mg/L. More than 100 mg/L sulphite can be achieved through high SO_2-producing strains. A comprehensive research study of Suzzi et al. (1985) verified that the majority of *S. cerevisiae* strains produced less than 10 mg/L SO_2. Considerable differences in the regulation of sulphur metabolism and the activity of sulphate permease, ATP-sulphurylase and sulphite reductase were noticed in high and low sulphite-forming yeast strains. These effects and findings were summarized in reviews of Pretorius (2000) and Rauhut (2017).

Excessive sulphite production can be the result of defects in sulphate uptake and reduction. These processes are normally regulated by certain metabolites, like methionyl-t-RNA and S-adenosylmethionine. For example, methionine is not repressing the sulphate-permease in high-sulphite-producing strains. High and low-sulphite-producing yeasts have no regulation of ATP-sulphurylase through S-containing metabolites. Low-sulphite producing yeasts develop a higher formation of NADPH-dependent sulphite reductase at exponential growth as high-sulphite producers (Henschke 1997). As mentioned above, sulphite production needs a lot of energy; therefore, the cellular metabolism of high SO_2-forming yeast strains is diminished, explaining a lower biomass and slow fermentation rate. Sulphite reductase can be reduced at higher concentrations of methionine and cysteine (Pretorius 2000; Rauhut 2017). Several studies demonstrated that sulphite formation by yeasts is also affected from nutrient composition and sulphate concentration in the must, its clarification, initial pH-value, temperature and other oenological conditions (Pretorius 2000; Ugliano and Henschke 2009; Rauhut 2017).

During alcoholic fermentation, bound SO_2 is mainly represented by the reaction of SO_2 with acetaldehyde, pyruvate and 2-ketoglutarate. The binding of sulphites is in addition increased by high glucose levels in musts of sweet wines, by molecules like keto-5-fructose, keto-2-gluconic acid, diketo-2,5-gluconic acid, which occur in healthy, ripe grapes. These substances are detected in large concentrations in grapes that are infected by *Botrytis cinerea* and acetic acid bacteria. Furthermore, elevated temperatures and pH-values in combination with aeration can favour the formation of these SO_2-binding compounds. On the opposite, numerous studies demonstrate that thiamine (vitamin B1) can reduce the production of pyruvate and 2-ketoglutarate and thus the need for SO_2 to stabilize the wine (Ribéreau-Gayon et al. 2006). Indeed, this vitamin is an essential part of the coenzyme of the pyruvate and 2-ketoglutarate decarboxylase, and, for this reason, its addition is allowed in concentrations up to 0.6 mg/L to grape must according to EU-regulations.

Consequently, important desirable characteristics of *Saccharomyces* and non-*Saccharomyces* wine yeasts are low sulphite formation and low production of sulphite binding compounds (like acetaldehyde, pyruvate and 2-ketoglutarate) (Comitini et al. 2017; Padilla et al. 2016).

The addition of sulphites will be also needed in future to reduce the risks of microbial spoilage of the wine during the production process and after bottling. No other preservative is available, which combines different properties to ensure that the product reaches the consumer in the best possible condition and quality. Health concerns and an unfavourable public perception of sulphite have led to demands for restriction of its use in all wine growing countries of the world. Due to extensive research studies several, physical, chemical and biological options are available for winemakers to minimize, but not to eliminate the application of sulphites (Guerrero and Cantos-Villar 2015). There are only a few treatments or additives, which are able to compensate one of the properties of SO_2 to a certain extent. Furthermore, it has to be taken into account that very low additions of SO_2 and the partial use of alternatives will probably change the ageing potential of wine and that also other wine characteristics and styles can be expected, in particular for white wines. No addition of SO_2 will be only possible for a few wines, which need specific oenological treatments to be produced and stabilized. Parameters, which are often investigated in SO_2 replacement research studies, are reviewed by Guerrero and Cantos-Villar (2015) and Santos et al. (2012).

An important tool to minimize the use of SO_2 and to lower its natural production is the selection of an adequate yeast strain with a low ability to produce SO_2 and with a low formation of SO_2-binding compounds. Furthermore, nutrient deficiencies in the grape musts have to be compensated by the addition of nutrient supplements, especially thiamine (vitamin B1), is required to diminish the production of SO_2-binding compounds like pyruvate and 2-ketoglutarate.

Climate change will lead to higher levels of sugar and increased pH-values in grapes and musts from certain vintages and wine-growing regions and consequently an increase in the need for sulphites will occur. More research is necessary in optimizing the conditions and perfecting the methods to apply sulphites during the wine making process. Furthermore, research should be focused more intensively on substitute products and on the selection of yeasts strains, the application of mixed yeast cultures (*Saccharomyces* and non-*Saccharomyces*) with suitable interactions that are better adapted to musts with specific requirements to minimize SO_2 levels in wine. In this connection, an evolution-based strategy was recently designed to screen novel yeast strains impaired in sulphate assimilation (De Vero et al. 2011). This strategy, that combines the sexual recombination of spores and application of a specific selective pressure, provides a rapid screening method to generate genetic variants and select improved wine yeast strains with an impaired metabolism regarding the production of sulphites and H_2S. Indeed, three strains with a low sulphite production ($SO_2 < 10$ mg/L) and with an impaired H_2S production in grape must without added sulphites were selected.

References

Alvarez, M. A., & Moreno-Arribas, M. V. (2014). The problem of biogenic amines in fermented foods and the use of potential biogenic amine-degrading microorganisms as a solution. *Trends in Food Science and Technology, 39*, 146–155.

Amézqueta, S., González-Peñas, E., Murillo-Arbizu, M., & López de Cerain, A. (2009). Ochratoxin A decontamination: A review. *Food Control, 20*, 326–333.

Anderson, M. E. (1998). Glutathione: An overview of biosynthesis and modulation. *Chemico-Biological Interactions, 111*, 1–14.

Angioni, A., Caboni, P., Garau, A., Farris, A., Orro, D., Budroni, M., & Cabras, P. (2007). In vitro interaction between ochratoxin A and different strains of *Saccharomyces cerevisiae* and *Kloeckera apiculata*. *Journal of Agricultural and Food Chemistry, 55*(5), 2043–2048.

Badea, G. A., & Antoce, A. O. (2015). Glutathione as a possible replacement of sulfur dioxide in winemaking technologies: A review. *Scientific Papers., Series B, Horticulture, LIX*, 123–139.

Bakalinsky, A. T. (1996). Sulfites, wine and health. In A. L. Waterhouse & R. M. Rantz (Eds.), *Wine in context: Nutrition, physiology, policy* (pp. 35–42). Davis: American Society for Enology and Viticulture.

Bassig, B. A., Lan, Q., Rothman, N., Zhang, Y., & Zheng, T. (2012). Current understanding of lifestyle and environmental factors and risk of non-hodgkin lymphoma: An epidemiological update. *Journal of Cancer Epidemiology*, 978930: 1–27.

Battilani, P., Giorni, P., & Pietri, A. (2003). Epidemiology of toxin-producing fungi and ochratoxin A occurrence in grape. *European Journal of Plant Pathology, 109*(7), 715–722.

Belda, I., Ruiz, J., Esteban-Fernández, A., Navascués, E., Marquina, D., Santos, A., & Moreno-Arribas, M. V. (2017). Microbial contribution to wine aroma and its intended use for wine quality improvement. *Molecules, 22*(189), 29.

Benito, Á., Jeffares, D., Palomero, F., Calderón, F., Bai, F.-Y., Bähler, J., & Benito, S. (2016). Selected *Schizosaccharomyces pombe* strains have characteristics that are beneficial for winemaking. *PLoS One, 11*, e0151102.

Bisquert, R., Muñiz-Calvo, S., & Guillamón, M. (2018). Protective role of intracellular melatonin against oxidative stress and UV radiation in *Saccharomyces cerevisiae*. *Frontiers in Microbiology, 9*, 318.

Brandolini, V., Fiore, C., Maietti, A., Tedeschi, P., & Romano, P. (2007). Influence of *Saccharomyces cerevisiae* strains on wine total antioxidant capacity evaluated by photochemiluminescence. *World Journal of Microbiology and Biotechnology, 23*, 581–586.

Brandolini, V., Maietti, A., Tedeschi, P., Capece, A., & Romano, P. (2011). Influence of *Saccharomyces cerevisiae* wine strains on total antioxidant capacity. *Annales de Microbiologie, 61*, 125–130.

Caruso, M., Fiore, C., Contursi, M., Salzano, G., Paparella, A., & Romano, P. (2002). Formation of biogenic amines as criteria for the selection of wine yeasts. *World Journal of Microbiology and Biotechnology, 18*, 159–163.

Cheynier, V., Souquet, J. M., & Moutounet, M. (1989). Glutathione content and glutathione to hydroxycinnamic acid ratio in *Vitis vinifera* grapes and musts. *American Journal of Enology and Viticulture, 40*, 320–324.

Comitini, F., Capece, A., Ciani, M., & Romano, P. (2017). New insights on the use of wine yeasts. *Current Opinion in Food Science, 13*, 44–49.

Costanigro, M., Appleby, C., & Menke, S. D. (2014). The wine headache: Consumer perceptions of sulfites and willingness to pay for non-sulfited wines. *Food Quality and Preference, 31*, 81–89.

Coulon, J., Husnik, J. I., Inglis, D. L., van der Merwe, G. K., Lonvaud, A., Erasmus, D. J., & van Vuuren, H. J. J. (2006). Metabolic engineering of *Saccharomyces cerevisiae* to minimize the production of ethyl carbamate in wine. *American Journal of Enology and Viticulture, 57*, 113–124.

Cubaiu, L., Abbas, H., Dobson, A. D., Budroni, M., & Migheli, Q. (2012). A *Saccharomyces cerevisiae* wine strain inhibits growth and decreases Ochratoxin A biosynthesis by *Aspergillus carbonarius* and *Aspergillus ochraceus*. *Toxins, 4*(12), 1468–1481.

Datta, S., Timson, D. J., & Annapurea, U. S. (2017). Antioxidant properties and global metabolite screening of the probiotic yeast *Saccharomyces cerevisiae* var. *boulardii*. *Journal of the Science of Food and Agriculture, 97*, 3039–3049.

De Vero, L., Solieri, L., & Giudici, P. (2011). Evolution-based strategy to generate non-genetically modified organisms *Saccharomyces cerevisiae* strains impaired in sulfate assimilation pathway. *Letters in Applied Microbiology, 53*(5), 572–575.

De Vero, L., Bonciani, T., Verspohl, A., Mezzetti, F., & Giudici, P. (2017). High-glutathione producing yeasts obtained by genetic improvement strategies: A focus on adaptive evolution approaches for novel wine strains. *AIMS Microbiology, 3*, 155–170.

Divol, B., du Toit, D., & Duckitt, E. (2012). Surviving in the presence of sulphur dioxide: Strategies developed by wine yeasts. *Applied Microbiology and Biotechnology, 95*, 601–613.

Duncan, W. S., & Derek, J. J. (1996). Glutathione is an important molecule in the yeast *Saccharomyces cerevisiae*. *FEMS Microbiology Letters, 141*, 207–212.

EFSA: European Food Safety Authority. (2011). Scientific opinion on risk based control of biogenic amine formation in fermented foods. *EFSA Journal, 9*, 1–93.

El Darra, N., Turk, M. F., Ducasse, M. A., Grimi, N., Maroun, R. G., Louka, N., & Vorobiev, E. (2016). Changes in polyphenol profiles and color composition of freshly fermented model wine due to pulsed electric field, enzymes and thermovinification pretreatments. *Food Chemistry, 194*, 944–950.

El Khoury, A., & Atoui, A. (2010). Ochratoxin a: General overview and actual molecular status. *Toxins, 2*(4), 461–493.

Elskens, M. T., Jaspers, C. H., & Penninckx, M. J. (1991). Glutathione as an endogenous sulphur source in the yeast *Saccharomyces cerevisiae*. *Journal of General Microbiology, 137*, 637–644.

Fenwick, E. K., Xie, J., Man, R. E. K., Lim, L. L., Flood, V. M., Finger, R. P., Wong, T. Y., & Lamoureux, E. L. (2015). Moderate consumption of white and fortified wine is associated with reduced odds of diabetic retinopathy. *Journal of Diabetes and its Complications, 29*, 1009–1014.

Fernandes, I., Pérez-Gregorio, R., Soares, S., Mateus, N., & De Freitas, V. (2017). In J. Frías, C. Martínez-Villaluenga, & E. Peñas (Eds.), *Fermented foods in health and disease prevention* (pp. 593–621). New York: Elsevier Inc.. ISBN 978-0-12-802309-9.

Fernández-Cruz, E., Álvarez-Fernández, M. A., Valero, E., Troncoso, A. M., & García-Parrilla, M. C. (2017). Melatonin and derived L-tryptophan metabolites produced during alcoholic fermentation by different wine yeast strains. *Food Chemistry, 217*, 431–437.

Fernández-Mar, M. I., Mateos, R., García-Parrilla, M. C., Puertas, B., & Cantos-Villar, E. (2012). Bioactive compounds in wine: Resveratrol, hydroxytyrosol and melatonin: A review. *Food Chemistry, 130*, 797–813.

Ferreira, V., Fernandes, F., Pinto-Carnide, O., Valentão, P., Falco, V., Martín, J. P., Ortiz, J. M., Arroyo-García, R., Andrade, P. B., & Castro, I. (2016). Identification of *Vitis vinifera* L. grape berry skin color mutants and polyphenolic profile. *Food Chemistry, 194*, 117–127.

Fleet, G. H. (2007). Yeasts in foods and beverages: Impact on product quality and safety. *Current Opinion in Biotechnology, 18*, 170–175.

Fragopoulou, E., Choleva, M., Antonopoulou, S., & Demopoulos, C. A. (2018). Wine and its metabolic effects. A comprehensive review of clinical trials. *Metabolism Clinical and Experimental, 83*, 102–119.

Gabrielli, M., Aleixandre-Tudo, J. L., Kilmartin, P. A., Sieczkowski, N., & du Toit, W. J. (2017). Additions of glutathione or specific glutathione-rich dry inactivated yeast preparation (DYP) to Sauvignon blanc must: Effect on wine chemical and sensory composition. *South African Journal of Enology and Viticulture, 38*, 18–28.

Gamero-Sandemetrio, E., Payá-Tormo, L., Gómez-Pastor, R., Aranda, A., & Matallana, E. (2018). Non-canonical regulation of glutathione and trehalose biosynthesis characterizes non-*Saccharomyces* wine yeasts with poor performance in active dry yeast production. *Microbial Cell, 5*, 184–197.

Garcia Moruno, E., Sanlorenzo, C., Boccaccino, B., & Di Stefano, R. (2005). Treatment with yeast to reduce the concentration of Ochratoxin A in Red Wine. *American Journal of Enology and Viticulture, 56*(1), 73–76.

Giovannini, C., Straface, E., Modesti, D., Coni, E., Cantafora, A., De Vincenzi, M., Malorni, W., & Masella, R. (1999). Tyrosol, the major olive oil biophenol, protects against oxidized-LDL-induced injury in Caco-2 cells. *The Journal of Nutrition, 129*, 1269–1277.

Granchi, L., Romano, P., Mangani, S., Guerrini, S., & Vincenzini, M. (2005). Production of biogenic amines by wine microorganisms. *Bulletin de l'O.I.V., 895*, 596–609.

Guaadaoui, A., Benaicha, S., Elmajdoub, N., Bellaoui, M., & Hamal, A. (2014). What is a bioactive compound? A combined definition for a preliminary consensus. *International Journal of Food Sciences and Nutrition, 3*, 174–179.

Guerrero, R. F., & Cantos-Villar, E. (2015). Demonstrating the efficiency of sulphur dioxide replacements in wine: A parameter review. *Trends in Food Science and Technology, 42*, 27–43.

Guerrini, S., Mangani, S., Romboli, Y., Luti, S., Pazzagli, L., & Granchi, L. (2018). Impact of *Saccharomyces cerevisiae* strains on health-promoting compounds in wine. *Fermentation, 4*, 14.

Guilford, J. M., & Pezzuto, J. M. (2011). Wine and health: A review. *American Journal of Enology and Viticulture, 62*, 471–486.

Henschke, P. A. (1997). Wine yeast. In F. K. Zimmermann & K.-D. Entian (Eds.), *Yeast sugar metabolism* (pp. 527–560). Lancaster: Technomic Publishing.

Henschke, P. A., & Jiranek, V. (1993). Yeast: Metabolism of nitrogen compounds. In G. H. Fleet (Ed.), *Wine microbiology and biotechnology* (pp. 77–164). GmbH, Chur-London-New York: Harwood Academic Publishers.

Herbert, P., Cabrita, M. J., Ratola, N., Laureano, O., & Alves, A. (2005). Free amino acid and biogenic amines in wines and musts from the Alentejo region. Evolution of amines during alcoholic fermentation and relationship with variety, sib-region and vintage. *Journal of Food Engineering, 66*, 315–322.

Hocking, A. D., Leong Su-lin, L., Kaz Benozir, A., Emmett, R. W., & Scott, E. S. (2007). Fungi and mycotoxins in vineyards and grape product. *International Journal of Food Microbiology, 119*, 84–88.

IARC. (2002). Monograph on the evaluation of carcinogenic risks to human: Some traditional herbal medicines, some mycotoxins, Naphtalene and styrene. IARC Monographs, 82: 68–128.

Ingledew, W. M., Magnus, C. A., & Patterson, J. R. (1987). Yeast foods and ethyl carbamate formation in wine. *American Journal of Enology and Viticulture, 38*, 332–335.

Jolly, N. P., Varela, C., & Pretorius, I. S. (2014). Not your ordinary yeast: Non-*Saccharomyces* yeasts in wine production uncovered. *FEMS Yeast Research, 14*, 215–237.

Kritzinger, E. C., Bauer, F. F., & du Toit, W. J. (2013a). Role of glutathione in winemaking: A review. *Journal of Agricultural and Food Chemistry, 61*, 269–277.

Kritzinger, E. C., Bauer, F. F., & du Toit, W. J. (2013b). Influence of yeast strain, extended lees contact and nitrogen supplementation on glutathione concentration in wine. *Australian Journal of Grape and Wine Research, 19*, 161–170.

Landete, J. M., Ferrer, S., & Pardo, I. (2007). Biogenic amine production by lactic acid bacteria acetic. *Food Control, 18*, 1569–1574.

Lavigne, V., Pons, A., & Dubourdieu, D. (2007). Assay of glutathione in must and wines using capillary electrophoresis and laser-induced fluorescence detection – changes in concentration in dry white wines during alcoholic fermentation and aging. *Journal of Chromatography. A, 1139*, 130–135.

Lavigne-Cruège, V., & Dubourdieu, D. (2002). Role of glutathione on development of aroma defects in dry white wines. 13th international enology symposium, management and wine marketing, montpellier, proceedings. In H. Trogus, J. Gafner, & A. Sütterlin (Eds.), *International association of enology, management and wine marketing* (pp. 331–347). Breisach Germany: TS Verlag, Neuenburg a. Rhein.

Lehtonen, P. (1996). Determination of amines and amino acids in wine – a review. *American Journal of Enology and Viticulture, 47*, 127–133.

Li, Y., Wei, G., & Chen, J. (2004). Glutathione: A review on biotechnological production. *Applied Microbiology and Biotechnology, 66*, 233–242.

Lindberg, M. L., & Amsterdam, E. A. (2008). Alcohol, wine, and cardiovascular health. *Clinical Cardiology, 31*, 347–351.

Luo, Y., Liu, X., & Li, J. (2018). Updating techniques on controlling mycotoxins – A review. *Food Control, 89*, 123–132.

Marques, A. P., Leitão, M. C., & San Romão, M. V. (2007). Biogenic amines in wines: Influence of oenological factors. *Food Chemistry, 107*, 853–860.

Martuscelli, M., Arfelli, G., Manetta, A. C., & Suzzi, G. (2013). Biogenic amines content as a measure of the quality of wines of Abruzzo (Italy). *Food Chemistry, 140*, 590–597.

Mas, A., Guillamon, J. M., Torija, M. J., Beltran, G., Cerezo, A. B., Troncoso, A. M., & Garcia-Parrilla M.C. (2014). Bioactive compounds derived from the yeast metabolism of aromatic amino acids during alcoholic fermentation. *BioMed Research International*, 898045: 1–7.

Medina, K., Boido, E., Farîna, L., Gioia, O., Gomez, M. E., Barquet, M., Gaggero, C., Dellacassa, E., & Carrau, F. (2013). Increased flavour diversity of Chardonnay wines by spontaneous fermentation and co-fermentation with *Hanseniaspora vineae*. *Food Chemistry, 141*, 2513–2521.

Mehdi, K., & Penninckx, M. J. (1997). An important role for glutathione and gamma-glutamylpeptidase in the supply of growth requirements during nitrogen starvation of the yeast *Saccharomyces cerevisiae*. *Microbiology, 143*, 1885–1889.

Mezzetti, F., De Vero, L., & Giudici, P. (2014). Evolved *Saccharomyces cerevisiae* wine strains with enhanced glutathione production obtained by an evolution-based strategy. *FEMS Yeast Research, 14*, 977–987.

Monagas, M., Bartolome, B., & Gomez-Cordoves, C. (2005). Updated knowledge about the presence of phenolic compounds in wine. *Critical Reviews in Food Science and Nutrition, 45*, 85–118.

Moss, M. (2002). Mycotoxin review – 1. *Aspergillus and Penicillium. Mycologist, 16*, 116–119.

Muller, C. J., & Fugelsang, K. C. (1997). Red wine but not white: The importance of fully characterizing wines used in health studies. *The American Journal of Clinical Nutrition, 66*, 447.

Nikolantonaki, M., Julien, P., Coelho, C., Roullier-Gall, C., Ballester, J., Schmitt-Kopplin, P., & Gougeon, R. D. (2018). Impact of glutathione on wines oxidative stability: A combined sensory and metabolomic study. *Frontiers in Chemistry, 6*, 182.

Nisamedtinov, I., Kevvai, K., Orumets, K., Arike, L., Sarand, I., Korhola, M., & Paalme, T. (2011). Metabolic changes underlying the higher accumulation of glutathione in *Saccharomyces cerevisiae* mutants. *Applied Microbiology and Biotechnology, 89*, 1029–1037.

OIV (Organisation of Vine and Wine). (2011). Resolution OIV-CST 369-2011. 9th General Assembly Porto, Portugal.

OIV (Organisation of Vine and Wine). (2015). Resolutions OIV-OENO 445-2015 and OIV-OENO 446-2015. 13th General Assembly, Mainz.

Otteneder, H., & Majerus, P. (2000). Occurrence of ochratoxin A (OTA) in wines: Influence of the type of wine and its geographical origin. *Food Additives and Contaminants, 17*, 793–798.

Ough, C. S., Crowell, E. A., & Gutlove, L. B. R. (1988). Carbamyl compound reactions with ethanol. *American Journal of Enology and Viticulture, 39*, 239–242.

Padilla, B., Gil, J. G., & Manzanares, P. (2016). Past and future of Non-*Saccharomyces* yeasts: From spoilage microorganisms to biotechnological tools for improving wine aroma complexity. *Frontiers in Microbiology, 7*(411), 20.

Park, S. K., Boulton, R. B., & Noble, A. C. (2000a). Formation of hydrogen sulphide and glutathione during fermentation of white grape must. *American Journal of Enology and Viticulture, 51*, 91–97.

Park, S. K., Boulton, R. B., & Noble, A. C. (2000b). Automated HPLC analysis of glutathione and other volatile thiols in grape musts and wine using pre-column derivatization with fluorescence detection. *Food Chemistry, 69*, 475–480.

Parker, M., Capone, D. L., Francis, I. L., & Herderich, M. J. (2017). Aroma precursors in grapes and wine: Flavor release during wine production and consumption. *Journal of Agricultural and Food Chemistry, 10*, 2281–2286.

Penninckx, M. J. (2000). A short review on the role of glutathione in the response of yeasts to nutritional, environmental, and oxidative stresses. *Enzyme and Microbial Technology, 26*, 737–742.

Penninckx, M. J. (2002). An overview on glutathione in *Saccharomyces versus* non-conventional yeasts. *FEMS Yeast Research, 2*, 295–305.

Petruzzi, L., Sinigaglia, M., Corbo, M. R., Campaniello, D., Speranza, B., & Bevilacqua, A. (2014). Decontamination of ochratoxin A by yeasts: possible approaches and factors leading to toxin removal in wine. *Applied Microbiology and Biotechnology, 98*, 6555–6567.

Petruzzi, L., Baiano, A., De Gianni, A., Sinigaglia, M., Corbo, M. R., & Bevilacqua, A. (2015). Differential adsorption of Ochratoxin A and anthocyanins by inactivated yeasts and yeast cell walls during simulation of wine aging. *Toxins, 7*, 4350–4365.

Petzinger, E., & Weidenbach, A. (2002). Mycotoxins in the food chain: The role of ochratoxins. *Livestock Production Science, 76*, 245–250.

Peyrot des Gachons, C., Tominaga, T., & Dubourdieu, D. (2002). Sulfur aroma precursor present in S-glutathione conjugate form: Identification of S-3-(hexan-1-ol)-glutathione in must from *Vitis vinifera* L. cv. Sauvignon Blanc. *Journal of Agricultural and Food Chemistry, 50*, 4076–4079.

Pfohl-Leszkowicz, A., & Manderville, R. A. (2007). Ochratoxin *A:* An overview on toxicity and carcinogenicity in animals and humans. *Molecular Nutrition & Food Research, 51*, 61–99.

Pierini, R., Gee, J. M., Belshaw, N. J., & Johnson, I. T. (2008). Flavonoids and intestinal cancers. *The British Journal of Nutrition, 99*, ES53–ES59.

Piotrowska, M., Nowak, A., & Czyzowska, A. (2013). Removal of ochratoxin A by wine *Saccharomyces* cerevisiae strains. *European Food Research and Technology, 236*, 441–447.

Poli, A., Marangoni, F., Avogaro, A., Barba, G., Bellentani, S., Bucci, M., Cambieri, R., Catapano, A. L., Costanzo, S., Cricelli, C., de Gaetano, G., Di Castelnuovo, A., Faggiano, P., Fattirolli, F., Fontana, L., Forlani, G., Frattini, S., Giacco, R., La Vecchia, C., Lazzaretto, L., Loffredo, L., Lucchin, L., Marelli, G., Marrocco, W., Minisola, S., Musicco, M., Novo, S., Nozzoli, C., Pelucchi, C., Perri, L., Pieralli, F., Rizzoni, D., Sterzi, R., Vettor, R., Violi, F., & Visioli, F. (2013). Moderate alcohol use and health: A consensus document. *Nutrition, Metabolism, and Cardiovascular Diseases, 23*, 487–504.

Pozo-Bayón, M. A., Andújar-Ortiz, I., & Moreno-Arribas, M. V. (2009). Scientific evidences beyond the application of inactive dry yeast preparations in winemaking. *Food Research International, 42*, 754–761.

Pretorius, I. S. (2000). Tailoring wine yeast for the new millennium: novel approaches to the ancient art of winemaking. *Yeast, 16*, 675–729.

Rauhut, D. (2003). Impact of volatile sulfur compounds on wine quality. In Hrsg, J.-C. Davidian, D. Grill, L. J. De Kok, I. Stulen, M. J. Hawkesford, E. Schnug, & H. Rennenberg (Eds.), *Sulfur transport and assimilation in plants. Regulation, Interaction and Signaling* (pp. 121–131). Leiden: Backhuys Publishers.

Rauhut, D. (2017). Usage and formation of sulfur compounds. In H. König, G. Unden, & J. Fröhlich (Eds.), *Biology of microorganisms on grapes, in must and in wine* (pp. 255–291). Cham: Springer International Publishing AG.

Rauhut, D., & Micheloni, C. (2010). Current issues in organic winemaking: Consumer expectations, producer attitudes and oenological innovation. Chapter 10. In A. G. Reynolds (Ed.), *Managing wine quality, Volume 2: Oenology and wine quality* (pp. 271–289). Woodhead Publishing Limited.

Renaud, S. D., & de Lorgeril, M. (1992). Wine, alcohol, platelets, and the French paradox for coronary heart disease. *Lancet, 339*, 1523–1526.

Renaud, S., & Gueguen, R. (1998). The French paradox and wine drinking. *Novartis Foundation Symposium, 216*, 208–217.

Restuccia, D., Loizzo, M. R., & Spizzirri, U. G. (2018). Accumulation of biogenic amines in wine: role of alcoholic and malolactic fermentation. *Fermentation, 4*, 6–18.

Ribéreau-Gayon, P., Glories, Y., Maujean, A., & Dubourdieu, D. (2000). Phenolic compounds. In P. Ribéreau-Gayon, Y. Glories, A. Maujean, & D. Dubourdieu (Eds.), *Handbook of enology, the chemistry of wine stabilization and treatments* (Vol. 2, 2nd ed., pp. 141–201). Chichester: Wiley.

Ribéreau-Gayon, P., Dubourdieu, D., Doneche, B., & Lonvaud, A. (2006). Handbook of enology. In *The microbiology of wine and vinifications* (Vol. 1, 2nd ed., pp. 193–220). The Atrium/Southern Gate/Chichester: Wiley.

Rodriguez-Naranjo, M. I., Gil-Izquierdo, A., Troncoso, A. M., Cantos, E., & Garcia-Parrilla, M. C. (2011). Melatonin: A new bioactive compound present in wine. *Journal of Food Composition and Analysis, 24*, 603–608.

Rodriguez-Naranjo, M. I., Torija, M. J., Mas, A., Cantos-Villar, E., & del Carmen Garcia-Parrilla, M. (2012). Production of melatonin by *Saccharomyces* strains under growth and fermentation conditions. *Journal of Pineal Research, 53*, 219–224.

Rodriguez-Naranjo, M. I., Ordóñez, J. L., Callejón, R. M., Cantos-Villar, E., & Garcia-Parrilla, M. C. (2013). Melatonin is formed during winemaking at safe levels of biogenic amines. *Food and Chemical Toxicology, 57*, 140–146.

Romano, P., & Suzzi, G. (1993). Sulfur dioxide and wine microorganisms. In G. H. Fleet (Ed.),. Chapter 13 *Wine microbiology and biochemistry* (pp. 373–393). Chur: Harward Academic Publishers.

Romano, P., Fiore, C., Paraggio, M., Caruso, M., & Capece, A. (2003). Function of yeast species and strains in wine flavour. *International Journal of Food Microbiology, 86*, 169–180.

Romboli, Y., Mangani, S., Buscioni, G., Granchi, L., & Vincenzini, M. (2015). Effect of *Saccharomyces cerevisiae* and *Candida zemplinina* on quercetin, vitisin A and hydroxytyrosol contents in Sangiovese wines. *World Journal of Microbiology and Biotechnology, 31*, 1137–1145.

Roussis, I. G., Lambropoulos, I., & Tzimas, P. (2007). Protection of volatiles in a wine with low sulfur dioxide by caffeic acid or glutathione. *American Journal of Enology and Viticulture, 58*, 274–278.

Samoticha, J., Wojdyło, A., Chmielewska, J., Politowicz, J., & Antoni, S. (2017). The effects of enzymatic pre-treatment and type of yeast on chemical properties of white wine. *LWT – Food Science and Technology, 79*, 445–453.

Samuel, S. M., Thirunavukkarasu, M., Penumathsa, S. V., Paul, D., & Maulik, N. (2008). Akt/FOXO3a/SIRT1-mediated cardioprotection by n-tyrosol against ischemic stress in rat in vivo model of myocardial infarction: Switching gears toward survival and longevity. *Journal of Agricultural and Food Chemistry, 56*, 9692–9698.

Santos, M. C., Nunes, C., Saraiva, J. A., & Coimbra, M. A. (2012). Chemical and physical methodologies for the replacement/reduction of sulfur dioxide use during winemaking: Review of their potentialities and limitations. *European Food Research and Technology, 234*, 1–12.

Schwartz, G. G. (2002). Hypothesis: Does ochratoxin A cause testicular cancer? *Cancer Causes & Control, 13*, 91–100.

Soufleros, E., Barrios, M., & Bertrand, A. (1998). Correlation between the content of biogenic amines and other wine compounds. *American Journal of Enology and Viticulture, 49*, 266–278.

Suzzi, G., Romano, P., & Zambonelli, C. (1985). *Saccharomyces* strain selection in minimizing SO_2 requirement during vinification. *American Journal of Enology and Viticulture, 36*, 199–202.

Thomas, D., & Surdin-Kerjan, Y. (1997). Metabolism of sulphur amino acids in *Saccharomyces cerevisiae*. *Microbiology and Molecular Biology Reviews, 61*, 503–532.

Til, H. P., Feron, V. J., & de Groot, A. P. (1972). The toxicity of sulphite. I. Long-term feeding and multigeneration studies in rats. *Food and Cosmetics Toxicology, 10*, 291–310.

Torrea, D., & Ancin, C. (2001). Influence of yeast strain on biogenic amines content in wines: Relationship with the utilization of amino acids during fermentation. *American Journal of Enology and Viticulture, 52*, 185–190.

Torrea, D., & Ancìn, C. (2002). Content of biogenic amines in a Chardonnay wine obtained through spontaneous and inoculated fermentations. *Journal of Agricultural and Food Chemistry, 50*, 4895–4899.

Uemura, T., Tomonari, Y., Kashiwagi, K., & Igarashi, K. (2004). Uptake of GABA and putrescine by UGA4 on the vacuolar membrane in *Saccharomyces cerevisiae*. *Biochemical and Biophysical Research Communications, 315*, 1082–1087.

Ugliano, M., & Henschke, P. A. (2009). Yeasts and wine flavour. Chapter 8D. In M. V. Moreno-Arribas & M. C. Polo (Eds.), *Wine chemistry and biochemistry* (pp. 313–392). New York: Springer Science Business Media, LLC.

Ugliano, M., Kwiatkowski, M., Vidal, S. P., Capone, D., Siebert, T., Dieval, J.-B., Aaagard, O., & Waters, E. J. (2011). Evolution of 3-mercaptohexanol, hydrogen sulfide, and methyl mercaptan during bottle storage of Sauvignon blanc wines. Effect of glutathione, copper, oxygen exposure, and closure-derived oxygen. *Journal of Agricultural and Food Chemistry, 59*, 2564–2572.

Vigentini, I., Romano, A., Compagno, C., Merico, A., Molinari, F., Tirelli, A., Foschino, R., & Volonterio, G. (2008). Physiological and oenological traits of different *Dekkera/Brettanomyces bruxellensis* strains under wine-model conditions. *FEMS Yeast Research, 8*, 1087–1096.

Vincenzini, M., Guerrini, S., Mangani, S., & Granchi, L. (2017). Amino acid metabolisms and production of biogenic amines and ethyl carbamate. In H. König, G. Unden, & J. Fröhlich (Eds.), *Biology of microorganisms on grapes, in must and in wine* (pp. 231–253). Cham: Springer International Publishing AG.

Wang, J., Ho, L., Zhao, Z., Seror, I., Humala, N., Dickstein, D. L., Thiyagarajan, M., Percival, S. S., Talcott, S. T., & Pasinetti, G. M. (2006). Moderate consumption of Cabernet Sauvignon attenuated beta-amyloid neuropathology in a mouse model of Alzheimer's disease. *The FASEB Journal, 20*, 2313–2320.

Webber, V., Dutra, S. V., Spinelli, F. R., Carnieli, G. J., Cardozo, A., & Vanderlinde, R. (2017). Effect of glutathione during bottle storage of sparkling wine. *Food Chemistry, 216*, 254–259.

Wegmann-Herr, P., Ullrich, S., Schmarr, H. G., & Durner, D. (2016). Use of glutathione during white wine production – impact on S-off-flavors and sensory production. *BIO Web of Conferences, 7*, 02031.

Wu, G., Fang, Y. Z., Yang, S., Lupton, J. R., & Turner, N. D. (2004). Glutathione metabolism and its implications for health. *The Journal of Nutrition, 134*(3), 489–492.

Xia, Q., Yang, C., Wu, C., Zhou, R., & Li, Y. (2018). Quantitative strategies for detecting different levels of ethyl carbamate (EC) in various fermented food matrices: An overview. *Food Control, 84*, 499–512.

Zechmann, B., Liou, L. C., Koffler, B. E., Horvat, L., Tomašić, A., Fulgosi, H., & Zhang, Z. (2011). Subcellular distribution of glutathione and its dynamic changes under oxidative stress in the yeast *Saccharomyces cerevisiae*. *FEMS Yeast Research, 11*, 631–642.

Zell, J. A., McEligot, A. J., Ziogas, A., Holcombe, R. F., & Anton-Culver, H. (2007). Differential effects of wine consumption on colorectal cancer outcomes based on family history of the disease. *Nutrition and Cancer, 59*, 36–45.

Zimmerli, B., & Dick, R. (1996). Ochratoxin A in table wine and grape-juice: Occurrence and risk assessment. *Food Additives and Contaminants, 13*, 655–668.

Chapter 12
Spoilage Yeasts in Wine Production

Manuel Malfeito-Ferreira and Ana Carla Silva

Contents

M. Malfeito-Ferreira (✉) · A. C. Silva
Linking Landscape Environment Agriculture and Food Research Centre (LEAF), Instituto Superior de Agronomia, University of Lisbon, Lisbon, Portugal
e-mail: mmalfeito@isa.ulisboa.pt; acsilva@isa.ulisboa.pt

P. Romano et al. (eds.), *Yeasts in the Production of Wine*,
https://doi.org/10.1007/978-1-4939-9782-4_12

1 Introduction

The yeast species *Saccharomyces cerevisiae* is the protagonist of wine fermentation justifying the epithet of "wine yeast" by excellence. However, several other species are particularly important mainly regarding spoilage. In fact, according to our experience in the industry, the central role of wine yeasts in enology is much higher by its negative effects and economical losses than by the claimed quality imparted by specific yeast starters. In other words, once appropriate conditions are given to fermenting yeasts they fulfil their role without meriting special attention. On the contrary, spoilage yeasts are a permanent concern due to their possible detrimental effects. In fact, presently, wine spoilage yeasts represent one of the most significant problems in modern enology, even if recent reviews emphasize possible beneficial activities (Padilla et al. 2016; Valera and Borneman 2017; Gschaedler 2017).

The description of yeast species is found in taxonomy handbooks, the last of which includes about 1500 species (Kurtzman et al. 2011). The number may double in the future due to the exploration of a wider range of sources, mainly within insects (Boekhout 2005). However, this increase is not expected to be observed in foods and beverages, where new specific names derive more from improved species delineation using molecular biology tools (Kurtzman and Robnett 2003) than from non-described species. As an example, *Starmerella bacillaris* is the current species replacing *C. zemplinina* which had been frequently identified as *C. stellata* (Valera and Borneman 2017). About a fifth of all contaminating species may be found in food and beverages but only about a dozen are really detrimental to food quality. Among contamination yeasts, those surviving in foods but without the ability to grow are called *adventitious*, *innocuous* or *innocent* yeasts (Fig. 12.1). Those responsible for unwanted modifications of the processed product – visual, textural

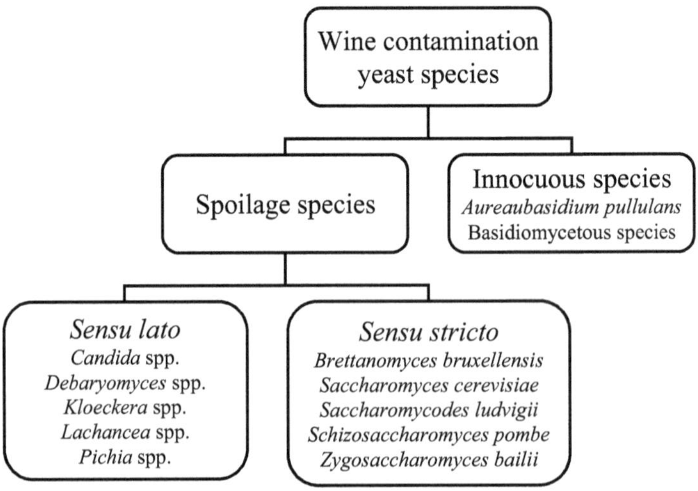

Fig. 12.1 Groups of most common contamination yeasts species with technological significance

or organoleptical (producing off-flavours or off-tastes) – are called *spoilage* yeasts. However, for technologists the concept of spoilage yeast is narrower and only fits to species able to adversely modify foods processed according to the standards of good manufacturing practices (GMPs) (Pitt and Hocking 1985). These are the sensu stricto spoilage yeasts and represent the most resistant species to the stresses provoked by food or beverage processing (Loureiro and Querol 1999; Loureiro 2000).

In wine industry, the definition of spoilage is not always obvious, because the microbial metabolites contribute to its flavour and aroma and their pleasantness is driven by many subjective factors (e.g. habits, fashions, opinion makers' choices) that persuade the consumer taste. Therefore the definition of spoilage as "the alteration of food recognised by the consumer" (Stratford 2006) is not fully satisfactory for wines. This situation is clearly demonstrated by the presence of volatile phenols in red wines, produced by the species *Brettanomyces bruxellensis*. While some consumers and opinion makers prefer wines tainted by volatile phenols, particularly before knowing its origin, others do consider that, even in low concentration, these compounds depreciate wine quality due to diminished flavour complexity (Malfeito-Ferreira 2018).

1.1 Yeast Species Significance and Diversity in Grapes and Wines

The detrimental effects caused by yeasts are the most frequent problems of microbial origin related with wine quality. The most common recognised symptoms of yeast spoilage are film formation in bulk wines, cloudiness, sediment formation and gas production in bottled wines, and off-flavour production during all processing and storing stages (Loureiro and Malfeito-Ferreira 2003a). The technological significance of wine or grape juice contamination yeasts has been thoroughly discussed in previous reviews (Thomas 1993; Kunkee and Bisson 1993; Loureiro and Malfeito-Ferreira 2003a; Fugelsang and Edwards 2007; Malfeito-Ferreira 2011; Zuehlke et al. 2013; Escott et al. 2017) which constitute the background for the following description. Table 12.1 lists the species recently identified as a result of the analytical activity of our laboratory showing that practically all have already been described in those reviews.

1.1.1 Grapes and Grape Juices

Common yeast contaminants of grapes and grape juices before fermentation include basidiomycetous ans ascomycetous genera like *Candida*, *Cryptococcus*, *Debaryomyces*, *Hansenula*, *Kloeckera/Hanseniaspora*, *Metchsnikowia*, *Pichia* and *Rhodotorula* (Fleet et al. 2002). The yeast like fungus *Aureaubasidium pullulans* is also frequent in grapes. These grape yeasts do not cause problems in wines bottled

Table 12.1 Contamination wine yeast species isolated in the ISA Microbiology laboratory from different sources during the last decade (unpublished data, species names resulting from DNA sequence matching and not updated to current synonyms)

Source	Species
Insects in the vineyard	*Candida albicans, C. apicola, C. parapsilosis, Debaryomyces hansenii, Hanseniaspora uvarum*
Damaged grapes	*C. fructus, C. stellimalicola, H. guilliermondii, H. uvarum, Issatchenkia orientalis, Pichia guilliermondii, Saccharomycopsis crataegensis, Zygoascus meyerae, Zygoascus hellenicus, Zygosaccharomyces bailii*
Grape juices	*C. diversa, C. zemplinina, H. uvarum, Issatchenkia terricola, Lachancea thermotolerans, P. guilliermondii, P. fermentans, P. kluyveri, Metschnikowia pulcherrima, Saccharomyces cerevisae, Starmerella* spp., *Z. bailii, Z. rouxii, Wickerhamomyces anomalus*
Concentrated and rectified grape juices	*C. apicola, S. cerevisiae, Z. bailii, Z. rouxii*
Winery equipment (tanks, hoses, pumps, buckets, purges, levels, presses)	*C. ishiwadae, H. uvarum, H. guilliermondii, Issatchenkia orientalis, M. pulcherrima, P. kudriazevii, P. membranifaciens, Rhodotorula mucilaginosa, W. anomalus, Yarrowia lypolitica, Zygoascus hellenicus, Z. meyerae*
Bottling equipment (tanks, filler, filters, hoses, capsules)	*C. amapae, C. boidinii, C. parapsilosis, Cryptococus magnus, P. guillermondii, Sporobolomyces roseus*
Base white wine for sparkling wine	*Dekkera bruxellensis*
Bulk and bottled red wine	*D. bruxellensis, Candida* spp., *C. cantarelli, C. oleophila, C. parapsilosis, Lodderomyces elongisporus, Pichia* spp., *P. manshurica, P. membranifaciens, S. cerevisiae, Saccharomycodes ludwigii, Schizosaccharomyces pombe, Trigonopsis californica, Z. bailii, Z. parabailii*
Bag in box red wine	*P. manshurica, Z. bailii*
Bottled white wine	*Cystobasidium sloofiae, P. kudriavzevii, P. guilliermondii, R. mucilaginosa, Z. bailii*
Rosé wine	*C. parapsilosis, Magnusiomyces capitatus, Schizosaccharomyces* spp., *Yarrowia lypolitica, Z. bailii, Z. parabailii*
Alchool free wine	*C. pararugosa, Magnosiomyces capitatus, Exophiala dermatiditis, Filobasidium uniguttulatum, Sporopachidermia cereana, Trichosporon mycotoxinivorans, T. montevideense, Y. lipolytica, Z. bailii*

according to good manufacturing practices (GMPs) but may be a cause of concern in early stages of wine production. Grape health is the main driver of yeast diversity and the influence of any factor changing microbial communities (e.g. grape variety, geographical location, climate, vineyard spraying, technological practices, processing stage and season) can only be properly studied if the damaged grapes are effectively avoided (Barata et al. 2012a), which has not been well understood by the scientific community (Valera and Borneman 2017). In particular sour rot is an infection where a wide number of yeast species may contribute to the high concentration of acetic acid and unwanted modifications of juice composition before winemaking (Guerzoni and Marchetti 1987; Barata et al. 2012b). The rotting process is not yet clear, being initiated by acetic acid bacteria and promoted by contamination and spoilage wine yeasts. The main measures to adopt in winemaking to overcome problems involve grape selection, increased sulphur dioxide usage and prompt inoculation of active starters.

In juices, given the short period before fermentation, spoilage events are rare. In white juices with long settling periods or with long skin contact and in long red pre-fermentative maceration, film-forming yeasts (e.g. *P. anomala*) or apiculate yeasts may grow very fast. These species are easily controlled by adequate winemaking measures (low temperature, sulphur dioxide, hygiene). In principle these species are inhibited during fermentation but even for a short period, due to their fast growth, they may produce unwanted amounts of metabolites like ethylacetate (vinegar smell) or acetaldehyde (oxidised taint) (Romano 2005), which can irremediably spoil the wine.

1.1.2 Wine Fermentation

Fermentation problems are related with the activity of fermenting yeasts (*S. cerevisiae* or *S. bayanus*). The production of off-flavours (sulphur reduced compounds) (Bell and Henschke 2005) and acetic acid is due to nutritional imbalances or deficient fermenting conditions (e. g. high temperature) that may lead to stuck fermentations (Bisson and Butzke 2000). These events are a result of the environment conditions and not of any particular spoilage characteristic and so the correct management of fermenting conditions overcomes the problem. The activity of other yeast species is limited or unknown, being stuck wines highly susceptible to bacterial spoilage.

1.1.3 Bulk and Bottled Wine

During bulk wine storage, film-forming yeasts (e.g. *Candida* spp., *Pichia* spp.) may form pellicles on wine surface and spoil wine by the production of odour active compounds. The absence of oxygen and proper sulphur dioxide usage together with appropriate hygienic measures prevent the emergence of films. These species are frequently encountered in bottled wine but, when in low number, tend to die, or to remain dormant, due to low resistance to the stress imposed by the bottled product.

The classical spoilage events that occur in bottled wine are due to a low number of species. Their spoiling abilities are related with bottled sweet wine refermentation or proliferation in bottled dry wines, leading to sediments or turbidity. The typical species are those capable of growing in bottled wines – *Z. bailii*, *S. cerevisiae*, *Schizosaccharomyces pombe* and *Saccharomycodes ludwigii*. Albeit not frequent grape or winery contaminants, the stress resistances of these yeasts enable their survival and proliferation under conditions that are not tolerated by the other species. According to our experience, refermentation problems have increased in the last years in red wines because of the addition of concentrated grape juices to make softer wines in accordance to modern market demands (Sena-Esteves et al. 2018). In bag in box wines this problem is easily recognised by swollen packages. In bottled wines, gas production and wine turbidity are also easily visible after bottle opening and pouring wine in the glass. In red wines the effect of sediment formation or haziness is less obvious, given the dark colour.

A particular mention should be made to *Dekkera bruxellensis*. The recognition of its role in red wine spoilage due to the production of volatile phenols in bulk or bottled wines ("horse sweat" taint) has revealed a new challenge to winemakers since the last decade (Loureiro and Malfeito-Ferreira 2006). Moreover, its effects are particularly notorious in premium red wines aged in costly oak barrels, which considerably increased the economical losses provoked by spoilage yeasts in the wine industry. Presently, this species is regarded as the main threat posed by yeasts to wine quality. The effect is not only direct, due to the production of volatile phenols, but also indirect due to the technological measures needed to control its activity and that may reduce wine sensory attributes. In particular, those winemakers stressed by having the slightest hint of "horse sweat" tend to adopt control and preventive measures that are reported as effective in research articles without validation under winery conditions (Malfeito-Ferreira 2018).

1.2 Factors Promoting the Dissemination of Spoilage Yeasts

The wine spoilage yeasts associated with wineries are disseminated by all surfaces with residues of nutrients where they can proliferate, namely wine residues in wooden barrels, valves, improperly cleaned tanks, hoses, fillers, corkers, filters, pumps, walls, floors, air, etc. They easily contaminate and grow in wine when environmental conditions are favourable. It is common knowledge that their incidence increases with low levels of hygiene, so that the best way to prevent their contamination is avoiding their propagation by sound hygiene procedures. However, the hygienic level is only a part of the problem. The primary source of spoilage yeasts are grapes, mainly those that are damaged. Yeasts may also be vectored by insects, air, mainly when the environment in the vicinity of the winery is dirty with winemaking residues. The additives used in vinification and/or in the preparation of sweet wines from dry wines, particularly those processed, such as concentrated juices, sulphited juice with sub-lethal preservative doses and sucrose are also sources of spoilage yeasts (Loureiro and Malfeito-Ferreira 2003a; Valero et al. 2005). Another source are finished wines from external suppliers that may harbour high contamination loads especially wines newly fermented.

The grapes are, probably, the main source for wine spoilage yeasts justifying the understanding of how yeasts survive along the year and reach the grapes in the vineyard.

It is generally accepted that damaged grapes bear a wider diversity of species and in numbers of 5–6 orders of magnitude higher (Fleet et al. 2002; Barata et al. 2012a). Therefore, it seems legitimate to assume that rotten grapes are one of the most important sources of food spoilage yeasts. Recent studies showed that sour rot grapes bear a high concentration of *Z. bailii* that survives along the fermentation and reaches high counts in the final wine (Barata et al. 2012b). Guerzoni and Marchetti (1987) also showed that *Brettanomyces/Dekkera* spp. were present in higher numbers in sour rot grapes and Renouf and Lonvaud-Funel (2007) recovered *D. bruxellensis* from sound grapes using an enrichment medium. Damage includes sour rot, grey rot, downy mildew, insects (*Lobesia botrana, Eupoecillia ambiguella,* honeydew producers, *Drosophila* spp., bees, wasps, ants, etc.), birds, hail, rain, and overripened/mummified grapes (Barata et al. 2012a).

Assuming that spoilage yeasts enter the winery and cannot be fully eradicated, we need to learn how to deal with them. In particular, the use of wood cooperage increases the risk of of *D. bruxellensis* (Chatonnet et al. 1993) which must be properly sanitised (Barata et al. 2013). Among the enological practices that disseminate yeasts are, naturally, all contaminated wine transfers, mainly when blends with wines aged in used barrels are required. Similarly, concentrated grape juices and sulphited juices with sub-lethal doses of sulphur dioxide are vectors and promote the dissemination of resistant strains of *Z. bailii* in the winery. These ingredients should be handled in restricted areas within the winery to decrease the incidence of infections. In addition, all enological practices that promote yeast growth like those leading to high levels of dissolved oxygen (e. g. micro-oxygenation, untopped tanks, racking with aeration), low sulphur dioxide or sorbate levels may enhance the proliferation of spoilage yeasts. Finally, improper management of wine residues (e.g. husks, lees, distillery pomaces) that lead to dirty winery vicinities may significantly contribute to yeast dissemination, mainly through insects (*Drosophila*) and air (Connel et al. 2002). Our experience tells us that wineries with dirty environments and without an efficient residue treatment system have, as a rule, higher incidence of spoilage yeasts. In some cases we observed that water in underground sheets have heavy yeast populations, suggesting their contamination by winery effluents.

The bottling line is the last circuit run by the wine in the winery. The filling operation is the last opportunity for the wine to be contaminated by spoilage yeasts. Usually, the contamination of wine in this phase results from lack of hygiene and, particularly, from deficient disinfection of all circuit, including surfaces contacting with the cork that closes the bottle. However, the lack of hygiene results most frequently from inadequate conditions to apply cleaning and disinfection programs – filler and corker design, microbiological quality of ambient air, equipment maintenance program and operator's training – than from the efficiency of sanitation programs. Thus, the specificities of each bottling line are determinant in the levels and kind of spoilage yeast contaminating it. From our experience, the main contamination sources are in the filler and in the corker and are more dangerous as equipment age increases and less maintenance is run (Loureiro and Malfeito-Ferreira 2003b).

When the fillers are poorly designed (or the sanitation procedures advised by the suppliers are not followed), their cleaning and disinfection may not reach all surfaces where there are wine residues, originating dangerous contamination sources, as self-levelling systems, surfaces protected by o-rings and valves of isobarometric fillers. The usage of steam disinfection is an alternative to the chemical disinfection given that is possible to kill microorganisms without directly contacting the cells. However, filler steaming may bring a problem frequently overlooked – the formation of a negative pressure in the filler interior during cooling – leading to air suction (with spoilage yeasts in suspension) and, as a rule, the immediate filler contamination. In these cases, the solution is to introduce sterile compressed air during cooling. The corker may also be an important contamination source when the surfaces contacting with the side part of the corks are heavily contaminated, as the transport tube, the corker jaws and the feeding bin, where the formation of water droplets condensation promotes the contamination and colonisation by yeasts suspended in the air. In modern machines heating the jaws up to lethal temperatures prevents this type of contamination. Other frequent contamination source is the bottle rinser or the rinsing water. When bottles arrive from the factory in good packaging conditions, the contamination appears after rinsing (Donnelly 1977b; Neradt 1982). As a final remark, in many companies, either the fillers or corkers are located in a closed room aiming to reduce contaminations from the winery ambient air. When they are correctly designed – with over pressure, wet air evacuation and correct hygiene – they are advantageous to proper bottling. However, in most cases they are poorly designed and become dangerous contamination sources, working as microbial incubators (Donnelly 1977a). In fact, bottle breakage is current during bottling, leading to wine dispersion through room surfaces and, consequently, to microbial growth, stimulated by the moist environmental conditions.

The few scientific studies on the ecology of bottling lines do not allow concluding on the most frequent type of contaminating yeasts. Film-forming species should prevail, but it is conceivable that dangerous contaminants are favoured by developing on wine residues and by resisting to sub-lethal doses of chemical or physical agents used in equipment cleaning and disinfection. From our experience, when concentrate grape juice is used for producing sweet wines, it is very frequent to detect yeasts of the genus *Zygosaccharomyces*, particularly *Z. bailii*, in the fillers.

2 Description of the Main Yeast Genera/Species Involved in Wine Spoilage

The species involved in wine spoilage are also known for affecting other food commodities and their taxonomical, physiological or technological properties have been described in excellent textbooks (Deak and Beuchat 1996; Boekhout and Robert 2003; Blackburn 2006; Querol and Fleet 2006). The description of the species presented below is mainly concerned with their wine relevant properties.

2.1 Apiculate Yeasts

Apiculate yeasts owe their denomination to lemon shaped form and include species of the genera *Kloeckera/Hanseniaspora*. They are particularly frequent on grape surfaces and in juices after grape crushing (Fleet et al. 2002). These species are easily controlled by adequate winemaking measures (low temperature, sulphur dioxide, hygiene) and are inhibited during fermentation. The production of unwanted amounts of metabolites like ethyl acetate (vinegar smell) (Romano 2005) may occur in white juices with long settling periods or with long skin contact and in long red pre-fermentative maceration. This spoilage activity is due to their fast growth but is not a great concern to winemakers because the preventive measures may be easily implemented.

2.2 Film-Forming Species

The denomination "film-forming yeasts" includes a group of species able to grow on the surface of wine developing pellicles. The species of the genus *Candida* and *Pichia* are regarded as the typical film-forming yeasts although *S. cerevisiae*, *D. bruxellensis* or *Z. bailii* may also be recovered from wines pellicles (Ibeas et al. 1996; Farris et al. 2002). In the case of *S. cerevisiae* it is even a desirable feature for the race *beticus*, which is one of the agents of sherry-type wine production (Suárez-Lepe and Iñigo-Leal 2004). The ability to form films by *Pichia* and *Candida* is probably explained by their aerobic nature and fast growth and so the other species are usually minor constituents of film microflora. In bulk wines they quickly cover wine surface when air has not been removed from the top of storage vessels. Although strains of *Candida* spp. or *Pichia* spp. are tolerant to preservatives, their control in wines is mainly due to their weak tolerance to low oxygen tensions, which enhances the inhibitory effect of ethanol or preservatives. In bottled wines they may cause cloudiness if the initial contamination load is high and so these species are regarded as indicators of poor GMPs. They may also produce, at the bottleneck, a film or a ring of cells adherent to the glass, if the closure does not prevent the diffusion of oxygen, the level of free sulphur dioxide is too low and the initial contamination is high.

2.3 Sensu Stricto Spoilage Yeasts

2.3.1 *Zygosaccharomyces bailii* and Related Species

The genus *Zygosaccharomyces* comprises some of the most feared species in the industries of high sugar and high acidic food products. *Z. rouxii* is mainly known for being highly osmotolerant while *Z. bailii* is notorious for its resistance to low pH, high concentration of organic acids, including preservatives, and high osmotolerance. *Z. bisporus* has been described as having intermediate features between the former two species (Pitt and Hocking 1985).

The most problematic species in wines is *Z. bailli* (Zuehlke et al. 2013). Its activity includes visible sediment formation, cloudiness or haziness in dry wines and refermentation in sweet wines. Given the visual nature of the spoiling effect it is a greater concern in white wines. It may also produce undesirable odour active metabolites (Fugelsang and Edwards 2007) but their relevance is lower that the former visual faults. One of the most relevant sources of *Z. bailii* are the grapes, particularly grapes damaged by sour rot (Barata et al. 2012a). These authors demonstrated that *Z. bailii* survived during wine fermentation being present in 10^5 CFU/ml in the fermented wine. Once established in the winery, the main risk is the contamination of wine after sterile filtration, due to improper sanitation, before bottling (Malfeito-Ferreira et al. 1997).

The species *Z. bisporus* has been isolated from grapes affected by honeydew and sour rotten grapes (Barata et al. 2012a) but it was not detected at the end of fermentation (Barata et al. 2012b), which may explain its lower incidence in wines than *Z. bailii.* In spite of the ability of *Z. bisporus*, isolated from sherry film, to resist to sorbic acid and sulphur dioxide (Splittstoesser et al. 1978) and to produce odorous acyloins in sherry wines (Neuser et al. 2000), we are not aware of spoilage events by this species (Loureiro and Malfeito-Ferreira 2003a).

The species *Z. rouxii* has been recovered from concentrated grapes juices. However, the importance of this yeast as bottled wine spoiler is much lower than that of *Z. bailii*, probably due to its weaker resistance to low pH and chemical preservatives.

A new species – *Z. lentus* – was recognized based on isolates from several food industries including one red wine. It is characterized by having similar stress tolerances to *Z. bailii* and *S. cerevisiae*, being distinguished by growing slowly at low temperature (4 °C) (Steels et al. 1999). Another species with physiological tests similar to *Z. bailii* was identified as *Z. parabailii* (Suh et al. 2013). We have already isolated this species from bottled wines (Table 12.1). It remains to be seen if these two species may be regarded as wine spoilers. Interestingly, a *Z. bailii*-derived interspecies hybrid strain ISA1307 with an unknown species, isolated from a continuous sparkling wine plant by our laboratory, was studied concerning its unusual acetic acid resistance (Mira et al. 2014).

Another species taxonomically closely related with *Z. bailii* is *Torulaspora delbrueckii* (Kurtzman et al. 2011). It contaminates concentrated grapes juices and its spoiling effects are related with growth in bottled wine, as described by Minarik (1983). We have detected this species, particularly in concentrated or sulphited grape juices, but in spoiled wines its incidence in bottled wines is much lower than that of *Z. bailii*, probably given its weaker resistance to preservatives.

The last taxonomical rearrangements have created the genus *Lachancea* that includes *L. thermotolerans* (ex *Kluyveromyces thermotolerans*), *L. waltii* (ex *K. waltii*), *L. cidri* (ex *Z. cidri*) and *L. fermentati* (ex *Z. fermentati*) (Kurtzman and James 2006). The latter species has been implicated in the production of odour active compounds in sherry-like medium (Freeman et al. 1977) but has not been implicated in wine spoilage.

2.3.2 *Saccharomyces cerevisiae* and Related Species

Yeast genome sequencing has provided a deeper understanding of the evolution of *S. cerevisiae* and related species like *S. uvarum*, *S. kudriavzevii* or *S. eubayanus* (Hittinger et al. 2015). The former species is the desired agent of wine fermentation. However, it may also be responsible for wine spoilage. During fermentation, the occasional nutritional imbalance of grape juice may lead to off-flavour production imparted by sulphur reduced compounds (Bell and Henschke 2005). Modern wine-making systems with juice pump over under anaerobic conditions tend to increase the problem, contrarily to old systems where juice aeration was present. If not treated in due time, these taints may persist during storage and in bottled wines. In finished wines the most frequent detrimental effects of this species are refermentation of sweet wines and sediment, cloudiness or haziness formation. These effects are similar to those provoked by *Z. bailii* and some *S. uvarum* (former *bayanus*) strains may be more dangerous given their higher tolerance to ethanol (Malfeito-Ferreira et al. 1990).

2.3.3 *Saccharomycodes ludwigii* and *Schizosaccharomyces pombe*

S. ludwigii and *S. pombe* are notorious agents of wine spoilage thanks to their high resistance to stress conditions. Despite this feature their overall incidence is much lower than that of *Z. bailii* and *S. cerevisiae*, for which there is no obvious explanation. Both species have been isolated from grapes in vineyard but with reduced incidence (Florenzano et al. 1977; Combina et al. 2005) and in grape juices (Pardo et al. 1989), being also rare contaminants of winery environments. We hypothesise that their natural contamination sources are more restricted and/or their ability to survive in winery environments is lower than those of *Z. bailii* or *S. cerevisiae*. Their common effects result from cell growth in bottled wine leading to sediment or turbidity formation and refermentation. We currently isolate *S. ludwigii* in bulk white Vinho Verde wine, particularly when an excess of sulphur dioxide is used, and in sparkling wine plants using the Charmat system where their growth may clog stainless steel pipes. *S. ludwigii* is particularly resistant to sulphur dioxide probably due to its ability to produce high amounts of acetaldehyde. *S. pombe* has been exploited for the reduction of malic acid (Delfini and Formica 2001) which, if adopted, must be followed by preventive measures to reduce the risk of post-treatment proliferation.

2.3.4 *Dekkera/Brettanomyces Bruxellensis*

Dekkera bruxellensis, or its anamorph, *Brettanomyces bruxellensis* is presently the most notorious wine spoilage yeast due to the production of ethylphenols in red wines (Loureiro and Malfeito-Ferreira 2006). This species is long known as an undesirable contaminant but not as a recognised producer of these metabolites. The present widespread use of oak barrels to age premium red wine, where the ability to produce ethylphenols overwhelms the presence of other contaminants, contributed significantly to its notoriety. In addition, the controversy about its influence on wine quality involving winemakers, journalists and consumers make this species as the most prominent microbial wine spoilage subject (Malfeito-Ferreira 2018).

D. bruxellensis is rather elusive yeast, being difficult to isolate from sources contaminated by other yeasts due to its low growth rates. Thus the use of selective media and long incubation periods are essential to its recovery. It has been rarely isolated from grapes (Guerzoni and Marchetti 1987; Renouf and Lonvaud-Funel 2007) and winery environments (Connel et al. 2002), being dominant in bottled red wines, as ethylphenols producers, or in sparkling wines, inducing cloudiness, when have no concurrence of the other yeasts (Loureiro and Malfeito-Ferreira 2006). In relative terms, it is not so tolerant to ethanol or preservatives, as *S. cerevisiae* or *Z. bailii* but has the ability to remain viable for long periods and proliferate when conditions become less severe (Renouf et al. 2007). Their occasional detection in sparkling wines may be related with their resistance to carbon dioxide as observed also for *D. naardenensis* in carbonated soft drinks (Esch 1992). However, it is seldom isolated from still white wines for which there is yet no satisfactory explanation (Barata et al. 2008). Another species, *P. guilliermondii*, is also able to produce ethylphenols in grape juices but not in wines (Barata et al. 2006) and so should not be regarded as a sensu stricto spoilage species.

3 Yeast Monitoring

3.1 Microbiological Control

The conservative attitude of wine industry and, namely, the absence of microbiological safety hazards, determine that the implementation of HACCP and self-control plans, mandatory in most food industries, is not dealt with the desirable strictness. In fact, the microbial stability of most dry table wines – white, rosé or red – attained when good winery practices are followed, leads to the absence of microbiological control by most producers. Exceptionally, commercial contracts with modern distributors (supermarket chains and others) or demanding clients may force the implementation of routine microbiological analysis. For this reason, microbiological control in wine industry is, as a rule, synonym of microbiological assessment (particularly yeasts) of bottled sweet wines processing where the risk of refermentation is high. However, the present microbiological hazards of wine industry should justify much more attention.

3.1.1 Grape and Grape Juice Monitoring

During wine fermentation it is neither easy nor justifiable to implement microbiological control plans to detect spoilage yeasts. Their influence in wine quality, as a rule, is irrelevant and possible corrective measures are practically absent. One of the few measures is to establish chemical indicators related with grape microbiological quality, already implemented in numerous wineries (particularly cooperatives or large companies to establish the price of grapes as a function of its health status), like laccase activity (indicator of grapes affected by grey rot) or volatile acidity and gluconic acid (indicator of grapes affected by sour rot). The utilisation of costly FTIR instruments makes these determinations readily available, thus providing the possibility of separate processing according to raw material quality. In smaller dimension wineries, grape selection enables the removal or separate processing of poor quality grapes.

3.1.2 Bulk Wine Monitoring

After wine fermentation, most wineries measure qualitative or quantitative chemical indicators to control the activity of lactic acid bacteria (malic acid assessment) and acetic acid bacteria (volatile acidity). It is not current practice to monitor the presence of spoilage yeasts. However, it would be useful to screen spoilage yeasts or its secondary metabolites, such as 4-ethylphenol and ethylacetate, namely in wines produced from poor sanitary quality grapes. In this case, the prevalence of such yeasts seems to be high and the wine resistance to microbial colonisation is reduced, creating conditions to product alteration. During this stage it is also important to monitor the presence of film-forming yeasts growing on wine surface, mainly in large volume vessels or untopped tanks where it is not easy to avoid the presence of oxygen required by these yeasts. Microbiological analysis is not a requirement but visual inspection of tank tops every 2 weeks is a simple and effective practice. In white wines, particularly those with residual sugar, the specific detection of *Z. bailii* or *S. cerevisiae* should be considered, because may cause refermentation problems during storage.

3.1.3 The Peculiar Case of *D. Bruxellensis*

The relatively low demanding microbiological control during bulk wine storage is no longer advisable concerning red wines, particularly those deserving appropriate aging. Presently, the detection of *D. bruxellensis* is a pre-requisite for wineries during all processing stages of premium red wines. In fact, it frequently appears in high levels just after the malolactic fermentation (Rodrigues et al. 2001) leading to premature "horse sweat" taint. During barrel ageing, irrespective of grape quality, it is essential to monitor *D. bruxellensis* periodically, mainly in used barrels, which are a well known ecological niche of these yeasts. We have established, for many Portuguese wineries, microbiological criteria that have been giving adequate results so far, and are given here only as guidelines. In the first case, for bulk-stored wines, it is satisfactory to detect *D. bruxellensis* monthly, bimonthly or even every 3 months (according to the type of wine and of container). The sample volumes are 1, 0.1, 0.01 and 0.001 ml, from a blend composed by wine from the interface air/liquid and from different depths of the container. When the result is positive for 1 ml, or less, and the level of 4-ethylphenol is higher than 150 μg/l, it is recommended a fine filtration immediately, accompanied by sulphite addition. In following analysis, after filtration, it is sufficient to monitor the level of 4-ethylphenol, as a rule. For wines before bottling, the criteria are more stringent, and detection should be made on 100, 10 and 1 ml of wine, sampled as described above. When the result is positive in 1 or 10 ml, it is recommended a very fine or sterilizing filtration. If positive detection is only obtained for 100 ml, it is acceptable to control viable cells only by addition of preservatives (e.g. 1 mg/l of molecular sulphite). In this case, bottling must be technically correct and dissolved oxygen should be lowered to practically zero. Otherwise, it is recommended a sterile filtration or, in alternative, a heat treatment of the wine to destroy viable cells.

The use of selective media for *Brettanomyces* may also give false positive results that are very rare because, in red wines, this is the single species that produces volatile phenols (positive responses include plate smelling). In these situations, microscopical examination is an easy alternative to check identity because cellular morphology is unique among wine contaminating yeasts. Moreover, periodical analysis using selective media is an efficient strategy to avoid the use of expensive molecular methods that may be an option when immediate results are required.

3.1.4 Wine Bottling

Wine bottling is the main stage of conventional microbiological control, if adopted by wineries. Common procedures including analysis of bottles, rinsing water, closures (corks, rip caps), bottling and corking machines and atmosphere. When properly applied this control enables the detection of contamination sources determining corrective measures. Most frequently, the contamination sources are located in the filling and corking machines (Loureiro and Malfeito-Ferreira 2003b). The final analysis concerns the evaluation of bottled wine contamination (Loureiro and Malfeito-Ferreira 2003a). Common microbial contaminants do not survive long time after bottling and if microbial counts are higher than the specifications, the product is retained until clearance is given.

3.2 Tools Used in Microbiological Control in the Wineries

As a rule, yeast detection and enumeration methodologies are based on growth on plates containing a general-purpose culture medium, after membrane filtration of wine samples or rinsing solutions (Loureiro et al. 2004). The use of Most Probable Number (MPN) technique is not common, but according to our practical experience would be useful in some situations, particularly when is desired an estimation of yeast contamination in bulk wine, or in wines with high percentage of suspended solids. Novel techniques based on direct molecular analysis, flow cytometry or biosensors are also available (Escott et al. 2017; Tubia et al. 2018) but their main limitation is the cost for most small and medium size companies. Even when cost is not an issue, our experience has shown that false positives due to the presence DNA fragments from dead cells may lead to exaggerated inactivation processes with occasional loss of wine quality. In our opinion, it is difficult for a winemaker to rely on techniques that frequently give false positives for microbial stable wines.

The utilisation of selective and/or differential culture media has only somewhat increased in the last years, mainly owing to the problems with *D. bruxellensis*. This situation shows a clear distinction from the typical bacterial control of other food industries, where bacterial indicators, based on the use of a wide variety of selective/differential media, play a central role. Likewise, we presented the concept of zymological (zymo = yeast) indicator to be applied to the wineries, in order to increase the utility of the routine microbiological control (Loureiro and Querol 1999).

The objective of using zymological indicators is to measure the hygienic quality of surfaces that contact with wine and the spoilage risks involved, given that the microbiological safety is not an issue. Taking in account the formerly defined wine yeast groups, the hygienic quality of the wine processing may be assessed by the detection of film-forming yeasts by the MPN technique, using a general-purpose culture medium. To detect sensu stricto spoilage yeasts selective/differential media have been developed directed to the most dangerous species – *Z. bailii* (Schuller et al. 2000) and *D. bruxellensis* (Rodrigues et al. 2001). *S. cerevisiae* could be indirectly estimated by the difference between counting on general purpose media in the absence or presence of cycloheximide or of lysine (Heard and Fleet 1986). The presumptive results obtained with culture media could be further confirmed, if necessary, using molecular biological identification. Currently, yeast identity is obtained by DNA sequencing, which are usually used by external laboratories and not by wineries, due to the degree of expertise and equipment required. In rare and special situations, particularly commercial conflicts, fine molecular typing techniques, adequate to source tracking (Giudici and Pulvirenti 2002), may be used for forensic studies of wine contamination. Additionally, chemical indicators can also be used to monitor yeast activity in an easy and fast way. Among the molecules produced by yeasts, 4-ethylphenol is currently the most common indicator of *D. bruxellensis* activity and should be used together with microbiological detection.

3.3 Acceptable Levels of Yeasts

The establishment of acceptable levels of microorganisms in the final product is a common concern to many food industries. In foods harbouring pathogenic microorganisms law regulates these levels and the technologist effort is addressed to their compliance. Regarding yeast spoilage legislation is practically absent and the aim of the technologist is to establish levels that are attainable under his industrial conditions and ensure product stability during its shelf life (Loureiro and Malfeito-Ferreira 2003a).

In the case of *Z. bailii*, one viable cell per bottle may cause spoilage (Davenport 1986; Deak and Reichart 1986; King et al. 1986; Thomas 1993) but such a strict limit is difficult to attain in winery practice and is not at all appropriate when yeasts counts belong to innocent contaminants. Occasionally specifications are established as a function of the sugar content, assuming that sweet wines are more vulnerable than dry ones. However, Deak and Reichart (1986) did not find differences in the microbial stability of white, red, semi-dry and semi-sweet wines and demonstrated that stability depends on the initial yeast population. Accordingly, Chandra et al. 2015) using a response surface methodology approach also demonstrated that residual sugar do not increase wine instability with varying levels of ethanol and sulphite. These authors speculated that sugar do not prevent the cell death process, however if cells are able to grow then sugar will promote yeast proliferation with consequent spoilage. The absence of inhibition by ethanol in alcohol free wine may explain the recovery of unusual contaminant species, not associated with spoilage, in such products (Table 12.1).

In the absence of sound scientific background to establish appropriate specifications, the industry empirically establishes its limits, which may be used for commercial purposes. A level of yeast counts as low as <1/500 ml or < 1/ml are currently regarded as the maximum acceptable level (Andrews 1992; Loureiro and Malfeito-Ferreira 2003a) reflecting the caution on the prevention of spoilage events. As a rule, the estimation of contaminating flora is obtained after growth on general media and so results reflect the total flora and not the spoiling one. If this may be wise for highly virulent species, in most of the cases these values are too strict for innocuous contamination ones. Most wineries, when levels are higher than acceptable, hold the product for enough time to meet specifications or to re-bottle it. This procedure gives an indication of the spoilage risk because the increase in yeast counts is a signal of contamination by spoiling yeasts. However, this may lead to long holding periods in the case of high initial yeast loads, which take time to be reduced to acceptable levels, even in the case of innocent contaminants. Their specific detection would give clearance to the final product sooner with obvious economical advantages.

References

Andrews, S. (1992). Specifications for yeasts in Australian beer, wine and fruit juice products. In R. A. Samson, A. D. Hocking, J. I. Pitt, & A. D. King (Eds.), *Modern methods in food mycology* (pp. 111–118). Amsterdam: Elsevier.

Barata, A., Correia, P., Nobre, A., Malfeito-Ferreira, M., & Loureiro, V. (2006). Growth and 4-ethylphenol production by the yeast *Pichia guilliermondii* in grape juices. *American Journal of Enology and Viticulture, 57*, 133–138.

Barata, A., Caldeira, J., Botelheiro, R., Pagliara, D., Malfeito-Ferreira, M., & Loureiro, V. (2008). Survival patterns of *Dekkera bruxellensis* in wines and inhibitory effect of Sulphur dioxide. *International Journal Food Microbiology, 121*, 201–207.

Barata, A., Malfeito-Ferreira, M., & Loureiro, V. (2012a). The microbial ecology of wine grape berries. *International Journal of Food Microbiology, 153*, 243–259.

Barata, A., Malfeito-Ferreira, M., & Loureiro, V. (2012b). Changes in sour rotten grape berry microbiota during ripening and wine fermentation. *International Journal of Food Microbiology, 154*, 152–161.

Barata, A., Laureano, P., D'Antuono, I., Martorell, P., Stender, H., Malfeito-Ferreira, M., Querol, A., & Loureiro, V. (2013). Enumeration and identification of 4-ethylphenol producing yeasts recovered from the wood of wine ageing barriques after different sanitation treatments. *Journal of Food Research, 2*, 140–149.

Bell, S.-J., & Henschke, P. A. (2005). Implications of nitrogen nutrition for grapes, fermentation and wine. *Australian Journal of Grape and Wine Research, 11*, 242–295.

Bisson, L. F., & Butzke, C. E. (2000). Diagnosis and rectification of stuck and sluggish fermentations. *American Journal of Enology and Viticulture, 51*, 168–177.

Blackburn, C. (Ed.). (2006). *Food spoilage microorganisms*. Cambridge: Woodhead Pub. Ltd..

Boekhout, T. (2005). Biodiversity: Gut feeling for yeasts. *Nature, 434*, 449–451.

Boekhout, T., & Robert, V. (Eds.). (2003). *Yeasts in food: Beneficial and detrimental aspects*. Hamburg: B. Behr's Verlag.

Chandra, M., Oro, I., Ferreira-Dias, S., & Malfeito-Ferreira, M. (2015). Effect of ethanol, sulfur dioxide and glucose on the growth of wine spoilage yeasts using response surface methodology. *PLoS One, 10*, e0128702. https://doi.org/10.1371/journal.pone.0128702.

Chatonnet, P., Boidron, J., & Dubourdieu, D. (1993). Influence des conditions d'élevage et de sulfitage des vins rouges en barriques sur le teneur en ácide acétique et en ethyl-phenols. *Journal International Sciences Vigne Vin, 27*, 277–298.

Combina, M., Mercado, L., Borgo, P., Elia, A., Jofré, V., Ganga, A., Martinez, C., & Catania, C. (2005). Yeasts associated to Malbec grape berries from Mendoza, Argentina. *Journal of Applied Microbiology, 98*, 1055–1061.

Connel, L., Stender, H., & Edwards, C. (2002). Rapid detection and identification of Brettanomyces from winery air samples based on peptide nucleic acid analysis. *American Journal of Enology and Viticulture, 53*, 322–324.

Davenport, R. (1986). Unnaceptable levels for yeasts. In A. King, J. Pitt, L. Beuchat, & J. Corry (Eds.), *Methods for the mycological examination of food* (pp. 214–215). New York: Plenum Press.

Deak, T., & Beuchat, L. (1996). *Handbook of food spoilage yeasts*. Boca Raton: CRC Press.

Deak, T., & Reichart, O. (1986). Unnaceptable levels of yeasts in bottled wine. In A. King, J. Pitt, L. Beuchat, & J. Corry (Eds.), *Methods for the mycological examination of food* (pp. 215–218). New York: Plenum Press.

Delfini, C., & Formica, J. (2001). *Wine microbiology: Science and technology*. New York: Marcel Dekker.

Donnelly, D. (1977a). Airborne microbial contamination in a winery bottling room. *American Journal of Enology and Viticulture, 28*, 176–181.

Donnelly, D. (1977b). Elimination from table wines of yeast contamination by filling machines. *American Journal of Enology and Viticulture, 28*, 182–184.

Esch, F. (1992). Yeast in soft drinks and concentrated fruit juices. *Brygmesteren, 4*, 9–20.

Escott, C., Loira, I., Morata, A., Bañuelos, M., & Suárez-Lepe, J. (2017). Wine spoilage yeasts: control strategy. In A. Morata (Ed.), *Yeast*. IntechOpen. https://doi.org/10.5772/intechopen.69942. Available from: https://www.intechopen.com/books/yeast-industrial-applications/wine-spoilage-yeasts-control-strategy.

Farris, G., Zara, S., Pinna, G., & Budroni, M. (2002). Genetic aspects of flor yeasts *Sardinian strains*, a case of study. In M. Ciani (Ed.), *Biodiversity and biotechnology of wine yeasts* (pp. 71–83). Kerala: Research Signpost.

Fleet, G., Prakitchaiwattana, C., Beh, A., & Heard, G. (2002). The yeast ecology of wine grapes. In M. Ciani (Ed.), *Biodiversity and Botechnology of wine yeasts* (pp. 1–17). Kerala: Research Signpost.

Florenzano, G., Balloni, W., & Materassi, R. (1977). Contributo alla ecologia dei lieviti *Schizosaccharomyces* sulle uve. *Vitis, 16*, 38–44.

Freeman, B., Muller, C., Kepner, R., & Webb, A. (1977). Some products of the metabolism of ethyl 4-oxobutanoate by *Saccharomyces fermentati* in the film form on the surface of simulated flor sherry. *American Journal of Enology and Viticulture, 28*, 119–122.
Fugelsang, K., & Edwards, C. (2007). *Wine microbiology*. Berlin: Springer Verlag.
Giudici, P., & Pulvirenti, A. (2002). Molecular methods for identification of wine yeasts. In M. Ciani (Ed.), *Biodiversity and biotechnology of wine yeasts* (pp. 35–52). Kerala: Research Signpost.
Gschaedler, A. (2017). Contribution of non-conventional yeasts in alcoholic beverages. *Current Opinion in Food Science, 13*, 73–77.
Guerzoni, E., & Marchetti, R. (1987). Analysis of yeast Flora associated with grape sour rot and of the chemical disease markers. *Applied and Environmental Microbiology, 53*, 571–576.
Heard, G., & Fleet, G. (1986). Evaluation of selective media for enumeration of yeasts during wine fermentation. *Journal of Applied Bacteriology, 60*, 477–481.
Hittinger, C., Rokas, A., Bai, F.-Y., Boekhout, T., Gonçalves, P., Jeffries, T., Kominek, J., Lachance, M.-A., Libkind, D., Rosa, C., Sampaio, J., & Kurtzman, C. (2015). Genomics and the making of yeast biodiversity. *Current Opinion in Genetics and Development, 35*, 100–109.
Ibeas, J. I., Lozano, I., Perdigones, F., & Jiménez, J. (1996). Detection of *Dekkera-Brettanomyces* strains in Sherry by a Nested PCR Method. *Applied and Environmental Microbiology, 62*, 998–1003.
King, A., Pitt, J., Beuchat, L., & Corry, J. (Eds.). (1986). *Methods for the mycological examination of food*. New York: Plenum Press.
Kunkee, R., & Bisson, L. (1993). Wine-making yeasts. In A. Rose & J. Harrison (Eds.), *The yeasts* (Vol. 5, 2nd ed., pp. 69–127). London: Academic.
Kurtzman, C., & James, S. (2006). *Zygosaccharomyces* and related genera. In C. Blackburn (Ed.), *Food spoilage microorganisms* (pp. 289–305). Cambridge: Woodhead Publishers.
Kurtzman, C., & Robnett, C. (2003). Phylogenetic relationships among yeasts of the '*Saccharomyces* complex' determined by multigene sequence analysis. *FEMS Yeast Research, 3*, 417–432.
Kurtzman, C., Fell, J., & Boekhout, T. (Eds.) (2011). The yeasts, 5th edition, *A taxonomic study*. Elsevier, Amsterdam.
Loureiro, V. (2000). Spoilage yeasts in food and beverages: Characterisation and ecology for improved diagnosis and control. *Food Research International, 33*, 247–256.
Loureiro, V., & Malfeito-Ferreira, M. (2003a). Spoilage yeasts in the wine industry. *International Journal of Food Microbiology, 86*, 23–50.
Loureiro, V., & Malfeito-Ferreira, M. (2003b). Yeasts in spoilage. In B. Caballero, L. Trugo, & P. Finglas (Eds.), *Encyclopedia of food sciences and nutrition* (2nd ed., pp. 5530–5536). London: Academic.
Loureiro, V., & Malfeito-Ferreira, M. (2006). Spoilage activities of *Dekkera/Brettanomyces* spp. In C. Blackburn (Ed.), *Food spoilage microorganisms* (pp. 354–398). Cambridge: Woodhead Publishers.
Loureiro, V., & Querol, A. (1999). The prevalence and control of spoilage yeasts in foods and beverages. *Trends Food Science and Technology, 10*(11), 356–365.
Loureiro, V., Malfeito-Ferreira, M., & Carreira, A. (2004). Detecting spoilage yeasts. In R. Steele (Ed.), *Understanding and measuring the shelf-life of food* (pp. 233–288). Cambridge: Woodhead Publishers.
Malfeito-Ferreira, M. (2011). Yeasts and wine off-flavours: A technological perspective. *Annals of Microbiology, 61*, 95–102.
Malfeito-Ferreira, M. (2018). Two decades of "horse sweat" taint and *Brettanomyces* yeasts in wines: Where do we stand now? *Beverages, 4*, 32. https://doi.org/10.3390/beverages4020032.
Malfeito-Ferreira, M., Lopes, J., & Loureiro, V. (1990). Characterization of spoilage yeasts in Portuguese bottled dry white wines. In P. Ribéreau-Gayon & A. Lonvaud (Eds.), *Actualités Oenologiques 89, Comptes rendus du 4eme Symposium International d'Oenologie, Bordeaux* (pp. 293–296). Paris: Dunod.
Malfeito-Ferreira, M., Tareco, M., & Loureiro, V. (1997). Fatty acid profiling: A feasible typing system to trace yeast contaminations in wine bottling plants. *International Journal of Food Microbiology, 38*, 143–155.

Minarik, E. (1983). Levures de contamination des vins embouteillés. *Bulletin OIV, 56*(628), 414–419.
Mira, N., Münsterkötter, M., Dias-Valada, F., Santos, J., Palma, M., Roque, F., Guerreiro, J., Rodrigues, F., Sousa, M., Leão, C., Güldener, U., & Sá-Correia, I. (2014). The genome sequence of the highly acetic acid-tolerant *Zygosaccharomyces bailii*-derived interspecies hybrid strain ISA1307, isolated from a sparkling wine plant. *DNA Research, 21*, 299–313.
Neradt, F. (1982). Sources of reinfections during cold-sterile bottling of wine. *American Journal of Enology and Viticulture, 33*, 140–144.
Neuser, F., Zorn, H., & Berger, R. (2000). Generation of odorous acyloins by yeast pyruvate decarboxylases and their occurrence in sherry and soy sauce. *Journal of Agriculture and Food Chemistry, 48*, 6191–6195.
Padilla, B., Gil, J., & Manzanares, P. (2016). Past and future of non-Saccharomyces yeasts: From spoilage microorganisms to biotechnological tools for improving wine aroma complexity. *Frontiers in Microbiology, 7*, 411.
Pardo, I., García, M. J., Zúniga, M., & Uruburu, F. (1989). Dynamics of microbial populations during fermentation of wines from the Utiel-Requena region of Spain. *Applied and Environmental Microbiology, 55*, 539–541.
Pitt, J., & Hocking, A. (1985). *Fungi and food spoilage*. Sydney: Academic.
Querol, A., & Fleet, G. (Eds.). (2006). *Yeasts in food and beverages*. Berlin: Springer-Verlag.
Renouf, V., & Lonvaud-Funel, A. (2007). Development of an enrichment medium to detect *Dekkera/Brettanomyces bruxellensis*, a spoilage wine yeast, on the surface of grape berries. *Microbiological Research, 162*, 154–157.
Renouf, V., Perello, M.-C., Revel, G., & Lonvaud-Funel, A. (2007). Survival of wine microorganisms in the bottle during storage. *American Journal of Enology and Viticulture, 58*, 379–386.
Rodrigues, N., Gonçalves, G., Pereira-da-Silva, S., Malfeito-Ferreira, M., & Loureiro, V. (2001). Development and use of a new medium to detect yeasts of the genera *Dekkera/Brettanomyces* spp. *Journal of Applied Microbiology, 90*, 588–599.
Romano, P. (2005). Proprietà technologiche e di qualità delle specie di lieviti vinari. In M. Vicenzini, P. Romano, & G. Farris (Eds.), *Microbiologia del vino* (pp. 101–131). Milano: Casa Editirice Ambrosiana.
Schuller, D., Côrte-Real, M., & Leão, C. (2000). A differential medium for the enumeration of the spoilage yeast *Zygosaccharomyces bailii* in wine. *Journal of Food Protection, 63*, 1570–1575.
Sena-Esteves, M., Mota, M., & Malfeito-Ferreira, M. (2018). Patterns of sweetness preference in red wine according to consumer characterization. *Food Research International, 106*, 38–44.
Splittstoesser, D. F., Queale, D. T., & Mattick, L. R. (1978). Growth of Saccharomyces bisporus var. bisporus, a yeast resistant to sorbic acid. *American Journal of Enology and Viticulture, 29*, 272–276.
Steels, H., James, S., Roberts, I., & Stratford, M. (1999). *Zygosaccharomyces lentus*: A significant new osmophilic, preservative-resistant spoilage yeast, capable of growth at low temperature. *Journal of Applied Microbiology, 87*, 520–527.
Stratford, M. (2006). Food and beverage spoilage Yeats. In A. Querol & G. Fleet (Eds.), *Yeasts in food and beverages* (pp. 335–379). Berlin: Springer-Verlag.
Suárez-Lepe, J., & Iñigo-Leal, B. (2004). *Microbiología Enológica: Fundamentos de Vinificación* (3rd ed.). Madrid: Ediciones Mundi-Prensa.
Suh, S.-O., Gujjari, P., Beres, C., Beck, B., & Zhou, J. (2013). Proposal of *Zygosaccharomyces parabailii* sp. nov. and *Zygosaccharomyces pseudobailii* sp. nov., novel species closely related to *Zygosaccharomyces bailii*. *International Journal of Systematic and Evolutionary Microbiology, 63*, 1922–1929.
Thomas, D. S. (1993). Yeasts as spoilage organisms in beverages. In A. Rose & J. Harrison (Eds.), *The yeasts* (Vol. 5, pp. 517–561). London: Academic.
Tubia, I., Prasad, K., Pérez-Lorenzo, E., Abadín, C., Zumárraga, M., Oyanguren, I., Barbero, F., Paredes, J., & Arana, S. (2018). Beverage spoilage yeast detection methods and control technologies: A review of *Brettanomyces*. *International Journal of Food Microbiology, 283*, 65–76.

Valera, C., & Borneman, A. (2017). Yeasts found in vineyards and wineries. *Yeast, 34*, 111–128.

Valero, E., Schuller, D., Cambon, B., Casal, M., & Dequin, S. (2005). Dissemination and survival of commercial wine yeast in the vineyard: A large-scale, three-years study. *FEMS Yeast Research, 5*, 959–969.

Zuehlke, J. M., Petrova, B., & Edwards, C. G. (2013). Advances in the control of wine spoilage by *Zygosaccharomyces* and *Dekkera/Brettanomyces*. *Annual Review of Food Science and Technology, 4*, 57–78.

Chapter 13
Yeasts and Sparkling Wine Production

Hervé Alexandre

Contents

H. Alexandre (✉)
Institut Universitaire de la Vigne et du Vin Jules Guyot, Université de Bourgogne, Dijon, France
e-mail: rvalex@u-bourgogne.fr; http://www.u-bourgogne.fr/IUVV/

P. Romano et al. (eds.), *Yeasts in the Production of Wine*,
https://doi.org/10.1007/978-1-4939-9782-4_13

1 Introduction

Sparkling wine can be produced by the Champenoise, bottle fermentation, transfer or Charmat processes. This chapter focuses principally on microbiological aspects of the traditional Champenoise method. Process details for the different methods of production are described in Howe (2003). All methods for sparkling wine production require two successive alcoholic fermentations, but the Champenoise method is characterized by a long period of aging the wine on yeast lees, which profoundly affects the flavor and quality of the final product.

The first fermentation in the champenoise method transforms grape must into base wine. The second fermentation is the key element of this method. It takes place in the bottle and increases alcohol content and internal bottle pressure (up to 5–7 atmospheres). After this second alcoholic fermentation, the wine is aged on the yeast lees for at least 9 months, depending on the legislation in the country where it is produced. Champagne is aged on the lees for 1 year, and vintage (millésimé) Champagne is aged for 3 years. Yeast autolysis occurs during this prolonged period of contact with the wine. This process is slow and involves cell death, with hydrolytic enzymes releasing cytoplasmic (peptides, fatty acids, nucleotides, amino acids) and cell wall (mannoproteins) compounds into the wine. Storage at low temperatures during aging may lead to slow cell death rates and slow enzymatic reactions, accounting for the long time required for the process. The organoleptic and foaming properties of the wine are modified during ageing on yeast lees, reflecting changes in wine composition.

2 Viticultural Parameters

Desired characteristics of the grape juice should be taken into account when considering cultivation of grapes for sparkling wine production. An example for the Champagne grape juice is given Table 13.1 (Tusseau 2004).

It is essential to check grape maturity before harvest to ensure that the appropriate analytical profile is obtained. Sparkling wines are generally white or rosé. They may be produced from either white or red grapes, but overripe red grapes are avoided to prevent coloration of the wine with red pigments. Different grape varieties are

Table 13.1 General properties of musts or juice used in Champagne production. (adapted from Tusseau 2004)

Potential alcohol (% v/v)	10.0
Sugar (g/l)	170
Total acidity (g/l H_2SO_4)	7.6
Ratio sugar/acidity	22.7
pH	3.10
Tartaric acid (g/l)	7.5
Malic acid (g/l)	6.5
Calcium (mg/l)	88
Total nitrogen (mg/l)	477
Ammoniacal nitrogen (mg/l)	127
Copper (mg/l)	1.8

Table 13.2 Grape varieties used for sparkling wine production in different countries (adapted from Buxaderas and Lopez-Tamames 2012)

Sparkling wine name	Country	Grape varieties
Crémant	France	Chardonnay, Chenin Blanc, Cabernet Franc, Aligoté, Gamay, …
Espumoso (Cava)	Spain	Macabeo, Parellada, Xarel-Lo, Chardonnay
Asti spumante	Italia	Moscato
Cap Classique	South Africa	Sauvignon blanc, Chenin blanc, …
Traditional method	Australia	Chardonnay, Pinot noir, Pinot meunier, Shiraz, occasionally, Cabernet-Sauvignon, Merlot
Traditional method	USA	Chardonnay, Pinot blanc and Pinot noir, Riesling and others
Champagne	France	Pinot noir, Pinot meunier, Chardonnay

used in order to produce different style of sparkling wine across the world. Table 13.2 summarized the main grape varieties used to produce the most known sparkling wines.

The chemical parameters of juices/musts used for sparkling wine production differ from those used for still wine production, with musts for still wine production having lower ripeness indices, lower total acidity, higher sugar content and lower nitrogen concentration (Maujean et al. 1987). In addition to these maturity criteria, the prevention of mold development on the grapes is important. *Botrytis cinerea* has highly detrimental effects on sparkling wine quality (Ribéreau-Gayon et al. 2004), modifying juice protein content and foaming properties (Marchal et al. 1999). Indeed, presence of *Botrytis* on grapes lead to oxidation and browning together with development of off-flavours production of β-glucans leading to clarification problems and above all *Botrytis* proteases alter effervescence, especially above 20% infestation (Cilindre et al. 2008; Buxaderas and Lopez-Tamames 2012).

3 Base Wine Production

Whole bunch pressing is the best method for pressing grapes for sparkling wine production, because the resulting juice is low in both phenolic compounds and sediment. Grapes are, therefore, best harvested by hand. The grape varieties used for sparkling wine production are highly diverse, resulting in very different products (Table 13.2).

3.1 Pressing

Mechanical harvesting is forbidden for grapes used for sparkling wine production in France. In contrast, mechanical harvesting is extensively used in Australia and elsewhere, making it possible to harvest at night, when it is cooler. Depending on the country, grapes may be destemmed after harvest, crushed or pressed as a whole (i.e. without prior removal of stems).

In the Champagne region of France, grapes for premium quality sparkling wines are transported for pressing as quickly as possible, and pressed without crushing or destemming, to limit maceration. Pressing is a key step in sparkling wine production, and serves to give a chemical fractionation of the juice released. The analytical composition of the juice changes during pressing, due to the heterogeneity of berry composition, especially in relation to the location of organic acids (Fig. 13.1).

The cells from the intermediate zones of the grape tissue are the most fragile, and release their contents first, followed by cells from the central zone and, finally, cells from the peripheral zone. Under appellation rules in France, 4000 kg of grapes can be pressed to create up to 25.5 hL of juice. The first 20.5 hL are known as the cuvée, and are obtained under a pressure of 0.2–0.6 bars. The remaining 5.0 hL are known as the taille, and are usually obtained under a standard pressure of 0.6–1.4 bars. Figure 13.2 illustrates the processes used to obtain the cuvée and the taille from grapes for the Champenoise method. The juice (0.75 hL) released during the loading of the grapes into the press (rebêches) is discarded. The 20.5 hL cuvée, plus 2–4% sediment, is then pumped into a racking tank, followed by the 5 hL taille plus

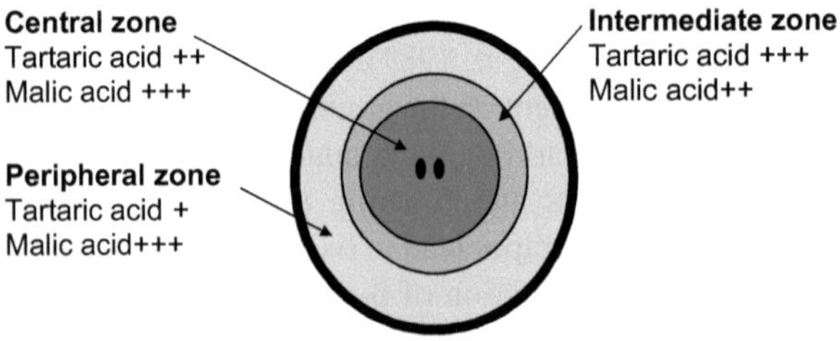

Fig. 13.1 Relative proportions of acids in different zones of grape berries used for wine production

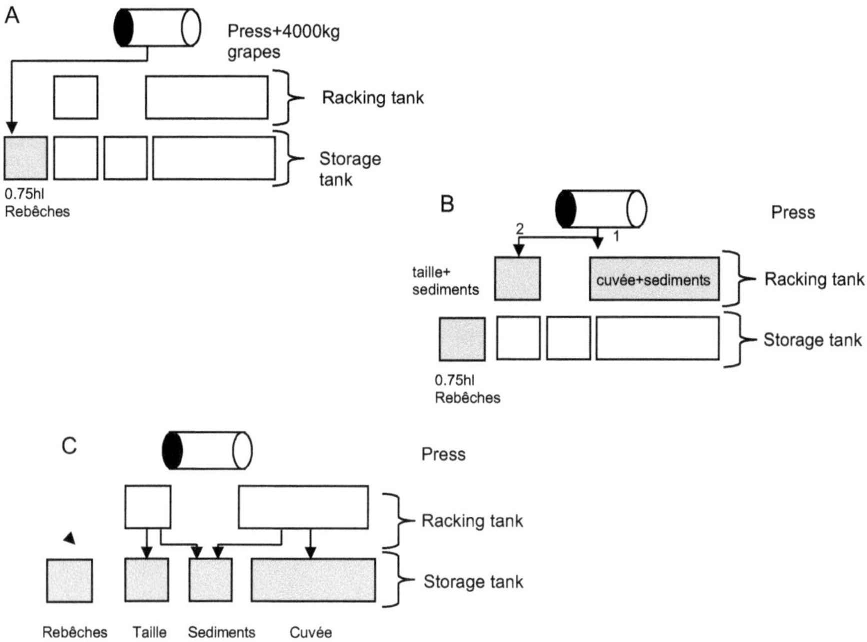

Fig. 13.2 The steps generating cuvée and taille during pressing of grapes for Champagne production. Step A, when the press is charged some juice is liberated and discarded (**A**). During pressing two fractions are obtained, cuvée and taille (**B**). Sediments from each fraction are separated from the juice after racking (**C**). (Kemp et al. 2015)

2–4% sediment which goes into a separate tank.. After racking, 4000 kg of grapes gives rise to 20.5 hL of cuvée, 5 hL taille, 0.75 hL rebêches and 0.5–1 hL sediment. Base wines are made from the cuvée and taille juices only, and, generally, they are fermented separately.

In summary, the composition of the juice changes during pressing, due to the heterogeneity of grape berry composition, the different grape berry zones and the different pressures applied. An example of changes in the acid constituents of the juice, pH and potassium content during pressing is shown in (Fig. 13.3a, b).

3.2 *Racking*

Racking is the operation used to clarify the juice after pressing. Various methods may be used, but all aim to remove solid/particulate material (Boulton et al. 1996). The process is a determinant of juice quality and consequently wine quality (Pernot 1999). Furthermore, the type of fining agent used to clarify the juices influences the quality of base and sparkling wines (Puig-Deu et al. 1999).

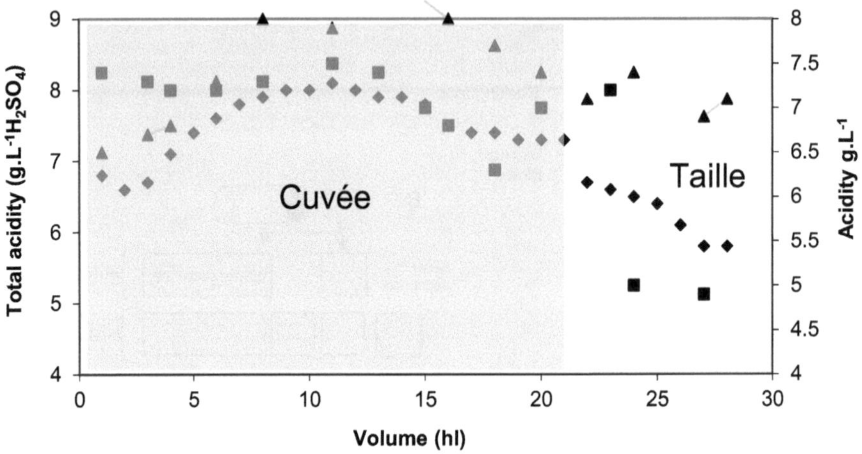

Fig. 13.3a Changes in acidity of juice obtained during pressing of grapes for Champagne production. (◆) Total acidity (g.L^{-1} H_2SO_4), (▲) malic acid (g.L^{-1}), (■) tartaric acid (g.L^{-1}). According to Valade and Blanck (1989)

Fig. 13.3b Changes in pH and potassium concentration of juice obtained during pressing of grapes for Champagne production. (◆) pH, (■) potassium. According to Valade and Blanck (1989)

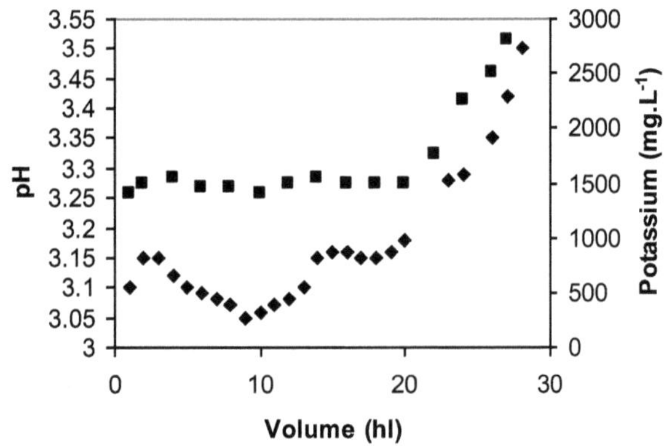

Sulfur dioxide is usually added to the must at a concentration of about 4 g.hL^{-1} to prevent oxidation and browning of the juice. Furthermore, sufur dioxide inhibits microflora activity which can occur during settling and clarification of the juice. Different methods such as cold settling, with or without the addition of pectinolytic enzymes, and centrifugation are used for racking. The settling technique may affect the foaming properties of the wine and the primary alcoholic fermentation. Musts treated with pectic enzymes give wines with lower levels of foam stability and foamability than untreated musts (Lao et al. 1999). The lower protein and polysaccharide contents of treated musts may account for the lower levels of foaming in later wines. Various fining agents may be added during racking to increase the efficiency of juice settling, and facilitate the precipitation of suspended solids. Bentonite and potassium caseinate or cellulose are often used, but these treatments may affect fermentation and foaming properties. Juices treated with caseinate and cellulose usually ferment more completely than juices treated with bentonite (Puig-Deu et al. 1999). After racking, juice turbidity usually ranges from 5 to 50 NTU.

4 First Alcoholic Fermentation

Surprisingly, very few studies have been published on the first alcoholic fermentation during sparkling wine production, probably because this fermentation closely resembles that of still wine. However, the yeasts involved in this fermentation may affect sparkling wine quality.

Traditionally, wine fermentation was spontaneous, mediated by indigenous yeasts present on the grapes at harvest, or introduced from the equipment and cellar during the vinification process (Fleet and Heard 1993). Many factors may affect the total yeast population, resulting in considerable variations in the quality and organoleptic characteristics of the wine produced. Most sparkling winemakers, therefore, now add selected yeast strains to the must. The technological traits on which yeasts are selected for base wine production are not specific to the sparkling wine process, and are similar to those for yeast strains used in still wine production.

Once the juice has settled, indigenous yeast density in the must is between 100 and 10,000 cells.mL^{-1}. Selected yeasts are added at a density of 1–3 × 10^6 cells. mL^{-1}. The must is generally fermented in stainless steel tanks in which the temperature can be controlled. Nitrogenous nutrients (diammonium-hydrogenphosphate,) are added to the must (5–10 g.hL^{-1}) before or at the start of alcoholic fermentation. The temperature is maintained at 18–20 °C during the alcoholic fermentation, giving the wines a more pronounced aromatic profile than fermentations at higher temperatures. Fermentation kinetics are usually consistent and predictable under these conditions. Sluggish or stuck fermentations are very rare, as the final alcohol concentration rarely exceeds 11% (v/v) in base wine production. The ecological profile of yeasts associated with these fermentations is expected to be the same as that for still wines. The early stages will involve contributions from various indigenous strains within genus of *Hanseniaspora*, *Metschnikowia*, *Candida*, *Pichia*, and *Kluyveromyces*, but inoculated strains of *Saccharomyces cerevisiae* or *Saccharomyces bayanus* will eventually dominate the fermentation (Fleet 2007, 2008).

Most of must to get base wine for sparkling wine production are inoculated with *Saccharomyces cerevisiae* (Pozo-Bayon et al. 2009). However more recently, trials have been done with non-*Saccharomyces* yeast in order to improve quality and complexity of wine, but also to improve the foaming properties of the base wine (González-Royo et al. 2015; Medina-trujillo et al. 2017). In the last study, the authors proved that the use of sequential inoculation for the production of base wine lead to a significant increase in higher maximum heights of foam compared to conventional inoculation. These results could reflect a higher release of low molecular weight proteins during the autolysis of *Torulaspora delbrueckii* (Medina-Trujillo et al. 2017).

5 Malolactic Fermentation

Malolactic fermentation generally takes place after alcoholic fermentation and is mediated by lactic acid bacteria, principally strains of *Oenococcus oeni.* This fermentation has a particularly important effect on the wine, changing its composition and organoleptic properties. Its main effect is to reduce the total acidity of the wine and increase its pH, due to the decarboxylation of L-malic acid to L-lactic acid. Bacterial activity also stabilizes the wine and enriches its aromatic composition (Henick-Kling 1993).

The main goal of sparkling wine production is to obtain a fine and fruity wine with a desirable acidity. The need for malolactic fermentation, therefore, remains unclear for such wines. In warm regions that produce wines of low acidity (e.g. regions in Spain, Australia, Greece), malolactic fermentation is not desirable. In cool regions, base wines are highly acidic (pH around 3.0) at the end of the alcoholic fermentation, and malolactic fermentation is needed to decrease this acidity. While malolactic fermentation is frequently an indigenous reaction, it can be carried out by inoculation with selected strains of *O.oeni* (Valade et al. 1987a, b).

After the alcoholic and malolactic fermentations, the wine is gradually cooled to 10–12 °C and clarified by racking and addition of fining agents. The wines are stabilized before bottling to prevent bitartrate precipitation. Indeed, during the secondary fermentation, alcohol concentration increases, and this decreases the solubility of potassium hydrogen tartrate. The presence of tartaric salts in the bottle leads to the release of large amounts of gas and wine when the bottle is opened. The prevention of bitartrate precipitation is, therefore, considered desirable. The wine is then filtered (Ribéreau-Gayon et al. 2004).

Blending is the final step before the secondary fermentation. It is often used to ensure that the base wine is balanced in terms of its composition and flavor. The blending of wines from different years, vineyards and grape varieties makes it possible to create a specific sparkling wine style.

6 Secondary Fermentation

The main goal of secondary fermentation is to obtain a sparkling wine with a pressure of about six bars (6 Kg/cm^2 or 6.10^5 pascal) at 10 °C. It is the unique step of sparkling wine production. Essentially, a mixture of yeast, sugar and fining agents (the *tirage* solution) is added to the base wine so that it contains about 1.5×10^6 yeast cells.mL^{-1} and 24 g.L^{-1} sugar. The tiraged wine is then transferred to bottles or tanks for the secondary fermentation.

During secondary fermentation, yeast cells grow and metabolize the sugar added with the tirage. The yeast inoculum and *tirage* solution (*liqueur de tirage)* must be carefully prepared. An example of a yeast inoculum or tirage preparation as described by Laurent and Valade (2007), is provided in Fig. 13.4.

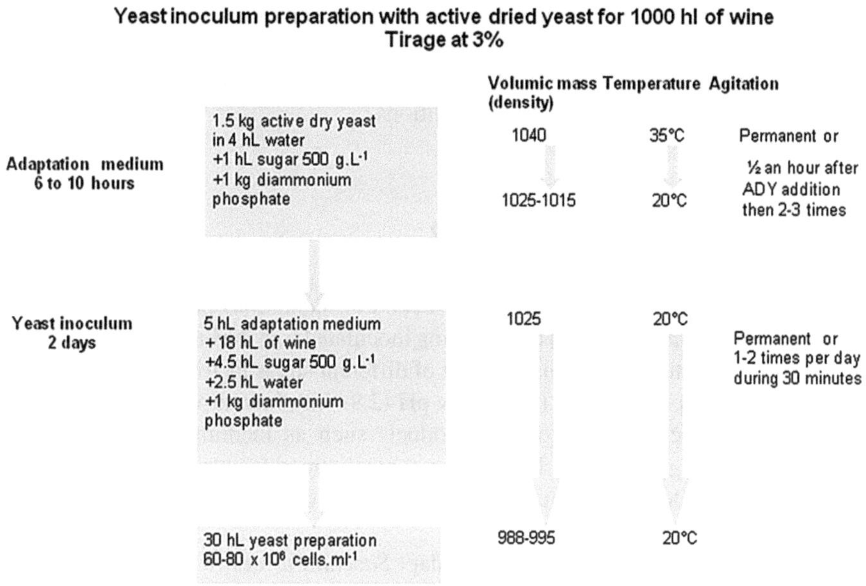

Fig. 13.4 Protocol for preparing tirage and yeast inoculum for sparkling wine production (Laurent and Valade 2007). The first step (adaptation medium) prepares the yeast in a rehydratation medium containing sugar and diammonium phosphate. During this step, the initial temperature is 35 °C and is gradually decreased to 20 °C. Diammonium phosphate addition during this step favors sugar consumption and even during the next step (yeast inoculum preparation). The second step allows adaptation of yeast to a wine medium, especially to alcohol. The temperature (20 °C) allows a better yeast survival than 35 °C. Agitation, decreases CO_2 concentration and leads to better yeast survival

Different commercial yeasts are generally used for the first and second alcoholic fermentations. The yeasts for the first fermentation are selected on the basis of high fermentation speed and low acid production, together with other desirable properties, whereas the yeasts for the second fermentation are selected on the basis of other technological properties (Martinez-Rodriguez et al. 2001a). These properties include the ability to grow at low temperatures and under pressure in a medium containing at least 10% (v/v) ethanol, and having desirable flocculating or agglutinating abilities. For sparkling wine production, the yeast strain is also selected on the basis of its autolytic capacity and foam quality.

Due to the extreme conditions (low pH, high ethanol, and a steady increase in CO_2 pressure), several weeks are required to complete this second fermentation. It was recently found that the second fermentation kinetics are significantly affected by the choice of the yeast strain with which the fermentation is conducted (Martí-Raga et al. 2015). Using a QTL mapping approach Martí-Raga et al. (2017) were able to identify 4 genes (*PMA1, PDR1*, *MSB2*, and *VMA13*) involved in the genetic determinism of second fermentation. These genes play a key role in maintaining intracellular pH regulation, cell detoxification, control of plasma membrane composition and the response to cold stress (Martí-Raga et al. 2017). According to these authors, the stressful conditions of second alcoholic fermentation have driven the selection of strains adapted to these conditions.

6.1 Yeast Inoculum Preparation

Before starting the "prise de mousse" (second fermentation) progressive acclimation of *S. cerevisiae* is required before being inoculated into the base wine (Fig. 13.4), taking into account the synergistic effect of different stress factor such as high ethanol concentration (10%–12% (v/v)), low pH (2.9–3.2), low nitrogen content, accumulation of toxic fermentation sub-products such as medium-chain fatty acids (C6-C12) and organic acids. In addition typical stress factors due to the second fermentation are low temperature (10 °C–15 °C), CO_2 overpressure and high total acidity (5-7gL1of H_2SO_4).

Different protocol exist in order to adapt *Saccharomyces* to this harsh environment (see below). Recently it has been shown that some physiological parameters could be monitored and played a decisive role with respect to reaching the maximum pressure of 6 bars in bottles. These parameters are higher glycogen and trehalose contents, lower ROS accumulation, better vacuolar activity and lower ratios of ergosterol/squalene and oleic acid/stearic acid (Borrull et al. 2016).

Active dry yeast (ADY) is generally added at a concentration of 3 $g.L^{-1}$. The addition of larger amounts of yeast generally prevents efficient cell multiplication and inhibits the recovery of activity (Laurent and Valade 2007). The yeast is rehydrated in water at 35–38 °C for 30 min. During the adaptation phase, the addition of nitrogen does not significantly increase growth rate (Mendes-Fereira et al. 2004), but sugar breakdown increases. The adaptation phase lasts 6–10 h — the time required to reach a volumic mass or density of 1.025–1.015 (Fig. 13.4). Yeast cell viability is greater if the mixture is stirred during the stages of yeast addition and adaptation, than in the absence of stirring. This is probably due to the lower CO_2 content in the presence of stirring. The preparation of the inoculum is a very important step and is a prerequisite for the success of the second fermentation. The second fermentation is realized by yeast under stressfull conditions including high ethanol content of the base wine, nitrogen starvation, low pH and low growth temperature (Penacho et al. 2012). These authors demonstrated that among the genes expressed during the second fermentation, there are those involved in respiratory metabolism, oxidative stress response, autophagy, and peroxisomal functions. Ethanol has been shown to be the main factor involved in the transcription profile, temperature also explain the observed transcription profile while nitrogen starvation does not seem to play a key role on this transcription profile (Penacho et al. 2012).

6.2 Fining Agents

Fining agents are added to the yeast inoculum to facilitate the shifting of yeast to the neck of the bottle during riddling (see Sect. 7). Three kinds of fining agents are used: bentonite, bentonite-kaolin, bentonite-alginate. All these agents facilitate riddling, but have a large effect on the foaming characteristics of sparkling wines. The electrostatic interaction of bentonite with proteins considerably decreases their concentration (Senee et al. 1998). However, proteins have a positive effect on foam stability (Martinez-Rodriguez et al. 2002; Todd et al. 2000). The addition of bentonite with other agents significantly modifies the protein and peptide composition of wines. It affects both foaming parameters and the visual, olfactory and gustatory properties of the wine, decreasing its sensory quality (Martinez-Rodriguez and Polo 2003; Vanrell et al. 2007). Despite these deleterious effects, fining agents are essential to eliminate the yeast in the subsequent riddling and disgorging operations (see Sect. 7).

6.3 Bottle Fermentation

Once the base wine is inoculated with the tirage solution, a small polyethylene cup known as a "bidule" is inserted into the top of the bottle. Deposits of yeast cells and other particulates collect in the bidule during riddling, facilitating disgorging. A crown cap is then placed over the bidule and the bottles are sealed and stored at 12–15 °C. Changes in sugar concentration, yeast populations or increase in pressure in the bottle (measured with an aphrometer) are used to monitor fermentation kinetics. Figure 13.5 shows an example of changes in yeast populations and bottle

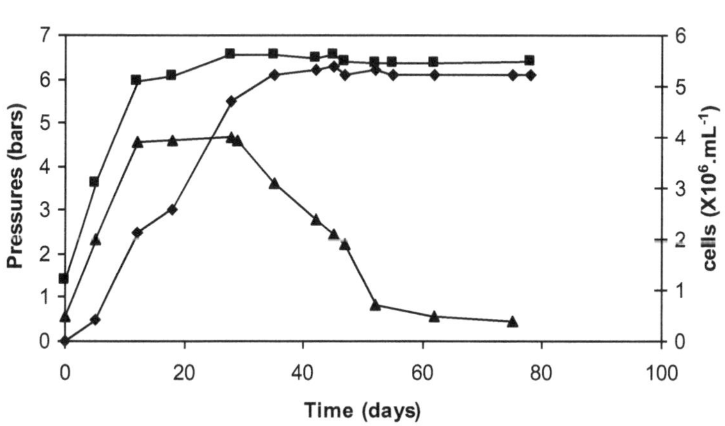

Fig. 13.5 Changes in pressure (◆), number of viable yeast cells (▲) and total population of yeast cells (■) during champenoise bottle fermentation. Valade and Laurent (1999)

pressure during champenoise bottle fermentation. Yeasts grow from the initial population of about 1–2 × 10^6 cfu.mL^{-1} to a maximum population of about 4-8×10^6 cfu.mL^{-1} within the first 20 days, using up all the fermentable sugar, and producing carbon dioxide to increase bottle pressure. Thereafter, they gradually lose viability and, usually, cannot be detected after 40–50 days. Yeasts are expected to perform under very stressful conditions (low pH, high initial ethanol content, increasing carbon dioxide concentration and pressure) during secondary fermentation, and the process must be properly managed to avoid stuck fermentations (Martinez-Rodriguez et al. 2001a).

6.3.1 Effect of Carbon Dioxide on Alcoholic Fermentation

Carbon dioxide (CO_2) is known to inhibit yeast growth at pressures of 3–5 atmospheres (3–5 Kg/cm^2), depending on the physical composition of the wine (Kunkee and Ough 1966). Base wine usually contains 0.5–1 g.L^{-1} CO_2, corresponding to an initial pressure of 0.3–0.5 atmospheres. Even at low concentration, CO_2 has an impact on fermentation kinetics. For base wines with an initial pressure of 0.6 atmospheres (1 g.L^{-1} CO_2) to 2 atmospheres (4 g.L^{-1} CO_2), the secondary fermentation can be incomplete (Valade and Laurent 1999).

6.3.2 Effect of the Physicochemical Parameters of Base Wine on Secondary Fermentation

In addition to CO_2, several other base wine properties, including pH, ethanol content, temperature and SO_2, affect growth and activity of the inoculated yeast. Sugar consumption is decreased from 0.3 g.L^{-1} per day at 16 °C to 0.2 g.L^{-1} per day at 13 °C. Fermentations at temperatures below 10 °C or above 20 °C should be avoided. Above 20 °C, the increase in pressure is faster and leads to premature secondary fermentation arrest (Valade and Laurent 1999). Ethanol concentration in the base wine affects the fermentation, with concentrations close to or exceeding 12% (v/v) limiting yeast growth and resulting in a sluggish, second fermentation. The alcohol content of the base wine increases by about 1.6% (v/v) during the secondary fermentation, and this factor needs to be taken into consideration.

The pH of the base wine should be kept above 2.9, as lower values result in sluggish fermentation. Sulfite content is one of the most important factors for a successful secondary fermentation. Free SO_2 concentration in the base wine should not exceed 10 mg.L^{-1}. Incomplete secondary fermentation, giving residual sugar, often results from the synergistic interaction of these factors and is illustrated in Table 13.3 for pH, SO_2 content and fermentation temperature (Charpentier et al. 1996; Valade and Laurent 1999).

Table 13.3 Combined effects of different physicochemical parameters of the base wine on residual sugar content of sparkling wine, 6 months after secondary bottle fermentation (Valade and Laurent 1999)

Trial	pH	Total SO_2/Free SO_2 (mg.L^{-1})	Yeast inoculum (%)	Residual sugar g.L^{-1} 15 °C	10 °C
A	3.20	80/15	5	0.2	0.6
B	2.90	80/15	5	0.9	7.2
C	3.20	64/5	5	0.2	0.4
D	2.90	64/5	5	0.9	4.2
E	3.20	80/15	1	0.2	0.2
F	2.90	80/15	1	2.8	9.3
G	3.20	64/5	1	0.2	2.1
H	2.90	64/5	1	1.2	8.3

6.3.3 Yeast Nutritional Parameters

During alcoholic fermentation, yeast growth is affected by deficiencies in nutrients, such as nitrogen or vitamins, or oxygen deficiency (Alexandre and Charpentier 1998). However, the addition of nitrogen (ammonium sulfate) or thiamine neither improves yeast growth nor affects the kinetics of secondary fermentation (Valade and Laurent 1999). These results obtained with the secondary fermentation of Champagne base wines may be specific, as such wines are already rich in nitrogen (300–400 mg.L^{-1} total nitrogen; 10–30 mg.L^{-1} ammoniacal nitrogen). Nevertheless, diammonium phosphate and thiamine are usually added to base wines at concentrations up to 0.2 g.L^{-1} and 1.2 mg.L^{-1}, respectively. While such addition may not affect the growth kinetics of the yeast, it may impact on wine flavour.

The production of higher alcohols, which is undesirable if they are produced at concentrations above 400 mg.L^{-1}, is affected by must nitrogen content, with total higher alcohol concentration being increased by decreasing assimilable nitrogen (Rapp and Versini 1991). Fatty acid esters are also of great importance to wine aroma, as they are the dominant esters formed, and usually impart pleasant odors. Ester formation is positively correlated with the nitrogen status of the must. The addition of nitrogen should, therefore, increase ester production during the secondary fermentation (Beltran et al. 2005; Vilanova et al. 2007).

Oxygen has been shown to have no effect on secondary fermentation kinetics (Valade and Laurent 1999), and this is probably due to the slower growth of the yeast under these conditions than during the primary alcoholic fermentation.

Secondary fermentation is followed by prolonged aging of the wine in contact with yeast cells (lees) during which time yeast autolysis occurs.

6.3.4 Volatile Compounds of Secondary Fermentation

The majority of studies involving volatile compounds and sparkling wines have been mainly focused on the aging stage. However, changes already occurred during the secondary fermentation compared to base wine (Torrens et al. 2010; Hidalgo et al. 2004).

7 Aging on Yeast Lees

Yeast autolysis in sparkling wine production has been the subject of many studies (for a review see, Alexandre and Guilloux-benatier 2006). The lees present in still wine during aging are composed of tartaric acid salts, organic residues and cells of various species of yeasts and bacteria, whereas sparkling wine lees are mainly composed of cells from a single species of yeast, along with technological co-adjuvants (fining agents) such as bentonite, which help to flocculate and eliminate the yeast lees at the end of the aging period. The aging of sparkling wine on the lees in the bottle is generally longer than still wine aging, and yeast autolysis occurs under CO_2 pressure (6 atmospheres).

In still wines, malolactic fermentation usually takes place during aging on yeast lees, whereas in sparkling wines, malolactic fermentation, if carried out, is completed before aging in the bottle.

7.1 Mechanisms of Yeast Autolysis

Yeast autolysis may be considered a lytic event. It is an irreversible process catalyzed by yeast intracellular enzymes. Autolysis generally takes place at the end of the stationary phase of growth, and is usually associated with cell death. The scientific basis of yeast autolysis has been the subject of numerous reviews (Babayan and Bezrukov 1985; Charpentier and Feuillat 1993; Fornairon-Bonnefond et al. 2001).

Babayan et al. (1981) suggested that yeast autolysis occurs in four stages:

- Degradation of internal cellular structures, leading to the release of vacuolar proteases into the cytoplasm.
- Inhibition of the released proteases by specific cytoplasmic inhibitors, followed by their activation due to the degradation of these inhibitors.
- Hydrolysis of intracellular polymer components, with the hydrolysis products accumulating in the space restricted by the cell wall.
- Release of hydrolysis products with molecular masses low enough to pass through the pores of the cell wall.

The autolysis of yeast cells may be induced by manipulating environmental conditions, or it can develop naturally, as occurs in wine production. While similar biochemical mechanisms may be involved in both circumstances, some differences also exist. Autolysis is often induced in industrial applications, such as in the production of yeast extract for use as a flavor enhancer or for the production of intracellular enzymes (Breddam and Beenfeldt 1991; Kollar et al. 1993; Zambonelli et al. 2000). Yeast autolysates are also added to growth culture media as they are rich in vitamins and amino acids. With industrial processes, yeast autolysis may be induced by physical factors (increase in temperature, alternate freezing and thawing, osmotic pressure), chemical factors (pH, detergents, and antibiotics), or biological factors (aeration and starvation). Under these conditions, it may be very rapid, taking 48–72 h, depending on the inducer.

Natural autolysis takes much longer. This is especially true in wines, in which the autolytic conditions, pH 3–4, aging temperature of 15 °C and presence of ethanol (12% v/v), are far from the ideal conditions of 45 °C at pH 5, often used to induce autolysis. These differences result in different autolysates (Charpentier and Feuillat 1993; Connew 1998).

In sparkling wines, yeast autolysis does not begin until 2–4 months after the completion of secondary fermentation (Charpentier and Feuillat 1993; Todd et al. 2000). Yeast autolysis can be promoted by using a mixture of "killer" and sensitive yeast for the secondary fermentation. In these conditions, the sensitive yeast cells rapidly die in the presence of the killer strains (Todd et al. 2000).

7.1.1 Biochemical and Morphological Changes

Hydrolytic enzymes play a major role in autolysis. Proteases are the most extensively studied of all the enzymes involved in autolysis. Lurton et al. (1989) used specific inhibitors to show that, in acidic conditions, protease A was the principal enzyme involved in proteolysis during autolysis in a model wine system, despite the presence of many other proteolytic enzymes in yeast. It was suggested that protease A activity may be responsible for 80% of the nitrogen released during autolysis under optimal conditions. Using a *Δpep4* mutant of *S. cerevisiae* lacking protease A, Alexandre et al. (2003) showed that protease A was responsible for 60% of the nitrogen release observed during autolysis in wine. These results suggest that other acidic proteases may also be involved in the proteolytic process. Consistent with this hypothesis, Komano et al. (1999) and Olsen et al. (1999) have identified other acidic proteases (Yapsin proteases).

In sparkling wines, proteolytic activity decreases during active bottle fermentation and in the following months, but after 9 months of fermentation and aging, it gradually increases (Feuillat and Charpentier 1982). Leroy et al. (1990) reported that proteolytic activity during Champagne aging may also depend on the yeast strain used.

The yeast cell wall is degraded during autolysis, but few studies have investigated the enzymes involved in this process during wine production. Cell wall degradation during autolysis has been observed by microscopy and in studies of cell wall composition. The cell wall of *S. cerevisiae* may account for 20–30% of cell dry mass. It consists principally of mannoproteins and β-glucans (85–90% of cell wall dry mass). The inner layer of the cell wall consists of glucans, with mannoproteins both embedded in and covering this glucan layer (Fleet 1991; Klis et al. 2002).

Glucanases are involved in yeast cell wall degradation (Arnold 1972; Charpentier and Freyssinet 1989). β-glucanases, classified as endo- and exoglucanases, hydrolyze the β-*O*-glycoside links of β-glucan chains, leading to the release of glucose, oligosaccharides and mannoproteins trapped in the cell wall or covalently bound to β-(1 → 6) and β-(1 → 3) glucans (Fleet 1991). The kinetics of β-glucanase activity during autolysis and the enological parameters affecting this activity in sparkling wine remain unknown. The action of these enzymes has been deduced from the compounds released. Cell wall degradation during autolysis results in the release of both amino acids and macromolecules, suggesting the action of both proteolytic and polysaccharide degrading enzymes (Hien and Fleet 1983).

Charpentier and Freyssinet (1989) and Feuillat et al. (1989) developed a model of cell wall degradation in which the glucans are first hydrolyzed by glucanases, releasing mannoproteins trapped within or covalently linked to the glucans. The glucans are then released, due to the activity of cell wall glucanases or solubilized glucanases in the medium. Finally, the protein fraction of the mannoproteins is degraded by proteolysis.

Microscopy has also been used to study the changes taking place in the yeast cell wall. Despite the action of proteases and glucanases on the cell wall, it does not break down completely (Vosti and Joslyn 1954; Avakyants 1982). The cell wall of yeasts grown in a synthetic medium for 24 h is thick and smooth and is easily distinguished from the plasma membrane. After autolysis, the yeast cells are much smaller and display wrinkles or folds and ridges (Avakyants 1982; Charpentier et al. 1986; Kollar et al. 1993; Hernawan and Fleet 1995). These wrinkles are thought to be due to plasmolysis, consistent with the increase in vacuole size due to solubilization of the cytoplasmic content (Martinez-Rodriguez et al. 2001b). In these studies, the structural and ultrastructural changes occurring in yeast cells during autolysis were compared in a model wine system and in sparkling wines. After 24 h of incubation in a model wine system, the yeast cells had lost most of their cytoplasmic content and had a large vacuole, whereas yeasts aged for 12 months in a sparkling wine still had most of their cytoplasmic content and had a small vacuole. These findings demonstrate that autolysis conditions are not optimal during the aging of sparkling wine or Champagne. Figure 13.6 schematically represents the various events that occur during yeast autolysis, but most of these processes are not completely understood.

Fig. 13.6 distributed throughout the vacuole. The cell wall is rough, small wrinkles or folds are observed. (**C**). After 12 months, the cell appears to have collapsed, accounting for its smaller size. The cell wall remains unbroken, with many ridges and folds. Yeast cells have lost much of their cytoplasmic content by this time. The fate of the plasma membrane during this process remains unclear (Alexandre and Guilloux-Benatier 2006). Nucleotides released during Champagne autolysis are ribonucleotides. The main nucleotides detected in Champagne wine extracts were 5′- and 3′- (or 2′-) UMP, 5′- and 3′- (or 2′-) GMP, 5′- and3′-AMP, 5′- and 3′-CMP, and 5′-IMP (Charpentier et al. 2005)

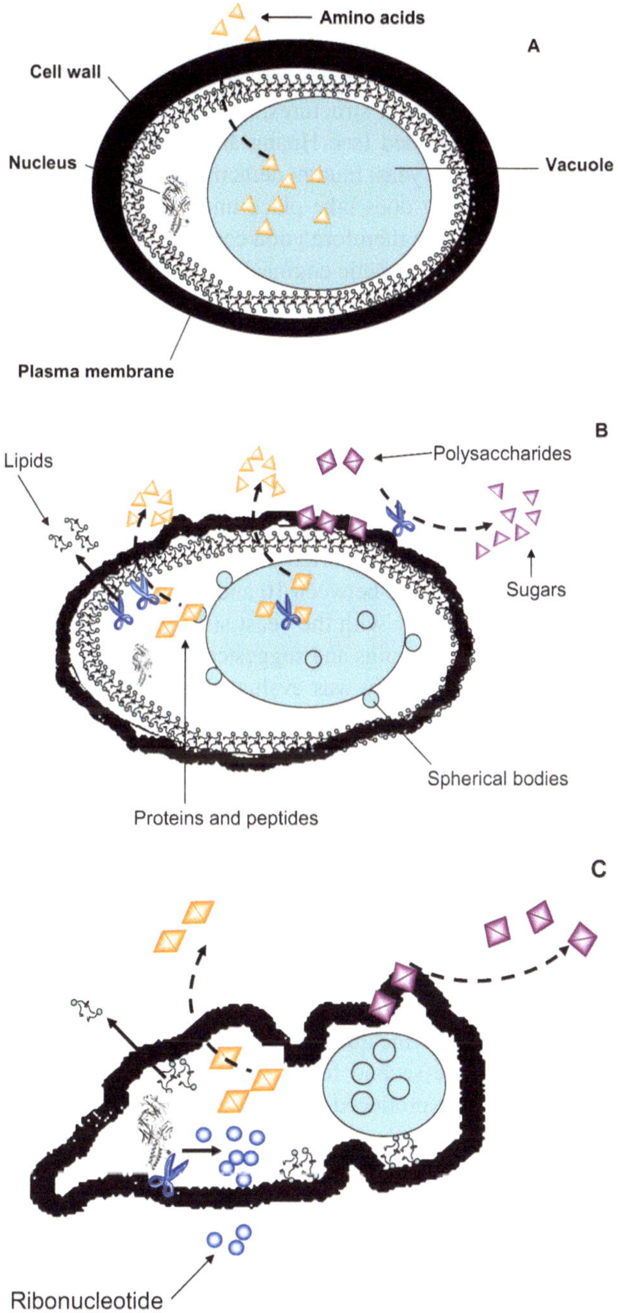

Fig. 13.6 Schematic representation of morphological and biochemical changes in yeast during autolysis in sparkling wine. (**A**) After the second alcoholic fermentation, yeast cells are elongated and ovoid. The cell wall is thick and smooth. Inside the cell, there is a large vacuole surrounded by spherical bodies. (**B**) After 6 months, the cell and vacuole are smaller in size. Spherical bodies are

Autophagy was shown to play a possible role in the release of yeast compounds into wines (Cebollero et al. 2005). Autophagy is a degradation pathway activated by nitrogen or carbon starvation. It is characterized by the formation of autophagosomes, which contain intracellular structures, including mitochondria, that are carried to the vacuole and degraded (see Huang and Klionsky 2002, for a review). Cebollero et al. (2005) used a yeast mutant defective in the autophagic or Cvt pathways to show that autophagy does take place under wine production conditions. Genes related to autophagy are therefore good candidates for studying the molecular basis of autolysis or for the genetic engineering of wine yeasts.

7.1.2 Factors Affecting Autolysis

The principal factors affecting yeast autolysis in wines are pH, temperature, ethanol concentration and the nature of the yeast strain. As pH and ethanol content cannot be changed in the wine, they will not be considered here.

High temperatures, up to 60 °C, have been reported to favor autolysis in a model wine system. Molnar et al. (1981) reported the optimal temperature for proteolysis in the Champenoise method to be between 10 and 12 °C.

Autolysis varies considerably with the yeast strain. Suzzi (1990) compared the autolytic capacity of different strains and suggested that this criterion could be used to select yeasts. Autolytic capacity was evaluated by measuring the amino acids released by the yeast at various temperatures, 10 days after fermentation. Significant differences were observed in the autolytic capacity of three strains. Thus, the amount of nitrogen released into the medium depends on the yeast strain used, opening up useful possibilities for modification in sparkling wine production (Martinez-Rodriguez et al. 2001c). Martinez-Rodriguez et al. (2001a) suggested that yeast strains with a high autolytic capacity would produce better quality sparkling wine than yeasts with a low autolytic capacity. They also suggested that autolytic capacity and foaming analysis should be used for selecting yeasts for sparkling wine production.

Nunez et al. (2005) recently confirmed the importance of the autolytic capacity of yeast strains for sparkling wine quality. A mutant strain with accelerated autolysis was used to conduct the second fermentation. The resulting wine had better foaming properties than that produced with a control strain. The aging period was also reduced from nine to 6 months with this mutant, potentially decreasing production costs.

7.2 *Yeast Autolysis Compounds and Their Impact on Sparkling Wine Quality*

The autolysis of yeast during aging results in the release of various compounds that modify the physical and organoleptic properties of sparkling wine (Alexandre 2011; Toressi et al. 2011).

7.2.1 Changes in the Nitrogen Compounds Present at Different Stages in the Champenoise Method

Numerous studies have investigated the changes in nitrogen composition that occur during the aging of wine on lees. Nitrogen release is thought to reflect the autolytic activity of the yeast — proteolytic activity in particular. In sparkling wine production, amino acids are released into the medium during bottle fermentation. Once the available glucose has been exhausted, amino acid levels in the wine increase (Feuillat and Charpentier 1982). This process, known as excretion (passive exorption), has been described by Morfaux and Dupuy (1966) and should not be confused with autolysis.

Yeast autolysis does not begin until 3–9 months after the end of the second fermentation. The lag time until the start of autolysis varies considerably and depends on base wine composition, aging temperature and yeast strain. Total amino acid concentration increases before the increase in free amino acid concentration. Thus, peptides are released into the medium and then broken down into amino acids. Moreno-Arribas et al. (1996) studied the changes in the various nitrogen fractions during sparkling wine aging according to the Champenoise method. Between 3 and 9 months after addition of the tirage solution, they observed no change in free amino-acid concentration, regardless of the grape variety used. Free amino-acid concentration increased after 9 months, indicating the start of autolysis. These results have been confirmed by Nunez et al. (2005).

Peptide content fluctuates, peaking after 12–15 months of aging on the lees, and decreasing thereafter. This behavior may reflect an initial release of peptides that are subsequently broken down. Moreno-Arribas et al. (1996) also showed that the distribution of free amino acids is very different from that of amino acids in peptides and proteins, as confirmed in other studies (Moreno-Arribas et al. 1998a; Guilloux-Benatier and Chassagne 2003).

The amount of peptides released by yeast autolysis during sparkling wine aging is variable, and depends on grape variety and aging time (Moreno-Arribas et al. 1998b). The nature of the peptides released also changes with aging, the length of the peptides released decreasing with increasing aging time (Martinez-Rodriguez and Polo 2000).

The amino-acid composition of the peptides present in sparkling wines has been investigated (Moreno-Arribas et al. 1996, 1998a, b). The peptides mostly originate from the break down of yeast proteins rather than grape juice proteins. Bartolomé et al. (1997) showed that different varieties of sparkling wines aged with the same yeast over 26 months, contained peptides with similar amino-acid compositions.

The high levels of threonine and serine in peptides from sparkling wine are consistent with their yeast origin, as these amino acids are not the predominant free amino acids in the proteins of base wine (Usseglio-Tomasset and Bosia 1990; Acedo et al. 1994; Moreno-Arribas et al. 1998a). Threonine and serine are involved in the glycosidic bonds between proteins and mannans in the yeast cell wall (Klis et al. 2002).

The concentration and composition of proteins that are released during autolysis in sparkling wine have been little studied, possibly because amino acids are considered as better markers for following autolysis. Changes in protein content during autolysis seem to depend on the yeast strain. Leroy et al. (1990) compared two different yeast strains and found that protein content remained stable during the first 9 months for one strain whereas it decreased greatly from the end of the second fermentation for the other strain. Nunez et al. (2005) reported an increase in protein and polypeptide levels during the first 3 months, followed by a decrease, attibuted to protease activity. Protein and peptide content then increased again after 6 months.

A total increase in protein content of 8–13% was reported by Todd et al. (2000) during the secondary fermentation and aging of sparkling wines. This release of protein by the yeast cells stabilized after 90 days, with only slight increases occurring thereafter.

7.2.2 Impact of Nitrogen Fractions on Sparkling Wine Quality

Amino-acid enrichment of the medium may improve the aromatic potential of sparkling wines, as amino acids are the precursors of aroma compounds. Aroma compounds may be generated by the deamination or decarboxylation of amino acids (Feuillat and Charpentier 1982). The levels of one lactone compound, 3-hydroxy-4,5-dimethyl-2(5*H*)-furanone, also known as sotolon (green nut, curry odor) gradually increase during the aging of sparkling wine. Pham et al. (1995) showed that sotolon is generated from threonine, which is transformed into α-ketobutyric acid that reacts with acetaldehyde. Vitispirane, a norisoprene-derived compound imparting eucalyptus odors, is synthesized from methionine and has been detected in aged Cava wine (Francioli et al. 2003).

The surfactant properties of peptides are thought to play a role in wine, and some peptides are also responsible for sweet and bitter tastes (Polo et al. 1992). Peptides probably play a similar role in foam stabilization in both sparkling wines and beer. Foam (bubble) properties are extremely important in sparkling wine. Usually, they are measured as height of foam, foam area, foam collar (Nunez et al. 2005). Positive correlations between polypeptide molecular mass, hydrophobicity and foam stabilizing activity have been found in beer (St John Coghlan et al. 1992; Onishi and Proudlove 1994). Moreno-Arribas et al. (1998a) suggested that the hydrophobicity of peptides might account for the foaming properties of sparkling wine.

Various studies have tried to identify the compounds affecting foam quality in wine (Brissonnet and Maujean 1991, Brissonnet and Maujean 1993; Malvy et al. 1994; Andres-Lacueva et al. 1997). Most studies have focused on base wine, and it is therefore difficult to extrapolate the results obtained to sparkling wines. The autolysis occurring in the Champenoise method may have a major effect. Moreno-Arribas et al. (2000) found that foam characteristics were positively correlated with the concentrations of most free amino acids and proteins, confirming the results of Malvy et al. (1994). However, no relationship was found between foam characteristics and the concentration of wine peptides.

7.2.3 Polysaccharides

Glucanases and proteases release polysaccharides from the yeast cell wall during autolysis in sparkling wines. These macromolecules contain mainly glucose (74%) and mannose (26%). Their mannose/glucose ratio increases during autolysis, possibly due to mannoprotein release after glucan degradation. Indeed, mannoproteins are trapped in the glucan layer of the wall, and the very low levels of mannosidase activity recorded cannot account for the increase in mannose concentration (Freyssinet et al. 1989). Polysaccharide concentrations in wines vary, and depend on the method used for their measurement. Charpentier (2000) reported an increase in polysaccharide content from 366 $mg.L^{-1}$ in the base wine to 602 $mg.L^{-1}$ after 9 months of aging of a sparkling wine. More recently, Martínez-Lapuente et al. (2018) confirmed an enrichment in mannose during the six first months of aging followed by a decrease in the content, which could be due to precipitation phenomena.

There is strong evidence to suggest that mannoproteins from the yeast cell wall play a key role in wine stability and in the organoleptic properties of sparkling wine. Mannoproteins have been shown to reduce haze formation (Ledoux et al. 1992; Dupin et al. 2000), presumably by competing with wine proteins for unknown factors. It has been suggested that, as the concentration of these factors decreases in the presence of mannoproteins, protein particle size and turbidity also decrease (Dupin et al. 2000).

Mannoproteins prevent the precipitation of tartaric salts (Lubbers et al. 1993; Gerbaud et al. 1997; Moine-Ledoux et al. 1997). They stick to the growth sites of the crystal, blocking growth of the crystal lattice (Gerbaud et al. 1997).

The effect of colloids (macromolecules) on foam quality has also been investigated (Brissonnet and Maujean 1991). Some of the compounds present in the foam precipitate in ethanol, consistent with the presence of macromolecules. Moreno-Arribas et al. (2000) showed that neutral polysaccharides are important for foam quality in sparkling wines. The optimum aging time for obtaining a high-quality, stable foam appears to be 18 months. However, foam quality decreases after 18 months, which coincides with an increase in the level of monomeric compounds, such as fructose, that probably arise from the hydrolysis of grape components by yeast enzymes released during autolysis (Andres-Lacueva et al. 1997). Finally, mannoproteins are thought to contribute to the mouthfeel of the wine. Bertuccioli and Ferrari (1999) developed an index for evaluating wine body, and showed that this index was increased by the presence of mannoproteins. Mannoproteins also influence the intensity and persistence of wine aromas (Lubbers et al. 1994; Feuillat 2003).

7.2.4 Lipids

Lipids are important components of sparkling wines because they are a major source of flavor compounds (Forss 1969) and affect foam stability. Changes in the lipid content of sparkling wine have been the focus of several studies.

Lipid content increases during the second fermentation (Troton et al. 1989). Lipid content increases and qualitative changes occur during aging in the bottle in contact with the lees (Piton et al. 1988). The concentration of polar lipids decreases whereas the concentration of neutral lipids increases (monoglycerides, diglycerides, and triglycerides). Experiments in a model wine system showed that the levels of triacylglycerols, 1,3-diacylglycerols, 2-monoacylglycerols, free fatty acids, sterol esters and sterols increase after 2 days of autolysis and then decrease, probably due to yeast hydrolytic enzymes (Pueyo et al. 2000). No phospholipids were released into the medium, confirming previous results from Hernawan and Fleet (1995), and it was suggested that any phospholipids present are degraded.

Conflicting results have been published concerning the influence of lipids on foam. Maujean et al. (1990) found that octanoic and decanoic fatty acids reduced foam stability, whereas Dussaud et al. (1994) reported that the addition of a lipid mixture did not affect the foam. Pueyo et al. (1995) noted that linolenic and palmitoleic acid levels were the best indicators of foam stability. The effects of fatty acids on the foaming properties of wine were investigated by Gallart et al. (2002). The C8, C10 and C12 acids had a negative effect on foam quality, whereas ethyl esters of hexanoic, octanoic and decanoic acids had a positive effect.

7.2.5 Nucleic Acids

Although the degradation of proteins during yeast autolysis has been extensively studied, the hydrolysis of RNA and DNA has received less attention. RNA and DNA comprise 5–15% and 0.1–1.5% of cell dry weight, respectively (Nagodawithana 1992).

The DNA of a brewing and baking strain of *Saccharomyces cerevisiae* was found to be almost completely degraded during autolysis (Hough and Maddox 1970; Suomalainen 1975). However, Trevelyan (1978) reported no decrease in DNA content during the autolysis of a baker's yeast strain. The extent of DNA degradation during autolysis appears to depend on the yeast strain (Hernawan and Fleet 1995). The very low levels of DNA detected in autolysates probably reflect its degradation by DNase activity. DNA degradation requires several active enzymes and leads to oligonucleotide, nucleotide and nucleoside degradation products. The predominance of deoxyribonucleotides in the autolysate indicates that endo- and exonucleases are principally responsible for the degradation process. Zhao and Fleet (2003) reported the degradation of up to 55% of total DNA in *S.cerevisiae* during autolysis, releasing 3′-and 5′ deoxyribonucleotides. Even under optimum autolytic conditions, some parts of the DNA were resistant to autolytic degradation. However, further studies of DNA degradation under enological conditions are required. The presence of ethanol, the lower pH of wine and low storage temperatures may give much lower levels of DNA degradation.

More than 95% of the nucleic acid in yeast cells consists of RNA. Zhao and Fleet (2005) suggested that RNA degradation is a key reaction in yeast autolysis. They showed, in different autolytic conditions, that up to 95% of cell RNA were degraded, releasing mainly 3′-, 5′- and 2′-ribonucleotides. Formation of the flavor-enhancing nucleotides 5'-AMP and 5'-GMP was favored at 50 °C (pH 7.0) and pH 4.0 (40 °C), respectively. These conditions are far from optimal for wine production, but the degradation of RNA and the release of nucleotides into sparkling wine during autolysis may affect its organoleptic properties (Courtis et al. 1998).

RNAse is active during autolysis in Champagne (Leroy et al. 1990), but data on the extent of nucleic acid degradation should be interpreted with caution because organic acids, phenolic compounds, peptides and other compounds in wine can interfere with the measurement of nucleotides. Monophosphate nucleotides in Champagne have been unequivocally identified (Aussenac et al. 2001; Charpentier et al. 2005). Three monophosphate nucleotides (5'-UMP, 5'-GMP and 5'-IMP) in Champagne aged on lees for 8 years were identified. Nucleotide monophosphate concentration ranged from 50–500 μgL^{-1}, considerably different from previously reported concentrations (Courtis et al. 1998).

Monophosphate nucleotides are used as flavorings in the food industry (Abbas 2006), but further studies are required to evaluate their impact on wine flavor.

7.2.6 Volatile Compounds

The release of volatile compounds during yeast autolysis has been studied less thoroughly than the release of non- aroma compounds. It is worth noting that reported changes of volatile compounds during aging are contradictory (Pozo-Bayon et al. 2009). The few studies carried out have shown that many compounds are released, some of which are detected by tasters at low concentrations. Chung (1986) reported that autolysis of *Saccharomyces cerevisae* at 15–20 °C or 35–40° in a model wine system (12% v/v ethanol, pH 3.5) released many different volatile compounds after 4–6 months. Esters comprise the major family of volatile compounds released during autolysis, both qualitatively and quantitatively. Small acyl chain esters (C3-C4) and medium acyl chain esters (C6-C12) with characteristic fruity odors are released at the start of yeast autolysis and are then broken down. Heavy acyl chain esters have also been identified in model and sparkling wines (Molnar et al. 1981). Pozo-Bayon et al. (2003) reported an important effect of ageing time on volatile compounds of Cava wines. They found that some esters (hexyl acetate, isopentyl acetate, ethyl butyrate, ethyl octanoate, diethyl succinate) are useful to discriminate Cava wines of different ages. However, contradictory results have been reported. Martinez-Garcia et al. (2017) observed a decrease in aliphatic esters of long chain fatty acids during the second fermentation process contrary to Welke et al. (2014) who obtained an increase in esters such as ethyl octanoate and ethyl decanoate during 9 months of lees contact. Ruiz-Moreno et al. (2017) reported that the profile ester was significantly influenced by ageing on lees. They observed a decrease of total esters during the first 3 months of ageing while no changes occurred between 3 and 9 months of ageing. However, different trends could be observed depending on the group of esters. For example, higher alcohol acetate decreased by 50%

during ageing while ethyl ester of branched acids increased during ageing (Ruiz-Moreno et al. 2017).

Regarding higher alcohol, Coelho et al. (2009) described an increase in 1-propanol, 2-methyl-1-propanol (isobutanol) and isoamyl alcohols after 24 months of 'dégorgement' (removal of yeast sediment from bottles) and Pozo-Bayón et al. (2010) the opposite. Recently Martinez-Garcia et al. (2017) show an increase for 2-phenyl-1-ethanol and 1-propanol and a decrease for isobutanol and isoamyl alcohols contents during the second fermentation. These differences could be attributed to complex balances of intracellular synthesis and extracellular adsorption-desorption processes in the cell walls during lees aging (Pozo-Bayón et al. 2010).

Terpenic alcohols and higher alcohols are also released during autolysis. Geraniol and α-terpineol, citronellol and farnesol have all been identified. These compounds are perceived by tasters at low concentrations, from 100 to 300 $\mu g.L^{-1}$. Molnar et al. (1981) suggested that farnesol contributes greatly to the aromatic quality of sparkling wine and Loyaux et al. (1981) suggested that nerolidol makes a similar contribution to Champagne. The rapid formation of two higher alcohols, isoamyl alcohol and phenyl ethyl alcohol (rose odor), has been observed during autolysis in a model wine system (Chung 1986).

About ten aldehydes have been identified in sparkling wines. Methyl 3-butanal is the most abundant, accounting for 40% of all aldehydes present, and may be formed through a mechanism involving isoamyl alcohol oxidation. Most of the aldehydes identified are present at levels close to or greater than the detection threshold of the human nose for aqueous solutions. Aldehydes are described as having a grassy odor with a negative effect on organoleptic properties, although this mostly disappears during aging (Chung 1986).

Francioli et al. (2003) characterized the volatile compounds released in sparkling wines during autolysis and suggested that they could be used as age markers. Acetates appeared to decrease during aging, whereas diethylsuccinate, vitispirane and TDN (1,1,6-trimethyl-1,2-dihydro naphthalene) levels increased over time. Hexanol and 2-phenylethanol were also released during autolysis. Compounds such as vitispirane, TDN and diethylsucinate may be good age markers and can discriminate between young and aged sparkling wines. Riu-Aumatell et al. (2006) obtained similar results, reporting that some high molecular weight acetates, and ethyl and isoamyl esters, are typical aroma compounds in young Cava (Spanish sparkling wine), whereas vitispirane, diethyl succinate, TDN, hexenol and ethyl acetate are typical aroma compounds in Cava aged over a prolonged period. The release of these aromas by yeast enzymes acting on glycoside precursors has been identified as a possible mechanism for their formation, with C-13 norisoprenoids and vitispirane being derived from glycoside-bound carotenoids and megastigma, respectively (Bosch-Fuste et al. 2007; Riu-Aumatell et al. 2006). TDN may be a direct degradation product of carotene (Rapp 1998), although precursors linked to a sugar molecule have also been reported (Winterhalter 1991).

Descriptive sensory analyses can be used to characterize the effects of the Champenoise process on aroma. In one study, the sensory properties of sparkling wines after 18 months of aging on lees could not be predicted from the profiles of the base wine (De La Presa-Owens et al. 1998). In this study, the authors demonstrated that descriptive analysis of the base wine can be used to distinguish between different grape varieties, such as *Chardonnay* or *Pinot noir*. However, after secondary fermentation, sparkling wines could not be differentiated on the basis of the grape variety or color of grape used. Thus, secondary fermentation and aging on the lees can profoundly modify the aromatic profile of the wine.

7.2.7 Phenolic Compounds

Contrary to expectation and the importance of phenolic compounds from a sensorial point of view, there are only few studies regarding the phenolic compounds composition of sparkling wines (Pozo-Bayon et al. 2003). Browning of sparkling wine during aging has been observed (Ibern-Gomez et al. 2000; Serra-Cayuela et al. 2013, 2014) and was attributed to two different mechanisms, oxidation of phenolic compounds released during yeast autolysis and the action of cytoplasmic enzymes. Resveratrol content has been shown to decrease during aging (Jeandet et al. 2006) which could be explained by the adsorption capacity of yeast cell wall (Mazauric and Salmon 2005, 2006). However, contradictory results are reported by Pozo-Bayon et al. (2003). In this study, the authors did not find changes in the low molecular weight phenolic composition of the sparkling wines aged for over 18 months with lees. These observations might be linked to the reduction conditions in the bottle during aging. This hypothesis is supported by two different studies. In the first one, it has be shown that antioxidant activity of a sparkling wine does not change during aging (Satue-Gracia et al. 1999). In the second one, the authors observed that the aldehyde level (oxidation marker) in sparkling wine is lower compared to other types of wines (Cullere et al. 2007). However, it has been reported that antioxidant potential of sparkling wines decreased during ageing on lees (Stefenon et al. 2014).

In a more in depth study it was shown that gallic acid, catechin and particularly GRP increased with time during ageing on lees while tartaric esterified hydroxycinnamic acids (cis-coutaric acid, caftaric acid and fertaric acid) tended to decrease (Serra-Cayuela et al. 2013). Other phenolic compounds such as trans-coutaric and p-coumaric increased significantly with time and was significantly correlated with browning (Serra-Cayuela et al. 2013). It is interesting to note that in this study the authors reported that 5-hydroymethylfurfural (5-HMF) concentration was significantly correlated with browning. Thus, they suggested the 5-HMF could be used as a quality marker (Serra-Cayuela et al. 2013).

8 Foaming Properties

Foaming properties of a sparkling wine is a very important criteria and is a major sensorial characteristic. Foaming properties depend to a great extent to the must and wine composition and has been reviewed (Martinez-Rodriguez and Pueyo 2009;

Pozo-Bayon et al. 2009; Kemp et al. 2015). Here we will only focus on the effect of lees aging on sparkling wine foaming properties. The main yeast component involved in foaming properties is the mannoproteins released during alcoholic fermentation and the process of autolysis. These amnnoproteins have been shown to increase the foaming properties of the sparkling wine (Nunez et al. 2005).

Based on this observation, some authors proposed to select sparkling wine yeast strain with high autolytic capacity (Martinez-Rodriguez et al. 2001a). However, according to Martínez-Lapuente et al. (2015) polysaccharides among which mannoproteins, did not correlate with foamability or foam stability. On the other hand, there were a positive correlation between polysaccharides and foam stability. These contradictory results highlight a need for future research.

9 Riddling, Disgorging and Dosage

9.1 *Riddling*

Riddling is carried out to move sediment (yeast lees, protein material, bitartrate) towards the neck of the bottle to facilitate its removal. It takes place between 6 and 36 months after tirage, depending on the legislation of the country concerned. Historically, riddling was carried out manually, with a riddling rack specially designed to allow the bottles to be gradually turned and brought from the horizontal to the vertical position. During this movement, the sediment is moved from the side of the bottle to the neck. This system is highly labor-intensive and takes about 20–30 days to complete. Many companies have replaced this method with the use of mechanical processes, using gyropallets. With these methods, large crates are electronically rotated and shaken from the horizontal to the vertical in a few days (Hardy 1993; Howe 2003).

9.2 *Disgorging*

Once the bottle has been inverted, the lees present in the neck of the bottle are removed, in a process known as disgorgement. The yeast sediment entrapped in the bidule is snap frozen by immersing it in a solution at around −20 °C. Once the extreme end of the neck is frozen, the bottle is placed the right way up, opened and the ice plug (containing the deposit) is ejected by a gas pressure corresponding to 2% of the bottle volume. As the wine is cold, gas loss from the wine is minimal during this process (1–2 atmospheres of carbon dioxide lost) (Lacoste 1983). After pressure release upon opening the bottle, spontaneous over-foaming called gushing can occur (Kupfer et al. 2017b). Specific proteins seem to be involved in the development of gushing. It has been reported that surface-active proteins produced by the grape-associated fungus *Penicillium oxalicum* are involved on the development of gushing (Kupfer et al. 2017b). Kupfer et al. (2017a) identified the highly glycosylated seripauperin 5 (PAU5) from *Saccharomyces cerevisiae* as a potential biomarker for gushing because the absence of the protein was associated with the occurrence of gushing in a high percentage of analyzed samples. The foam stabilizing property of PAU5 was recently demonstrated (Kupfer et al. 2017b). The authors suggested that PAU5 might play a role not only indirectly as a marker for gushing but also have a direct impact as a protein that stabilizes sparkling wine against gushing-inducing factors.

9.3 Dosage

After disgorging, the bottle is then topped up by addition of a small volume of a reserved wine, mixed with or without sugar, to give the final sparkling wine its desired balance. This process is known as *dosage*. Even dry sparkling wines ("brut") are sweetened, and contain 5–12 gL^{-1} residual sugar. A cork is inserted into the bottle and fixed in place with a wire. The bottle is then labeled for the market (Lacoste 1983).

10 Secondary Fermentation with Flocculent or Immobilized Yeast

10.1 Flocculent Yeasts

Flocculation is useful property of industrial *Saccharomyces* strains for winemaking (Shinohara et al. 1997). The flocculation phenomenon is complex, and is controlled by dominant and recessive genes *FLO1, FLO5, FLO8, FLO9, FLO10, FLO11, flo3, flo6, flo7, fsu1, fsu2, tup1* (Verstrepen et al. 2003). Flocculent wine yeasts are rare, and the frequency of strains displaying flocculation among wild *Saccharomyces* yeasts isolated from grape musts is low (Speers et al. 1992). Flocculent-bond formation in *Saccharomyces cerevisiae* involves a protein-carbohydrate interaction between specific proteins on the surface of flocculent cells and mannan present on the cell surface. Flocculent yeasts form aggregates (flocks), which sediment rapidly after fermentation.

Some winemakers use flocculent strains for the secondary fermentation, to facilitate riddling. Flocculent strains decrease the time and cost required for riddling, provided that there is a good interaction between the yeast and the fining agents (Bidan et al. 1986). Riddling takes a few days with flocculent strains, rather than several weeks.

As flocculation properties depend on the genetic constitution of the strain, several studies have focused on the construction of flocculating enological strains (Romano et al. 1985; Thornton 1985; Vezinhet et al. 1992; Shinohara et al. 1997). These studies demonstrated that hybrid strains had useful flocculating and wine-making properties. Coloretti et al. (2006) characterized interspecific hybrids of *Saccharomyces*. They reported that flocculent hybrid strains of *Saccharomyces cerevisiae* and the cold-fermentation strain *Saccharomyces uvarum* had interesting characteristics. The hybrid strains carried out fermentation at low temperature and formed a compact deposit that sedimented much faster than the non flocculent *Saccharomyces cerevisiae* strain. The aim when using such strains is not to shorten the aging time but to facilitate and accelerate the ridling process. Indeed, aging still requires several months since flocculant strains do not necessarily give faster autolysis.

10.2 Immobilized Sparkling Wine Yeasts

Use of immobilized yeast cells to conduct fermentations has several technical and economic advantages over the use of free cells (Stewart and Russell 1986; Strehaiano et al. 2006). Immobilized yeast technology was applied to sparkling wine production and the Champenoise method as early as the 1960s and 1980s, respectively (Martynenko and Gracheva 2003). The chief benefit of this system is elimination of the riddling step, which is time consuming. Production costs are, therefore, lower with this technique.

In the wine industry, yeasts are immobilized by adsorption or adhesion to various supports or by incorporation into a carrier gel (calcium alginate, carrageenan etc.). Carrier gels are widely used because a high density of yeast cells may be entrapped in such systems, and have a high level of operational functionality (see Chap. 9, Strehaiano et al. 2006).

Immobilized yeasts were first used for sparkling wine production by the Champenoise method by Bidan et al. (1980). Bottle fermentation is conducted by adding around 200 yeast- gel beads per bottle which gave 10^6–10^7 cells.mL^{-1} and 6 pKa pressure after 20–30 days at 15 °C. As in the traditional process, beads are left in the bottle for aging for at least 1 year. According to Yokotsuka et al. (1997), similar changes in the nitrogen fractions due to autolysis were observed. Furthermore, there were no distinct differences in taste and smell between the sparkling wine made using free and immobilized yeasts. Indeed, in another study no chemical or organoleptic differences have been found between Champagne produced with free or immobilized cells (Martynenko and Gracheva 2003). Calcium alginate is now the major carrier for immobilized yeasts used for bottle fermentation. Calcium alginate beads are easy to prepare and have no effect on wine aroma. The secondary fermentation process is also unaffected by the presence of beads. However, there are several serious drawbacks of this method. Yeast cells may leak from the gel beads into the bottle, increasing turbidity (Fumi et al. 1987, 1988). A double immobilization technique has been used to overcome this problem, with the beads being coated with a gel layer without microorganisms (Klein and Wagner 1986; Duteurtre et al. 1987; Strehaiano et al. 2006). However, this seems to decrease mass exchange in the biocatalyst, resulting in partial cell release. Martynenko et al. (2004) tackled this problem by treating the biocatalysts with an aromatic alcohol (factor d1) that inhibited yeast growth and proliferation without affecting the fermentation.

Although, all the studies demonstrated that entrapped cells in double-layered alginate beads led to wines similar to those produced by the traditional method, this technology has not received wide application, except for its use by the Moët et Chandon winery (Strehaiano et al. 2006).

11 Conclusion

Major changes in the sparkling wine production process have occurred over the last 20 years. Grape and must quality have benefited from viticultural progress (clonal selection, grape yield, maturity). Technological advances in pressing have made it possible to obtain juice of higher quality. Active dry yeasts are widely used, making it possible to avoid stuck or sluggish fermentations, and making it easier to control the quality of the base wine. Similar improvements have also been observed with selection of the lactic acid bacteria used for malolactic fermentation. Advances in secondary fermentation techniques reflect a greater knowledge in microbiological aspects of this fermentation and a better understanding of the physicochemical factors influencing its kinetics. Moreover, control of the yeast inoculum has also been improved, increasing the likelihood of successful secondary fermentation.

In microbiological terms, future progress is likely to be less spectacular. Indigenous yeasts are still widely used for still wines. Investigations are currently underway to determine the impact of these yeasts on aromatic profile and the control of such fermentations. Progress in this field will also probably benefit sparkling wine production. Indigenous strains could be used during the primary alcoholic fermentation, to improve the aromatic quality of the base wine. Yeast cocktails could be used as an alternative, mimicking the natural process (Fleet 2008).

Finally, genetic engineering could potentially be used to improve sparkling wine production. Many targets for the improvement of wine yeast and malolactic bacteria have been identified. For example, the overexpression of alcohol acetyl transferase genes (*ATF1* and *ATF2*) in *Saccharomyces cerevisiae* has been shown to have a marked effect on ester formation in wine. A number of genetically modified yeasts already exist, but their commercial use will depend on consumer acceptance. However, genetic modification could constitute an interesting strategy for obtaining yeast strains carrying out accelerated autolysis. Indeed, accelerated autolysis is highly desirable since it can reduce production costs. As stated above, increasing the aging temperature or addition of autolysate have been considered with various results. Thus, this aspect of the process needs to be improved in the future.

References

Abbas, C. A. (2006). Production of antioxidants, aromas, colours, flavours, and vitamins by yeasts. In A. Querol & G. H. Fleet (Eds.), *Yeasts in food and beverages* (pp. 285–334). Berlin: Springer.

Acedo, M. I., Pueyo, E., & Polo, M. C. (1994). Preliminary studies on peptides in wine by HPLC. *American Journal of Enology and Viticulture, 45*, 167–172.

Alexandre, H. (2011). Downstream processing and product recovery: Autolysis of yeasts. In M. Moo-Young (Ed.), *Comprehensive biotechnology* (Vol. 2, 2nd ed., pp. 641–649). Elsevier.

Alexandre, H., & Charpentier, C. (1998). Biochemical aspects of stuck and sluggish fermentation in grape must. *Journal of Industrial Microbiology and Biotechnology, 20*, 20–27.

Alexandre, H., & Guilloux-Benatier, M. (2006). Yeast autolysis in sparkling wine. *Australian Journal of Grape and Wine Research, 12*, 119–127.

Alexandre, H., Guilloux-Benatier, M., Chassagne, D., Charpentier, C., & Feuillat, M. (2003). Les peptides du vin: Origine et impact. In A. Lonvaud-Funel, G. de Revel, & P. Darriet (Eds.),. Œnologie 2003 *Tec & Doc* (pp. 512–514). Paris: Lavoisier.

Andres-Lacueva, C., Lamuela-Raventos, R. M., Buxaderas, S., & del Carmen de la Torre-Boronat, M. (1997). Influence of variety and aging on foaming properties of Cava (sparkling wine).2. *Journal of Agricultural and Food Chemistry, 45*, 2520–2525.

Arnold, W. N. (1972). The structure of the yeast cell wall. Solubilization of a marker enzyme, β-fructofuranosidase, by the autolytic system. *Journal of Biological Chemistry, 4*, 1161–1169.

Aussenac, J., Chassagne, D., Claparol, C., Charpentier, M., Duteurtre, B., Feuillat, M., & Charpentier, C. (2001). Purification method for the isolation of monophosphate nucleotides from Champagne wine and their identification by mass spectrometry. *Journal of Chromatography, 907*, 155–164.

Avakyants, S. P. (1982). Etudes au microscope électronique de l'autolyse des levures de vin. *Vinodelie i vinogradstvo (USSR), 2*, 55–59.

Babayan, T. L., & Bezrukov, M. G. (1985). Autolysis in yeast. *Acta Biotechnology, 5*, 129–136.

Babayan, T. L., Bezrukov, M. G., Latov, V., Belikov, V. M., Belatseva, E. M., & Titova, E. F. (1981). Induced autolysis of *Saccharomyces cerevisiae*. Morphological effects, rheological effects and dynamics of accumulation of extracellular hydrolysis products. *Current Microbiology, 5*, 163–168.

Bartolomé, B., Moreno-Arribas, V., Pueyo, E., & Polo, M. C. (1997). Online HPLC photodiode array detection and OPA derivation for partial identification of small peptides from white wine. *Journal of Agricultural and Food Chemistry, 45*, 3374–3381.

Beltran, G., Esteve-Zarzoz, B., Rozès, N., Mas, A., & Guillamon, J. M. (2005). Influence of the timing of nitrogen additions during synthetic grape must fermentations on fermentation kinetics and nitrogen consumption. *Journal of Agricultural and Food Chemistry, 53*, 996–1002.

Bertuccioli, M., & Ferrari, S. (1999). Laboratory experience on the influence of yeast in mouthfeel. Proceedings Les entretiens scientifiques de Lallemand. Montreal.

Bidan, P., Divies, C., & Dupuy, P. (1980). France patent n°. 2432045.

Bidan, P., Feuillat, M., & Moulin, J. P. (1986). Les vins mousseux. *Bulletin de l'OIV, 59*, 563–626.

Borrull, A., Lopez-Martinez, G., Miro-Abella, E., Salvado, Z., Poblet, M., & Cordero-Otero, R. (2016). New insights into the physiological state of Saccharomyces cerevisiae during ethanol acclimatation for producing sparkling wines. *Food Microbiology, 54*, 20–29.

Bosch-Fuste, J., Riu-Aumatell, M., Guadayol, J. M., Calxach, J., Lopez-Tamames, E., & Buxaderas, S. (2007). Volatile profiles of sparkling wines obtained by three extraction methods and gas chromatography mass spectrometry (GC-MS) analysis. *Food Chemistry, 105*, 428–435.

Boulton, R. B., Singleton, V. L., Bisson, L. F., & Kunkee, R. E. (1996). *Principles and practices of winemaking*. New York: Chapman & Hall.

Breddam, K., & Beenfeldt, T. (1991). Acceleration of yeast autolysis by chemical methods for production of intracellular enzymes. *Applied Microbiology and Biotechnology, 35*, 323–329.

Brissonnet, F., & Maujean, A. (1991). Identification of some foam active compounds in Champagne base wines. *American Journal of Viticulture and Enology, 42*, 97–102.

Brissonnet, F., & Maujean, A. (1993). Characterization of foaming proteins in a Champagne base wine. *American Journal of Viticulture and Enology, 44*, 297–301.
Buxaderas, S., & Lopez-Tamames, E. (2012). Sparkling wines: Features and trends from tradition. *Advances in Food Nutrition and research, 66*, 1–45.
Cebollero, E., Carrascosa, A. V., & Gonzalez, R. (2005). Evidence for yeast autophagy during simulation of sparkling wine aging: A reappraisal of the mechanism of yeast autolysis in wine. *Biotechnology Progress, 21*, 614–616.
Charpentier, C. (2000). Yeast autolysis and yeast macromolecules? Their contribution to wine flavour and stability. *American Journal of Viticulture and Enology, 51*, 271–275.
Charpentier, C., & Feuillat, F. (1993). Yeast autolysis. In G. H. Fleet (Ed.), *Wine microbiology and biotechnology* (pp. 243–264). Chur: Harwood Academic Publishers.
Charpentier, C., & Freyssinet, M. (1989). The mechanism of yeast autolysis in wine. *Yeasts, 5*, S181–S186.
Charpentier, C., Nguyen Van Long, T., Bonaly, R., & Feuillat, M. (1986). Alteration of cell wall structure in *Saccharomyces cerevisiae* and *Saccharomyces bayanus* during autolysis. *Applied Microbiology and Biotechnology, 24*, 405–413.
Charpentier, M., Bondet, V., Duteurtre, B., Meirland, S., Virion, M.-C., & Valade, M. (1996). Enqête sur les fins de prise de mousse difficiles. *Le Vigneron Champenois, 3*, 6–15.
Charpentier, C., Aussenac, J., Charpentier, M., Prome, J. C., Duteurtre, B., & Feuillat, M. (2005). Release of nucleotides and nucleosides during yeast autolysis: kinetics and potential impact on flavor. *Journal of Agriculture and Food Chemistry, 53*, 3000–3007.
Chung, S. (1986). Contribution à l'étude de la formation des composés volatils au cours de l'autolyse des levures de vinification. Thesis, Université de Bourgogne.
Cilindre, C., Jegou, S., Hovasse, A., Schaeffer, C., Castro, A. J., Clement, C., Van Dorsselaer, A., Jeandet, P., & Marchal, R. (2008). Proteomic approach to identify champagne wine proteins as modified by Botrytis cinerea infection. *Journal of Proteome Research, 7*, 1199–1208.
Coelho, E., Coimbra, M. A., Nogueira, J. M. F., & Rocha, S. M. (2009). Quantification approach for assessment of sparkling wine volatiles from different soils, ripening stages, and varieties by stir bar sorptive extraction with liquid desorption. *Analytica Chimica Acta, 635*, 214–221.
Coloretti, F., Zambonelli, C., & Tini, V. (2006). Characterization of flocculent *Saccharomyces* interspecific hybrids for the production of sparkling wines. *Food Microbiology, 23*, 672–676.
Connew, S. J. (1998). Yeast autolysis: A review of current research. *Australian and New Zealand Wine Industry Journal, 13*, 31–64.
Courtis, K., Todd, B., & Zhao, J. (1998). The potential role of nucleotides in wine flavour. *Australian Grapegrower and Winemaker, 409*, 51–53.
Cullere, L., Cacho, J., & Ferreira, V. (2007). An assessment of the role played by some oxidation-related aldehydes in wine aroma. *Journal of Agricultural and Food Chemistry, 55*, 876–881.
De La Presa-Owens, C., Schlich, P., Davies, H. D., & Noble, A. C. (1998). Effect of Method Champenoise process on aroma flavour of four *V. vinifera* varieties. *American Journal of Enology and Viticulture, 49*, 289–294.
Dupin, I. V., McKinnon, B. M., Ryan, C., Boulay, M., Markides, A. J., Jones, G. P., Williams, P. J., & Waters, E. J. (2000). *Saccharomyces cerevisiae* mannoproteins that protect wine from protein haze: their release during fermentation and lees contact and a proposal for their mechanism of action. *Journal of Agricultural and Food Chemistry, 48*, 3098–3105.
Dussaud, A., Robillard, B., Carles, B., & Duteurtre, B. (1994). Exogenous lipids and ethanol influences on the foam behaviour of sparkling base wines. *Journal of Food Science, 59*, 148–151.
Duteurtre, B., Ors, P., & Hennequin, D. (1987). Les levures incluses. Développement semi-industriel. *Vigneron Champenois, 108*, 595–602.
Feuillat, M. (2003). Yeast macromolecules: origin, composition and enological interest. *American Journal of Enology and Viticulture, 54*, 211–213.
Feuillat, M., & Charpentier, C. (1982). Autolysis of yeasts in Champagne. *American Journal of Viticulture and Enology, 33*, 6–13.

Feuillat, M., Freyssinet, M., & Charpentier, C. (1989). L'élevage sur lies des vins blancs de Bourgogne II. Evolution des macromolécules (polysaccharides et protéines). *Vitis, 28*, 161–176.

Fleet, G. H. (1991). Cell walls. In A. H. Rose & J. S. Harrison (Eds.), *The Yeasts* (Vol. 4, 2nd ed., pp. 199–277). London: Academic.

Fleet, G. H. (2007). Wine. In M. P. Doyle & L. R. Beuchat (Eds.), *Food microbiology fundamentals and frontiers* (3rd ed., pp. 863–890). Washington DC: ASM Press.

Fleet, G. H. (2008). Wine yeasts for the future. *FEMS Yeast Research, 8*, 979–995.

Fleet, G. H., & Heard, G. M. (1993). Yeast-growth during fermentation. In G. H. Fleet (Ed.), *Wine microbiology and biotechnology* (pp. 27–54). Chur: Harwood Academic Publishers.

Fornairon-Bonnefond, C., Camarasa, C., Moutounet, M., & Salmon, J. M. (2001). Etat des connaissances scientifiques actuelles sur le phénomène d'autolyse des levures et l'élevage des vins sur lies. *Journal International des Sciences de la Vigne et du Vin, 35*, 57–78.

Forss, D. A. (1969). Role of lipids in flavours. *Journal of Agricultural and Food Chemistry, 17*, 681–685.

Francioli, S., Torrens, J., Riu-Aumatell, M., Lopez-Tamanez, E., & Buxaderas, S. (2003). Volatile compounds by SPME-GC as age markers of sparkling wines. *American Journal of Viticulture and Enology, 54*, 158–162.

Freyssinet, M., Feuillat, M., & Charpentier, C. (1989). Rôle de la paroi cellulaire dans l'autolyse des levures. Applications oenologiques. In *Actualités oenologiques 89* (pp. 160–168). Paris: Masson.

Fumi, M. F., Trioli, G., & Colagrande, O. (1987). Preliminary assessment on the use of immobilized yeast cells in sodium alginate for sparkling wine processes. *Biotechnology Letters, 9*, 339–342.

Fumi, M. F., Trioli, G., & Colagrande, O. (1988). Immobilization of *Saccharomyces cerevisae* in calcium alginate gel and its application to bottle-fermented sparkling wine production. *American Journal of Enology and Viticulture, 39*, 267–272.

Gallart, M., Lopez-Tamames, E., Suberbiola, G., & Buxaderas, S. (2002). Influence of fatty acids on wine foaming. *Journal of Agricultural and Food Chemistry, 50*, 7042–7045.

Gerbaud, V., Gabas, N., Blouin, J., Pellerin, P., & Moutounet, M. (1997). Influence of wine polysaccharides and polyphenols on the crystallisation of potassium hydrogen tartrate. *Journal International des Sciences de la Vigne et du Vin, 31*, 65–83.

González-Royo, E., Pascual, O., Kontoudakis, N., Esteruelas, M., Esteve-Zarzoso, B., Mas, A., Canals, J. M., & Zamora, F. (2015). Oenological consequences of sequential inoculation with non-*Saccharomyces* yeasts (*Torulaspora delbrueckii* or *Metschnikowia pulcherrima*) and *Saccharomyces cerevisiae* in base wine for sparkling wine production. *European Food Research and Technology, 240*, 999–1012.

Guilloux-Benatier, M., & Chassagne, D. (2003). Comparison of components released by fermented or active dried yeasts after aging on lees in a model wine. *Journal of Agricultural and Food Chemistry, 51*, 746–751.

Hardy, F. (1993). La clarification des vins en bouteilles en method champenoise. Le remuage (2ème partie). *Revue des Œnologues, 68*, 21–26.

Henick-Kling, T. (1993). Malolactic fermentation. In G. H. Fleet (Ed.), *Wine microbiology and biotechnology* (pp. 289–326). Chur: Harwood Academic Publishers.

Hernawan, T., & Fleet, G. H. (1995). Chemical and cytological changes during the autolysis of yeasts. *Journal of Industrial Microbiology, 14*, 440–450.

Hidalgo, P., Pueyo, E., Pozo-Bayón, M. A., Martínez-Rodríguez, A. J., Martín-Álvarez, P., & Polo, M. C. (2004). Sensory and analytical study of rosé sparkling wines manufactured by second fermentation in the bottle. *Journal of Agricultural and Food Chemistry, 52*, 6640–6645.

Hien, N. H., & Fleet, G. H. (1983). Variation of (1→3)-β-glucanases in *Saccharomyces cerevisiae* during vegetative growth, conjugation and sporulation. *Journal of Bacteriology, 156*, 1214–1221.

Hough, M. D., & Maddox, I. S. (1970). Yeast autolysis. *Process Biochemistry, 5*, 50–52.

Howe, P. (2003). Sparkling wines. In A. G. H. Lea & J. R. Piggot (Eds.), *Fermented beverage production* (2nd ed., pp. 139–155). New York: Kluwer Academic.
Huang, W. P., & Klionsky, D. J. (2002). Autophagy in yeast: A review of the molecular machinery. *Cell Structure and Function, 27*, 409–420.
Ibern-Gomez, M., Andres-Lacueva, C., Lamuela-Ravento, R. M., Buxaderas, S., Singleton, V. L., & de la Torre-Boronat, M. C. (2000). Browning of cava (sparkling wine) during aging in contact with lees due to the phenolic composition. *American Journal of Enology and Viticulture, 51*, 29–36.
Jeandet, P., Chaudruc, D., Robillard, B., Peters, F., Tusseau, D., Conreux, A., et al. (2006). Determination of the trans-resveratrol content of Champagne wines by reversed-phase HPLC. *Journal International Des Sciences De La Vigne Et Du Vin, 40*, 117–119.
Kemp, B., Alexandre, H., Robillard, B., & Marchal, R. (2015). Review: Effect of production phases on bottle fermented sparkling wine quality. *Journal of Agricultural and Food Chemistry, 63*, 19–38.
Klein, J., & Wagner, F. (1986). Methods for the immobilization of microbial cells. *Applied Biochemistry and Bioengineering, 4*, 11–51.
Klis, F. M., Mol, P., Hellingwerf, K., & Brul, S. (2002). Dynamics of cell wall structure in *Saccharomyces cerevisiae*. *FEMS Microbiological Review, 26*, 239–256.
Kollar, R., Sturdik, E., & Sablaturova, E. (1993). Biochemical, morphological and cytochemical studies of enhanced autolysis of *Saccharomyces cerevisiae*. *Folia Microbiologica, 38*, 473–478.
Komano, H., Rockwell, N., Wang, G., & Krafft, G. (1999). Purification and characterization of the yeast GPI-anchored, mono-basic aspartyl protease yapsin 2 (Mkc7p). *Journal of Biological Chemistry, 274*, 24421–24437.
Kunkee, R. E., & Ough, C. S. (1966). Multiplication and fermentation of *Saccharomyces cerevisiae* under carbon dioxide pressure in wine. *Applied Microbiology, 4*, 643–648.
Kupfer, V. M., Vogt, E. I., Ziegler, T., Vogel, R. F., & Niessen, L. (2017a). Comparative proteomic protein profile analysis of wines made from Botrytis cinerea infected and healthy grapes reveals a novel biomarker for gushing in sparkling wine. *Food Research International, 99*, 501–509.
Kupfer, V. M., Vogt, E. I., Siebert, A. K., Meyer, M. L., Vogel, R. F., & Niessen, L. (2017b). Foam-stabilizing properties of the yeast protein PAU5 and evaluation of factors that can influence its concentration in must and wine. *Food Research International, 102*, 111–118.
Lacoste, J. (1983). Elaboration des vins effervescents (méthode champenoise). *Vigne et Vin, 325*, 16–19.
Lao, C., Santamaria, A., Lopez-Tamanes, E., Bujan, J., Buxaderas, S., & Torre-Boronat, M. C. (1999). Effect of grape pectic enzyme treatment on foaming properties of white musts and wines. *Food Chemistry, 65*, 169–173.
Laurent, M., & Valade, M. (2007). La préparation du levain de tirage à partir des levures sèches actives. *Le Vigneron Champenois, 3*, 81–90.
Ledoux, V., Dulau, L., & Dubourdieu, D. (1992). Interprétation de l'amélioration de la stabilité protéique des vins au cours de l'élevage sur lies. *Journal International des Sciences de la Vigne et du Vin, 26*, 239–251.
Leroy, M. J., Charpentier, M., Duteurtre, B., Feuillat, M., & Charpentier, C. (1990). Yeast autolysis during Champagne aging. *American Journal of Viticulture and Enology, 41*, 21–28.
Loyaux, D., Roger, S., & Adda, J. (1981). The evolution of champagne volatiles during ageing. *Journal of the Science of Food and Agriculture, 32*, 1254–1258.
Lubbers, S., Leger, B., Charpentier, C., & Feuillat, M. (1993). Effet colloïdes-protecteur d'extraits de parois de levures sur la stabilité tartrique d'une solution hydro-alcoolique modèle. *Journal International des Sciences de la Vigne et du Vin, 27*, 13–22.
Lubbers, S., Voilley, A., Feuillat, M., & Charpentier, C. (1994). Influence of mannoproteins from yeast on the aroma intensity of a model wine. *Lebensmittelwissenschaft und Technologie, 27*, 108–114.
Lurton, L., Segain, J. P., & Feuillat, M. (1989). Etude de la protéolyse au cours de l'autolyse de levures en milieu acide. *Sciences des Aliments, 9*, 111–124.

Malvy, J., Robillard, B., & Duteurtre, B. (1994). Influence des protéines sur le comportement de la mousse des vins de Champagne. *Sciences des Aliments, 14*, 87–98.

Marchal, R., Brunet, S., Marchal-Delahaut, L., Jeandet, P., Maujean, A., & Legendre, L. (1999). Conséquences de l'infection des raisins par Botrytis cinerea sur la composition protéique des moûts champenois. *Revue Française d'œnologie, 177*, 23–27.

Martinez-Garcia, R., Garcia-Martinez, T., Puig-Pujol, A., Mauricio, J. C., & Moreno, J. (2017). Changes in sparkling wine aroma during the second fermentation under CO_2 bottle in sealed Bottle. *Food Chemistry, 237*, 1030–1040.

Martínez-Lapuente, L., Guadalupe, Z., AyestarÁna, B., & Pérez-Magariño, S. (2015). Role of major wine constituents in the foam properties of white and rosé sparkling wines. *Food Chemistry, 174*, 330–338.

Martínez-Lapuente, L., Apolinar-Valiente, R., Guadalupe, Z., Ayestarán, B., Pérez-Magariño, S., Williams, P., et al. (2018). Polysaccharides, oligosaccharides and nitrogenous compounds change during the ageing of Tempranillo and Verdejo sparkling wines. *Journal of the Science of Food and Agriculture, 98*, 291–303.

Martinez-Rodriguez, A. J., & Polo, M. C. (2000). Characterization of the nitrogen compounds released during yeast autolysis in a model wine system. *Journal of Agricultural and Food Chemistry, 48*, 1081–1085.

Martinez-Rodriguez, A. J., & Polo, M. C. (2003). Effect of the addition of bentonite to the tirage solution on the nitrogen composition and sensory quality of sparkling wines. *Food Chemistry, 81*, 383–388.

Martinez-Rodriguez, A. J., & Pueyo, E. (2009). Sparkling wines and yeast autolysis. In V. Moreno-Arribas & C. Polo (Eds.), *Wine chemistry and biochemistry* (pp. 61–80). New York: Springer.

Martinez-Rodriguez, A. J., Carrascosa, A. V., Barcenilla, J. M., Pozo-Bayon, M. A., & Polo, M. C. (2001a). Autolytic capacity and foam analysis as additional criteria for the selection of yeast strains for sparkling wine production. *Food Microbiology, 18*, 183–191.

Martinez-Rodriguez, A. J., Polo, M. C., & Carrascosa, A. V. (2001b). Structural and ultrastructural changes in yeast cells during autolysis in a model wine system and in sparkling wines. *International Journal in Food Microbiology, 71*, 45–51.

Martinez-Rodriguez, A. J., Carrascosa, A. V., & Polo, M. C. (2001c). Release of nitrogen compounds to the extracellular medium by three strains of *Saccharomyces cerevisiae* during induced autolysis in a model wine system. *International Journal in Food Microbiology, 68*, 155–160.

Martinez-Rodriguez, A. J., Carrascosa, A. V., Martin-Alvarez, P. J., Moreno-Arribas, V., & Polo, M. C. (2002). Influence of the yeast strain on the changes of the amino acids, peptides and proteins during sparkling wine production by the traditional method. *Journal of Industrial Microbiology and Biotechnology, 29*, 314–322.

Martí-Raga, M., Sancho, M., Guillamón, J. M., Mas, A., & Beltran, G. (2015). The effect of nitrogen addition on the fermentative performance during sparkling wine production. *Food Research International, 67*, 126–135.

Martí-Raga, M., Peltier, E., Mas, A., Beltran, G., & Marullo, P. (2017). Genetic causes of phenotypic adaptation to the second fermentation of sparkling wines in Saccharomyces cerevisiae. *G3: Genes, Genomes, Genetics, 7*, 399–412.

Martynenko, N. N., & Gracheva, I. M. (2003). Physiological and biochemical characteristics of immobilized Champagne yeasts and their participation in champagnizing process: A review. *Applied Biochemistry and Microbiology, 39*, 439–445.

Martynenko, N. N., Gracheva, I. M., Sarishvili, N. G., Zubov, A. L., El'Registan, G. I., & Lozinsky, V. I. (2004). Immobilization of Champagne yeasts by inclusion into cryogels of polyvinyl alcohol: Means of preventing cell release from the carrier matrix. *Applied Biochemistry and Microbiology, 40*, 158–164.

Maujean, A., Brun, O., Veselle, G., & Sugot, O. (1987). Study of the maturity of Champagne varieties. *Bulletin de l'OIV, 60*, 27–35.

Maujean, A., Poinsaut, P., Dantan, H., Brissonnet, F., & Cossiez, E. (1990). Etude de la tenue et de la qualité de la mousse des vins effervescents. *Bulletin de l'OIV, 63*, 405–427.

Mazauric, J. P., & Salmon, J. M. (2005). Interactions between yeast lees and wine polyphenols during simulation of wine aging: I. Analysis of remnant polyphenolic compounds in the resulting wines. *Journal of Agricultural and Food Chemistry, 53*, 5647–5653.

Mazauric, J. P., & Salmon, J. M. (2006). Interactions between yeast lees and wine polyphenols during simulation of wine aging: II. Analysis of desorbed polyphenol compounds from yeast lees. *Journal of Agricultural and Food Chemistry, 54*, 3876–3881.

Medina-Trujillo, L., Gonzalez-Royo, E., Sieczkowski, N., Heras, J., Canals, J. M., & Zamora, F. (2017). Effect of sequential inoculation (Torulaspora delbrueckii/Saccharomyces cerevisiae) in the first fermentation on the foaming properties of sparkling wine. *European Food Research Technology, 243*, 681–688.

Mendes-Fereira, A., Mendes-Faia, A., & Leao, C. (2004). Growth and fermentation patterns of *Saccharomyces cerevisiae* under different ammonium concentration and its implication in winemaking industry. *Journal of Applied Microbiology, 97*, 540–545.

Moine-Ledoux, V., Perrin, A., Paladin, I., & Dubourdieu, D. (1997). First results of tartaric acid stabilisation by adding mannoproteins (Mannostab™). *Journal International des Sciences de la Vigne et du Vin, 31*, 23–29.

Molnar, I., Oura, E., & Suomalainen, H. (1981). Study of the volatile substances produced during the autolysis of Champagne yeasts. *Acta Alimentaria, 10*, 27–36.

Moreno-Arribas, V., Pueyo, E., & Polo, M. C. (1996). Peptides in musts and wines. Changes during the manufacture of Cavas (Sparkling Wines). *Journal of Agricultural and Food Chemistry, 44*, 3783–3788.

Moreno-Arribas, M. V., Bartolomé, B., Pueyo, E., & Polo, M. C. (1998a). Isolation and characterization of individual peptides from wine. *Journal of Agricultural and Food Chemistry, 46*, 3422–3425.

Moreno-Arribas, V., Pueyo, E., Polo, M. C., & Martin-Alvarez, P. J. (1998b). Changes in the amino acid composition of the different nitrogenous fractions during the aging of wine with yeasts. *Journal of Agricultural and Food Chemistry, 46*, 4042–4051.

Moreno-Arribas, V., Pueyo, E., Nieto, F. J., Martin-Alvarez, P. J., & Polo, M. C. (2000). Influence of the polysaccharides and the nitrogen compounds on foaming properties of sparkling wines. *Food Chemistry, 70*, 309–317.

Morfaux, J. N., & Dupuy, P. (1966). Comparaison de l'exorption des acides aminés par une souche de *Saccharomyces cerevisiae* et un mutant résistant à la canavanine. *Compte Rendu de l' Académie des Sciences, 263*, 1224.

Nagodawithana, T. (1992). Yeast-derived flavors and flavor enhancers and their probable mode of action. *Food Technology, 46*, 138–144.

Nunez, Y. P., Carrascosa, A. V., Gonzalez, R., Polo, M. C., & Martinez-Rodriguez, A. J. (2005). Effect of accelerated autolysis of yeast on the composition and foaming properties of sparkling wines elaborated by a Champenoise method. *Journal of Agricultural and Food Chemistry, 53*, 7232–7237.

Olsen, V., Cawley, N. X., Brandt, J., Egel-Mitani, M., & Loh, Y. P. (1999). Identification and characterization of *Saccharomyces cerevisiae* yapsin 3, a new member of the yapsin family of aspartic proteases encoded by the YPS 3 gene. *Biochemistry Journal, 339*, 407–411.

Onishi, A., & Proudlove, M. O. (1994). Isolation of beer foam polypeptides by hydrophobic interaction chromatography and their partial characterization. *Journal of Science and Food Agriculture, 65*, 233–240.

Penacho, V., Valero, E., & Gonzalez, R. (2012). Transcription profiling of sparkling wine second fermentation. *International Journal of Food Microbiology, 153*, 172–186.

Pernot, N. (1999). Le débourbage des moûts en Champagne. *Le Vigneron Champenois, 8*, 43–60.

Pham, T. T., Guichard, E., Schlich, P., & Charpentier, C. (1995). Optimal conditions for the formation of sotolon from α-cetobutyric acid in the French "Vin Jaune". *Journal of Agricultural and Food Chemistry, 43*, 2616–2619.

Piton, F., Charpentier, M., & Troton, D. (1988). Cell wall and lipid changes in *Saccharomyces cerevisiae* during ageing of Champagne wine. *American Journal of Viticulture and Enology, 39*, 221–226.

Polo, M. C., Gonzalez de Llano, M. D., & Ramos, M. (1992). Derivatization and liquid chromatographic separation of peptides. In M. L. Nollet (Ed.), *Food analysis by HPLC* (pp. 117–140). New York: Dekker.

Pozo-Bayon, M. A., Hernandez, M. T., Martin-Alvarez, P. J., & Polo, M. C. (2003). Study of low molecular weight phenolic compounds during the aging of sparkling wines manufactured with red and white grape varieties. *Journal of Agricultural and Food Chemistry, 51*, 2089–2095.

Pozo-Bayon, M. A., Martınez-Rodrıguez, A., Pueyo, E., & Moreno-Arribas, M. V. (2009). Chemical and biochemical features involved in sparkling wine production: From a traditional to an improved winemaking technology. *Trends Food Sciences and Technology, 20*, 289–299.

Pozo-Bay6n, M. A., Martín-Álvarez, P. J., Moreno-Arribas, M. V., Andújar-Ortiz, I., & Pueyo, E. (2010). Impact of using Trepat and Monastrell red grapes varieties on volatile and nitrogen composition during the manufacture of rosé Cava sparkling wines. *LWT-Food Science and Technology, 43*, 1526–1532.

Pueyo, E., Martin-Alvarez, P., & Polo, M. C. (1995). Relationship between foam characteristics and chemical composition in wines and cavas (sparkling wine). *American Journal of Viticulture and Enology, 46*, 518–524.

Pueyo, E., Martinez-Rodriguez, A., Polo, M. C., Santa-Maria, G., & Bartolome, B. (2000). Release of lipids during yeast autolysis in a model wine system. *Journal of Agricultural and Food Chemistry, 46*, 116–122.

Puig-Deu, M., Lopez-Tamanes, E., Buxaderas, S., & Torre-Boronat, M. C. (1999). Quality of base and sparkling wines as inflenced by the type of fining agent added pre-fermentation. *Food Chemistry, 66*, 35–42.

Rapp, A. (1998). Volatile flavour of wine: Correlation between instrumental analysis and sensory perception. *Nahrung, 42*, 351–363.

Rapp, A., & Versini, G. (1991). Influence of nitrogen compounds in grapes on aroma compounds of wines. In J. M. Rantz (Ed.), *Proceedings of the international symposium on nitrogen in grapes and wine* (pp. 156–164). Washington: Seattle.

Ribéreau-Gayon, P., Dubourdieu, D., Donèche, B., & Lonvaud, A. (2004). Traité d'Oenologie.1.microbiologie du vin, 5ème ed Dunod, France.

Riu-Aumatell, M., Bosch-Fusté, J., Lopez-Tamames, E., & Buxaderas, S. (2006). Development of volatile compounds of cava (Spanish sparkling wine) during long ageing time in contact with lees. *Food Chemistry, 95*, 237–242.

Romano, P., Soli, M. G., Suzzi, G., Grazia, L., & Zambonelli, C. (1985). Improvement of a wine *Saccharomyces cerevisae* strain by a breeding program. *Applied and Environmental Microbiology, 50*, 1064–1067.

Ruiz-Moreno, M. J., Munoz-Redondo, J. M., Cuevas, F. J., Marrufo-Curtido, A., Leon, J. M., Ramirez, P., et al. (2017). The influence of pre-fermentative maceration and ageing on ester profile and marker determination of Pedro Ximenez sparkling wine. *Food Chemistry, 230*, 697–704.

Satue-Gracia, M. T., Andres-Lacueva, C., Lamuela-Raventos, R. M., & Frankel, E. N. (1999). Spanish sparkling wines (cavas) as inhibitors of in vitro human low-density lipoprotein oxidation. *Journal of Agricultural and Food Chemistry, 47*, 2198–2202.

Senee, J., Viaux, L., Robillar, B., Duteurtre, B., & Vignes-Adler, M. (1998). The endogenous particles of a sparkling wine and their influence on the foaming behaviour. *Food Hydrocolloids, 12*, 217–226.

Serra-Cayuela, A., Aguilera-Curiel, M. A., Riu-Aumatel, M., Buxaderas, S., & López-Tamames, E. (2013). Browning during biological aging and commercial storage of Cava sparkling wine and the use of 5-HMF as a quality marker. *Food Research International, 53*, 226–231.

Serra-Cayuela, A., Jourdes, M., Riu-Aumatell, M., Buxaderas, S., Teisseidre, P. L., & López-Tamames, E. (2014). Kinetics of browning, phenolics, and 5-hydroxymethylfurfural in commercial sparkling wines. *Journal of Agricultural and Food Chemistry, 62*, 1159–1166.

Shinohara, T., Mamiya, S., & Yanagida, F. (1997). Introduction of flocculation property into wine yeasts (*Saccharomyces cerevisiae*) by hybridization. *Journal of Fermentation and Bioengineering, 83*, 96–101.

Speers, R. A., Tung, M. A., Durance, T. D., & Stewart, G. G. (1992). Biochemical aspects of yeast flocculation and its measurement. *Journal of the Institute of Brewing, 98*, 293–300.
St John Coghlan, D., Woodrow, C. W., Bamforth, C. W., & Hinchliffe, E. (1992). Polypeptides with enhanced foam potential. *Journal of the Institute Brewing, 98*, 207–213.
Stefenon, C. A., Bonesi, C. D. M., Marzarotto, V., Barnabé, D., Spinelli, F. R., Webber, V., et al. (2014). Phenolic composition and antioxidant activity in sparkling wines: Modulation by the ageing on lees. *Food Chemistry, 145*, 292–299.
Stewart, G. G., & Russell, I. (1986). One hundred years of yeast research and development in the brewing industry. *Journal of the Institute of Brewing, 92*, 537–558.
Strehaiano, P., Ramon-Portugal, F., & Taillandier, P. (2006). Yeasts as biocatalysts. In A. Querol & G. H. Fleet (Eds.), *Yeasts in food and beverages* (pp. 243–284). Berlin: Springer.
Suomalainen, H. (1975). Some enzymological factors influencing the leavening capacity and keeping quality of baker's yeast. *European Journal of Applied Microbiology, 1*, 1–12.
Suzzi, G. (1990). La capacita autolitica come caracterre di selezione in *Saccharomyces cerevisiae. Industrie delle Bevande, 19*, 318–321.
Thornton, R. J. (1985). The introduction of flocculation into homothallic wine yeast. *American Journal of Enology and Viticulture, 36*, 47–49.
Todd, B., Fleet, G. H., & Henschke, P. A. (2000). Promotion of autolysis through the interaction of killer and sensitive yeasts: Potential application in sparkling wine production. *American Journal of Enology and Viticulture, 51*, 65–72.
Toressi, S., Frangipane, M. T., & Anelli, G. (2011). Biotechnologies in sparkling wine production. Interesting approaches for quality improvement: A review. *Food Chemistry, 129*, 1232–1241.
Torrens, J., Riu-Aumatell, M., Vichi, S., López-Tamames, E., & Buxaderas, S. (2010). Assessment of volatile and sensory profiles between base and sparkling wines. *Journal of Agricultural and Food Chemistry, 58*, 2455–2461.
Trevelyan, W. E. (1978). Effect of procedures for the reduction of the nucleic acid content of SCP on the DNA content of *Saccharomyces cerevisiae. Journal of Science and Food Agriculture, 29*, 903–908.
Troton, D., Charpentier, M., Robillard, B., Calvayrac, R., & Duteurtre, B. (1989). Evolution of the lipid contents of Champagne wine during the second fermentation of *Saccharomyces cerevisiae. American Journal of Enology and Viticulture, 40*, 175–182.
Tusseau, D. (2004). Les moûts 2003. *Le Vigneron champenois, 1*, 42–69.
Usseglio-Tomasset, L., & Bosia, P. D. (1990). Evolution des acides aminés et des oligopeptides du moût au vin. *Bulletin de l'OIV, 63*, 21–46.
Valade, M., & Blanck, G. (1989). Evolution des paramètres analytiques au cours du pressurage en Champagne. *Revue Française d'œnologie, 118*, 23–27.
Valade, M., & Laurent, M. (1999). La prise de mousse: Les phénomènes microbiologiques. *Le Vigneron Champenois, 6*, 67–89.
Valade, M., Laurent, M., Nail, T. H., Delgenes, P. H., & Putier, F. (1987a). La fermentation malolactique en Champagne, son ensemencement par l'intermédiaire de bactéries lyophilisées sélectionnées. Première partie. *Le Vigneron Champenois, 108*, 399–414.
Valade, M., Laurent, M., Nail, T. H., Delgenes, P. H., & Putier, F. (1987b). La fermentation malolactique en Champagne, son ensemencement par l'intermédiaire de bactéries lyophilisées sélectionnées. Deuxième partie. *Le Vigneron Champenois, 108*, 455–471.
Vanrell, G., Canals, R., Esteruelas, M., Fort, F., Canals, J. M., & Zamora, F. (2007). Influence of the use of bentonite as a riddling agent on foam quality and protein fraction of sparkling wines (Cava). *Food Chemistry, 104*, 148–155.
Verstrepen, K. J., Derdelinckx, G., Verachtert, H., & Delvaux, F. R. (2003). Yeast flocculation: What brewers should know. *Applied Microbiology and Biotechnology, 61*, 197–205.
Vezinhet, F., Barre, P., Laurent, M., & Valade, M. (1992). Introduction of flocculation into an industrial yeast strain by transfer of a single chromosome. *Journal of the Institute of Brewing, 98*, 315–319.

Vilanova, M., Ugliano, M., Varela, C., Siebert, T., Pretorius, I. S., & Hensckke, P. A. (2007). Assimilable nitrogen utilisation and production of volatile and non-volatile compounds in chemically defined medium by Saccharomyces cerevisiae wine yeasts. *Applied Microbiology and Biotechnology, 77*, 145–157.

Vosti, D. C., & Joslyn, M. A. (1954). Autolysis of baker's yeast. *Applied Microbiology, 2*, 70–78.

Welke, J. E., Zanus, M., Lazzarotto, M., Hepp-Pulgati, F., & Alcaraz-Zini, C. (2014). Main differences between volatiles of sparkling and base wines accessed through comprehensive two dimensional gas chromatography with time-of-flight spectrometric detection and chemometric tools. *Food Chemistry, 164*, 427–437.

Winterhalter, P. (1991). 1,1,6-trimethyl-1,2-dihydronaphthalene (TDN) formation in wine. 1. Studies on the hydrolysis of 2,6,10,10-tetramethyl-1-oxaspiro[4.5]dec-6-ene-2,8-diol rationalizing the origin of TDN and related C13 norisoprenoids in Riesling wine. *Journal of Agricultural and Food Chemistry, 39*, 1825–1829.

Yokotsuka, K., Yajima, M., & Matsudo, T. (1997). Production of bottle-fermented sparkling wine using yeast immobilized in double-layer gel beads or strands. *American Journal of Enology and Viticulture, 48*, 471–481.

Zambonelli, C., Rainieri, S., Chiavari, C., Montanari, G., Benevelli, M., & Grazia, L. (2000). Autolysis of yeasts and bacteria in fermented foods. *Italian Journal of Food Science, 1*, 9–21.

Zhao, J., & Fleet, G. H. (2003). Degradation of DNA during the autolysis of *Saccharomyces cerevisiae*. *Journal of Industrial Microbiology and Biotechnology, 30*, 175–182.

Zhao, J., & Fleet, G. H. (2005). Degradation of RNA during the autolysis of *Saccharomyces cerevisiae* produces predominantly ribonucleotides. *Journal of Industrial Microbiology and Biotechnology, 32*(9), 415–423.

Chapter 14
Yeasts Associated With Biological Ageing of Fortified Wines

Tahía Benítez, Ana M. Rincón, and Antonio C. Codón

Contents

1 Introduction: Fortified Wines Subjected to Biological Ageing: Sherry and Sherry-Like Wines

Fortified wines are produced from fully fermented, partially fermented or unfermented grape juice, by fortification with wine derived ethanol. Historically, these wines were created in response to technical problems encountered in warm climatic regions: sugar-rich grapes and high temperatures frequently led to stuck fermentations producing unstable wines easily susceptible to subsequent acidification and spoilage. The addition of alcohol stopped fermentation, stabilized the wine, and

T. Benítez • A. C. Codón
Department of Genetics, Faculty of Biology, University of Seville, Sevilla, Spain
e-mail: tahia@us.es; accodon@us.es

A. M. Rincón (✉)
Department of Genetics, Centro Andaluz de Biología Molecular y Medicina Regenerativa (CABIMER), University of Seville, Sevilla, Spain
e-mail: amrincon@us.es

P. Romano et al. (eds.), *Yeasts in the Production of Wine*,
https://doi.org/10.1007/978-1-4939-9782-4_14

resulted in an alcoholic, sweet wine product. Generally, these wines have final ethanol concentrations between 15 and 23% (v/v). General descriptions of the history and production of these types of wines can be found in Moreira and Guedes de Pinho (2011); Perestrelo et al. (2011); Pozo-Bayon and Moreno-Arribas (2011); Reader and Dominguez (2003) and Ribereau-Gayon et al. (2006).

Fortified wines are classified according to their sugar levels as semi-dry (up to 20 $g.L^{-1}$), sweet (up to 36 $g.L^{-1}$) and syrupy sweet (liquoreux) (above 36 $g.L^{-1}$). For semi-dry and sweet wines, alcoholic fermentation is usually stopped before completion by adding ethanol from fermented grape distillations; syrupy sweet wines require higher sugar concentrations that are not attained by conventional winemaking procedures. In these cases, grapes can be dried naturally by the sun, frozen on the vine (e.g. producing icewine), or subjected to noble rot (e.g. Botrytized wines, see Chap. 7) to increase their initial sugar contents to levels such as 400 $g.L^{-1}$.

These wines vary according to grape variety and ageing process, which may be biological (e.g. as for Sherry wines) or physico-chemical (e.g. as for Port and Madeira wines).

Sherry wines from Spain are among the most prestigious fortified wines (Parish and Fleet 2013; Ribereau-Gayon et al. 2006) and are the types discussed in this Chapter. In 2017, global production of Sherry wines, made exclusively from white grapes, was about 326.000 HL (https://www.sherry.wine/sites/default/files/memoria_2017.pdf).

Sherries are mainly produced from grapes cultivated in Spain in areas around Jerez de la Frontera, Cádiz. Their production is regulated by the "Consejo Regulador de la Denominación de Origen". The vineyards are situated in alkaline soils called *albariza*, characterized by their capacity for water retention (Reader and Dominguez 2003). Over 95% grape vines are the Palomino Fino variety and the rest are mostly Moscatel and Pedro Ximénez varieties. As for other Spanish regions, Sherry-like wines are mainly produced in the Montilla-Moriles area, near Córdoba, where temperatures are higher than in Cádiz, and wines are produced almost exclusively with Pedro Ximénez grapes that are more resistant to those climatic conditions (Table 14.1) (Benitez et al. 2011; Moreno-Garcia et al. 2015a, b; Pozo-Bayon and Moreno-Arribas 2011; Suarez Lepe and Iñigo Leal 2011).

Sherry and Sherry-like wines possess alcohol content from 15 to 18%, which derives from adding alcohol to fully fermented musts that usually have ethanol concentrations of 10–12%. Sherry and Sherry-like wines are classified as *fino, amontillado* or *oloroso*, based on the process of ageing (Table 14.1). They are all aged by the *soleras* and *criaderas* method (Fig. 14.1). The butts selected for *fino* and *amontillado* production are filled up to 5/6 of volume; those selected for *oloroso* are filled completely. The ageing of *fino*, and the first period of *amontillado*, are conducted in the coolest part of cellars where wines are aged biologically by flor yeasts that form a film (velum) on the wine surface; *olorosos* are maintained in a warmer place and are aged chemically in the presence of oxygen (Reader and Dominguez 2003).

There are also sweet, fortified Sherry wines that possess sugar concentration over 50 $g.L^{-1}$; the grape variety is Pedro Ximénez. Berries are sun-dehydrated and an incomplete fermentation is stopped by adding brandy up to 18% ethanol, thus

Table 14.1 Grape varieties, type of wines and ageing procedures carried out in the areas of Jerez and Montilla-Moriles

D. O.	Jerez-Xerez-Sherry and Manzanilla-Sanlúcar		Montilla-Moriles	
Regulations				
Ministry order:	O.M. 2-V-77		O.M. 12-XII-85	
	B.O.E. 12-V-77		B.O.E. 27-XII-85	
Grape Varieties				
	Whites:		Whites:	
	Palomino de Jerez		Airén	
	Palomino Fino		Baladí-Verdejo = Jaén blanco	
	Pedro Ximénez		Moscatel	
	Moscatel		Pedro Ximénez	
Type of wine				
	Fortified wines:	Ethanol (v/v)	Fortified wines:	Ethanol (v/v)
	Fino	15°	Fino	14–17.5°
	Amontillado	16–18°	Amontillado	16–22°
	Oloroso	18–20°	Oloroso	16–20°
	Palo Cortado and Raya	18–20°	Palo Cortado	16–18°
	Manzanilla	15°	Raya	16–20°
	Sweet and syrupy wines	15–20°	Sweet and syrupy wines	15–20°
			Pedro Ximénez	
			Whites:	
			Without ageing	10–12°
			With ageing	Mín. 13°
			Young wines	
			Ruedos	Mín. 14°
Ageing				
	Procedure of criaderas and soleras in oak butts. Wines aged for at least 3 years		Procedure of criaderas and soleras in oak butts. Wines aged for at least 2 years	

producing very sweet wines which are then submitted to oxidative ageing by the *soleras* and *criaderas* procedure (Campo et al. 2008; Suarez-Lepe 1997). Other Sherry wines marketed as medium dry, medium cream or pale cream Sherries usually contain *oloroso* wine and also small amounts of *fino* -to lighten the color- and of *amontillado*, together with variable amounts of sweetening material. *Finos* elaborated in Sanlúcar de Barrameda are called *manzanillas*; Sanlúcar de Barrameda is located on the Atlantic coast, and so *manzanillas* have an intense and characteristic iodine flavor (Table 10.1). Wines from some butts subjected to physico-chemical ageing develop characteristics from both *amontillado* (aroma) and *oloroso* (taste) and are called *palo cortado*; in addition, fortified wines recovered from fermentation lees and subjected to physico-chemical ageing but with less intense aroma than that of *olorosos* are classified as *raya* (Table 14.1).

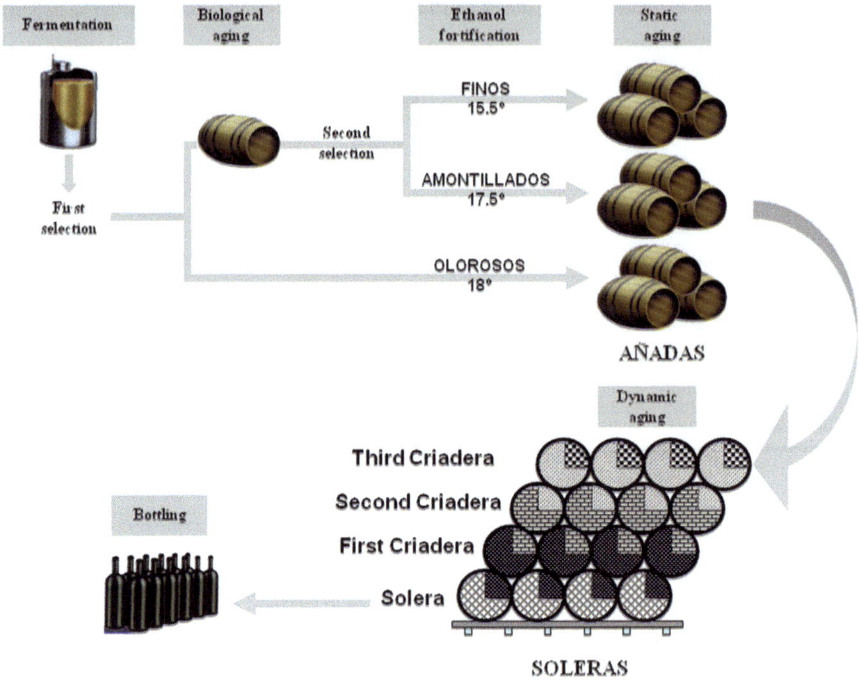

Fig. 14.1 The elaboration scheme (fermentation and ageing) of Sherry wines. Musts are fermented producing dry wines with 10–12% ethanol. A first classification will select the coarser wines that will be fortifed to 18% ethanol and will be subjected to physico-chemical ageing for *OLOROSOS* elaboration. The paler and more delicate wines will be fortified to 15.5% ethanol and subjected to biological ageing. After a second classification, some butts will be selected for *FINOS* and will maintain their biological ageing whereas other butts will be fortified to 17.5% ethanol for *AMONTILLADOS* and will be subjected to physico-chemical ageing. Wines, unblended, are then allowed to undergo static ageing (*AÑADAS*) for 1–3 years. *Finos*, *amontillados* and *olorosos* are further subjected to a dynamic process of continuous blending (*SOLERAS*). Portions of wine from the oldest butts (*Solera*) are periodically blended and bottled. The volume removed is replaced with a similar amount of younger wine (*First Criadera*), the process being repeated until wines from the youngest *Criadera* are replaced with wine from the *AÑADAS*

2 Yeasts Associated With the Ageing of Sherry Wines

Yeasts have two roles in the process of Sherry production. First, they conduct alcoholic fermentation in the grape must. The wine so produced is then subject to a specific storage process where yeasts have a second role in contributing to wine maturation and ageing. After must fermentation, lees are removed, and alcohol is added to the wine. Increase in ethanol concentration to 15.5% induces flor yeast growth aerobically forming a white, rough layer on the surface of the wine, called the velum (Campo et al. 2008).

Fino wine under velum is submitted to some special conditions as a result of yeast oxidative metabolism and the reducing environment created in the wine, since dissolved oxygen is consumed by the flor yeasts and the velum prevents wine from uptaking new oxygen from the air (Mauricio and Ortega 1997; Pozo-Bayon and Moreno-Arribas 2011). A combination of both processes causes transformation in the substances contained in *fino* wines that result in the formation of different and unique organoleptic flavors and aromas (Berlanga et al. 2004; Mauricio and Ortega 1997; Moreno-Garcia et al. 2015a) (Table 14.2).

For wines subjected to biological ageing, velum develops in each of the butts, either *añadas*, *criaderas* or *soleras*. During the transfer of wine from one stage of the *solera* to the next, velum is periodically broken and newly formed. This phenomenon has important implications in flor yeasts ecology and population frequencies.

S. cerevisiae strains, responsible for must fermentation, are present in vineyards and fermentation tanks only for a few weeks a year, whereas yeast strains involved in Sherry ageing are always present in cellars (Benitez et al. 2011; Martinez et al. 1995; Infante et al. 2003b). The matter of whether flor yeasts, responsible for Sherry maturation, were in fact fermenting yeasts adapted to the conditions of the ageing wine (Suarez Lepe and Iñigo Leal 2011), or different strains or species present at very low concentration in the fermenting must and further selected, has long been debated (Capece et al. 2013).

Table 14.2 Suitable *añada* base wine for flor maturation and characteristics and average values of Sherry wine components from Jerez area

	Wines[c]			
Characteristics	Añadas[d]	Finos	Olorosos	Pedro Ximénez
Soil	Albariza	Albariza	Albariza	Albariza
Grape variety	Palomino	Palomino	Palomino	Pedro Ximénez
Alcohol content	14.8–15.3	15.5	18	15.5
Ageing	Biological	Biological	Physico-chemical	Physico-chemical
Years of ageing	1–3	3–5	8–10	8–25
Analytical values				
pH	3.0–3.25	2.9–3.3	3.1–3.5	3.6–4.1
alcohol	14.8–15.3	15.5–17.0	18–21	15.5
acidity[a] (g L^{-1})	–	3.7–5.2	4.5–6.0	5.2–7.1
volatile acidity[b] (g L^{-1})	<0.65	<0.3	<0.8–1.2	<0.8–1.3
acetaldehyde (mg L^{-1})	–	200–400	60–80	150–200
glycerol (g L^{-1})	6.7–7.2	<1.0	5–8	3–5
malic acid (mg L^{-1})	<0.15	134–268	335–603	2500
lactic acid (mg L^{-1})	<1.15	<900	<720	<400
polyphenols (mg L^{-1})	<250	250	275–350	500

[a]In g L^{-1} tartaric acid
[b]In g L^{-1} acetic acid
[c]Modified from Garcia Maiquez E. (1995)
[d]Modified from Reader and Dominguez (2003)

2.1 Taxonomic Characterization and Ecology of Flor Yeasts

The velum is in fact a heterogeneous mixture of microorganisms. They are mostly yeasts and over 95% of them are formed by different strains of *S. cerevisiae* (Kurtzman et al. 2011). The remaining yeasts are species of *Debaryomyces*, *Pichia*, *Hansenula* and *Candida* (Alexandre 2013; Marin-Menguiano et al. 2017; Martinez et al. 1995, 1997a, b, c).

Taxonomic descriptions of flor yeasts are complex and confusing. Flor yeasts within *Saccharomyces* have been identified by several authors and considered to represent a variety of species, including *Saccharomyces beticus* and *Saccharomyces cheresiensis* (synonymous with *S. cerevisiae*), *Saccharomyces montuliensis* (now *T. delbrueckii*) and *Saccharomyces rouxii* (now *Zygosaccharomyces rouxii*) (Alexandre 2013; Barnett et al. 1990; Reader and Dominguez 2003; Toledano et al. 1991).

Merida et al. (2005) described the predominance of specific varieties according to *country region*: In most cases, there is a taxonomic linkage between specific varieties and strain properties: *T. delbruckii* strains (*S. cerevisae* var. *montuliensis*) have low acetogenic power; *S. cerevisiae* var. *cerevisiae* strains possess the highest alcoholigenic power (Toledano et al. 1991), var. *ellipsoideus* have higher ethanol tolerance, vars. *capensis* and *bayanus* give higher color reduction capacity, var. *bayanus* has higher tolerance to CO_2, var. *capensis* gives intensive flavour, var. *rosei* renders low volatile acidity, vars. *beticus* and *cheresiensis* have strong oxidative capabilities and film forming potential, and vars. *montuliensis* and *rouxii* produce high concentrations of acetaldehyde production and tolerance. However, all varieties belong to the same species, *S. cerevisiae* (Alexandre 2013; Barnett et al. 1990; Moreno et al. 1991; Marin-Menguiano et al. 2017; Martinez et al. 1997a, b, c; Merida et al. 2005).

Because of their diversity, *S. cerevisiae* flor strains have been classified according to "races" by other authors, depending on their capacity to ferment and assimilate different sugars (Legras et al. 2014; Marin-Menguiano et al. 2017; Sancho et al. 1986; Suarez Lepe and Iñigo Leal 2011). Of the different races of *S. cerevisiae*, race *montuliensis* is able to assimilate only glucose; race *beticus* assimilates glucose and sucrose; race *cheresiensis*, glucose, sucrose and maltose; race *rouxii* assimilates glucose and maltose; race *aceti* glucose, sucrose and trehalose, and race *chevalieri*, glucose, sucrose and galactose. A sugar fermentation and assimilation test is routinely used and the strains are classified according to races for quality control and rapid identification of *S. cerevisiae* strains in biological ageing (Marin-Menguiano et al. 2017; Sancho et al. 1986; Suarez-Lepe 1997). Species and races are cited in this Chapter as they appeared in the original papers.

The presence of different races, in diverse proportions affects the characteristics of wines, and so classifications according to races is still maintained in cellars. Velum from Sherry-like wines from Montilla-Moriles was formed by *S. cerevisiae* races *capensis*, *chevalieri* and *aceti* (Guijo et al. 1986). Other authors found that *S. beticus* (ca. 85%) and *S. montuliensis* (ca. 15%) were the predominant races isolated in different wineries of this region (Marin-Menguiano et al. 2017).

Variable flor yeast populations have also been reported in *different cellars within the same area*. In some cellars from Jerez, as also occurs in those of Montilla-Moriles (Marin-Menguiano et al. 2017) 95% of the velum population was formed by *S. cerevisae* strains, with two main races, *beticus* and *montuliensis* (Martinez et al. 1995). Two races, *beticus* and *cheresiensis*, and a small population of *Pichia* species (Mesa et al. 1999), a single race, *beticus* (Esteve-Zarzoso et al. 2001), a single population of *beticus* per butt, together with species of *Brettanomyces* and *Dekkera* as a minority population (Ibeas et al. 1996), have also been reported in different cellars. In consequence, the organoleptic characteristics of the Sherries elaborated are also very different, and specific for each cellar (Esteve-Zarzoso et al. 2001; Marin-Menguiano et al. 2017).

Characterization of flor yeasts in añadas, the youngest Sherries, indicated the presence of *S. cerevisae* races *beticus* and *cheresiensis* (Mesa et al. 2000). The presence of *montuliensis* and *rouxii* races has only been described in the final blending stages of relatively old Sherries. These differences in yeast populations have been attributed to differences in sensitivity to compounds such as acetaldehyde (Martinez et al. 1997b), or to the presence of a killer factor (Ibeas and Jimenez 1996; Mesa et al. 1999) that would select the most tolerant populations, or to differences in the time needed for velum formation (Martinez et al. 1995). Furthermore, within a single race, phenotypes with low, medium and high acetaldehyde production have been described (Martinez et al. 1997a, b, c; Marin-Menguiano et al. 2017). *S. cerevisiae* races *beticus* and *cheresiensis* are faster at forming velum, phenotypic differences being also found within the same race. For this reason, *beticus* and *cheresiensis* races are predominant in younger Sherries, in which acetaldehyde concentration is never higher than 300 mg.L^{-1}. During biological ageing, all wines undergo an increase in acetaldehyde concentration. *S. cerevisiae* races *montuliensis* and *rouxii* are more resistant to this compound and so the latter strains are progressively selected in the velum against the former races. Disruptions in the dynamic system, for instance every time wine is removed and the butts refilled, produce loss of velum that needs to be formed again, thus favoring the development of *beticus* and *cheresiensis* races (Alexandre 2013; Marin-Menguiano et al. 2017; Martinez et al. 1997a, b, c).

Some reports have aimed at *correlating the enological and genetic characteristics* of flor and fermenting yeasts (Martinez et al. 1995, 1997a, b, c, 1998), after using molecular and physiological techniques to identify them. Different isolates belonging to the same yeast race displayed great variability in the RFLP of their mtDNA but rendered little diversity in karyotype. However, after comparison of physiological and molecular characteristics based on the 5.8S rRNA gene, internal transcribed regions (5.8S-ITS) and phylogenetic analyses (Esteve-Zarzoso et al. 2004) no correlation was found between flor races, strains isolated from culture collections and those from natural environments (Capece et al. 2013). It is interesting to note that two flor yeast groups have been identified in Vins Jaunes. Yeasts isolated physically from different areas and characterized by the production of a thin velum had similar "interdelta" genetic profile, different from that of yeasts producing a thick velum (Alexandre 2013).

By applying molecular and physiological analyses, other authors (Esteve-Zarzoso et al. 2001; Granchi et al. 2002; Ibeas et al. 1997a), identified flor strains, which differed in their molecular karyotype but yielded identical mtDNA restriction patterns, whereas others showed distinguishable mitochondrial genomes but exhibited identical karyotype. The interpretation was that the different strains identified were very closely related. Colonies isolated from individual barrels surprisingly corresponded in most cases to a single karyotype, and that was still the case over a period of at least 2 years, thus indicating that the genotype of the dominant strain was quite stable (Ibeas et al. 1997a).

2.2 *Metabolic and Physiological Characteristics of* **Saccharomyces** *Flor Yeasts Involved in Biological Ageing*

The physiological properties of flor yeasts are very different from those of fermenting yeasts, and very different among the various flor isolates (Esteve-Zarzoso et al. 2001, 2004; Marin-Menguiano et al. 2017; Martinez et al. 1997a, b, c, 1998) (Table 14.3A). These flor yeasts are metabolically very active, to the extent that ethanol consumption could be of 7.5–9 L per butt (500 L) per year (Benitez et al. 2011).

Comparative genomics, proteomics and metabolomics of flor and wine yeasts have revealed the extent of proteome remodelling imposed by the biofilm life-style (Legras et al. 2014). Moreno-Garcia et al. (2014, 2015a, b, 2016, 2017) performed a proteome analysis during biofilm formation to elucidate the role of the mitochondria and detected several mitochondrion-located proteins that were highlighted and were involved in carbohydrate oxidative metabolism, velum formation, apoptosis, stress responses to ethanol, acetaldehyde reactive-oxygen species, metabolism of non-fermentable carbon uptake, respiration, aroma formation and other phenomena that were particularly expressed in yeasts forming velum as compared to fermentative conditions. As a result, production of specific compounds take place under velum formation and wine ageing (Legras et al. 2016).

Sherry *finos*, *amontillados* and *olorosos* processes extract tanins, phenols (ferulic acid, syringaldehydes) and other compounds from the oak butts (Martinez de la Ossa et al. 1987a, b); in *olorosos*, oxidation of polyphenols from the grapes gives rise to organic acids such as caffeic, cumaric and gentisic (Martinez de la Ossa et al. 1987a, b). For *finos* and *amontillados*, oxidative metabolism of different strains of *S. cerevisiae* is mainly responsible for the characteristics of the final products (Marin-Menguiano et al. 2017; Martinez et al. 1997a, b, c; Pozo-Bayon and Moreno-Arribas 2011). These wines are all relatively rich in 3-methylbutanal, phenylacetaldehyde, methional, ethyl esters of methylpentanoic acids and sotolon (Fig. 14.2), the latter contributing to the typical aged aroma of Sherries (Collin et al. 2012; Martin et al. 1992). Nevertheless, *finos* are particularly rich in acetaldehyde,

Table 14.3A Concentrations of volatile and other compounds produced by different species of wine yeasts[a]

Volatile compounds concentration (µg mL^{-1})							
Yeast species	Propanol	Isobutanol	Isoamyl alcohol	2-Phenyl etanol	Ethyl acetate	Isoamyl acetate	Acetoin
Saccharomyces cerevisiae	0.4–170	5–666	17–769	5–83	10–205	0.1–16	0–29
Kloeckera apiculata	4–25	3–60		10–35	40–870	0.04–1.1	56–187
Candida stellata	4–8	13–21		6–11	7–25	0.1–0.4	35–254
Hansenula anomala[b]	3–15	18–29	11–25	27	137–2150	1–11	
Metschinikowia pulcherrima	1–43	37–123	21–243	22	150–382	0.1–0.8	
Zygosaccharomyces bailii	18–25	20–30	48–85	13–22	23–53	0.1–0.5	17–24
Pichia membranaefaciens	<1	1–9	0.5–9.5		16–21	1–6	
Brettanomyces bruxellensis	2–3	19	46	26	36–860	<1	

Other compounds					
	Ethanol (%)	Glycerol (mg mLl^{-1})	Acetaldehyde (µg mL^{-1})	Acetic acid (mg mL^{-1})	Succinic acid (mg mL^{-1})
Saccharomyces cerevisiae	6–23	3.7–6.8	15–30	0.1–2.0	0.6–1.7
Kloeckera apiculata	2–7	5.5–8.2	8–54	0.1–1.2	0.3
Candida stellata	6–7			1.08–1.3	
Hansenula anomala[3]	0.5–5	0.2–2.2	3.2–8.1	1.6	0.2
Metschinikowia pulcherrima	2–4	2.7–4.2	23–40	0.1–0.2	
Zygosaccharomyces bailii	5–13			0.1–0.3	1.6
Pichia membranaefaciens	0.1–0.5	4.1–5.4	2.9	0.3	
Brettanomyces bruxellensis	9–12			1–7	

[a]Modified from Fleet (2002)
[b]Nowadays *Pichia anomala* (Kurtzman et al. 2011)

diacetyl, ethyl esters of branched aliphatic acids and 4-ethylguaicol (Campo et al. 2008; Marin-Menguiano et al. 2017; Pozo-Bayon and Moreno-Arribas 2011).

Flor yeasts play two central roles in wine maturation. They use and transform various substances such as ethanol and glycerol and produce acetaldehyde, acetic acid, acetoin and intermediate compounds that are oxidized via reactions in which they act as electron donors (Berlanga et al. 2004). Changes in ethanol, glycerol, acetaldehyde, acetic acid, nitrogen compounds, including amino acids, urea and ammonium and major volatile compounds are continuously ocurring as the Sherries are growing older (Cortes et al. 1998) (Table 14.3B). An active consumption of oxygen and the isolating effect of velum prevent oxidation of the wine, thus

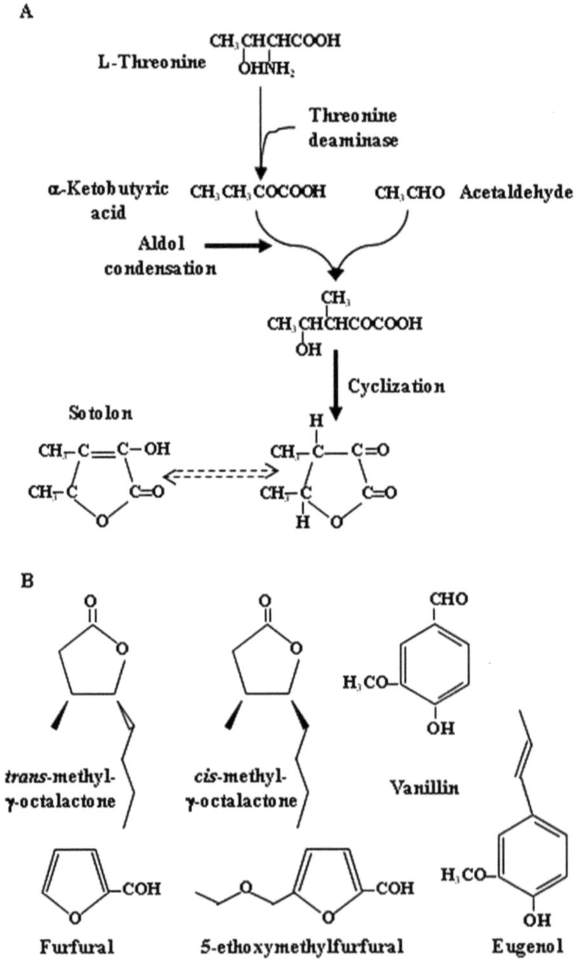

Fig. 14.2 (**A**) Formation of sotolon. Sotolon contributes to the typical aged aroma of Port, Vins Jaunes, Vins Doux Naturels, Botrytized and Sherries. α-ketobutyric acid may be formed enzymatically by deamination of threonine with threonine deaminase from *S. cerevisiae* or during wine physico-chemical maturation. Sotolon would be formed by a chemical reaction between α-ketobutyric acid and acetaldehyde produced by yeasts, in Sherries and French Sherry-type wines. Sotolon is formed as a result of condensation and further cyclization of two carbonyl compounds in Port wines and wines subjected to physico-chemical ageing (from Pham et al. 1995). (**B**) Structures of volatile compounds different from sotolon, formed during physico-chemical ageing of fortified wines. Their development increases with ageing and they are reliable indicators of ageing in oak, having a major impact on the aromas of fortified wines (from Cutzach et al. 2000)

providing a highly reductive medium (Marin-Menguiano et al. 2017; Martinez et al. 1995, 1998).

Ethanol is rapidly oxidized to acetaldehyde and acetate at the time of velum formation during biological ageing (Gutierrez et al. 2010). Velum formation requires energy consumption, its main carbon source available being ethanol (Martinez et al. 1998; Mauricio et al. 2001). Once velum is completed, *glycerol* is consumed, together with ethanol, organic acids (acetate, lactate, citrate and succinate) and amino acids, mostly proline (Gutierrez et al. 2010). Glycerol (about 7 $g.L^{-1}$) is steadily consumed, reaching about 0.3 $g.L^{-1}$ in the oldest Sherries (Martinez et al. 1998).

Amino acid concentration diminishes progressively in wine. Proline stands for over 75% total amino acid concentration and so utilization of proline reflects overall amino acid consumption. Utilization of proline is not restricted, since flor yeasts undergo oxidative metabolism. Under oxygen limitation conditions, flor yeasts may liberate amino acids such as threonine, methionine, cysteine and tryptophan, synthesized from ethanol, to balance their redox potential. These amino acids act as electron acceptors to oxidize excess of NADH (Martinez et al. 1998; Mauricio et al. 2001). Higher alcohols (isobutanol and isoamylic), acetaldehyde and acetoin are produced (Berlanga et al. 2001). Isobutanol that derives from valine increases steadily after velum formation, and then stays almost unalterable, whereas isoamylic alcohols that derive from isoleucine and leucine increase later in the process; their increase corresponds to a decrease in the corresponding wine amino acids (Martinez et al. 1998).

Acetaldehyde is mainly responsible for the typical flavor of Sherries. Together with acetaldehyde, diethylethane and acetoin in high concentrations are typical of aged Sherries, thus contributing strongly to their sensory properties (Mauricio and Ortega 1997).

Malo-lactic fermentation is carried out by bacteria present in the velum, which also consume gluconic, malic and lactic acid (Bravo-Abad 1986; Martinez et al. 1998; Pozo-Bayon and Moreno-Arribas 2011), and takes place only during the first period of ageing (5th and 4th *criaderas*) (Bravo-Abad 1986). Lactic bacteria are absent in younger wines (*sobretablas*) because the SO_2 concentration is too high for them to survive (50–60 $mg.L^{-1}$). During the dynamic phase of ageing process, this concentration is reduced to 30–40 $mg.L^{-1}$, and SO_2 is mostly bound to acetaldehyde (Martinez et al. 1998).

Gluconic acid is almost never found in the juice of sound grapes, and so it is an indicator of *Botrytis* infection, although is also formed by acetic acid bacteria (Magyar 2011). This compound is not metabolized by fermenting yeasts. Lactic bacteria undergo fermentation of gluconic acid and produce lactic and above all acetic acid (volatile acidity) during biological ageing of wine obtained from grapes infected with *B. cinerea*, affecting the quality of the wine. Both lactic and acetic acids can be metabolized by flor yeasts when concentration of gluconic acid is under 600 $mg.L^{-1}$(Perez et al. 1991). Among the flor yeasts, *S. cerevisiae* race *capensis* is the most efficient flor race at using gluconic acid during biological

Table 14.3B Concentrations of volatile compounds present in Palomino grape musts fermented with different species of wine yeasts[a]

	Volatile compounds concentration (μg mL^{-1})							
	n-Propanol	Isobutanol	2-Methyl butanol	3-Methyl butanol	Ethyl acetate	Ethyl caproate	Ethyl caprilate	Ethyl caprate
S. cerevisiae	24	56	22	178	10	0.7	1.1	2.8
Saccharomyces								
(flor strain)	24	95	27	207	-	0.4	0.9	3.3
K. apiculata	4	10	1	10	5	0.4	-	-
Candida sp.	21	19	6	47	21	0.1	0.4	0.4
Pichia fermentans	4	6	2	7	500	0.2	0.1	-

[a]Modified from Garcia Maiquez (1995)

ageing (Peinado et al. 2004a, b, c). Yeasts of the races *beticus* and *cheresiensis* are more efficient at reducing volatile acidity (Martinez de la Ossa et al. 1987a, b; Martinez et al. 1998).

2.3 Genetic Characteristics of Flor Yeasts

Molecular techniques have allowed for a much better understanding of the taxonomic identities of yeasts pecies and strains associated with Sherry production and have revealed significant variability in nuclear and mitochondrial genomes of *Saccharomyces* flor yeasts.

Spanish and Jura flor strains of *S. cerevisiae* possess *specific genomic patterns* due to the occurrence of a 24 bp deletion and a G insertion respectively in the *ITS1* region (Legras et al. 2014). Restriction analysis of this region from over 150 flor strains indicated that such deletion is found only in flor yeast populations (Alexandre 2013; Esteve-Zarzoso et al. 2001, 2004; Marin-Menguiano et al. 2017). That deletion was present in sequences of three other *Saccharomyces* strains, described as *Saccharomyces aceti* and *Saccharomyces gaditensis* (isolated from velum of Sherry wines), and *S. prostoserdovii* (isolated from Vernaccia di Oristano wine velum) (Fernadez-Espinar et al. 2000). Flor yeast populations from Hungary (Tokay), France (Jura), Italy (Sardinia) and Spain (Jerez and Cordoba) have recently been compared by applying different molecular techniques (microsatellite markers, CGH on array, polymorphism of *FLO11* gene and others) (Legras et al. 2014, 2016). Results reveal that most flor strains share the same unique origin, possibly Mesopotamia, where wine domestication originated.

Flor strains show little variability both in number and size of chromosomes (Legras et al. 2014; Martinez et al. 1995; Valero et al. 2007). The severe selective conditions of Sherry production have very likely favored an almost unique chromosomal pattern, similar in all flor yeasts, but different from that of fermenting yeasts. This lack of karyotype polymorphism may also be related to the scant presence of Ty1 elements (Ibeas and Jimenez 1996). However, Mesa et al. (1999, 2000) found high polymorphism of nuclear and mitochondrial genome in flor yeasts, although there were preferential nucleus-mitochondria associations of specific patterns, and associations of these patterns with ageing phases.

Aneuploidy is also very common in flor strains (Legras et al. 2014). Polysomy of chromosome XIII was observed in all flor strains examined by Guijo et al. (1997). Strains of the races *beticus* and *montuliensis* showed aneuploidies of chromosomes I, III and VI (*beticus*) and X and XII (*montuliensis*) as well as differences in copy number in 38% ORFs of the total genome. Those ORFs were very often associated with the presence of Ty and LTR regions (Codon et al. 1998; Storchova et al. 2006).

Most flor strains studied by Martinez et al. (1995) were unable to sporulate or, when sporulating, they produced non viable spores. The presence of lethal recessive mutations in heterozygosis in flor yeasts (Jimenez and Benitez 1988) supports this view, since the sporulation frequency is lower than that of mutation. This sexual isolation prevented a random distribution of taxonomic characters.

Several mutations in genes involved in mating type switching have been detected in *S. cerevisiae* flor yeasts isolated from Sardinian Sherry-like wines (Pirino et al. 2004; Zara et al. 2008). When flor and non-flor (fermenting) *S. cerevisiae* wine yeasts were compared, non-flor strains were all homothallic whereas almost 80% flor strains were semi-homothallic (Budroni et al. 2000). Observed semi-homothallic behavior derived from several mutations and transpositions occurring in flor yeasts genome, as was further verified in other flor strains by Infante et al. (2003a). Ristow et al. (1995) demonstrated that ethanol induced DNA single-strand breaks in *S. cerevisiae*, that acetaldehyde had a deleterious effect on chromosomal DNA in cells as well as on isolated DNA, and that ethanol is mutagenic via its metabolite, acetaldehyde. This suggestion has proved to be true on mtDNA of wine yeasts (Castrejon et al. 2002; Codon et al. 1998; Ibeas and Jimenez 1996).

A combination of methods revealed several differentiated regions between wine and flor yeasts with sites positively selected in flor yeasts such as the high-affinity transporter gene *ZRT1*, the hexose transporter gene *HXT7*, the yapsin gene *YPS6*, the membrane protein gene *MTS27* and above all the flocculine gene *FLO11*. Therefore there are several genomic regions in flor yeasts clearly adapted to ageing conditions such as cell-cell adhesion, zinc or hexose transport or signaling pathways (Coi et al. 2017; Marsit and Dequin 2015). Microsatellite genotyping also reveals that flor yeasts are close to fermenting wine strains (Legras et al. 2016). Comparative genome hybridization indicated amplification of a few genes, but single nucleotide polymorphisms are very frequent (Legras et al. 2016).

Hybridization experiments using DNA microarrays have allowed for a comparison of the complete genome of flor yeasts with that of yeasts from other sources. Specific amplifications or deletions of some genome regions explain the high genetic variability of flor yeasts and some of their enological features (Backhus et al. 2001; Infante et al. 2003a).

Most examples of genome reorganizations in industrial yeasts refer to carbon source metabolism (Codon et al. 1998; Alves Jr. et al. 2008). Velum forming ability in flor yeasts seems to be positively correlated to resistance to oxidative stress, copper, acetaldehyde and ethanol and inversely correlated to utilization of fermentable carbon sources other than glucose such as maltose, galactose, sucrose or raffinose (Budroni et al. 2005; Capece et al. 2013). The ability to assimilate different sugars and all characteristics related to flor metabolism, such as alcohol dehydrogenases or superoxide dismutases, were checked in *S. cerevisiae* strains, races *montuliensis*, *beticus*, *cheresiensis* and *rouxii* and compared with the presence or absence of genes encoding the corresponding enzymes. Amplification of genes encoding a putative monocarboxylic acid transporter and endocytosis trafficking have also been described (Legras et al. 2014, 2016). Genes encoding *ADH1* and *ADH2* enzymes had several locations in chromosomes; genes related to sugar metabolism such as *AGT1*, *MAL23*, *SUC2* or *GAL1* were also present, even when flor yeasts were unable to utilize the corresponding sugar (Fierro-Risco 2011; Legras et al. 2014).

Sherry wine lacks fermentable carbon sources, and "petites" are bound to disappear in cellars during biological ageing. However, they can transitorily be detected. Ibeas et al. (1997b) reported a synergistic effect of ethanol and temperature on flor yeasts, and so at 26 °C and 15.5% ethanol, 20–30% cells lacked functional mitochondria. Esteve-Zarzoso et al. (2001) also found a percentage of "petite" mutants in Sherry wines whose mitochondria displayed RFLP patterns different from those of "grande" cells, with functional mitochondria; the percentage of "petite" mutants was increased in butts with higher acetaldehyde concentrations. When mtDNA is repaired, DNA polymerase introduces mistakes accounting for the observed changes in the RFLP. Since mitochondrial DNA polymerase lacks proof-reading capability whereas nuclear DNA polymerase possesses such capacity, the preferential effect of ethanol on mitochondrial rather than nuclear genome could be the result of a better system to repair breaks in nucleus (Castrejon et al. 2002).

3 Velum Formation

Velum is considered an adaptive structure made up by yeasts that modify their cell shape, size and hydrophobicity (Legras et al. 2016; Martinez et al. 1995). During velum formation, cells increase in size, and become elongate; their walls seem to be thinner than those of yeasts at fermenting phase (Martinez et al. 1997b, c). Increase in hydrophobicity results from the synthesis of hydrophobic proteins (Martinez et al. 1997b, c) that causes cells to aggregate, and those aggregates retain gas bubbles, which originate in respiratory process. As a result, cells create a multilayered floating film (Martinez et al. 1997b).

3.1 External Conditions that Modulate Velum Formation

Cell changes during velum formation, that resulted from the lipogenic activity of yeasts were reported (Bravo-Abad 1986; Valero et al. 2002). However, other authors found that addition of oleic acid or ergosterol did not affect velum formation, whereas *addition of proteases* decreased cell hydrophobicity and eliminated cell aggregates (Martinez et al. 1995, 1997a), thus indicating that the process depended on hydrophobic proteins present on cell surfaces (Douglas et al. 2007; Van Mulders et al. 2009). Adherence is a property conferred by molecules called adhesins, which have a central role in formation of fungal biofilms. Biofilms are resistant to multiple hostile conditions, including ethanol, acetaldehyde or antimicrobial drugs. Velum also protects against additional extreme conditions, including low pH, oxidative stress, hydric stress and the presence of high metal concentrations such as copper (Martinez et al. 1997a, b, c, 1998). Adhesin encoding genes of yeasts are thus activated by a series of *adverse environmental effects*, such as carbon or nitrogen starvation, and the presence of ethanol or acetaldehyde. Biofilm also develops on non-fermentable carbon sources other than ethanol, such as glycerol or ethyl acetate (Alexandre 2013; Zara et al. 2010).

Synergy between temperature, low pH, ethanol, acetaldehyde, oxidative stress and other adverse effects may slow down velum development, may increase the time necessary for velum completion or may contribute to velum loss (Ibeas and Jimenez 1997; Martinez et al. 1997b). Velum does not develop in the presence of fermentable carbon sources or ammonium, whereas proline and ethanol seem to activate the process (Martinez et al. 1997b). The presence of polyphenols and biotin has also been described as activating velum formation (Bravo-Abad 1986; Budroni et al. 2005; Martinez et al. 1997b).

Amino acids can induce yeast cell adhesion; as it seems that amino acids transporters are regulators of *FLO11* and invasive growth (Torbensen et al. 2012). Among them, glutamine transporter genes *DIP5* and *GNP1* were essential for biofilm formation in a protein kinase dependent manner (Torbensen et al. 2012). Tpk3p, a protein kinase A, also seems to function specifically in biofilm formation, since biofilm development and cell-cell adhesion was absent in *tpk3* mutants (Bojsen et al. 2012; Scherz et al. 2014). The presence of *FOT* genes that encode oligopeptide transporters in several flor strains favor their adaptation to nitrogen-limited conditions during wine ageing (Legras et al. 2016). In addition, flor strains cannot metabolize dipeptides containing L-histidine, and so L-histidine reduces dramatically velum formation (Bou Zeidan et al. 2014; Legras et al. 2016). For the same reason, L-carnosine and L-histidine-containing dipeptide also inhibits velum formation (Legras et al. 2016).

3.2 Genes Induced During Velum Formation

Velum formation has also been considered to be a modified form of flocculation, with cells acquiring a pseudohyphal shape (Budroni et al. 2000; Palecek et al. 2000). *FLO genes*, responsible for flocculation, constitute subtelomeric gene families which encode proteins located in the cell surface and anchored in the cell membrane by a GPI residue (Teunissen and Steensma 1995; Bony et al. 1998; Legras et al. 2014; Verstrepen et al. 2004, 2005). One of these genes, *FLO11* (also called *MUC1*), encodes the Flo11p adhesin which plays an important role in determination of cell surface hydrophobicity, pseudophyphal development and invasive growth (Lo and Dranginis 1998; Ma et al. 2007; Tamaki et al. 2000; van Dyk et al. 2005), and also in the ability of cells to form biofilms (Reynolds and Fink 2001; Zara et al. 2005), similar to the velum formed on Sherry wines (Guo et al. 2000).

Expression of *FLO11* increased considerably in flor yeasts during velum formation induced by high ethanol and acetaldehyde concentration, low pH and oxidative stress, which suggested that this gene is a suitable candidate responsible for velum development (Infante et al. 2003a; Purevdorj-Gage et al. 2007; Barrales et al. 2008). A comparison of flor and non-flor yeasts indicated that the ability to form velum exclusively depended on the *FLO11* gene. Furthermore, a genome screening identified 71 genes essential for velum formation, half of which were necessary for *FLO11* transcription (Scherz et al. 2014).

Polymorphism of *FLO11* is a key feature in flor strains since their hydrophobicity is regulated by the level of *FLO11* expression and Flo11p length (Barua et al. 2016; Zara et al. 2005; Legras et al. 2014). Flor yeasts possess a deletion of 0.1 Kb within a repression region of the *FLO11* gene promoter, that increased *FLO11* gene expression, and rearrangements within the tandem repeats of the coding region, which yielded more hydrophobic Flo11p variants (Fig. 14.3) (Alexandre 2013; Fidalgo et al. 2008). Analysis of *FLO11* gene sequence carried out on 20 different flor yeast strains identified thirteen different alleles whose sizes varied from 3 to 6.1 Kb (Alexandre 2013). Halme et al. (2004) found that the *FLO* gene family was regulated genetically and epigenetically. Regulatory regions varied according to the presence or absence of the 0.1 Kb repression sequence, two short repeats sequences and single point mutations, which were targets for at least the pH, the MAPK pathways, the cAMP cascade and the Gcn4p pathways (Alexandre 2013; Wang et al. 2015).

Apart from the *FLO11* gene, there are other genes related to ethanol metabolism, glycerol assimilation, oxidative stress and oxido-reduction potential that increase their expression during velum formation (Alexandre 2013; Espinazo-Romeu et al. 2008; Infante et al. 2003a; Martineau et al. 2007); among them, the *heat shock proteins genes HSP12, HSP26, HSP82, HSP104* (Alexandre 2013). Expression of *ALD2, ALD3* and *ALS4* genes that encode aldehyde dehydrogenases, *NRG1*, a negative regulator of glucose-repressed genes, and *BTN2* that encodes a protein involved in intracellular protein trafficking, are also induced (Alexandre 2013; Aranda and del Olmo 2003; Aranda et al. 2002). A truncated form of Nrg1p affects *FLO11* repression whereas deletion of *BTN2* affects velum formation (Alexandre 2013). It seems that the role of Btn2p in amino acid transport is linked to ethanol resistance (Alexandre 2013). Zara et al. (2002) reported that the *HSP12* gene is strongly induced in the absence of glucose and in the presence of ethanol; its lack of expression resulted in the lack of velum formation (Alexandre 2013). Furthermore, overexpression of *HSP12* increased the levels of glutathione peroxidase and reductase activities. As a result, a higher intracellular glutathione content and a reduced peroxided lipid concentration, that resulted in a higher resistance to oxidative stress conditions, faster growth of velum and higher survival of flor yeast transformants overexpressing *HSP12*, was observed (Fierro-Risco et al. 2013).

Other *genes that encode cell surface proteins* have also been related to velum formation. Alexandre et al. (2000) identified a 49 kDa hydrophobic mannoprotein present in flor yeasts, and associated it with velum formation. Reynolds and Fink (2001) described cell surface adhesine-like glycoproteins anchored to the cell membrane; the genes encoding these proteins were expressed under carbon or nitrogen starvation conditions (Lambrechts et al. 1996a, b; Lo and Dranginis 1998); the proteins allowed cell to cell adherence or adherence of cells to inert solid surfaces. One of those proteins, Sed1p, regulated by Msn2p and Msn4p, conferred resistance to 2.4 D and to β-1.3-glucanase and controlled pH gradient through the plasma membrane, under hostile conditions (Puig and Perez-Ortin 2000).

Most *S. cerevisiae* genes containing intragenic repeats, encode cell wall proteins (Pir) such as adhesins. Recombination events among internal repeats of adhesin

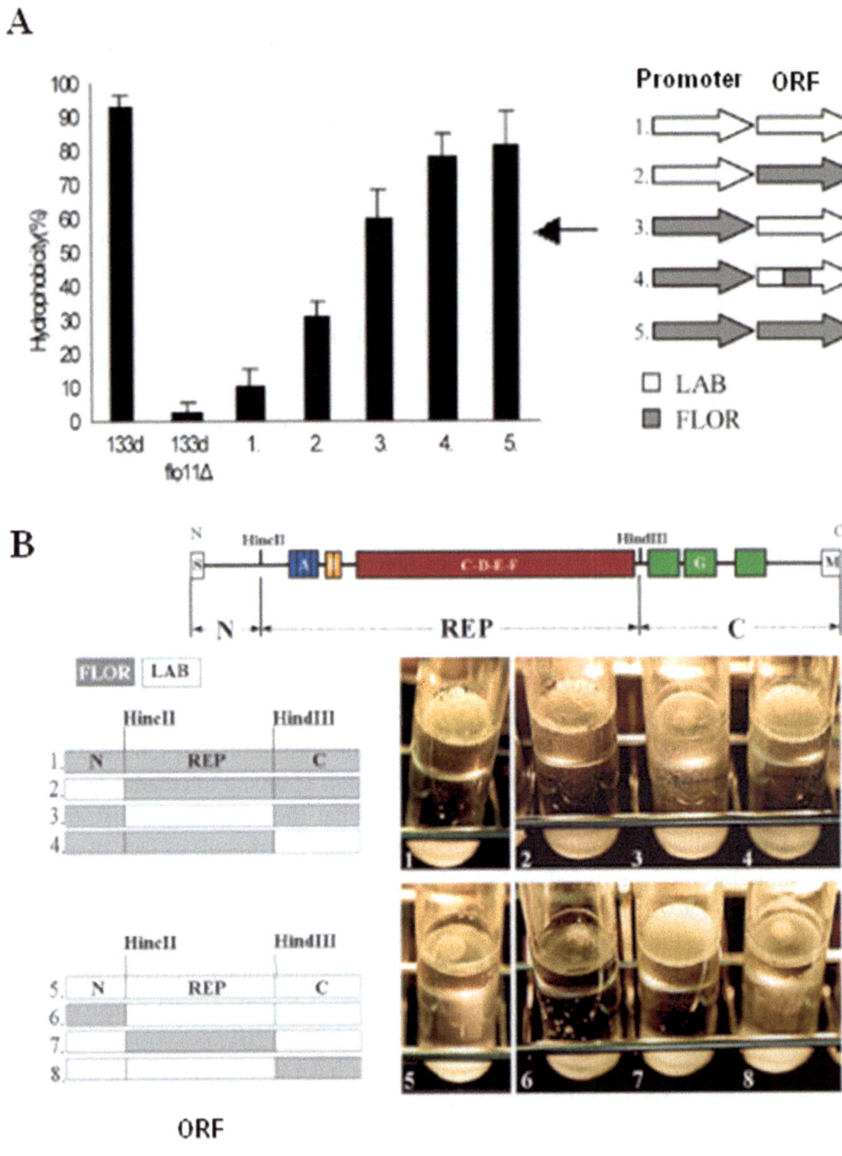

Fig. 14.3 Increased hydrophobicity and velum formation depend on two modifications of *FLO11* gene of flor yeasts: deletion of a repression region in the promoter, and rearrrangements of the tandem repeats of the ORF that yield more hydrophobic Flo11 proteins. (**A**) Increase in hydrophobicity when promoters and ORFs from flor (FLOR) and laboratory (LAB) genes or different combinations of both were cloned in a plasmid and yeast strain 133dflo11Δ, lacking *FLO11* gene, was transformed with the constructions. 133dflo11Δ strain derived from strain 133, a meiotic product of a flor/laboratory hybrid able to form flor. Over a certain hydrophobicity limit (arrow) hydrophobicity permits velum formation. (**B**) Velum formation after chimeric *FLO11* genes that combined the different domain repeats (C, D, E, F) were cloned in a plasmid, and those constructions were used to transform strain 133dflo11Δ. Velum formation was assessed in the transformants (from Fidalgo et al. 2006)

genes also cause expansion and contraction of gene size (Verstrepen and Klis 2006). Thus, coupled with epigenetic changes in gene expression, these genetic rearrangements provide a reservoir of cell surface molecules with new functions, and/or create alterations in phenotypes such as adhesion, flocculation and biofilm formation, or functional diversity of cell surface antigens that elude the host immune system in pathogens (Verstrepen et al. 2004, 2005). The *CCW7/HSP150* gene has been shown to encode a cell surface protein, one of the members of the Pir protein family (Moukadiri and Zueco 2001a, b). This protein possesses an average of eleven repeat units, as occurs with other adhesins. Kovacs et al. (2008) compared the characteristics of Hsp150p in numerous fermenting and flor yeasts isolated from several geographical regions, including Sherry and Sherry-like wines, as well as Botrytized biological aged Hungarian wine regions. All the analyzed *S. cerevisiae* flor yeasts, regardless of their geographical regions, have a deletion of three repeat units. Non-flor wine strains show pronounced polymorphism of the *CCW7* locus.

Ethanol and acetaldehyde toxicity are related to the formation of reactive oxygen species (ROS) in the mitochondria and to mutations and deletions ocurring mostly in the mtDNA (Lloyd et al. 2003). Glutathione metabolism plays an important role in oxidative stress tolerance in cells, growing as a biofilm (Gales et al. 2008). In addition, superoxide dismutases encoded by *SOD1* and *SOD2* genes protect against ROS effects (Costas et al. 1997; Longo et al. 1996). *SOD1* is strongly expressed during stationary phase and during velum formation in flor yeasts (Infante et al. 2003a), and the protective role of Sod1p in ethanol tolerance has been amply demonstrated (Fig. 14.4) (Costas et al. 1997). Overexpression of genes involved in oxidation stress resistance in flor yeasts allows for faster velum formation; *SOD1* and *SOD2* overexpression increased superoxide dismutase activity as well as activities of catalase, glutathione reductase and glutathione peroxidase (Fierro-Risco et al. 2013).

4 New Technologies in Sherry Production and Application of Flor Yeasts

With regards to Sherry wine making the following new technology trends are now envisaged:

Reduction in Sherry ageing time. It has been attemped by increasing the surface-volume ratio, using stainless steel tanks or increasing aireation with stirrers or with pulses of short micro areation periods (Pozo-Bayon and Moreno-Arribas 2011). Other methods used submerged cultures of *S. cerevisiae* var. *capensis* (Peinado et al. 2004a, b, c; Pozo-Bayon and Moreno-Arribas 2011). Addition of quercetin -a major flavonol found in several fruits that strongly protects yeast cells against oxidative stress- may also contribute to increase flor yeasts viability and velum stability during biological Sherry ageing (Belinha et al. 2007). Applications of flor yeasts might also be widened by immobilization within a fungal hyphal framework (Legras

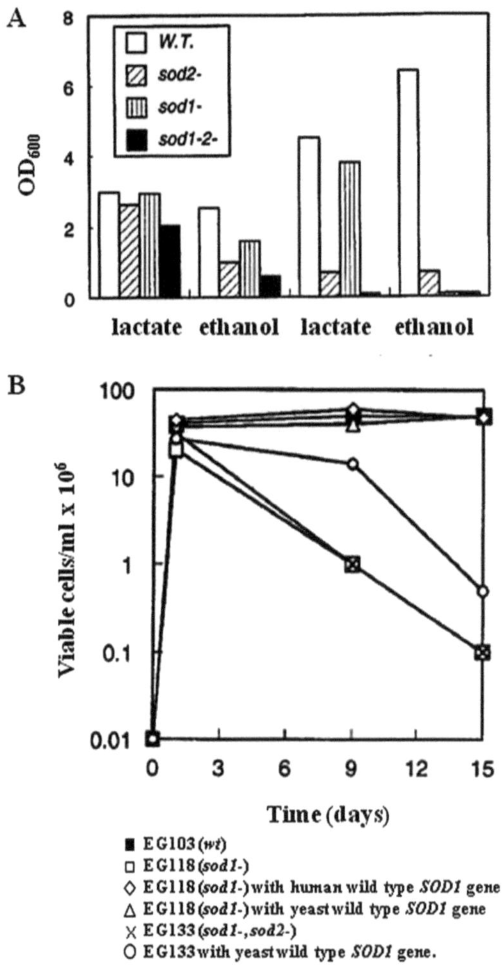

Fig. 14.4 Oxygen as well as respirable carbon sources such as ethanol and lactate favor oxidative stress and formation of reactive oxygen species (ROS). Superoxide dismutases Sod1p and Sod2p encoded by *SOD1* and *SOD2* genes protect against ROS effects. (**A**) Growth (O.D. at 660 nm) of wild type yeast strain EG103 and deficient SOD mutants in minimal medium with non-fermentable carbon sources at high (left) or low (right) aeration. ROS are produced even under low aeration. (**B**) Transformation with wild type yeast or human SOD gene restored survival to sod$^-$ mutants. Viability of transformed sod1$^-$ and sod1$^-$ sod2$^-$ yeast strains with either yeast or human *SOD1* gene (from Longo et al. 1996)

et al. 2016). Different fungus-yeast combinations could be further extended for application in different biotechnological purposes.

Improving sweet Sherry production by acceleration during grape drying. Alternative processes involve the use of hot air driers, facilitating rapid water loss from harvested grapes (Pozo-Bayon and Moreno-Arribas 2011).

Sherry from organic grapes. There have been attemps to produce Sherries from grapes cultivated ecologically (Pozo-Bayon and Moreno-Arribas 2011).

To introduce these new technologies it would be necessary to add the advances in cell inmobilization described above. These new methods may facilitate both the establishment of more stable velum and the shortage of ageing time.

5 Conclusion

The addition of neutral spirits or brandy to grape must fermentation to avoid wine spoilage has given rise to excellent fortified wines, mainly due to the introduction of sophisticated maturation methods, such as the *solera* system in the production of Sherry wines, and to very restrictive conditions in the vineyards with regards to grape varieties and cultivation procedures. Blending procedures mostly in Port and Sherry wines allow formation of very high quality wines. From high quality Palomino grape *añadas* (low polyphenol and sugar content), *fino* Sherry wines are obtained after subjecting them to biological ageing. During biological ageing, a velum is formed on the wine surface. That, together with the respiratory metabolism of flor yeasts creates a reducing atmosphere in the wine, and this condition is responsible for its pale color and many of its delicate organoleptic properties.

Neverthless, because of the blending during the winemaking process and external or metabolic derived stress conditions, velum ocasionally undergoes disruption and flocculation. Molecular technics have allowed the identification of many genes responsible for flor formation and stability as long as stress resitance. Manipulation of these genes encoding sugar transporters, alcohol-acyltransferases, flocculin-like proteins and heat shock proteins will allow the isolation of new yeasts with improved properties for fermentation and maturation of these types of wines.

Finally, there is more a more knowledge about the link between intracellular proteins and metabolites excreted by flor yeasts that are related to sensorial properties and that may result in information useful for innovative wine making.

Acknowledgements This research was supported by the CICYT project numbers AGL2006-03947 and PETRI-95-1010.90.01, and Junta de Andalucía PAI CVI-107 PO6-CVI-01546.

References

Alexandre, H. (2013). Flor yeasts of Saccharomyces cerevisiae—their ecology, genetics and metabolism. *International Journal of Food Microbiology, 167*(2), 269–275. https://doi.org/10.1016/j.ijfoodmicro.2013.08.021.

Alexandre, H., Blanchet, S., & Charpentier, C. (2000). Identification of a 49-kDa hydrophobic cell wall mannoprotein present in velum yeast which may be implicated in velum formation. *FEMS Microbiology Letters, 185*(2), 147–150. https://doi.org/10.1126/science.1131969.

Alves, S. L. Jr., Herberts, R. A., Hollatz, C., Trichez, D., Miletti, L. C., de Araujo, P. S., & Stambuk, B. U. (2008). Molecular analysis of maltotriose active transport and fermentation by Saccharomyces cerevisiae reveals a determinant role for the AGT1 permease. *Applied and Environmental Microbiology 74* (5): 1494–1501. AEM.02570-07 [pii] https://doi.org/10.1128/AEM.02570-07.

Aranda, A., & del Olmo, M. (2003). Response to acetaldehyde stress in the yeast Saccharomyces cerevisiae involves a strain-dependent regulation of several ALD genes and is mediated by the general stress response pathway. *Yeast, 20*(8), 747–759.

Aranda, A., Querol, A., & del Olmo, M. (2002). Correlation between acetaldehyde and ethanol resistance and expression of HSP genes in yeast strains isolated during the biological aging of sherry wines. *Archives of Microbiology, 177*(4), 304–312.

Backhus, L. E., DeRisi, J., & Bisson, L. F. (2001). Functional genomic analysis of a commercial wine strain of Saccharomyces cerevisiae under differing nitrogen conditions. *FEMS Yeast Research, 1*(2), 111–125. https://doi.org/10.1111/j.1567-1364.2008.00399.x.

Barnett, J. A., Payne, R. W., & Yarrow, D. (1990). *Yeast: Characteristics and identification* (2nd ed.). Cambridge: Cambridge University Press.

Barrales, R. R., Jimenez, J., & Ibeas, J. I. (2008). Identification of novel activation mechanisms for FLO11 regulation in Saccharomyces cerevisiae. *Genetics, 178*(1), 145–156. https://doi.org/10.1534/genetics.107.081315.

Barua, S., Li, L., Lipke, P. N., & Dranginis, A. M. (2016). Molecular basis for strain variation in the Saccharomyces cerevisiae Adhesin Flo11p. *mSphere, 1*(4). https://doi.org/10.1128/mSphere.00129-16mSphere00129-16.

Belinha, I., Amorim, M. A., Rodrigues, P., de Freitas, V., Moradas-Ferreira, P., Mateus, N., & Costa, V. (2007). Quercetin increases oxidative stress resistance and longevity in *Saccharomycescerevisiae. Journal of Agricultural and Food Chemistry, 55*(6), 2446–2451. https://doi.org/10.1016/j.ijfoodmicro.2007.11.083.

Benitez, T., Rincon, A. M., & Codon, A. C. (2011). Yeasts used in biologically aged wines. In A. V. Carrascosa, R. Muñoz, & R. Gozalez (Eds.), *Molecular wine microbiology* (pp. 51–84). Elseiver: Academic.

Berlanga, T. M., Atanasio, C., Mauricio, J. C., & Ortega, J. M. (2001). Influence of aeration on the physiological activity of flor yeasts. *Journal of Agricultural and Food Chemistry, 49*(7), 3378–3384.

Berlanga, T. M., Peinado, R., Millan, C., Mauricio, J. C., & Ortega, J. M. (2004). Influence of blending on the content of different compounds in the biological aging of sherry dry wines. *Journal of Agricultural and Food Chemistry, 52*(9), 2577–2581. https://doi.org/10.1534/genetics.106.057836.

Bojsen, R. K., Andersen, K. S., & Regenberg, B. (2012). Saccharomyces cerevisiae—a model to uncover molecular mechanisms for yeast biofilm biology. *FEMS Immunology and Medical Microbiology, 65*(2), 169–182. https://doi.org/10.1111/j.1574-695X.2012.00943.x.

Bony, M., Barre, P., & Blondin, B. (1998). Distribution of the flocculation protein, flop, at the cell surface during yeast growth: The availability of flop determines the flocculation level. *Yeast, 14*(1), 25–35. https://doi.org/10.1111/j.1567-1364.2008.00434.x.

Bou Zeidan, M., Zara, G., Viti, C., Decorosi, F., Mannazzu, I., Budroni, M., Giovannetti, L., & Zara, S. (2014). L-histidine inhibits biofilm formation and FLO11-associated phenotypes in Saccharomyces cerevisiae flor yeasts. *PLoS One, 9*(11), e112141. https://doi.org/10.1371/journal.pone.0112141.

Bravo-Abad, F. (1986). Crianza biológica de vino: Procedimiento tradicional de vinos finos de D.O. Jerez y D. O. Montilla-Moriles. Enología y Enotecnia: 15–19.

Budroni, M., Giordano, G., Pinna, G., & Farris, G. A. (2000). A genetic study of natural flor strains of Saccharomyces cerevisiae isolated during biological ageing from Sardinian wines. *Journal of Applied Microbiology, 89*(4), 657–662.

Budroni, M., Zara, S., Zara, G., Pirino, G., & Mannazzu, I. (2005). Peculiarities of flor strains adapted to Sardinian sherry-like wine ageing conditions. *FEMS Yeast Research, 5*(10), 951–958.

Campo, E., Cacho, J., & Ferreira, V. (2008). The chemical characterization of the aroma of dessert and sparkling white wines (Pedro Ximenez, Fino, Sauternes, and Cava) by gas chromatography-olfactometry and chemical quantitative analysis. *Journal of Agricultural and Food Chemistry, 56*(7), 2477–2484. https://doi.org/10.1021/jf072968l.

Capece, A., Siesto, G., Poeta, C., Pietrafesa, R., & Romano, P. (2013). Indigenous yeast population from Georgian aged wines produced by traditional "Kakhetian" method. *Food Microbiology, 36*(2), 447–455. https://doi.org/10.1016/j.fm.2013.07.008.

Castrejon, F., Codon, A. C., Cubero, B., & Benitez, T. (2002). Acetaldehyde and ethanol are responsible for mitochondrial DNA (mtDNA) restriction fragment length polymorphism (RFLP) in flor yeasts. *Systematic and Applied Microbiology, 25*(3), 462–467.

Codon, A. C., Benitez, T., & Korhola, M. (1998). Chromosomal polymorphism and adaptation to specific industrial environments of Saccharomyces strains. *Applied Microbiology and Biotechnology, 49*(2), 154–163.

Coi, A. L., Bigey, F., Mallet, S., Marsit, S., Zara, G., Gladieux, P., Galeote, V., Budroni, M., Dequin, S., & Legras, J. L. (2017). Genomic signatures of adaptation to wine biological ageing conditions in biofilm-forming flor yeasts. *Molecular Ecology, 26*(7), 2150–2166. https://doi.org/10.1111/mec.14053.

Collin, S., Nizet, S., Claeys Bouuaert, T., & Despatures, P.-M. (2012). Main odorants in Jura flor-Sherry wines. Relative contributions of Sotolon, Abhexon, and Theaspirane-derived compounds. *Journal of Agricultural and Food Chemistry, 60*(1), 380–387. https://doi.org/10.1021/jf203832c.

Cortes, M. B., Moreno, J., Zea, L., Moyano, L., & Medina, M. (1998). Changes in aroma compounds of Sherry wines during their biological aging carried out by Saccharomyces cerevisiae races bayanus and capensis. *Journal of Agricultural and Food Chemistry, 46*(6), 2389–2394. https://doi.org/10.1021/jf970903k.

Costas, V., Amorim, M. A., Reis, E., Quintanilha, A., & Moradas-Ferreira, P. (1997). Mitochondrial superoxide dismutase is essential for ethanol tolerance of Saccharomyces cerevisiae in the post-diauxic phase. *Microbiology, 143*, 1649–1656. https://doi.org/10.1016/j.fm.2013.06.008.

Cutzach, I., Chatonnet, P., & Dubourdieu, D. (2000). Influence of storage conditions on the formation of some volatile compounds in white fortified wines (Vins doux naturels) during the aging process. *Journal of Agricultural and Food Chemistry, 48*(6), 2340–2345. https://doi.org/10.1128/AEM.71.1.255-260.2005.

Douglas, L. M., Li, L., Yang, Y., & Dranginis, A. M. (2007). Expression and characterization of the flocculin Flo11/Muc1, a Saccharomyces cerevisiae mannoprotein with homotypic properties of adhesion. *Eukaryotic Cell, 6*(12), 2214–2221. https://doi.org/10.1128/EC.00284-06.

Espinazo-Romeu, M., Cantoral, J. M., Matallana, E., & Aranda, A. (2008). Btn2p is involved in ethanol tolerance and biofilm formation in flor yeast. *FEMS Yeast Research, 8*(7), 1127–1136. https://doi.org/10.1111/j.1567-1364.2008.00397.x.

Esteve-Zarzoso, B., Peris-Toran, M. J., Garcia-Maiquez, E., Uruburu, F., & Querol, A. (2001). Yeast population dynamics during the fermentation and biological aging of sherry wines. *Applied and Environmental Microbiology, 67*(5), 2056–2061.

Esteve-Zarzoso, B., Fernandez-Espinar, M. T., & Querol, A. (2004). Authentication and identification of Saccharomyces cerevisiae 'flor' yeast races involved in sherry ageing. *Antonie Van Leeuwenhoek, 85*(2), 151–158.

Fernadez-Espinar, M. T., Esteve-Zarzoso, B., Querol, A., & Barrio, E. (2000). RFLP analysis of the ribosomal internal transcribed spacers and the 5.8S rRNA gene region of the genus *Saccharomyces*: A fast method for species identification and the differentiation of flor yeasts. *Antonie Van Leeuwenhoek, 78*(1), 87–97.

Fidalgo, M., Barrales, R. R., Ibeas, J. I., & Jimenez, J. (2006). Adaptive evolution by mutations in the FLO11 gene. *Proceedings of the National Academy of Sciences of the United States of America, 103*(30), 11228–11233.

Fidalgo, M., Barrales, R. R., & Jimenez, J. (2008). Coding repeat instability in the FLO11 gene of Saccharomyces yeasts. *Yeast, 25*(12), 879–889. https://doi.org/10.1002/yea.1642.

Fierro-Risco, J. (2011). *Mejora genética de levaduras implicadas en la elaboración de vinos tipo Fino*. Seville: Seville.

Fierro-Risco, J., Rincon, A. M., Benitez, T., & Codon, A. C. (2013). Overexpression of stress-related genes enhances cell viability and velum formation in Sherry wine yeasts. *Applied Microbiology and Biotechnology, 97*(15), 6867–6881. https://doi.org/10.1007/s00253-013-4850-9.

Fleet, G. H. (2002). *Wine microbiology and biotechnology*. London: Taylor & Francis. https://doi.org/10.1099/mic.0.041673-0.

Gales, G., Penninckx, M., Block, J. C., & Leroy, P. (2008). Role of glutathione metabolism status in the definition of some cellular parameters and oxidative stress tolerance of Saccharomyces cerevisiae cells growing as biofilms. *FEMS Yeast Research, 8*(5), 667–675. https://doi.org/10.1111/j.1567-1364.2008.00401.x.

Garcia Maiquez, E. (1995). Sherry wine microorganisms. *Microbiología, 11*(1), 51–58. https://doi.org/10.1111/j.1567-1364.2006.00126.x.

Granchi, L., Ganucci, D., Messini, A., & Vincenzini, M. (2002). Oenological properties of Hanseniaspora osmophila and Kloeckera corticis from wines produced by spontaneous fermentations of normal and dried grapes. *FEMS Yeast Research, 2*(3), 403–407.

Guijo, S., Millan, C., & Ortega, J. M. (1986). Fermentative features of vinification and maturation yeasts isolated in the Montilla-Moriles region of southern Spain, 3. *Food Microbiology, 2*, 133–142. https://doi.org/10.1016/S0740-0020(86)80036-3.

Guijo, S., Mauricio, J. C., Salmon, J. M., & Ortega, J. M. (1997). Determination of the relative ploidy in different Saccharomyces cerevisiae strains used for fermentation and 'flor' film ageing of dry sherry-type wines. *Yeast, 13*(2), 101–117.

Guo, B., Styles, C. A., Feng, Q., & Fink, G. R. (2000). A Saccharomyces gene family involved in invasive growth, cell-cell adhesion, and mating. *Proceedings of the National Academy of Sciences of the United States of America, 97*(22), 12158–12163.

Gutierrez, P., Roldan, A., Caro, I., & Perez, L. (2010). Kinetic study of the velum formation by Saccharomyces cerevisiae (beticus ssp.) during the biological aging of wines. *Process Biochemistry, 45*(4), 493–499. https://doi.org/10.1016/j.procbio.2009.11.005.

Halme, A., Bumgarner, S., Styles, C., & Fink, G. R. (2004). Genetic and epigenetic regulation of the FLO gene family generates cell-surface variation in yeast. *Cell, 116*(3), 405–415.

Ibeas, J. I., & Jimenez, J. (1996). Genomic complexity and chromosomal rearrangements in wine-laboratory yeast hybrids. *Current Genetics, 30*(5), 410–416.

Ibeas, J. I., & Jimenez, J. (1997). Mitochondrial DNA loss caused by ethanol in Saccharomyces flor yeasts. *Applied and Environmental Microbiology, 63*(1), 7–12.

Ibeas, J. I., Lozano, I., Perdigones, F., & Jimenez, J. (1996). Detection of Dekkera-Brettanomyces strains in sherry by a nested PCR method. *Applied and Environmental Microbiology, 62*(3), 998–1003.

Ibeas, J. I., Lozano, I., Perdigones, F., & Jimenez, J. (1997a). Dynamics of flor yeast populations during the biological aging of Sherry wines. *American Journal of Enology and Viticulture, 48*, 75–79.

Ibeas, J. I., Lozano, I., Perdigones, F., & Jimenez, J. (1997b). Effects of ethanol and temperature on the biological aging of Sherry wines. *American Journal of Enology and Viticulture, 48*, 71–74.

Infante, J. J., Dombek, K. M., Rebordinos, L., Cantoral, J. M., & Young, E. T. (2003a). Genome-wide amplifications caused by chromosomal rearrangements play a major role in the adaptive evolution of natural yeast. *Genetics, 165*(4), 1745–1759.

Infante, J. J., Rodriguez, M. E., Rebordinos, L., & Cantoral, J. M. (2003b). Industrial yeast for wine production. New methods for control and improvement. In J. L. Barredo (Ed.), *Microorganismos for health care, food and enzyme production*. Kerala.

Jimenez, J., & Benitez, T. (1988). Selection of ethanol-tolerant yeast hybrids in pH-regulated continuous culture. *Applied and Environmental Microbiology, 54*(4), 917–922. https://doi.org/10.1021/jf802476w.

Kovacs, M., Stuparevic, I., Mrsa, V., & Maraz, A. (2008). Characterization of Ccw7p cell wall proteins and the encoding genes of Saccharomyces cerevisiae wine yeast strains: Relevance for flor formation. *FEMS Yeast Research, 8*(7), 1115–1126. https://doi.org/10.1111/j.1567-1364.2008.00413.x.

Kurtzman, C., Fell, J. W., & Boekhout, T. (2011). The yeasts. In *A taxonomic study* (5th ed.). San Diego: Elsevier Science.

Lambrechts, M. G., Bauer, F. F., Marmur, J., & Pretorius, I. S. (1996a). Muc1, a mucin-like protein that is regulated by Mss10, is critical for pseudohyphal differentiation in yeast. *Proceedings of the National Academy of Sciences of the United States of America, 93*(16), 8419–8424.

Lambrechts, M. G., Sollitti, P., Marmur, J., & Pretorius, I. S. (1996b). A multicopy suppressor gene, MSS10, restores STA2 expression in Saccharomyces cerevisiae strains containing the STA10 repressor gene. *Current Genetics, 29*(6), 523–529.

Legras, J. L., Erny, C., & Charpentier, C. (2014). Population structure and comparative genome hybridization of European flor yeast reveal a unique group of Saccharomyces cerevisiae

strains with few gene duplications in their genome. *PLoS One, 9*(10), e108089. https://doi.org/10.1371/journal.pone.0108089.
Legras, J. L., Moreno-Garcia, J., Zara, S., Zara, G., Garcia-Martinez, T., Mauricio, J. C., Mannazzu, I., Coi, A. L., Bou Zeidan, M., Dequin, S., Moreno, J., & Budroni, M. (2016). Flor yeast: New perspectives beyond wine aging. *Front Microbiology, 7*, 503. https://doi.org/10.1016/j.copbio.2013.03.005.
Lloyd, D., Lemar, K. M., Salgado, L. E., Gould, T. M., & Murray, D. B. (2003). Respiratory oscillations in yeast: Mitochondrial reactive oxygen species, apoptosis and time; a hypothesis. *FEMS Yeast Research, 3*(4), 333–339.
Lo, W. S., & Dranginis, A. M. (1998). The cell surface flocculin Flo11 is required for pseudohyphae formation and invasion by Saccharomyces cerevisiae. *Molecular Biology of the Cell, 9*(1), 161–171.
Longo, V. D., Gralla, E. B., & Valentine, J. S. (1996). Superoxide dismutase activity is essential for stationary phase survival in Saccharomyces cerevisiae. Mitochondrial production of toxic oxygen species in vivo. *The Journal of Biological Chemistry, 271*(21), 12275–12280.
Ma, J., Jin, R., Jia, X., Dobry, C. J., Wang, L., Reggiori, F., Zhu, J., & Kumar, A. (2007). An interrelationship between autophagy and filamentous growth in budding yeast. *Genetics, 177*(1), 205–214. https://doi.org/10.1534/genetics.107.076596.
Magyar, I. (2011). Botrytized wines. *Advances in Food and Nutrition Research, 63*, 147–206. https://doi.org/10.1016/j.ijfoodmicro.2007.11.003.
Marin-Menguiano, M., Romero-Sanchez, S., Barrales, R. R., & Ibeas, J. I. (2017). Population analysis of biofilm yeasts during fino sherry wine aging in the Montilla-Moriles D.O. region. *International Journal of Food Microbiology, 244*, 67–73. https://doi.org/10.1016/j.ijfoodmicro.2016.12.019.
Marsit, S., & Dequin, S. (2015). Diversity and adaptive evolution of Saccharomyces wine yeast: A review. *FEMS Yeast Research, 15*(7). https://doi.org/10.1093/femsyr/fov067. 10.1093/molbev/msv057.
Martin, B., Etievant, P. X., Le Quere, J. L., & Schlich, P. (1992). More clues about sensory impact of sotolon in some flor Sherry wines. *Journal of Agricultural and Food Chemistry, 40*, 475–478.
Martineau, C. N., Beckerich, J. M., & Kabani, M. (2007). Flo11p-independent control of "mat" formation by hsp70 molecular chaperones and nucleotide exchange factors in yeast. *Genetics, 177*(3), 1679–1689. https://doi.org/10.1534/genetics.107.081141.
Martinez de la Ossa E, Caro, I., Bonat, M., Perez, L., & Domecq, B. (1987a). Dry extract in sherry and its evolution in the aging process. *American Journal of Enology and Viticulture, 38*(4), 321–325.
Martinez de la Ossa E, Perez, L., & Caro, I. (1987b). Variation of the major volatiles trough aging of sherry. *American Journal of Enology and Viticulture, 38*(4), 293–297.
Martinez, P., Codon, A. C., Perez, L., & Benitez, T. (1995). Physiological and molecular characterization of flor yeasts: Polymorphism of flor yeast populations. *Yeast, 11*(14), 1399–1411.
Martinez, P., Perez, L., & Benitez, T. (1997a). Evolution of flor yeast population during the biological aging of fino sherry wine. *American Journal of Enology and Viticulture, 48*, 160–168.
Martinez, P., Perez, L., & Benitez, T. (1997b). Factor which affect velum formation by flor yeast isolated from sherry wine. *Systematic and Applied Microbiology, 20*(1), 154–157.
Martinez, P., Perez, L., & Benitez, T. (1997c). Velum formation by flor yeasts isolated from sherry wine. *American Journal of Enology and Viticulture, 48*, 55–62.
Martinez, P., Valcarcel, M. J., Perez, L., & Benitez, T. (1998). Metabolism of *Saccharomyces cerevisiae* flor yeasts during fermentation and bilogical aging of fino sherry: By-products and aroma compounds. *American Journal of Enology and Viticulture, 49*, 240–250.
Mauricio, J. C., & Ortega, J. M. (1997). Nitrogen compounds in wine during its biological aging by two flor film yeasts: An approach to accelerated biological aging of dry sherry-type wines. *Biotechnology and Bioengineering, 53*(2), 159–167. https://doi.org/10.1002/(SICI)1097-0290(19970120)53:2<159::AID-BIT5>3.0.CO;2-N.
Mauricio, J. C., Valero, E., Millan, C., & Ortega, J. M. (2001). Changes in nitrogen compounds in must and wine during fermentation and biological aging by flor yeasts. *Journal of Agricultural and Food Chemistry, 49*(7), 3310–3315.

Merida, J., Lopez-Toledano, A., Marquez, T., Millan, C., Ortega, J. M., & Medina, M. (2005). Retention of browning compounds by yeasts involved in the winemaking of sherry type wines. *Biotechnology Letters, 27*(20), 1565–1570.

Mesa, J. J., Infante, J. J., Rebordinos, L., & Cantoral, J. M. (1999). Characterization of yeasts involved in the biological aging of sherry wines. *Lebensmittel-Wissenschaft und Technologie, 32*(2), 114–120.

Mesa, J. J., Infante, J. J., Rebordinos, L., Sanchez, J. A., & Cantoral, J. M. (2000). Influence of the yeast genotypes on enological characteristics of sherry wines. *American Journal of Enology and Viticulture, 51*(1), 15–21.

Moreira, N., & Guedes de Pinho, P. (2011). Port Wine. *Advances in Food and Nutrition Research, 63*, 119–146.

Moreno, J. J., Millan, C., Ortega, J. M., & Medina, M. (1991). Analytical differentiation of wine fermentations using pure and mixed yeast cultures. *Journal of Industrial Microbiology, 7*(3), 181–189. https://doi.org/10.1007/bf01575881.

Moreno-Garcia, J., Garcia-Martinez, T., Moreno, J., Millan, M. C., & Mauricio, J. C. (2014). A proteomic and metabolomic approach for understanding the role of the flor yeast mitochondria in the velum formation. *International Journal of Food Microbiology, 172*, 21–29. https://doi.org/10.1016/j.ijfoodmicro.2013.11.030.

Moreno-Garcia, J., Garcia-Martinez, T., Millan, M. C., Mauricio, J. C., & Moreno, J. (2015a). Proteins involved in wine aroma compounds metabolism by a Saccharomyces cerevisiae flor-velum yeast strain grown in two conditions. *Food Microbiology, 51*, 1–9. https://doi.org/10.1016/j.fm.2015.04.005.

Moreno-Garcia, J., Garcia-Martinez, T., Moreno, J., & Mauricio, J. C. (2015b). Proteins involved in flor yeast carbon metabolism under biofilm formation conditions. *Food Microbiology, 46*, 25–33. https://doi.org/10.1016/j.fm.2014.07.001.

Moreno-Garcia, J., Mauricio, J. C., Moreno, J., & Garcia-Martinez, T. (2016). Functional analysis of stress protein data in a flor yeast subjected to a biofilm forming condition. *Data in Brief, 7*, 1021–1023. https://doi.org/10.1016/j.dib.2016.03.072.

Moreno-Garcia, J., Mauricio, J. C., Moreno, J., & Garcia-Martinez, T. (2017). Differential proteome analysis of a flor yeast strain under biofilm formation. *International Journal of Molecular Sciences, 18*(4). https://doi.org/10.3390/ijms18040720.

Moukadiri, I., & Zueco, J. (2001a). Evidence for the attachment of Hsp150/Pir2 to the cell wall of Saccharomyces cerevisiae through disulfide bridges. *FEMS Yeast Research, 1*(3), 241–245.

Moukadiri, I., & Zueco, J. (2001b). YJL159w does encode Pir2/Hsp150. *Yeast, 18*(4), 323–324. https://doi.org/10.1534/genetics.107.082602.

Palecek, S. P., Parikh, A. S., & Kron, S. J. (2000). Genetic analysis reveals that FLO11 upregulation and cell polarization independently regulate invasive growth in Saccharomyces cerevisiae. *Genetics, 156*(3), 1005–1023.

Parish, M. E., & Fleet, G. H. (2013). Wine. In M. P. Doyle & R. L. Buchanan (Eds.), *Food microbiology-fundamentals and frontiers* (pp. 915–947). Washington DC: American Society for Microbiology.

Peinado, R. A., Mauricio, J. C., Medina, M., & Moreno, J. J. (2004a). Effect of Schizosaccharomyces pombe on aromatic compounds in dry sherry wines containing high levels of gluconic acid. *Journal of Agricultural and Food Chemistry, 52*(14), 4529–4534.

Peinado, R. A., Moreno, J., Medina, M., & Mauricio, J. C. (2004b). Changes in volatile compounds and aromatic series in sherry wine with high gluconic acid levels subjected to aging by submerged flor yeast cultures. *Biotechnology Letters, 26*(9), 757–762.

Peinado, R. A., Moreno, J. J., Maestre, O., Ortega, J. M., Medina, M., & Mauricio, J. C. (2004c). Gluconic acid consumption in wines by Schizosaccharomyces pombe and its effect on the concentrations of major volatile compounds and polyols. *Journal of Agricultural and Food Chemistry, 52*(3), 493–497. https://doi.org/10.1016/j.foodchem.2014.04.039.

Perestrelo, R., Albuquerque, F., Rocha, S. M., & Camara, J. S. (2011). Distinctive characteristics of Madeira wine regarding its traditional winemaking and modern analytical methodologies. *Advances in Food and Nutrition Research, 63*, 207–249.

Perez, L., Valcarcel, M. J., Gonzalez, P., & Domecq, B. (1991). F influence of Botrytis infection of the grapes on the biological aging process of Fino Sherry. *American Journal of Enology and Viticulture, 42*(1), 58–62.

Pham, T. T., Guichard, E., Schlich, P., & Charpentier, C. (1995). Optimal conditions for the formation of sotolon from alpha-ketobutyric acid in the French "Vin Jaune". *Journal of Agricultural and Food Chemistry, 43*, 2616–2619.

Pirino, G., Zara, S., Pinna, G., Farris, G. A., & Budroni, M. (2004). Diversity of Y region at HML locus in a Saccharomyces cerevisiae strain isolated from a Sardinian wine. *Antonie Van Leeuwenhoek, 85*(1), 29–36.

Pozo-Bayon, M. A., & Moreno-Arribas, M. V. (2011). Sherry wines. *Advances in Food and Nutrition Research, 63*, 17–40.

Puig, S., & Perez-Ortin, J. E. (2000). Stress response and expression patterns in wine fermentations of yeast genes induced at the diauxic shift. *Yeast, 16*(2), 139–148.

Purevdorj-Gage, B., Orr, M. E., Stoodley, P., Sheehan, K. B., & Hyman, L. E. (2007). The role of FLO11 in Saccharomyces cerevisiae biofilm development in a laboratory based flow-cell system. *FEMS Yeast Research, 7*(3), 372–379.

Reader, H. P., & Dominguez, M. (2003). Fortified wines: Sherry, Port and Madeira. In A. G. H. Lea & J. R. Piggott (Eds.), *Fermented beverage production* (pp. 157–194). New York: Springer.

Reynolds, T. B., & Fink, G. R. (2001). Bakers' yeast, a model for fungal biofilm formation. *Science, 291*(5505), 878–881.

Riberereau-Gayon, P., Dubordieu, D., Donèche, B., & Lonvaud, A. (2006). Handbook of enology. In *The microbiologyof wine and vinifications* (Vol. 1). Chichester: Willey.

Ristow, H., Seyfarth, A., & Lochmann, E. R. (1995). Chromosomal damages by ethanol and acetaldehyde in Saccharomyces cerevisiae as studied by pulsed field gel electrophoresis. *Mutation Research, 326*(2), 165–170. https://doi.org/10.1016/j.fm.2011.10.010.

Sancho, E. D., Hernandez, E., & Rodriguez-Navarro, A. (1986). Presumed sexual isolation in yeast populations during production of Sherrylike wine. *Applied and Environmental Microbiology, 51*(2), 395–397.

Scherz, K., Andersen, Bojsen, R., Gro, L., Rejkjaer, Sorensen, Weiss, M., Nielsen, Lisby, M., Folkesson, A., & Regenberg, B. (2014). Genetic basis for Saccharomyces cerevisiae biofilm in liquid medium. *G3 (Bethesda), 4*(9), 1671–1680. https://doi.org/10.1534/g3.114.010892.

Storchova, Z., Breneman, A., Cande, J., Dunn, J., Burbank, K., O'Toole, E., & Pellman, D. (2006). Genome-wide genetic analysis of polyploidy in yeast. *Nature, 443*(7111), 541–547.

Suarez Lepe, J. A., & Iñigo Leal, B. (2011). *Microbiología enológica: Fundamentos de vinificación*. Madrid: Ediciones Mundi-Prensa.

Suarez-Lepe, J. A. (1997). *Levaduras vínicas: Funcionalidad y uso en bodega*. Madrid: Mundi-prensa.

Tamaki, H., Miwa, T., Shinozaki, M., Saito, M., Yun, C. W., Yamamoto, K., et al. (2000). GPR1 regulates filamentous growth through FLO11 in yeast Saccharomyces cerevisiae. *Biochemical and Biophysical Research Communications, 267*(1), 164–168. https://doi.org/10.1006/bbrc.1999.1914.

Teunissen, A. W., & Steensma, H. Y. (1995). Review: The dominant flocculation genes of Saccharomyces cerevisiae constitute a new subtelomeric gene family. *Yeast, 11*(11), 1001–1013. https://doi.org/10.1111/j.1567-1364.2008.00381.x.

Toledano, F., Martinez, J., Moreno, J., Ortega, J. M., & Medina, M. (1991). Utilisation sequentielle de deux levures poour la fermentation de mouts destines a la production de vins du type sherry. *Cerevisiae and Biotechnol, 1*, 35–41.

Torbensen, R., Moller, H. D., Gresham, D., Alizadeh, S., Ochmann, D., Boles, E., & Regenberg, B. (2012). Amino acid transporter genes are essential for FLO11-dependent and FLO11-independent biofilm formation and invasive growth in Saccharomyces cerevisiae. *PLoS One, 7*(7), e41272. https://doi.org/10.1371/journal.pone.0041272. PONE-D-11-08927 [pii].

Valero, E., Millan, C., & Ortega, J. M. (2002). Changes in the lipid composition of *Saccharomyces cerevisiae* race *capensis* (G-1) during alcoholic fermentation and flor film formation. *Lebensmittel-Wissenschaft und Technologie, 35*, 593–599.

Valero, E., Cambon, B., Schuller, D., Casal, M., & Dequin, S. (2007). Biodiversity of Saccharomyces yeast strains from grape berries of wine-producing areas using starter commercial yeasts. *FEMS Yeast Research, 7*(2), 317–329.

van Dyk, D., Pretorius, I. S., & Bauer, F. F. (2005). Mss11p is a central element of the regulatory network that controls FLO11 expression and invasive growth in Saccharomyces cerevisiae. *Genetics, 169*(1), 91–106. https://doi.org/10.1534/genetics.104.033704.

Van Mulders, S. E., Christianen, E., Saerens, S. M., Daenen, L., Verbelen, P. J., Willaert, R., Verstrepen, K. J., & Delvaux, F. R. (2009). Phenotypic diversity of Flo protein family-mediated adhesion in Saccharomyces cerevisiae. *FEMS Yeast Research, 9*(2), 178–190. https://doi.org/10.1111/j.1567-1364.2008.00462.x.

Verstrepen, K. J., & Klis, F. M. (2006). Flocculation, adhesion and biofilm formation in yeasts. *Molecular Microbiology, 60*(1), 5–15. https://doi.org/10.1111/j.1365-2958.2006.05072.x.

Verstrepen, K. J., Reynolds, T. B., & Fink, G. R. (2004). Origins of variation in the fungal cell surface. *Nature Reviews Microbiology, 2*(7), 533–540. https://doi.org/10.1038/nrmicro927.

Verstrepen, K. J., Jansen, A., Lewitter, F., & Fink, G. R. (2005). Intragenic tandem repeats generate functional variability. *Nature Genetics, 37*(9), 986–990. https://doi.org/10.1038/ng1618.

Wang, L. C., Montalvo-Munoz, F., Tsai, Y. C., Liang, C. Y., Chang, C. C., & Lo, W. S. (2015). The histone acetyltransferase Gcn5 regulates ncRNA-ICR1 and FLO11 expression during Pseudohyphal development in Saccharomyces cerevisiae. *BioMed Research International, 2015*, 284692. https://doi.org/10.1155/2015/284692.

Zara, S., Antonio Farris, G., Budroni, M., & Bakalinsky, A. T. (2002). HSP12 is essential for biofilm formation by a Sardinian wine strain of S. cerevisiae. *Yeast, 19*(3), 269–276.

Zara, S., Bakalinsky, A. T., Zara, G., Pirino, G., Demontis, M. A., & Budroni, M. (2005). FLO11-based model for air-liquid interfacial biofilm formation by Saccharomyces cerevisiae. *Applied and Environmental Microbiology, 71*(6), 2934–2939.

Zara, G., Mannazzu, I., Sanna, M. L., Orro, D., Farris, G. A., & Budroni, M. (2008). Exploitation of the semi-homothallic life cycle of Saccharomyces cerevisiae for the development of breeding strategies. *FEMS Yeast Research, 8*(7), 1147–1154. https://doi.org/10.1111/j.1567-1364.2008.00393.x.

Zara, S., Gross, M. K., Zara, G., Budroni, M., & Bakalinsky, A. T. (2010). Ethanol-independent biofilm formation by a flor wine yeast strain of Saccharomyces cerevisiae. *Applied and Environmental Microbiology, 76*(12), 4089–4091. https://doi.org/10.1128/AEM.00111-10.

Chapter 15
Yeasts in Fruit Wine Fermentation

Charoen Charoenchai

Contents

1 Introduction

It has been recognized internationally that wine is the alcoholic beverage resulting exclusively from the fermentation of grape or grape must. This is stated clearly in the definition of wine in the International Code of Oenological Practices issued by International Organization of Vine and Wine with its 47 member states. Various wine producing countries have also implemented regulations controlling the use of grape varieties and the amounts of harvest that can be used to produce particular wines. Thus, it is apparent that alcoholic beverages fermented from other fruits are not recognized as wine by international convention. By trials and errors since antiquity, grape was found most suitable and is solely utilized as raw material for winemaking. All modern winemakers still place great care in the vineyards to obtain perfectly sound grapes for winemaking and they understand that winery operations can only enhance the already good quality grapes (Rankine 2004). These facts indicate that fermented beverages of grape origin only are accepted as wine.

C. Charoenchai (✉)
Department of Food Science and Technology, Faculty of Agricultural Technology, Rajamangala University of Technology Thanyaburi, Thanyaburi, Pathumthani, Thailand
e-mail: c_charoen@rmutt.ac.th

P. Romano et al. (eds.), *Yeasts in the Production of Wine*,
https://doi.org/10.1007/978-1-4939-9782-4_15

However, other fruits and sugary substrates have been fermented to obtain various alcoholic beverages but they were assigned specific names corresponding to the type of raw materials employed. These include cider from apple, mead from honey and sake from rice.

Due to the lack of international agreement on production practices for fruit wines, alcoholic beverages from fruits other than grape and apple are called fruit wine with a particular fruit name, for example, pineapple wine, peach wine and strawberry wine. These names have confused consumers with wine (from grape) causing them to expect similar quality and flavor as that of table wine. But most fruits have different compositions and microflora from grape and the resulting beverages are usually significantly different in quality.

In order to produce high quality fruit wines, these differences must be taken into account and utilized to optimize the style of the beverage that reflects the nature of the fruit as well as the production process to obtain a suitable product which could be much different from conventional wines, with unique characteristics and benefits. In this respect, the roles of yeasts during fermentation are one of the most important factors to be considered in the development of alcoholic beverages from fruits.

In wine made from grape, the roles of yeasts that contribute to overall wine quality are well documented and already discussed in previous chapters of this book. These include the following:

1. Yeasts metabolize sugars and nitrogen compounds into alcohols and other flavor compounds, impacting wine quality;
2. Ethanol produced by yeasts not only affects organoleptic property but also acts as solvent extracting color and flavor, such as tannins from grape into wine;
3. Enzymatic activities resulting from yeast growth can transform flavor precursors present in grape must into flavor active compounds, such as glycosidase enzymes;
4. Yeast fermentation increases the concentrations of some organic acids but decreases some acids and some phenolic compounds are transformed into volatile compounds;
5. In cases of lees contact, dead yeast cells will decompose by autolytic enzymes and will impart specific characteristics known to be associated with autolysis;
6. Yeast cell walls can absorb certain wine constituents, resulting in decreasing amounts of these compounds.

These yeast contributions to wine quality in different fruit systems could differ from those occurring in grape and, therefore, must be thoroughly investigated before optimized production process can be designed for a certain fruit.

2 Opportunities for Fruit Wines

Throughout the world, many different kinds of fruits are cultivated and commercialized. Some fruits are only available during specific seasons and can attain high value when retailed as fresh products such as durian, lychee and mangosteen but others are produced in large quantities during longer period of time, causing their values to decline significantly. Even high-valued fruits may undergo similar problem with over production in certain years due to favorable climatic conditions. Furthermore, cultivation areas for popular fruits are expanding, replacing original crops, causing over supply. In order to gain some revenues from overproduced fruits, they are normally processed into preserves, jams, canned fruits and other products but these are too common for consumers and do not increase much higher revenue for producers. An interesting alternative of producing value added products from over-supplied fruits is fruit-winemaking. They can be marketed at higher values and attain much higher profits due to the nature of alcoholic beverages which tends to be marketed as high value products indicating social status.

A new trend in beverage market is the consumption of low alcohol and this trend has been determined by many factors, such as health awareness of consumers and the enforcement of restrictions on alcohol consumption by excise taxes and alcohol control laws in many countries. These restrictions aim to reduce alcohol related illnesses and road accidents caused by irresponsible drivers under the influence of alcohol intoxication. Fruit juices usually contain lower amounts of fermentable sugars than wine grapes and therefore could yield new products with low concentrations of ethanol. International Code of Oenological Practices indicates that wine must contain over 8.5% alcohol but fruit wine products can have lower alcohol content without violating any international standard.

Modern consumers are always searching for new products. Alcoholic beverages from different kinds of fruits with their unique flavor could cater for this market. Alcoholic beverages from exotic fruits can deliver their special characteristics to consumers in distance market without the need to transport fresh products over long distance in refrigerated containers. Even commonly known product in one country could be a new product in another. For example, sake or rice wine from Japan has been popular in the United States where as many brands of American cider are distributed widely in Japan. Thus, wine made from tropical fruits could be appreciated by consumers in temperate zones and vice versa.

All over the world, health and functional foods have gained increasing interests by producers and consumers alike. Fruit wines offer various health benefits such as antioxidant activity from phenolic compounds found in many fruits. Total phenols and antioxidant activities had been investigated by Kalkan Yildirim (2006) in different fruit wines made from black mulberry, blackberry, quince, apple, apricot, melon, red raspberry, bilberry, sour cherry and strawberry. The highest value of antioxidant activities and total phenolic contents were determined in bilberry, blackberry and black mulberry wines (61.80%, 1161 mg/l gallic acid equivalents; 60.00%, 1232 mg/l gallic acid equivalents; 58.10%, 1081 mg/l gallic acid equivalents), respectively. These results showed a potential as natural antioxidants of bilberry, blackberry and black mulberry wines.

Ortiz et al. (2013) determined antioxidant activity of wines made in Ecuador from Andean blackberries (*Rubus glaucus* Benth.) and bluberries (*Vaccinium floribundum* Kunth.) and Golden Reinette apples and found that Andean blackberries had the highest total phenolic content (1265 ± 91 mg/L) and antioxidant activity (12 ± 1 mM).

They concluded that these berries are suitable raw materials to produce wines with an in vitro antioxidant capacity that is comparable to red grape wines.

Commercially available fruit wines from blackberry, cherry, raspberry, blackcurrant, strawberry and apple produced in Croatia were analyzed by Ljevar et al. (2016). The results showed that blackberry, cherry and blackcurrant wines contained the highest amount of total phenolic compounds, while cherry and blackcurrant wines also contained the highest amount of total anthocyanins. Blackberry, followed by cherry, raspberry and blackcurrant wines also had a significantly higher antioxidant capacity than strawberry and apple wines. Fruit wines inhibited the growth of human cancer cells in vitro in a dose-dependent manner with differing susceptibility among tested cancer cells.

Phenolic contents and antioxidant activities found in fruit wines might not correlate with actual health benefits, so many researchers had investigated the effects of fruit wines as antioxidant agents in vivo. Srikanta et al. (2016) found that feeding mulberry and jamun wines to streptozotocin-induced diabetic rats increased antioxidant enzymes and hepatic glutathione contents but decreased non-esterified fatty acids and lipid peroxidation. These findings suggest that fruit wines can be beneficial as antioxidants. Escudero-López et al. (2018) evaluated the potential effect on cardiovascular risk factors of the regular consumption by healthy humans of a beverage obtained by alcoholic fermentation and pasteurization of orange juice. It was found that fermented orange beverage intake significantly increased oxygen radical absorbance capacity (43.9%) and reduced uric acid (−8.9%), catalase (CAT) (−23.2%), thiobarbituric acid reactive substances (TBARS) (−30.2%) and C-reactive protein (−2.1%). They concluded that regular consumption of orange could result in protecting the cardiovascular system in healthy humans and be considered a novel functional beverage.

Fruits are important sources of vitamins. Alcoholic fermentation affects vitamin contents in grape wines, resulting in decreased vitamin contents (Rankine 2004). However, in fruit wines the effect varies, depending on the fruit matrixes. Escudero-López et al. (2018) found that ascorbic acid contents in fermented orange beverage did not undergo a significant change. In another study (Cerrillo et al. 2014), significant increases in carotenoid content and provitamin A value of fermented orange juice were observed from day 0 (5.37 mg/L and 75.32 RAEs/L, respectively) until day 15 (6.65 mg/L and 90.57 RAEs/L, respectively). The authors suggested that the increases were probably due to a better extractability of the carotenoids from the food matrix as a result of fermentation.

3 Challenges in Fruit Wine Fermentation

In order to fully utilize the potential benefits of fruit wines, information on various aspects of production process and the nature of the fruits including the following topics must be investigated.

1. Fruit juice extraction and process to obtain suitable musts for fermentation
2. Ecology of yeasts and factors affecting their presence in fruit juices and wines

3. Microbiology of fermentation, starter culture and fermentation conditions tailored for particular fermentation
4. Contributions of fermentation to beverage quality

Figure 15.1 illustrates various aspects and approaches in the development of fruit wines that need to be chosen according to different fruits and style of beverage desired.

Fig. 15.1 Different approaches in the development of fruit wines

It can be seen that at each stage of production, a set of alternatives is available. These factors influence the outcome of the finished wines as in conventional winemaking and have to be taken into account with detailed scientific studies.

4 Fruit Juice Pre-Treatments

Unlike grape, many fruits contain skins that must be peeled off and the pulps cut in to small pieces for subsequent juice extraction. During the peeling and cutting processes, indigenous yeast flora on the surface of the peels could be transferred into the extracted juice. Fruits without skin also have yeast microflora on the surface that can carry out alcoholic fermentation. Furthermore, depending on the equipment and method being employed, yeasts colonizing in the environment of the processing plant can also contaminate into the musts and influence the overall fermentation.

Some fruits, especially those with low water contents, require further extraction processes in order to obtain sufficient amount of juice to be fermented. Heat treatment, such as pasteurization, is used to reduce microbial contamination and facilitate juice extraction of some fruits. However, heat can change the composition of the juice and affect nutrient availability for yeast fermentation. Ascorbic acid, total flavanones, total carotenoids and provitamin A values were found to decrease after pasteurization (Escudero-López et al. 2016). On the other hand, Cerrillo et al. (2015) found that, although pasteurization of orange juice produced partial amino acid degradation, the total amino acid content was higher in the final product than in the original juice. These changes could affect the growth of yeasts, resulting in modified flavor of finished beverages.

Higher acidity of the musts can affect organoleptic characteristics of fruit wines. This problem was rectified by dilution of fruit juices to obtain acceptable ranges of acidity (Akubor 1996). However, this practice also reduces concentrations of fruit sugars as well as nitrogen compounds necessary for yeast growth. Most fruits, even without acid dilution, contain less sugars than grape and therefore chaptalization or the additions of sucrose into the musts can be employed in order to increase alcohol contents of the finished wines (Akubor 1996; Jitjaroen 2007; Duarte et al. 2010; Won et al. 2015; Satora et al. 2018). Nutrient addition commonly practiced in the wine industry is also necessary in fruit juices pre-treatment, especially those with acid dilution. Jitjaroen (2007) studied the effects of nutrient supplements in banana, santol and ma-mao juices. It was found that banana juice supplemented with 1000 mg/L diammonium phosphate (DAP) and santol juice supplemented with 1000 mg/L DAP and 0.6 mg/L thiamine showed less sulfur binding capacities, resulting in perceptible dryness.

These pre-treatment practices affect yeast growth and their metabolic activities, resulting in the final quality of the finished wines.

5 Ecology of Yeasts in Fruit Juices and Wines

It is well documented that indigenous yeasts belonging to the genera *Candida, Kloeckera/Hanseniaspora, Pichia, Metschnikowia, Kluyveromyces* and other non-*Saccharomyces* grow during the early stages of alcoholic fermentation of grape juice into wine, although strains of *Saccharomyces cerevisiae* predominate mid to later stages of fermentation (Fleet and Heard 1993; Fleet 2003). Indigenous yeasts represent the microflora of the grapes and *S. cerevisiae* is present in low numbers on grapes and can also originate from winery equipment. Similar occurrences were observed in other fermentation systems such as those in palm wines (Stringini et al. 2009; Ouoba et al. 2012) and traditional rice wine in South East Asia where *Saccharomycopsis fibuligera* occurred during saccharification of rice (Thanh et al. 2016; Farh et al. 2017). Therefore, it can be expected that in many fruit wine fermentation systems, succession of yeast domination phases could also occur. Indigenous yeasts from the fruits would grow during the early stages of fermentation and then die out, leaving *S. cerevisiae* as the main fermenting yeast to complete the alcoholic fermentation.

Among the fruit wines, cider is one of the oldest known beverages and is produced in more than 25 countries around the world and the highest production and consumption is mainly in European countries (about 70–80%). It must be underlined that the appreciation of this fermented beverage is linked to the recognition of its "territoriality". Indigenous yeasts can actively contribute to the expression of cider typicality and significantly influence the sensory profile of cider.

Thus, Morrissey et al. (2004) examined the role of the indigenous yeast flora in traditional Irish cider fermentations, finding *Hanseniaspora uvarum* to predominate in the initial phase of fermentation. Thereafter *Saccharomyces cerevisiae* dominated in the alcoholic fermentation phase. Coton et al. (2016) isolated from unpasteurized ciders and different cider musts 15 yeast species, the dominant being *Saccharomyces bayanus* (about 35%), followed by *S. cerevisiae* and non-*Saccharomyces* yeasts, such as *Lachancea cidri*, *Dekkera anomala* and *Hanseniaspora valbyensis*.

The roles of yeasts in apple or cider fermentation have been reviewed by Cousin et al. (2017).

Fruit juices under sterile condition without contact with winery equipment such as the juice press, fermentation tanks and pumps do not yield adequate alcohol concentrations for stable beverage products without pasteurization. This is due to the lack of fermentative *Saccharomyces* yeasts from winery equipment. Freshly pressed pineapple juices obtained from Thailand, Australia and Angola were allowed to ferment spontaneously by native microflora (Chanprasartsuk et al. 2010; Dellacassa et al. 2017). The yeasts *Hanseniaspora uvarum* and *Pichia guilliermondii* predominated all fermentations yielding low alcohol contents (2–4% by volume). *S. cerevisiae* was not found in any of the fermentations. When pineapple juice was inoculated with selected strains of *S. cerevisiae*, the fermentable sugars were completely consumed from initial value of 24 °Brix (Baidya et al. 2016).

6 Yeast Starter Cultures and Their Contributions to Fruit Wine Quality

Since relying on naturally occurring *Saccharomyces* yeasts from winery environment can result in uncontrollable fermentations with inconsistent wine qualities, inoculation with desired strains of *S. cerevisiae* has become standard practice in most wineries (Reed and Nagodawithana 1988). However, Fleet and Heard (1993) had proposed that not only the inoculated *S. cerevisiae* that determine the characteristics of the resulting wines but also indigenous yeasts that grow during the early stages of fermentation. Subsequent studies have confirmed the contributions of indigenous yeasts to overall wine flavor profiles and the use of alternative yeasts as co-starter with *S. cerevisiae* or as single cultures in grape wine fermentations have been reported and these have been reviewed by Fleet (2008).

For fruit wine fermentations, many studies regarding yeast fermentation have been carried out for different fruit juices and these have been reviewed by Prakitchaiwattana and Tananuwong (2011) and Chanprasartsuk and Prakitchaiwattana (2015). Species of *Saccharomyces* have been examined for their potential as starter cultures. Sixteen different strains of *Saccharomyces cerevisiae* and *Saccharomyces bayanus* were evaluated in the production of raspberry fruit wine (Duarte et al. 2010). Various kinetic parameters were determined and compared. One strain of *S. cerevisiae* was recommended for the fermentation of raspberry juice which produced a fruit wine with low concentrations of acids and high concentrations of acetates, higher alcohols and ethyl esters.

Three commercial *S. cerevisiae* yeast strains were evaluated for the production of pomegranate wine (Berenguer et al. 2016). The same fermentation patterns were observed for pH, titratable acidity, density, sugar consumption, and ethanol and glycerol production. A high ethanol concentration (10.91 ± 0.27% v/v) in combination with 1.49 g/L glycerol was achieved. Citric acid concentration increased rapidly at 31.7%, malic acid disappeared as a result of malolactic fermentation and the lactic acid levels reached values between 0.40 and 0.96 g/L. The analysis of total anthocyanin content highlighted a lower degradation of monomeric anthocyanins during winemaking with Viniferm PDM yeast. The resulting wine retained 34.5% of total anthocyanin content of pomegranate juice blend.

Saccharomyces uvarum strains were isolated from traditional fermentations of apple chicha in Patagonia, a region covering Argentina and Chile (Rodríguez et al. 2017). This research group also studied the physiological characteristics of *S. uvarum* and *Saccharomyces eubayanus* strains recovered from natural habitats and traditional fermentations. The yeast *S. uvarum* produced high glycerol levels, low acetic acid and increased production of the higher alcohol 2-phenylethanol and 2-phenylethyl acetate. Similar properties were observed for *S. eubayanus*. The combination of these strains can be used as a starter culture in cidermaking (González et al. 2017).

Non-*Saccharomyces* yeast species possess various characteristics which could be beneficial to fruit wine fermentations and these yeasts were investigated for their potential as single, sequential or co-starter cultures with *S. cerevisiae*. Volatile compounds formed during fermentation of papaya juice using a mixed culture of *Saccharomyces cerevisiae* and *Williopsis saturnus* were analyzed by Lee et al. (2010). Different volatile compounds were produced during fermentation including fatty acids, alcohols, aldehydes and esters but some volatile compounds, including those initially present in the juice, were utilized. The mixed culture fermentation by *S. cerevisiae* and *W. saturnus* benefited from the presence of both yeasts, with more esters being produced than the *S. cerevisiae* monoculture fermentation and more alcohols formed than the product fermented with *W. saturnus* alone. It was suggested that papaya juice fermentation with a mixed culture of *S. cerevisiae* and *W. saturnus* may be able to result in the formation of more complex aroma compounds and higher ethanol level than those using single yeasts. They also found that the yeast ratio of *W. saturnus* and *S. cerevisiae* in sequential fermentation of papaya wine was an important factor affecting fermentation performances (Lee et al. 2013).

Rodríguez-Lerma et al. (2011) studied the microbial ecology of spontaneous fermentation to select a starter culture for prickly pear wine production. Results showed that a mixed starter inoculum containing *Pichia fermentans* and *S. cerevisiae* yielded a fermented product that contained 8.37% alcohol (v/v). Analysis of volatile compounds revealed the presence of 9 major alcohols and esters (isobutanol, isopentanol, ethyl acetate, isoamyl acetate, ethyl octanoate, ethyl decanoate, ethyl 9-decanoate, β-phenylethyl acetate, and phenylethyl alcohol) that contributed to fruity, aromatic notes essential for desirable wine quality. It was concluded that combinations of *Saccharomyces* and non-*Saccharomyces* strains could be used to obtain high-quality fermented beverages from prickly pear juice.

A mixed culture of *Saccharomyces cerevisiae* and *Williopsis saturnus* var. *mrakii* was used to ferment three varieties of mango juices (Li et al. 2012). Both yeasts grew well and fructose, glucose and sucrose were consumed to trace levels in all juices. But since only one ratio of yeast mixture was used, comparison between volatile constituents of mango wines could be distinguished between different mango varieties.

Strains of *S. cerevisiae*, *Pichia kudriavzevii*, *Pichia fabianii* and *Saccharomycopsis fibuligera* were isolated from masau fruits and their traditionally fermented fruit pulp in Zimbabwe and tested for production of flavor compounds during fermentation of masau wines (Nyanga et al. 2013). It was found that *S. cerevisiae* strains produced higher amounts of ethanol and flavor compounds as compared to the other species, especially fatty acid ethyl esters that provide the major aroma impact of freshly fermented wines.

Sun et al. 2014 examined the effect of mixed fermentation of non-*Saccharomyces* (*Torulaspora delbrueckii* and *Metschnikowia pulcherrima*) and *Saccharomyces cerevisiae* on the production of cherry wines. Mixed culture of *S. cerevisiae*/*M. pulcherrima* was found to significantly enhance the production of higher alcohols, esters, acids and terpenes; while the characteristic of *S. cerevisiae*/*T. delbrueckii* pair was an increase in fruity esters, higher alcohols and decrease in acid production. The differences in the aromatic composition of the cherry wines were confirmed by the sensory evaluation.

Satora et al. (2014) investigated the influence of *Wickerhamomyces anomalus* killer yeast on the fermentation and chemical composition of apple wines. The yeast was inoculated together with *S. cerevisiae* strain as a mixed culture. It was found that the addition of *W. anomalus* killer strains to the unpasteurized must significantly decreased volatile acidity, while increased the amount of higher alcohols and titratable acidity. It was concluded that the use of *W. anomalus* strains together with *S. cerevisiae* as a mixed culture positively influenced the chemical composition and sensory characteristics of apple wines.

Ye et al. (2014) studied the effects of sequential mixed cultures of *Wickerhamomyces anomalus* and *S. cerevisiae* on apple cider fermentation. The results showed that growth of *W. anomalus* and *S. cerevisiae* was affected by each other during co-fermentation process. All the mixed cultures produced statistically the same level of ethanol as *S. cerevisiae* monoculture. The mixed fermentation produced more variability and higher amounts of acetate esters, ethyl esters, higher alcohols, aldehydes, and ketones. Sensory evaluation showed that ciders obtained from co-fermentation with *W. anomalus* obtained higher scores than ciders fermented by pure culture of *S. cerevisiae.*

Chen et al. (2015) evaluated the performance of *Torulaspora delbrueckii, Williopsis saturnus*, and *Kluyveromyces lactis* in lychee wine fermentation. It was found that *T. delbrueckii* had the fastest growth rate and high sugar consumption, producing the highest level of ethanol (7.6% v/v), while *K. lactis* and *W. saturnus* produced lower amounts (3.4% v/v and 0.8% v/v, respectively). Furthermore, *K. lactis* and *W. saturnus* over-produced ethyl acetate which was considered detrimental to wine quality. The yeast *T. delbrueckii* produced high levels of isoamyl alcohol, 2-phenylethyl alcohol, ethyl octanoate, and ethyl decanoate and retained high aroma-character compounds. It was concluded that this yeast could be a promising non-*Saccharomyces* yeast for lychee wine fermentation.

Minnaar et al. (2017) studied the effect of using *Schizosaccharomyces pombe* and *Saccharomyces cerevisiae* yeasts in sequential fermentations on phenolic acids of fermented Kei-apple (*Dovyalis caffra* L.) juice. Kei-apple wines obtained by sequential cultures of *S. pombe* and *S. cerevisiae* showed substantially lower concentrations of L-malic acid and had lower phenolic acid concentrations than Kei-apple wines produced with *S. cerevisiae* only. However, wine judges described the Kei-apple wines produced with the two yeast combination as having noticeable off-odors, while those produced with *S. cerevisiae* were described as fresh and fruity.

Influence of selected *Saccharomyces* and *Schizosaccharomyces* strains and their mixed cultures on chemical composition of apple wines was investigated by Satora et al. (2018). It was found that *S. bayanus* strain increased the level of malic acid and carbonyl compounds in apple wines, while *Sch. pombe* highly deacidified it and produced the most amounts of glycerol, esters, and acetic acid. The wines obtained with these species gained the best and the worse notes, respectively, during sensory analysis. Mixed yeast cultures produce higher amounts of ethanol, methanol, and volatile esters compared to pure cultures.

7 Factors Affecting the Growth and Metabolites Production of Yeasts in Fruit Wines

Various factors influence yeast growth and their metabolic activities, resulting in differing flavor profiles of finished products, depending on the compositions of fruit juices and other environmental factors. These factors such as sugar concentrations, nitrogen and other nutrients availability, pH, temperature and the addition of sulfur dioxide are important parameters in controlling fermentation of grape wines and have been well documented (Fleet and Heard 1993). In fruit juice fermentations, these parameters are also important in determining the final quality of fruit wines. Most studies concerning fruit wines mainly focused on yeast fermentation. There were fewer studies carried out on different fermentation conditions as affected by environmental factors.

Temperature of fermentation is one of the most important factors determining the production of volatile compounds by yeasts but it is also the most difficult parameter to control in fruit wine fermentations, especially in the tropical regions. Many studies on fruit wine fermentations were carried out at higher temperatures than those used for grape wine (Jitjaroen 2007; Peng et al. 2015; Lu et al. 2017a, b). This is probably due practical reasons since cooling requirements in these climates involve high running costs. Reddy and Reddy (2011) examined the effect of fermentation conditions on yeast growth and volatile composition of mango wine. Temperature had important effect on yeast growth and on the levels of volatile compounds. It was observed that the final concentrations of ethyl acetate and some of the higher alcohols decreased when fermentation temperature increased to 25 °C. Sulfur dioxide stimulated the yeast growth up to certain levels and in excess it inhibited the yeast metabolism. Ethanol concentration slightly increased with the addition of 100 ppm SO_2 and decreased with increase in concentration of SO_2. Aeration by shaking increased the viable cell count but decreased the ethanol productivity.

Temperature effect during fermentation of apple wine on the key aroma compounds and sensory properties were investigated by Peng et al. (2015). The concentration of nine key aroma compounds (ethyl acetate, isobutyl acetate, isopentyl acetate, ethyl caprylate, ethyl 4-hydroxybutanoate, isobutyl alcohol, isopentyl alcohol, 3-methylthio-1-propanol, and benzeneethanol) in apple wine significantly increased with the increase of fermentation temperature from 17 to 20 °C, and then eight out of the nine key aroma compounds, with an exception of ethyl 4-hydroxybutanoate, decreased when the temperature increased from 20 to 26 °C. The results showed that temperature control is of great importance in fruit winemaking.

Lu et al. (2017b) investigated the effects of temperature (20 and 30 °C) and pH (pH 3.1, 3.9) on the changes in chemical constituents of durian wine fermented with *S. cerevisiae*. Temperature significantly affected growth of *S. cerevisiae* EC-1118 regardless of pH with a higher temperature leading to a faster cell death. The pH had a more significant effect on ethanol production than temperature. However, relatively higher levels of isobutyl alcohol and isoamyl alcohol were produced at pH 3.1 than at pH 3.9 regardless of temperature. In contrast, production of esters was more affected by temperature than pH, where levels of ethyl esters and acetate esters were significantly higher at 20 °C than at 30 °C. Higher temperature improved the reduction of volatile sulfur compounds. The authors concluded that temperature control would be a more effective tool than pH in modulating the resulting aroma compound profile of durian wine.

Sulfur dioxide is widely used in the wine industry to control oxidation and for microbial stability. It is also used in industrial fermentation of other fruits, such as apple juice into alcoholic cider. Herrero et al. (2003) studied the effect of sulfur dioxide on the production of flavor volatiles during industrial cider fermentation. Addition of 100 mg/L SO_2, which is the level frequently used in industrial cider-making, induced higher acetaldehyde production by yeast than that obtained without SO_2. These amounts of acetaldehyde could prevent the occurrence of simultaneous alcoholic and malolactic fermentation, which is desirable in reducing malic acid in apple cider.

Lu et al. (2017a) investigated the effect of initial sugar concentrations (17, 23 and 30 °Brix) on mango wine composition fermented by *Saccharomyces cerevisiae* MERIT.ferm. It was found that growth rate and maximum cell population were inversely correlated with initial sugar levels, with the fastest growth rate and largest cell population in the low sugar fermentation. However, the cell population in the low and medium sugar fermentation declined significantly relative to the high sugar fermentation in which cell populations remained stable upon reaching the stationary phase. Glycerol production increased with increasing sugar content in low, medium and high sugar fermentation. In addition, high sugar fermentation had a negative impact on volatile production with significantly lower amounts of acetate esters but more acetic acid compared to the low and medium sugar fermentations. The authors concluded that mango wines fermented with different levels of sugars would have different flavor and aromas.

8 Conclusions

Fruit juices present opportunities for the development of new alcoholic beverages with their original color, flavor, antioxidants and other bioactive compounds with potential health benefits. The development of alcoholic beverages from fruits must take into account various aspects and approaches available at each stage of the production process. Pre-treatment of the juices can provide suitable medium for yeast fermentation but some practices could be detrimental to wine quality due to their effect on chemical compositions of the mash. Indigenous yeasts already present in the pressed musts could survive during fruit-winemaking process and influence chemical and sensorial properties of the resulting wine. These yeasts have different fermentation profiles than *Saccharomyces cerevisiae* wine strains and their roles in spontaneous or inoculated fermentations must be elaborated in each particular juice.

Factors such as temperature, pH, sugar concentrations, and the presence of sulphur dioxide affect growth and fermentation activities of yeasts at varying degrees and a suitable combination of these factors must be obtained by systematic investigations. Thorough understandings is necessary of metabolic behavior and characteristics of *Saccharomyces* and non-*Saccharomyces* yeasts that can offer unique characteristics and can be employed to obtain desirable styles of fruit wines.

9 Dedication

This chapter is based on a lecture given by Prof. Graham H. Fleet during the Thai Fruit Wine Seminar held in conjunction with the Third International Symposium on Tropical Wine held during 12–18 November 2011 in Chiang Mai, Thailand. The author wishes to dedicate this chapter to his memory.

References

Akubor, P. I. (1996). The suitability of African bush mango juice for wine production. *Plant Foods for Human Nutrition, 49*(3), 213–219.

Baidya, D., Chakraborty, I., & Saha, J. (2016). Table wine from tropical fruits utilizing natural yeast isolates. *Journal of Food Science and Technology, 53*(3), 1663–1669.

Berenguer, M., Vegara, S., Barrajón, E., Saura, D., Valero, M., & Martí, N. (2016). Physicochemical characterization of pomegranate wines fermented with three different *Saccharomyces cerevisiae* yeast strains. *Food Chemistry, 190*, 848–855.

Cerrillo, I., Escudero-López, B., Hornero-Méndez, D., Martín, F., & Fernández-Pachón, M. S. (2014). Effect of alcoholic fermentation on the carotenoid composition and provitamin A content of orange juice. *Journal of Agricultural and Food Chemistry, 62*(4), 842–849.

Cerrillo, I., Fernández-Pachón, M. S., Collado-González, J., Escudero-López, B., Berná, G., Herrero-Martín, G., Martín, F., Ferreres, F., & Gil-Izquierdo, A. (2015). Effect of fermentation and subsequent pasteurization processes on amino acids composition of orange juice. *Plant Foods for Human Nutrition, 70*(2), 153–159.

Chanprasartsuk, O., & Prakitchaiwattana, C. (2015). Impacts of Allochthonous and autochthonous yeast starters: Case studies in fruit wine fermentations. In E. Perkins (Ed.), *Food microbiology: Fundamentals, challenges and health implications* (pp. 117–160). New York: Nova Publisher.

Chanprasartsuk, O., Prakitchaiwattana, C., Sanguandeekul, R., & Fleet, G. H. (2010). Autochthonous yeasts associated with mature pineapple fruits, freshly crushed juice and their ferments; and the chemical changes during natural fermentation. *Bioresource Technology, 101*(19), 7500–7509.

Chen, D., Yap, Z. Y., & Liu, S. Q. (2015). Evaluation of the performance of *Torulaspora delbrueckii, Williopsis saturnus*, and *Kluyveromyces lactis* in lychee wine fermentation. *International Journal of Food Microbiology, 206*, 45–50.

Coton, E., Coton, M., Levert, D., Casaregola, S., & Sohier, D. (2016). Yeast ecology in French cider and black olive natural fermentations. *International Journal of Food Microbiology, 108*, 130–135.

Cousin, F. J., Le Guellec, R., Schlusselhuber, M., Dalmasso, M., Laplace, J. M., & Cretenet, M. (2017). Microorganisms in fermented apple beverages: Current knowledge and future directions. *Microorganisms, 5*(3), 39.

Dellacassa, E., Trenchs, O., Fariña, L., Debernardis, F., Perez, G., Boido, E., & Carrau, F. (2017). Pineapple (*Ananas comosus* L. Merr.) wine production in Angola: Characterisation of volatile aroma compounds and yeast native flora. *International Journal of Food Microbiology, 241*, 161–167.

Duarte, W. F., Dragone, G., Dias, D. R., Oliveira, J. M., Teixeira, J. A., Silva, J. B., & Schwan, R. F. (2010). Fermentative behavior of *Saccharomyces* strains during microvinification of raspberry juice (*Rubus idaeus* L.). *International Journal of Food Microbiology, 143*(3), 173–182.

Escudero-López, B., Cerrillo, I., Gil-Izquierdo, Á., Hornero-Méndez, D., Herrero-Martín, G., Berná, G., Medina, S., Ferreres, F., Martín, F., & Fernández-Pachón, M. S. (2016). Effect of thermal processing on the profile of bioactive compounds and antioxidant capacity of fermented orange juice. *International Journal of Food Sciences and Nutrition, 67*(7), 779–788.

Escudero-López, B., Ortega, Á., Cerrillo, I., Rodríguez-Griñolo, M. R., Muñoz-Hernández, R., Macher, H. C., Martín, F., Hornero-Méndez, D., Mena, P., Del Rio, D., & Fernández-Pachón, M. S. (2018). Consumption of orange fermented beverage improves antioxidant status and reduces peroxidation lipid and inflammatory markers in healthy humans. *Journal of Science and Food Agriculture, 98*(7), 2777–2786.

Farh, M. E., Cho, Y., Lim, J. Y., & Seo, J. A. (2017). A diversity study of *Saccharomycopsis fibuligera* in rice wine starter nuruk, reveals the evolutionary process associated with its interspecies hybrid. *Journal of Microbiology, 55*(5), 337–343.

Fleet, G. H. (2003). Yeast interactions and wine flavour. *International Journal of Food Microbiology, 86*(1–2), 11–22.

Fleet, G. H. (2008). Wine yeasts for the future. *FEMS Yeast Research, 8*(7), 979–995.

Fleet, G. H., & Heard, G. (1993). Yeasts-growth during fermentation. In G. H. Fleet (Ed.), *Wine microbiology and biotechnology* (pp. 27–54). Chur: Harwood Academic Publishers.

González Flores, M., Rodríguez, M. E., Oteiza, J. M., Barbagelata, R. J., & Lopes, C. A. (2017). Physiological characterization of *Saccharomyces uvarum* and *Saccharomyces eubayanus* from Patagonia and their potential for cidermaking. *International Journal of Food Microbiology, 249*, 9–17.

Herrero, M., García, L. A., & Díaz, M. (2003). The effect of SO_2 on the production of ethanol, acetaldehyde, organic acids, and flavor volatiles during industrial cider fermentation. *Journal of Agricultural and Food Chemistry, 51*(11), 3455–3459.

Jitjaroen, W. (2007). *Influence of yeast strains and nutritive supplements on enological characteristics of tropical fruit wines*. Göttingen: Cuvillier Verlag.

Kalkan Yildirim, H. (2006). Evaluation of colour parameters and antioxidant activities of fruit wines. *International Journal of Food Sciences and Nutrition, 57*(1–2), 47–63.

Lee, P. R., Ong, Y. L., Yu, B., Curran, P., & Liu, S. Q. (2010). Profile of volatile compounds during papaya juice fermentation by a mixed culture of *Saccharomyces cerevisiae* and *Williopsis saturnus*. *Food Microbiology, 27*(7), 853–861.

Lee, P. R., Kho, S. H., Yu, B., Curran, P., & Liu, S. Q. (2013). Yeast ratio is a critical factor for sequential fermentation of papaya wine by *Williopsis saturnus* and *Saccharomyces cerevisiae*. *Microbial Biotechnology, 6*(4), 385–393.

Li, X., Chan, L. J., Yu, B., Curran, P., & Liu, S. Q. (2012). Fermentation of three varieties of mango juices with a mixture of *Saccharomyces cerevisiae* and *Williopsis saturnus* var. mrakii. *International Journal of Food Microbiology, 158*(1), 28–35.

Ljevar, A., Ćurko, N., Tomašević, M., Radošević, K., Srček, V. G., & Ganić, K. K. (2016). Phenolic composition, antioxidant capacity and in vitro cytotoxicity assessment of fruit wines. *Food Technology and Biotechnology, 54*(2), 145–155.

Lu, Y., Chan, L., Li, X., & Liu, S. (2017a). Effects of sugar concentration on mango wine composition fermented by Saccharomyces cerevisiae MERIT.Ferm. *Food Science and Technology, 53*(1), 199–208.

Lu, Y., Voon, M. K., Huang, D., Lee, P. R., & Liu, S. Q. (2017b). Combined effects of fermentation temperature and pH on kinetic changes of chemical constituents of durian wine fermented with *Saccharomyces cerevisiae*. *Applied Microbiology and Biotechnology, 101*(7), 3005–3014.

Minnaar, P. P., Jolly, N. P., Paulsen, V., Du Plessis, H. W., & Van Der Rijst, M. (2017). *Schizosaccharomyces pombe* and *Saccharomyces cerevisiae* yeasts in sequential fermentations: Effect on phenolic acids of fermented Kei-apple (*Dovyalis caffra* L.) juice. *International Journal of Food Microbiology, 257*, 232–237.

Morrissey, W. F., Davenport, B., Querol, A., & Dobson, A. D. (2004). The role of indigenous yeasts in traditional Irish cider fermentations. *Journal of Applied Microbiology, 97*(3), 647–655.

Nyanga, L. K., Nout, M. J., Smid, E. J., Boekhout, T., & Zwietering, M. H. (2013). Fermentation characteristics of yeasts isolated from traditionally fermented masau (*Ziziphus mauritiana*) fruits. *International Journal of Food Microbiology, 166*(3), 426–432.

Ortiz, J., Marín-Arroyo, M. R., Noriega-Domínguez, M. J., Navarro, M., & Arozarena, I. (2013). Color, phenolics, and antioxidant activity of blackberry (*Rubus glaucus* Benth.), blueberry (*Vaccinium floribundum* Kunth.), and apple wines from Ecuador. *Journal of Food Science, 78*(7), C985–C993.

Ouoba, L. I., Kando, C., Parkouda, C., Sawadogo-Lingani, H., Diawara, B., & Sutherland, J. P. (2012). The microbiology of Bandji, palm wine of *Borassus akeassii* from Burkina Faso: Identification and genotypic diversity of yeasts, lactic acid and acetic acid bacteria. *Journal of Applied Microbiology, 113*(6), 1428–1441.

Peng, B., Li, F., Cui, L., & Guo, Y. (2015). Effects of fermentation temperature on key aroma compounds and sensory properties of apple wine. *Journal of Food Science, 80*(12), S2937–S2943.

Prakitchaiwattana, C., & Tananuwong, K. (2011). Wines from tropical plants: Processing, microbiological, chemical and health aspects. In A. S. Peeters (Ed.), *Wine type, production and health* (pp. 1–5). New York: Nova Science Publishers.

Rankine, B. C. (2004). *Making good wine* (p. 318). Sydney: Pan Macmillan.

Reddy, L. V. A., & Reddy, O. V. S. (2011). Effect of fermentation conditions on yeast growth and volatile composition of wine produced from mango (*Mangifera indica* L.) fruit juice. *Food and Bioproducts Processing, 89*(4), 487–491.

Reed, G., & Nagodawithana, T. W. (1988). Technology of yeast usage in winemaking. *American Journal of Enology and Viticulture, 39*, 83–90.

Rodríguez, M. E., Pérez-Través, L., Sangorrín, M. P., Barrio, E., Querol, A., & Lopes, C. A. (2017). *Saccharomyces uvarum* is responsible for the traditional fermentation of apple chicha in Patagonia. *FEMS Yeast Research, 17*(1).

Rodríguez-Lerma, G. K., Gutiérrez-Moreno, K., Cárdenas-Manríquez, M., Botello-Álvarez, E., Jiménez-Islas, H., Rico-Martínez, R., & Navarrete-Bolaños, J. L. (2011). Microbial ecology studies of spontaneous fermentation: Starter culture selection for prickly pear wine production. *Journal of Food Science, 76*(6), M346–M352.

Satora, P., Tarko, T., Sroka, P., & Blaszczyk, U. (2014). The influence of *Wickerhamomyces anomalus* killer yeast on the fermentation and chemical composition of apple wines. *FEMS Yeast Research, 14*(5), 729–740.

Satora, P., Semik-Szczurak, D., Tarko, T., & Bułdys, A. (2018). Influence of selected *Saccharomyces* and *Schizosaccharomyces* strains and their mixed cultures on chemical composition of apple wines. *Journal of Food Science, 83*(2), 424–431.

Srikanta, A. H., Kumar, A., Sukhdeo, S. V., Peddha, M. S., & Govindaswamy, V. (2016). The antioxidant effect of mulberry and jamun fruit wines by ameliorating oxidative stress in streptozotocin-induced diabetic Wistar rats. *Food & Function, 7*(10), 4422–4431.

Stringini, M., Comitini, F., Taccari, M., & Ciani, M. (2009). Yeast diversity during tapping and fermentation of palm wine from Cameroon. *Food Microbiology, 26*(4), 415–420.

Sun, S. Y., Gong, H. S., Jiang, X. M., & Zhao, Y. P. (2014). Selected non-*Saccharomyces* wine yeasts in controlled multistarter fermentations with *Saccharomyces cerevisiae* on alcoholic fermentation behaviour and wine aroma of cherry wines. *Food Microbiology, 44*, 15–23.

Thanh, V. N., Thuy, N. T., Chi, N. T., Hien, D. D., Ha, B. T., Luong, D. T., Ngoc, P. D., & Ty, P. V. (2016). New insight into microbial diversity and functions in traditional Vietnamese alcoholic fermentation. *International Journal of Food Microbiology, 232*, 15–21.

Won, S. Y., Seo, J. S., Kwak, H. S., Lee, Y., Kim, M., Shim, H. S., & Jeong, Y. (2015). Quality characteristics and quantification of acetaldehyde and methanol in apple wine fermentation by various pre-treatments of mash. *Preventive Nutrition and Food Science, 20*(4), 292–297.

Ye, M., Yue, T., & Yuan, Y. (2014). Effects of sequential mixed cultures of *Wickerhamomyces anomalus* and *Saccharomyces cerevisiae* on apple cider fermentation. *FEMS Yeast Research, 14*(6), 873–882.

Chapter 16
Yeasts Associated With the Production of Distilled Alcoholic Beverages

Graeme M. Walker, Patricia Lappe-Oliveras, Rubén Moreno-Terrazas C., Manuel Kirchmayr, Melchor Arellano-Plaza, and Anne Christine Gschaedler-Mathis

Contents

G. M. Walker (✉)
School of Applied Sciences, Abertay University, Dundee, Scotland, UK
e-mail: g.walker@abertay.ac.uk

P. Lappe-Oliveras
Departamento de Botánica, Instituto de Biología, Universidad Nacional Autónoma de México, Cd. de México, México
e-mail: lappe@ib.unam.mx

R. Moreno-Terrazas C.
Departamento de Ingenierías Química, Industrial y de Alimentos, Universidad Iberoamericana, Cd. de Mexico, México
e-mail: ruben.moreno@ibero.mx

M. Kirchmayr · M. Arellano-Plaza · A. C. Gschaedler-Mathis
Unidad de Biotecnología Industrial, Centro de Investigación y Asistencia en Tecnología y Diseño del Estado de Jalisco A.C. (CIATEJ), Zapopan, Jalisco, México
e-mail: mkichmayr@ciatej.mx; marellano@ciatej.mx; agschaedler@ciatej.mx

P. Romano et al. (eds.), *Yeasts in the Production of Wine*,
https://doi.org/10.1007/978-1-4939-9782-4_16

1 Introduction

The production of alcoholic beverages from fermentable carbon sources by yeast is the oldest and most economically important of all biotechnologies. Yeast, in particular the species *Saccharomyces cerevisiae*, plays a vital role in the production of all alcoholic beverages (see Fig. 16.1) and the selection of suitable yeast strains is essential not only to maximise alcohol yield, but also to enhance beverage sensory quality.

The yeast species that dominates in the production of worldwide distilled spirits is *S. cerevisiae*, and the specific strains of this species employed in fermentation exert a profound influence on spirit flavour and aroma characteristics. For large-scale beverage fermentations, as in brewing, winemaking and distilled spirit production, pure cultures of selected strains of *S. cerevisiae* are typically employed (Walker 1998). These strains are either sourced and cultivated *in house* or supplied for direct inoculation from yeast producing companies. In smaller-scale (artisanal) processes, spontaneous fermentations may be allowed to occur that rely on indigenous microorganisms (wild yeasts and bacteria) present in the raw materials and in the production facility. For example, this would be typical in small distilleries in México (for Tequila and Mezcal production) and in Brazil (for Cachaça production). In some types of alcoholic beverage fermentations, non-*S. cerevisiae* yeasts may be employed either as starter cultures, or occur naturally. For example,

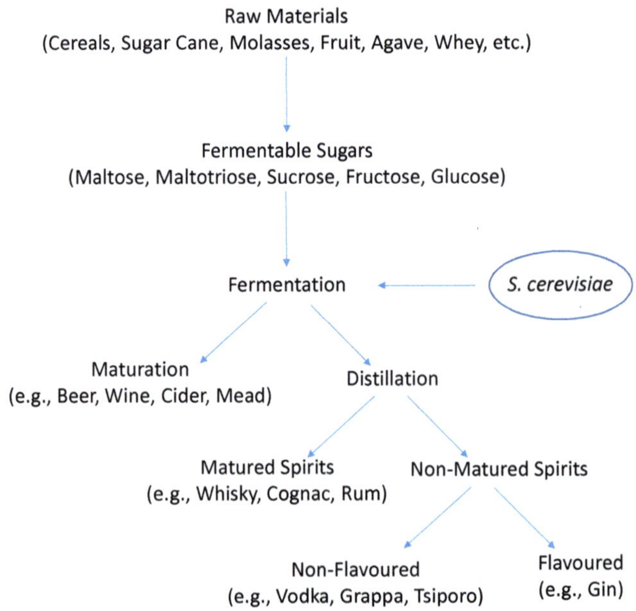

Fig. 16.1 The key role of *Saccharomyces cerevisiae* in production of fermented beverages. Adapted from Walker and Stewart (2016)

Table 16.1 Yeasts used in the production of selected distilled alcoholic beverages. Adapted from Walker and Stewart (2016)

Beverage	Yeast involved	Comments
Whisk(e)y	*Saccharomyces cerevisiae*	Scotch whisky producers currently use selected distilling strains of *S. cerevisiae* in three main formats, cream yeast, pressed (cake) and dried yeast. Malt whisky distilleries traditionally use pressed yeast, but larger grain distillers have now adopted cream yeast. Dried yeasts are not as prevalent as pressed and cream formats in whisky fermentations.
Rum	*Saccharomyces cerevisiae* and *Schizosaccharomyces pombe*	*Saccharomyces cerevisiae* strains in rum fermentations are developed as starter cultures and provide faster fermentation with more higher alcohols and fatty acids, but less esters resulting in lighter style rums.
		Schizosaccharomyces pombe in rum fermentations provides slower fermentations leading to less higher alcohols and fatty acids, but more esters resulting in heavy, strong aroma rums. Growth of *Schiz. pombe* is favoured by low pH, higher sugar concentrations.
Tequila, Mezcal, Bacanora	Natural yeasts in artisanal *Agave* fermentations	Various yeasts have been isolated from such processes: *S. cerevisiae*, *Kluyveromyces marxianus*, *Pichia* spp., *Brettanomyces* spp., *Rhodotorula* spp., etc.
Brandies, Gin, Vodka, etc.	*Saccharomyces cerevisiae*	For brandies, cognac, etc. the base wine is produced by pure starter cultures of *S. cerevisiae*. For gin, vodka, etc. selected distilling strains of *S. cerevisiae* will be used.
Cheese whey-derived beverages	*Kluyveromyces marxianus*	Lactose-fermenting yeast to produce ethanol destined for gin, vodka and cream liqueurs, etc.

Schizosaccharomyces pombe is found in molasses fermentations for rum production, and *Kluyveromyces marxianus* strains are employed in cheese whey fermentations for the production of white spirits such as vodka and gin. Table 16.1 summarizes different yeast species encountered in alcoholic beverage fermentations.

This Chapter will focus on the zymology aspects pertaining to distilled spirit production and will compare and contrast fermentation processes for both wine and spirits.

2 Yeasts in Production of Cereal-Based Spirits

Distilled spirits that employ cereals as their starting raw materials include: whisky (e.g. Scotch whisky), whiskey (e.g. Irish and American), vodka, gin and shochu (see Table 16.1). The cereals in question are predominantly barley, wheat, rye, maize, rice and sorghum. The starting carbohydrate in all cases is starch which cannot be fermented directly by *S. cerevisiae*. This glucose polysaccharide requires pre-hydrolysis to simple sugars prior to yeast fermentation, and this contrasts markedly from winemaking where fermentable sugars (glucose and fructose) are readily available in the grape berries and in the subsequent must. Other salient differences exist between wine and distilled cereal spirits. For example, the alcohol content of bottled spirits is approximately 3–four fold higher than in finished wines (e.g. a typical whisky would have an alcohol concentration of 40% v/v, whereas a typical table wine would be 12% v/v). Yeast strains employed for spirits and for wine are also different and the following discussion covers fermentation aspects of distilled spirits from cereals, with a special emphasis on whisky production processes. Figure 16.2 outlines the major categories of global whiskies.

One of the best selling spirit drinks in the world is Scotch whisky which is produced by fermentation of an infusion of malted barley and other cereals with strains of *S. cerevisiae* and matured over time in oak barrels (Russell and Stewart 2014; Walker and Hill 2016). There are two main types of Scotch whisky: malt whisky and grain whisky. Blended Scotch whisky is a mix of these types. Malt whisky is produced using malted barley as the cereal and enzyme source, and the fermented wash is distilled in copper pot stills. Grain whisky is produced using wheat or maize as the predominant cereal, with a small proportion (e.g. 15%) of malted barley as a source of amylolytic enzymes, and the fermented wash is distilled continuously in large "Patent" or "Coffey" stills. In the UK, Scotch Whisky has had a legal definition since 1909 (recognised by the EC in 1989) and the current (2009) Scotch Whisky Regulations define five categories of Scotch Whisky: *Single Malt, Single Grain, Blended, Blended Malt* and *Blended Grain.* Blended Scotch whisky is typically a mix of malt and grain whiskies, with some blends having a much a 50 individual malt and grain whiskies. These Regulations state that "Scotch Whisky":

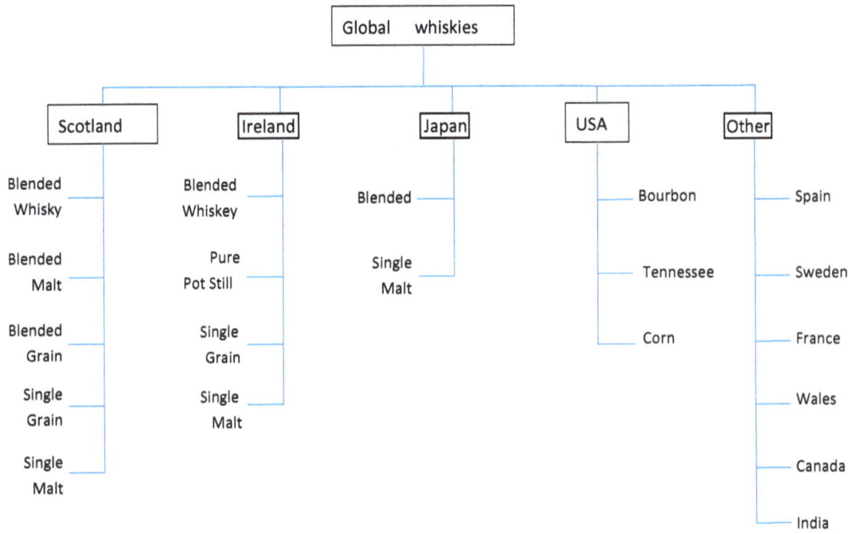

Fig. 16.2 Categories of the main global whiskies. Adapted from Walker and Hill (2016)

a. Has been distilled at a distillery in Scotland from water and malted barley (to which only whole grains of other cereals have been added) all of which have been -

   (i) processed at that distillery into a mash;
   (ii) converted at that distillery into a fermentable substrate only by endogenous enzyme systems; and
   (iii) fermented at that distillery only by the addition of yeast;

b. Has been distilled at an alcoholic strength by volume of less than 94.8%.
c. Has been matured in oak casks not exceeding 700 L for a period not less than 3 years.

Whisky processes involve the production of a sugary solution called *wort*, which is generated following the enzymatic extraction of maltose and other sugars from an aqueous mash of barley malt grist (as in the case of malt whisky) or an aqueous mash of malt and other cereals (as in the case of grain whisky). The saccharification of cereal starch is accomplished by amylolytic enzymes present in malted barley. For Scotch whisky production, exogenous (commercial) enzymes are not permitted, but for grain neutral spirit (GNS) destined for vodka or gin production, application of such enzymes is permitted. After cooling, the wort is then fermented with selected strains of *S. cerevisiae* to produce wash at around 8% v/v ethanol. For malt whisky, fresh alcohol distillate is produced following two batch distillations as shown in Fig. 16.3.

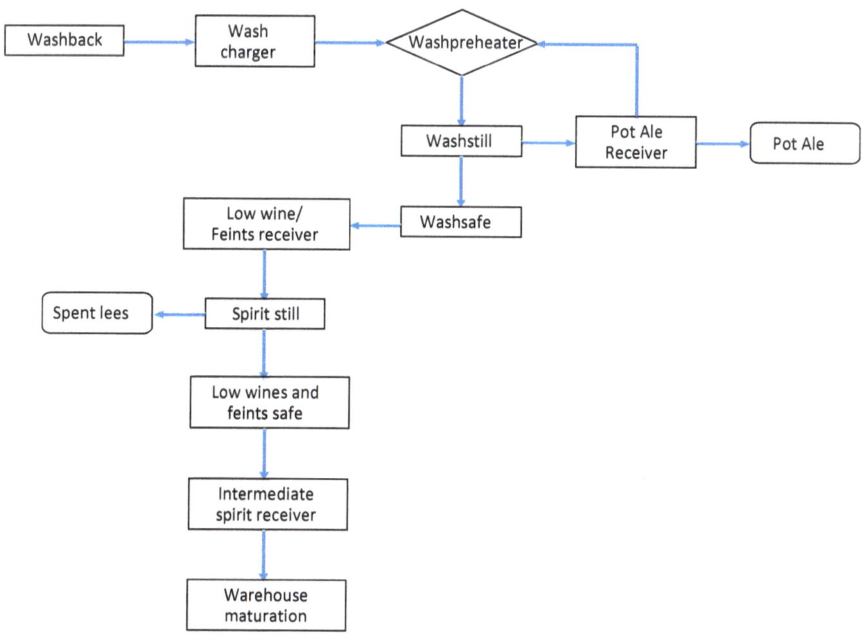

Fig. 16.3 Summary of the main processes involved in Scotch malt whisky production. Adapted from Walker and Hill (2016)

The spirit fraction typically has an alcohol concentration of 63–70% v/v and is matured in oak barrels for a minimum of 3 years (but often for 10–15 years) to impart characteristic flavour, aroma and taste to the spirit (Russell and Stewart 2014; Bryce and Stewart 2004; Murray 2017). "Single malt" Scotch whisky is such whisky produced only from malted barley and from a single distillery. Grain whisky distillations employ continuous Coffey (or Patent) stills comprising a rectifier and an analyser to produce spirit at a strength of 94.5% v/v alcohol. Grain whiskies are mainly used for blending with malt whiskies (Lea and Piggott 2003).

Regarding the yeasts used for whisky production, fermentations are conducted by specific strains of *S. cerevisiae* which convert mash sugars into ethanol, carbon dioxide and numerous secondary fermentation metabolites that collectively act as flavour congeners in the final spirit (Walker and Hill 2016). Yeast strain selection for whisky production is therefore critically important in dictating the organoleptic qualities of the final product. The same is true for wine yeasts. The fermentable sugars extracted following cereal mashing are predominantly maltose and maltotriose, in contrast to glucose, fructose and sucrose in wine musts. An important distinction between beer and whisky production is that in whisky wort preparation, because the wort is not boiled, starch degradation processes do not stop when the wort leaves the mashing vessel. Consequently, residual malt enzymes continue their amylolytic activity in fermentation. This has similarities with the Simultaneous Saccharification and Fermentation (SSF) processes typically found in fuel alcohol plants that process maize (Walker 2011).

In Scotch whisky processes, where no exogenous enzymes are allowed, maltodextrin molecules (small branched oligosaccharides) in the wort may be utilised by some whisky yeast strains. For example, a widely used Scotch whisky yeast strain, named "M type" (thought to be a hybrid of *S. cerevisiae* and *S. cerevisiae* var. *diastaticus*) possesses limited starch-debranching amylolytic activity (Watson 1993). Whisky yeast strains have been described as "*maltose + oligosaccharide type*" to reflect their properties in rapidly fermenting maltose, maltotriose and other oligosaccharides. In contrast, other beverage yeasts have been described as "*sucrose + maltose type*" to reflect fermentation of sucrose, glucose and maltose (Jones 1998). The ability to efficiently and completely ferment maltotriose is an important distinguishing characteristic of Scotch whisky yeasts. Table 16.2 outlines the main desired attributes of distiller's yeast, compared with wine yeasts.

Spontaneous fermentations are no longer conducted in modern whisky distilleries that use freshly propagated or commercially supplied pure-cultured strains of *S. cerevisiae*. This yeast may be supplied in compressed (cake) form, liquid (cream) yeast, or in dried form (Walker and Hill 2016). The pitching rate (inoculum) is generally $0.5–2 \times 10^7$ cells/mL. Unlike breweries or wineries, whisky fermentations are typically allowed to proceed for 2–3 days without precise temperature control. Wort pH in a whisky fermentation will start at pH 5–5.5 and will fall to pH 4.2–4.5 at the end of fermentation. Yeast viability at the end of fermentation is very low due to the combination of low pH, temperatures >30 °C, and high final ethanol concentrations. These factors exert considerable physiological stress on yeast (Walker and van Dijck 2006).

Table 16.2 Attributes of distiller's yeasts compared with wine yeasts

Attribute	Distillers yeast	Wine yeasts
Genetics	Some whisky yeasts are natural hybrids between *S. cerevisiae* and *S. cerevisiae* var. *diastaticus*, but generally have complex ploidy (e.g. polyploid). Some Scotch whisky is made using mixtures of distiller's and brewer's yeast cultures. *S. cerevisiae* strains found in agave juice fermentations show high genetic diversity.	Wine yeasts are generally homothallic diploids, but some strains may be polyploid or aneuploid.
Metabolism	Rapid and complete fermentation of cereal sugars. Most distillers yeasts will ferment glucose, maltose, matotriose and some limited fermentation of maltodextrins (glucoamylase activity). *Agave* juice yeasts ferment fructose and glucose. Production of desirable flavour congeners (esters, organic acids, aldehydes, higher alcohols etc.). In Scotch whisky and *Agave* fermentations, there is also a contribution to spirit flavour made by lactic acid bacteria (see text).	Vigorous fermentation desired, with correct volatile acidity, aromatic character (esters, succinic acid) and viscosity (glycerol). Low acetaldehyde and correct balance of sulphur compound production.
Stress physiology	Stress-tolerance to temperature, ethanol, osmotic pressure and competitive microbes (lactic acid bacteria) desired.	In addition to tolerance to variable temperature, osmotic pressure and pH, wine yeasts should be SO_2 tolerant.

Whisky-distillers (and oenologists), unlike brewers, do not recycle yeast. Fermented wash including yeast is distilled resulting with concomitant destruction of yeast cells. This necessitates the supply of freshly propagated yeast from separate commercial yeast organisations (Jones 1998). The specifications for Scotch whisky yeast strains include yeast cell viability, bacterial count and moisture content (Korhola et al. 1989). Some Scotch whisky distilleries formerly supplemented their distiller's yeast with a small proportion of spent brewer's yeast. The presence of brewers' yeast provides flavour benefits in terms of final spirit quality (Korhola et al. 1989) and in controlling fermentation pH due to pyruvic acid uptake (McGill 1990).

In addition to the selected distillers strain of yeast to initiate fermentation, various wild yeasts such as non-distilling strains of *S. cerevisiae, Pichia membranifaciens, Torulaspora delbrueckii* and *Candida* species may also be present in whisky fermentations. Although such yeasts are potentially problematic in affecting fermentation progress, their levels are usually kept low due to the dominance of the main production yeast strain of *S. cerevisiae.* Another wild yeast, *Dekkera bruxellensis* (anamorph *Brettanomyces bruxellensis*), which is a common contaminant in wine fermentations, may play an important role in cereal fermentations for distilled spirits production. For example, Passoth et al. (2007) found that *D. bruxellensis* dominated wheat-based distillery fermentations following out-competition of the *S. cerevisiae* starter cultures.

Due to the lack of a pre-fermentation boiling stage (as in brewing) and due to the non-aseptic processes employed, other microorganisms, notably bacteria, are present during cereal fermentations. For example, the importance of lactic acid bacteria has been shown in grain whisky processes (van Beek and Priest 2002); in wheat-based processes (Passoth et al. 2007); in corn mash-processes (Thomas et al. 2001; Smith 2017) and in rice-based processes (Watanabe et al. 2007). In a wheat distillery, Passoth et al. (2007) proposed that *Lactobacillus vini* contributed, with yeast, to an "ethanol-producing consortium". In Scotch malt whisky fermentations, limited bacterial activity plays an important role in flavour development of the final spirit. In particular, whisky producers recognise positive influences of lactic acid bacteria including *Lactobacillus plantarum* and *L. fermentum* during the latter stages of fermentation. Lactic acid bacteria (LAB) can produce lactic and acetic acids, which become esterified to ethyl lactate and ethyl acetate, respectively. These compounds impart sweet, fruity, creamy, and pineapple-like flavours to the spirit. *Lactobacillus* spp. can also produce γ-dodecalactone which imparts a "sweet and fatty" characteristic to the spirit. In other cereal-based distilled spirits such as "sour-mash" bourbons, lactic acid bacterial growth is encouraged during certain process steps to depress wort pH and impart desired flavour congeners to the final distillate (Smith 2017).

The choice of the distillers strain, together with contributions from other microorganisms, will play an important role in dictating the final flavour and aroma characteristics of cereal-based distilled spirits. In addition to the main fermentation metabolites, ethanol and carbon dioxide, numerous secondary fermentation compounds will be produced that act as important flavour congeners in the final spirit. Table 16.3 summarises the principal flavour congeners in distilled spirits, and these include: fatty acids and their esters (e.g. ethyl caprate, ethyl laurate), organic acids (e.g. succinic acid), higher alcohols or "fusel oils" (e.g. n-propanol, isoamylalcohol), aldehydes (e.g. acetaldehyde) and vicinal diketones (diacetyl).

Table 16.3 Important yeast-derived flavour congeners in distilled spirits

Compound class	Example	Flavour/Aroma
Higher alcohols	*n*-propanol Isobutanol	Alcoholic
	Iso-amyl alcohol	Pharmacy
	(3-methylbutan-1-ol)	Fusel, alcoholic, fruity, banana
	Phenylethanol	Roses, perfume
Esters	Ethyl acetate	Solvent, acetone
	Ethyl butyrate	Pineapple, banana, mango
	Ethyl caproate	Apple, aniseed
	Ethyl caprylate	Apple
	Ethyl hexanoate	Pineapple, unripe banana
	Ethyl lactate	Butter/cream
	Ethyl octanoate	Sour apple, apricot
	Iso-amyl acetate	Banana, fruity
Aldehydes	Acetaldehyde	Green apple
Vicinal diketones	Diacetyl	Butter, butterscotch
Phenolics	4-Vinyl guaiacol	Clove-like
S-Compounds	Hydrogen sulphide	Rotten eggs

Many flavour-active compounds are produced by yeast in reactions between alcohols and acyl CoA molecules but some esters, notably ethyl lactate, are linked to the presence of lactic acid bacteria. Ester production during fermentation is linked to the relative abundance of the corresponding alcohols and acyl CoAs, with ethyl acetate being the predominant ester produced, with isoamyl acetate at lower concentrations. The latter has a much lower flavour threshold than ethyl acetate and contributes a fruity (banana) aroma to beverages. Another yeast-derived group of flavour congeners important in distilled spirits (and in wine) are the fusel oils. Their concentration levels in beverages are linked to the levels of corresponding amino acids in the fermentation medium (for example, phenylalanine stimulates phenylethanol production leading to a rose-like aroma). In distilled spirits these compounds, if controlled within certain limits, are beneficial contributors to the aroma characteristics of distillates.

Further information on the origin of different flavour congeners in distilled spirits can be found in: Lea and Piggott (2003); Walker and Hughes (2010); Russell and Stewart (2014); Goodall et al. (2015) and Walker et al. (2017).

3 Yeasts in the Production of *Agave* and *Dasylirion*-Based Spirits

Agaves or *magueyes* (*Agave*) and *sotoles* (*Dasylirion*) are succulent plants that have great biological, ecological and economic importance. They belong to the family Asparagaceae, subfamilies Agavoideae and Nolinoideae, respectively, which are endemic to America. During pre-Hispanic times the integral use of these plants was so important, that the survival and cultural development of several Mesoamerican civilizations could not be explained without their existence (Fiore et al. 2005; Lappe-Oliveras et al. 2008). Different *Agave* and *Dasylirion* species have long been exploited for the production of distilled alcoholic beverages, which are produced by distillation of the fermented must of several species, and are generically known with the names *mezcal* or *sotol*, respectively (Lappe-Oliveras et al. 2008).

In México, a great diversity of agaves (175 species) exists with around 50 species being used in the production of mezcal. From these, 28 are frequently employed in artisanal production and, 14 in larger-scale commercial processes. The final product may come from the exploitation of a single agave species, or from the combination of several ones (Aguirre and Eguiarte 2013; Torres et al. 2015). Thus, the distinctive characteristics of mezcal spirit beverages depends on: the agave species or mixture of species used; the growth conditions and maturity of the plants, as well as from the cooking, fermentation, distillation and aging processes. Together, these factors give rise to a great diversity of mezcales. Some of them, such as tequila, bacanora and mezcal, are widely known, have national and international recognition and are protected by their appellation of origin (AO). However, most *mezcales* are regional products, not known outside their place of origin and do not have an AO that protects them.

The genus *Dasylirion* consists of 20 species, 16 of them endemic, which grow in several states of México (Reyes-Valdés et al. 2012). *Sotol* is produced in the states of Chihuahua, Coahuila and Durango from several *Dasylirion* species, mainly *D. durangense.*

3.1 Distilled Agave Spirits: Tequila, Mezcal, Bacanora, Raicilla and Sotol

Tequila, Mezcal, Bacanora, Raicilla and sotol are specific names of distilled beverages obtained from different *Agave* and *Dasylirion* species (Table 16.4). Tequila is the famous spirit beverage classically associated with México. The word *tequila* derives from the Nahuatl *tequillan*, from *tequitl* = tribute, work or employment and *tlan* = place, meaning a place of tribute or in which work is done. However, this word may also be associated with the *Ticuilas* a tribe who lived in the hillside of the extinct volcano Tequila, in Jalisco. The word *mezcal, mescal or mexcal* derives from the Nahuatl words *metl* = maguey or agave, and *ixcaloa* = to roast, and means roasted agave. In pre-Hispanic México the agave plants had several uses (food, fodder, medicine, construction material, textiles, and soap, among others) and for centuries they have been used to produce alcoholic beverages.

With the introduction by the Spaniards of the Philippine and Arab stills, in the second half of the sixteenth and the beginning of the seventeenth centuries, respectively, the distilled agave beverages originated and over time took their present form (Cedeño 2003). However, recently some researchers based on strong archaeological and ethnographic evidence have proposed that distillation is not of mestizo origin, but pre-hispanic, which has caused great controversy that remains to be elucidated.

The following contribution will present the different processes of elaboration of agave spirits with special emphasis on yeasts species and their contribution to the characteristics of the final product.

3.2 Tequila Categories, Types and Production Process

The most famous agave spirit from México is undoubtedly Tequila. This alcoholic beverage is obtained from distillation of the fermented must of *Agave tequilana* Weber var. azul; it is produced in the territory of AO which includes all the state of Jalisco and some regions of the states of Guanajuato, Michoacán, Nayarit and Tamaulipas (Table 16.4). The principal characteristics of this agave are: a high concentration of a complex mixture of highly branched fructooligosaccharides containing principally β (2–1) linkages (Mancilla-Margalli and López 2006) which are used as reserve carbohydrate by the plant, low fiber content, and the presence of some chemical compounds like terpenes, which contribute to the flavour and taste

Table 16.4 Bacteria and yeasts isolated from the must fermentation process of Mexican distilled *Agave* and *Dasylirion* spirits

Beverage	*Agave* species	States of production	Microbiota[1]	References
Tequila	*A. tequilana* var. azul	Jalisco, regions of the states of Guanajuato, Michoacán, Nayarit and Tamaulipas	LAB, non-*Saccharomyces* (*Candida* spp., *C. intermedia*, *C. magnoliae*, *Dekkera anomala*, *D. bruxellensis*, *Hanseniaspora* spp., *H. uvarum*, *H. vineae*, *Issatchenkia orientalis*, *Kazachstania humilis*, *Kluyveromyces lactis*, *K. marxianus*, *Meyerozyma guilliermondii*, *Pichia membranifaciens*, *Torulaspora delbrueckii*, *Zygosaccharomyces bailii*) and *Saccharomyces* (*S. cerevisiae*) yeasts In industrialized process selected *S. cerevisiae* strains	Lachance (1995) Cedeño (2003) Gschaedler et al. (2004)
Mezcal	*A. fourcroydes*	Yucatán	Non-*Saccharomyces* (*Candida parapsilosis*, *Clavispora lusitaniae*, *Debaryomyces hansenii*, *I. orientalis*, *K. marxianus*, *Meyerozyma caribbica*, *M. guilliermondii*, *Millerozyma farinosa*, *Ogataea angusta*, *P. membranifaciens*, *T. delbrueckii*, *Wickerhamomyces anomalus*) Basidiomycetous yeasts (*Rhodotorula* spp., *Rhodotorula mucilaginosa*)	Lappe et al. (2004)
Mezcal	*A. angustifolia*	Oaxaca	LAB, AAB; non-*Saccharomyces* (*Candida* spp., *C. apicola*, *C. boidinii*, *C. parapsilosis*, *C. zemplinina*, *Citeromyces matritensis*, *Cl. lusitaniae*, *D. hansenii*, *D. anomala*, *Diutina rugosa*, *Hanseniaspora* spp., *H. guilliermondii*, *H. osmophila*, *I. orientalis*, *K. lactis*, *K. marxianus*, *M. guilliermondii*, *Pichia fermentans*, *P. mandshurica*, *P. membranifaciens*, *Schizosaccharomyces pombe*, *T. delbrueckii*, *W. anomalus*, *Z. bailii*, *Z. bisporus*, *Z. rouxii*) and *Saccharomyces* (*S. cerevisiae*) yeasts Basidiomycetous yeasts (*Cryptococcus uniguttulatus*, *Naganishia albida*, *Pseudozyma prolifica*, *Rhodosporidiobolus fluvialis*, *Rhodotorula glutinis*, *Rh. mucilaginosa*, *Sporidiobolus salmonicolor*)	Andrade Meneses and Ruiz Terán (2004) Kirchmayr et al. (2017)

(continued)

Table 16.4 (continued)

Beverage	*Agave* species	States of production	Microbiota[1]	References
Mezcal	*A. salmiana* subsp. *crassispina*	San Luis Potosí	LAB, *Zymomonas mobilis*; non-*Saccharomyces* (*C. ethanolica, Cl. lusitaniae, I. orientalis, Kazachstania exigua, K. marxianus. P. fermentans, Pichia kluyveri, T. delbrueckii, Z. bailii*) and *Saccharomyces* (*S. cerevisiae*) yeasts	Escalante-Minakata et al. (2008) Verdugo Valdez et al. (2011)
Mezcal	*A. angustifolia* *A. lechuguilla* *A. montium-sancticaroli*	Tamaulipas	LAB, Non-*Saccharomyces* (*C. parapsilosis, Cl. lusitaniae, K. marxianus, M. guilliermondii, P. kluyveri, T. delbrueckii, Yamadazyma mexicana, Z. bailii*) and *Saccharomyces* (*S. cerevisiae*) yeasts Basidiomycetous yeasts (*Rh. mucilaginosa*)	Arratia Mireles (2009) Narváez-Zapata et al. (2010)
Mezcal	*A. durangensis*	Durango	Non-*Saccharomyces* (*H. uvarum, K. marxianus, T. delbrueckii, P. fermentans, Saturnispora diversa*) and *Saccharomyces* (*S. cerevisiae*) yeasts	Páez-Lerma et al. (2013)
Mezcal	*A. cupreata* *A. angustifolia*	Guerrero	Non-*Saccharomyces* (*Kz. exigua, K. marxianus, M. guilliermondii, P. kluyveri, T. delbrueckii and Z. rouxii*) and *Saccharomyces* (*S. cerevisiae*) yeasts	Kirchmayr et al. (2014)
Mezcal	*A. cupreata* *A. inaequidens* *A. angustifolia* *A. americana* *A. hookerii*	Michoacán	Non-*Saccharomyces* (*K. lactis, K. marxianus*) and *Saccharomyces* (*S. cerevisiae*) yeasts	Gallardo-Valdez et al (2008), Pérez et al. (2013)
Bacanora	*A. angustifolia*	Sonora	Non-*Saccharomyces* (*Candida blankii, C. silvatica, D. bruxellensis, Mi. farinosa, M. guilliermondii, Ogataea polymorpha, T. delbrueckii*) and *Saccharomyces* (*S. cerevisiae*) yeasts. Basidiomycetous yeasts (*Rhodotorula* sp.)	Vallejo-Córdoba et al. (2005) Zamora-Quiñonez (2006) Álvarez-Ainza et al. (2015)
Raicilla	*A. angustifolia* *A. inaequidens* *A. maximiliana*	Jalisco	Non-*Saccharomyces* (*Cl. lusitaniae, K. marxianus*) and *Saccharomyces* (*S. cerevisiae*) yeasts	Arrizon et al. (2007)
Sotol	*Dasylirion* spp.	Chihuahua Durango Coahuila	Non-*Saccharomyces* (*Dekkera* sp., *Kloeckera* sp., *Hanseniaspora* sp., *Kluyveromyces* spp., *K. marxianus, Zygosaccharomyces* sp.), and *Saccharomyces* (*S. cerevisiae*) yeasts	De la Garza-Toledo et al. (2008)

[a]Microbiota in spontaneous fermentation. LAB = Lactic acid bacteria, AAB = Acetic acid bacteria

of the final product. The elaboration process is subject to the Mexican Official Regulation NOM-006-SCFI-2012, which recognizes only two tequila categories: "*Tequila* 100%" obtained exclusively from sugars of *A. tequilana* var. azul, and "*Tequila*", produced using 51% of agave sugars and 49% from other sugar sources (sugar cane, molasses or hydrolyzed corn syrup). In each category there are five types of tequila: *tequila blanco* (silver) without maturation; *tequila joven* or *oro* (gold) containing permitted additives (oak extract, glycerin, sugar syrup) and colours (generally caramel colour); *tequila reposado* (aged) matured at least 2 months in white oak barrels (this being the most popular kind of tequila); *tequila añejo* (extra aged) and *tequila extra añejo* (ultra aged) matured for 1 or 3 years in white oak barrels, respectively. According to the Consejo Regulador del Tequila (CRT) (Tequila Regulatory Council) in 2017 the global production of tequila was 271.4 million liters. "*Tequila* 100%" represented more than 56% of the total production, and more than 80% of the annual production was exported. By mid-2018 there were 1450 registered tequila brands produced in 235 Mexican distilleries certified by the CRT, which is the council that verifies the compliance of the Official Mexican Regulation.

The process of elaboration begins with the harvest of 7–9 year old agave plants (Cedeño 2003). As with many other agave plants, *A. tequilana* var. azul is rich in carbohydrates, mainly highly branched fructans and neo-fructans (Mancilla-Margalli and López 2006; Waleckx et al. 2008). The complete plant is cut down, the leaves are removed leaving the stem and the leaf bases, what is called the head or *piña*. The heads are cooked in brick ovens heated by steam injection for 36–48 h at 100 °C; then the steam injection is suspended and the agave pieces are left in the oven for two more days to complete the cooking process. Nowadays, in most distilleries, brick ovens have been replaced by steel autoclaves to increase efficiency. The main objectives of the cooking process are: (1) To accomplish hydrolysis of fructans into simple sugars, mainly fructose, glucose and sucrose, which are easily fermented by yeasts; (2) To facilitate the milling operation and the extraction of sugars, since during cooking the agave plants acquire a soft texture (3) To generate some important chemical compounds (e.g. fusel oils) which determine the sensorial characteristics of the final product; some of these are produced by caramelisation and Maillard reactions (Cedeño 2003).

The cooked agave is milled to extract a sweet must which contains a high concentration of fructose and other fermentable sugars. In some distilleries, the milling process is still done with rudimentary mills (*tahona*), such as those used in the elaboration of mezcal. Nowadays, the mills used in the tequila industry are similar to those used in the sugarcane industry. In the last 15 years a new milling/cooking technology has been developed to extract fructans or fructose with hot water using diffusors. This technology is applied to crude crushed agave or to wash the agave fibers improving the efficiency of the sugar extraction.

The fermentation process will be discussed in detail in the Sect. 3.4. Most distilleries use stainless steel fermentation tanks (whose capacity ranges from 2000 to 120,000 L) although some still employ wooden tanks. The fermentation wort of "*Tequila* 100%" is comprised solely of agave must, with an initial sugar concentration between 4–10% v/v, depending on the amount of water added during milling.

For "*Tequila*" other sugars are added, which are previously dissolved and mixed with the agave must to obtain an initial sugar concentration of 8–16%, depending on the sugar tolerance of the yeast strain that will be used in the fermentation process. The pH of agave must oscillates around 4.5, needing no adjustment. Wort formulation is based on the composition of raw materials and the nutritional requirements for yeast growth and fermentation. Generally, a nitrogen source (urea, ammonium sulphate, ammonium phosphate) and salts (magnesium sulphate) can be added. The fermentation wort may be left to ferment spontaneously or in some distilleries selected *S. cerevisiae* strains are employed (Gschaedler et al. 2004).

Distillation involves the separation and concentration of the alcohol from the fermented wort. Tequila is obtained after two consecutive differential distillations in copper or stainless steel stills. Some distilleries also use rectification columns. During the first distillation the fermented must is split into three different products: a light product (head), a tail product or vinasses which is discarded and a slope cut product called *ordinario* (with an ethanol content of 20–30% v/v) which is subjected to a second distillation to obtain a distillate or spirit with around 55% v/v ethanol (Prado-Ramírez et al. 2005).

Finally, tequila can be matured in different ways, depending on the type of tequila to be obtained. According to the norm, this process can only be carried out in the region of AO. The regulation specifies that the maturation tanks have to be made of oak or holm oak wood with a maximum capacity of 600 L. Prior to bottling, tequila is filtered. In the case of "Tequila 100%" the product has to be bottled in the region of AO, "*Tequila*" can be exported in bulk outside of México, but when bottled, it must be labelled with the legend Made in México or Product of México.

3.3 *Mezcal, Bacanora and Sotol Categories, Types and Production Process*

For the elaboration of others spirits obtained from *Agave* and *Dasylirion* species, the general stages of the processes are the same as in tequila. The main differences between tequila and the different types of mezcal are the species of agave used as raw material, and that the elaboration process of mezcal is more artisanal than the tequila process. It is important to highlight that the word *mezcal* is the generic name of all the agave distilled beverages, which are produced in 26 of the 32 states of Mexico (Aguirre and Eguiarte 2013; Torres et al. 2015). This word refers also to a spirit with AO whose elaboration process is subjected to the Mexican Official Regulation NOM-070-SCFI-2016). The region of AO includes the states of Durango, Guerrero, Oaxaca, San Luis Potosí, Zacatecas and some regions of Guanajuato, Michoacán, Puebla and Tamaulipas (Table 16.4). The regulation stipulates that mezcal is a distilled alcoholic beverage, 100% maguey or agave obtained by distillation of the fermented juices, extracted from mature cooked heads of different agave species (*A. angustifolia, A. cupreata, A. durangensis, A. inaequidens, A. maximiliana, A. potatorum, A. salmiana*, among others) (Fig. 16.4a) harvested in the territory of

Fig. 16.4 Mezcal elaboration process. (**a**) *Agave angustifolia.* (**b**) Raw agave heads or *piñas.* (**c**) Pit oven filed with stones. (**d**) Brick oven. (**e**) Cooked agave heads or piñas. (**f**) Rudimentary mill or *tahona.* (**g**) Fermentation of cooked agave must with fiber in wooden vats. (**h**) Fermentation of agave must without fiber in a brick tank. (**i**) Distillation of fermented agave must in a rudimentary still. (**j**) Arab type still. (**k**) Mezcal aging in oak barrels (optional)

AO. Three categories are recognized *Mezcal*, *Mezcal Artesanal* (*Artisanal Mezcal*) and *Mezcal Ancestral* (*Ancestral Mezcal*); each one with six classes.

i) *Blanco o joven* (white or young), without any further processing
ii) *Madurado en vidrio* (matured in glass); Mezcal stabilized in a glass container for more than 12 months
iii) *Reposado* (aged); matured from 2 to 12 months in wooden containers of any size, form and capacity
iv) *Añejo* (extra aged); matured more than 12 months in wooden containers with less than 1000 L capacity
v) *Abocado con* (doomed with); added with authorized ingredients to give flavour, such as maguey's worm, damiana, lemon, honey, orange, among others
vi) *Destilado con* (distilled with); Mezcal must be distilled with ingredients to incorporate flavours, such as turkey or chicken breast, rabbit, *mole*, fruits, among others

In 2017 the global production of mezcal reported by the Consejo Regulador del Mezcal was 3,986,221 L, 88% of *Mezcal Artesanal*, with Oaxaca being the largest producer (87.0%). In recent years Mezcal has achieved national and international recognition, and is the second most consumed agave distilled beverage in the country after Tequila.

Bacanora and sotol are other agave spirits with government official recognition and AO. The Mexican Official Regulation NOM-168-SCFI-2004 specifies that bacanora is only produced in the state of Sonora from *A. angustifolia*. Sotol is produced in the states of Chihuahua, Coahuila and Durango. The Mexican regulation NOM-159-SCFI-2004 recognized two categories: "*Sotol 100% puro*", obtained exclusively from sugars of the *Dasylirion* spp. and "*Sotol 51%* or *sotol*", produced using 51% of *Dasylirion* sugars and 49% from other sugar sources. Four types of bacanora and sotol are allowed: *blanco* (silver), *joven* u *oro* (gold), *reposado* (aged) and *añejo* (extra aged) with the same description as the different tequila types.

Outside these areas with AO other agave spirits are produced. Raicilla is elaborated in the Western region of Jalisco with different agave species; another spirit is produced in Southern Jalisco with different varieties of *A. angustifolia.*

After harvest of the raw material (wild or cultivated) the mezcal elaboration process is similar to the one described for tequila, although there are modifications since in general it is a more artisanal process (Fig. 16.4). The raw agave heads (Fig. 16.4b) are usually cooked in pit ovens filled with stones, heated with wood and covered with earth, to impart a smoked flavour to the product, or it can be done in brick ovens or steel autoclaves, as in some distilleries of San Luis Potosí and Zacatecas (Fig. 16.4c, d). The milling of the cooked agave heads is commonly done in a rudimentary mill or *tahona* (Fig. 16.4e, f). In San Luis Potosí during the milling process water is added, and the juices are collected by gravity. In Oaxaca and Guerrero all the crushed agave is used in the fermentation process added with some water. In some parts of Guerrero and Michoacán milling is still carried out with wood or steel mallets and the juices are collected in a *canoa*, a hollow-log fermentation container (Gallardo-Valdez et al. 2008; Kirchmayr et al. 2014).

A wide variety of fermentation vessels are used: round holes carved directly in the ground, rectangular wooden crates buried in the ground, wooden vats with 1000 L capacity (Fig. 16.4g), rectangular stone or brick tanks with 3000–10,000 L capacity (Fig. 16.4h), or stainless steel tanks as those used in the tequila industry. In general, the fermentation process is carried out spontaneously with the microbiota present in the must, and lasts 1–10 days depending on the temperature, region and weather conditions. The fermented must is distillated twice in rudimentary equipment as Philippine-type stills (whose advantage is that they allow the production of spirits from a small amount of *a Agave*, and that they can be disassembled and transported quickly) or metal stills (Fig. 16.4i). In others distilleries the classic Arab-type still with serpentine is used (Fig. 16.4j).

3.4 *Yeasts and Fermentation Aspects*

3.4.1 Yeasts Identified in Natural Fermentations

Few research papers deal with the identification and characterization of the yeasts involved in the fermentation process of the different agave spirits. Lachance (1995) firstly reported the yeast communities present in a tequila distillery, where a natural fermentation was undertaken. Due to the cooking step carried out to hydrolyse fructans into fermentable sugars, the yeasts found on fresh agave plants (*Clavispora lusitaniae, Kluyveromyces marxianus, Metschnikowia agaves*, and *Pichia membranifaciens*) differed from those found on cooked agave, fresh must and crushing equipment (*Candida* spp., *Hanseniaspora vineae, P. membranifaciens, S. cerevisiae* and *Torulaspora delbrueckii*). During fermentation, a succession of different yeast species was observed. During earlier fermentation stages a rich mixture of species was detected including *Dekkera bruxellensis, Hanseniaspora guilliermondii, H. vineae, K. marxianus, P. membranifaciens, T. delbrueckii* as secondary yeasts and, *S. cerevisiae* (three biotypes) as dominant species. During the progression of the fermentation the heterogeneity of species diminished and, at the end of the fermentation the maltose-positive non-flocculent *S. cerevisiae* biotype became dominant (Table 16.4). Gschaedler et al. (2004) reported the isolation of different yeast species in 13 tequila distilleries: *Candida magnoliae, Hanseniaspora uvarum, H. vineae Issatchenkia orientalis* and *Kluyveromyces lactis* (Table 16.4). Karyotype analysis of seven *S.cerevisae* isolates showed six different profiles indicating the existence of a wide genetic heterogeneity within this species (Table 16.3).

For mezcal, studies have been published dealing with the microbiota present during the natural fermentation stage in different production regions of México. A great diversity of yeasts was found in all studies, especially in the initial fermentation stages, where several common genera and species were recognized (Table 16.4). For example, in mezcal from Yucatán, Lappe et al. (2004) reported *Candida parapsilosis, Cl. lusitaniae, Debaryomyces hansenii, I. orientalis, K. marxianus, Millierozyma farinosa, Meyerozyma caribbica, My. guilliermondii, Ogataea angusta*, *P. membranifaciens*, *Rhodotorula* spp., *Rh. mucilaginosa*, *T. delbrueckii* and *Wickerhanomyces anomalus* at the beginning of the fermentation. As fermentation progressed they also observed a dramatic reduction in yeast heterogeneity, probably due to the fermentation conditions, until the end of the fermentation, stage in which *K. marxianus* was the dominant species. During fermentation of mezcal from Oaxaca, Andrade-Meneses and Ruiz-Terán (2004) identified species of the genera *Candida, Hanseniaspora, Rhodotorula* and *Saccharomyces.* In two other mezcal factories from the same region, Kirchmayr et al. (2017) reported a great diversity of non-*Saccharomyces* species, stating great differences in yeast diversity between factories and production years (Table 16.4). In this study, the coexistence of non-*Saccharomyces* populations (*Kluyveromyces*, *Torulaspora* and *Zygosaccharomyces*) with *S. cerevisiae* until the end of fermentation was described.

In two studies of mezcal from San Luis Potosí produced with *A. salmiana* subsp. *crassispina*, a low yeast diversity was detected during fermentation. Escalante-Minakata et al. (2008) identified the yeasts *Cl. lusitaniae, K. marxianus* and *Pichia fermentans*; and the bacteria *Zymomonas mobilis* subsp. *mobilis*, *Z. mobilis* subsp. *pomaceae*, *Weissella cibaria*, *W. paramesenteroides*, *Lactobacillus farraginis*, *L. kefiri*, *L. plantarum* and *L. pontis*, highlighting the participation of a mixed microbial culture in mezcal fermentations. Verdugo Valdez et al. (2011) described species of four additional yeast genera (*Candida, Kazachstania, Saccharomyces, Torulaspora* and *Zygosaccharomyces*) (Table 16.4). As the presence of growth inhibitors (e.g. saponines) has been reported in the agave species used in this region, the raw material might have a direct impact on the presence and survival of some yeast species.

In mezcal from Tamaulipas, species of non-*Saccharomyces* (*Candida*, *Clavispora*, *Kluyveromyces*, *Meyerozyma*, *Pichia*, *Torulaspora*, *Yamadazyma*, *Zygosaccharomyces*) and *Saccharomyces* (*S. cerevisiae*) were detected in the early stages of the fermentation (Table 16.4). Only *T. delbrueckii* and *S. cerevisiae* persisted until the end of the process.

In two mezcal distilleries from Durango, Páez-Lerma et al. (2013) identified the same yeast biota (*H. uvarum*, *K. marxianus*, *P. fermentans, S. cerevisiae, Saturnispora diversa* and *T. delbrueckii*) at the beginning of the fermentation. As fermentation progressed the yeast diversity decreased and at the end of both processes only *S. cerevisiae* or *S. cerevisiae* and *T. delbrueckii* were recovered, respectively.

Kirchmayr et al. (2014) described the microbial consortia found in three mezcal distilleries in Guerrero. Besides species of the genera *Kazachstania, Kluyveromyces, Meyerozyma, Pichia, Saccharomyces, Torulaspora* and *Zygosaccharomyces,* which changed clearly between factories the authors also described the presence of high populations of lactic acid bacteria during these fermentations (Table 16.4).

In fact, the high abundance of bacterial populations has been detected in several of the mentioned studies, e.g. tequila and mezcal from San Luis Potosí, Oaxaca, Guerrero and Tamaulipas, in which different species of *Lactobacillus*, *Leuconostoc, Oenococcus* and *Weissella,* as well as acetic acid bacteria have been reported. Acid fermentation has been observed parallel to the alcoholic fermentation carried out mainly by yeasts and in some cases also by *Zymomonas mobilis* (Escalante-Minakata et al. 2008; Narváez-Zapata et al. 2010; Kirchmayr et al. 2017).

For bacanora, *S. cerevisiae* was reported as the predominant yeast during natural fermentation processes carried out in different municipalities, although several non-*Saccharomyces* yeasts species of the genera *Candida, Dekkera, Meyerozyma, Millerozyma, Ogataea, Torulaspora* and *Rhodotorula* were also identified (Table 16.4). For raicilla and sotol, few studies have been conducted to characterize the microbial diversity present during fermentation. In both beverages, *K. marxianus* and *S. cerevisiae* have been reported, in addition to *Cl. lusitania* for the former and, *Dekkera* spp., *Kloeckera* sp., *Hanseniaspora* sp., *Zygosaccharomyces* spp. for the latter Vallejo-Cordoba et al. 2005; Zamora-Quinonez 2006; Alvarez-Ainza et al 2015; De La Garza Toledo et al. 2008).

It is clear that during the initial phases of the fermentation of agave distillates from different regions of México, a great yeasts diversity is present that influences the quality and sensory properties of the final product. Yeast diversity tends to diminish towards the end, as also shown in other spirits and in wine fermentations, although several non-*Saccharomyces* yeasts persist during the whole process. Besides *S. cerevisiae*, the dominant and persistent species, *K. marxianus* and *T. delbrueckii* were reported in most of the aforementioned studies; while *Cl. lusitaniae, M. guilliermondii, P. membranifaciens* and *Z. bailii* only in some The remaining yeast genera and species were only sporadically isolated. It is clear that the yeast species involved in agave must spontaneous fermentations are variable, and no mezcal has the same microbial diversity profile, which makes these beverages unique.

3.4.2 Fermentation Development

The general practice in the production of mezcal, raicilla, bacanora and sotol is the spontaneous or natural fermentation of the must by the microbiota present in the substrate. In the tequila industry, few companies maintain the natural fermentation because the microbial consortium produces a wide diversity of volatile compounds that contribute to the sprit flavour and bouquet, despite the lower productivity of ethanol. In some distilleries, mainly in the tequila industry, the wort is inoculated with *S. cerevisiae* commercial strains (fresh baker's yeasts or dried yeast for wine, beer, or rum production). This practice is not the most appropriate because these yeasts are adapted to other substrates different from agave must and could have a negative impact on the sensory profiles of the final product. Another option is to use yeast strains isolated from natural fermentations which is deemed the most appropriate practice (Gschaedler et al. 2015). When a starter culture inoculum is used, the selected yeast strain (maintained on agar slants, lyophilized or in liquid nitrogen) is propagated in a medium with the same composition as the agave wort. The inoculum is scaled up with continuous aeration to produce enough volume to inoculate 5 or 10% of the final volume of the fermentation tank. Depending on the yeast strain, populations from 100–300 × 10^6 cells mL^{-1} are normally achieved at the onset of fermentation (Cedeño 2003). Tequila fermentation starts when the formulated must is poured in the fermentation tanks with or without yeast inoculation. Normally the temperature of the wort at the beginning of the fermentation is around 30 °C, increasing during the process and can exceed 40 °C. This has a negative effect on the yeasts, so the strains employed in the process require to be temperature tolerant. Using yeast inocula, the fermentation time ranges from 12 h in the fastest process to 3 or 4 days in the slowest. Without inocula the fermentation lasts from 2 days to 1 week. The time of fermentation has an important impact on the generation of the volatile compounds – these are lower in a fast fermentation compared with a slow fermentation. The rate of the fermentation depends mainly on the yeast strain, medium composition (sugar concentration at the beginning of the fermentation, nutriment supplementation), operating and weather conditions. Ethanol production can be detected from the beginning of the fermentation and, depending on the yeasts involved and the initial sugar concentration, ethanol concentrations reach 4–9% v/v at the end of the process (Gschaedler et al. 2015). In order to increase fermentation yields, the use of enzymes to convert residual agave polymers into fermentable sugars has been reported (Cedeño 2003). Arrizon and Gschaedler (2002) showed the possibility to achieve high fermentation efficiency (above 90%) at high initial sugar concentration or when an additional nitrogen source is added during the exponential growth phase of the yeast.

Several studies have focused on the use of non-*Saccharomyces* yeasts (alone or in co-culture with *Saccharomyces*). Díaz-Montaño et al. (2008) compared, at laboratory scale, the fermenting behaviour in *A. tequilana* juice of 5 yeasts strains: 3 of *S. cerevisiae* and 2 of *Hanseniaspora*. This study highlighted major differences between the three *S. cerevisiae* strains, especially in the production of volatile compounds, and the disadvantage of *Hanseniaspora* strains to achieve complete alcoholic fermentation, although these strains produced large amounts of esters. Valle-Rodríguez et al. (2012) showed that supplementation of the agave juice with certain specific amino acids allowed *H. vineae* to complete the fermentation. González-Robles et al. (2015) explored the use of mixed cultures of *Saccharomyces*/*Hanseniaspora* observing the contribution of the latter in aromatic profiles of tequila. Amaya-Delgado et al. (2013) carried out fermentations of *A. tequilana* juice at an industrial scale using two non-*Saccharomyces* yeasts (*P. kluyveri* and *K. marxianus*) with a fermentation efficiency higher than 85% and an interesting production of volatile compounds, mainly esters. This behaviour was confirmed at laboratory scale by Segura-García et al. (2015).

Tequila fermentations are generally carried out in open tanks, allowing evaporation of alcohol and carbon dioxide to alleviate yeast stress. Nowadays, some distilleries use cooling systems to reduce alcohol evaporation and to keep fermentation temperature tolerable for yeasts (Gschaedler et al. 2015).

Finally, it is important to mention that during tequila fermentation lactic acid bacteria from the genus *Lactobacillus*, *Lactococcus*, *Leuconostoc* and *Pediococcus* are also present and, *Acetobacter*, an acetic acid bacterium may appear in old fermented worts (Cedeño 2003). These bacteria may have a positive impact on the generation of volatile compounds, as has been reported in whiskey, cider, and wine fermentations. However, if the bacterial populations are too large, they could affect the ethanol yield and produce some undesirable compounds.

In the case of mezcal, bacanora and raicilla fermentations, these are generally carried out with the agave fibers present (except for mezcal from San Luis Potosí), increasing the fermentation time. A general characteristic of these processes is that they are completely spontaneous, without any control of the key parameters of fermentation, such as sugars concentration or temperature. In San Luis Potosí, the fermentation of *A. salmiana* subsp. *crassispina* juice is carried out without the agave fibers, with low initial sugar concentration and is generally induced by the addition of a mixed inoculum containing low counts ($>1 \times 10^6$ cells mL^{-1}) of non-*Saccharomyces* yeasts and bacteria. Verdugo Valdez et al. (2011) observed a short fermentation (around 30 h), with low residual sugars concentration and yeast population between 10–15 $\times 10^6$ cells mL^{-1}. The amount of ethanol and volatile compounds were relatively low and were directly related to the initial sugar content of the agave must. From Oaxaca, where spontaneous fermentation is generally carried out with agave fiber, Kirchmayr et al. (2017) published a study about mezcal production process in two different distilleries, in two consecutive years. Fermentation kinetics and volatile compound generation differed markedly between the studied processes (all of them spontaneous), mainly due to the lack of control during the fermentation and changes in the ambient condition (mainly temperature). This study pointed out great differences in the initial sugar concentration and total yeast populations (that ranged from 12 to 180 $\times 10^6$ cells mL^{-1}). Low fermentation efficiencies were observed due to high concentration of residual sugars at the end of the fermentation. Nevertheless, this study also demonstrated that the employment of wild inocula in the fermentation of cooked agave juice may be a good practice in order to increase alcoholic fermentation efficiency.

There are very few *in situ* fermentation studies of agave distillates. However, in spontaneous fermentations, in which non-*Saccharomyces*, *Saccharomyces*, lactic and acetic acid bacteria participate, although they sometimes have low conversion rates of sugars to ethanol, they do result in beverages with complex sensory profiles that are currently sought by a new type of consumer. Some of the yeasts isolated from these spontaneous fermentations display interesting characteristics, compared with wine strains. Fiore et al. (2005) evaluated parameters of technological interest, such as SO_2 and copper resistance, ethanol tolerance and enzymatic activities in *Candida krusei*, *C. magnolia*, *Kloeckera africana*, *K.. apiculata*, and *S. cerevisiae* strains isolated from agave, sotol and grape must. All agave strains were more resistant to SO_2 than wine strains, and non-*Saccharomyces* agave yeasts were more tolerant to ethanol. This behaviour is not common in non-*Saccharomyces* associated with alcoholic beverages. In contrast, all the non-*Saccharomyces* strains showed similarities in β-glucosidase and β-xylosidase activities, except *C. krusei*. Remarkable characteristics were the β-glucosidase activity of a *S. cerevisiae* strain and β-xylosidase activity of a *C. krusei* strain, both isolated during the fermentation of sotol carried out with must and fiber. This special condition could have caused the adaptation of the yeasts that would explain these activities.

3.4.3 Production of Aromatic Compounds

Regarding the process of elaboration of these spirits, the production of volatile aromatic compounds in such beverages is influenced by several factors. These include raw materials, fructan hydrolysis processes (oven, acidic or enzymatic hydrolysis), spontaneous or directed fermentations, distilling systems (still or column) and finally aging (Table 16.5).

Raw materials used in agave distilled beverages play an important role in the development of the sensory characteristics of the final product. Compounds including terpenes, alcohols, esters, and acids derive from the raw materials (Prado-Jaramillo et al. 2015). Peña-Álvarez et al. (2004) identified diverse terpenes in different agave species used in mezcal (7 in *A. angustifolia*, 9 in *A. salmiana*) and tequila production (32 in *A. tequilana*). Arrizon et al. (2007) determined the volatile compounds in tequila and raicilla. Sixteen terpenes were found in both beverages mainly: β-myrcene, isocineole, linalool, 1-terpineol, 4-terpineol, citronellol, nerol, geraniol and nerolidol; 15 more were detected only in raicilla, predominantly α-pinene, camphene, limonene, γ-terpinene, ρ-cymene, myrcenol, neryl acetate, geranyl acetate, α-eudesmol. From these, geraniol, linalool, limonene, myrcenol, γ-terpinene, 4-terpineol, were among others are found in the final product. This highlights the impact of the raw material on the sensory profiles of agave distillates (Table 16.5).

During production of agave distillates, it is necessary to carry out the hydrolysis of fructans, which can be accomplished by a cooking process. During this process agave heads and/or agave juice are exposed to temperatures between 95 °C and 121 °C, causing caramelisation and Maillard reactions. Some compounds that have been identified in cooked agave juice are: 5-hydroxymethylfurfural, methyl-2-furonoate and 2, 3-dihydroxy-3,5-dihydro-6-methyl-4 (H) -pyran-4-one (Mancilla-Margalli and López 2002). Prado-Jaramillo et al. (2015) reported the presence of acids, aldehydes, ketones, esters, and mainly furans (1-furan-2-yl-ethanone, 5-acetoxymethyl-furfural, 5-methyl-furfural, furfural and hydroxymethylfurfural) and terpenes (α-terpineol, Δ-cadinene, linalool, nerolidol, ocimene and γ-terpinene). These are compounds that persist in the distilled and aged beverage (Tables 16.5).

Another compound generated during cooking or the agave fructans hydrolysis stage is methanol, whose concentration is stipulated in the regulations of tequila and mezcal. Due to the cooking conditions (high temperatures and low pH) methoxyl groups are released from the pectin present in the agave heads, and methanol is formed in the presence of water. This alcohol is partially separated during distillation (Prado-Ramírez et al. 2005).

The most studied stage in agave distillates production is fermentation. Usually, at the beginning of the process a high yeast diversity is present, mainly non-*Saccharomyces* species (of the genera *Candida*, *Hanseniaspora*, *Meyerozyma*, *Kluyveromyces*, *Pichia*, *Torulaspora*, *Zygosaccharomyces*, etc.) which produce a large amount of volatile compounds and flavour congeners that impact in the sensory quality of the agave spirits (Table 16.5). The generation of these compounds depends on different factors, including: agave species; preparation of the must with

Table 16.5 Volatile compounds production along the elaboration process of agave distilled beverages

Process stage	Beverage	Agave used	Volatile compounds	References
Raw material	Tequila	*A. tequilana* var. azul	Terpenes, alcohols, furans, acids, aldehydes, esters, hydrocarbures	Prado-Jaramillo et al. (2015)
	Mezcal, Tequila	*A. angustifolia, A. salmiana, A. tequilana* var azul	Acids, esters, terpenes	Peña-Álvarez et al. (2006)
	Mezcal	*A. salmiana* subsp. *crassispina, A. salmiana* var. *salmiana, A. angustifolia, A. cupreata, A. karwinskii*	Acids, lipids	Martínez-Aguilar and Peña-Álvarez (2009)
Cooking/ hydrolysis	Tequila	*A. tequilana* var. azul	Furans, pyrans, aldehydes, nitrogen and sulfur compounds terpenes, alcohols, furans, acids, aldehydes, esters, hydrocarbures	Mancilla-Margalli and López (2002)
				Prado-Jaramillo et al. (2015)
	Mezcal	*A. salmiana*	Furans	Garcia-Soto et al. (2011)
Fermentation	Tequila	*A. tequilana* var. azul	Terpenes, alcohols, higher alcohols, furans, acids, esters, aldehydes, hydrocarbures	Arrizon et al. (2006)
				Arellano et al. (2008)
				Díaz-Montaño et al. (2008) Pinal et al. (2009) Morán-Marroquín et al. (2011) Valle-Rodríguez et al. (2012) Amaya-Delgado et al. (2013) González-Robles et al. (2015) Prado-Jaramillo et al. (2015) Segura-García et al. (2015)
	Mezcal	*A. angustifolia*	Esters, higher alcohols, aldehydes	Kirchmayr et al. (2017)
	Mezcal	*A. salmiana* subsp. *crassispina*	Esters, higher alcohols, aldehydes	Verdugo Valdez et al. (2011) De León-Rodríguez et al. (2008)
	Mezcal	*A. angustifolia, A. potatorum*	Alcohols, esters, ketones, acids, furans	Vera-Guzmán et al. (2009), (2012)

(continued)

Table 16.5 (continued)

Process stage	Beverage	Agave used	Volatile compounds	References
	Tequila, Raicilla	*A. tequilana* var. azul, *A. maximilana*, *A. inaequidens*	Alcohols, esters, aldehydes	Arrizon et al. (2007)
	Mezcal	*A. durangensis*	Alcohols, higher alchohols, esters, acids, acetals, ketones, aldehydes, esters, terpenes	Soto-García et al. (2009) Rutiaga-Quiñones et al. (2012) De los Ríos-Deras et al. (2015) Núñez-Guerrero et al. (2016)
	Mezcal	*Agave* spp.	Acetals, acids, alcohols, esters, ketones, aldehydes, phenols, terpenes	Molina-Guerrero et al. (2007)
	Mezcal	*A. angustifolia, A. lechugilla* *A. americana*	Acids	Narváez-Zapata et al. (2010)
	Bacanora	*A. angustifolia*	Alcohols, aldehydes, esters,	Álvarez-Ainza et al. (2013)
	Mezcal	Definied wort medium	Alcohols, aldehydes, esters	Arrizon et al. (2006)
Distillation	Mezcal	*A. salmiana*	Esters, higher alcohols, furans, acids, aldehydes	De León-Rodríguez et al. (2006)
	Mezcal	*A. angustifolia, A. potatorum*	Alcohols, esters, ketones, acids, furans, terpens	Vera-Guzmán et al. (2009), (2018)
	Tequila, Raicilla	*A. tequilana* var. azul, *A. maximiliana*, *A. inaequidens*	Terpenes	Arrizon et al. (2007)
	Raicilla, Sisal, Tequila, Mezcal, Bacanora, Sotol	*A. maximiliana,A. sisalana*, *A. tequilana* var. azul, *A. salmiana*, *A. potatorum, A.angustifolia*, *A. durangensis, A. angustifolia* *Dasylirion* spp.	Esters, higher alchohols, furans, acids, aldehydes	De León-Rodríguez et al. (2008)
	Tequila, Mezcal, Bacanora, Sotol	*A. tequilana* var. azul, *A. angustifolia*, *Dasylirion* spp.	Alcohols, esters, aldehydes	Lachenmeier et al. (2006)
	Sotol	*Dasylirion* spp.	Alcohols, esters, acids, furfural, ketones	De la Garza et al. (2010)

	Mezcal	*A. durangensis*	Esters, higher alchohols, acids aldehydes, tepens,, ketones	De los Ríos-Deras et al. (2015)
	Tequila	*A. tequilana* var. azul	Alcohols, esters, aldehydes, furans, terpenes, acids, hydrocarbures	Prado-Ramírez et al. (2005) Ceballos-Magaña et al. (2013) Prado-Jaramillo et al. (2015)
Aging	Tequila	*A. tequilana* var. azul	Acetals, terpenes, aldehydes, esters, ketones, furans phenols, acids, hydrocarburs, alcohols	López-Ramírez et al. (2013) Ceballos-Magaña et al. (2013) Prado-Jaramillo et al. (2015) González-Robles and Cook (2016)
	Tequila, Mezcal	*A. tequilana* var. azul, *A. angustifolia*	Furans	Muñoz-Muñoz et al. 2010)

hydrolyzed agave juice, with or without bagasse fibers; spontaneous or induced fermentation; yeast strain; fermentation conditions (temperature, dissolved oxygen, pH, nitrogen concentration), among others, (Arrizon and Gschaedler 2002, 2006; Arellano et al. 2008; Díaz-Montaño et al. 2008; Verdugo Valdez et al. 2011; López-Álvarez et al. 2012; Morán-Marroquín et al. 2011; Amaya-Delgado et al. 2013; Segura-García et al. 2015; González-Robles et al. 2015; Kirchmayr et al. 2017) (Tables 16.5).

During agave juice fermentation higher alcohols are the most important aromatic compounds produced, representing between 50–90% of the total of volatile compounds produced during this stage. These include, mainly: amyl alcohol, 1-propanol, isobutanol and furfuryl alcohol (Gschaedler et al. 2015). Factors that influence their production are the yeast strain (Arellano et al. 2008; Díaz-Montaño et al. 2008; Segura et al. 2015), high fermentation temperature and low nitrogen concentration (Arellano et al. 2008; Arrizon et al. 2006). Esters are the most diverse compounds found, and their production depends mainly on the yeast strain. Non-*Saccharomyces* species as *K. marxianus*, *P. kluyveri* and *T. delbrueckii* have been reported to produce more esters than *S. cerevisiae* (Amaya-Delgado et al. 2013; Segura-García et al. 2015; Núñez-Guerrero et al. 2016). However, anaerobic conditions (Morán-Marroquín et al. 2011) as well as high nitrogen concentrations, can influence the production of these compounds in *A. tequilana* and *A. durangensis* juices (Arrizon and Gschaedler 2002; Rutiaga-Quiñones et al. 2012). At the end of tequila fermentation, Prado-Jaramillo et al. (2015) detected 71 compounds belonging to alcohols, acids, terpenes, furans, aldehydes, ketones and esters (Table 16.5).

Other volatile compounds produced during agave must fermentation are: acetals, aldehydes, acids, ketones, phenols, hydrocarbons, sulfur, etc., although they have been identified, have not been quantified, and the parameters that influence their production are still unknown (Prado-Jaramillo et al. 2015) (Table 16.5).

It is important to highlight that agave species, place of origin, age, year season, spontaneous fermentation, and fermentation conditions can change the profile of volatile compounds produced (Pinal et al. 2009). The different yeast and bacteria species that participate in the spontaneous fermentation of agave-spirits can also impact the production of non-volatile compounds such as: ethyl acetate, ethyl lactate, acetic acid and lactic acid, that modify the titratable acidity of the must (Kirchmayr et al. 2017; Escalante-Minakata et al. 2008; Narváez-Zapata et al. 2010; Páez-Lerma et al. 2013).

There are few reports regarding the production and recovery of volatile compounds during the distillation of agave spirits (Prado-Ramírez et al. 2005). In these beverages more than 200 compounds have reported mainly: alcohols, acids, terpenes, furans, aldehydes, ketones and esters, whose profile varies depending on the species of agave used, its region of origin, the fermentation process and distillation conditions (Table 16.5) (Álvarez-Ainza et al. 2013; Prado-Jaramillo et al. 2015).

Agave distillates can be aged in wooden (oak, white oak, holm oak) barrels for months or years, depending of the type of distillate to be obtain (Fig 16.4k). In aged agave spirits more 327 compounds have been identified (Prado-Jaramillo et al. 2015). During tequila aging the main compounds detected are: higher alcohols, methanol, esters, furans, gallic acid, vanillin, syringaldehyde, synapinaldehyde, coniferaldehyde, syringic acid, ferulic acid, esculetin, scopoletin, cis/trans whisky lactones, guaiacol, 4-ethyl guaiacol and, vainillin (González-Robles and Cook 2016).

3.5 *Recent Developments and Future Prospects*

The great number of agave species and their many uses are a natural and cultural symbol of México. For more than 10,000 years these plants have been used for different purposes (nutritional, medicinal, fibers extraction, production of alcoholic beverages, construction, etc.). This emphasises why the diversity of agaves must be protected to maintain their ecology, ethnobotany and evolution, ensuring their availability to support the economic and cultural needs of the Mexican people.

The production processes of *Agave* and *Dasylirion* spirits have to be understood and precisely described. The genetic diversity of the plants used as raw material as well as the production and composition of the must have to be determined. The microbiota present in the must that participates in the complex fermentation process has to be phenotypically and genotypically characterized, and their potential as starter cultures. Finally, the precise roles of microbes during spontaneous fermentations in the production of ethanol and aromatic compounds remains to be evaluated.

4 Yeasts in Production of Miscellaneous Spirits

This section covers yeast and fermentation aspects of selected miscellaneous distilled spirits. It is outwith the scope of this Chapter to provide a comprehensive coverage of these topics and the reader is directed to several publications and websites for more detailed information (e.g. for fruit-based spirits: http://www.pediacognac.com/en/; for sugar/molasses-based spirits: Piggot (2003); and for whey-based spirits: O'Shea (2003).

4.1 *Yeasts in Production of Fruit-Based Spirits*

Various spirits are produced following the distillation of fermented fruit sugars. For example, brandies are distilled wines and eau-de-vies are distilled fruit beverages. The fruits in question are those from which fermentable sugars, principally glucose and fructose, can be easily extracted and include grapes, apples, plums, apricots and several others.

For brandies, including the best-known example Cognac, the starting material for the distillation is wine. Yeast and fermentation aspects of wine production have been well covered in previous chapters in this book and the reader is also directly to the recent elegant paper on wine yeasts by Pretorius (2016).

Several naturally-occurring and selected yeasts are involved in grape fermentations for spirit production, principally including: *S. cerevisiae*, *Candida famata* and *K. apiculata*. For cognac production, the Bureau National Interprofessionnel du Cognac (BNIC, http://www.bnic.fr/cognac/) in France recommends 8 strains of yeast (Dr L. Lurton, BNIC, personal communication) which will produce different concentrations of secondary fermentation metabolites that act as flavour congeners in cognac spirits (e.g. ethyl acetate, isoamyl acetate, higher alcohols, 2-phenyethanol, ethyl laurate etc.). Ester formation by cognac yeasts is a desirable attribute as these compounds impart characteristic fruity and floral aromas to the spirit. The choice of yeast strain is therefore critical in defining spirit quality.

A typical composition of a freshly-distilled wine spirit like cognac would comprise: ~72% v/v ethanol; ~28% v/v water and < 1% volatile substances. Of the latter, 80 olfactive zones have been characterized by gas chromatography-olfactometry and many of these have been identified (e.g. in the following chemical classes: alcohol, esters, aldehydes, norisoprenoïds, terpenes, etc.) in concentrations ranging from nanogram/l to several hundred milligrams/l. Although cognac aroma originates from the grapes, distillation and ageing, the role of yeast and progress of fermentation is of paramount importance for cognac sensory characteristics. The BNIC have developed a cognac aroma wheel that depicts the cycle of the seasons (Fig. 16.5).

Cognac wine fermentations are complete within 4–8 days, depending on temperature which also influences volatile flavour and aroma compounds in the spirit. For example, increasing temperature (e.g. 18–22 °C) results in elevated levels of higher alcohols, but decreasing levels of isoamyl acetate. In order to have some control over congener profiles and also off-flavours, the yeasts employed for cognac fermentations (at an inoculation rate of around 1×10^6 cells/mL) must be able to initiate alcoholic fermentation of grape must rapidly to prevent growth of contaminant microbes. This is especially important since SO_2 is not allowed in cognac production processes. In addition, the level of utilisable nitrogen in the grape must will dictate the kinetics and extent of yeast growth and nitrogen deficiency can be addressed by supplementation (within defined limits) with ammonium salts. Rapid yeast growth also exerts some degree of control over bacterial malolactic fermentation. Generally speaking, distillers producing wine-based spirits such as cognac must select the most suitable strains of *S. cerevisiae* to dominate the fermentation and to liberate the correct balance of congeners into the spirit. In other words, the sensory profile of a wine spirit, as it is with other alcoholic beverages, is highly yeast strain-dependent.

4.2 Yeasts in Production of Sugarcane-Based Spirits

Various distilled spirits can be produced from sugar cane. Those emanating from the fermentation of the raw sugar cane juice are spirits such as Rhum Agricole (from Réunion and Martinique) and Cachaça (from Brazil), and those emanating from

Fig. 16.5 Cognac aroma depicting the cycles of the seasons From: http://www.bnic.fr/cognac/ © BNIC / Gérard MARTRON

sugar cane molasses are spirits collectively called rum. There are several styles of rum, classed as white rum (having no colour, generally unaged in oak wood barrels); amber rum (with a golden colour, typically aged in oak barrels for between 1–3 years) and dark rum (typically aged 3 or more years). Molasses (or sugar cane juice) fermentations can either be spontaneous, relying on natural microorganisms (wild yeasts and bacteria) indigenous to the raw materials and distillery environments, or can be specially selected *S. cerevisiae* strains to be used as starter cultures.

Fermentations can either be batch, continuous or semi-continuous processes, but in all cases, the main fermentable sugar available to yeast is sucrose. Therefore, yeasts with a high invertase activity are employed to facilitate sucrose hydrolysis to glucose and fructose. In addition to *S. cerevisiae,* other yeasts play important roles in rum fermentations such as *Schizosaccharomyces pombe* (Pech et al. 1984; Fahrasmane et al. 1988).1., This fission yeast can impart desirable flavour and aroma characteristics, particularly in dark, heavy rum styles. Compared with *S. cerevisiae*, *Schizo. pombe* conducts slower fermentations, produces less higher alcohols and fatty acids, but more esters. This yeast is favoured by low pH and higher sugar concentrations in molasses. *Schizo. pombe* also has interesting enological properties such as malo-ethanolic fermentation (Benito et al. 2012).

4.3 *Yeasts in Production of Whey-Based Spirits*

There are a variety of beverages, both alcoholic and non-alcoholic, derived from cheese whey, a by-product of the cheese-making process (Jeličić et al. 2008). In addition to whey "beer" and whey "wine", distilled spirits are also made from cheese whey (O'Shea 2003). In all cases, the fermentation of lactose in the whey is conducted by lactose-fermenting yeasts, notably *K. marxianus* (Walker and O'Neill 1990). The spirits include whey-based vodka and such beverages are produced on a large scale in several countries (e.g. Ireland, New Zealand, Turkey). Because whey only contains around 5% w/v lactose, it has to be concentrated (following ultrafiltration to remove whey protein) using reverse osmosis to increase the final alcohol concentration in the beer prior to distillation. The alcohol levels produced in whey fermentations are also restricted by the relatively low ethanol tolerance of *Kluyveromyces* spp. employed and this is an area for further yeast research.

5 Conclusions and Future Prospects

The diversity of distilled spirits is linked to the diverse starting raw materials and the main available sugars for yeast fermentation, which are: maltose (for whiskies), glucose and fructose (for brandies), sucrose (for rums), lactose (for cheese whey-based spirits) and fructose (for tequila and mezcal). The choice of yeast strain for such spirits must therefore reflect specific sugar fermentability. For example, a fructophilic yeast would not be appropriate for malt wort fermentations for whisky.

Many developments have taken place aimed at improving the efficiency of sugar conversion to alcohol, and in selecting new strains of *S. cerevisiae* to impart desirable sensory characteristics to spirit beverages. Walker et al. (2012) have identified desirable characteristics for Scotch whisky distilling yeast strains.

Although scientific advances in knowledge of yeast genetics and molecular biology as applied to brewing and winemaking have been made in recent times this has proved of very limited practical value to the distiller. In particular, strain engineering employing recombinant DNA techniques has not found favour for production of spirit beverages, mainly due to public perception issues. In other words, the constraints in employing GM yeast strains in this field are primarily sociological, rather than technological. *Self-cloned* yeast strains, involving yeast-yeast genetic modification may be more attractive to consumers than yeast transformation with non-yeast genes and may also be more favourable from a regulatory standpoint (Argyros and Stonehouse 2017). Where such genetically manipulated (GM) strains are being successfully exploited is in the fuel ethanol sector (see Walker and Walker 2018). Nevertheless, some molecular biological techniques applied to beverage spirit yeast strains have proved useful. For example, proteomic analysis of whisky yeast strains during fermentation has provided insight into stress responses of yeast during fermentation of industrial grain mashes (Hansen et al. 2006). Molecular genetics and bioinformatics applied to industrial strains of *S. cerevisiae* represent powerful tools in monitoring and control of fermentations for beer, wine and spirit production (Bond and Blomberg 2006). Future research into yeast physiology and genetics will lead to a deeper understanding of *S. cerevisiae* strains exploited for spirit fermentations. In turn, novel yeast strains with interesting new attributes with improved fermentation performance and flavour quality are on the horizon.

References

Aguirre, D., & Eguiarte, L. (2013). Genetic diversity, conservation and sustainable use of wild *Agave cupreata* and *Agave potatorum* extracted for mezcal production in Mexico. *Journal of Arid Environments, 90*, 36–44.

Álvarez-Ainza, M. L., González-Ríos, H., González-León, A., Ojeda-Contreras, A. J., Valenzuela-Quintanar, A. I., & Acedo-Félix, E. (2013). Quantification of mayor volatile compounds from artisanal agave distilled: Bacanora. *American Journal of Analytical Chemistry, 4*, 683–688.

Álvarez-Ainza, M. L., Zamora-Quiñonez, K. A., Moreno-Ibarra, G. M., & Acedo-Félix, E. (2015). Genomic diversity of *Saccharomyces cerevisiae* yeasts associated with alcoholic fermentation of Bacanora produced by artisanal methods. *Applied Biochemistry and Biotechnology, 175*(5), 2668–2676.

Amaya-Delgado, L., Herrera-López, E. J., Arrizon, J., Arellano-Plaza, M., & Gschaedler, A. (2013). Performance evaluation of *Pichia kluyveri, Kluyveromyces marxianus* and *Saccharomyces cerevisiae* in industrial tequila fermentation. *World Journal of Microbiology and Biotechnology, 29*(5), 875–881.

Andrade Meneses, O. E., & Ruiz Terán, F. (2004, August). Study of yeast populations in a mezcal fermentation. In *Abstracts of the 11th international congress on yeasts*. Yeasts in Science and Technology, The Quest for Sustainable Development., Rio de Janeiro, 15–20.

Arellano, M., Pelayo, C., Ramírez, J., & Rodríguez, I. (2008). Characterization of kinetic parameters and the formation of volatile compounds during the tequila fermentation by wild yeasts isolated from agave juice. *Journal of Industrial Microbiology & Biotechnology, 35*(8), 835–841.

Argyros, D. A. & Stonehouse, E. A. (2017). Yeast strain development for alcohol production. In: G. M. Walker, C. Abbas, W. M. Ingledew & C. Pilgrim (Eds.), *The alcohol textbook* (6th ed.). (pp. 287–297). Duluth: Ethanol Technology Institute.

Arratia Mireles, J. M. (2009). Diversidad genética de levadura de mezcal tamaulipeco. M Sc Dissertation, Centro de Biotecnología Genómica, Instituto Politécnico Nacional. Tamaulipas.

Arrizon, J., & Gschaedler, A. (2002). Increasing fermentation efficiency at high sugars concentrations by supplementing an additional source of nitrogen during the exponential phase of the tequila fermentation process. *Canadian Journal of Microbiology, 48*, 965–970.

Arrizon, J., Fiore, C., Acosta, G., Romano, P., & Gschaedler, A. (2006). Fermentation behaviour and volatile compound production by agave and grape must yeasts in high sugar *Agave tequilana* and grape must fermentations. *Antonie Van Leeuwenhoek, 89*(1), 181–189.

Arrizon, J., Arizaga, J. J., Hernández, R. E., Estarrón, M., & Gschaedler, A. (2007). Production of volatile compounds in tequila and raicilla musts by different yeasts isolated from Mexican agave beverages. In M. H. Tunick & E. González de Mejía (Eds.), *Hispanic foods chemistry and flavor* (pp. 167–177). Washington: American Chemical Society Symposium Series 946.

Benito, S., Palomero, F., Morata, A., Claderon, F., & Suarez-Lepe, J. A. (2012). New applications for *Schizosaccharomyces pombe* in the alcoholic fermentation of red wine. *International Journal of Food Science & Technology*. https://doi.org/10.1111/j.1365-2621.2012.03076.

Bond, U., & Blomberg, A. (2006). Principles and applications of genomics and proteomics in the analysis of industrial yeast strains. In A. Querol & G. H. Fleet (Eds.), *The yeast handbook. Yeasts in food and beverages* (pp. 175–213). Berlin & Heidelberg: Springer-Verlag.

Bryce, J. H., & Stewart, G. G. (2004). *Distilled spirits: Tradition and innovation*. Nottingham: Nottingham University Press.

Ceballos-Magaña, S. G., de Pablos, F., Jurado, J. M., Martín, M. J., Alcázar, Á., Muñiz-Valencia, R., Gonzalo-Lumbreras, R., & Izquierdo-Hornillos, R. (2013). Characterisation of tequila according to their major volatile composition using multilayer perceptron neural networks. *Food Chemistry, 136*(3–4), 1309–1315.

Cedeño, M. (2003). Tequila production from agave: historical influences and contemporary processes. In K. A. Jacques, T. P. Lyons, & D. R. Kelsall (Eds.), *The alcohol textbook* (4th ed., pp. 223–245). Nottingham: Nottingham University Press.

De la Garza, H., Buenrostro, J., Reyes-Vega, M., Rodríguez, R., Martínes, D. G., & Aguilar, C. N. (2010). Chemical profile of sotol analyzed by solid phase microextraction-gas chromatography. *American Journal of Agricultural and Biological Science, 5*(3), 261–268.

De la Garza-Toledo, H., Martínez, M., Lara, L., Rodríguez-Herrera, R., Rodríguez-Martínez, J., & Aguilar, C. N. (2008). Production of a Mexican alcoholic beverage: Sotol. *Research Journal of Biological Sciences, 3*(6), 566–571.

De León-Rodríguez, A., González-Hernández, L., de la Barba, R. A., Escalante-Minakata, P., & López, M. G. (2006). Characterization of volatile compounds of mezcal, an ethnic alcoholic beverage obtained from *Agave salmiana*. *Journal of Agricultural and Food Chemistry, 54*, 1337–1341.

De León-Rodríguez, A., Escalante-Minakata, P., de la Barba, R. A. P., & Blaschek, H. P. (2008). Optimization of fermentation conditions for the production of the mezcal from *Agave salmiana* using response surface methodology. *Chemical Engineering & Processing: Process Intensification, 47*(1), 76–82.

De los Ríos-Deras, G. C., Rutiaga-Quiñones, O. M., López-Miranda, J., Páez-Lerma, J. B., López, M., & Soto-Cruz, N. O. (2015). Improving *Agave duranguensis* must for enhanced fermentation. c/n ratio effects on mezcal composition and sensory properties. *Revista Mexicana de Ingeniería Química, 14*(2), 363–371.

Díaz-Montaño, D. M., Délia, M. L., Estarrón-Espinosa, M., & Strehaiano, P. (2008). Fermentative capability and aroma compound production by yeast strains isolated from *Agave tequilana* Weber juice. *Enzyme and Microbial Technology, 42*(7), 608–616.

Escalante-Minakata, P., Blaschek, H. P., De la Barba, R. A. P., Santos, L., & De León-Rodríguez, A. (2008). Identification of yeasts and bacteria involved in the mezcal fermentation of *Agave salmiana*. *Letters in Applied Microbiology, 46*(6), 626–630.

Fahrasmane, L., Ganou-Parfait, B., & Parfait, A. (1988). Yeast flora of Haitian rum distilleries. *MIRCEN Journal of Applied Microbiology and Biotechnology, 4*, 239–241.

Fiore, C., Arrizon, J., Gschaedler, A., Flores, J., & Romano, P. (2005). Comparison between yeasts from grape and agave musts for traits of technological interest. *World Journal of Microbiology and Biotechnology, 21*, 1141–1147.

Gallardo-Valdez, J., Gshaedler-Mathis, A., Chazaro-Basañes, M. J., Rodríguez Domínguez, J. M., Tapia-Campos, E., Villanueva-Rodríguez, S., Salado-Ponce, J. H., Villegas-García, E., Medina-Niño, R., Aguirre-Ochoa, M., Vallejo-Pedraza, M. (2008). *La producción de mezcal en el Estado de Michoacán*. Gobierno del Estado de Michoacán, Guadalajara, México: Centro de Investigación y Asistencia en Tecnología y Diseño del Estado de Jalisco AC (CIATEJ).

Garcia-Soto, M. J., Jimenez-Islas, H., Navarrete-Bolaños, J. L., Rico-Martínez, R., Miranda-López, R., & Botello-Álvarez, J. E. (2011). Kinetic study of the thermal hydrolysis of *Agave salmiana* for mezcal production. *Journal of Agricultural and Food Chemistry, 59*(13), 7333–7340.

González-Robles, I. W., & Cook, D. J. (2016). The impact of maturation on concentrations key odour active compounds which determine the aroma of tequila. *Journal of the Institute of Brewing, 122*(3), 369–380.

González-Robles, I. W., Estarrón-Espinosa, M., & Díaz-Montaño, D. M. (2015). Fermentative capabilities and volatile compounds produced by *Kloeckera/Hanseniaspora* and *Saccharomyces* yeast strains in pure and mixed cultures during *Agave tequilana* juice fermentation. *Antonie Van Leeuwenhoek, 108*(3), 525–536.

Goodall, I., Fotheringham, R., Murray, D., Speers, R. A., & Walker, G. M. (2015). *Distilled spirits: Future challenges, new solutions*. Packington: Context Products Ltd..

Gschaedler Mathis, A., Ramírez Cordova, J., Diaz Montaño, D., Herrera-López, J. E., Arellano-Plaza, M., Arrizon-Gaviño, J., & Pinal-Zuazo, L. (2004). Fermentación: Etapa clave en la elaboración del tequila. In A. C. Gschaedler Mathis & J. A. Ruiz González (Eds.), *Ciencia y Tecnología del Tequila: Avances y perspectivas* (pp. 62–12). Guadalajara, México: Centro de Investigación y Asistencia en Tecnológica y Diseño del Estado de Jalisco AC (CIATEJ).

Gschaedler Mathis, A., Amaya-Delgado, L., Arellano-Plaza, M., Arrizón-Gabiño, J., Herrera-López, E., Iñiguez-Muñoz, L., & Kirchmayr, M. (2015). La fermentación: Etapa clave en la elaboración del tequila. In A. C. Gschaedler Mathis, B. Rodríguez-Garay, R. Prado-Ramírez, & J. L. Flores-Montaño (Eds.), *Ciencia y Tecnología del Tequila, avances y perspectivas* (pp. 97–170). Guadalajara, México: Centro de Investigación y Asistencia en Tecnología y Diseño del Estado de Jalisco AC (CIATEJ).

Hansen, R., Pearson, S. Y., Brosnan, J. M., Meaden, P. G., & Jamieson, D. J. (2006). Proteomic analysis of a distilling strain of *Saccharomyces cerevisiae* during industrial grain fermentation. *Applied Microbiology and Biotechnology, 72*, 116–125.

Jeličić, I., Božani, R., & Tratnik, L. (2008). Whey-based beverages- a new generation of diary products. *Mljekarstvo, 58*, 257–274.

Jones, R. C. (1998). Distillery yeast technology. In I. Campbell & F. G. Priest (Eds.), *Proceedings of the 5th Aviemore conference on malting, brewing & distilling* (pp. 65–77). London: Institute of Brewing.

Kirchmayr, M. R., Arellano Plaza, M., Estarrón-Espinosa, M., Gallardo Valdez, J., Gschaedler Mathis, A. C., López Ramírez, J. E., Navarro Hurtado, A., Prado Ramírez, R., & Ramírez Romo. (2014). *Manual para la estandarización de los procesos de producción de mezcal guerrerense*. Guadalajar, México: Centro de Investigación y Asistencia en Tecnología y Diseño del Estado de Jalisco AC (CIATEJ).

Kirchmayr, M., Segura, L. E., Lappe-Oliveras, P., Moreno-Terrazas, R., de la Rosa, M., & Gschaedler Mathis, A. (2017). Impact of environmental conditions and process modifications on microbial diversity, fermentation efficiency and chemical profile during the fermentation of Mezcal in Oaxaca. *LWT – Food Science and Technology, 79*, 160–169.

Korhola, M., Harju, L., & Lehtonen, M. (1989). Fermentation. In J. R. Piggott, R. Sharp, & R. E. B. Duncan (Eds.), *The science and technology of whiskies* (pp. 89–117). Harlow: Longman Scientific and Technical.

Lachance, M. A. (1995). Yeast communities in a natural tequila fermentation. *Antonie Van Leeuwenhoek, 68*, 151–160.

Lachenmeier, D. W., Sohnius, E. M., Attig, R., et al. (2006). Quantification of selected volatile constituents and anions in Mexican *Agave* spirits (Tequila, Mezcal, Sotol, Bacanora). *Journal of Agricultural and Food Chemistry, 54*(11), 3911–3915.

Lappe, P., Ulloa, M., Arce Rocha, G., Cáceres Farfán, M., Tapia-Tussell, R., Pérez-Brito, D., & Larque, A. (2004, August). Isolation and identification of the mycobiota present in Agave

fourcroydes. In *Abstracts of the 11th International congress on yeasts* (pp. 15–20). Rio de Janeiro: Yeasts in Science and Technology, The Quest for Sustainable Development..
Lappe-Oliveras, P., Moreno-Terrazas, R., Arrizon-Gaviño, J., Herrera-Suárez, T., García-Mendoza, A., & Gschaedler-Mathis, A. (2008). Yeasts associated with the production of Mexican alcoholic non distilled and distilled *Agave* beverages. *FEMS Yeast Research, 8*, 1037–1052.
Lea, A. G. H., & Piggott, J. R. P. (2003). *Fermented beverage production*. (2nd ed.). Springer.
López-Álvarez, A., Díaz-Pérez, A. L., & Sosa-Aguirre, C. (2012). Ethanol yield and volatile compound content in fermentation of agave must by *Kluyveromyces marxianus* UMPe-1 comparing with *Saccharomyces cerevisiae* baker's yeast used in tequila production. *Journal of Bioscience and Bioengineering, 113*(5), 614–618.
López-Ramírez, J. E., Martín-del-Campo, S. T., Escalona-Buendía, H., García-Fajardo, J. A., & Estarrón-Espinosa, M. (2013). Physicochemical quality of tequila during barrel maturation. A preliminary study. *CyTA Journal of Food, 11*(3), 223–233.
Mancilla-Margalli, N., & López, M. (2002). Generation of Maillard compounds from inulin during the thermal processing of *Agave tequilana* Weber var. Azul. *Journal of Agricultural and Food Chemistry, 50*, 806–812.
Mancilla-Margalli, N., & López, M. (2006). Water-soluble carbohydrates and fructan structure patterns from *Agave* and *Dasylirion* species. *Journal of Agricultural and Food Chemistry, 54*, 7832–7839.
Martínez-Aguilar, J. F., & Peña-Álvarez, A. (2009). Characterization of five typical agave plants used to produce mezcal through their simple lipid composition analysis by gas chromatography. *Journal of Agricultural and Food Chemistry, 57*(5), 1933–1939.
McGill, D. J. (1990). Brewers yeast effect. In I. Campbell (Ed.), *Proceedings of the 3rd Aviemore conference on malting, brewing & distilling* (pp. 407–412). London: Institute of Brewing.
Molina-Guerrero, J. A., Botello-Álvarez, J. E., Estrada-Baltazar, A., Navarrete-Bolaños, J. L., Jiménez-Islas, H., Cárdenas-Manríquez, M., & Rico-Martínez, R. (2007). Compuestos volátiles en el mezcal. *Revista Mexicana de Ingenieria Quimica, 6*(1), 41–50.
Morán-Marroquín, G. A., Cordova, J., Valle-Rodríguez, J. O., Estarrón-Espinosa, M., & Díaz-Montaño, D. M. (2011). Effect of dilution rate and nutrients addition on the fermentative capability and synthesis of aromatic compounds of two indigenous strains of *Saccharomyces cerevisiae* in continuous cultures fed with *Agave tequilana* juice. *International Journal of Food Microbiology, 151*(1), 87–92.
Muñoz-Muñoz, A. C., Pichardo-Molina, J. L., Ramos-Ortíz, G., Barbosa-García, J. L., Maldonado, M. A., Meneses, N. E., Ornelas, A., Escobedo, P., & López de Alba, L. (2010). Identification and quantification of furanic compounds in tequila and mezcal using spectroscopy and chemometric methods. *Journal of the Brazilian Chemical Society, 21*(6), 1077–1087.
Murray, D. M. (2017). Distillation practices for beverage alcohol and impact on flavour. In: G. M. Walker, C. Abbas, W. M. Ingledew & C. Pilgrim (Eds.), *The alcohol textbook* (6th ed.). (pp. 455–469). Duluth: Ethanol Technology Institute.
Narváez-Zapata, J. A., Rojas-Herrera, R. A., Rodríguez-Luna, I. C., & Larralde-Corona, C. P. (2010). Culture-independent analysis of lactic acid bacteria diversity associated with mezcal fermentation. *Current Microbiology, 61*(5), 444–450.
Núñez-Guerrero, M. E., Páez-Lerma, J. B., Rutiaga-Quiñones, O. M., González-Herrera, S. M., & Soto-Cruz, N. O. (2016). Performance of mixtures of *Saccharomyces* and non-*Saccharomyces* native yeasts during alcoholic fermentation of *Agave duranguensis* juice. *Food Microbiology, 54*, 91–97.
O'Shea, J. (2003). Whey alcohol – a viable outlet for whey? *The alcohol textbook* 4th Edition. Eds K. A. Jacques, T. P. Lyons & D. R. Kelsall. Nottingham University Press. pp 65–73.
Páez-Lerma, J. B., Arias-García, A., Rutiaga-Quiñones, O. M., Barrio, E., & Soto-Cruz, N. (2013). Yeasts isolated from the alcoholic fermentation of *Agave duranguensis* during mezcal production. *Food Biotechnology, 27*(4), 342–356.
Passoth, V., Blomqvist, J., & Schnurer, J. (2007). *Dekkera bruxellensis* and *Lactobacillus vini* form a stable ethanol-producing consortium in a commercial alcohol production process. *Applied and Environmental Microbiology, 73*, 4354–4356.

Pech, B., Lavoue, G., Parfait, A., & Belin, J. M. (1984). Rum fermentation: Suitability of strains of *Schizosaccharomyces pombe*. *Science des Aliments, 4*, 67–72.

Peña-Álvarez, A., Díaz, L., Medina, A., Labastida, C., Capella, S., & Vera, L. E. (2004). Characterization of three *Agave* species by gas chromatography and solid-phase microextraction-gas chromatography-mass spectrometry. *Journal of Chromatography. A, 1027*, 131–136.

Peña-Álvarez, A., Capella, S., Juárez, R., & Labastida, C. (2006). Determination of terpenes in tequila by solid phase microextraction-gas chromatography–mass spectrometry. *Journal of Chromatography. A, 1134*(1–2), 291–297.

Pérez E., González-Hernández1 J.C., Chávez-Parga M.C., Cortés-Penagos C. (2013). Caracterización fermentativa de levaduras productoras de etanol a partir de jugo de *Agave cupreata* en la elaboración de mezcal. *Revista Mexicana de Ingeniería Química 12*(3), 451–461.

Piggot, R. (2003). Production of heavy and light rums: Fermentation and maturation. In K. A. Jacques, T. P. Lyons & D. R. Kelsall (Eds.), *The alcohol textbook* (4th ed.). (pp. 246–253). Nottingham University Press.

Pinal, L., Cornejo, E., Arellano, M., Herrera, E., Nuñez, L., Arrizon, J., & Gschaedler, A. (2009). Effect of *Agave tequilana* age, cultivation field location and yeast strain on tequila fermentation process. *Journal of Industrial Microbiology & Biotechnology, 36*(5), 655–661.

Prado-Jaramillo, N., Estarrón-Espinosa, M., Escalona-Buendía, H., Cosío-Ramírez, R., & Martín-del-Campo, S. T. (2015). Volatile compounds generation during different stages of the tequila production process. A preliminary study. *LWT – Food Science and Technology, 61*(2), 471–483.

Prado-Ramírez, R., Gonzáles-Alvarez, V., Pelayo-Ortiz, C., Casillas, N., Estarrón-Espinosa, M., & Gómez-Hernández, H. E. (2005). The role of distillation on the quality of tequila. *International Journal of Food Science and Technology, 40*, 701–708.

Pretorius, I. S. (2016). Conducting wine symphonics with the aid of Yeast Genomics. *Beverages, 2*(4), 36. https://doi.org/10.3390/beverages2040036.

Reyes-Valdés, M. H., Benavides-Mendoza, A., & Ramírez-Rodríguez, H. (2012). Biología e importancia del sotol (*Dasylirion* spp). Parte I: Sistemática, genética y reproducción. *Planta, 7*(14), 11–14.

Russell, I., & Stewart, G. G. (2014). *Whisky: Technology, production and marketing*, (2nd ed.). Boston: Academic Press/Elsevier.

Rutiaga-Quiñones, O. M., Córdova, É., Martell-Nevárez, M. A., Guillamón, J. M., Rozès, N., & Páez, J. (2012). Volatile compound production in agave duranguensis juice fermentation using four native yeasts and NH4Cl supplementation. *European Food Research and Technology, 235*(1), 29–35.

Segura-García, L. E., Taillandier, P., Brandam, C., & Gschaedler, A. (2015). Fermentative capacity of *Saccharomyces* and non-*Saccharomyces* in agave juice and semi-synthetic medium. *LWT – Food Science and Technology, 60*(1), 284–291.

Smith, K. B. (2017). Yeast practices in the production of American whiskies. In: *The alcohol textbook* 6th Edition. pp 335–361. Eds. Walker, G. M., Abbas, C., Ingledew, W. M., & Pilgrim, C. Ethanol Technology Institute, Duluth.

Soto-García, E., Rutiaga-Quiñones, M., López-Miranda, J., Montoya-Ayón, L., & Soto-Cruz, O. (2009). Effect of fermentation temperature and must processing on process productivity and product quality in mescal fermentation. *Food Control, 20*, 307–309.

Thomas, K. C., Hynes, S. H., & Ingeldew, W. M. (2001). Effect of lactobacilli on yeast growth, viability and batch and semi-continuous alcoholic fermentation of corn mash. *Journal of Applied Microbiology, 90*, 819–828.

Torres, I., Blancas, J., León, A., & Casas, A. (2015). TEK, local perceptions of risk, and diversity of management practices of *Agave inaequidens* in Michoacán. México. *Journal of Ethnobiology and Ethnomedicine, 11*(61), 1–20.

Vallejo-Córdoba, A., Gonzalez-Córdoba, M., & Estrada-Montoya. (2005, March 13–17). The latest advantages in the characterization of Mexican distilled agave beverage: Tequila, Mezcal and Bacanora. In *Abstracts of the 229th National meeting of the American chemical society*, San Diego.

Valle-Rodríguez, J. O., Hernández-Cortés, G., Cordova-López, J., Estarrón-Espinosa, M., & Díaz-Montaño, D. M. (2012). Fermentation of *Agave tequilana* juice by *Kloeckera africana*: Influence of amino-acid supplementations. *Antonie Van Leeuwenhoek, 101*, 195–204.

van Beek, S., & Priest, F. G. (2002). Evolution of the lactic acid bacterial community during malt whisky fermentation: A polyphasic study. *Applied and Environmental Microbiology, 68*, 297–305.

Vera-Guzmán, A. M., Santiago-García, P. A., & López, M. G. (2009). Compuestos volátiles aromáticos generados durante la elaboración de mezcal de *Agave angustifolia* y *Agave potatorum*. *Revista Fitotecnia Mexicana, 32*(4), 273–279.

Vera-Guzmán, A. M., López, M. G., & Chávez-Servia, J. L. (2012). Chemical composition and volatile compounds in the artisanal fermentation of mezcal in Oaxaca, Mexico. *African Journal of Biotechnology, 11*(78), 14344–14353.

Vera-Guzmán, A. M., Guzmán-Gerónimo, R. I., López, M. G., & Chávez-Servia, J. L. (2018). Volatile compound profiles in Mezcal spirits as influenced by agave species and production processes. *Beverages, 4*, 1–10.

Verdugo Valdez, A., Segura García, L., Kirchmayr, M., Ramírez Rodríguez, P., González Esquinca, A., Coria, R., & Gschaedler Mathis, A. (2011). Yeast communities associated with artisanal mezcal fermentations from *Agave salmiana*. *Antonie Van Leeuwenhoek, 100*(4), 497–506.

Waleckx, E., Gschaedler, A., Colonna-Ceccaldi, B., & Monsan, P. (2008). Hydrolysis of fructans from *Agave tequilana* Weber var. azul during the cooking step in a traditional tequila elaboration process. *Food Chemistry, 108*, 40–48.

Walker, G. M. (1998). *Yeast physiology and biotechnology*. Chichester and New York: Wiley.

Walker, G. M. (2011). Fuel alcohol: Current production and future challenges. *Journal of the Institute of Brewing, 117*, 3–22.

Walker, G., & Hill, A. (2016). *Saccharomyces* cerevisiae in the production of whisky. In: G. M. Walker & G. G. Stewart (Eds.), *Beverages. Special edition on Saccharomyces cerevisiae* (Vol. 2(4), p. 38). Basel: MDPI AG Publishers. https://doi.org/10.3390/beverages2040038. Beverages

Walker, G. M., & Hughes, P. S. (2010). *Distilled spirits. New horizons: Energy, environment and enlightenment*. Nottingham: Nottingham University Press.

Walker, G. M., & O'Neill, J. D. (1990). Morphological and metabolic changes in *Kluyveromyces marxianus* var. *marxianus* NNRLy 2415 during fermentation of lactose. *Journal of Chemical Technology & Biotechnology, 49*, 75–89.

Walker, G. M., & Stewart, G. G. (2016). *Saccharomyces* cerevisiae in the production of fermented beverages. In: G. M Walker, & G. G. Stewart (Eds.), *Special edition on Saccharomyces cerevisiae* (Vol. 2(4), p. 38). Basel: MDPI AG Publishers. https://doi.org/10.3390/beverages2040038. Beverages.

Walker, G. M., & van Dijck, P. (2006). Physiological and molecular responses of yeasts to the environment. In A. Querol & G. H. Fleet (Eds.), *The yeast handbook. Yeasts in food and beverages* (pp. 111–152). Berlin/Heidelberg: Springer-Verlag.

Walker, G. M., & Walker, R. S. K. (2018). Enhancing yeast alcoholic fermentations. *Advances in Applied Microbiology, 105*. https://doi.org/10.1016/bs.aambs.2018.05.003. ISSN 0065-2164.

Walker, G. M., Bringhurst, T., Brosnan, J., & Jack, F. (2012). Selecting new distilling yeasts for improved fermentation and for sustainability. In G. M. Walker, I. Goodall, R. Fotheringham, & D. Murray (Eds.), *Distilled spirits: Science and sustainability* (pp. 127–136). Nottingham: Nottingham University Press.

Walker, G. M., Abbas, C., Ingledew, W. M., & Pilgrim, C. (2017). *The alcohol textbook*. 6th Edition. Ethanol Technology Institute, Duluth.592pp.

Watanabe, S., Kanauchi, M., Kakuta, T., & Koizumi, T. (2007). Isolation and characteristics of lactic acid bacteria in Japanese spirit Awamori mash. *Journal of the Institute of Brewing, 65*, 197–201.

Watson, D. C. (1993). Yeasts in distilled alcoholic beverage production. In A. H. Rose & J. S. Harrison (Eds.), *The yeasts* (Vol. 5, 2nd ed., pp. 215–244). London: Academic.

Zamora-Quiñonez, K. (2006). Caracterización fenotípica de levaduras presentes durante el proceso de fermentación del Bacanora. BSc Dissertation. Universidad de Sonora. Hermosillo.

Index

P. Romano et al. (eds.), *Yeasts in the Production of Wine*,
https://doi.org/10.1007/978-1-4939-9782-4

MIX
Papier aus verantwortungsvollen Quellen
Paper from responsible sources
FSC® C105338

If you have any concerns about our products,
you can contact us on
ProductSafety@springernature.com

In case Publisher is established outside the EU,
the EU authorized representative is:
Springer Nature Customer Service Center GmbH
Europaplatz 3, 69115 Heidelberg, Germany

Printed by Libri Plureos GmbH
in Hamburg, Germany